Foundations of Algebraic Analysis

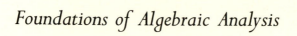

PRINCETON MATHEMATICAL SERIES

Editors: WU-CHUNG HSIANG, ROBERT P. LANGLANDS, JOHN D. MILNOR, and ELIAS M. STEIN

1. The Classical Groups, By HERMAN WEYL
3. An Introduction to Differential Geometry, By LUTHER PFAHLER EISENHART
4. Dimension Theory, By W. HUREWICZ and H. WALLMAN
6. The Laplace Transform, By D.V. WIDDER
8. Theory of Lie Groups: 1, By C. CHEVALLEY
9. Mathematical Methods of Statistics, By HARALD CRAMÉR
10. Several Complex Variables, By S. BOCHNER and W.T. MARTIN
11. Introduction to Topology, By S. LEFSCHETZ
12. Algebraic Geometry and Topology, edited by R.H. FOX, D.C. SPENCER, and A.W. TUCKER
14. The Topology of Fibre Bundles, By NORMAN STEENROD
15. Foundations of Algebraic Topology, By SAMUEL EILENBERG and NORMAN STEENROD
16. Functionals of Finite Riemann Surfaces, By MENAHEM SCHIFFER and DONALD C. SPENCER.
17. Introduction to Mathematical Logic, Vol. 1, By ALONZO CHURCH
19. Homological Algebra, By H. CARTAN and S. EILENBERG
20. The Convolution Transform, By I.I. HIRSCHMAN and D.V. WIDDER
21. Geometric Integration Theory, By. H. WHITNEY
22. Qualitative Theory of Differential Equations, By V.V. NEMYTSKII and V.V. STEPANOV
23. Topological Analysis, By GORDON T. WHYBURN (revised 1964)
24. Analytic Functions, By AHLFORS, BEHNKE and GRAUERT, BERS, et al.
25. Continuous Geometry, By JOHN VON NEUMANN
26. RIEMANN Surfaces, By L. AHLFORS and L. SARIO
27. Differential and Combinatorial Topology, edited By S.S. CAIRNS
28. Convex Analysis, By R.T. ROCKAFELLAR
29. Global Analysis, edited by D.C. SPENCER and S. IYANAGA
30. Singular Integrals and Differentiability Properties of Functions, By E.M. STEIN
31. Problems in Analysis, edited By R.C. GUNNING
32. Introduction to Fourier Analysis on Euclidean Spaces, By E.M. STEIN and G. WEISS
33. Étale Cohomology, By J.S. MILNE
34. Pseudodifferential Operators, By MICHAEL E. TAYLOR
36. Representation Theory of Semisimple Groups: An Overview Based on Examples, By ANTHONY W. KNAPP
37. Foundations of Algebraic Analysis, By MASAKI KASHIWARA, TAKAHIRO KAWAI, and TATSUO KIMURA, translated by Goro Kato

Foundations of Algebraic Analysis

By

Masaki Kashiwara,

Takahiro Kawai, and

Tatsuo Kimura

Translated by Goro Kato

PRINCETON UNIVERSITY PRESS

PRINCETON, NEW JERSEY

Published by Princeton University Press, 41 William Street, Princeton, New Jersey 08540
In the United Kingdom: Princeton University Press, Guildford, Surrey

This book was originally published by Kinokuniya Company Ltd. under
the title *Daisūkaisekigaku no kiso*

Library of Congress Cataloging in Publication Data will be
found on the last printed page of this book
ISBN 0-691-08413-0

This book has been composed in Linotron Times Roman
Clothbound editions of Princeton University Press books
are printed on acid-free paper, and binding materials are
chosen for strength and durability.
Printed in the United States of America by
Princeton University Press
Princeton, New Jersey

Contents

Preface

Prior to its founding in 1963, the Research Institute for Mathematical Sciences (to which we are gratefully indebted for support) was the focus of divers discussions concerning goals. One of the more modest goals was to set up an institution that would create a "Courant-Hilbert" for a new age.[1] Indeed, our intention here—even though this book is small in scale and only the opening chapter of our utopian "Treatise of Analysis"—is to write just such a "Courant-Hilbert" for the new generation. Each researcher in this field may have his own definition of "algebraic analysis," a term included in the title of this book. On the other hand, algebraic analysts may well share a common attitude toward the study of analysis: the essential use of algebraic methods such as cohomology theory. This characterization is, of course, too vague: one can observe such common trends whenever analysis has made serious reformations. Professor K. Oka, for example, once spoke of the "victory of abstract algebra" in regard to his theory of ideals of undetermined domains.[2] Furthermore, even Leibniz's main interest, in the early days of analysis, seems to have been in the algebraization of infinitesimal calculus. As used in the title of our book, however, "algebraic analysis" has a more special meaning, after Professor M. Sato: it is that analysis which holds onto substance and survives the shifts of fashion in the field of analysis, as Euler's mathematics, for example, has done. In this book, as the most fruitful result of our philosophy, we pay particular attention to the microlocal theory of linear partial differential equations, i.e. the new thinking on the local analysis on contangent bundles. We hope that the fundamental ideas that appear in this book will in the near future become the conventional wisdom

[1] R. Courant and D. Hilbert, *Methods of Mathematical Physics*, vols. 1 and 2 (Interscience, 1953 and 1962). These two volumes seem to reflect the strong influence of the Courant Institute; the countervailing influence must be strong as well.
[2] Quoted by Professor Y. Akizuki in *Sūgaku 12* (1960), 159. A general theory of ideals of undetermined domains has been reorganized by H. Cartan and Serre and is now called the theory of coherent sheaves (see Hitotumatu [1]).

among analysts and theoretical physicists, just as the Courant-Hilbert treatise did.

Despite our initial determination and sense of purpose, the task of writing was a heavy burden for us. It has been a time-consuming project, while our first priority has been to be at the front of the daily rapid progress in this field. Thus, we cannot deny the existence of minor areas that do not yet meet with our full satisfaction. Still, a proverb says, "Striving for the best is an enemy of the good." We are content, then, to publish our book in this form, hoping that the intelligent reader will benefit despite several defects, and expecting that this will become the first part of our "Treatise of Analysis." We would also like to emphasize that our comparison of this book with "Courant-Hilbert" is only a goal, and that we do not pretend to equate the maturity of this book with that of Courant and Hilbert's. Theirs is the crystallization of the great scholar Courant's extended effort. Therefore, we would appreciate hearing the critical reader's opinions on the content of this book, for the purpose of improvement.

Let us turn to the content of each chapter. In Chapter I, §1, a review of cohomology theory is given, with which we define the sheaf of hyperfunctions. Since students of analysis nowadays seem to be given little opportunity to learn cohomology theory, despite its importance, we have prepared a rather comprehensive treatment of sheaf cohomology theory as an introduction to notions and notations used in later chapters. One may skip this material if it is familiar. The main purpose of Chapter I, §2, is to present the mathematical formulation, via the Čech cohomology group, of the idea that "hyperfunctions are boundary values of holomorphic functions." The reader can then obtain the explicit description of a hyperfunction by combining this with the results in §3 of Chapter II.

In Chapter II, §1, the sheaf of microfunctions is constructed on a cotangent bundle, by which the stage for our main theme, microlocal analysis, is established. After some preparation of the theory of holomorphic functions of several complex variables, in §2, the properties of microfunctions will be studied in detail in §3. Furthermore, in §4, specific examples will be treated.

In §1 and §2 of Chapter III, where we basically followed Sato, Kawai, and Kashiwara [1] (hereafter SKK [1]), fundamental operations on microfunctions are discussed. However, the approach taken in SKK [1] may not be suited to the novice; hence the method of description has been changed. There it was necessary to prove a certain lemma (Proposition 3.1.1) directly, which is technical and intricate and could be tiresome for the reader. Because of the introductory nature of this book, therefore, we decided to treat this lemma as an "axiom," so to speak, and to proceed to what follows from it. In §4 through §6, elliptic and hyperbolic differential equations are treated explicitly to show how effectively microfunction theory applies to the theory of linear partial differential equations. These three sections also serve as preparation for the theory of microdifferential equations considered in Chapter IV. Prior to these three sections, we discuss (in §3) the analyticity of Feynman integrals. This section has a somewhat different flavor than other sections; it is intended as an invitation to a new trend in mathematical physics: namely, the study of theoretical physics through methods of algebraic analysis. We also thought that it might be a good exercise to go through the operations on microfunctions. In §7, we prove the flabbiness of the microfunction sheaf; and, in §8, a hyperfunction containing holomorphic parameters is discussed. The last two sections are intended to take into account some important properties of microfunctions not covered by the previous sections.

In Chapter IV, we discuss the theory of microdifferential equations, the most effective application of microfunction theory. In §1, we define a microdifferential operator, and the fundamental properties are given. "Qauntized contact transformations" of microdifferential operators are treated in §2. A quantized contact transformation is an extremely important notion, one that revolutionized the theory of linear differential equations. The reader may be astonished to see how easily one can obtain profound results with the structures of solutions of linear (micro)-differential equations by combining microfunction theory with the theory of quantized contact transformations. This point should be considered as the quintessence of microlocal analysis. As in Chapter III, we proceed in Chapter IV in a manner accesible to the reader rather than in the most logical order, which may be less accessible. For example, in §1 we chose the plane-wave decomposition of the δ-function as a starting point for the introduction of microdifferential operators, and in §2 we restricted our discussion to those contact transformations which have generating functions. We decided not to present our more "algebro-analytic" treatments of the above topics until we write a treatise on microdifferential equations centered around the theory of holonomic systems. Likewise, so that the essence of the theory might be plain to the reader, we did not aim at full generality in §3.

As we close this preface, we would like to express our most sincere gratitude to our teacher Professor Mikio Sato, who indeed provided almost all the essential ideas this book contains. We hope that this book will succeed in imparting the emanation of Professor Sato's throbbing mathematics. It is quite fortunate that authors Kashiwara and Kawai, just at the point when they were choosing their specialities, were able to attend Professor Hikosaburo Komatsu's introductory lectures in hyperfunction theory.[3] This book might be thought of as a report to Professor Komatsu ten years later. Furthermore, activity centered around Professor Sato and the authors' works has received warm encouragement and support from Professors Kōsaku Yosida and Yasuo Akizuki. Two graduate students at Kyoto University, Mr. Kimio Ueno and Mr. Akiyoshi Yonemura, have read our manuscript and have given beneficial advice. Mr. Yonemura and a graduate student at Sophia University, Mr. Masatoshi Noumi, helped us read the proofs; we would like to take this opportunity to offer our sincere thanks. During the preparation of this book, one or another of us was affiliated with the Research Institute for Mathematical Sciences, Kyoto University; the Department of Mathematics, Nagoya University; the Miller Institute for Basic Research in Science, University of California–Berkeley; the Mathematics Department, Harvard University; the Institute for Advanced Study, Princeton; the Department of Mathematics, Université Paris–Nord; and the Department of Mathematics, Massachusetts Institute of Technology. We thank these institutions and their members for their hospitality during our stay. Last, but not least, we would like to express our profound gratitude to Professor Seizō Itō, who not only gave us the opportunity to write this book, but also kept us from proceeding too slowly. We would again like to apologize to Professor Itō for our delay. Without his warm encouragement, in fact, it is doubtful that this book could ever have been published.

August of the coming-of-age year [1978] of hyperfunction theory[4]

The Authors

[3] *Sato's Hyperfunction Theory and Linear Partial Differential Equations with Constant Coefficients*, Seminar Notes 22 (University of Tokyo). At the time (1968), the above lecture note was at the highest level in the field, rather than at the introductory level.

[4] It was in 1958 that Professor Sato published his outline of hyperfunction theory.

Notations

(I) Sheaf Theory ($\mathscr{F}$ and $\mathscr{G}$ denote sheaves)

(II) Manifold Theory (M and N denote manifolds; however X and Y are sometimes used instead of M and N)

$$DM$$

$$\sqrt{-1}SM \qquad\qquad \sqrt{-1}S^*M \qquad\qquad 40$$

$$M$$

(with arrows π, τ from DM and τ, π to M)

$$N \underset{M}{\times} \sqrt{-1}S^*M - \sqrt{-1}S_N^*M \xrightarrow{\tilde{\omega}_f} \sqrt{-1}S^*M$$

$$\Big\downarrow \rho_f \qquad\qquad\qquad 105$$

$$\sqrt{-1}S^*N$$

where $f : N \to M$

(III) Hyperfunction Theory

\mathscr{A}	39	$b(\varphi)$	28, 78, 80
\mathscr{A}^*	81	sp	50
\mathscr{B}	19	S.S.	71
\mathscr{C}	40	$\widehat{\text{S.S.}}$	102
$\widehat{\mathscr{C}}$	179	$\delta(x)$	85, 86
\mathscr{O}	19	$Y(x)$	84, 85
\mathscr{D}	39	x_+^λ	83, 85
$\varphi(x + \sqrt{-1}v0)$	81		

(IV) Microdifferential Operator Theory

$\mathscr{E}_{(\lambda)}^\infty$	195	\mathscr{L}	139
$\mathscr{E}_{(\lambda)}$	195	$N_l^\omega(P; t)$	212
$\mathscr{E}(\lambda)$	195	$\sigma(P)$	139, 208
\mathscr{E}^∞	195	D^α	139
\mathscr{E}	195		

(V) Others

$H^n(K)$	9	Z°	79
$H^k(A^\bullet \xrightarrow{f} B^\bullet)$	61	$A(t) \ll B(t)$	213

Foundations of Algebraic Analysis

CHAPTER I

Hyperfunctions

§1. Sheaf Theory

Recall some of the basic concepts from sheaf theory.

Definition 1.1.1. *A presheaf $\mathscr{F}$ over a topological space X associates with each open set U of X an abelian group $\mathscr{F}(U)$, such that there exists an abelian group homomorphism $\rho_{V,U}:\mathscr{F}(U) \to \mathscr{F}(V)$ for open sets $U \supset V$ with the following axioms:*

(1) $\rho_{U,U} = \mathrm{id}_U$ (= *the identity map on* $\mathscr{F}(U)$)

(2) *For $V_1 \subset V_2 \subset V_3$, open sets of X, we have*

$$\rho_{V_1,V_2} \circ \rho_{V_2,V_3} = \rho_{V_1,V_3}.$$

The homomorphism $\rho_{V,U}$ is called the restriction map, and for $s \in \mathscr{F}(U)$ $\rho_{V,U}(s)$ is often denoted by $s|_V$.

Definition 1.1.2. *Let $\mathscr{F}$ be a presheaf over X. The stalk of the presheaf $\mathscr{F}$ at $x \in X$ is defined as $\mathscr{F}_x = \varinjlim_{x \in U} \mathscr{F}(U)$, where $\varinjlim$ denotes the inductive limit, where an equivalence relation "$\sim$" on $\bigcup_{x \in U} \mathscr{F}(U)$ is defined as follows: $s_1 \sim s_2$, for $s_1 \in F(U)$ and $s_2 \in F(V)$, if and only if there exists a sufficiently small open set $W \subset U \cap V$ such that $s_1|_W = s_2|_W$. Therefore a canonical map is induced: $\mathscr{F}(U) \to \mathscr{F}_x$ for $x \in U$. The image of $s \in \mathscr{F}(U)$ under the canonical map is denoted by s_x. Hence we have $(s_1)_x = (s_2)_x$ if and only if there exists an open set V such that $x \in V \subset U$ and such that $s_1|_V = s_2|_V$.*

Definition 1.1.3. *A presheaf $\mathscr{F}$ over X is said to be a sheaf if the following axioms are satisfied: it is given an open covering $\{U_i\}_{i \in I}$ of U in X, $U = \bigcup_{i \in I} U_i$.*

(a) *Let $s \in \mathscr{F}(U)$. If $s|_{U_i} = 0$ for each $i \in I$, then $s = 0$.*

(b) *Suppose that for each $i \in I$ there exists $s_i \in \mathscr{F}(U_i)$ such that $s_i|_{U_i \cap U_j} = s_j|_{U_i \cap U_j}$ for $i, j \in I$. Then there exists $s \in \mathscr{F}(U)$ such that $s|_{U_i} = s_i$ for each $i \in I$.*

Definition 1.1.4. *Suppose that $\mathscr{F}$ and $\mathscr{G}$ are presheaves. Then $f:\mathscr{F} \to \mathscr{G}$ is said to be a morphism if for each open set U the morphism $f(U):\mathscr{F}(U) \to \mathscr{G}(U)$ is an abelian group homomorphism and if open sets U and V are given such that $U \supset V$, then the following diagram commutes.*

$$\begin{array}{ccc} \mathscr{F}(U) & \xrightarrow{f(U)} & \mathscr{G}(U) \\ {\scriptstyle \rho_{V,U}}\downarrow & & \downarrow{\scriptstyle \rho_{V,U}} \\ \mathscr{F}(V) & \xrightarrow{f(V)} & \mathscr{G}(V) \end{array}$$

Hence there is induced a homomorphism on each stalk, $f_x:\mathscr{F}_x \to \mathscr{G}_x$.

Definition 1.1.5. *Let $\mathscr{F}$ be a presheaf over X. A sheaf $\mathscr{F}'$ is said to be the sheaf associated to the presheaf $\mathscr{F}$ (or $\mathscr{F}'$ is the sheafication of $\mathscr{F}$, or $\mathscr{F}'$ is the induced sheaf from the presheaf $\mathscr{F}$) if for each open set U of X the presheaf $\mathscr{F}'(U)$ (which is actually a sheaf) associates all the maps: $U \xrightarrow{s} \bigcup_{x \in U} \mathscr{F}_x$ such for each $X \in U$ there exists a neighborhood U' of x and $s' \in \mathscr{F}(U')$ such that $s(x') = s'_{x'}$ is true for any x' in U'.*

For a given morphism from a presheaf $\mathscr{F}$ into a sheaf $\mathscr{G}$ there is induced a unique morphism from $\mathscr{F}'$ into $\mathscr{G}$. Note that $\mathscr{F}$ and $\mathscr{F}'$ are isomorphic on each stalk.

Definition 1.1.6. *Let $\mathscr{F}'$, $\mathscr{F}$, and $\mathscr{F}''$ be sheaves over a topological space X. A sequence $\mathscr{F}' \xrightarrow{f'} \mathscr{F} \xrightarrow{f} \mathscr{F}''$ is said to be exact if $\mathscr{F}'_x \xrightarrow{f'_x} \mathscr{F}_x \xrightarrow{f_x} \mathscr{F}''_x$ is an exact sequence, i.e.* Ker f_x = Im f'_x, *at each $x \in X$.*

Let $\mathscr{F}$ and $\mathscr{G}$ be sheaves over X, and let $f:\mathscr{F} \to \mathscr{G}$ be a morphism. Then the presheaf assignment of U, an open subset, to $\mathrm{Ker}(\mathscr{F}(U) \xrightarrow{f(U)} \mathscr{G}(U))$ *is a sheaf, denoted by* Ker(f). *One also has the presheaf* $\mathrm{Coker}(\mathscr{F}(U) \xrightarrow{f(U)} \mathscr{G}(U)) = \mathscr{G}(U)/\mathrm{Im}\, f(U)$. *This presheaf is not a sheaf in general. The sheaf associated to this presheaf is denoted by* Coker(f). *Then, by definition, we have the exact sequence of sheaves*

$$0 \to \mathrm{Ker}(f) \to \mathscr{F} \xrightarrow{f} \mathscr{G} \to \mathrm{Coker}(f) \to 0.$$

In the case where Ker(f) = 0, *we often write $\mathscr{G}/\mathscr{F}$ instead of* Coker(f) *by identifying $\mathscr{F}$ with* Im f.

Definition 1.1.7. *Let X and Y be topological spaces, and let $f:X \to Y$ be a continuous map. For a sheaf $\mathscr{F}$ over X the presheaf assignment of an open subset U of Y to $\mathscr{F}(f^{-1}(U))$ is a sheaf over Y. This sheaf is called the direct image of $\mathscr{F}$ under the continuous map f, denoted by $f_*(F)$. For a sheaf $\mathscr{G}$ on Y there can be defined the presheaf $\varinjlim_{V \supset f(U)} \mathscr{G}(V)$ for an open set U of X. Generally this presheaf is not a sheaf. The associated sheaf is called the inverse image of $\mathscr{G}$ under f, denoted by $f^{-1}(\mathscr{G})$. Suppose that S is an arbitrary subset of X, and let $j_S:S \to X$ be the imbedding map. Then* *the inverse image $j_S^{-1}(\mathscr{F})$ of the sheaf $\mathscr{F}$ is called the restriction of $\mathscr{F}$ to S, and we often denote it by $\mathscr{F}|_S$.*

If $\mathscr{G}$ is a sheaf over Y, then there exists a natural morphism $\mathscr{G} \to f_*(f^{-1}(\mathscr{G}))$. Notice that $(f^{-1}\mathscr{G})_x = \mathscr{G}_{f(x)}$ and that giving a morphism $\mathscr{G} \to f_*\mathscr{F}$ for a sheaf $\mathscr{F}$ over X is equivalent to giving a morphism $f^{-1}\mathscr{G} \to \mathscr{F}$.

Definition 1.1.8. *Let $\mathscr{F}$ be a sheaf over a topological space X, and let U be an open subset of X. The subset $\{x \in U \mid s_x \neq 0\}$ for $s \in \mathscr{F}(U)$ is called the support of s, denoted by* $\operatorname{supp}(s)$. *Note that* $\operatorname{supp}(s)$ *is closed in U.*

Definition 1.1.9. *Let $\mathscr{F}$ be a sheaf over a topological space X, and let S be a locally closed subset of X; i.e. it is the intersection of an open set and a closed set in X. Then define $\Gamma_S(X, \mathscr{F}) = \{s \in \mathscr{F}(U) \mid \operatorname{supp}(s) \subset S\}$, where U is an open set in X such that S is closed in U.*

The definition above is independent of the choice of U.

Proof. Let U_1 and U_2 be such open sets; then $U_1 \cap U_2$ contains S as a closed subset. Therefore one can assume that $S \subset U_1 \subset U_2 \subset X$ and that S is closed in U_1 and U_2. Define a map φ from $\{s \in \mathscr{F}(U_2) \mid \operatorname{supp}(s) \subset S\}$ to $\{s \in \mathscr{F}(U_1) \mid \operatorname{supp}(s) \subset S\}$ by $\varphi(s) = s|_{U_1}$. Then φ is bijective. Therefore $\Gamma_S(X, \mathscr{F})$ is independent of the choice of U.

In the case that $S = X$, we denote $\Gamma_S(X, \mathscr{F})$ with $\Gamma(X, \mathscr{F})$, whose elements are called the global sections of $\mathscr{F}$, i.e. $\Gamma(X, \mathscr{F}) = \mathscr{F}(X)$. Generally we also denote $\mathscr{F}(U)$ with $\Gamma(U, \mathscr{F})$ for an open set U in X, whose elements are called the sections of $\mathscr{F}$ over U.

Definition 1.1.10. *Let $\mathscr{F}$ be a sheaf over a topological space X, and let S be a locally closed subset of X. We denote the sheaf associated to a presheaf $\Gamma_{S \cap U}(U, \mathscr{F}) = \{s \in \mathscr{F}(U) \mid \operatorname{supp}(s) \subset S \cap U\}$, for an open set U of X, by $\Gamma_S(\mathscr{F})$.*

Definition 1.1.11. *A sheaf $\mathscr{F}$ over a topological space X is said to be flabby if for an arbitrary open set U the homomorphism $\rho_{U,X}: \mathscr{F}(X) \to \mathscr{F}(U)$ is an epimorphism. Therefore, for a flabby sheaf $\mathscr{F}$ any section of $\mathscr{F}$ over U can be extended to a section over X.*

Proposition 1.1.1. *Let $\mathscr{F}$ be a flabby sheaf over a topological space X, and let S be a locally closed subset of X. Then $\Gamma_S(\mathscr{F})$ is a flabby sheaf.*

Proof. Let U_1 be an open set such that S is closed in U_1. Then, for any open set U of X, the set $U_1 \cap U$ is open in X and contains $S \cap U$ as a closed set. Let s be an element of $\Gamma_S(\mathscr{F})(U) = \Gamma_{S \cap U}(U, \mathscr{F})$; then $s \in \mathscr{F}(U_1 \cap U)$ and $\operatorname{supp}(s) \subset S \cap U$. Therefore we have $s|_{(U_1 - S) \cap (U_1 \cap U)} = 0$. Then there exists a unique $s' \in \mathscr{F}((U_1 - S) \cup (U_1 \cap U))$ such that $s'|_{(U_1 - S)} = 0$ and $s'|_{U_1 \cap U} = s$. Since the sheaf $\mathscr{F}$ is flabby, s' can be extended to a section $\tilde{s} \in \mathscr{F}(U_1)$. Then we have $\tilde{s}|_{(U_1 - S)} = 0$. Hence $\operatorname{supp}(\tilde{s}) \subset S$, i.e. $\tilde{s} \in \Gamma_S(X, \mathscr{F}) = \Gamma_S(\mathscr{F})(X)$.

Proposition 1.1.2. *Let* $\mathscr{F}'$, $\mathscr{F}$, and $\mathscr{F}''$ *be sheaves over a topological space* X, *let* U *be an open set, and let* S *be a locally closed set in* X.

(1) *If* $0 \to \mathscr{F}' \xrightarrow{f'} \mathscr{F} \xrightarrow{f} \mathscr{F}''$ *is an exact sequence of sheaves, then*
  (i) $0 \to \mathscr{F}'(U) \xrightarrow{f'(U)} \mathscr{F}(U) \xrightarrow{f(U)} \mathscr{F}''(U)$ *and*
  (ii) $0 \to \Gamma_S(X, \mathscr{F}') \to \Gamma_S(X, \mathscr{F}) \to \Gamma_S(X, \mathscr{F}'')$ *are exact.*

(2) If $0 \to \mathscr{F}' \xrightarrow{f'} \mathscr{F} \xrightarrow{f} \mathscr{F}'' \to 0$ *is an exact sequence of sheaves, and if* $\mathscr{F}'$ *is a flabby sheaf, then*
  (i) $0 \to \mathscr{F}'(U) \xrightarrow{f'(U)} \mathscr{F}(U) \xrightarrow{f(U)} \mathscr{F}''(U) \to 0$ *and*
  (ii) $0 \to \Gamma_S(X, \mathscr{F}') \to \Gamma_S(X, \mathscr{F}) \to \Gamma_S(X, \mathscr{F}'') \to 0$ *are exact.*

Proof. (1.i) First we will show that $f'(U)$ is a monomorphism. Suppose that $f'(U)s' = 0$ for $s' \in \mathscr{F}'(U)$. Then $f'_x s'_x = 0$ for each x in U. Therefore $s'_x = 0$; i.e. there exists a neighborhood $V(x)$ of x such that $s'|_{V(x)} = 0$. By the definition of a sheaf, we know that $s' = 0$. Therefore $f'(U)$ is monomorphic. Next we will prove that $\operatorname{Im} f'(U) \subset \operatorname{Ker} f(U)$. Since $(f_x \circ f'_x)s'_x = 0$ for $s' \in \mathscr{F}(U)$, for each x one can find a neighborhood $V(x)$ of x such that $f(U)f'(U)s'|_{V(x)} = 0$. Therefore, since $\mathscr{F}''$ is a sheaf we have $f(U)f'(U)s' = 0$. It remains to be proved that $\operatorname{Im} f'(U) \supset \operatorname{Ker} f(U)$. Let $s \in \mathscr{F}(U)$ such that $f(U)s = 0$. Then, for each $x \in U$, $f_x s_x = 0$ holds. By the exactness there exists $s'_x \in \mathscr{F}'_x$ such that $f'_x s'_x = s_x$. This implies that $f'(V(x))s'(x) = s|_{V(x)}$ for some $s'(x) \in \mathscr{F}(V(x))$ in some neighborhood $V(x)$ of x such that $V(x) \subset U$. Since $f'(V(x))$ is a monomorphism, $s'(x)$ is unique. Therefore we have $s'(x)|_{V(x) \cap V(y)} = s'(y)|_{V(x) \cap V(y)}$. By the sheaf axiom, we have $s' \in \mathscr{F}'(U)$ and $s'|_{V(x)} = s'(x)$. Then $f'(U)s' = s$.

Next we will give a proof of (1.ii). Let U be an open set in X such that S is closed in U. It is to be shown that $\operatorname{supp}(s') \subset S$ for the s', as in the above, provided that $\operatorname{supp}(s) \subset S$ for an $s \in \mathscr{F}(U)$. Note that $f'(U - S)s'|_{(U-S)} = s|_{(U-S)} = 0$ and that $f'(U - S)$ is a monomorphism. Therefore $s'|_{(U-S)} = 0$, i.e. $\operatorname{supp}(s') \subset S$.

It suffices to show that $f(U)$ is an epimorphism to prove (2.i). Let $s'' \in \mathscr{F}''(U)$, and let $\mathscr{M} = \{(s, V) \mid V$ is an open subset of U, $s \in \mathscr{F}(V)$ and $f(V)s = s''|_V\}$. Then define an order relation, denoted with $\succ$, in $\mathscr{M}$ as follows: let (s_1, V_1) and (s_2, V_2) be elements of $\mathscr{M}$. The expression $(s_1, V_1) \succ (s_2, V_2)$ holds if and only if $V_1 \supset V_2$ and $s_1|_{V_2} = s_2$. Then $\mathscr{M}$ is a non-empty, inductively ordered set. Therefore there exists a maximal element in $\mathscr{M}$ by Zorn's lemma. Let (s, V) be a maximal element. $V = U$ is left to be proved. Suppose $V \neq U$, and let $x \in U - V$. Then there exists a neighborhood $V(x)$ and $s(x) \in \mathscr{F}(V(x))$ such that $(s(x), V(x)) \in \mathscr{M}$. Then notice that $f(V \cap V(x))(s - s(x))|_{V \cap V(x)} = 0$. One can then find $s' \in \mathscr{F}'(V \cap V(x))$ such that $f'(V \cap V(x))s' = (s - s(x))|_{V \cap V(x)}$ by (1.i). The flabbiness of $\mathscr{F}'$ implies that there exists $\tilde{s}' \in \mathscr{F}'(V(x))$ such that $\tilde{s}'|_{V \cap V(x)} = s'$. Define $\tilde{s} \in \mathscr{F}(V \cup V(x))$ as $\tilde{s}|_V = s$ and $\tilde{s}|_{V(x)} = s(x) + f'(V(x))\tilde{s}'$. Then $(\tilde{s}, V \cup V(x)) \in \mathscr{M}$, which contradicts the maximality of (s, V) in $\mathscr{M}$. Therefore $V = U$; that is, $f(U)$ is an epimorphism. (2.ii) can be proved similarly. Let $\mathscr{M}' = \{(s, U) | s \in \Gamma_{S \cap U}(U, \mathscr{F}), U$ is an open set such that $f(U)s = s''|_U\}$. Then take a maximal element of $\mathscr{M}'$ to be (s, V) satisfying that $(s, V) \succ (0, X - \operatorname{supp}(s''))$.

Remark. Conversely, if $0 \to \mathscr{F}'(U) \to \mathscr{F}(U) \to \mathscr{F}''(U)(\to 0)$ is exact for any open set U, then it is plain that $0 \to \mathscr{F}' \to \mathscr{F} \to \mathscr{F}''(\to 0)$ is exact.

Corollary 1. *Let* $0 \to \mathscr{F}' \xrightarrow{f'} \mathscr{F} \xrightarrow{f} \mathscr{F}'' \to 0$ *be an exact sequence of sheaves.*

(1) *If* $\mathscr{F}'$ *and* $\mathscr{F}$ *are flabby sheaves, then* $\mathscr{F}''$ *is a flabby sheaf.*

(2) *If* $\mathscr{F}'$ *and* $\mathscr{F}''$ *are flabby sheaves, then* $\mathscr{F}$ *is a flabby sheaf.*

Proof. Since $\mathscr{F}'$ is a flabby sheaf, we have the commutative diagram

$$\begin{array}{ccccccccc} 0 & \longrightarrow & \mathscr{F}'(X) & \xrightarrow{f'(X)} & \mathscr{F}(X) & \xrightarrow{f(X)} & \mathscr{F}''(X) & \longrightarrow & 0 \\ & & \downarrow{\scriptstyle \rho'_{U,X}} & & \downarrow{\scriptstyle \rho_{U,X}} & & \downarrow{\scriptstyle \rho''_{U,X}} & & \\ 0 & \longrightarrow & \mathscr{F}'(U) & \xrightarrow{f'(U)} & \mathscr{F}(U) & \xrightarrow{f(U)} & \mathscr{F}''(U) & \longrightarrow & 0 \end{array}$$

with exact rows, where U is any open set in X. To prove (1), first notice that $\rho_{U,X}$ is an epimorphism since $\mathscr{F}$ is flabby. On the other hand, $f(U)$ is an epimorphism. Therefore, for any $s'' \in \mathscr{F}''(U)$, there exists an $s \in \mathscr{F}(X)$ such that $s'' = f(U)\rho_{U,X}(s) = \rho''_{U,X}(f(X)s)$. This implies that $\rho''_{U,X}$ is an epimorphism; i.e. $\mathscr{F}''$ is a flabby sheaf. Next we will prove (2). Let $s \in \mathscr{F}(U)$. Since $\rho''_{U,X}$ and $f(X)$ are both epimorphisms, one can find $\tilde{s} \in \mathscr{F}(X)$ such that $\rho''_{U,X}f(X)\tilde{s} = f(U)s$. By commutativity we have $\rho''_{U,X}f(X)\tilde{s} = f(U)\rho_{U,X}\tilde{s}$. Therefore $f(U)(s - \rho_{U,X}\tilde{s}) = 0$. Then note that $\rho'_{U,X}$ is an epimorphism. Hence there exists $\tilde{s}' \in \mathscr{F}'(X)$ such that $s - \rho_{U,X}\tilde{s} = f'(U)\rho'_{U,X}\tilde{s}' = \rho_{U,X}f'(X)\tilde{s}'$. That is, $s = \rho_{U,X}(\tilde{s} + f'(X)\tilde{s}')$, showing that $\rho_{U,X}$ is an epimorphism. Therefore $\mathscr{F}$ is a flabby sheaf.

Corollary 2. *Suppose that* $0 \to \mathscr{F}^0 \to \mathscr{F}^1 \to \cdots \to \mathscr{F}^r \to \mathscr{G} \to 0$ *is an exact sequence of sheaves and that each* $\mathscr{F}^i$, $i = 0, 1, \ldots, r$, *is a flabby sheaf. Then* $\mathscr{G}$ *is a flabby sheaf, and the following sequences are exact:*

$$0 \to \Gamma_S(X, \mathscr{F}^0) \to \Gamma_S(X, \mathscr{F}^1) \to \cdots \to \Gamma_S(X, \mathscr{F}^r) \to \Gamma_S(X, \mathscr{G}) \to 0$$

and

$$0 \to \mathscr{F}^0(U) \to \mathscr{F}^1(U) \to \cdots \to \mathscr{F}^r(U) \to \mathscr{G}(U) \to 0,$$

where S is a locally closed subset of X, and where U is an open subset of X.

Proof. First split the given long exact sequence into short exact sequences, as follows:

$$
\begin{array}{cccccccccccccccc}
 & & & & & 0 \quad 0 & & & & 0 \quad 0 & & & & & & 0 \\
 & & & & & \searrow \ \nearrow & & & & \searrow \ \nearrow & & & & & & \nearrow \\
 & & & & & \mathscr{G}^1 & & & & \mathscr{G}^3 & & & & & & \mathscr{G} \\
 & & & & & \nearrow \ \searrow & & & & \nearrow \ \searrow & & & & & & \nearrow \\
0 & \longrightarrow & \mathscr{F}^0 & \longrightarrow & \mathscr{F}^1 & \longrightarrow & \mathscr{F}^2 & \longrightarrow & \mathscr{F}^3 & \longrightarrow & \cdots & \longrightarrow & \mathscr{F}^{r-1} & \longrightarrow & \mathscr{F}^r & \\
 & & & & & & & \searrow \ \nearrow & & & & & & \searrow \ \nearrow & & \\
 & & & & & & & \mathscr{G}^2 & & & & & & \mathscr{G}^{r-1} & & \\
 & & & & & & & \nearrow \ \searrow & & & & & & \nearrow \ \searrow & & \\
 & & & & & & & 0 \quad 0 & & & & & & 0 \quad 0 & &
\end{array}
$$

By Corollary 1, $\mathscr{G}^1$ is a flabby sheaf and, since $\mathscr{F}^2$ is flabby, $\mathscr{G}^2$ is a flabby sheaf. Therefore, by repeating this process, we can conclude that $\mathscr{G}$ is a flabby sheaf. These short exact sequences provide the short exact sequences

$$0 \to \Gamma_S(X, \mathscr{F}^0) \to \Gamma_S(X, \mathscr{F}^1) \to \Gamma_S(X, \mathscr{G}^1) \to 0,$$

$$0 \to \Gamma_S(X, \mathscr{G}^i) \to \Gamma_S(X, \mathscr{F}^{i+1}) \to \Gamma_S(X, \mathscr{G}^{i+1}) \to 0 \qquad \text{for } 1 \leqq i \leqq r-2,$$

and

$$0 \to \Gamma_S(X, \mathscr{G}^{r-1}) \to \Gamma_S(X, \mathscr{F}^r) \to \Gamma_S(X, \mathscr{G}) \to 0.$$

The latter assertion follows plainly from these short exact sequences.

Definition 1.1.12. *An exact sequence* $0 \to \mathscr{F} \to \mathscr{L}^0 \to \mathscr{L}^1 \to \cdots$ *is said to be a flabby resolution of a sheaf* $\mathscr{F}$ *if each* $\mathscr{L}^i$, $i = 0, 1, \ldots,$ *is a flabby sheaf*.

Proposition 1.1.3. *For an arbitrary sheaf* $\mathscr{F}$ *over a topological space* X *there exists a flabby sheaf* $\mathscr{L}$ *such that* $0 \to \mathscr{F} \to \mathscr{L}$ *is an exact sequence. Therefore there exists a flabby resolution of* $\mathscr{F}$.

Proof. Consider a sheaf $\mathscr{C}^0F$ which associates with each open set U of X an abelian group $\left\{ s \,|\, U \xrightarrow{s} \bigcup_{x \in U} \mathscr{F}_x,\right.$ where s is an arbitrary mapping such that $s(x) \in \mathscr{F}_x$ holds$\left.\right\}$. It is plain that $\mathscr{C}^0(\mathscr{F})$ is a flabby sheaf, and then the sequence $0 \to \mathscr{F} \to \mathscr{C}^0(\mathscr{F})$ is exact. Define a sheaf $\mathscr{Z}^0(\mathscr{F})$ so that the sequence $0 \to \mathscr{F} \to \mathscr{C}^0(\mathscr{F}) \to \mathscr{Z}^0(\mathscr{F}) \to 0$ is exact. Similarly, we also have the following exact sequence: $0 \to \mathscr{Z}^0(\mathscr{F}) \to \mathscr{C}^0(\mathscr{Z}^0(\mathscr{F})) \to \mathscr{Z}^1(\mathscr{F})$. If one defines $\mathscr{C}^n(\mathscr{F})$ as $\mathscr{C}^0(\mathscr{Z}^{n-1}(\mathscr{F}))$, the sequence $0 \to \mathscr{F} \to \mathscr{C}^0(\mathscr{F}) \to \mathscr{C}^1(\mathscr{F}) \to \cdots$ gives a flabby resolution of $\mathscr{F}$.

Note. The flabby resolution constructed above is said to be the canonical flabby resolution of $\mathscr{F}$.

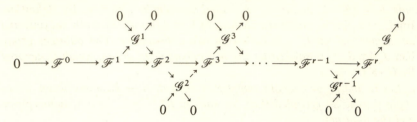

Definition 1.1.13. *Let K^n be an abelian group, and let $d_n: K^n \to K^{n+1}$ be a homomorphism such that $d_{n+1} \circ d_n = 0$ for each $n = 0, 1, 2, \ldots$. Then the pair $K = \{K^n, d_n\}_{n=0,1,2,\ldots}$ is called a cochain complex. By definition,* $\operatorname{Im} d_{n-1} \subset \operatorname{Ker} d_n$ *holds. An element of* $\operatorname{Ker} d_n$ *is called an* nth *cocycle, and an element of* $\operatorname{Im} d_{n-1}$ *is called an* nth *coboundary. The quotient group* $\operatorname{Ker} d_n/\operatorname{Im} d_{n-1}$ *is said to be the* nth *cohomology group, which is denoted by $H^n(K)$.*

Let K and K′ be cochain complexes. We call $K \xrightarrow{f} K'$ a morphism of cochain complexes provided that, for each n, $K^n \xrightarrow{f_n} K'^n$ is a homomorphism and the diagram

$$\begin{array}{ccccccccccccc}
0 \longrightarrow & K^0 & \xrightarrow{d_0} & K^1 & \xrightarrow{d_1} & \cdots & \longrightarrow & K^{n-1} & \xrightarrow{d_{n-1}} & K^n & \xrightarrow{d_n} & K^{n+1} & \xrightarrow{d_{n+1}} \cdots \\
 & \downarrow f_0 & & \downarrow f_1 & & f_{n-1} & & \downarrow & & \downarrow f_n & & \downarrow f_{n+1} & \\
0 \longrightarrow & K'^0 & \xrightarrow{d'_0} & K'^1 & \xrightarrow{d'_1} & \cdots & \longrightarrow & K'^{n-1} & \xrightarrow{d'_{n-1}} & K'^n & \xrightarrow{d'_n} & K'^{n+1} & \xrightarrow{d'_{n+1}} \cdots
\end{array}$$

is commutative. Let K', K, and *K'' be cochain complexes. A sequence*

$$0 \to K' \xrightarrow{f'} K \xrightarrow{f} K'' \to 0$$

is said to be exact if, for each n,

$$0 \to K'^n \xrightarrow{f'_n} K^n \xrightarrow{f_n} K''^n \to 0$$

is an exact sequence. That is, all the vertical sequences are exact in the diagram

$$\begin{array}{ccccccccc}
 & 0 & & 0 & & 0 & & 0 & \\
 & \downarrow & & \downarrow & & \downarrow & & \downarrow & \\
\cdots \longrightarrow & K'^{n-2} & \xrightarrow{d'_{n-2}} & K'^{n-1} & \xrightarrow{d'_{n-1}} & K'^n & \xrightarrow{d'_n} & K'^{n+1} & \longrightarrow \cdots \\
 & \downarrow f'_{n-2} & & \downarrow f'_{n-1} & & \downarrow f'_n & & \downarrow f'_{n+1} & \\
\cdots \longrightarrow & K^{n-2} & \xrightarrow{d_{n-2}} & K^{n-1} & \xrightarrow{d_{n-1}} & K^n & \xrightarrow{d_n} & K^{n+1} & \xrightarrow{d_{n+1}} \cdots \\
 & \downarrow f_{n-2} & & \downarrow f_{n-1} & & \downarrow f_n & & \downarrow f_{n+1} & \\
\cdots \longrightarrow & K''^{n-2} & \xrightarrow{d''_{n-2}} & K''^{n-1} & \xrightarrow{d''_{n-1}} & K''^n & \xrightarrow{d''_n} & K''^{n+1} & \xrightarrow{d''_{n+1}} \cdots \\
 & \downarrow & & \downarrow & & \downarrow & & \downarrow & \\
 & 0 & & 0 & & 0 & & 0 &
\end{array}$$

Notice that the nth cocycles of K' are mapped into the nth cocycles of K and similarly for the nth coboundaries. Therefore f'_n induces a homomorphism f'^*_n on the cohomology groups $H^n(K') \to H^n(K)$. Likewise there is induced a homomorphism $f^*_n: H^n(K) \to H^n(K'')$. Furthermore, since f_n is an epimorphism, for an nth cocycle z''_n of K'' there exists $z_n \in K^n$ such that $f_n(z_n) = z''_n$. Therefore one has $f_{n+1}(d_n z_n) = d''_n f_n(z_n) = d''_n z''_n = 0$, which implies that $f'_{n+1}(z'_{n+1}) = d_n z_n$ for some z'_{n+1} in K'^{n+1}. Notice that z'_{n+1} is uniquely determined modulo $\operatorname{Im} d'_n$, the $(n + 1)$-coboundaries for the given

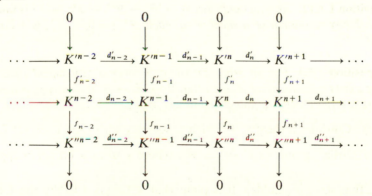

z_n''. Therefore one has a well-defined homomorphism $h_n^*: H^n(K'') \to H^{n+1}(K')$. Then we have the following important proposition.

Proposition 1.1.4. *Let* $0 \to K' \to K \to K'' \to 0$ *be an exact sequence of cochain complexes. Then the induced sequence of cohomology groups* $0 \to H^0(K') \to H^0(K) \to H^0(K'') \to H^1(K') \to \cdots$ *is exact.*

One can plainly prove this assertion from definitions.

Corollary (Nine Lemma). *Suppose that in the commutative diagram below the three vertical sequences are exact and L is a cochain complex. If any two of the horizontal sequences are exact, then the remaining horizontal sequence is exact.*

$$
\begin{array}{ccccccccc}
 & & 0 & & 0 & & 0 & & \\
 & & \downarrow & & \downarrow & & \downarrow & & \\
0 & \to & K^0 & \to & K^1 & \to & K^2 & \to & 0 \\
 & & \downarrow & & \downarrow & & \downarrow & & \\
0 & \to & L^0 & \to & L^1 & \to & L^2 & \to & 0 \\
 & & \downarrow & & \downarrow & & \downarrow & & \\
0 & \to & M^0 & \to & M^1 & \to & M^2 & \to & 0 \\
 & & \downarrow & & \downarrow & & \downarrow & & \\
 & & 0 & & 0 & & 0 & &
\end{array}
$$

Proof. Since the second horizontal sequence is a cochain complex, all three of the horizontal sequences are cochain complexes. By Proposition 1.1.4 we have the long exact sequence $0 \to H^0(K) \to H^0(L) \to H^0(M) \to H^1(K) \to \cdots \to H^2(M) \to 0$. For instance, if the first and second horizontal sequences are exact, then one has $H^i(K) = H^i(L) = 0$, $i = 0, 1, 2$. Then $H^j(M)$ must be trivial for $j = 0, 1, 2$; i.e. the third sequence is exact.

Remark. The assumption that L is a cochain complex in Proposition 1.1.4 is necessary for this claim. In fact, one can have additive groups with the property $a + a = b + b = 0$ and the commutative diagram

$$
\begin{array}{ccccccccc}
 & & 0 & & 0 & & 0 & & \\
 & & \downarrow & & \downarrow & & \downarrow & & \\
0 & \to & 0 & \longrightarrow & \{0, a\} & \longrightarrow & \{0, a\} & \to & 0 \\
 & & \downarrow & & \downarrow & & \downarrow & & \\
0 & \to & \{0, b\} & \to & \{0, b, a, a+b\} & \xrightarrow{\varphi} & \{0, a\} & \to & 0 \\
 & & \downarrow & & \downarrow & & \downarrow & & \\
0 & \to & \{0, b\} & \to & \{0, b\} & \longrightarrow & 0 & \longrightarrow & 0 \\
 & & \downarrow & & \downarrow & & \downarrow & & \\
 & & 0 & & 0 & & 0 & &
\end{array}
$$

where φ is defined as $\varphi(b) = \varphi(a) = a$, $\varphi(0) = \varphi(a + b) = 0$. Then φ is a homomorphism, but the second horizontal sequence is not a cochain complex.

The following lemma is also well-known.

Five Lemma.

$$\begin{array}{ccccccccc} K_1 & \to & K_2 & \to & K_3 & \to & K_4 & \to & K_5 \\ \downarrow h_1 & & \downarrow h_2 & & \downarrow h_3 & & \downarrow h_4 & & \downarrow h_5 \\ L_1 & \to & L_2 & \to & L_3 & \to & L_4 & \to & L_5 \end{array}$$

Suppose that the horizontal sequences are exact in the above diagram. If h_2 and h_4 are isomorphisms, h_1 is an epimorphism, and h_5 is a monomorphism, then h_3 is an isomorphism.

Proving this lemma is left to the reader.

Definition 1.1.14. *Let $\mathscr{F}$ be a sheaf over a topological space X, and let S be a locally closed subset of X. The* jth *relative cohomology of $\mathscr{F}$ with supports in S, denoted by $H_S^j(X, \mathscr{F})$, is defined as follows. Let $0 \to \mathscr{F} \to \mathscr{L}^0 \to \mathscr{L}^1 \to \cdots$ be a flabby resolution of $\mathscr{F}$. Then one obtains the complex $\{\Gamma_S(X, \mathscr{L}^*)\}$. Define*

$$H_S^j(X, \mathscr{F}) = \frac{\operatorname{Ker}(\Gamma_S(X, \mathscr{L}^j) \to \Gamma_S(X, \mathscr{L}^{j+1}))}{\operatorname{Im}(\Gamma_S(X, \mathscr{L}^{j-1}) \to \Gamma_S(X, \mathscr{L}^j))}.$$

Note that this definition is independent of the choice of the flabby resolution $\{L^*\}$ by Theorem 1.1.1 below.

Theorem 1.1.1. *The $H_S^j(X, \mathscr{F})$ is determined canonically by any flabby resolution of $\mathscr{F}$.*

We begin with lemmas.

Lemma 1. *Let $f: \mathscr{F} \to \mathscr{G}$ be a morphism, and let $0 \to \mathscr{F} \to \mathscr{L}^0 \xrightarrow{g_0} \mathscr{L}^1 \xrightarrow{g_1} \cdots$ be an exact sequence of sheaves. Then there exists a flabby resolution of $\mathscr{G}$, $0 \to \mathscr{G} \to \mathscr{M}^0 \to \cdots$, such that for each $j \geqq 0$, there exists a morphism $f_j: \mathscr{L}^j \to \mathscr{M}^j$ so that the diagram*

$$\begin{array}{ccccccccc} 0 & \to & \mathscr{F} & \to & \mathscr{L}^0 & \xrightarrow{g_0} & \mathscr{L}^1 & \xrightarrow{g_1} & \cdots \\ & & \downarrow f & & \downarrow f_0 & & \downarrow f_1 & & \\ 0 & \to & \mathscr{G} & \to & \mathscr{M}^0 & \longrightarrow & \mathscr{M}^1 & \longrightarrow & \cdots \end{array}$$

is commutative.

Proof. We prove this lemma inductively on j. Suppose that we have constructed $\mathscr{M}^j$ and f_j for $0 \leqq j \leqq k$ as claimed.

$$\begin{array}{cccccccccccc} 0 & \to & \mathscr{F} & \to & \mathscr{L}^0 & \to & \mathscr{L}^1 & \to \cdots \to & \mathscr{L}^k & \xrightarrow{g_k} & \mathscr{L}^{k+1} & \to \\ & & \downarrow f & & \downarrow f_0 & & \downarrow f_1 & & \downarrow f_k & & & \\ 0 & \to & \mathscr{G} & \to & \mathscr{M}^0 & \to & \mathscr{M}^1 & \to \cdots \to & \mathscr{M}^k & & & \end{array}$$

Let $\mathscr{Z}$ be a sheaf so that the sequence $0 \to \mathscr{G} \to \mathscr{M}^0 \to \cdots \to \mathscr{M}^k \xrightarrow{h} \mathscr{Z} \to 0$ is exact. Then define a morphism $\mathscr{L}^k \to \mathscr{Z} \oplus \mathscr{L}^{k+1}$ by $x \mapsto ((h \circ f_k)(x), -g_k(x))$. Then we have the exact sequence, $0 \to \mathscr{Z} \xrightarrow{\iota} \operatorname{Coker}(\mathscr{L}^k \to \mathscr{Z} \oplus \mathscr{L}^{k+1})$. We also have the commutative diagram

$$\begin{array}{ccc} \mathscr{L}^k & \xrightarrow{g_k} & \mathscr{L}^{k+1} \\ \downarrow{\scriptstyle f_k} & & \downarrow{\scriptstyle f'_{k+1}} \\ \mathscr{M}^k & \xrightarrow{\iota \circ h} & \operatorname{Coker}(\mathscr{L}^k \to \mathscr{Z} \oplus \mathscr{L}^{k+1}) \end{array}$$

where $f'_{k+1}: \mathscr{L}^{k+1} \to \mathscr{Z} \oplus \mathscr{L}^{k+1}$ is the canonical morphism defined by $f'_{k+1}(x) = (0, x)$. Let $\mathscr{M}^{k+1}$ be a flabby sheaf such that $0 \to \operatorname{Coker}(\mathscr{L}^k \to \mathscr{Z} \oplus \mathscr{L}^{k+1}) \xrightarrow{\iota'} \mathscr{M}^{k+1}$ is an exact sequence. Then we obtain the commutative diagram

$$\begin{array}{ccc} \mathscr{L}^k & \xrightarrow{g_k} & \mathscr{L}^{k+1} \\ \downarrow{\scriptstyle f_k} & & \downarrow{\scriptstyle \iota' \circ f'_{k+1} = f_{k+1}} \\ \mathscr{M}^k & \xrightarrow[\iota' \circ \iota \circ h]{} & \mathscr{M}^{k+1} \end{array}$$

Lemma 2. *Suppose that* $0 \to \mathscr{F} \to \mathscr{L}^0 \to \mathscr{L}^1 \to \cdots$ *and* $0 \to \mathscr{F} \to \mathscr{L}'^0 \to L'^1 \to \cdots$ *are flabby resolutions of a sheaf* $\mathscr{F}$. *Then there exists a flabby resolution of* $\mathscr{F}$, $0 \to \mathscr{F} \to \mathscr{L}''^0 \to \mathscr{L}''^1 \to \cdots$, *such that the diagrams*

$$\begin{array}{ccccccccc} 0 & \to & \mathscr{F} & \to & \mathscr{L}^0 & \to & \mathscr{L}^1 & \to & \cdots \\ & & \downarrow{\scriptstyle \mathrm{id}} & & \downarrow & & \downarrow & & \\ 0 & \to & \mathscr{F} & \to & \mathscr{L}''^0 & \to & \mathscr{L}''^1 & \to & \cdots \end{array} \quad \text{and} \quad \begin{array}{ccccccccc} 0 & \to & \mathscr{F} & \to & \mathscr{L}'^0 & \to & \mathscr{L}'^1 & \to & \cdots \\ & & \downarrow{\scriptstyle \mathrm{id}} & & \downarrow & & \downarrow & & \\ 0 & \to & \mathscr{F} & \to & \mathscr{L}''^0 & \to & \mathscr{L}''^1 & \to & \cdots \end{array}$$

are commutative.

Proof. Apply Lemma 1 to the case where $f: \mathscr{F} \oplus \mathscr{F} \to \mathscr{F}$, defined by $f(x, y) = x + y$, and the flabby resolution of $\mathscr{F} \oplus \mathscr{F}$, $0 \to \mathscr{F} \oplus \mathscr{F} \to \mathscr{L}^0 \oplus \mathscr{L}'^0 \to \cdots$. Then take a flabby resolution of $\mathscr{F}$ such that

$$\begin{array}{ccccccccc} 0 & \to & \mathscr{F} \oplus \mathscr{F} & \to & \mathscr{L}^0 \oplus \mathscr{L}'^0 & \to & \mathscr{L}' \oplus \mathscr{L}'^1 & \to & \cdots \\ & & \downarrow{\scriptstyle f} & & \downarrow & & \downarrow & & \\ 0 & \longrightarrow & \mathscr{F} & \longrightarrow & \mathscr{L}''^0 & \longrightarrow & \mathscr{L}''^1 & \longrightarrow & \cdots \end{array}$$

is a commutative diagram.

Proof of Theorem 1.1.1. It is to be proved that $H^j(\Gamma_S(X, \mathscr{L}^*))$ and $H^j(\Gamma_S(X, \mathscr{L}'^*))$ are isomorphic to $H^j(\Gamma_S(X, \mathscr{L}''^*))$ in order to claim $H^j(\Gamma_S(X, \mathscr{L}^*)) \cong H^j(\Gamma_S(X, \mathscr{L}'^*))$. Therefore one can assume that there exists h_i, $i = 0, 1, 2, \ldots$, such that the diagram

$$\begin{array}{ccccccccccccccc}
0 \to \mathscr{F} \to & \mathscr{L}^0 & \xrightarrow{f_0} & \mathscr{L}^1 & \to \cdots \to & \mathscr{L}^{i-1} & \xrightarrow{f_{i-1}} & \mathscr{L}^i & \xrightarrow{f_i} & \mathscr{L}^{i+1} & \to \cdots \\
\| \qquad & \downarrow{h_0} & & \downarrow{h_1} & & \downarrow{h_{i-1}} & & \downarrow{h_i} & & \downarrow{h_{i+1}} & \\
0 \to \mathscr{F} \to & \mathscr{L}'^0 & \xrightarrow{f'_0} & \mathscr{L}'^1 & \to \cdots \to & \mathscr{L}'^{i-1} & \xrightarrow{f'_{i-1}} & \mathscr{L}'^i & \xrightarrow{f'_i} & \mathscr{L}'^{i+1} & \to \cdots
\end{array}$$

is commutative. Then define a morphism $\mathscr{L}^i \oplus \mathscr{L}'^{i-1} \to \mathscr{L}^{i+1} \oplus \mathscr{L}'^i$ by $(x, y) \mapsto (f_i(x), f'_{i-1}(y) + (-1)^i h_i(x))$. Then we have the exact sequence of flabby sheaves, $0 \to \mathscr{L}^0 \to \mathscr{L}^1 \oplus \mathscr{L}'^0 \to \mathscr{L}^2 \oplus \mathscr{L}'^1 \to \cdots$. Furthermore, $0 \to \mathscr{L}'^i \to \mathscr{L}^{i+1} \oplus \mathscr{L}'^i \to \mathscr{L}^{i+1} \to 0$, $i = 0, 1, \ldots$, is exact. Therefore, by Proposition 1.1.2 and its Corollary 2, all the vertical sequences are exact and the second horizontal sequence is exact in the diagram

$$\begin{array}{ccccc}
 & 0 & & 0 & \\
 & \downarrow & & \downarrow & \\
0 \to & 0 & \longrightarrow & \Gamma(X, \mathscr{L}'^0) & \longrightarrow \cdots \\
 & \downarrow & & \downarrow & \\
0 \to & \Gamma_S(X, \mathscr{L}^0) & \to & \Gamma_S(X, \mathscr{L}^1 \oplus \mathscr{L}'^0) & \to \cdots \\
 & \downarrow & & \downarrow & \\
0 \to & \Gamma_S(X, \mathscr{L}^0) & \to & \Gamma_S(X, \mathscr{L}^1) & \longrightarrow \cdots \\
 & \downarrow & & \downarrow & \\
 & 0 & & 0 &
\end{array}$$

Hence we have, by Proposition 1.1.4, the exact sequence

$$0 \to H^0(\Gamma_S(X, \mathscr{L}^*)) \to H^0(\Gamma_S(X, \mathscr{L}'^*)) \to 0 \to$$
$$H^1(\Gamma_S(X, \mathscr{L}^*)) \to H^1(\Gamma_S(X, \mathscr{L}'^*)) \to 0 \to \cdots.$$

Thus, we conclude, $H^j(\Gamma_S(X, \mathscr{L}^*)) \cong H^j(\Gamma_S(X, \mathscr{L}'^*))$.

Definition 1.1.15. *Let $\mathscr{F}$ be a sheaf over a topological space X, and let S be a locally closed subset of X. A sheaf $\mathscr{H}^j_S(\mathscr{F})$, $j = 0, 1, 2, \ldots$, over X is defined as follows*:

$$\mathscr{H}^j_S(\mathscr{F}) = \mathscr{H}^j(\Gamma_S(\mathscr{L}^*)) = \frac{\operatorname{Ker}(\Gamma_S(\mathscr{L}^j) \to \Gamma_S(\mathscr{L}^{j+1}))}{\operatorname{Im}(\Gamma_S(\mathscr{L}^{j-1}) \to \Gamma_S(\mathscr{L}^j))}.$$

The sheaf $\mathscr{H}^j_S(\mathscr{F})$ *is said to be the* jth *derived sheaf of $\mathscr{F}$ with support in S.*

Note. For $x \notin S$ the stalk $\mathscr{H}^j_S(\mathscr{F})_x = 0$. Therefore the sheaf $\mathscr{H}^j_S(\mathscr{F})$ is concentrated on the set S. In view of this, $\mathscr{H}^j_S(\mathscr{F})|_S$ can be denoted simply by $\mathscr{H}^j_S(\mathscr{F})$.

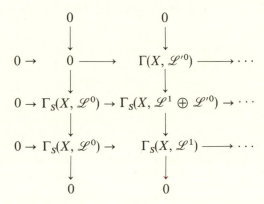

Proposition 1.1.5. *If* $\mathscr{H}^j_S(\mathscr{F}) = 0$ *for* $j < k$, *then* $\mathscr{H}^k_S(\mathscr{F})(U) = H^k_{S\cap U}(U, \mathscr{F})$.

Proof. Let $0 \to \mathscr{F} \to \mathscr{L}^0 \to \mathscr{L}^1 \to \cdots$ be a flabby resolution of $\mathscr{F}$. By hypothesis, the sequence $0 \to \Gamma_S(\mathscr{L}^0) \to \Gamma_S(\mathscr{L}^1) \to \cdots \to \Gamma_S(\mathscr{L}^{k-1}) \to \mathscr{I} \to 0$ is exact, where $\mathscr{I} = \operatorname{Im}(\Gamma_S(\mathscr{L}^{k-1}) \to \Gamma_S(\mathscr{L}^k))$. Generally we note the following. Let $\mathscr{F}_1 \xrightarrow{\varphi} \mathscr{F}_2$ be a morphism of sheaves, and let U be an open set of X. Let $\operatorname{Im}\varphi$ be the sheaf associated to the presheaf $\operatorname{Im}(\mathscr{F}_1(U) \xrightarrow{\varphi(U)} \mathscr{F}_2(U)) = \operatorname{Im}(\varphi(U))$. Then one plainly has $(\operatorname{Im}\varphi)(U) \supset \operatorname{Im}(\varphi(U))$. Hence, in particular, $\mathscr{I}(U) \supset \operatorname{Im}(\Gamma_{S\cap U}(U, \mathscr{L}^{k-1}) \to \Gamma_{S\cap U}(U, \mathscr{L}^k))$. By Proposition 1.1.1, $\Gamma_S(\mathscr{L}^k)$ is a flabby sheaf. Therefore, by Corollary 2 of Proposition 1.1.2, $\mathscr{I}$ is flabby. Then $\Gamma_S(\mathscr{L}^{k-1})(U) = \Gamma_{S\cap U}(U, \mathscr{L}^{k-1}) \to \mathscr{I}(U) \to 0$ is an exact sequence. Hence we have $\mathscr{I}(U) = \operatorname{Im}(\Gamma_{S\cap U}(U, \mathscr{L}^{k-1}) \to \Gamma_{S\cap U}(U, \mathscr{L}^k))$. On the other hand, let $0 \to \mathscr{Z} \to \Gamma_S(\mathscr{L}^k) \to \Gamma_S(\mathscr{L}^{k+1})$ be an exact sequence. Then, by Proposition 1.1.2(1), $\mathscr{Z}(U) = \operatorname{Ker}(\Gamma_{S\cap U}(U, \mathscr{L}^k) \to \Gamma_{S\cap U}(U, \mathscr{L}^{k+1}))$. Therefore, $H^k_{S\cap U}(U, \mathscr{F}) = \mathscr{Z}(U)/\mathscr{I}(U)$. By definition, $0 \to \mathscr{I} \to \mathscr{Z} \to \mathscr{H}^k_S(\mathscr{F}) \to 0$ is exact. Since $\mathscr{I}$ is flabby, the sequence $0 \to \mathscr{I}(U) \to \mathscr{Z}(U) \to \mathscr{H}^k_S(\mathscr{F})(U) \to 0$ is also exact. That is, $\mathscr{H}^k_S(\mathscr{F})(U) = \mathscr{Z}(U)/\mathscr{I}(U) = H^k_{S\cap U}(U, \mathscr{F})$.

Theorem 1.1.2. *Let* $\mathscr{F}$, $\mathscr{F}'$, *and* $\mathscr{F}''$ *be sheaves over a topological space* X; *let* U *be an open set of* X; *let* S *be a locally closed subset of* X; *and let* S' *be a closed subset of* S.

(1) *Suppose* $S \subset U \subset X$ *holds, then we have*

$$H^j_S(X, \mathscr{F}) = H^j_S(U, \mathscr{F}), \qquad j = 0, 1, 2, \ldots.$$

(2) *We have* $H^0_S(X, \mathscr{F}) = \Gamma_S(X, \mathscr{F})$.

(3) *The sequence*

$$0 \to H^0_{S'}(X, \mathscr{F}) \to H^0_S(X, \mathscr{F}) \to H^0_{S-S'}(X, \mathscr{F}) \to H^1_{S'}(X, \mathscr{F}) \to \cdots$$

is exact. In particular, for a closed set Z *of* X, *we have the exact sequence*

$$0 \to H^0_Z(X, \mathscr{F}) \to H^0(X, \mathscr{F}) \to H^0(X - Z, \mathscr{F}) \to H^1_Z(X, \mathscr{F}) \to \cdots.$$

(4) *If* $0 \to \mathscr{F}' \to \mathscr{F} \to \mathscr{F}'' \to 0$ *is an exact sequence of sheaves, we have the induced long sequences* (i) *and* (ii):
(i) $0 \to H^0_S(X, \mathscr{F}') \to H^0_S(X, \mathscr{F}) \to H^0_S(X, \mathscr{F}'') \to H^1_S(X, \mathscr{F}) \to \cdots$
(ii) $0 \to \mathscr{H}^0_S(\mathscr{F}') \to \mathscr{H}^0_S(\mathscr{F}) \to \mathscr{H}^0_S(\mathscr{F}'') \to \mathscr{H}^1_S(\mathscr{F}') \to \cdots$.

Proof.

(1) is plainly true by Definition 1.1.9.

(2) Let $0 \to \mathscr{F} \to \mathscr{L}^0 \to \mathscr{L}^1 \to \cdots$ be a flabby resolution of $\mathscr{F}$. Since $\Gamma_S(X, \cdot)$ is a left exact functor, it follows that $H^0_S(X, \mathscr{F}) = \Gamma_S(X, \mathscr{F})$.

(3) Consider the commutative diagram

$$\begin{array}{ccccc}
 & 0 & & 0 & \\
 & \downarrow & & \downarrow & \\
0 \to & \Gamma_{S'}(U, \mathscr{L}^0) & \longrightarrow & \Gamma_{S'}(U, \mathscr{L}^1) & \longrightarrow \cdots \\
 & \downarrow & & \downarrow & \\
0 \to & \Gamma_S(X, \mathscr{L}^0) & \longrightarrow & \Gamma_S(U, \mathscr{L}^1) & \longrightarrow \cdots \\
 & \downarrow & & \downarrow & \\
0 \to & \Gamma_{S-S'}(U, \mathscr{L}^0) & \to & \Gamma_{S-S'}(U, \mathscr{L}^1) & \to \cdots \\
 & \downarrow & & \downarrow & \\
 & 0 & & 0 &
\end{array}$$

where U is an open set of X so that S is closed in the relative topology of U. Since the vertical sequences are exact, one obtains the long exact sequence induced by taking cohomologies.

(4) Let $\mathscr{C}^*(\mathscr{F}')$, $\mathscr{C}^*(\mathscr{F})$, and $\mathscr{C}^*(\mathscr{F}'')$ be canonical flabby resolutions of $\mathscr{F}'$, $\mathscr{F}$, and $\mathscr{F}''$ respectively. For each open set U we have the sequence

$$0 \to \mathscr{C}^0(\mathscr{F}')(U) \to \mathscr{C}^0(\mathscr{F})(U) \to \mathscr{C}^0(\mathscr{F}'')(U) \to 0$$

which is exact by Proposition 1.1.2. By the Remark that follows Proposition 1.1.2, the first and the second horizontal sequences are exact in the commutative diagram

$$\begin{array}{ccccccc}
 & 0 & & 0 & & 0 & \\
 & \downarrow & & \downarrow & & \downarrow & \\
0 \to & \mathscr{F}' & \longrightarrow & \mathscr{F} & \longrightarrow & \mathscr{F}'' & \longrightarrow 0 \\
 & \downarrow & & \downarrow & & \downarrow & \\
0 \to & \mathscr{C}^0(\mathscr{F}') & \to & \mathscr{C}^0(\mathscr{F}) & \to & \mathscr{C}^0(\mathscr{F}'') & \to 0 \\
 & \downarrow & & \downarrow & & \downarrow & \\
0 \to & \mathscr{Z}^0(\mathscr{F}') & \to & \mathscr{Z}^0(\mathscr{F}) & \to & \mathscr{Z}^0(\mathscr{F}'') & \to 0 \\
 & \downarrow & & \downarrow & & \downarrow & \\
 & 0 & & 0 & & 0 &
\end{array}$$

Therefore, Nine Lemma implies that the third sequence is also exact. By repeated use of this argument, one has the exact sequence, $0 \to \mathscr{C}^n(\mathscr{F}') \to \mathscr{C}^n(\mathscr{F}) \to \mathscr{C}^n(\mathscr{F}'') \to 0$, for each n. Since $\mathscr{C}^n(\mathscr{F}')$ is flabby, by Proposition 1.1.2,

$$0 \to \Gamma_S(X, \mathscr{C}^n(\mathscr{F}')) \to \Gamma_S(X, \mathscr{C}^n(\mathscr{F})) \to \Gamma_S(X, \mathscr{C}^n(\mathscr{F}'')) \to 0$$

is an exact sequence for each n.

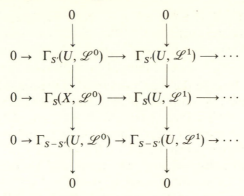

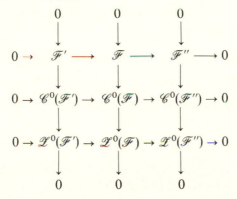

All the vertical sequences are exact in the commutative diagram

$$\begin{array}{ccccc}
 & 0 & & 0 & \\
 & \downarrow & & \downarrow & \\
0 \to & \Gamma_S(X, \mathscr{C}^0(\mathscr{F}')) & \to & \Gamma_S(X, \mathscr{C}^1(\mathscr{F}')) & \to \cdots \\
 & \downarrow & & \downarrow & \\
0 \to & \Gamma_S(X, \mathscr{C}^0(\mathscr{F})) & \to & \Gamma_S(X, \mathscr{C}^1(\mathscr{F})) & \to \cdots \\
 & \downarrow & & \downarrow & \\
0 \to & \Gamma_S(X, \mathscr{C}^0(\mathscr{F}'')) & \to & \Gamma_S(X, \mathscr{C}^1(\mathscr{F}'')) & \to \cdots \\
 & \downarrow & & \downarrow & \\
 & 0 & & 0 &
\end{array}$$

Therefore one obtains (4.i). Since $\mathscr{H}^j_S(\mathscr{F})_x = \varinjlim_{x \in U} H^j_{S \cap U}(U, \mathscr{F})$ and $\varinjlim$ is an exact functor, (4.ii) follows immediately from (4.i).

Definition 1.1.16. *A sheaf $\mathscr{F}$ is said to be of flabby dimension $\leqq r$, denoted by flabby dim $\mathscr{F} \leqq r$, if there exists an exact sequence*

$$0 \to \mathscr{F} \to \mathscr{L}^0 \to \mathscr{L}^1 \to \cdots \to \mathscr{L}^r \to 0$$

such that each $\mathscr{L}^i$, $i = 0, 1, \ldots, r$, is a flabby sheaf.

Note. Observe that $\mathscr{F}$ is a flabby sheaf if and only if flabby dim $\mathscr{F} \leqq 0$. When flabby dim $\mathscr{F} \leqq r$ holds, we sometimes say that $\mathscr{F}$ has a flabby resolution of length r.

Theorem 1.1.3. *Let $\mathscr{F}$ be a sheaf over a topological space X. Then the following statements are equivalent:*

(1) flabby dim $\mathscr{F} \leqq r$

(2) $H^{r+1}_S(X, \mathscr{F}) = 0$ *for an arbitrary closed set S in X*

(2′) $H^j_S(X, \mathscr{F}) = 0$ *for $j > r$ and an arbitrary, locally closed subset S in X*

(3) $\mathscr{H}^{r+1}_S(\mathscr{F}) = 0$ *for an arbitrary closed subset S in X*

(4) $H^r(X, \mathscr{F}) \to H^r(U, \mathscr{F})$ *is an epimorphism for any open set U of X.*

Proof. By definitions, (1) → (2′) → (2) and (1) → (3) follow plainly. We will show (2) → (4) → (1) and (3) → (1).

(2) → (4): Let $S = X$ and let $S' = X - U$ in (3) of Theorem 1.1.2. Then the long exact sequence becomes

$$\cdots \to H^r(X, \mathscr{F}) \to H^r(U, \mathscr{F}) \to H^{r+1}_{X-U}(X, \mathscr{F}) \to \cdots.$$

The third term $H^{r+1}_{X-U}(X, \mathscr{F}) = 0$ by (2). Therefore one concludes (4).

(4) → (1): Take a flabby resolution of $\mathscr{F}$, $0 \to \mathscr{F} \to \mathscr{L}^0 \xrightarrow{f_0} \mathscr{L}^1 \xrightarrow{f_1} \cdots \to \mathscr{L}^{r-1} \xrightarrow{f_{r-1}} \mathscr{L}^r \to \cdots$. Let $\mathscr{G} = \operatorname{Im} f_{r-1}$. Then the sequence, $0 \to \mathscr{F} \to \mathscr{L}^0 \to \mathscr{L}^1 \to \cdots \to \mathscr{L}^{r-1} \to \mathscr{G} \to 0$, is exact. It suffices to prove that $\mathscr{G}$ is a flabby sheaf. Notice that $0 \to \mathscr{G} \to \mathscr{L}^r \to \mathscr{L}^{r+1} \to \cdots$ is an exact sequence. Therefore

$$H^r(U, \mathscr{F}) = \frac{\operatorname{Ker}(\mathscr{L}^r(U) \to \mathscr{L}^{r+1}(U))}{\operatorname{Im}(\mathscr{L}^{r-1}(U) \to \mathscr{L}^r(U))} = \frac{\mathscr{G}(U)}{f_{r-1}(U)\mathscr{L}^{r-1}(U)}.$$

Statement (4) implies that

$$\frac{\mathscr{G}(X)}{f_{r-1}(X)\mathscr{L}^{r-1}(X)} \to \frac{\mathscr{G}(U)}{f_{r-1}(U)\mathscr{L}^{r-1}(U)}$$

is an epimorphism. Hence, for $u \in \mathscr{G}(U)$ there exist $\tilde{u} \in \mathscr{G}(X)$ and $s \in \mathscr{L}^{r-1}(U)$ such that $u = \tilde{u}|_U + f_{r-1}(U)s$. Since $\mathscr{L}^{r-1}$ is flabby, one can find $\tilde{s} \in \mathscr{L}^{r-1}(X)$ so that $f_{r-1}(U)s = f_{r-1}(U)(\tilde{s}|_U) = f_{r-1}(X)\tilde{s}|_U$. Then $u = (\tilde{u} + f(X)\tilde{s})|_U$ holds; i.e. $\mathscr{G}$ is a flabby sheaf.

(3) → (1): We will give a proof by induction on r. First, when $r = 0$, one must show that $\mathscr{F}$ is a flabby sheaf provided that $\mathscr{H}^1_S(\mathscr{F}) = 0$. Let U_0 be an open set in X, and let $s_0 \in \mathscr{F}(U_0)$. Define an order relation $\succ$ in the set $\mathscr{M} = \{(U', s') \mid U' \supset U_0, s' \in \mathscr{F}(U'), s|_{U_0} = s_0\}$, where $(U, s) \succ (U', s')$ if and only if $U \supset U'$ and $s \mid U' = s'$. Then $\mathscr{M}$ is an inductively ordered set. By Zorn's lemma, there exists a maximal element (U_1, s_1). We claim $U_1 = X$. Suppose $U_1 \neq X$, and then let $x \in X - U_1$. For the closed set $S = X - U_1$, we have $\mathscr{H}^1_S(\mathscr{F})_x = \varinjlim_{x \in V} H^1_{S \cap V}(V, \mathscr{F}) = 0$ by the hypothesis. By (3) in Theorem 1.1.2, for any open set V containing x, the sequence

$$0 \to H^0_{S \cap V}(V, \mathscr{F}) \to H^0(V, \mathscr{F}) \to H^0(V - S, \mathscr{F}) \to H^1_{S \cap V}(V, \mathscr{F})$$

is exact. Taking the inductive limit of each term in this exact sequence, there exists $\tilde{s} \in H^0(V, \mathscr{F}) = \mathscr{F}(V)$ so that $\tilde{s}|_{(V-S)} = s_1|_{V \cap U_1}$ for a sufficiently small V. Therefore, if one lets $s_2|_V = \tilde{s}$ and $s_2|_{U_1} = s_1$, then $(U_1 \cup V, s_2) \in \mathscr{M}$. This contradicts the maximality of (U_1, s_1). Hence one has $s_1 \in \mathscr{F}(X)$ and $s_1|_{U_0} = s_0$, which shows the flabbiness of $\mathscr{F}$.

Next, suppose the case $r = r_0$ is true; then we will prove the statement (1) for $r = r_0 + 1$. Let $\mathscr{L}^0$ be a flabby sheaf such that the sequence, $0 \to \mathscr{F} \to \mathscr{L}^0 \to \mathscr{G} \to 0$ is exact. Then, by (4) in Theorem 1.1.2, the sequence

$$0 = \mathscr{H}^{r_0}_S(\mathscr{L}^0) \to \mathscr{H}^{r_0}_S(\mathscr{G}) \to \mathscr{H}^{r_0+1}_S(\mathscr{F}) \to \mathscr{H}^{r_0+1}_S(\mathscr{L}^0) = 0$$

is exact. Therefore $\mathscr{H}^{r_0+1}_S(\mathscr{F}) = 0$ implies $\mathscr{H}^{r_0}_S(\mathscr{G}) = 0$. Then, by the inductive assumption, one has a flabby resolution

$$0 \to \mathscr{G} \to \mathscr{L}^1 \to \mathscr{L}^2 \to \cdots \to \mathscr{L}^{r_0+1} \to 0.$$

Therefore $\mathscr{F}$ has the flabby resolution of length $r_0 + 1$

$$0 \to \mathscr{F} \to \mathscr{L}^0 \to \mathscr{L}^1 \to \cdots \to \mathscr{L}^{r_0+1} \to 0.$$

Definition 1.1.17. *Let $\mathscr{F}$ be a sheaf over a topological space X, and let S be a closed subset of X. Then S is said to be purely r-codimensional with respect to $\mathscr{F}$ if*

$$\mathscr{H}^j_S(\mathscr{F}) = 0 \qquad \text{for } j \neq r.$$

Definition 1.1.18. *Let X be a topological space, and let A be an additive group. For an open subset U of X, consider the presheaf $A_X(U) = \{$mappings from U to A which are locally constant, i.e. all the continuous mappings for the discrete topology in $A\}$. The sheaf A_X associated to this presheaf is called the constant sheaf. Notice that for each $x \in X$ one has $(A_X)_x = A$.*

Definition 1.1.19. *Let $\mathscr{F}$ and $\mathscr{G}$ be sheaves over a topological space X, and let U be an open subset of X. We denote the sheaf associated to the presheaf $\mathscr{F}(U) \underset{\mathbf{Z}}{\otimes} \mathscr{G}(U)$ by $\mathscr{F} \underset{\mathbf{Z}}{\otimes} \mathscr{G}$. Then we have $(\mathscr{F} \underset{\mathbf{Z}}{\otimes} \mathscr{G})_x = \mathscr{F}_x \underset{\mathbf{Z}}{\otimes} \mathscr{G}_x$. If $\mathscr{F}$ and $\mathscr{G}$ are sheaves of vector spaces over $\mathbf{C}$, $\mathscr{F} \underset{\mathbf{C}}{\otimes} \mathscr{G}$ is defined similarly. Also note that $(\mathscr{F} \underset{\mathbf{C}}{\otimes} \mathscr{G})_x = \mathscr{F}_x \underset{\mathbf{C}}{\otimes} \mathscr{G}_x$. We may denote $\underset{\mathbf{Z}}{\otimes}$ and $\underset{\mathbf{C}}{\otimes}$ simply by $\otimes$ when there is no fear of confusion.*

We are now ready to define the sheaf $\mathscr{B}_M$ of hyperfunctions on a real analytic manifold M. We begin with preliminary notions.

Definition 1.1.20. *Let M be an n-dimensional real analytic manifold. If X is a complex manifold of dimension n containing M such that locally $M \cong \mathbf{R}^n \hookrightarrow \mathbf{C}^n \cong X$, then X is said to be a complexification of M. That is, there exists a neighborhood Ω of each point $x \in X$ and an injective holomorphic map $f: \Omega \to \mathbf{C}^n$ such that $\Omega \cap M = f^{-1}(\mathbf{R}^n)$. Equivalently, M is a real analytic submanifold of X such that, for $x \in M$, $T_xX = T_xM \oplus \sqrt{-1}T_xM$ holds, where we denote the tangent vector space at $x \in M$ by T_xM.*

Remark. If M is paracompact, then the complexification X is unique in the following sense: suppose X_1 and X_2 are complexifications of M; then there exists a complexification X_3 of M such that X_3 is an open subset in X_1 and X_2 (see Bruhat and Whitney [1]).

Definition 1.1.21. *Let $\mathbf{Z}_X$ be the constant sheaf on X such that each stalk is $\mathbf{Z}$, with $\mathbf{Z}$ being the ring of rational integers. The orientation sheaf ω_M over an n-dimensional (real analytic) manifold M is defined as $\mathscr{H}^n_M(\mathbf{Z}_X)$. If an open subset U is oriented, then $\omega_M(U) \cong \mathbf{Z}_M(U)$; see §2. The sign in this isomorphism depends upon the orientation on M. Note that giving a section of ω_M is equivalent to giving an orientation.*

Definition 1.1.22. *Let X be a complex manifold. We denote the sheaf of holomorphic functions on X by $\mathcal{O}_X$. That is, for an open subset U of X, it is the sheaf associated to the presheaf* $\mathcal{O}(U) = \{$holomorphic functions defined on $U\}$.

Definition 1.1.23. *Let M be an n-dimensional real analytic manifold; let X be a complexification of M; let $\mathcal{O}_X$ be the sheaf of holomorphic functions on M; and let ω_M be the orientation sheaf on M. Then we define the sheaf $\mathcal{B}_M$ of hyperfunctions on M by $\mathcal{B}_M = \mathcal{H}^n_M(\mathcal{O}_X) \otimes_{\mathbf{Z}} \omega_M$, and the sections of $\mathcal{B}_M$ are called hyperfunctions.*

In the case when M is oriented one has $\omega_M = \mathbf{Z}_M$. Therefore $\mathcal{B}_M = \mathcal{H}^n_M(\mathcal{O}_X) \otimes \omega_M = \mathcal{H}^n_M(\mathcal{O}_X)$. Let the open set X_U of X contain $U \subset M$ as a closed set. Then the sections of $\mathcal{B}_M$ over U can be written as $H^n_U(X_U, \mathcal{O}_X)$, by Proposition 1.1.5, provided that M is purely n-codimensional; see Theorem 2.2.1 for proof. In the next section we will examine $H^n_U(X_U, \mathcal{O}_X)$ via Cech cohomology theory. Using this theory, one can write hyperfunctions as sums of boundary values of holomorphic functions.

We will close this section with the proof of the flabbiness of the sheaf $\mathcal{B}_M$. We begin with a generalization of Proposition 1.1.5.

Proposition 1.1.6. *Let $\mathcal{F}$ be a sheaf over a topological space X, and let S and Z be closed sets of X such that $Z \subset S$. Suppose S is purely k-codimensional with respect to $\mathcal{F}$, i.e. $\mathcal{H}^j_S(\mathcal{F}) = 0$ for $j \neq k$; then*

$$H^j_Z(X, \mathcal{H}^k_S(\mathcal{F})) = H^{j+k}_Z(X, \mathcal{F})$$

holds for $j = 0, 1, 2, \ldots$.

Proof. First notice that one has $\Gamma_Z(X, \Gamma_S(\mathcal{L}^i)) = \Gamma_Z(X, \mathcal{L}^i)$, which is the case $j = 0$. Notations being the same as in Proposition 1.1.5, since the sequence $0 \to \mathcal{I} \to \mathcal{Z} \to \mathcal{H}^k_S(\mathcal{F}) \to 0$ is exact and $\mathcal{I}$ is flabby, together with (4) in Theorem 1.1.2, one obtains the exact sequence $0 = H^1_Z(X, \mathcal{I}) \to H^1_Z(X, \mathcal{Z}) \to H^1_Z(X, \mathcal{H}^k_S(\mathcal{F})) \to H^2_Z(X, \mathcal{I}) = 0 \to \cdots$. Hence $H^j_Z(X, \mathcal{H}^k_S(\mathcal{F})) = H^j_Z(X, \mathcal{Z})$ holds for $j \geqq 1$. On the other hand, by the assumption, for $j > k$, $H^j_S(\mathcal{F}) = 0$ holds. Therefore,

$$0 \to \mathcal{Z} \to \Gamma_S(\mathcal{L}^k) \to \Gamma_S(\mathcal{L}^{k+1}) \to \cdots$$

is a flabby resolution of $\mathcal{Z}$. Then one has

$$H^j_Z(X, \mathcal{Z}) = \frac{\operatorname{Ker}(\Gamma_Z(X, \mathcal{L}^{j+k}) \to \Gamma_Z(X, \mathcal{L}^{j+k+1}))}{\operatorname{Im}(\Gamma_Z(X, \mathcal{L}^{j+k-1}) \to \Gamma_Z(X, \mathcal{L}^{j+k}))} = H^{j+k}_Z(X, \mathcal{F}).$$

Therefore $H^j_Z(X, \mathcal{H}^k_S(\mathcal{F})) = H^{j+k}_Z(X, \mathcal{F})$ holds for $j = 0, 1, 2, \ldots$.

We quote a fundamental theorem from the theory of holomorphic functions of several variables.

Theorem 1.1.4 (Malgrange [1]). *Let X be a complex manifold, and let $\mathcal{O}_X$ be the sheaf of holomorphic functions on X. Then*

$$\text{flabby dim } \mathcal{O}_X \leqq \dim X.$$

Note. In fact we have the equality flabby dim $\mathcal{O}_X = \dim X$, as will be seen in §2.

Proposition 1.1.6 and the fact that an n-dimensional real analytic manifold M is purely n-codimensional with respect to $\mathcal{O}_X$ provide $H^1_Z(X, \mathcal{H}^n_M(\mathcal{O}_X)) = H^{n+1}_Z(X, \mathcal{O}_X)$ for any closed set Z of M. Theorem 1.1.3 and Theorem 1.1.4 imply $H^1_Z(X, \mathcal{H}^n_M(\mathcal{O}_X)) = H^{n+1}_Z(X, \mathcal{O}_X) = 0$. Hence, again by Theorem 1.1.3, one concludes that flabby dim $\mathcal{H}^n_M(\mathcal{O}_X) \leqq 0$; i.e. $\mathcal{H}^n_M(\mathcal{O}_X)$ is a flabby sheaf. Let $M = \bigcup_{i \in I} U_i$ be an open covering, where U_i is an orientable open set; then $\mathcal{B}_M|_{U_i} = \mathcal{H}^n_M(\mathcal{O}_X)|_{U_i}$, as we noted before. In general, if $\mathcal{F}$ is a sheaf on M and if $M = \bigcup_{i \in I} U_i$ is an open covering, then $\mathcal{F}|_{U_i}$ being flabby for each $i \in I$ is equivalent to $\mathcal{F}$ being a flabby sheaf, by Theorem 1.1.3. Because, by Theorem 1.1.3, flabby dim $\mathcal{F} \leqq r$ and the triviality of $\mathcal{H}^{r+1}_S(\mathcal{F})$ for an arbitrary closed S are equivalent statements. Then note that the triviality of $\mathcal{H}^{r+1}_S(\mathcal{F})$ is a local property, since $\mathcal{H}^{r+1}_S(\mathcal{F})$ is a sheaf. That is, $\mathcal{H}^{r+1}_S(\mathcal{F}) = 0$ if and only if $\mathcal{H}^{r+1}_S(\mathcal{F})|_{U_i} = 0$ for each U_i.

Hence we now conclude the following theorem.

Theorem 1.1.5. *The sheaf $\mathcal{B}_M$ of hyperfunctions on a real analytic manifold M is a flabby sheaf.*

This theorem, together with the general theory of systems of linear differential equations, provides us with explicit flabby resolutions of the constant sheaf $\mathbf{C}_X$ and the sheaf $\mathcal{O}_X$ of holomorphic functions via differential operators (see Komatsu [1]). This is an important and interesting result from the analyst's point of view.

Exercise 1. The presheaf of all the continuous and bounded functions on each open set of $\mathbf{R}^n$ is not a sheaf. Find the sheaf associated to this presheaf.

Exercise 2. Consider the presheaf of all the real-valued, locally constant functions on each open set of $\mathbf{R}^n$ modulo constant functions on $\mathbf{R}^n$. This presheaf is not a sheaf. Find the sheaf associated to the presheaf.

Exercise 3. Let X and Y be topological spaces, and let $f: X \to Y$ be a continuous map.

(1) Suppose $0 \to \mathcal{F}' \to \mathcal{F} \to \mathcal{F}'' \to 0$ is an exact sequence of sheaves; then prove that $0 \to f_*\mathcal{F}' \to f_*\mathcal{F} \to f_*\mathcal{F}''$ is exact.

(2) Suppose $\mathscr{F}$ is a flabby sheaf over X; then prove that $f_*\mathscr{F}$ is a flabby sheaf over Y.

(3) Suppose $\mathscr{G}' \to \mathscr{G} \to \mathscr{G}''$ is an exact sequence of sheaves over Y; then prove that $f^{-1}\mathscr{G}' \to f^{-1}\mathscr{G} \to f^{-1}\mathscr{G}''$ is exact over X.

Exercise 4. Suppose that, in the commutative diagram below, all the vertical sequences are exact and that the second and third horizontal sequences are exact. Then prove that there are induced maps $A' \to A \to A''$ and $C' \to C \to C''$ such that they are both exact.

$$\begin{array}{ccccccc}
 & & 0 & & 0 & & 0 \\
 & & \downarrow & & \downarrow & & \downarrow \\
 & & A' & & A & & A'' \\
 & & \downarrow & & \downarrow & & \downarrow \\
 & & K' & \to & K & \to & K'' \to 0 \\
 & & \downarrow & & \downarrow & & \downarrow \\
0 & \to & L' & \to & L & \to & L'' \\
 & & \downarrow & & \downarrow & & \downarrow \\
 & & C' & & C & & C'' \\
 & & \downarrow & & \downarrow & & \downarrow \\
 & & 0 & & 0 & & 0
\end{array}$$

§2. Hyperfunctions as Boundary Values of Holomorphic Functions

We will examine the orientation sheaf $\omega_M = \mathscr{H}^n_M(\mathbf{Z}_X)$ and the sheaf of hyperfunctions $\mathscr{B}_M = \mathscr{H}^n_M(\mathscr{O}_X) \otimes \omega_M$ via the Čech cohomology theory. The Čech theory is useful for the explicit presentation of hyperfunctions. In fact, M. Sato seems to have been "naturally" led to the notion of relative cohomology groups expressed in terms of covering, i.e. Čech cohomology groups—independently of Grothendieck—when he tried to find the correct formulation of the idea that a hyperfunction of several variables is a tensor product of hyperfunctions of one variable.

Let X be a topological space, and let Z be a closed subset of X. Suppose $X = \bigcup_{\lambda \in \Lambda} U_\lambda$ and $X - Z = \bigcup_{\lambda \in \Lambda'} U_\lambda$, where $\Lambda' \subset \Lambda$, are open coverings of X and $X - Z$. We will denote these open coverings by $\mathscr{U} = \{U_\lambda\}_{\lambda \in \Lambda}$ and $\mathscr{U}' = \{U_\lambda\}_{\lambda \in \Lambda'}$ respectively. For a sheaf $\mathscr{F}$ over X the Čech cohomology groups are defined as follows. First define

$$C^k(\mathscr{U}, \mathscr{F}) = \{\varphi = \{\varphi_{\lambda_0, \ldots, \lambda_k}\}_{(\lambda_0, \ldots, \lambda_k) \in \Lambda^{k+1}} \mid \varphi_{\lambda_0, \ldots, \lambda_k} \in \Gamma(U_{\lambda_0} \cap \cdots \cap U_{\lambda_k}, \mathscr{F}) \text{ and}$$
$$\varphi_{\lambda_0, \ldots, \lambda_k} = -\varphi_{\lambda_0, \ldots, \lambda_{i+1}, \lambda_i, \ldots \lambda_k}\}.$$

Then notice $C^k(\mathscr{U}, \mathscr{F}) \subset \bigoplus_{(\lambda_0, \ldots, \lambda_k) \in \Lambda^{k+1}} \Gamma(U_{\lambda_0} \cap \cdots \cap U_{\lambda_k}, \mathscr{F})$. Furthermore we define

$$C^k(\mathscr{U} \bmod \mathscr{U}', \mathscr{F}) = \{\varphi = \{\varphi_{\lambda_0, \ldots, \lambda_k}\}_{(\lambda_0, \ldots, \lambda_k) \in \Lambda^{k+1}} \in C^k(\mathscr{U}, \mathscr{F}) | \text{ if } (\lambda_0, \ldots, \lambda_k) \in \Lambda'^{k+1}, \text{ then } \varphi_{\lambda_0, \ldots, \lambda_k} = 0\}.$$

Then define a map $\delta: C^k(\mathscr{U}, \mathscr{F}) \to C^{k+1}(\mathscr{U}, \mathscr{F})$ by

$$\delta(\{\varphi_{\lambda_0, \ldots, \lambda_k}\}) = \{\psi_{\lambda_0, \ldots, \lambda_{k+1}}\}_{(\lambda_0, \ldots, \lambda_{k+1}) \in \Lambda^{k+2}},$$

where

$$\psi_{\lambda_0, \ldots, \lambda_{k+1}} = \sum_{i=0}^{k+1} (-1)^i \varphi_{\lambda_0, \ldots, \lambda_{i-1}, \lambda_{i+1}, \lambda_{k+1}}|_{U_{\lambda_0} \cap \cdots \cap U_{\lambda_{k+1}}}.$$

Note that one has $\delta^2 = 0$. Hence there is induced a cochain complex

$$0 \to C^0(\mathscr{U}, \mathscr{F}) \xrightarrow{\delta} C^1(\mathscr{U}, \mathscr{F}) \to \cdots.$$

Definition 1.2.1. *The* kth *Čech cohomology groups,* $H^k(\mathscr{U}, \mathscr{F})$ *and* $H^k(\mathscr{U} \bmod \mathscr{U}', \mathscr{F})$, *are defined as*

$$H^k(\mathscr{U}, \mathscr{F}) = H^k(C^*(\mathscr{U}, \mathscr{F})) = \frac{\operatorname{Ker}(C^k(\mathscr{U}, \mathscr{F}) \to C^{k+1}(\mathscr{U}, \mathscr{F}))}{\operatorname{Im}(C^{k-1}(\mathscr{U}, \mathscr{F}) \to C^k(\mathscr{U}, \mathscr{F}))}$$

and

$$H^k(\mathscr{U} \bmod \mathscr{U}', \mathscr{F}) = H^k(C^*(\mathscr{U} \bmod \mathscr{U}', \mathscr{F})) = \frac{\operatorname{Ker}(C^k(\mathscr{U} \bmod \mathscr{U}', \mathscr{F}) \to C^{k+1}(\mathscr{U} \bmod \mathscr{U}', \mathscr{F}))}{\operatorname{Im}(C^{k-1}(\mathscr{U} \bmod \mathscr{U}', \mathscr{F}) \to C^k(\mathscr{U} \bmod \mathscr{U}', \mathscr{F}))}.$$

For each $k \geqq 0$ one has the exact sequence

$$0 \to C^k(\mathscr{U} \bmod \mathscr{U}', \mathscr{F}) \to C^k(\mathscr{U}, \mathscr{F}) \to C^k(\mathscr{U}'; \mathscr{F}) \to 0.$$

Therefore the long exact sequence

$$0 \to H^0(\mathscr{U} \bmod \mathscr{U}', \mathscr{F}) \to H^0(\mathscr{U}, \mathscr{F}) \to H^0(\mathscr{U}', \mathscr{F}) \to H^1(\mathscr{U} \bmod \mathscr{U}', \mathscr{F}) \to \cdots$$

is induced.

Next we will prove

$$H_Z^k(X, \mathscr{F}) \cong H^k(\mathscr{U} \bmod \mathscr{U}', \mathscr{F}),$$

provided that for each $k \geqq 0$ and arbitrary $\lambda_0, \ldots, \lambda_r$

$$H^k(U_{\lambda_0} \cap \cdots \cap U_{\lambda_r}, \mathscr{F}) = 0$$

By this theorem of Leray, one can present explicitly the sections of the hyperfunction shear $\mathscr{B}_M$. We will begin with lemmas.

Lemma 1. $H^0(\mathscr{U} \bmod \mathscr{U}', \mathscr{F}) = \Gamma_Z(X, \mathscr{F})$.

Proof. By definition one has

$$H^0(\mathscr{U} \bmod \mathscr{U}', \mathscr{F}) = \{\{\varphi_\lambda\}_{\lambda \in \Lambda} \in C^0(\mathscr{U} \bmod \mathscr{U}', \mathscr{F}) | \delta(\{\varphi_\lambda\}_{\lambda \in \Lambda}) = 0\},$$

where $\delta(\{\varphi_\lambda\}_{\lambda \in \Lambda}) = \{(\varphi_\mu|_{U_\lambda \cap U_\mu} - \varphi_\lambda|_{U_\lambda \cap U_\mu})\}_{(\lambda,\mu) \in \Lambda^2}$. Therefore, by the definition of sheaf, there exists a unique $\varphi \in \Gamma(X, \mathscr{F})$ such that $\varphi|_{U_\lambda} = \varphi_\lambda$. Since $\{\varphi_\lambda\}_{\lambda \in \Lambda} \in C^0(\mathscr{U} \bmod \mathscr{U}', \mathscr{F})$, $\varphi|_{(X-Z)} = 0$ holds. This implies $\varphi \in \Gamma_Z(X, \mathscr{F})$. Conversely, if φ belongs to $\Gamma_Z(X, \mathscr{F})$, then clearly $\varphi \in H^0(\mathscr{U} \bmod \mathscr{U}', \mathscr{F})$.

Lemma 2. *If $\mathscr{F}$ is a flabby sheaf, then $H^k(\mathscr{U} \bmod \mathscr{U}', \mathscr{F}) = 0$ holds for any integer $k > 0$.*

Proof. In Theorem 1.1.2, (2) and (3) imply that the sequence

$$0 \to \Gamma_Z(X, \mathscr{F}) \to \Gamma(X, \mathscr{F}) \to \Gamma(X - Z, \mathscr{F}) \to H^1_Z(X, \mathscr{F})$$

is exact. Furthermore, $H^1_Z(X, \mathscr{F}) = 0$ since $\mathscr{F}$ is flabby by Theorem 1.1.3. Lemma 1 implies that

$$\begin{aligned} 0 &\to \Gamma_Z(X, \mathscr{F}) \to \Gamma(X, \mathscr{F}) \to \Gamma(X - Z, \mathscr{F}) \to H^1(\mathscr{U} \bmod \mathscr{U}', \mathscr{F}) \\ &\to H^1(\mathscr{U}, \mathscr{F}) \to H^1(\mathscr{U}', F) \to H^2(\mathscr{U} \bmod \mathscr{U}', \mathscr{F}) \to \cdots \end{aligned}$$

is an exact sequence. As we will show below, $H^1(\mathscr{U}, \mathscr{F}) = 0$; then $H^1(\mathscr{U} \bmod \mathscr{U}', \mathscr{F}) = 0$ is true. For the case $k \geqq 2$, $H^k(\mathscr{U} \bmod \mathscr{U}', \mathscr{F}) = 0$ follows simply from $H^k(\mathscr{U}, \mathscr{F}) = 0$, and $H^k(\mathscr{U}', \mathscr{F}) = 0$ for $k \geqq 1$. It suffices to prove that $H^k(\mathscr{U}, \mathscr{F})$ and therefore $H^k(\mathscr{U}', \mathscr{F})$ vanish for $k \geqq 1$. Let $f = \{f_{\lambda_0, \ldots, \lambda_k}\} \in C^k(\mathscr{U}, \mathscr{F})$ such that $\delta f = 0$. Consider the set $\mathscr{M} = \{(g, U) | U$ is an open set in X, $g \in C^{k-1}(\mathscr{U} \cap U, \mathscr{F})$ such that $\delta g = f|_U$, where $\mathscr{U} \cap U = \{U_\lambda \cap U\}$ and $f|_U$ is the image under the map $C^k(\mathscr{U}, \mathscr{F}) \to C^k(\mathscr{U} \cap U, \mathscr{F})\}$. First of all, $\mathscr{M} \neq \varnothing$ will be shown. If one lets $h_{\lambda_0, \ldots, \lambda_{k-1}} = f_{\lambda, \lambda_0, \ldots, \lambda_{k-1}} \in \Gamma(U_\lambda \cap U_{\lambda_0} \cap \cdots \cap U_{\lambda_{k-1}}, \mathscr{F})$, i.e. $h = \{h_{\lambda_0, \ldots, \lambda_{k-1}}\}_{(\lambda_0, \ldots, \lambda_{k-1}) \in \Lambda^k} \in C^{k-1}(\mathscr{U} \cap U_\lambda, \mathscr{F})$, then $(\delta h)_{\lambda_0, \ldots, \lambda_k} = \sum_{i=0}^{k} (-1)^i h_{\lambda_0, \ldots, \lambda_{i-1}, \lambda_{i+1}, \ldots, \lambda_k} = \sum (-1)^i f_{\lambda, \lambda_0, \ldots, \lambda_{i-1}, \lambda_{i+1}, \ldots, \lambda_k}$. By the assumption, one has $(\delta f)_{\lambda, \lambda_0, \ldots, \lambda_k} = f_{\lambda_0, \ldots, \lambda_k} - \sum_{i=0}^{k} (-1)^i \times f_{\lambda, \lambda_0, \ldots, \lambda_{i-1}, \lambda_{i+1}, \ldots, \lambda_k} = 0$. Therefore $\delta h = f|_{U_\lambda}$, i.e. $(h, U_\lambda) \in \mathscr{M}$. So $\mathscr{M} \neq \varnothing$. Next we will define an order relation in $\mathscr{M}$. Define $(g_1, U_1) \succ (g_2, U_2)$ if and only if $U_1 \supset U_2$ and $g_1|_{U_2} = g_2$. Then $\mathscr{M}$ is an inductively ordered set. Therefore there exists a maximal element in $\mathscr{M}$ by Zorn's lemma. Let (g, U) be a maximal element; then one is to show $U = X$. Suppose $U \neq X$, and let $x \in X - U$. Since $X = \bigcup_{\lambda \in \Lambda} U_\lambda$, there exists U_λ such that $x \in U_\lambda$. Then, as it was described before, $(h, U_\lambda) \in \mathscr{M}$. Therefore one has $\delta h|_{U \cap U_\lambda} = f|_{U \cap U_\lambda} = \delta g|_{U \cap U_\lambda}$. Then $\delta(h|_{U \cap U_\lambda} - g|_{U \cap U_\lambda}) = 0$. Note

$h|_{U\cap U_\lambda} - g|_{U\cap U_\lambda} \in C^{k-1}(\mathscr{U} \cap U \cap U_\lambda, \mathscr{F})$. When $k = 1$, by letting $Z = X$ in Lemma 1, $h|_{U\cap U_\lambda} - g|_{U\cap U_\lambda} \in H^0(\mathscr{U} \cap U \cap U_\lambda, \mathscr{F}) = \Gamma(U \cap U_\lambda, \mathscr{F})$. Since $\mathscr{F}$ is a flabby sheaf, there exists $s \in \Gamma(U_\lambda, \mathscr{F})$ such that $s|_{U\cap U_\lambda} = h|_{U\cap U_\lambda} - g|_{U\cap U_\lambda}$. Let $g'|_U = g$ and $g'|_{U_\lambda} = h - s$; then $\delta(h - s) = \delta h = f|_{U_\lambda}$. Therefore $(g', U \cup U_\lambda) \in \mathscr{M}$, which contradicts the choice of (g, U) in $\mathscr{M}$. Hence $X = U$. When $k > 1$, we give a proof by induction on k. The inductive assumption $H^{k-1}(\mathscr{U} \cap U \cap U_\lambda, \mathscr{F}) = 0$ implies that there exists $s \in C^{k-2}(\mathscr{U} \cap U \cap U_\lambda, \mathscr{F})$ such that $\delta s = h|_{U\cap U_\lambda} - g|_{U\cap U_\lambda}$. The flabbiness of $\mathscr{F}$ implies that there exists $\tilde{s} \in C^{k-2}(\mathscr{U} \cap U_\lambda, \mathscr{F})$ such that $\tilde{s}|_{U\cap U_\lambda} = s$. Define $g' \in C^{k-1}(\mathscr{U} \cap (U \cap U_\lambda), \mathscr{F})$ to be $g'|_U = g$ and $g'|_{U_\lambda} = h - \delta\tilde{s}$. Then $g|_{U\cap U_\lambda} = (h - \delta\tilde{s})|_{U\cap U_\lambda}$. Therefore g' is well defined and $\delta g'|_U = \delta g|_U = f|_U$ and $\delta g'|_{U_\lambda} = \delta(h - \delta\tilde{s})|_{U_\lambda} = \delta h|_{U_\lambda} = f|_{U_\lambda}$. Then $(g', U \cup U_\lambda) \in \mathscr{M}$, contradicting the maximality of (g, U). Hence one has $g \in C^{k-1}(\mathscr{U}, \mathscr{F})$ such that $\delta g = f$.

We are now ready to prove the theorem of Leray which is fundamental to our theory.

Theorem 1.2.1 (Leray). *Suppose* $H^j(U_{\lambda_0} \cap \cdots \cap U_{\lambda_r}, \mathscr{F}) = 0$ *for an arbitrary integer* $j > 0$ *and arbitrary* $\lambda_0, \ldots, \lambda_r$. *Then*

$$H^k(\mathscr{U} \bmod \mathscr{U}', \mathscr{F}) = H^k_Z(X, \mathscr{F}).$$

Remark. The covering that satisfies the condition of this theorem is sometimes called a Leray covering.

Proof. Lemma 1 is the case when $k = 0$ in this theorem. Therefore, let $k \geqq 1$. Let $0 \to \mathscr{F} \to \mathscr{L} \to \mathscr{G} \to 0$ be an exact sequence of sheaves such that $\mathscr{L}$ is a flabby sheaf. Then there is induced the long exact sequence

$$\begin{aligned} 0 &\to \Gamma(U_{\lambda_0} \cap \cdots \cap U_{\lambda_r}, \mathscr{F}) \to \Gamma(U_{\lambda_0} \cap \cdots \cap U_{\lambda_r}, \mathscr{L}) \\ &\to \Gamma(U_{\lambda_0} \cap \cdots \cap U_{\lambda_r}, \mathscr{G}) \to H^1(U_{\lambda_0} \cap \cdots \cap U_{\lambda_r}, \mathscr{F}) \to \cdots. \end{aligned}$$

By the assumption, the sequence

$$0 \to C^k(\mathscr{U} \bmod \mathscr{U}', \mathscr{F}) \to C^k(\mathscr{U} \bmod \mathscr{U}', \mathscr{L}) \to C^k(\mathscr{U} \bmod \mathscr{U}', \mathscr{G}) \to 0$$

is exact. On the other hand, Lemma 1 implies that the sequence

$$\begin{aligned} 0 &\to \Gamma_Z(X, \mathscr{F}) \to \Gamma_Z(X, \mathscr{L}) \to \Gamma_Z(X, \mathscr{G}) \to H^1(\mathscr{U} \bmod \mathscr{U}', \mathscr{F}) \\ &\to H^1(\mathscr{U} \bmod \mathscr{U}', \mathscr{L}) \to H^1(\mathscr{U} \bmod \mathscr{U}', \mathscr{G}) \to H^2(\mathscr{U} \bmod \mathscr{U}', \mathscr{F}) \\ &\to H^2(\mathscr{U} \bmod \mathscr{U}', \mathscr{L}) \to \cdots \end{aligned}$$

is exact. Note that $H^1(\mathscr{U} \bmod \mathscr{U}', \mathscr{L})$ and $H^2(\mathscr{U} \bmod \mathscr{U}', \mathscr{L})$ are both trivial by Lemma 2. Therefore one has the isomorphisms

$$H^k(\mathscr{U} \bmod \mathscr{U}', \mathscr{F}) \cong \begin{cases} \operatorname{Coker}(\Gamma_Z(X, \mathscr{L}) \to \Gamma_Z(X, \mathscr{G})) & \text{for } k = 1 \\ H^{k-1}(\mathscr{U} \bmod \mathscr{U}', \mathscr{G}) & \text{for } k > 1 \end{cases}.$$

Similarly, by Theorem 1.1.3,

$$H^k_Z(X, \mathscr{F}) = \begin{cases} \operatorname{Coker}(\Gamma_Z(X, \mathscr{L}) \to \Gamma_Z(X, \mathscr{G}) & \text{for } k = 1 \\ H^{k-1}_Z(X, \mathscr{G}) & \text{for } k > 1 \end{cases}$$

Hence $H^1(\mathscr{U} \bmod \mathscr{U}', \mathscr{F}) = H^1_Z(X, \mathscr{F})$ holds. When $k > 1$, we prove the assertion by induction on k. Let $j > 0$ and r be arbitrary. One has the exact sequence

$$\begin{aligned} H^j(U_{\lambda_0} \cap \cdots \cap U_{\lambda_r}, \mathscr{L}) &\to H^j(U_{\lambda_0} \cap \cdots \cap U_{\lambda_r}, \mathscr{G}) \\ &\to H^{j+1}(U_{\lambda_0} \cap \cdots \cap U_{\lambda_r}, \mathscr{F}). \end{aligned}$$

Note that $H^j(U_{\lambda_0} \cap \cdots \cap U_{\lambda_r}, \mathscr{L}) = 0$ since $\mathscr{L}$ is flabby and that $H^{j+1}(U_{\lambda_0} \cap \cdots \cap U_{\lambda_r}, \mathscr{F}) = 0$ by the assumption. Therefore one concludes $H^j(U_{\lambda_0} \cap \cdots \cap U_{\lambda_r}, \mathscr{G}) = 0$ for arbitrary $j > 0$ and $\lambda_0, \ldots, \lambda_r$. Now by the inductive assumption one has

$$H^{k-1}(\mathscr{U} \bmod \mathscr{U}', \mathscr{G}) = H^{k-1}_Z(X, \mathscr{G}).$$

Therefore one finally obtains

$$H^k(U \bmod \mathscr{U}', \mathscr{F}) = H^k_Z(X, \mathscr{F}).$$

We will apply this theorem to our theory of hyperfunctions. First we will recall some of the most fundamental results from the theory of holomorphic functions of several variables in order to prove that there is a covering satisfying the assumption of Theorem 1.2.1.

Definition 1.2.2. *Let X be a paracompact complex manifold. Then X is said to be a Stein manifold if* (i) *and* (ii) *are satisfied.*

(i) *Holomorphically convex condition: let K be a compact set in X; then $\hat{K} = \{p \in X \,|\, |f(p)| \leqq \sup_{x \in K} |f(x)|$ for any $f \in \Gamma(X, \mathscr{O}_X)\}$ is a compact set.*

(ii) *Holomorphically separable condition: for any distinct points p and q in X, there exists $f \in \Gamma(X, \mathscr{O}_X)$ such that $f(p) \neq f(q)$.*

Examples. The space $\mathbf{C}^n$ is a Stein manifold. For a Stein manifold X and holomorphic function $f(x)$ on X, $\{x \in X \,|\, \operatorname{Im} f(x) > 0\}$ and $\{x \in X \,|\, |f(x)| > 1\}$ are both Stein manifolds. The direct product and the intersection of two Stein manifolds are also Stein manifolds. A closed analytic submanifold of a Stein manifold is a Stein manifold. A 1-dimensional complex manifold without compact components is a Stein manifold.

The following theorems are crucial. In particular, Theorem 1.2.2 seems to be one of the most profound results in the field of analysis in this century.

Theorem 1.2.2 (Oka-Cartan). *If X is a Stein manifold, then $H^i(X, \mathscr{O}_X) = 0$ for any integer $i > 0$.*

Theorem 1.2.3 (Grauert [1]). *A paracompact real analytic manifold has complex neighborhoods which are Stein manifolds.*

Consult, for example, Hitotumatu [1] for proofs. Let M be an n-dimensional real analytic manifold, and let X be a complexification. If an open subset U of M is oriented, then $\mathscr{B}_M \cong \mathscr{H}^n_M(\mathcal{O}_X)$ holds over U since $\omega_M|_U = \mathbf{Z}_U$. Let Ω be an open set of X containing U as a closed set. Then one has $\mathscr{B}_M(U) = H^n_U(\Omega, \mathcal{O}_X)$ as we showed in §1. Furthermore, by Theorem 1.2.3, the open set Ω of X can be taken to be a Stein manifold. Suppose that real analytic functions $f_0, f_1, \ldots, f_n$ on M satisfy the conditions:

(1) f_j is real-valued on M for each j, $0 \leqq j \leq n$, and

(2) for each point $x \in M$ the convex hull of $\{df_0(x), \ldots, df_n(x)\} \subset T^*_xM$, i.e. $\left\{\sum_{i=0}^n t_i\, df_i(x) \,\middle|\, \sum_{i=0}^n t_i = 1, t_i \geqq 0\right\}$, is a neighborhood of the origin in T^*_xM. In the case $M = \mathbf{R}^n$, if the convex hull of $\{\xi_0, \ldots, \xi_n\} \subset M$ is a neighborhood of the origin, then one can take $f_j = \langle \xi_j, x\rangle$. Let $V_j = \{z \in \Omega \,|\, \operatorname{Im} f_j > 0\}$ for $j = 0, 1, \ldots, n$. Then V_j is a Stein manifold. Since $\operatorname{Im} f_j = 0$ on U, one has $\left(\bigcup_{j=0}^n V_j\right) \cap U = \varnothing$.

Lemma 1. $\left(\bigcup_{j=0}^n V_j\right) \cup U$ *is a neighborhood of* U.

Proof. Let $x_0 \in U$, and let $x + \sqrt{-1}y$ be sufficiently near the point x_0. When $x + \sqrt{-1}y$ does not belong to $\bigcup_{j=0}^n V_j$, we must show $y = 0$, i.e. $x \in U$. Consider the Taylor expansion $f_j(x + \sqrt{-1}y) = f_j(x) + \sqrt{-1}\langle y, df_j(x)\rangle$ + (terms of degree greater than 2 in y). Since y is sufficiently small and $x + \sqrt{-1}y \notin V_j$, one has $\langle y, df_j(x)\rangle \leqq 0$ for each j. On the other hand, the convex hull of df_j, $j = 0, \ldots, n$, is a neighborhood of the origin. Therefore, if $y \neq 0$, $\left\langle y, \sum_{j=0}^n t_j\, df_j(x)\right\rangle > 0$ holds for some $t_j > 0, j = 0, \ldots, n$, $\sum_{j=0}^n t_j = 1$. Then $\left\langle y, \sum_{j=0}^n t_j\, df_j(x)\right\rangle = \sum_{j=0}^n t_j\langle y, df_j(x)\rangle \leqq 0$ is contradictory.

From Theorem 1.2.3, U has a fundamental neighborhood system consisting of Stein manifolds. Such a neighborhood of U is called a Stein neighborhood. Therefore there exists a Stein neighborhood Ω' such that $\left(\bigcup_{j=0}^n V_j\right) \cup U \supset \Omega'$ holds. Replacing V_j by $V_j \cap \Omega'$ and Ω' by Ω, one has $\Omega = \left(\bigcup_{j=0}^n V_j\right) \cup U$. Then $\mathscr{V} = \{\Omega, V_0, \ldots, V_n\}$ and $\mathscr{V}' = \{V_0, V_1, \ldots, V_n\}$ are Stein open coverings of Ω and $\Omega - U$ respectively. Then, by Oka-

Cartan Theorem 1.2.2, for any integer $j > 0$ and arbitrary $\lambda_0, \dots, \lambda_r$, $H^j(V_{\lambda_0} \cap \cdots \cap V_{\lambda_r}, \mathcal{O}_X) = 0$ holds. Therefore Leray Theorem 1.2.1 implies $\mathscr{B}_M(U) = H^n_U(\Omega, \mathcal{O}_X) = H^n(\mathscr{V} \bmod \mathscr{V}', \mathcal{O}_X)$.

Recall the following definition:

$$H^n(\mathscr{V} \bmod \mathscr{V}', \mathcal{O}_X) = \frac{\operatorname{Ker}(C^n(V \bmod \mathscr{V}', \mathcal{O}_X) \to C^{n+1}(\mathscr{V} \bmod \mathscr{V}', \mathcal{O}_X))}{\operatorname{Im}(C^{n-1}(\mathscr{V} \bmod \mathscr{V}', \mathcal{O}_X) \to C^n(\mathscr{V} \bmod \mathscr{V}', \mathcal{O}_X))}.$$

Lemma 2. $C^{n+1}(\mathscr{V} \bmod \mathscr{V}', \mathcal{O}_X) = 0$.

Proof. As $C^{n+1}(\mathscr{V} \bmod \mathscr{V}', \mathcal{O}_X) \to C^{n+1}(\mathscr{V}, \mathcal{O}_X)$ is a monomorphism, it is sufficient to show $C^{n+1}(\mathscr{V}, \mathcal{O}_X) = 0$. $C^{n+1}(\mathscr{V}, \mathcal{O}_X) = \{\varphi \in \Gamma(\Omega \cap V_0 \cap \cdots \cap V_n, \mathcal{O}_X)\} = \{\varphi \in \Gamma(V_0 \cap \cdots \cap V_n, \mathcal{O}_X)\}$. When Ω is a sufficiently small neighborhood of U, we will prove $V_0 \cap \cdots \cap V_n = \varnothing$. Let $x + \sqrt{-1}y \in V_0 \cap \cdots \cap V_n$. Then one obtains $\langle y, df_j(x)\rangle > 0$ as in the proof of Lemma 1. Since $y \neq 0$, there exists $t_j > 0$, $j = 0, 1, \dots, n$, $\sum_{j=0}^{n} t_j = 1$ such that $\left\langle y, \sum_{j=0}^{n} t_j\, df_j(x)\right\rangle \leqq 0$. Then one has $\left\langle y, \sum_{j=0}^{n} t_j\, df_j(x)\right\rangle = \sum_{j=0}^{n} t_j \langle y, df_j(x)\rangle > 0$, which is a contradiction. Hence we conclude $C^{n+1}(\mathscr{V}, \mathcal{O}_X) = 0$.

Therefore from this lemma we have

$$\mathscr{B}_M(U) = \frac{C^n(\mathscr{V} \bmod \mathscr{V}', \mathcal{O}_X)}{\operatorname{Im}(C^{n-1}(\mathscr{V} \bmod \mathscr{V}', \mathcal{O}_X) \to C^n(\mathscr{V} \bmod \mathscr{V}', \mathcal{O}_X))}.$$

We now compute $C^n(\mathscr{V} \bmod \mathscr{V}', \mathcal{O}_X)$.

$$\begin{aligned} C^n(\mathscr{V} \bmod \mathscr{V}', \mathcal{O}_X) &= \Big\{(\varphi_{0,1,\dots,n}, \varphi_{\Omega,1,\dots,n}, \dots, \varphi_{\Omega,0,1,\dots,n-1}) \\ &\qquad \in \mathcal{O}_X\Big(\bigcap_{j=0}^{n} V_j\Big) \oplus \bigoplus_{j=0}^{n} \mathcal{O}_X\Big(\Omega \cap \bigcap_{k \neq j} V_k\Big) \Big| \varphi_{0,1,\dots,n} = 0\Big\} \\ &= \bigoplus_{j=0}^{n} \mathcal{O}_X\Big(\Omega \cap \bigcap_{k \neq j} V_k\Big) = \bigoplus_{j=0}^{n} \mathcal{O}_X\Big(\bigcap_{k \neq j} V_k\Big). \end{aligned}$$

Denote $W_{\hat{j}} = \bigcap_{k \neq j} V_k$. Then $C^n(\mathscr{V} \bmod \mathscr{V}', \mathcal{O}_X) = \bigoplus_{j=0}^{n} \mathcal{O}_X(W_{\hat{j}})$. We also abbreviate $\varphi_{\Omega, j_1, \dots, j_n} \in \mathcal{O}(W_{\hat{j}_0})$ as $\varphi_{\hat{j}_0}$, where the sign of the permutation

$$\begin{pmatrix} 0, & 1, & \dots, & n \\ j_0, & j_1, & \dots, & j_n \end{pmatrix} \text{is } +1.$$

In the last place we will compute $C^{n-1}(\mathscr{V} \bmod \mathscr{V}', \mathcal{O}_X)$. Let $\{\varphi_{\lambda_0, \dots, \lambda_{n-1}}\} \in C^{n-1}(\mathscr{V} \bmod \mathscr{V}', \mathcal{O}_X)$. If $\varphi_{\lambda_0, \dots, \lambda_{n-1}} \in \Gamma(V_{\lambda_0} \cap \cdots \cap V_{\lambda_{n-1}}, \mathcal{O}_X)$, then $\varphi_{\lambda_0, \dots, \lambda_{n-1}} = 0$ by the definition of "mod $\mathscr{V}'$." Therefore we only need to consider the type $\varphi_{\Omega,i_0,\ldots,i_{n-2}} \in \Gamma(\Omega \cap V_{i_0} \cap \cdots \cap V_{i_{n-2}}, \mathcal{O}_X)$. We abbreviate this as $\varphi_{\widehat{i_{n-1},i_n}}$, where sign

$$\begin{pmatrix} 0, & 1, & \ldots, & n \\ i_0, & i_1, & \ldots, & i_n \end{pmatrix} = +1.$$

Then $\varphi_{\widehat{j,k}} \in \mathcal{O}_X\left(\Omega \cap \bigcap_{l \neq j,k} V_l\right) = \mathcal{O}_X\left(\bigcap_{l \neq j,k} V_l\right) = \mathcal{O}_X(W_{\widehat{j,k}})$, where $W_{\widehat{j,k}} = \bigcap_{l \neq j,k} V_l$, satisfying $\varphi_{\widehat{j,k}} = -\varphi_{\widehat{k,j}}$. That is, $C^{n-1}(\mathscr{V} \bmod \mathscr{V}', \mathcal{O}_X) = \bigoplus'_{\substack{j,k \\ j \neq k}} \mathcal{O}_X(W_{\widehat{j,k}})$, where $\bigoplus'$ denotes the alternating sum; i.e. $\{\varphi_{\widehat{j,k}}\} \in \bigoplus' \mathcal{O}_X(W_{\widehat{j,k}})$ if and only if $\varphi_{\widehat{j,k}} = -\varphi_{\widehat{k,j}}$ holds for arbitrary j and k.

Finally we have the isomorphism

$$\mathscr{B}_M(U) \cong \frac{\bigoplus_{j=0}^{n} \mathcal{O}_X(W_{\hat{j}})}{\operatorname{Im}\left(\bigoplus'_{j,k} \mathcal{O}_X(W_{\widehat{j,k}}) \xrightarrow{\delta} \bigoplus_{j=0}^{n} \mathcal{O}_X(W_{\hat{j}})\right)}.$$

Next we will compute the image of δ, i.e. $(\delta\varphi)_{\hat{j}}$ for $\varphi = \{\varphi_{\widehat{j,k}}\} \in \bigoplus'_{j,k} \mathcal{O}_X(W_{\widehat{j,k}})$. By definition, $(\delta\varphi)_{\hat{j}_n} = (\delta\varphi)_{\Omega,j_0,\ldots,j_{n-1}} = \varphi_{j_0,\ldots,j_{n-1}} - \varphi_{\Omega,j_1,\ldots,j_{n-1}} + \varphi_{\Omega,j_0,j_2,\ldots,j_{n-1}} - \cdots$. As before, $\varphi_{j_0,\ldots,j_{n-1}} = 0$. Let $-\varphi_{\Omega,j_1,\ldots,j_{n-1}} = \varphi_{\widehat{j_0,j_n}}$, where sign $\begin{pmatrix} 0, & 1, & \ldots, & n \\ j_n, & j_0, & \ldots, & j_{n-1} \end{pmatrix} = 1$, $\varphi_{\Omega,j_0,j_2,\ldots,j_{n-1}} = \varphi_{\widehat{j_1,j_n}}$ and so on. Then $(\delta\varphi)_{\hat{j}_n} = \varphi_{\widehat{j_0,j_n}} + \varphi_{\widehat{j_1,j_n}} + \cdots + \varphi_{\widehat{j_{n-1},j_n}} (+\varphi_{\widehat{j_n,j_n}}(=0))$. Therefore $(\delta\varphi)_{\hat{j}} = \sum_{k=0}^{n} \varphi_{\widehat{k,j}}$; i.e. $\operatorname{Im}\delta = \delta(\{\varphi_{\widehat{j,k}}\}) = \left\{\sum_{k=0}^{n} \varphi_{\widehat{k,j}}\right\}_j$. For $\varphi \in \mathcal{O}_X(W_{\hat{j}})$ the boundary value $b(\varphi)$ of φ is defined by the image of the composite map

$$\begin{array}{ccccc} \mathcal{O}_X(W_{\hat{j}}) & \longrightarrow & \bigoplus_{l=0}^{n} \mathcal{O}_X(W_{\hat{l}}) & \longrightarrow & \dfrac{\bigoplus_{l=0}^{n} \mathcal{O}_X(W_{\hat{l}})}{\operatorname{Im}\left(\bigoplus'_{l,k} \mathcal{O}_X(W_{\widehat{l,k}}) \rightarrow \bigoplus_{l=0}^{n} \mathcal{O}_X(W_{\hat{l}})\right)} = \mathscr{B}_M(U). \\ \Cup & & \Cup & & \Cup \\ \varphi & \longmapsto & (0, \ldots, 0, \phi_{\hat{l}}, 0, \ldots 0) & \longmapsto & b(\varphi) \end{array}$$

Therefore $\mathscr{B}_M(U) \cong \sum_{j=0}^{n} b(\mathcal{O}_X(W_{\hat{j}}))$.

Note. We will investigate further the notion of boundary values of holomorphic functions in §3 of the next chapter.

If U is an oriented open set, then the hyperfunctions on U can be expressed as the sum of boundary values of holomorphic functions which are defined on $(n + 1)$ angular domains (see Figure 1.2.1) when $n = 2$.

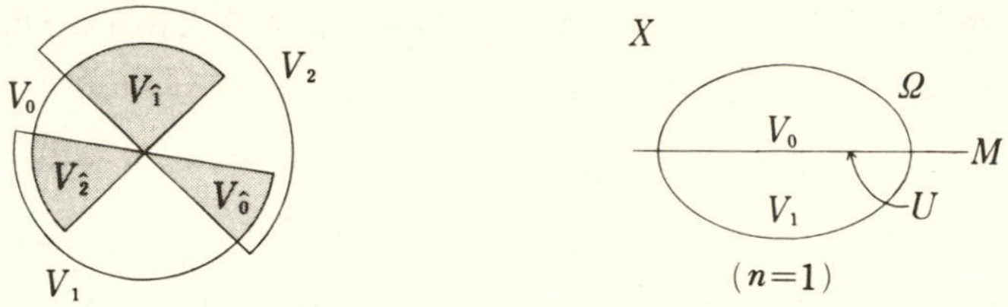

Figure 1.2.1

Notice that for $g_{\hat{j}} \in \mathcal{O}_X(W_{\hat{j}})$, $0 \leqq j \leqq n$, $\sum_{j=0}^{n} b(g_{\hat{j}}) = 0$ holds if and only if $g_{\hat{j}} = \sum_{k=0}^{n} f_{\widehat{j,k}}$, $0 \leqq j \leqq n$, for $f_{\widehat{j,k}} \in \mathcal{O}(W_{\widehat{j,k}})$ such that $f_{\widehat{j,k}} = -f_{\widehat{k,j}}$. That is, two boundary values of holomorphic functions define the same hyperfunction if and only if the difference of the boundary values is a coboundary. That was the reason for introducing the notion of relative cohomology.

For example, when $n = 1$, let $M = \mathbf{R}$, $X = \mathbf{C}$, and $f_0(x) = x$, $f_1(x) = -x$. From the definition

$$V_0 = \{z \in \Omega \mid \operatorname{Im} z > 0\}$$

and

$$V_1 = \{z \in \Omega \mid \operatorname{Im} z < 0\}.$$

Then $W_{\hat{0}} = V_1$, $W_{\hat{1}} = V_0$ (see Figure 1.2.1). Note that $C^{n-1}(\mathscr{V} \bmod \mathscr{V}', \mathcal{O}_X) = C^0(\mathscr{V} \bmod \mathscr{V}', \mathcal{O}_X) = \{(\varphi_\Omega, \varphi_0, \varphi_1) \in \mathcal{O}_X(\Omega) \oplus \mathcal{O}(V_0) \oplus \mathcal{O}(V_1) \mid \varphi_0 = \varphi_1 = 0\} = \mathcal{O}_X(\Omega)$. Therefore

$$\mathscr{B}_M(U) = \frac{\mathcal{O}_X(W_{\hat{0}}) \oplus \mathcal{O}_X(W_{\hat{1}})}{\operatorname{Im}(\mathcal{O}_X(\Omega) \to \mathcal{O}_X(W_{\hat{0}}) \oplus \mathcal{O}(W_{\hat{1}}))}$$
$$= \frac{\mathcal{O}_X(V_1) \oplus \mathcal{O}_X(V_0)}{\operatorname{Im}(\mathcal{O}_X(\Omega) \to \mathcal{O}_X(V_1) \oplus \mathcal{O}_X(V_0))}.$$

Suppose that $g_+ \in \mathcal{O}(V_0)$ and $g_- \in \mathcal{O}(V_1)$ have the same boundary value, i.e. $b(g_+) = b(g_-)$, then $g_+ - g_- = (g_{\hat{0}}, g_{\hat{1}}) \in \mathcal{O}(W_{\hat{0}}) \oplus \mathcal{O}(W_{\hat{1}})$, where $g_{\hat{0}} = -g_-$, $g_{\hat{1}} = g_+$, is a coboundary. Therefore, there exists $f = f_{\widehat{0,1}} = -f_{\widehat{1,0}} \in \mathcal{O}_X(\Omega)$ such that $g_{\hat{0}} = f_{\widehat{0,0}} + f_{\widehat{1,0}} = -f$ and $g_{\hat{1}} = f_{\widehat{0,1}} + f_{\widehat{1,1}} = f$. This means that there is $f \in \mathcal{O}_X(\Omega)$ such that $g_+ = f|_{V_0}$ and $g_- = f|_{V_1}$ hold. Hence one obtains $\mathscr{B}_M(U) = \mathcal{O}_X(\Omega - U)/\mathcal{O}_X(\Omega)$.

Remark. We will compute $\mathscr{B}_{\mathbf{R}^n} = \mathscr{H}^n_{\mathbf{R}^n}(\mathcal{O}_{\mathbf{C}^n})$, hyperfunctions on $\mathbf{R}^n$, expressing hyperfunctions as the sum of 2^n boundary values rather than as the sum of boundary values from $(n+1)$ angular domains.

Let $U_{k,+} = \{z \in \mathbf{C}^n \mid \operatorname{Im} z_k > 0\}$, and let $U_{k,-} = \{z \in \mathbf{C}^n \mid -\operatorname{Im} z_k > 0\}$ for $k = 1, 2, \ldots, n$. Then $\{U_{k,\pm}\}_{k=1,2,\ldots,n}$ is a Stein covering of $\mathbf{C}^n - \mathbf{R}^n$.

Denote $\mathscr{V}' = \{U_{k,\pm}\}_{k=1,2,\ldots,n}$ and $\mathscr{V} = \mathscr{V}' \cup \{\mathbf{C}^n\}$. Then we have

$$H^n_{\mathbf{R}^n}(\mathbf{C}^n, \mathcal{O}) = H^n(\mathscr{V} \bmod \mathscr{V}', \mathcal{O})$$
$$= \frac{\operatorname{Ker}(C^n(\mathscr{V} \bmod \mathscr{V}', \mathcal{O}) \xrightarrow{\delta} C^{n+1}(\mathscr{V} \bmod \mathscr{V}', \mathcal{O}))}{\operatorname{Im}(C^{n-1}(\mathscr{V} \bmod \mathscr{V}', \mathcal{O}) \xrightarrow{\delta} C^n(\mathscr{V} \bmod \mathscr{V}', \mathcal{O}))}.$$

One has $C^{n+1}(\mathscr{V} \bmod \mathscr{V}', \mathcal{O}) = 0$ as before, $C^n(\mathscr{V} \bmod \mathscr{V}', \mathcal{O}) = \bigoplus_{\epsilon_1 = \pm, \ldots, \epsilon_n = \pm} \mathcal{O}(U_{1\epsilon_1} \cap \cdots \cap U_{n\epsilon_n})$, $C^{n-1}(\mathscr{V} \bmod \mathscr{V}', \mathcal{O}) = \bigoplus_{k,\epsilon_k} \mathcal{O}(\hat{k}, \epsilon_{\hat{k}})$, where $\mathcal{O}(\hat{k}, \epsilon_{\hat{k}}) = \mathcal{O}(U_{1,\epsilon_1} \cap \cdots \cap U_{k-1,\epsilon_{k-1}} \cap U_{k+1,\epsilon_{k+1}} \cap \cdots \cap U_{n,\epsilon_n})$, $\epsilon_{\hat{k}} = (\epsilon_1, \ldots, \epsilon_{k-1}, \epsilon_{k+1}, \ldots, \epsilon_n)$.

Next we compute the coboundary. Let

$$\psi = \{\psi(\hat{k}, \epsilon_{\hat{k}})\} \in C^{n-1}(\mathscr{V} \bmod \mathscr{V}', \mathcal{O}).$$

Then $(\delta\psi)_{\epsilon_1,\ldots,\epsilon_n} = \sum_{k=1}^{n} (-1)^k \psi(\hat{k}, \epsilon_{\hat{k}})$, $(\epsilon_{\hat{k}} = (\epsilon_1, \ldots, \epsilon_{k-1}, \epsilon_{k+1}, \ldots, \epsilon_n))$. We denote the element $\{\psi_{\epsilon_1,\ldots,\epsilon_n}\} \bmod \{\sum (-1)^k \psi(\hat{k}, \epsilon_{\hat{k}})\}$ of $H^n_{\mathbf{R}^n}(\mathbf{C}^n, \mathcal{O})$ by $\sum_\epsilon \epsilon_1 \cdots \epsilon_n b(\psi_{\epsilon_1,\ldots,\epsilon_n}) = \sum_\epsilon b(g_\epsilon)$. Then notice that if $\sum_\epsilon b(g_\epsilon) = 0$, then $\epsilon_1 \cdots \epsilon_n g_\epsilon = \sum_k (-1)^k \varphi_{k,\epsilon_1 \cdots \epsilon_{k-1}\epsilon_{k+1} \cdots \epsilon_n}$ for $\varphi(\hat{k}, \epsilon_{\hat{k}}) \in \mathcal{O}(\hat{k}, \epsilon_{\hat{k}})$. Therefore $g_\epsilon = \sum_k (-1)^k \epsilon_1 \cdots \epsilon_n \varphi_{k,\epsilon_1 \cdots \epsilon_{k-1}\epsilon_{k+1} \cdots \epsilon_n}$. If one lets

$$h_{\epsilon_{\hat{k}}} = \epsilon_1 \cdots \epsilon_{k-1}\epsilon_{k+1} \cdots \epsilon_n \varphi_{k,\epsilon_1 \cdots \epsilon_{k-1}\epsilon_{k+1} \cdots \epsilon_n},$$

then one has $g_\epsilon = \sum_k (-1)^k \epsilon_k h_{\epsilon_{\hat{k}}}$. Therefore, if $\sum_\epsilon b(g_\epsilon) = 0$, then $g_\epsilon = \sum_k (-1)^k \epsilon_k h_{\epsilon_{\hat{k}}}$ holds. This is fundamental when one expresses hyperfunctions explicitly. Hyperfunctions defined over an open set in $\mathbf{R}^n$ can be treated similarly.

We will treat the orientation sheaf $\omega_M = \mathscr{H}^n_M(\mathbf{Z}_X)$ via covering cohomology as we did for the sheaf of hyperfunctions $\mathscr{H}^n_M(\mathcal{O}_X)$. Then we will show that ω_M is locally isomorphic to $\mathbf{Z}_M$. This implies that the hyperfunction sheaf $\mathscr{B}_M$ is locally isomorphic to $\mathscr{H}^n_M(\mathcal{O}_X)$ provided that M is oriented.

Some preliminary notions are necessary.

Definition 1.2.3. *Let X and Y be topological spaces, and let $f_i: X \to Y$ be continuous maps for $i = 0, 1$. Then f_0 and f_1 are said to be homotopic if there exists a continuous map $F: X \times I \to Y$, where $I = [0, 1]$, such that $F(x, 0) = f_0(x)$ and $F(x, 1) = f_1(x)$. We call the continuous map F a homotopy between f_0 and f_1, denoted by $f_0 \simeq f_1$.*

Two topological spaces X and Y are said to have the same homotopy type if there exist continuous maps $f: X \to Y$ and *$g: Y \to X$ such that $g \circ f \simeq 1_X$ and $f \circ g \simeq 1_Y$.*

A topological space is said to be contractible to a point x_0 if there exists a continuous map $F: X \times I \to X$ such that, for any point $x \in X$ and any $t \in I$, $F(x, 0) = x$, $F(x, 1) = x_0$ and $F(x_0, t) = x_0$.

Remark. If X is contractible to a point, then X has the same homotopy type as a point.

The next theorem indicates that cohomology groups with coefficients in a constant sheaf are homotopy invariant.

Theorem 1.2.4. *Let X and Y be topological spaces, and let M be an additive group. If X and Y have the same homotopy type, then $H^k(X, M) \cong H^k(Y, M)$ for any integer $k \geqq 0$. In particular, when X is contractible, one has*

$$H^k(X, M) = \begin{cases} 0 & \text{for } k \neq 0 \\ M & \text{for } k = 0 \end{cases}$$

where M is regarded as a constant sheaf in this theorem.

The proof of this theorem will be given in §2, Chapter II.

Theorem 1.2.5. *Let M be an additive group. Then*

$$\mathscr{H}^k_{\mathbf{R}^n \times \{0\}}(M_{\mathbf{R}^{n+l}}) = \begin{cases} M_{\mathbf{R}^n} & \text{for } k = l \\ 0 & \text{for } k \neq l \end{cases}$$

holds. In particular, when $k \neq n$, $\mathscr{H}^k_M(\mathbf{Z}_X) = 0$.

Proof. Let $A_{n,l}$ be the statement of Theorem 1.2.5. Then we will give a proof by induction on n and l. We also denote $A_{n,l} \Rightarrow A_{n',l'}$ when the statement $A_{n,l}$ implies the statement $A_{n',l'}$. Let U and V be open balls in $\mathbf{R}^n$ and $\mathbf{R}^l$ respectively. Then we must show

$$H^k_{U \times \{0\}}(U \times V, M) = \begin{cases} M & \text{for } k = l \\ 0 & \text{for } k \neq l \end{cases}.$$

Since U is contractible, Theorem 1.2.4 implies the isomorphisms $H^k(U \times V, M) \cong H^k(V, M)$ and $H^k(U \times (V - \{0\}), M) \cong H^k(V - \{0\}, M)$. Therefore one obtains

$$\begin{array}{ccccccc}
\cdots \to & H^{k-1}(U \times V, M) & \to & H^{k-1}(U \times (V - \{0\}), M) & \to & H^k_{U \times \{0\}}(U \times V, M) & \to \\
& \| & & \| & & \uparrow & \\
\cdots \to & H^{k-1}(V, M) & \longrightarrow & H^{k-1}(V - \{0\}, M) & \longrightarrow & H^k_{\{0\}}(V, M) & \longrightarrow
\end{array}$$

$$\begin{array}{cccc}
H^k(U \times V, M) & \to & H^k(U \times (V - \{0\}), M) & \to \cdots. \\
\| & & \| & \\
H^k(V, M) & \longrightarrow & H^k(V - \{0\}, M) & \longrightarrow \cdots.
\end{array}$$

By Five Lemma, one concludes the isomorphism $H^k_{U\times\{0\}}(U \times V, M) \cong H^k_{\{0\}}(V, M)$. Hence, for an arbitrary l, the implication $A_{0,l} \Rightarrow A_{n,l}$ is true. Next we will prove the implication $A_{0,1} \Rightarrow A_{0,l}$. Then we need to show the implication $A_{0,l-1} \Rightarrow A_{0,l}$. But since $A_{0,l-1} \Rightarrow A_{1,l-1}$ is true, it suffices to prove the implication $A_{1,l-1} \Rightarrow A_{0,l}$. Since $\mathbf{R} \times \{0\}$ is purely $(l-1)$-codimensional with respect to $M_{\mathbf{R}^l}$, $A_{1,l-1}$ implies

$$H^k_{\{0\}}(\mathbf{R}^l, M) = H^{k-(l-1)}_{\{0\}}(\mathbf{R} \times \{0\}, \mathscr{H}^{l-1}_{\mathbf{R}\times\{0\}}(M)) =$$

$$H^{k-l+1}_{\{0\}}(\mathbf{R}, M) = \begin{cases} M & \text{for } k-l+1 = 1, \text{ i.e. } k = l \\ 0 & \text{for } k-l+1 \neq 1, \text{ i.e. } k \neq l \end{cases}$$

by Proposition 1.1.6. This proves $A_{0,l}$, provided that the last equality is true, i.e. $A_{0,1}$ (which remains to be proved). The exact sequence, since $\mathbf{R}$, $\mathbf{R}^+$, and $\mathbf{R}^-$ are contractible,

$$\cdots \to H^{k-1}(\mathbf{R} - \{0\}, M) \to H^k_{\{0\}}(\mathbf{R}, M) \to H^k(\mathbf{R}, M) \to \cdots$$

gives

$$H^k(\mathbf{R}, M) = \begin{cases} M & \text{for } k = 0 \\ 0 & \text{for } k \neq 0 \end{cases}$$

and

$$H^{k-1}(\mathbf{R} - \{0\}, M) = H^{k-1}(\mathbf{R}^+, M) \oplus H^{k-1}(\mathbf{R}^-, M) = \begin{cases} M^2 & \text{for } k = 1 \\ 0 & \text{for } k \neq 1 \end{cases}.$$

Therefore one has $H^k_{\{0\}}(\mathbf{R}, M) = 0$ for $k \neq 0, 1$. We now treat the case where $k = 0, 1$. Consider the exact sequence

$$\begin{array}{ccccccccc} 0 \to H^0_{\{0\}}(\mathbf{R}, M) \to & H^0(\mathbf{R}, M) & \xrightarrow{\varphi} & H^0(\mathbf{R} - \{0\}, M) & \to H^1_{\{0\}}(\mathbf{R}, M) \to 0 \\ & \| & & \| & \\ & M & & M \oplus M & \end{array}$$

where $\varphi(x) = (x, x)$. Then $H^0_{\{0\}}(\mathbf{R}, M) = 0$ since φ is a monomorphism, and therefore $H^1_{\{0\}}(\mathbf{R}, M) = (M \oplus M)/M \cong M$.

We will examine the sections of the sheaf $\omega_M = \mathscr{H}^n_M(\mathbf{Z}_X)$, utilizing Theorems 1.2.4 and 1.2.5. Let $\xi_0, \ldots, \xi_n$ be $(n+1)$ vectors in $\mathbf{R}^n$ such that their convex hull is a neighborhood of the origin, and let $\{U, x = (x_1, \ldots, x_n)\}$ be a connected local-coordinate system of $P \in M$. Let $f_j(x) = \langle \xi_j, x \rangle$ for $0 \leq j \leq n$. Then $f_0(x), \ldots, f_n(x)$ are real-valued analytic functions on M, and for each $x \in M$ the convex hull of $df_0(x), \ldots, df_n(x)$ is a neighborhood of the origin in T^*_xM. Let Ω be a contractible neighborhood in X containing U as a closed set. Then the set $V_j = \{z \in \Omega \mid \operatorname{Im} f_j(z) > 0\}$ for each j, $0 \leqq j \leqq n$, is also contractible. Hence, by Theorem 1.2.4, $\mathscr{V} = \{\Omega, V_0, \ldots, V_n\}$ and $\mathscr{V}' = \{V_0, \ldots, V_n\}$ are Leray coverings of Ω and $\Omega - U$ respectively; i.e. they satisfy the condition of Theorem 1.2.1. Then

Theorem 1.2.5, Proposition 1.1.5, and Theorem 1.2.1 imply

$$\omega_M(U) = H^n_{M\cap U}(\Omega, \mathbf{Z}_X) = \frac{\bigoplus_{j=0}^{n} \Gamma(W_{\hat{j}}, \mathbf{Z}_X)}{\delta\left(\bigoplus'_{j,k} \Gamma(W_{\widehat{j,k}}, \mathbf{Z}_X)\right)},$$

where $W_{\hat{j}} = \bigcap_{k \neq j} V_k$, $W_{\widehat{j,k}} = \bigcap_{l \neq j,k} V_l$ and $\bigoplus'$ is the alternative sum. Since $W_{\hat{j}}$ and $W_{\widehat{j,k}}$ are connected, $\Gamma(W_{\hat{j}}, \mathbf{Z}_X) \cong \mathbf{Z}$ and $\Gamma(W_{\widehat{j,k}}, \mathbf{Z}_X) \cong \mathbf{Z}$ hold. Consider the map $\varphi: \bigoplus_{j=0}^{n} \Gamma(W_{\hat{j}}, \mathbf{Z}_X) \cong \mathbf{Z}^{n+1} \ni (s_i) \mapsto \sum_{i=0}^{n} s_i \in \mathbf{Z}$. Notice that φ is an epimorphism. We will show next that the kernel of φ is $\delta\left(\bigoplus'_{j,k} \Gamma(W_{\widehat{j,k}}, \mathbf{Z}_X)\right)$.

Lemma. *Suppose that integers* $s_0, \ldots, s_n$ *satisfy* $\sum_{i=0}^{n} s_i = 0$. *Then there exist integers* $s_{\hat{j},k}$, *where* $j, k = 0, \ldots, n$, *such that* $s_{j,k} = -s_{k,j}$ *and* $s_i = \sum_{k=0}^{n} s_{k,i}$ *for* $0 \leqq i \leqq n$. *Note that the converse is also true.*

Proof. When $n = 1$, we have $s_0 + s_1 = 0$. Let $s_{1,0} = -s_{0,1} = s_0$. Then $s_0 = s_{0,0} + s_{1,0}$ and $s_1 = s_{0,1} + s_{1,1}$. Notice that $s_{0,0} = s_{1,1} = 0$. Next, assume that $\sum_{i=0}^{n} s_i = 0$ holds. Let $s'_0 = s_0 + s_n$, $s'_1 = s_1, \ldots, s'_{n-1} = s_{n-1}$. Then $\sum_{i=0}^{n-1} s'_i = 0$ holds. Therefore, by the inductive assumption, there exist integers $s_{j,k}$ where $j, k = 0, \ldots, n-1$, such that $s_{j,k} = -s_{k,j}$ and $s'_i = \sum_{k=0}^{n-1} s_{k,i}$ for $i = 0, 1, \ldots, n-1$. Let $s_{0,n} = -s_{n,0} = s_n$ and $s_{n,j} = -s_{j,n} = 0$ for $j = 1, \ldots, n$. Then $s_i = \sum_{k=0}^{n} s_{k,i}$ for $0 \leqq i \leqq n$. The converse is plainly true.

This lemma shows that the kernel of φ is $\delta\left(\bigoplus'_{j,k} \Gamma(W_{\widehat{j,k}}, \mathbf{Z}_X)\right)$. Therefore we conclude $H^n_U(\Omega, \mathbf{Z}_X) \cong \mathbf{Z}$; i.e. if U is a connected coordinate neighborhood, $\omega_M(U) \cong \mathbf{Z}$ holds. Let $(y_1, \ldots, y_n)$ be another local coordinate system; then there is induced an isomorphism $\mathbf{Z} \to \mathbf{Z}$. Notice that this isomorphism is either $1_{\mathbf{Z}}$ or $-1_{\mathbf{Z}}$, since the only automorphisms : $\mathbf{Z} \to \mathbf{Z}$ are $\pm 1_{\mathbf{Z}}$ when $\det(\partial(y_i)/\partial(x_j)) > 0$ and $\det(\partial(y_i)/\partial(x_j)) < 0$ respectively.

Let U be a connected open set of M. If there is an $s \in \omega_M(U)$ such that $s \neq 0$, then s is not zero at each point x in U; i.e. $s_x \neq 0$. Let $U = \bigcup_{i \in I} U_i$, where U_i is a connected coordinate neighborhood; let

$(x_i^1, \ldots, x_i^n)$ be the local coordinate system of U_i; and let the orientation $\mathscr{or}$ be such that $(x_i^1, \ldots, x_i^n)$ is a positive local coordinate system with respect to $\mathscr{or}$. Then $s|_{U_i}$ defines an element $\varphi_i(s)$ in $\mathbf{Z}$. Suppose that $U_i \cap U_j \neq \varnothing$; then $\varphi_i(s) = \pm\varphi_j(s)$, since ± 1 are the only automorphisms of $\mathbf{Z}$. Since U is connected, $\varphi_i(s) = \epsilon_i c$ for some $c \in \mathbf{Z}$, where $\epsilon_i = \pm 1$. So $s_x = c \neq 0$. Let $\widetilde{\mathscr{or}}_{U_i}$ be another orientation such that $\widetilde{\mathscr{or}}_{U_i} = \epsilon_i \mathscr{or}_{U_i}$. Then one has $\widetilde{\mathscr{or}}_{U_i}|_{U_i \cap U_j} = \widetilde{\mathscr{or}}_{U_j}|_{U_i \cap U_j}$. Let $(x_i^1, \ldots, x_i^n)$ be a positive local coordinate system with respect to $\widetilde{\mathscr{or}}_{U_i}$. Then $\det(\partial(x_i^l)/\partial(x_j^k))_{1 \leqq l,k \leqq n} > 0$ holds, provided $U_i \cap U_j \neq 0$. Therefore U is orientable. Furthermore, if U is a paracompact space, one can show that there is a continuous n-form on U, using a partition of unity on U. On the other hand, if U is non-orientable, then $\omega_M(U) = 0$. That is, giving a non-zero section of ω_M over U is equivalent to giving an orientation on U.

Hence, so long as a local coordinate system is fixed, or M is orientable, or one considers locally, then there exist isomorphisms $\mathscr{B}_M = \mathscr{H}^n_M(\mathscr{O}_X) \underset{\mathbf{Z}_M}{\oplus} \omega_M \cong \mathscr{H}^n_M(\mathscr{O}_X)$ and $\mathscr{B}_M(U) \cong H^n_U(X, \mathscr{O}_X)$.

CHAPTER II

Microfunctions

§1. Definition of Microfunctions

Let M be a manifold, and let N be a submanifold of M. We always assume that a submanifold is regular; i.e. its topology is provided with the topology as a subspace. Equivalently, $N \cap U_j = \{x_j^1 = \cdots = x_j^l = 0\}$ for a local coordinate system $\{U_j, (x_j^1, \ldots, x_j^m)\}$ of M; see, for example, Matsushima [1].

Definition 2.1.1.

(a) *For $x \in M$ we denote the tangent space of M at x by T_xM. The tangent bundle of M is denoted by $TM = \bigcup_{x \in M} T_xM$.*

(b) *Let $f: X \to Y$ and $g: X' \to Y$ be mappings. The fibre product, denoted by $X \underset{Y}{\times} X'$, of X and X' is defined by*

$$X \underset{Y}{\times} X' = \{(x, x') \in X \times X' \,|\, f(x) = g(x')\}.$$

(c) *Let N be a submanifold of M. Then*

$$\mathrm{Coker}(TN \to N \underset{M}{\times} TM) = \bigcup_{x \in N} T_xM/T_xN$$

is called the normal bundle of N in M, denoted by T_NM.

(d) *The cotangent space T_x^*M is the dual vector space of T_xM, and $T^*M = \bigcup_{x \in M} T_x^*M$ is called the cotangent bundle of M.*

(e) $\mathrm{Ker}(N \underset{M}{\times} T^*M \to T^*N) = \bigcup_{x \in N} \{\eta \in T_x^*M \,|\, \langle \eta, T_xN \rangle = 0\}$ *is called the conormal bundle on N with respect to M, denoted by T_N^*M.*

(f) *Let $\mathbf{R}_+^\times$ be the multiplicative group of positive real numbers. The tangent sphere bundle of M, denoted by SM, is defined by*

$$(TM - M)/\mathbf{R}_+^\times = \bigcup_{x \in M} (T_xM - \{0\})/\mathbf{R}_+^\times.$$

Similarly, the cotangent sphere bundle S^*M is defined as

$$(T^*M - M)/\mathbf{R}_+^\times = \bigcup_{x \in M} (T_x^*M - \{0\})/\mathbf{R}_+^\times.$$

Furthermore, $(T_N M - N)/\mathbf{R}_+^\times = \bigcup_{x \in N} ((T_x M/T_x N) - \{0\})/\mathbf{R}_+^\times$ *is said to be the normal sphere bundle on* N *with respect to* M *and is denoted by* $S_N M$. *The conormal sphere bundle, denoted by* $S_N^* M$, *is defined as*

$$(T_N^* M - N)\mathbf{R}_+^\times = \bigcup_{x \in N} \{\eta \in T_x^* M \,|\, \langle \eta, T_x N \rangle = 0\}/\mathbf{R}_+^\times.$$

Note. M can be identified with zero sections of TM and T^*M, $\{(x, \xi) \in TM \,|\, \xi = 0\}$ and $\{(x, \eta) \in T^*M \,|\, \eta = 0\}$ respectively. Therefore, we often denote M for zero sections of TM or T^*M.

We will define the notion of real monoidal transform of M with center N. Suppose that $M = \bigcup_{j \in J} U_j$ and the local coordinates $(x_j^1, \ldots, x_j^m)$ of U_j satisfy $U_j \cap N = \{x_j \,|\, x_j^1 = \cdots = x_j^l = 0\}$, where $l = \operatorname{codim}_M N$. One can assume that the equations of a coordinate transformation

$$\left.\begin{aligned} x_j^\nu &= f_{jk}^\nu(x_k), && \nu = l + 1, \ldots, m \\ x_j^\nu &= \sum_{\mu=1}^{l} x_k^\mu g_{jk,\mu}^\nu(x_k), && \nu = 1, \ldots, l \end{aligned}\right\} \tag{2.1.1}$$

hold between local coordinates. Then define $U_j' = \{(x_j, \xi_j) = (x_j^1, \ldots, x_j^m; \xi_j^1, \ldots, \xi_j^l) \in U_j \times (\mathbf{R}^l - \{0\}) \,|\, x_j^\nu \xi_j^\mu = x_j^\mu \xi_j^\nu$ and $x_j^\nu \xi_j^\nu \geqq 0$ for $\mu, \nu = 1, \ldots, l\}$. Notice that, if $(x_j, \xi_j) \in U_j'$ and $x_j \notin N \cap U_j$, then there exists a positive real number t such that $\xi_j = (tx_j^1, tx_j^2, \ldots, tx_j^l)$; and if $x_j \in N \cap U_j$, then no condition is needed on $\xi_j \in \mathbf{R}^l - \{0\}$. If one defines a map $\mathbf{R}_+^\times \times U_j' \to U_j'$ by $((x_j, \xi_j), t) \mapsto (x_j, t\xi_j)$, then $\mathbf{R}_+^\times$ acts on U_j'. Then the transitivity of this action defines an equivalence relation in U_j' and we denote the quotient by $\tilde{U}_j$; i.e. $\tilde{U}_j = U_j'/\mathbf{R}_+^\times$. We will paste together $\tilde{U}_j$ as follows: points $(x_j, \xi_j) \in \tilde{U}_j$ and $(x_k, \xi_k) \in \tilde{U}_k$ are identified if and only if x_j and x_k satisfy the equations in (2.1.1) and $\xi_j^\nu = \sum_{\mu=1}^{l} \xi_k^\mu g_{jk,\mu}^\nu(x_k)$ for $\nu = 1, \ldots, l$. The manifold obtained in this manner is denoted by $\widetilde{{}^N M}$, which is called the real monoidal transform of M with center N. We remark that, as a set, $\widetilde{{}^N M} = (M - N) \sqcup S_N M$, where $\sqcup$ indicates a disjoint union.

Let M be a real analytic manifold, and let X be a complexification of M. Then $TX|_M = TM \oplus \sqrt{-1}TM$ holds; i.e. $T_x X = T_x M \oplus \sqrt{-1}T_x M$ for $x \in M$. Therefore, one has $T_M X = \sqrt{-1}TM\ (\cong TM)$, which implies $S_M X \cong \sqrt{-1}SM$. Let $x \in M$ and $v \in T_x M - \{0\}$; then one may consider $\sqrt{-1}v \in (T_M X)_x$. Let $x + \sqrt{-1}v0$ be the point in $\sqrt{-1}SM\ (= S_M X)$ which corresponds to $\sqrt{-1}v \in (T_M X)_x$. Note that we have $\widetilde{{}^M X} = (X - M) \sqcup \sqrt{-1}SM$ as a set. In particular, if $M = \mathbf{R}^n$ and $X = \mathbf{C}^n$, then $\widetilde{{}^M X} = (\mathbf{C}^n - \mathbf{R}^n) \sqcup \sqrt{-1}S^{n-1} \times \mathbf{R}^n$. Therefore, if $x \in M = \mathbf{R}^n$, $t \in \mathbf{R}_+^\times$, and $v \in \mathbf{R}^n - \{0\}$, then $x + \sqrt{-1}tv$ is a point in $\widetilde{{}^M X}$ and $\lim_{t \to +0} (x + \sqrt{-1}tv) = x + \sqrt{-1}v0 \in \widetilde{{}^M X}$. This indicates that the real monoidal transform $\widetilde{{}^M X}$

consists of all the possible directions toward M from $X - M$. Let

$$A_\epsilon = \{x + \sqrt{-1}v0 \in \sqrt{-1}SM \,|\, |x - x_0| < \epsilon \text{ and } |v - v_0| < \epsilon\}$$

and

$$B_\epsilon = \{x + \sqrt{-1}tv \in {}^{M}\widetilde{X} - \sqrt{-1}SM = X - M \,|\, 0 < t < \epsilon, |x - x_0| < \epsilon \text{ and } |v - v_0| < \epsilon\}.$$

Then one can take $\{A_\epsilon \cup B_\epsilon\}$ as a base of $x_0 + \sqrt{-1}v_0 0 \in {}^{M}\widetilde{X}$. See Figure 2.1.1, below.

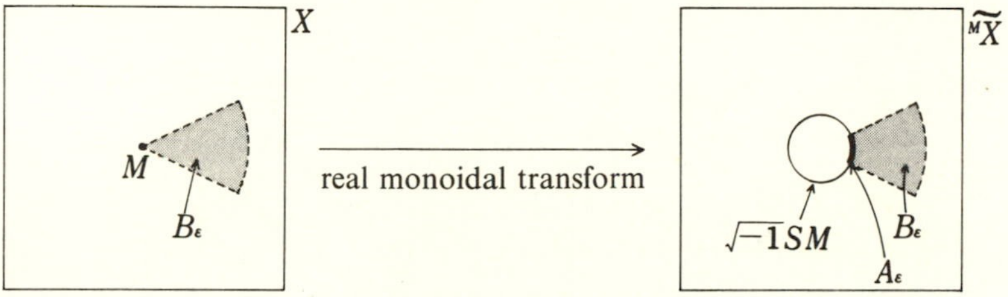

Figure 2.1.1

Definition 2.1.2. *Denote the natural embeddings $X - M \hookrightarrow X$ and ${}^{M}\widetilde{X} - \sqrt{-1}SM \hookrightarrow {}^{M}\widetilde{X}$ by ϵ and $\tilde{\epsilon}$ respectively. Let τ be the projection map from ${}^{M}\widetilde{X}$ onto X (see Figure 2.1.2), and let $\tilde{\tau}$ be the natural embedding $\sqrt{-1}SM \hookrightarrow {}^{M}\widetilde{X}$.*

$$\begin{array}{ccc} {}^{M}\widetilde{X} & \overset{\tilde{\epsilon}}{\hookleftarrow} & {}^{M}\widetilde{X} - \sqrt{-1}SM \\ \downarrow \tau & & \| \\ X & \overset{\epsilon}{\hookleftarrow} & X - M \end{array}$$

Figure 2.1.2

Then sheaves $\tilde{\mathcal{O}}$ over ${}^{M}\widetilde{X}$ and $\tilde{\mathcal{A}}$ over $\sqrt{-1}SM$ are defined by $\tilde{\mathcal{O}} = \tilde{\epsilon}_* \epsilon^{-1}\mathcal{O}_X$ and $\tilde{\mathcal{A}} = \tilde{\mathcal{O}}|_{\sqrt{-1}SM} = \tilde{\tau}^{-1}\tilde{\mathcal{O}}$ respectively, where $\mathcal{O}_X$ is the sheaf of holomorphic functions on X.

Remark. Let $x + \sqrt{-1}v0 \in \sqrt{-1}SM$; then

$$\tilde{\mathcal{A}}_{x+\sqrt{-1}v0} = \tilde{\mathcal{O}}_{x+\sqrt{-1}v0} = \varinjlim_{x+\sqrt{-1}v0 \in \tilde{U}} \tilde{\mathcal{O}}(\tilde{U}) = \varinjlim_{x+\sqrt{-1}v0 \in \tilde{U}} (\epsilon^{-1}\mathcal{O}_X)(\tilde{\epsilon}^{-1}(\tilde{U}))$$

$$= \varinjlim_{x+\sqrt{-1}v0 \in \tilde{U}} \mathcal{O}_X(\tilde{U} - \sqrt{-1}SM).$$

Therefore $\tilde{\mathscr{A}}$ may be considered as the sheaf of boundary values of holomorphic functions, where v indicates the direction along which the boundary value is taken.

Proposition 2.1.1. *$\sqrt{-1}SM$ is purely 1-codimensional with respect to $\tau^{-1}\mathcal{O}_X$; i.e. $\mathscr{H}^k_{\sqrt{-1}SM}(\tau^{-1}\mathcal{O}_X) = 0$ for $k \neq 1$.*

Proof. Since the assertion is local in nature, one can assume $M = \mathbf{R}^n$, $X = \mathbf{C}^n$, and $x_0 = 0$. Therefore, we must show $\mathscr{H}^k_{\sqrt{-1}SM}(\tau^{-1}\mathcal{O}_X)_{0+\sqrt{-1}v_00} = 0$ for $k \neq 1$, where $v_0 \in \mathbf{R}^n - \{0\}$. First note

$$\mathscr{H}^k_{\sqrt{-1}SM}(\tau^{-1}\mathcal{O}_X)_{0+\sqrt{-1}v_00} = \varinjlim_{0+\sqrt{-1}v_00 \in \tilde{U}} H^k_{\sqrt{-1}SM \cap \tilde{U}}(\tilde{U}, \tau^{-1}\mathcal{O}_X).$$

There is induced the long exact sequence, by Theorem 1.1.2 (3),

$$\begin{aligned} \cdots &\to H^{k-1}(\tilde{U}, \tau^{-1}\mathcal{O}_X) \to H^{k-1}(\tilde{U} - \sqrt{-1}SM, \tau^{-1}\mathcal{O}_X) \\ &\to H^k_{\sqrt{-1}SM \cap \tilde{U}}(\tilde{U}, \tau^{-1}\mathcal{O}_X) \to H^k(\tilde{U}, \tau^{-1}\mathcal{O}_X) \\ &\to H^k(\tilde{U} - \sqrt{-1}SM, \tau^{-1}\mathcal{O}_X) \to \cdots. \end{aligned} \tag{2.1.2}$$

If one lets $A_\epsilon = \{x + \sqrt{-1}v0 \,|\, |x| < \epsilon$ and $|v - v_0| < \epsilon\}$, $B_\epsilon = \{x + \sqrt{-1}tv \,|\, 0 < t < \epsilon, |v - v_0| < \epsilon$ and $|x| < \epsilon\}$, and $\tilde{U}_\epsilon = A_\epsilon \cup B_\epsilon$, then $\{\tilde{U}_\epsilon\}$ is a fundamental neighborhood system. Since $\tilde{U}_\epsilon - \sqrt{-1}SM = B_\epsilon$ is a convex set in $\mathbf{C}^n$, it in particular is a Stein manifold. Therefore, $H^k(\tilde{U}_\epsilon - \sqrt{-1}SM, \tau^{-1}\mathcal{O}_X) = 0$ for $k \neq 0$ by the Oka-Cartan theorem (Theorem 1.2.2). We need the following lemma to complete the proof of Proposition 2.1.1.

Lemma. *Let $\mathscr{F}$ be a sheaf over a topological space X, and let $x \in X$. Then one has*

$$\varinjlim_{x \in U} H^k(U, \mathscr{F}) = \begin{cases} 0 & \text{for } k \neq 0 \\ \mathscr{F}_x & \text{for } k = 0. \end{cases}$$

Proof of Lemma. Let $0 \to \mathscr{F} \to \mathscr{L}^0 \to \mathscr{L}^1 \to \cdots$ be a flabby resolution of $\mathscr{F}$. Then $H^k(U, \mathscr{F}) = H^k(\Gamma(U, \mathscr{L}^*))$. Note that taking cohomology of the cochain complex commutes with the direct limit. This implies

$$\varinjlim_{x \in U} H^k(U, \mathscr{F}) = H^k\left(\varinjlim_{x \in U} \Gamma(U, \mathscr{L}^*)\right) = H^k(\mathscr{L}^*_x) = \begin{cases} 0, & k \neq 0 \\ \mathscr{F}_x, & k = 0. \end{cases}$$

This lemma shows $\varinjlim_{0+\sqrt{-1}v_00 \in \tilde{U}} H^k(\tilde{U}, \tau^{-1}\mathcal{O}_X) = 0$ for $k > 0$, and then $\varinjlim_{0+\sqrt{-1}v_00 \in \tilde{U}} H^0(\tilde{U}, \tau^{-1}\mathcal{O}_X) = (\tau^{-1}\mathcal{O}_X)_{0+\sqrt{-1}v_00} = \mathcal{O}_{X,0}$. Taking the direct limit as $\epsilon \to 0$ in (2.1.2) and replacing $\tilde{U}$ by $\tilde{U}_\epsilon$ provides the following exact sequences

$$\left.\begin{aligned} &0 \to \mathscr{H}^0_{\sqrt{-1}SM}(\tau^{-1}\mathcal{O}_X)_{0+\sqrt{-1}v_0 0} \to \mathcal{O}_{X,0} \\ &\to \varinjlim_{0+\sqrt{-1}v_0 0 \in \tilde{U}} \mathcal{O}_X(\tilde{U} - \sqrt{-1}SM) \\ &\to \mathscr{H}^1_{\sqrt{-1}SM}(\tau^{-1}\mathcal{O}_X)_{0+\sqrt{-1}v_0 0} \to 0 \\ &\text{and} \\ &0 \to \mathscr{H}^k_{\sqrt{-1}SM}(\tau^{-1}\mathcal{O}_X) \to 0 \qquad (\text{for } k \geqq 2). \end{aligned}\right\} \tag{2.1.3}$$

Hence $\mathscr{H}^k_{\sqrt{-1}SM}(\tau^{-1}\mathcal{O}_X) = 0$ for $k \geqq 2$. By the uniqueness of the continuation of holomorphic functions, one has the monomorphism $\mathcal{O}_{X,0} \to \varinjlim_{0+\sqrt{-1}v_0 0 \in \tilde{U}} \mathcal{O}_X(\tilde{U} - \sqrt{-1}SM)$, which implies $\mathscr{H}^0_{\sqrt{-1}SM}(\tau^{-1}\mathcal{O}_X) = 0$.

Definition 2.1.3. *Define the sheaf $\mathscr{Q}$ over $\sqrt{-1}SM$ by $\mathscr{Q} = \mathscr{H}^1_{\sqrt{-1}SM}(\tau^{-1}\mathcal{O}_X)$.* Denote the sheaf of real analytic functions on M by $\mathscr{A}$ (or $\mathscr{A}_M$). I.e., $\mathscr{A} = \mathcal{O}_X|_M$.

Proposition 2.1.2. *The sequence of sheaves over $\sqrt{-1}SM$,*

$$0 \to \tau^{-1}\mathscr{A} \to \tilde{\mathscr{A}} \to \mathscr{Q} \to 0$$

is exact.

Proof. Notice that $\mathcal{O}_{X,0} = \mathscr{A}_{M,0} = (\tau^{-1}\mathscr{A})_{0+\sqrt{-1}v_0 0}$ and

$$\varinjlim_{0+\sqrt{-1}v_0 0 \in \tilde{U}} \mathcal{O}_X(\tilde{U} - \sqrt{-1}SM) = \tilde{\mathscr{A}}_{0+\sqrt{-1}v_0 0};$$

see the remark following Definition 2.1.2. Then the exactness of this proposition follows from the exact sequence in (2.1.3).

Remark. The exact sequence in Proposition 2.1.2 indicates that the sheaf $\mathscr{Q}$ may be considered to reflect the irregularity of the sheaf $\tilde{\mathscr{A}}$ of boundary values.

Since $T^*X|_M = T^*M \oplus \sqrt{-1}T^*M$, i.e. for $x \in M$, $T^*_xX = T^*_xM \oplus \sqrt{-1}T^*_xM$, one has $T^*_MX = \sqrt{-1}T^*M$ and $S^*_MX = \sqrt{-1}S^*M$. Let $x \in M$, and let $\eta \in T^*_xM - \{0\}$; then denote the point by $(x, \sqrt{-1}\eta\infty)$ in $\sqrt{-1}S^*M$ corresponding to the point $\sqrt{-1}\eta$. The symbol ∞ is used to suggest that the point $(x, \sqrt{-1}\eta\infty)$ is dual to $(x, \sqrt{-1}\xi) \in \sqrt{-1}TM$, which is denoted by $x + \sqrt{-1}\xi 0$. When we emphasize the point as being on a cotangent bundle, we write $(x, \sqrt{-1}\langle\eta, dx\rangle\infty)$ instead of $(x, \sqrt{-1}\eta\infty)$.

Definition 2.1.4. *Denote* $DM = \{(x+\sqrt{-1}v0, (x, \sqrt{-1}\eta\infty)) \in \sqrt{-1}SM \underset{M}{\times} \sqrt{-1}S^*M \mid \langle v, \eta\rangle \leqq 0\}$.

Definition 2.1.5. *The antipodal mapping* $a:\sqrt{-1}S^*M \to \sqrt{-1}S^*M$ *is defined by* $a(x, \sqrt{-1}\eta\infty) = (x, -\sqrt{-1}\eta\infty)$. *If* $\mathscr{F}$ *is a sheaf over* $\sqrt{-1}S^*M$, *we define* $\mathscr{F}^a = a_*\mathscr{F}$ $(=a^{-1}\mathscr{F})$. *The sheaf* $\mathscr{F}^a$ *is called the antipodal image of* $\mathscr{F}$. Note that $\mathscr{F}^a_{(x,\sqrt{-1}\eta\infty)} = \mathscr{F}_{(x,-\sqrt{-1}\eta\infty)}$.

Definition 2.1.6. *Let X and Y be topological spaces, and let* $f:X \to Y$ *be a continuous map. For a sheaf* $\mathscr{F}$ *over X and an open set U in Y, the assignment* $H^k(f^{-1}(U), \mathscr{F})$ *is a presheaf. The associated sheaf is denoted by* $R^kf_*(\mathscr{F})$.

Remark. When $k = 0$, $R^0f_*(\mathscr{F})$ is isomorphic to $f_*\mathscr{F}$.

Definition 2.1.7. *Let X and Y be topological spaces. A continuous map* $f:X \to Y$ *is said to be purely r-dimensional with respect to a sheaf* $\mathscr{F}$ *over X if* $R^jf_*\mathscr{F} = 0$ *for* $j \neq r$.

Proposition 2.1.2′. *The map* $\tau:DM \to \sqrt{-1}S^*M$ *is purely* (*n*-1)*-dimensional with respect to* $\pi^{-1}\mathscr{Q}$; *i.e.* $R^k\tau_*\pi^{-1}\mathscr{Q} = 0$ *for* $k \neq n-1$ (*see Figure* 2.1.3).

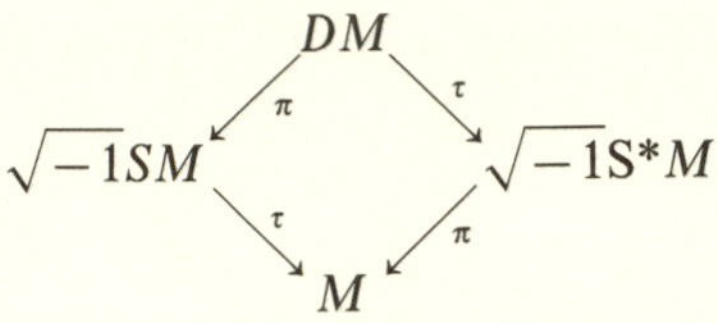

Figure 2.1.3

The proof of Proposition 2.1.2′ is lengthy. In this section we will prove Proposition 2.1.2′ by momentarily accepting a theorem on the triviality of cohomology groups. See the remark following Proposition 2.1.6. The proofs for the needed theorem and the purely *n*-codimensionality of *M* with respect to $\mathscr{O}_X$ will be given in the next section.

We are now in the position to give the definition of microfunctions.

Definition 2.1.8. *The sheaf* $\mathscr{C}_M$ *is defined as follows:*

$$\mathscr{C}_M = (R^{n-1}\tau_*\pi^{-1}\mathscr{Q})^a \otimes \pi^{-1}\omega_M.$$

The sections of the sheaf $\mathscr{C}_M$ *are called microfunctions. The stalk* $\mathscr{C}_{M,(x,\sqrt{-1}\xi\infty)} = (R^{n-1}\tau_*\pi^{-1}Q)_{(x,-\sqrt{-1}\xi\infty)} \otimes \omega_{M,x}$ (*see Figure* 2.1.3 *for the maps* τ *and* π).

We begin with preliminaries for the proof of Proposition 2.1.2′.

Definition 2.1.9. *Let* $\mathscr{F}$ *be a sheaf over a topological space X. Define a topology on the set* $F = \bigcup_{x \in X} \mathscr{F}_x$ *with the fundamental neighborhood system*

$\{s_x | x \in U\}$ for open sets, where U is an open set in X and $s \in \Gamma(U, \mathscr{F})$. Then, F is called the sheaf space of $\mathscr{F}$. Define a map $\varpi: F \to X$ by $\varpi(x') = x$ for $x' \in \mathscr{F}_x$. Then it is plain to see that ϖ is continuous.

Now we have the following facts.

(a) The map ϖ is a locally homeomorphism; i.e., for $y \in F$ there exists a neighborhood U of y such that, for some neighborhood V of $\varpi(y)$, $\varpi: U \to V$ is a homeomorphism. This statement (a) follows immediately from the definition of the topology on F.

(b) *Let* $\mathscr{G}(U) = \{f | f: U \to F$ is continuous and $\varpi \circ f = \mathrm{id}_U\}$. Then $\mathscr{G}(U) \cong \mathscr{F}(U)$ holds, where the map is given by $f(x) = s_X$ for $s \in \mathscr{F}(U)$. In fact, since f is continuous, there exists a neighborhood U' of x such that for $x' \in U'$ one has $f(x') = s'_{x'}$ for some $s' \in \mathscr{F}(U')$.

Therefore $\mathscr{F}_x = \varinjlim_{x \in U} \mathscr{G}(U)$; i.e. the sequence $0 \to \mathscr{F} \to \mathscr{G} \to 0$ is exact. Then, $0 \to \mathscr{F}(U) \to \mathscr{G}(U) \to 0$ is exact (see Proposition 1.1.2 (1.i)).

Conversely, let F and X be topological spaces, and let $\varpi: F \to X$ be a locally homeomorphism which is onto. For an open set U in X, the assignment $\{f | f: U \to F$ is continuous such that $\varpi \circ f = \mathrm{id}_U\}$ induces a sheaf $\mathscr{F}$. Notice that the sheaf space of $\mathscr{F}$ is then F. Therefore there is a one-to-one correspondence between sheaf spaces and sheaves. In particular, one can define the sections over an arbitrary subset A by $\{f | f: A \to F$ is continuous and $\varpi \circ f = \mathrm{id}_U\}$.

Let $f: X \to Y$ be a continuous map, let $\mathscr{F}$ be a sheaf over Y, and let F be the sheaf space of $\mathscr{F}$. Then the sheaf space of $f^{-1}\mathscr{F}$ is given as the fibre product $X \underset{Y}{\times} F = \{(x, x') \in X \times F | f(x) = \varpi(x')\}$. See Figure 2.1.4., below:

$$\begin{array}{ccc} X \underset{Y}{\times} F & \xrightarrow{\tilde{f}} & F \\ \downarrow{\scriptstyle \varpi} & & \downarrow{\scriptstyle \varpi} \\ X & \xrightarrow{f} & Y \end{array}$$

Figure 2.1.4

Let Z be a closed set in Y, and let $s \in \Gamma_Z(Y, \mathscr{F})$. Then, $s: Y \to F$ is continuous such that, for $y \notin Z$, $s(y) = 0$ in $\mathscr{F}_y$. Define a map $\tilde{s}: X \to X \times F$ by $\tilde{s}(x) = (x, (s \circ f)(x))$; then $\tilde{s}$ is continuous and $\varpi \circ \tilde{s} = \mathrm{id}_X$. Therefore, $\tilde{s} \in \Gamma_{f^{-1}(Z)}(X, f^{-1}\mathscr{F})$. That is, there is a canonical map $\Gamma_Z(Y, \mathscr{F}) \to \Gamma_{f^{-1}(Z)}(X, f^{-1}\mathscr{F})$. Let $0 \to \mathscr{F} \to \mathscr{L}^0 \to \mathscr{L}^1 \to \cdots$ be a flabby resolution of $\mathscr{F}$; then the sequence $0 \to f^{-1}\mathscr{F} \to f^{-1}\mathscr{L}^0 \to f^{-1}\mathscr{L}^1 \to \cdots$ is exact. Note that $f^{-1}\mathscr{L}^i$, $i = 0, 1, \ldots,$ is not a flabby sheaf, in general. From Lemma 1 in the proof of Theorem 1.1.1, there exists a flabby resolution

$\mathscr{M}^*$ of $f^{-1}\mathscr{F}$ such that the diagram

$$\begin{array}{ccccccc} 0 \to f^{-1}\mathscr{F} & \to & f^{-1}\mathscr{L}^0 & \to & f^{-1}\mathscr{L}^1 & \to & \cdots \\ \| & & \downarrow & & \downarrow & & \\ 0 \to f^{-1}\mathscr{F} & \to & \mathscr{M}^0 & \to & \mathscr{M}^1 & \to & \cdots \end{array}$$

is commutative. Therefore the commutative diagram

$$\begin{array}{ccccccc} 0 \longrightarrow \Gamma_Z(Y, \mathscr{L}^0) & \longrightarrow & \Gamma_Z(Y, \mathscr{L}^1) & \longrightarrow & \Gamma_Z(Y, \mathscr{L}^2) & \longrightarrow & \cdots \\ \downarrow & & \downarrow & & \downarrow & & \\ 0 \to \Gamma_{f^{-1}(Z)}(X, f^{-1}\mathscr{L}^0) & \to & \Gamma_{f^{-1}(Z)}(X, f^{-1}\mathscr{L}^1) & \to & \Gamma_{f^{-1}(Z)}(X, f^{-1}\mathscr{L}^2) & \to & \cdots \\ \downarrow & & \downarrow & & \downarrow & & \\ 0 \longrightarrow \Gamma_{f^{-1}(Z)}(X, \mathscr{M}^0) & \longrightarrow & \Gamma_{f^{-1}(Z)}(X, \mathscr{M}^1) & \longrightarrow & \Gamma_{f^{-1}(Z)}(X, \mathscr{M}^2) & \longrightarrow & \cdots \end{array}$$

is induced. Hence there is induced a map from $H^k_Z(Y, \mathscr{F})$ to $H^k_{f^{-1}(Z)}(X, f^{-1}\mathscr{F})$. Then there are the long exact sequences such that the diagram

$$\begin{array}{ccccccc} \cdots \longrightarrow H^k_Z(Y, \mathscr{F}) & \longrightarrow & H^k(Y, \mathscr{F}) & \longrightarrow & H^k(Y - Z, \mathscr{F}) & \longrightarrow & \\ \downarrow & & \downarrow & & \downarrow & & \\ \cdots \to H^k_{f^{-1}(Z)}(X, f^{-1}\mathscr{F}) & \to & H^k(X, f^{-1}\mathscr{F}) & \to & H^k(X - f^{-1}(Z), f^{-1}\mathscr{F}) & \to & \end{array}$$

$$\begin{array}{c} H^{k+1}_Z(Y, \mathscr{F}) \to \cdots \\ \downarrow \\ H^{k+1}_{f^{-1}(Z)}(X, f^{-1}\mathscr{F}) \to \cdots \end{array}$$

commutes. This proves (1) of Proposition 2.1.3, below.

Proposition 2.1.3. *Let X and Y be topological spaces, and let $f: X \to Y$ be a continuous map. Suppose that $\mathscr{F}$ is a sheaf over Y, and that Z is a closed set of Y.*

(1) *Then there exists a map $H^k_Z(Y, \mathscr{F}) \to H^k_{f^{-1}(Z)}(X, f^{-1}\mathscr{F})$ such that the diagram*

$$\begin{array}{ccccccc} \cdots \longrightarrow H^k_Z(Y, \mathscr{F}) & \longrightarrow & H^k(Y, \mathscr{F}) & \longrightarrow & H^k(Y - Z, \mathscr{F}) & \longrightarrow & \cdots \\ \downarrow & & \downarrow & & \downarrow & & \\ \cdots \to H^k_{f^{-1}(Z)}(X, f^{-1}\mathscr{F}) & \to & H^k(X, f^{-1}\mathscr{F}) & \to & H^k(X - f^{-1}(Z), f^{-1}\mathscr{F}) & \to & \cdots \end{array}$$

is commutative.

(2) *If f is an open map which is onto and if each fibre $f^{-1}(y)$ for $y \in Y$ is connected, then $\Gamma_Z(Y, \mathscr{F})$ and $\Gamma_{f^{-1}(Z)}(X, f^{-1}\mathscr{F})$ are isomorphic.*

(3) *In the case where f is a homeomorphism, $H^k_Z(Y, \mathscr{F}) = H^k_{f^{-1}(Z)}(X, f^{-1}\mathscr{F})$ holds for $k = 0, 1, 2, \ldots$.*

Proof. Let $\tilde{s}$ be the image of $s \in \Gamma_Z(Y, \mathscr{F})$ under the above-constructed canonical map $\Gamma_Z(Y, \mathscr{F}) \to \Gamma_{f^{-1}(Z)}(X, f^{-1}\mathscr{F})$. Denote the sheaf space of $\mathscr{F}$ by F. Then recall that $\tilde{s}: X \to X \underset{Y}{\times} F$ is defined by $\tilde{s}(x) = (x, (s \circ f)(x))$ for $s: Y \to F$. Therefore one may regard $\tilde{s}: X \to F$. Since f is onto for an arbitrary $y \in Y$, there exists $x \in X$ such that $y = f(x)$. Suppose $\tilde{s}_1 = \tilde{s}_2$; then $s_1(y) = (s_1 \circ f)(x) = \tilde{s}_1(x) = \tilde{s}_2(x) = (s_2 \circ f)(x) = s_2(y)$ holds, i.e. $s_1 = s_2$. Therefore the map $s \mapsto \tilde{s}$ is a monomorphism. Let $\tilde{s}: X \to F$. Then define $s: Y \to F$ by $s(y) = \tilde{s}(x)$, where $y = f(x)$. Next we will show that this definition does not depend upon the choice of x. Fix $x_0 \in X$ such that $f(x_0) = y$. Then the set $\{x \in f^{-1}(y) \mid \tilde{s}(x) = \tilde{s}(x_0)\} = \tilde{s}^{-1}(\tilde{s}(x_0))$ is a closed set (see Figure 2.1.5),

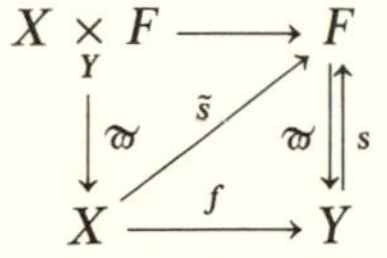

Figure 2.1.5

since $\tilde{s}$ is continuous. Since ϖ is a local homeomorphism, $f^{-1}(y) \xrightarrow{\tilde{s}} \varpi^{-1}(y)$ is continuous and $\varpi^{-1}(y)$ is discrete, therefore the inverse image $\tilde{s}^{-1}(\tilde{s}(x_0))$ of a point in $\varpi^{-1}(y)$ is open. By the assumption, $f^{-1}(y)$ is connected; thus $f^{-1}(y) = \tilde{s}^{-1}(\tilde{s}(x_0))$. Hence $s: Y \to F$ is uniquely determined. We will show next that s is continuous. Let U be an open set in F. Then one has $s^{-1}(U) = f(f^{-1}(s^{-1}(U))) = f(\tilde{s}^{-1}(U))$, and $\tilde{s}^{-1}(U)$ is open. Since f is an open map by assumption, the set $s^{-1}(U) = f(\tilde{s}^{-1}(U))$ is open. Therefore s is continuous.

Lastly we will prove (3). Let $0 \to \mathscr{F} \to \mathscr{L}^0 \to \mathscr{L}^1 \to \cdots$ be a flabby resolution of $\mathscr{F}$. Since f is a homeomorphism, $0 \to f^{-1}\mathscr{F} \to f^{-1}\mathscr{L}^0 \to f^{-1}\mathscr{L}^1 \to \cdots$ is a flabby resolution of $f^{-1}\mathscr{F}$. From (2), one has the commutative diagram

$$\begin{array}{ccccccccc}
 & & 0 & & 0 & & 0 & & \\
 & & \downarrow & & \downarrow & & \downarrow & & \\
0 & \longrightarrow & \Gamma_Z(Y, \mathscr{L}^0) & \longrightarrow & \Gamma_Z(Y, \mathscr{L}^1) & \longrightarrow & \Gamma_Z(Y, \mathscr{L}^2) & \longrightarrow & \cdots \\
 & & \downarrow & & \downarrow & & \downarrow & & \\
0 & \to & \Gamma_{f^{-1}(Z)}(X, f^{-1}\mathscr{L}^0) & \to & \Gamma_{f^{-1}(Z)}(X, f^{-1}\mathscr{L}^1) & \to & \Gamma_{f^{-1}(Z)}(X, f^{-1}\mathscr{L}^2) & \to & \cdots \\
 & & \downarrow & & \downarrow & & \downarrow & & \\
0 & \longrightarrow & 0 & \longrightarrow & 0 & \longrightarrow & 0 & \longrightarrow &
\end{array}$$

Therefore $H^k_Z(Y, \mathscr{F}) = H^k_{f^{-1}(Z)}(X, f^{-1}\mathscr{F})$ for $k = 0, 1, 2, \ldots$.

Proposition 2.1.4. *Let X be a topological space, and assume that all the open subsets are paracompact and Hausdorff. Then, for an arbitrary subset S of X, there exists the isomorphism* $H^k(S, \mathscr{F}|_S) \cong \varinjlim_{S \subset U} H^k(U, \mathscr{F})$.

Note. This proposition still holds if one assumes that X is paracompact and that S is closed in X. For example, if X is a metric space, the hypotheses of this proposition are satisfied.

Proof. As the first step, we will prove $\Gamma(S, \mathscr{F}|_S) \cong \varinjlim_{S \subset W} \Gamma(W, \mathscr{F})$. Let $s_1 \in \Gamma(W_1, \mathscr{F})$ and $s_2 \in \Gamma(W_2, \mathscr{F})$ such that $s_1|_S = s_2|_S$ in $\Gamma(S, \mathscr{F}|_S)$. Denote the embedding $S \hookrightarrow X$ by j. Then, for any $x \in S$, one has $(\mathscr{F}|_S)_x = (j^{-1}\mathscr{F})_x = \mathscr{F}_x$. Therefore $(s_1)_x = (s_2)_x$ holds for $x \in S$. This implies that there exists a neighborhood of x, $W(x)(\subset W_1 \cap W_2)$, so that $s_1|_{W(x)} = s_2|_{W(x)}$ is true. Then $W = \bigcup_{x \in S} W(x)$ is a neighborhood of S such that $s_1|_W = s_2|_W$ and $W \subset W_1 \cap W_2$. Therefore the map $\varinjlim_{S \subset W} \Gamma(W, \mathscr{F}) \to \Gamma(S, \mathscr{F}|_S)$ is monomorphic. Next we will show that this map is an epimorphism. Let $s \in \Gamma(S, \mathscr{F}|_S)$ and $x \in S$; then $s_x \in (j^{-1}\mathscr{F})_x = \mathscr{F}_x$. So there is a neighborhood of x, $W(x)$, such that for $x \in S \cap W(x)$ one has $\tilde{s}_x = s_x$ for some $\tilde{s} \in \Gamma(W(x), \mathscr{F})$. Then $S \subset \bigcup_{x \in S} W(x)$ is a covering of S. Therefore one obtains a locally finite refinement $\{U_i\}_{i \in I}$. Then there exists $s_i \in \Gamma(U_i, \mathscr{F})$ so that $(s_i)_x = s_x$ holds for $x \in S \cap U_i$. Furthermore, one may assume $X = \bigcup_i U_i$ without loss of generality. Since X is paracompact and Hausdorff, X is a normal space. Therefore there can be found a covering $\{V_i\}$ such that $\bar{V}_i \subset U_i$ and $X = \bigcup_i V_i$, since $\{U_i\}$ is a locally finite covering of X. Define $W = \{x \in X \mid (s_i)_x = (s_j)_x \text{ for } x \in \bar{V}_i \cap \bar{V}_j\}$. We will show that W is a neighborhood of S. The assumption of being locally finite implies that one finds a neighborhood U of x and $(i_1, \ldots, i_N)$ with the property $U \cap \bar{V}_i = \varnothing$ for $i \neq i_1, \ldots, i_N$. In the case where $x \notin \bar{V}_i$, replace U by $U - \bar{V}_i$ so that one may have $x \in U \cap \bar{V}_i$ for $i = i_1, \ldots, i_N$. For a sufficiently small neighborhood V of x, $s_{i_\mu}|_V = s_{i_\nu}|_V$ for $\mu, \nu = 1, \ldots, N$. Then there exists $\tilde{s} \in \Gamma(\Omega, \mathscr{F})$ for some open set Ω, $S \subset \Omega \subset W$, such that $\tilde{s}|_{V_i \cap \Omega} = s_i|_{V_i \cap \Omega}$ and $\tilde{s}_x = s_x$ for $x \in S$. Hence, we have proved $\Gamma(S, \mathscr{F}|_S) \cong \varinjlim_{S \subset W} \Gamma(W, \mathscr{F})$.

We will prove the case of higher cohomology groups next. Let $0 \to \mathscr{F} \to \mathscr{L}^0 \to \mathscr{L}^1 \to \cdots$ be a flabby resolution of $\mathscr{F}$. Let V be an arbitrary open set in X. Then $\Gamma(V \cap S, \mathscr{L}^i|_S) = \varinjlim_{V \cap S \subset W} \Gamma(W, \mathscr{L}^i)$. Therefore, $s \in \Gamma(V \cap S, \mathscr{L}^i|_S)$ can be extended to a section of $\mathscr{L}^i$ over an open set W containing $V \cap S$; and since $\mathscr{L}^i$ is flabby, it can be extended to a section over X, whose restriction to $V \cap S$ of the image in $\Gamma(S, \mathscr{L}^i|_S)$ is s. Therefore, $0 \to \mathscr{F}|_S \to \mathscr{L}^0|_S \to \mathscr{L}^1|_S \to \cdots$ gives a flabby resolution of $\mathscr{F}|_S$. Hence one has isomorphisms $\varinjlim_{S \subset U} H^k(U, \mathscr{F}) = \varinjlim_{S \subset U} H^k(\Gamma(U, \mathscr{L}^*)) \cong H^k\left(\varinjlim_{S \subset U} \Gamma(U, \mathscr{L}^*)\right) \cong H^k(\Gamma(S, \mathscr{L}^*|_S)) = H^k(S, \mathscr{F}|_S)$ for $k = 0, 1, 2, \ldots$.

Corollary. *Let X be a topological space whose open subsets are paracompact and Hausdorff. Let $\mathscr{F}$ be a sheaf over X, and let $f: X \to Y$ be a closed continuous map. Then for each $y \in Y$, $R^k f_*(\mathscr{F})_y = H^k(f^{-1}(y), \mathscr{F}|_{f^{-1}(y)})$ for $k = 0, 1, 2, \ldots$.*

Proof. Let U be an open neighborhood of $f^{-1}(y)$. Since f is a closed map, $V = Y - f(X - U)$ is an open neighborhood of y such that $f^{-1}(V) \subset U$ holds. Therefore, by Proposition 2.1.4, $H^k(f^{-1}(y), \mathscr{F}|_{f^{-1}(y)}) \cong \varinjlim_{f^{-1}(y) \subset U} H^k(U, \mathscr{F}) = \varinjlim_{y \in V} H^k(f^{-1}(V), \mathscr{F}) = R^k f_*(\mathscr{F})_y$.

Definition 2.1.10. *A continuous map $f: X \to Y$ is said to be proper if conditions* (1) *and* (2) *are satisfied:*

(1) *For each $y \in Y$ the fibre $f^{-1}(y)$ is compact.*

(2) *f is a closed map; i.e. the image of any closed set in X under f is always a closed set in Y.*

Remark. The above corollary holds if X and Y are Hausdorff and if $f: X \to Y$ is a proper map.

Proposition 2.1.5. *The map $\tau: DM \to \sqrt{-1}S^*M$ is proper.*

In order to prove this proposition we need several lemmas.

Lemma 1. *Let X and Y be topological spaces, and let $f: X \to Y$ be a continuous map. Then the following* (1) *and* (2) *are equivalent:*

(1) *f is a closed map.*

(2) *For each $y \in Y$, let U be a neighborhood of $f^{-1}(y)$. Then there exists a neighborhood V of y such that $f^{-1}(V) \subset U$. That is, $\{f^{-1}(V) \mid V$ is an element of a fundamental neighborhood system $V \ni y\}$ is a fundamental neighborhood system for $f^{-1}(y)$.*

Proof. Let $V = Y - f(X - U)$; then (2) follows plainly from (1). Let A be a closed set in X. If $y \notin f(A)$, then $A \cap f^{-1}(y) = \varnothing$ and $(X - A) \supset f^{-1}(y)$. Therefore, there exists a neighborhood V of y so that $f^{-1}(V) \subset X - A$, i.e. $V \cap f(A) = \varnothing$, which means that $f(A)$ is closed.

Lemma 2. *In the Figure* 2.1.6, *if f is a proper map, then f' is proper, where $X \times_Y Y'$ indicates the fibre product of $f: X \to Y$ and $g: Y' \to Y$, where g' and f' are defined as $g'(x, y') = x$ and $f'(x, y') = y'$, respectively, for $(x, y') \in X \times_Y Y'$.*

$$\begin{array}{ccc} X & \xleftarrow{g'} & X \underset{Y}{\times} Y' \\ \Big\downarrow{\scriptstyle f} & & \Big\downarrow{\scriptstyle f'} \\ Y & \xleftarrow{g} & Y' \end{array}$$

Figure 2.1.6

Proof. For $y' \in Y'$ one has the isomorphism $f'^{-1}(y') = \{(x, y') \in X \underset{Y}{\times} Y' \mid f(x) = g(y')\} \cong f^{-1}(g(y'))$. Since f is a proper map, $f'^{-1}(y')$ is compact. Hence it suffices to prove f' as a closed map. Let $y' \in Y'$, and let U be a neighborhood of $f'^{-1}(y')$. Then, by Lemma 1, it is sufficient to prove that $f'^{-1}(U') \subset U$ for some neighborhood U' of y'. Let $x' \in f'^{-1}(y')$; then let $x = g'(x')$ and $y = g(y')$. Then one can find neighborhoods $W(x')$ of $x = g'(x')$ and $V(x')$ of y' such that $g'^{-1}(W(x')) \cap f'^{-1}(V(x')) \subset U$. Note $\bigcup_{x' \in f'^{-1}(y')} W(x') \supset f^{-1}(y)$. Since $f^{-1}(y)$ is compact, there are finitely many $x_1, \ldots, x_N \in f'^{-1}(y')$ such that $W = \bigcup_{i=1}^{N} W(x_i) \supset f^{-1}(y)$. Then define $V = \bigcap_{i=1}^{N} V(x_i)$. Notice that V is also a neighborhood of y' and that $f'^{-1}(y') \subset g'^{-1}(W) \cap f'^{-1}(V) \subset U$. Since f is a closed map, there exists a neighborhood V' of y such that $f^{-1}(y) \subset f^{-1}(V') \subset W$. Therefore $g'^{-1}(W) \supset g'^{-1}f^{-1}(V') = f'^{-1}g^{-1}(V')$. Let $U' = V \cap g^{-1}(V')$. Then U' is a neighborhood of y' with the property $f'^{-1}(U') \subset U$.

Lemma 3. *Let $f: X \to Y$ be a proper map and Y be compact. Then X is compact.*

Proof. Let $X = \bigcup_{i \in I} U_i$ be an arbitrary open covering of X. For each $y \in Y$, $f^{-1}(y)$ is compact and $f^{-1}(y) \subset \bigcup_{i \in I} U_i$. Therefore there exists a finite subset $I(y)$ of I such that $f^{-1}(y) \subset \bigcup_{i \in I(y)} U_i$. Since f is a closed map, by Lemma 1 there is a neighborhood $V(y)$ of y so that $f^{-1}(V(y)) \subset \bigcup_{i \in I(y)} U_i$. Since $\bigcup_{y \in Y} V(y)$ is an open covering of Y and Y is compact, $Y = \bigcup_{i=1}^{N} V(y_i)$ for finitely many points $y_1, y_2, \ldots, y_N$ in Y. Then $X = \bigcup_{i \in \left(\bigcup_{i=1}^{N} I(y_i)\right)} U_i$ is a finite covering of X.

Lemma 4. *If $f: X \to Y$ and $g: Y \to Z$ are proper maps, then $g \circ f: X \to Z$ is proper.*

Proof. If f and g are closed maps, then the composition $g \circ f$ is a closed map. Therefore it suffices to prove that $f^{-1}(g^{-1}(z))$ is compact for each $z \in Z$. By Lemma 3, it is enough to show that $f : f^{-1}(g^{-1}(z)) \to g^{-1}(z)$ is proper. This is plainly so from Lemma 2.

Lemma 5. *$\sqrt{-1}SM \xrightarrow{\tau} M$ is a proper map.*

Proof. From Lemma 1, the map τ being proper is local in nature. For a continuous map $f : X \to Y$ and an open covering of Y, $Y = \bigcup_{i \in I} U_i$, it suffices to prove that $f : f^{-1}(U_i) \to U_i$ is proper for each $i \in I$. Hence, from the beginning, one may assume $\sqrt{-1}SM \cong S^{n-1} \times M$ to prove this assertion. For a point x in M, the map $S^{n-1} \to x$ is a proper map. Therefore, by Lemma 2, τ is a proper map.

We are now ready to prove Proposition 2.1.5. Consider the commutative diagram

$$
\begin{array}{ccccc}
DM & \overset{\iota}{\hookrightarrow} & \sqrt{-1}SM \underset{M}{\times} \sqrt{-1}S^*M & \to & \sqrt{-1}SM \\
 & \searrow^{\tau} & \downarrow p_1 & & \downarrow p_2 \\
 & & \sqrt{-1}S^*M & \xrightarrow{\pi} & M
\end{array}
$$

In this commutative diagram, p_2 is a proper map by Lemma 5. Therefore, by Lemma 2, p_1 is a proper map. Then, by Lemma 4, $\tau = p_1 \circ \iota$ is proper, since ι is proper.

Combining the corollary of Proposition 2.1.4 and Proposition 2.1.5, we have

$$(R^k\tau_*\pi^{-1}\mathscr{Q})_{(x_0, \sqrt{-1}\xi_0\infty)} = H^k(\tau^{-1}(x_0, \sqrt{-1}\xi_0\infty), \pi^{-1}\mathscr{Q}|_{\tau^{-1}(x_0, \sqrt{-1}\xi_0\infty)}).$$

The paracompactness condition on M, below, is satisfied at least locally. We will rephrase this cohomology group into the terms of holomorphic functions (see Proposition 2.1.6, below).

First note that one has the homeomorphism $\pi : \tau^{-1}(x_0, \sqrt{-1}\xi_0\infty) \xrightarrow{\sim} \pi\tau^{-1}(x_0, \sqrt{-1}\xi_0\infty)$. Therefore, by (3) of Proposition 2.1.3, $H^k(\tau^{-1}(x_0, \sqrt{-1}\xi_0\infty), \pi^{-1}\mathscr{Q}|_{\tau^{-1}(x_0, \sqrt{-1}\xi_0\infty)}) = H^k(F, \mathscr{Q}|_F)$, where $F = \pi\tau^{-1}(x_0, \sqrt{-1}\xi_0\infty) = \{x + \sqrt{-1}v0 \in \sqrt{-1}SM \mid \langle v, \xi_0 \rangle \leqq 0\}$. On the other hand, Proposition 2.1.4 implies $H^k(F, \mathscr{Q}) = \varinjlim_{F \subset \tilde{U} \subset \widetilde{M}X} H^k(\tilde{U}, \mathscr{H}^1_{\sqrt{-1}SM}(\tau^{-1}\mathscr{O}_X))$, which can also be written as $\varinjlim_{F \subset \tilde{U}} H^{k+1}_{\sqrt{-1}SM \cap \tilde{U}}(\tilde{U}, \tau^{-1}\mathscr{O}_X)$ by Propositions 2.1.1 and 1.1.6. Since the question is local in nature, one may assume $M = \mathbf{R}^n$ without loss of generality. We will construct a neighborhood $\tilde{U}_{(\xi,\epsilon)}$ of F. Let $\xi_1, \xi_2, \ldots, \xi_n$ be n real-vectors so that the convex hull of $\xi_0, \xi_1, \ldots, \xi_n$ contains a neighborhood of the origin. For $\xi = (\xi_1, \ldots, \xi_n)$ and $\varepsilon > 0$, define $\tilde{U}_{(\xi,\epsilon)} = \{x + \sqrt{-1}y \in X \mid |x - x_0| < \epsilon,\ |y| < \epsilon$ and

$\langle y, \xi_j \rangle > 0$ for some j, $1 \leqq j \leqq n\} \cup \{x + \sqrt{-1}v0 \in \sqrt{-1}SM \,|\, |x - x_0| < \epsilon$ and $\langle v, \xi_j \rangle > 0$ for some j, $1 \leqq j \leqq n\}$. Then $\tilde{U}_{(\xi,\epsilon)}$ is a neighborhood of F. When one takes $\xi_1, \ldots, \xi_n$ in a neighborhood of $-\xi_0$ and ϵ in a neighborhood of zero, $\{\tilde{U}_{(\xi,\epsilon)}\}$ forms a fundamental neighborhood system of F (see Figure 2.1.7 for $n = 2$). Note that when $\xi_j \to -\xi_0$ and $\varepsilon \to 0$, $\tilde{U}_{(\xi,\epsilon)}$ goes to F.

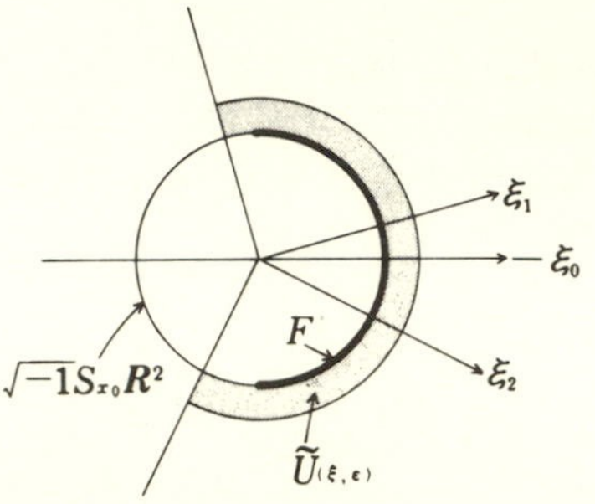

Figure 2.1.7. Cross-section of ${}^{R^2}\widetilde{C}^2$ at x_0

Lemma. *Let* $U_\epsilon = \{x + \sqrt{-1}y \in X \,|\, |y| < \epsilon, |x - x_0| < \epsilon\}$, *and let* $Z_{(\xi,\epsilon)} = \{x + \sqrt{-1}y \in U_\epsilon \,|\, \langle y, \xi_j \rangle \leqq 0$ *for any* j, $1 \leqq j \leqq n\}$. *Then* $\tilde{U}_{(\xi,\epsilon)} \cap \tau^{-1}(Z_{(\xi,\epsilon)}) = \tilde{U}_{(\xi,\epsilon)} \cap \sqrt{-1}SM$ *holds, where* $\tau: \widetilde{{}^M X} \to X$.

Proof. Let $x + \sqrt{-1}v0 \in \tilde{U}_{(\xi,\epsilon)} \cap \sqrt{-1}SM$. Then $x \in Z_{(\xi,\epsilon)}$. Therefore $x + \sqrt{-1}v0 \in (\sqrt{-1}SM)_x = \tau^{-1}(x) \subset \tau^{-1}(Z_{(\xi,\epsilon)})$. That is, $\tilde{U}_{(\xi,\epsilon)} \cap \sqrt{-1}SM \subset \tilde{U}_{(\xi,\epsilon)} \cap \tau^{-1}(Z_{(\xi,\epsilon)})$ holds. Suppose $x + \sqrt{-1}y \notin \sqrt{-1}SM$ and $x + \sqrt{-1}y \in \tilde{U}_{(\xi,\epsilon)} \cap \tau^{-1}(Z_{(\xi,\epsilon)})$. Then it would imply that $\langle y, \xi_j \rangle > 0$ for some j and $\langle y, \xi_j \rangle \leqq 0$ for all j, which is a contradiction.

Let $X = \tilde{U}_{(\xi,\epsilon)}$, $Y = U_\epsilon$, $f = \tau|_{\tilde{U}_{(\xi,\epsilon)}}$, $Z = Z_{(\xi,\epsilon)}$, and let $\mathscr{F} = \mathcal{O}_X$ in (1) of Proposition 2.1.3. Note $f^{-1}(Z) = \sqrt{-1}SM \cap \tilde{U}_{(\xi,\epsilon)}$ from the above lemma. Then we have the following commutative diagram:

$$\begin{array}{ccccc}
\cdots \longrightarrow H^{k-1}(U_\epsilon, \mathcal{O}_X) & \longrightarrow & H^{k-1}(U_\epsilon - Z_{(\xi,\epsilon)}, \mathcal{O}_X) & \longrightarrow & \\
\downarrow & & \downarrow & & \\
\cdots \to H^{k-1}(\tilde{U}_{(\xi,\epsilon)}, \tau^{-1}\mathcal{O}_X) & \to & H^{k-1}(\tilde{U}_{(\xi,\epsilon)} - \sqrt{-1}SM, \tau^{-1}\mathcal{O}_X) & \to & \\
H^k_{Z(\xi,\epsilon)}(U_\epsilon, \mathcal{O}_X) \longrightarrow & H^k(U_\epsilon, \mathcal{O}_X) & \longrightarrow & H^k(U_\epsilon - Z_{(\xi,\epsilon)}, \mathcal{O}_X) & \longrightarrow \cdots \\
\downarrow & \downarrow & & \downarrow & \\
H^k_{\sqrt{-1}SM \cap \tilde{U}(\xi,\epsilon)}(\tilde{U}_{(\xi,\epsilon)}, \tau^{-1}\mathcal{O}_X) \to & H^k(\tilde{U}_{(\xi,\epsilon)}, \tau^{-1}\mathcal{O}_X) & \to & H^k(\tilde{U}_{(\xi,\epsilon)} - \sqrt{-1}SM, \tau^{-1}\mathcal{O}_X) & \to \cdots
\end{array} \tag{2.1.4}$$

Since $\tilde{U}_{(\xi,\epsilon)} - \sqrt{-1}SM \cong U_\epsilon - Z_{(\xi,\epsilon)}$, from (3) of Proposition 2.1.3, one has

$$H^k(U_\epsilon - Z_{(\xi,\epsilon)}, \mathcal{O}_X) = H^k(\tilde{U}_{(\xi,\epsilon)} - \sqrt{-1}SM, \tau^{-1}\mathcal{O}_X).$$

Furthermore, one obtains

$$\varinjlim_{x_0 \in U_\epsilon} H^k(U_\epsilon, \mathcal{O}_X) = \begin{cases} 0 & \text{for } k \neq 0 \\ \mathcal{O}_{X,x_0} & \text{for } k = 0 \end{cases}$$

by the lemma in the proof of Proposition 2.1.1. Notice $\varinjlim_{F \subset \tilde{U}} H^k(\tilde{U}, \tau^{-1}\mathcal{O}_X) = H^k(F, \tau^{-1}\mathcal{O}_X|_F)$ and $\tau(F) = x_0$. Therefore, $\tau^{-1}\mathcal{O}_X|_F$ is a constant sheaf on F. Since F is homeomorphic to a demisphere, F is contractible. Hence, by Theorem 1.2.4,

$$H^k(F, \tau^{-1}\mathcal{O}_X|_F) = \begin{cases} 0 & \text{for } k \neq 0 \\ \mathcal{O}_{X,x_0} & \text{for } k = 0 \end{cases}$$

holds. Taking the direct limit in the commutative diagram (2.1.4), Five Lemma implies

$$\varinjlim_{F \subset \tilde{U}} H^k_{\tilde{U} \cap \sqrt{-1}SM}(\tilde{U}, \tau^{-1}\mathcal{O}_X) \cong \varinjlim_{\xi,\epsilon} H^k_{Z(\xi,\epsilon)}(U_\epsilon, \mathcal{O}_X) = \varinjlim_{\xi} \mathscr{H}^k_{Z_\xi}(\mathcal{O}_X)_x,$$

where $Z_\xi = \{x + \sqrt{-1}y \in X \mid \langle y, \xi_j \rangle \leqq 0 \text{ for } j = 1, 2, \ldots, n\}$. Notice $Z_{(\xi,\epsilon)} = Z_\xi \cap U_\epsilon$. Therefore we could rewrite $(R^k\tau_*\pi^{-1}\mathscr{Q})_{(x_0, \sqrt{-1}\xi\infty)}$ in terms of the cohomology group with coefficients in $\mathcal{O}_X$. This clarifies the relationship to hyperfunctions. More precisely, we have the following.

Proposition 2.1.6.

(1) *For an arbitrary k*

$$(R^k\tau_*\pi^{-1}\mathscr{Q})_{(x_0, \sqrt{-1}\xi_0\infty)} = \varinjlim_{\xi} \mathscr{H}^{k+1}_{Z_\xi}(\mathcal{O}_X)_{x_0}$$

holds.

(2) *A canonical epimorphism* $sp: \mathscr{B}_{M,x_0} \to \mathscr{C}_{M,(x_0, -\sqrt{-1}\xi_0\infty)}$ *exists.*

Remark. We will prove $\mathscr{H}^k_{Z_\xi}(\mathcal{O}_X)_{x_0} = 0$ for $k \neq n$ in the next section. This statement, with (1), proves Proposition 2.1.2′.

Proof. (1) has been proved; we will prove (2) here. It is sufficient to prove that the composite map

$$\mathscr{B}_{M,x_0} = \mathscr{H}^n_M(\mathcal{O}_X)_{x_0} \to \mathscr{H}^n_{Z_\xi}(\mathcal{O}_X)_{x_0} \to \varinjlim_{\xi} \mathscr{H}^n_{Z_\xi}(\mathcal{O}_X)_{x_0} = \mathscr{C}_{M,(x_0, -\sqrt{-1}\xi_0\infty)}$$

is an epimorphism. Notice that

$$\mathscr{H}^n_M(\mathcal{O}_X)_{x_0} = \varinjlim_{\epsilon} H^n_{M \cap U_\epsilon}(U_\epsilon, \mathcal{O}_X)$$

and

$$\mathscr{H}^n_{Z_\xi}(\mathscr{O}_X)_{x_0} = \varinjlim_{\epsilon} H^n_{Z_\xi \cap U_\epsilon}(U_\epsilon, \mathscr{O}_X)$$

hold. We first consider the map before taking the direct limit. Let $U_j = \{x + \sqrt{-1}y \in X \mid \langle y, \xi_j \rangle > 0\}$ for $j = 0, 1, \ldots, n$. Let the open coverings $\mathscr{U}$, $\mathscr{U}'$, and $\mathscr{U}''$ of U_ϵ, $U_\epsilon - U_\epsilon \cap M$, and $U_\epsilon - U_\epsilon \cap Z_\xi$ be as follows: $\mathscr{U} = \{U_\epsilon, U_\epsilon \cap U_0, U_\epsilon \cap U_1, \ldots, U_\epsilon \cap U_n\}$; $\mathscr{U}' = \{U_\epsilon \cap U_0, U_\epsilon \cap U_1, \ldots, U_\epsilon \cap U_n\}$; and $\mathscr{U}'' = \{U_\epsilon \cap U_1, \ldots, U_\epsilon \cap U_n\}$. Similarly as in §2, Chapter I, one has

$$C^{n+1}(\mathscr{U} \bmod \mathscr{U}', \mathscr{O}_X) = C^{n+1}(\mathscr{U} \bmod \mathscr{U}'', \mathscr{O}_X) = 0,$$

$$C^n(\mathscr{U} \bmod \mathscr{U}', \mathscr{O}_X) = C^n(\mathscr{U} \bmod \mathscr{U}'', \mathscr{O}_X) = \bigoplus_{j=0}^{n} \mathscr{O}_X(U_{\hat{j}}),$$

$$C^{n-1}(\mathscr{U} \bmod \mathscr{U}', \mathscr{O}_X) = \bigoplus_{j,k}{}' \mathscr{O}_X(U_{\widehat{j,k}}),$$

and

$$C^{n-1}(\mathscr{U} \bmod \mathscr{U}'', \mathscr{O}_X) = \bigoplus_{j,k}{}' \mathscr{O}_X(U_{\widehat{j,k}}) \oplus \bigoplus_{j=1}^{n} \mathscr{O}_X(U_{\hat{j}})$$

where $U_{\hat{j}} = \bigcap_{k \neq j} U_k$ and $U_{\widehat{j,k}} = \bigcap_{l \neq j,k} U_l$. Therefore the canonical epimorphism

$$H^n_{M \cap U_\epsilon}(U_\epsilon, \mathscr{O}_X) = \frac{\bigoplus_{j=0}^{n} \mathscr{O}_X(U_{\hat{j}})}{\bigoplus_{j,k}{}' \mathscr{O}_X(U_{\widehat{j,k}})} \to \frac{\bigoplus_{j=0}^{n} \mathscr{O}_X(U_{\hat{j}})}{\bigoplus_{j,k}{}' \mathscr{O}_X(U_{\widehat{j,k}}) \oplus \bigoplus_{j=1}^{n} \mathscr{O}_X(U_{\hat{j}})}$$
$$= H^n_{Z_\xi \cap U_\varsigma}(U_\epsilon, \mathscr{O}_X)$$

is induced. Therefore the map $\mathscr{B}_{M,x_0} \to \mathscr{C}_{M,(x_0, -\sqrt{-1}\xi_0 \infty)}$, obtained by taking the direct limit of the above epimorphism, is an epimorphism.

Definition 2.1.11. *The epimorphism sp is called the spectrum map (see Definition 2.3.3).*

Note. For $u \in \mathscr{B}_{M,x_0}$ we have

$$\mathrm{sp}(u)_{(x_0, -\sqrt{-1}\xi_0 \infty)} = 0 \quad \text{if and only if} \quad u = \sum_{j=1}^{n} b(\varphi_j), \tag{2.1.5}$$

where $\varphi_{\hat{j}} \in \mathscr{O}_X(U_{\hat{j}})$.

§2. Vanishing Theorems of Relative Cohomology Groups, Pure n-Codimensionality of $\mathbf{R}^n$ with respect to $\mathcal{O}_{\mathbf{C}^n}$ etc.

We will prove in this section that a real analytic manifold M is purely n-codimensional with respect to the sheaf $\mathcal{O}_X$ of holomorphic functions on a complexification X of M. That is, we show $\mathcal{H}^k_M(\mathcal{O}_X) = 0$ for $k \neq n$. We will also prove $\mathcal{H}^k_{Z_\xi}(\mathcal{O}_X)_{x_0} = 0$ for $k \neq n$, which was assumed in the proof of Proposition 2.1.6. These complete the proof of Proposition 2.1.2′ and Theorem 1.2.4.

Proposition 2.2.1. *Let Y be a complex manifold, and let Z be a closed set in Y. If $H^k_{V \times Z}(V \times Y, \mathcal{O}_{V \times Y}) = 0$ for an arbitrary complex manifold V and for $k < r$, then for any compact set $K \subset \mathbf{C}$, $H^k_{W \times K \times Z}(W \times \mathbf{C} \times Y, \mathcal{O}_{W \times \mathbf{C} \times Y)} = 0$ holds for an arbitrary complex manifold W and for $k < r + 1$.*

Proof. We will prove that, for $k < r + 1$, $H^k_{K \times Z}(\mathbf{C} \times Y, \mathcal{O}_{\mathbf{C} \times Y}) = 0$. As we will point out later in the proof, the general case can be reduced to this special case. The long exact sequence (3) in Theorem 1.1.2 implies that

$$\cdots \to H^{k-1}_{\mathbf{C} \times Z}(\mathbf{C} \times Y, \mathcal{O}_{\mathbf{C} \times Y}) \to H^{k-1}_{(\mathbf{C}-K) \times Z}(\mathbf{C} \times Y, \mathcal{O}_{\mathbf{C} \times Y})$$
$$\to H^k_{K \times Z}(\mathbf{C} \times Y, \mathcal{O}_{\mathbf{C} \times Y}) \to H^k_{\mathbf{C} \times Z}(\mathbf{C} \times Y, \mathcal{O}_{\mathbf{C} \times Y}) \to \cdots$$

is exact. Notice that $H^{k-1}_{(\mathbf{C}-K) \times Z}(\mathbf{C} \times Y, \mathcal{O}_{\mathbf{C} \times Y}) \cong H^{k-1}_{(\mathbf{C}-K) \times Z}((\mathbf{C} - K) \times Y, \mathcal{O}_{(\mathbf{C}-K) \times Y})$ by (1) of Theorem 1.1.2. Therefore for $k < r$ one obtains $H^k_{K \times Z}(\mathbf{C} \times Y, \mathcal{O}_{\mathbf{C} \times Y}) = 0$ from this long exact sequence.

Next we will prove the case when $k = r$. Let $\mathbf{P}^1 = \mathbf{C} \cup \{\infty\}$, a projective line, and let $\mathcal{I} = \{\varphi \in \mathcal{O}_{\mathbf{P}^1 \times Y} | \varphi(\infty, y) = 0\}$. Then one has $\mathcal{I}|_{\mathbf{C} \times Y} = \mathcal{O}_{\mathbf{C} \times Y}$ and $\mathcal{I}|_{(\mathbf{P}^1 - \{0\}) \times Y} \cong \mathcal{O}_{(\mathbf{P}^1 - \{0\}) \times Y}$ with respect to the origin $\{0\}$ via the map $\varphi(x, y) \mapsto x\varphi(x, y)$ (the inverse map being $(1/x)\psi(x, y) \leftarrow\!\shortmid \psi(x, y)$). We identify $\mathcal{I}|_{(\mathbf{P}^1 - \{0\}) \times Y}$ with $\mathcal{O}_{(\mathbf{P}^1 - \{0\}) \times Y}$ under the assumption $0 \in K$ so that $\mathcal{I}|_{(\mathbf{P}^1 - K) \times Y} = \mathcal{O}_{(\mathbf{P}^1 - K) \times Y}$, $\{0\} \subset K \subset \mathbf{P}^1$. Then, one has $H^{r-1}_{(\mathbf{P}^1 - K) \times Z}(\mathbf{P}^1 \times Y, \mathcal{I}) = H^{r-1}_{(\mathbf{P}^1 - K) \times Z}((\mathbf{P}^1 - K) \times Y, \mathcal{O}_{(\mathbf{P}^1 - K) \times Y}) = 0$. On the other hand, $H^r_{K \times Z}(\mathbf{C} \times Y, \mathcal{O}_{\mathbf{C} \times Y}) = H^r_{K \times Z}(\mathbf{C} \times Y, \mathcal{I}|_{\mathbf{C} \times Y}) = H^r_{K \times Z}(\mathbf{P}^1 \times Y, \mathcal{I})$ holds. Furthermore, the long exact sequence

$$\cdots \to H^{r-1}_{(\mathbf{P}^1 - K) \times Z}(\mathbf{P}^1 \times Y, \mathcal{I}) \to H^r_{K \times Z}(\mathbf{P}^1 \times Y, \mathcal{I})$$
$$\to H^r_{\mathbf{P}^1 \times Z}(\mathbf{P}^1 \times Y, \mathcal{I}) \to \cdots$$

implies that $H^r_{K \times Z}(\mathbf{C} \times Y, \mathcal{O}_{\mathbf{C} \times Y}) \to H^r_{\mathbf{P}^1 \times Z}(\mathbf{P}^1 \times Y, \mathcal{I})$ is a monomorphism. Hence, it suffices to prove $H^r_{\mathbf{P}^1 \times Z}(\mathbf{P}^1 \times Y, \mathcal{I}) = 0$ to complete the proof of Proposition 2.2.1. We need two lemmas.

Lemma 1. *Let $f: \mathbf{P}^1 \times Y \to Y$ be the projection map. Then $R^k f_* \mathcal{I} = 0$ for $k \geqq 0$.*

Proof. It is sufficient to prove $R^k f_*(\mathscr{I})_y = \varinjlim_{y \in U} H^k(\mathbf{P}^1 \times U, \mathscr{I}) = 0$ for each $y \in Y$. Since one can take a base of y consisting of Stein manifolds, one is to prove that, for any Stein manifold Y, $H^k(\mathbf{P}^1 \times Y, \mathscr{I}) = 0$ for $k \geqq 0$. Let $U_0 = (\mathbf{P}^1 - \{\infty\}) \times Y$, and let $U_1 = (\mathbf{P}^1 - \{0\}) \times Y$. Then $\mathbf{P}^1 \times Y = U_0 \cup U_1$ holds. Since U_0 and U_1 are Stein manifolds, Oka and Cartan's theorem (Theorem 1.2.2) says that $\mathscr{U} = \{U_0, U_1\}$ is a Leray covering for the sheaf $\mathscr{I}$. We can compute the cohomology groups $H^k(\mathbf{P}^1 \times Y, \mathscr{I})$ using the covering $\mathscr{U} = \{U_0, U_1\}$ by Leray's theorem (Theorem 1.2.1).

Note that $C^0(\mathscr{U}, \mathscr{I}) = \Gamma(U_0, \mathscr{I}) \oplus \Gamma(U_1, \mathscr{I})$, $C^1(\mathscr{U}, \mathscr{I}) = \Gamma(U_0 \cap U_1, \mathscr{I})$ and that $C^k(\mathscr{U}, \mathscr{I}) = 0$ for $k \geqq 2$. So $H^k(\mathbf{P}^1 \times Y, \mathscr{I}) = 0$ for $k \geqq 2$. Next we will prove $H^0(\mathbf{P}^1 \times Y, \mathscr{I}) = \Gamma(\mathbf{P}^1 \times Y, \mathscr{I}) = 0$. Let $\varphi(x, y) \in \Gamma(\mathbf{P}^1 \times Y, \mathscr{I})$, and let y be fixed. Then φ is a holomorphic function on the compact manifold $\mathbf{P}^1$. Therefore φ must be a constant function in x by the maximum principle. Since $\varphi(\infty, y) = 0$, $\varphi(x, y) = 0$ holds. So we have $H^0(\mathbf{P}^1 \times Y, \mathscr{I}) = 0$. Lastly we will show $H^1(\mathbf{P}^1 \times Y, \mathscr{I}) = \Gamma(U_0 \cap U_1, \mathscr{I})/[\delta(\Gamma(U_0, \mathscr{I}) \oplus \Gamma(U_1, \mathscr{I}))] = 0$. Let $\varphi(x, y) \in \Gamma(U_0 \cap U_1, \mathscr{I})$. Then φ is holomorphic in $\{(x, y) | 0 < |x| < \infty, y \in Y\}$. Let the Laurent expansion of $\varphi(x, y)$ be $\sum_{n=-\infty}^{\infty} a_n(y)x^n$. Notice that $\varphi_0(x, y) = \sum_{n \geqq 0} a_n(y)x^n$ is holomorphic in U_0 and that $\varphi_1(x, y) = \sum_{n<0} a_n(y)x^n$ is holomorphic in U_1. Furthermore, $\varphi(x, y) = \varphi_0(x, y) + \varphi_1(x, y)$ holds; i.e. $\varphi(x, y) \in \delta(\Gamma(U_0, \mathscr{I}) \oplus \Gamma(U_1, \mathscr{I}))$. This impiles $H^1(\mathbf{P}^1 \times Y, \mathscr{I}) = 0$.

Lemma 2. *Let X and Y be topological spaces, and let $\mathscr{F}$ be a sheaf over X. If a continuous map $f: X \to Y$ is purely r-dimensional with respect to $\mathscr{F}$, i.e. $R^k f_* \mathscr{F} = 0$ for $k \neq r$, then $H^k_{f^{-1}(Z)}(X, \mathscr{F}) = H^{k-r}_Z(Y, R^r f_*(\mathscr{F}))$ for an arbitrary locally closed subset Z of Y.*

Proof. Let $0 \to \mathscr{F} \to \mathscr{L}^0 \to \mathscr{L}^1 \to \cdots$ be a flabby resolution of $\mathscr{F}$. Recall that the sheaf $R^k f_* \mathscr{F}$ is associated to the presheaf $H^k(f^{-1}(U), \mathscr{F}) = H^k(\Gamma(f^{-1}(U), \mathscr{L}^\bullet)) = H^k(\Gamma(U, f_* \mathscr{L}^\bullet))$ for an open set U in Y. Therefore, one has

$$\mathscr{H}^k(f_* \mathscr{L}^\bullet) = \frac{\operatorname{Ker}(f_* \mathscr{L}^k \to f_* \mathscr{L}^{k+1})}{\operatorname{Im}(f_* \mathscr{L}^{k-1} \to f_* \mathscr{L}^k)}.$$

Let $f_* \mathscr{L}^{r-1} \xrightarrow{\varphi} f_* \mathscr{L}^r \xrightarrow{\psi} f_* \mathscr{L}^{r+1}$, and let $\mathscr{I} = \operatorname{Im} \varphi$ and $\mathscr{Z} = \operatorname{Ker} \psi$. Then

$$0 \to f_* \mathscr{L}^0 \to f_* \mathscr{L}^1 \to \cdots \to f_* \mathscr{L}^{r-1} \to \mathscr{I} \to 0$$

is an exact sequence of flabby sheaves. Noticing that $\Gamma_Z(Y, f_* \mathscr{L}^k) = \Gamma_{f^{-1}(Z)}(X, \mathscr{L}^k)$, one obtains the exact sequence

$$0 \to \Gamma_{f^{-1}(Z)}(X, \mathscr{L}^0) \to \cdots \to \Gamma_{f^{-1}(Z)}(X, \mathscr{L}^{r-1}) \to \Gamma_Z(Y, \mathscr{I}) \to 0.$$

From this exact sequence, $H^k_{f^{-1}(Z)}(X, \mathscr{F}) = 0$ for $k \leqq r - 1$ holds. In the case where $k \geqq r$, one has the exact sequence $0 \to \mathscr{I} \to \mathscr{Z} \to R^r f_*(\mathscr{F}) \to 0$ and the flabby resolution of $\mathscr{Z}$

$$0 \to \mathscr{Z} \to f_*\mathscr{L}^r \to f_*\mathscr{L}^{r+1} \to \cdots.$$

Then the methods in the proof of Proposition 1.1.5 and in the one of Proposition 1.1.6 can be applied to complete the proof.

By Lemmas 1 and 2, for an arbitrary k, one obtains

$$H^k_{\mathbf{P}^1 \times Z}(\mathbf{P}^1 \times Y, \mathscr{I}) = H^{k-r}_Z(Y, R^r f_*(\mathscr{I})) = H^{k-r}_Z(Y, 0) = 0.$$

This implies $H^r_{k \times Z}(\mathbf{C} \times Y, \mathscr{O}_{\mathbf{C} \times Y}) = 0$, as we noted previously. Having proved the above, one obtains

$$H^k_{K \times Z}(\mathbf{C} \times Y, \mathscr{O}_{\mathbf{C} \times Y}) = 0 \qquad \text{for } k < r + 1.$$

By replacing Y with $Y \times W$, and Z with $W \times Z$, Proposition 2.2.1 follows.

Corollary. *Let Ω be a connected open set in $\mathbf{C}$; let Z be a closed set in Ω, $Z \subsetneqq \Omega$; and let K_j be a compact set in $\mathbf{C}$ for each j, $1 \leqq j \leqq n$. Then, for $k \neq n + 1$, $H^k_{Z \times K_1 \times \cdots \times K_n}(\Omega \times \mathbf{C}^n, \mathscr{O}_{\mathbf{C}^{n+1}}) = 0$ holds.*

Proof. If a holomorphic function φ in Ω is zero on the open set $\Omega - Z$, then φ is zero in Ω by analytic continuation. Let V be any complex manifold. If a holomorphic function $\varphi(x, y)$ in $V \times \Omega$, where $x \in V$ and $y \in \Omega$, has a value of zero in $V \times (\Omega - Z)$, then $\varphi(x, y) = 0$ for each $x \in V$. Therefore, $\varphi = 0$ in $V \times \Omega$. That is, $H^0_{V \times Z}(V \times \Omega, \mathscr{O}_{V \times \Omega}) = \Gamma_{V \times Z}(V \times \Omega, \mathscr{O}_{V \times \Omega}) = 0$. Then, by Proposition 2.2.1, $H^k_{W \times K_1 \times Z}(W \times \mathbf{C} \times Y, \mathscr{O}_{W \times \mathbf{C} \times Y}) = 0$ is true for $k < 2$. Replacing $\mathbf{C} \times Y$ by Y and $K_1 \times Z$ by Z as the initial step, Proposition 2.2.1 can be applied inductively. Hence one has $H^k_{Z \times K_1 \times \cdots \times K_n}(\Omega \times \mathbf{C}^n, \mathscr{O}_{\mathbf{C}^{n+1}}) = 0$ for $k < n + 1$. When $k > n + 1$, Malgrange's theorem (Theorem 1.1.4) implies flabby dim $\mathscr{O}_{\mathbf{C}^{n+1}} \leqq n + 1$. This completes the proof.

Definition 2.2.1. *A subset K of $\mathbf{C}^n$ is said to be an analytic polyhedron if there exists $f_1, \ldots, f_N \in \mathscr{O}(\mathbf{C}^n)$ such that $K = \{z \in \mathbf{C}^n \mid |f_1(z)| \leqq 1, \ldots, |f_N(z)| \leqq 1\}$.*

Proposition 2.2.2. *Let K_1 and K_2 be compact analytic polyhedrons in $\mathbf{C}^n$. Then $H^k_{K_1 - K_2}(\mathbf{C}^n, \mathscr{O}_{\mathbf{C}^n}) = 0$ for $k \neq n$.*

Proof. Since one can assume $K_1 \supset K_2$ without loss of generality, one may let $K_1 = \{z \in \mathbf{C}^n \mid |f_j(z)| \leqq 1$ for $j = 1, \ldots, N\}$ and $K_2 = \{z \in \mathbf{C}^n \mid |f_j(z)| \leqq 1$ for $j = 1, \ldots, N, \ldots, N + N'\}$. Define $K'_j = \{z \in \mathbf{C}^n \mid |f_l(z)| \leqq 1$ for $l = 1, \ldots, N + j\}$, where $1 \leqq j \leqq N'$. By the long exact sequence (see (3) in Theorem 1.1.2)

$$\cdots \to H^k_{K'_j - K'_{j+1}}(\mathbf{C}^n, \mathscr{O}_{\mathbf{C}^n}) \to H^k_{K_1 - K'_{j+1}}(\mathbf{C}^n, \mathscr{O}_{\mathbf{C}^n}) \to H^k_{K_1 - K'_j}(\mathbf{C}^n, \mathscr{O}_{\mathbf{C}^n}) \to \cdots,$$

it is sufficient to prove the case where $N' = 1$. The general case follows inductively from this long exact sequence. Therefore we may now let $K_1 = \{z \in \mathbf{C}^n \| f_j(z)| \leqq 1 \text{ for } j = 1, \ldots, N-1\}$ and $K_2 = \{z \in \mathbf{C}^n \| f_j(z)| \leqq 1 \text{ for } j = 1, \ldots, N\}$. Furthermore, since K_1 and K_2 are compact, one may let $f_j(z) = z_j$, $1 \leqq j \leqq n \leqq N - 1$.

Define a map $F: X = \mathbf{C}^n \to Y = \mathbf{C}^N$ by $F(x) = (x_1, \ldots, x_n, f_{n+1}(x), \ldots, f_N(x))$, where $x = (x_1, \ldots, x_n) \in \mathbf{C}^n$. Choose a real number a so that $\max_{z \in K_1} (|f_N(z)|, 1) < a$. Then define $\tilde{K}_1$ and $\tilde{K}_2$, compact sets in Y, as follows:

$$\tilde{K}_1 = \{y \in Y \| y_j| \leqq 1 \text{ for } j = 1, \ldots, N-1 \text{ and } |y_N| \leqq a\}$$
$$\tilde{K}_2 = \{y \in Y \| y_j| \leqq 1 \text{ for } j = 1, \ldots, N\}.$$

Note that $F^{-1}(\tilde{K}_1) = K_1$ and $F^{-1}(\tilde{K}_2) = K_2$ hold.

Generally speaking, let X and Y be topological spaces, and let a closed continuous map $F: X \to Y$ be injective. Then, for a locally closed set S in Y and a sheaf $\mathscr{F}$ over X,

$$H^k_{F^{-1}(S)}(X, \mathscr{F}) \cong H^k_S(Y, F_*\mathscr{F}) \text{ holds for } k = 0, 1, \ldots. \tag{*}$$

Note that $R^k F_*(\mathscr{F})_y = \varinjlim_{y \in U} H^k(F^{-1}(U), \mathscr{F})$ for $y \in Y$. Since F is a closed map, $\varinjlim_{y \in U} H^k(F^{-1}(U), \mathscr{F}) = \varinjlim_{F^{-1}(y) \subset V} H^k(V, \mathscr{F})$. But F is injective. Therefore $F^{-1}(y)$ is either a point or an empty set. Either of these cases implies that the map F is purely 0-dimensional with respect to $\mathscr{F}$. Therefore the isomorphism in (*) exists by Lemma 2 in the proof of Proposition 2.2.1. Hence, in our case, one has the isomorphism $H^k_{K_1 - K_2}(\mathbf{C}^n, \mathcal{O}_{\mathbf{C}^n}) \cong H^k_{\tilde{K}_1 - \tilde{K}_2}(Y, F_*\mathcal{O}_X)$. It is sufficient to prove that $H^k_{\tilde{K}_1 - \tilde{K}_2}(Y, F_*\mathcal{O}_X) = 0$ holds for $k \neq n$.

Lemma 1. *Let $\mathcal{O}^l_Y$ be the direct sum of l copies of the sheaf $\mathcal{O}_Y$ of holomorphic functions in $Y = \mathbf{C}^N$. Then there exists an exact sequence*

$$0 \leftarrow F_*\mathcal{O}_X \leftarrow \mathcal{O}_Y \leftarrow \mathcal{O}_Y^{N-n} \leftarrow \mathcal{O}_Y^{\binom{N-n}{2}} \leftarrow \cdots \leftarrow \mathcal{O}_Y^{\binom{N-n}{N-n}} \leftarrow 0.$$

Proof. For the coordinate system $(y_1, \ldots, y_N)$ of Y, define a new coordinate system $(w_1, \ldots, w_N)$ by $w_j = y_j$ for $1 \leqq j \leqq n$, and $w_j = y_j - f_j(y_1, \ldots, y_n)$ for $n + 1 \leqq j \leqq N$. Then $F(X) = \{(w_1, \ldots, w_N) \in \mathbf{C}^N | w_{n+1} = \cdots = w_N = 0\}$. If $y \notin F(X)$, then $(F_*\mathcal{O}_X)_y = 0$; and if $y \in F(X)$, then $(F_*\mathcal{O}_X)_y = (\mathcal{O}_{F(X)})_y$. Associate $\varphi(w_1, \ldots, w_N) \in \mathcal{O}_Y$ with $\varphi|_{w_{n+1} = \cdots = w_N = 0} \in F_*\mathcal{O}_X$. Therefore one sees the exactness of $0 \leftarrow F_*\mathcal{O}_X \leftarrow \mathcal{O}_Y$. Next consider the map $\{\varphi_j\}_{n+1 \leqq j \leqq N} \mapsto \sum_{j=n+1}^{N} \varphi_j w_j$. This defines a map from $\mathcal{O}_Y^{N-n}$ to $\mathcal{O}_Y$, inducing the exact sequence $F_*\mathcal{O}_X \leftarrow \mathcal{O}_Y \leftarrow \mathcal{O}_Y^{N-n}$. Note that an arbitrary element in $\mathcal{O}_Y^{\binom{N-n}{k}}$ can be described as

$\{\varphi_{i_1,\ldots,i_k}\}_{n+1\leqq i_1,\ldots,i_k\leqq N}$ with an alternating index; i.e. $\varphi_{i_{\sigma(1)},\ldots,i_{\sigma(k)}} = (\operatorname{sgn}\sigma)\varphi_{i_1,\ldots,i_k}$ for a permutation σ of k letters $i_1,\ldots,i_k$. We will define a map δ

$$\mathcal{O}_Y^{\binom{N-n}{k}} \xrightarrow{\delta} \mathcal{O}_Y^{\binom{N-n}{k-1}}.$$

Let $(\delta\varphi)_{i_1,\ldots,i_{k-1}} = \sum_{j=n+1}^{N} \varphi_{i_1,\ldots,i_{k-1},j}w_j$ and $\delta(\{\varphi_{i_1,\ldots,i_k}\}) = \{(\delta\varphi)_{i_1,\ldots,i_{k-1}}\}$. Since $\varphi_{i_1,\ldots,i_{k-2},j,l} + \varphi_{i_1,\ldots,i_{k-2},l,j} = 0$, one has $(\delta^2\varphi)_{i_1,\ldots,i_{k-2}} = \sum_{j=n+1}^{N} (\delta\varphi)_{i_1,\ldots,i_{k-2},j}w_j = \sum_{j,l} \varphi_{i_1,\ldots,i_{k-2},j,l}w_jw_l = 0$. That is, $\delta^2 = 0$. Next, suppose $\delta\varphi = 0$; i.e. $\sum_{j=n+1}^{N} \varphi_{i_1,\ldots,i_{k-1},j}w_j = 0$. We will show that there exists h such that $\delta h = \varphi$. If $y \notin F(X)$, a direct computation can be applied. We will prove the case when $y \in F(X)$.

We will give a proof by induction on N. If $N = n + 1$, i.e. $k = 1$, then $\varphi_{n+1}w_{n+1} = 0$ holds. Then $\varphi_{n+1} = 0$ on the open set $\{w \mid w_{n+1} \neq 0\}$. By the analytic continuation, $\varphi_{n+1} = 0$, securing the exact sequence $\mathcal{O}_Y \xleftarrow{\delta} \mathcal{O}_Y \leftarrow 0$. Next assume that the claim is true for the case $N - 1$.

$$\sum_{j=n+1}^{N} \varphi_{i_1,\ldots,i_{k-1},j}w_j = 0$$

implies, in particular, $\sum_{j=n+1}^{N-1} (\varphi_{i_1,\ldots,i_{k-1},j}w_j|_{\{w_N=0\}}) = 0$. From the inductive assumption, one obtains

$$\varphi_{i_1,\ldots,i_{k-1},i_k} = \sum_{j=n+1}^{N-1} \psi_{i_1,\ldots,i_k,j}w_j + w_N g_{i_1,\ldots,i_k},$$

where $i_1,\ldots,i_k < N$, for some $\psi_{i_1,\ldots,i_k,j}$ and $g_{i_1,\ldots,i_k}$. Therefore, $0 = \sum_{j=n+1}^{N} \varphi_{i_1,\ldots,i_{k-1},j}w_j = \varphi_{i_1,\ldots,i_{k-1},N}w_N + w_N \sum_{l=n+1}^{N-1} g_{i_1,\ldots,i_{k-1},l}w_l$ enables one to get $\varphi_{i_1,\ldots,i_{k-1},N} = -\sum_{j=n+1}^{N-1} g_{i_1,\ldots,i_{k-1},j}w_j$. Then define $h_{i_1,\ldots,i_{k+1}}$, $i_1 < \cdots < i_{k+1}$, to be $\psi_{i_1,\ldots,i_{k+1}}$ if $i_{k+1} < N$, and to be $g_{i_1,\ldots,i_{k-1},i_k}$ if $i_{k+1} = N$. Hence, for $i_1 < \cdots < i_k < N$, $(\delta h)_{i_1,\ldots,i_k} = \sum_{j=n+1}^{N} h_{i_1,\ldots,i_k,j}w_j = \sum_{j=n+1}^{N-1} \psi_{i_1,\ldots,i_k,j}w_j + w_N g_{i_1,\ldots,i_k} = \varphi_{i_1,\ldots,i_k}$ holds; and also, for $i_k = N$,

$$\begin{aligned}(\delta h)_{i_1,\ldots,i_{k-1},N} &= \sum_{j=n+1}^{N-1} h_{i_1,\ldots,i_k,j}w_j = -\sum_{j=n+1}^{N-1} h_{i_1,\ldots,i_{k-1},j,N}w_j \\ &= -\sum_{j=n+1}^{N-1} g_{i_1,\ldots,i_{k-1},j}w_j = \varphi_{i_1,\ldots,i_{k-1},N}\end{aligned}$$

holds. This proves $\delta h = \varphi$; therefore one obtains the exactness of the sequence

$$\mathcal{O}_Y^{\binom{N-n}{k-1}} \xleftarrow{\delta} \mathcal{O}_Y^{\binom{N-n}{k}} \xleftarrow{\delta} \mathcal{O}_Y^{\binom{N-n}{k+1}}.$$

Lemma 2. *Let $\mathscr{F}$ be a sheaf over $Y = \mathbf{C}^N$. If the sequence*

$$0 \leftarrow \mathscr{F} \leftarrow \mathcal{O}_Y^{l_0} \leftarrow \mathcal{O}_Y^{l_1} \leftarrow \cdots \leftarrow \mathcal{O}_Y^{l_r} \leftarrow 0$$

is exact, then for $k < N - r$ one has

$$H^k_{\tilde{K}_1 - \tilde{K}_2}(Y, \mathscr{F}) = 0.$$

Proof. First we will prove $H^k_{\tilde{K}_1 - \tilde{K}_2}(Y, \mathcal{O}_Y) = 0$ for $k \neq N$. Let $\Omega = \{y_N \in \mathbf{C} \mid |y_N| > 1\}$, let $Z = \{y_N \in \mathbf{C} \mid 1 < |y_N| \leqq a\}$, and let $K_j = \{y_j \in \mathbf{C} \mid |y_j| \leqq 1\}$. Then $\tilde{K}_1 - \tilde{K}_2 = Z \times K_1 \times \cdots \times K_{N-1}$ holds. By Corollary of Proposition 2.2.1,

$$H^k_{\tilde{K}_1 - \tilde{K}_2}(Y, \mathcal{O}_Y) = H^k_{Z \times K_1 \times \cdots \times K_{N-1}}(\Omega \times \mathbf{C}^{N-1}, \mathcal{O}_{\mathbf{C}^N}) = 0 \qquad \text{for } k \neq N.$$

Therefore, one obtains $H^k_{\tilde{K}_1 - \tilde{K}_2}(Y, \mathcal{O}^{l_0}) = 0$ for $k \neq N$, proving the case where $r = 0$ in this lemma. We will prove the general case by induction on r. The given exact sequence implies that the sequences

$$0 \leftarrow \mathscr{F} \leftarrow \mathcal{O}_Y^{l_0} \leftarrow \mathscr{G} \leftarrow 0 \quad \text{and} \quad 0 \leftarrow \mathscr{G} \leftarrow \mathcal{O}_Y^{l_1} \leftarrow \cdots \leftarrow \mathcal{O}_Y^{l_r} \leftarrow 0$$

are exact. By the hypothesis, for $k < N - (r - 1) = N - r + 1$,

$$H^k_{\tilde{K}_1 - \tilde{K}_2}(Y, \mathscr{G}) = 0$$

holds. Therefore, for $k < N - r$, the exact sequence

$$H^k_{\tilde{K}_1 - \tilde{K}_2}(Y, \mathcal{O}_Y^{l_0}) \to H^k_{\tilde{K}_1 - \tilde{K}_2}(Y, \mathscr{F}) \to H^{k+1}_{\tilde{K}_1 - \tilde{K}_2}(Y, \mathscr{G})$$

implies that $H^k_{\tilde{K}_1 - \tilde{K}_2}(Y, \mathscr{F}) = 0$ holds.

Combining these two lemmas, one concludes that $H^k_{\tilde{K}_1 - \tilde{K}_2}(\mathbf{C}^n, \mathcal{O}_{\mathbf{C}^n}) = H^k_{\tilde{K}_1 - \tilde{K}_2}(Y, F_* \mathcal{O}_X) = 0$ for $k < n$. On the other hand, Malgrange's theorem (Theorem 1.1.4) implies flabby dim $\mathcal{O}_{\mathbf{C}^n} \leqq n$. From this, one has

$$H^k_{K_1 - K_2}(\mathbf{C}^n, \mathcal{O}_{\mathbf{C}^n}) = 0 \qquad \text{for } k > n.$$

Finally, $H^k_{K_1 - K_2}(\mathbf{C}^n, \mathcal{O}_{\mathbf{C}^n}) = 0$ for $k \neq n$ is acquired.

Remark. If $f_{n+1}(z) = 2$ in Proposition 2.2.2, then $K_2 = \varnothing$. Hence, for K_1 satisfying the condition of Proposition 2.2.2,

$$H^k_{K_1}(\mathbf{C}^n, \mathcal{O}_{\mathbf{C}^n}) = 0$$

holds for $k \neq n$.

Now we come to the main theorems.

Theorem 2.2.1. *Let M be a real analytic manifold, and let X be a complexification of M. Then M is purely n-codimensional with respect to $\mathcal{O}_X$.*

Proof. The assertion being local in nature, one may assume $M = \mathbf{R}^n$ and $X = \mathbf{C}^n$ without loss of generality. It is sufficient to prove that at the origin

$$\mathcal{H}^k_{\mathbf{R}^n}(\mathcal{O}_{\mathbf{C}^n})_0 = 0$$

holds for $k \neq n$. Note that $\operatorname{Im} f \geqq 0$ for $f \in \mathcal{O}_{\mathbf{C}^n}$ is equivalent to the condition $|e^{\sqrt{-1}f(z)}| \leqq 1$. Therefore $\mathbf{R} = \{z \in \mathbf{C} \,|\, |e^{\sqrt{-1}z}| \leqq 1 \text{ and } |e^{-\sqrt{-1}z}| \leqq 1\}$. Then notice that for an arbitrary polynomial $\varphi(z)$ the sets $K_1 = \mathbf{R}^n \cap \{z \in \mathbf{C}^n \,|\, |z_1| \leqq 1, \ldots, |z_n| \leqq 1\}$ and $K_2 = K_1 \cap \{z \in \mathbf{C}^n \,|\, \operatorname{Im} \varphi(z) \leqq 0\}$ are both compact analytic polyhedrons. Therefore, by Proposition 2.2.2, one has

$$H^k_{K_1 - K_2}(\mathbf{C}^n, \mathcal{O}_{\mathbf{C}^n}) = 0$$

for $k \neq n$. Let the polynomial φ be $\varphi(z) = z_1 + \sqrt{-1} - 2\sqrt{-1}(z_1^2 + \cdots + z_n^2)$. Then $K_1 - K_2 = \mathbf{R}^n \cap \{z \in \mathbf{C}^n \,|\, \operatorname{Im} \varphi > 0\}$. This is so because if, for $z \in \mathbf{R}^n$, $\operatorname{Im} \varphi(z) = 1 - 2(z_1^2 + \cdots + z_n^2) > 0$, then particularly $|z_j| < 1/\sqrt{2}$ for $j = 1, \ldots, n$. Consequently, $z \in K_1$. Define $\Omega = \{z \in \mathbf{C}^n \,|\, |z_1| < 1, \ldots, |z_n| < 1 \text{ and } \operatorname{Im} \varphi > 0\}$. Then $K_1 - K_2 = \Omega \cap \mathbf{R}^n$ holds. Hence one has

$$H^k_{\mathbf{R}^n \cap \Omega}(\Omega, \mathcal{O}_{\mathbf{C}^n}) = H^k_{K_1 - K_2}(\mathbf{C}^n, \mathcal{O}_{\mathbf{C}^n}) = 0$$

for $k \neq n$. By definition

$$\mathcal{H}^k_{\mathbf{R}^n}(\mathcal{O}_{\mathbf{C}^n})_0 = \varinjlim_{0 \in U} H^k_{\mathbf{R}^n \cap U}(U, \mathcal{O}_{\mathbf{C}^n}).$$

Let $U = a\Omega$ for $a > 0$. Then for $k \neq n$

$$H^k_{\mathbf{R}^n \cap U}(U, \mathcal{O}_{\mathbf{C}^n}) = 0.$$

Since $\{a\Omega\}_{a > 0}$ is a fundamental neighborhood system at the origin, one obtains

$$\mathcal{H}^k_{\mathbf{R}^n}(\mathcal{O}_{\mathbf{C}^n}) = 0 \qquad \text{for } k \neq n$$

as $a \to 0$.

Theorem 2.2.2. *Let $Z = \{z \in \mathbf{C}^n \,|\, \operatorname{Im} z_i \geqq 0 \text{ for } 1 \leqq i \leqq n\}$. Then*

$$\mathcal{H}^k_Z(\mathcal{O}_{\mathbf{C}^n})_0 = 0$$

holds for $k \neq n$.

Proof. Let $\Omega \subset \mathbf{C}^n$ be a sufficiently small neighborhood of the origin 0. Then it is sufficient to prove

$$H^k_{Z \cap \Omega}(\Omega, \mathcal{O}_{\mathbf{C}^n}) = 0 \qquad \text{for } k \neq n.$$

Define $\varphi(z) = -(z_1 + \cdots + z_n) + \sqrt{-1}a - b\sqrt{-1}(z_1^2 + \cdots + z_n^2)$. Let $K_1 = Z \cap \{z \in \mathbf{C}^n \mid |z_j| \leqq 1 \text{ for } j = 1, \ldots, n\}$, and let $K_2 = K_1 \cap \{z \in \mathbf{C}^n \mid \operatorname{Im} \varphi(z) \leqq 0\}$. Notice that K_1 and K_2 are compact analytic polyhedrons. Therefore, by Proposition 2.2.2, one has

$$H^k_{K_1 - K_2}(\mathbf{C}^n, \mathcal{O}_{\mathbf{C}^n}) = 0 \qquad \text{for } k \neq n.$$

Let $\Omega = \{z \in \mathbf{C}^n \mid \operatorname{Im} \varphi(z) > 0, |z_j| < 2\sqrt{a} \text{ for } j = 1, \ldots, n\}$. Then $(K_1 - K_2) \supset \Omega \cap Z$ holds for sufficiently small $a > 0$. Next we will show that one can choose a and b so that $K_1 - K_2 = \Omega \cap Z$. It is sufficient to prove that $|z_j| < 2\sqrt{a}$ holds for $j = 1, \ldots, n$ on the set $K_1 - K_2$. Note that

$$\operatorname{Im} \varphi(z) = -b(x_1^2 + \cdots + x_n^2) + a - \{y_1(1 - by_1) + \cdots + y_n(1 - by_n)\},$$

where $z_j = x_j + \sqrt{-1}y_j$, $1 \leqq j \leqq n$. Let $b = 1/2$. Then

$$0 < \operatorname{Im} \varphi(z) \leqq a - \tfrac{1}{2}(x_1^2 + \cdots + x_n^2) - \tfrac{1}{2}(y_1 + \cdots + y_n).$$

Therefore, one obtains

$$\tfrac{1}{2}(x_1^2 + \cdots + x_n^2) + \tfrac{1}{2}(y_1 + \cdots + y_n) \leqq a.$$

By the assumption, $\operatorname{Im} z_j = y_j \geqq 0$ for $j = 1, \ldots, n$. Then $x_j^2 \leqq 2a$ and $0 \leqq y_j \leqq 2a$, $1 \leqq j \leqq n$, hold. Therefore, for a sufficiently small real number $a > 0$, one has $|z_j| < 2\sqrt{a}$ for $j = 1, \ldots, n$. Note that Ω forms a fundamental neighborhood system of the origin as $a \to 0$. Hence, if $k \neq n$, then

$$H^k_{\Omega \cap U}(\Omega, \mathcal{O}_{\mathbf{C}^n}) = H^k_{K_1 - K_2}(\mathbf{C}^n, \mathcal{O}_{\mathbf{C}^n}) = 0.$$

Taking the direct limit over Ω as $a \to 0$, one has

$$\mathscr{H}^k_Z(\mathcal{O}_{\mathbf{C}^n}) = 0 \qquad \text{for } k \neq n.$$

Proposition 2.2.3. *Let the sets K_1 and K_2 be the same as in Proposition 2.2.2. Then, for an arbitrary complex manifold W,*

$$H^k_{(K_1 - K_2) \times W}(\mathbf{C}^n \times W, \mathcal{O}_{\mathbf{C}^n \times W}) = 0 \qquad \text{for } k < n.$$

Proof. The proof of Proposition 2.2.3 is quite similar to the one of Proposition 2.2.2. Replace $H^k_{K'_j - K'_{j+1}}(\mathbf{C}^n, \mathcal{O}_{\mathbf{C}^n})$ with $H^k_{(K'_j - K'_{j+1}) \times W}(\mathbf{C}^n \times W, \mathcal{O}_{\mathbf{C}^n \times W})$; $X = \mathbf{C}^n$ with $\mathbf{C}^n \times W$; and $Y = \mathbf{C}^N$ with $\mathbf{C}^N \times W$. The details are left for the reader.

As a byproduct of the discussion in this section, we will give a proof of Theorem 1.2.4 in Chapter I, i.e. the homotopy invariance of the cohomology groups with coefficient in a constant sheaf. Let X and Y be topological spaces, and let M be an additive group. Proposition 2.1.3 implies that for continuous maps $f_\nu: X \to Y$, $\nu = 0, 1$, the maps $f_\nu^*: H^k(Y, M_Y) \to H^k(X, f_\nu^{-1}M_Y) = H^k(X, M_X)$, $\nu = 0, 1$, are induced. Then we will prove that $f_0^* = f_1^*$ holds if f_0 and f_1 are homotopic. There exists a continuous map $F:X \times I \to Y$, where $I = [0, 1]$ such that $F(x, v) = f_v(x)$ for $v = 0, 1$. Define a map $\iota_v:X \to X \times I$ by $\iota_v(x) = (x, v)$, $v = 0, 1$. Then $f_v = F \circ \iota_v$. This implies that $f_v^* = \iota_v^* \circ F^*$, where

$$H^k(Y, M_Y) \xrightarrow{F^*} H^k(X \times I, M_{X \times I}) \underset{\iota_1^*}{\overset{\iota_0^*}{\rightrightarrows}} H^k(X, M_X).$$

In order to claim $f_0^* = f_1^*$, it is sufficient to prove $\iota_0^* = \iota_1^*$. Let p be the projection: $X \times I \to X$. Then, notice that $p \circ \iota_v$, $v = 0, 1$, are identity maps on X. Therefore one has $\iota_v^* \circ p^* = \mathrm{id}$ for $v = 0, 1$. If p^* is an isomorphism, then one obtains $\iota_0^* = \iota_1^*$. But it is sufficient to show

$$R^k p_*(M_{X \times I}) = \begin{cases} M_X & \text{for } k = 0 \\ 0 & \text{for } k \neq 0 \end{cases}$$

to claim that p^* is an isomorphism. This is because, from Lemma 2 in the proof of Proposition 2.2.1, one has

$$H^k(X \times I, M_{X \times I}) \cong H^{k-0}(X, R^0 p_*(M_{X \times I})) = H^k(X, M_X).$$

We will compute $R^k p_*(M_{X \times I})$ below. Since p is a proper map, the Corollary of Proposition 2.1.4 can be applied. Hence, one has for $x \in X$

$$R^k p_*(M_{X \times I})_x = H^k(I, M) \qquad \text{for } k = 0, 1, \ldots .$$

Consequently, it suffices to prove the following lemma.

Lemma. *Let I and M be as the above. Then*

$$H^k(I, M) = \begin{cases} M & \text{for } k = 0 \\ 0 & \text{for } k \neq 0 \end{cases}$$

holds.

Proof. Since I is connected, $H^0(I, M) = M$. Next, when $k > 1$, assume that $s \neq 0$ for $s \in H^k(I, M)$. Let the set $\mathcal{M} = \{[a, b] \subset I \mid s|_{[a,b]} \neq 0\}$. Define an order relation $\succ$ in $\mathcal{M}$ as follows: $[a, b] \succ [a', b']$ if and only if $[a, b] \subset [a', b']$. Note that $\mathcal{M}$ is an inductively ordered set. Let $\{F_j = [a_j, b_j]\}$ be a totally ordered subset of $\mathcal{M}$. Then $F = \bigcap F_j$ is a closed interval in I. Since $H^k(F, M) = \varinjlim_j H^k(F_j, M)$, $s|_F \neq 0$ holds. Therefore there exists a maximal element $[a, b]$ (i.e. a minimal interval in our case) by Zorn's lemma. One can choose a real number c such that $a < c < b$, since $a = b$ would imply $s|_{[a,b]} = 0$. Then the sequence

$$0 \to \Gamma([a, b], \mathcal{L}) \xrightarrow{i} \Gamma([a, c], \mathcal{L}) \oplus \Gamma([c, b], \mathcal{L})$$

and

$$\Gamma([a, c], \mathcal{L}) \oplus \Gamma([c, b], \mathcal{L}) \xrightarrow{j} \Gamma(\{c\}, \mathcal{L}) \to 0$$

are exact if $\mathscr{L}$ is either a constant sheaf or a flabby sheaf, where $i(s) = (s|_{[a,c]}, s|_{[c,b]})$ and $j(s_1, s_2) = (s_1|_c - s_2|_c)$. Consequently, there is induced the long exact sequence

$$\cdots \to H^{k-1}(\{c\}, M) \to H^k([a, b], M) \to H^k([a, c], M) \oplus H^k([c, b], M]$$
$$\to H^k(\{c\}, M) \to \cdots.$$

In the case where $k > 1$, one has $H^{k-1}(\{c\}, M) = H^k(\{c\}, M) = 0$. Then there exists the isomorphism

$$\begin{array}{ccc} H^k([a, b], M) \cong & H^k([a, c], M) \oplus H^k([c, b], M) \\ \Cup & \Cup \\ s|_{[a,b]} \longmapsto & s|_{[a,c]} \oplus s|_{[c,b]} \end{array}$$

for each $k > 1$. Since $[a, b]$ is maximal, one must have

$$s|_{[a,b]} \neq 0 \quad \text{and} \quad s|_{[a,c]} = s|_{[c,b]} = 0,$$

contradicting the above isomorphism. Lastly, if $k = 1$, one obtains $H^1([a, b], M) = 0$ from the exact sequence

$$0 \to H^0([a, b], M) \to H^0([a, c], M) \oplus H^0([c, b], M) \to H^0(\{c\}, M) \to 0.$$

Let X and Y have the same homotopy type. That is, there are continuous maps $f: X \to Y$ and $g: Y \to X$ such that $g \circ f \simeq 1_X$ and $f \circ g \cong 1_Y$ hold. By the above result, we have $f^* \circ g^* = 1$ and $g^* \circ f^* = 1$, where $H^k(X, M) \underset{f^*}{\overset{g^*}{\rightleftarrows}} H^k(Y, M)$ such that g^* and f^* are inverse maps to each other. Therefore $H^k(X, M) \cong H^k(Y, M)$ is true for $k = 0, 1, 2, \ldots$. In particular, if X is contractible to a point x_0,

$$H^k(X, M) \cong H^k(x_0, M) \qquad \text{holds for } k \geqq 0.$$

Since any sheaf over a point is by definition a flabby sheaf, $H^k(x_0, M) = 0$ for $k > 0$ and $H^0(x_0, M) = M$. This completes the proof of Theorem 1.2.4.

§3. Fundamental Exact Sequences

We will fix a real analytic manifold M throughout this section. Therefore we will simply use $\mathscr{A}$, $\mathscr{B}$, and $\mathscr{C}$ instead of $\mathscr{A}_M$, $\mathscr{B}_M$, and $\mathscr{C}_M$. The most fundamental exact sequence on M

$$0 \to \mathscr{A} \to \mathscr{B} \to \pi_* \mathscr{C} \to 0$$

in the theory of microfunctions will be established in this section. This exact sequence tells us that the "sheaf $\mathscr{B}/\mathscr{A}$ of the irregularity" is isomorphic to $\pi_*\mathscr{C}$. In other words, the structure of $\mathscr{B}/\mathscr{A}$ can be analyzed sharply on $\sqrt{-1}S^*M$. It will be proved that there exists the exact sequence

$$0 \to \tilde{\mathscr{A}} \to \tau^{-1}\mathscr{B} \to \pi_*\tau^{-1}\mathscr{C} \to 0$$

on $\sqrt{-1}SM$. This exact sequence clarifies the inner relationship between the sheaf $\mathscr{C}$ and the notion of the "boundary values of holomorphic functions." Important for the applications are the theorems on singularity spectrum (Theorems 2.3.4 and 2.3.5) which will be proved as its consequence.

We will begin with the notion of generalized relative cohomology groups $H^k(A^\bullet \xrightarrow{f} B^\bullet)$.

Definition 2.3.1. *Let* $A^\bullet = \{\cdots \to A^n \xrightarrow{d_A^n} A^{n+1} \to \cdots\}$ *and* $B^\bullet = \{\cdots \to B^n \xrightarrow{d_A^n} B^{n+1} \to \cdots\}$ *be cochain complexes of abelian groups. Let* $f: A^\bullet \to B^\bullet$ *be a morphism, i.e.* $f = \{f_n\}$, *where* $f_n: A^n \to B^n$ *is a homomorphism with the property* $f_{n+1} \circ d_A^n = d_B^n \circ f_n$ *for any integer n.*

Define $C^\bullet = C(A^\bullet \to B^\bullet)$ *by* $C^n = A^n \oplus B^{n-1}$, *and define* $d_C^n: A^n \oplus B^{n-1} \to A^{n+1} \oplus B^n$ *by* $d_C^n(x, y) = (d_A^n x, d_B^{n-1} y + (-1)^n f_n(x))$ *for each n. Then* $d_C^{n+1} \circ d_C^n = 0$ *holds. Therefore* $C^\bullet = \{\cdots \to C^n \xrightarrow{d_C^n} C^{n+1} \to \cdots\}$ *is a cochain complex. The cohomology groups* $H^k(C^\bullet)$ *of the cochain complex* $C^\bullet$ *are called the generalized relative cohomology groups.*

Note. We often denote $H^k(C^\bullet)$ by $H^k(A^\bullet \to B^\bullet)$. Propositions 2.3.1 and 2.3.2 (particularly (1) of Proposition 2.3.2) explain why $H^k(A^\bullet \to B^\bullet)$ is called a generalized relative cohomology group.

Proposition 2.3.1.

(1) *There is induced the long exact sequence*

$$\cdots \to H^k(A^\bullet \xrightarrow{f} B^\bullet) \to H^k(A^\bullet) \to H^k(B^\bullet) \to H^{k+1}(A^\bullet \xrightarrow{f} B^\bullet) \to \cdots.$$

(2) *For* $A^\bullet \xrightarrow{f} B^\bullet \xrightarrow{g} C^\bullet$, *the sequence*

$$\cdots \to H^k(A^\bullet \xrightarrow{f} B^\bullet) \to H^k(A^\bullet \xrightarrow{g \circ f} C^\bullet) \to H^k(B^\bullet \xrightarrow{g} C^\bullet) \to H^{k+1}(A^\bullet \xrightarrow{f} B^\bullet) \to \cdots$$

is exact.

Proof. Define the cochain complex $B^\bullet[k]$ by $(B^\bullet[k])^n = B^{n+k}$. Then one has the exact sequence

$$0 \to B^\bullet[-1] \to C(A^\bullet \to B^\bullet) \to A^\bullet \to 0.$$

The long exact sequence of cohomology groups induced from this exact sequence of cochain complexes is the one sought.

Next we will prove (2). Let $X^\bullet = C(A^\bullet \xrightarrow{g \circ f} C^\bullet)$, and let $Y^\bullet = C(B^\bullet \xrightarrow{g} C^\bullet)$. Then define a morphism $H: X^\bullet \to Y^\bullet$ as $h_n(x, y) = (f_n(x), y) \in B^n \oplus C^{n-1} = Y^n$ for $(x, y) \in A^n \oplus C^{n-1} = X^n$. If one lets $Z^\bullet = C(X^\bullet \xrightarrow{h} Y^\bullet)$, then from (1) one obtains the exact sequence

$$\cdots \to H^k(X^\bullet \to Y^\bullet) \to H^k(A^\bullet \to C^\bullet) \to H^k(B^\bullet \to C^\bullet) \to \cdots.$$

Therefore one must show $H^k(X^\bullet \xrightarrow{h} Y^\bullet) = H^k(A^\bullet \to B^\bullet)$ to complete the proof.

Let $((x, y), (x', y')) \in Z^n = (A^n \oplus C^{n-1}) \oplus (B^{n-1} \oplus C^{n-2})$. Then $d_Z^n((x, y), (x', y')) = ((d_A^n x, d_C^{n-1} y + (-1)^n (g_n \circ f_n)(x)), (d_B^{n-1} x' + (-1)^n f_n(x), (-1)^n y + d_C^{n-2} y' + (-1)^{n-1} g_{n-1}(x')))$. Letting $W^\bullet = C(A^\bullet \to B^\bullet)$, one defines $F_n : Z_n \to W_n$ by $F_n(((x, y), (x', y'))) = (x, x')$. Then $F : Z^\bullet \to W^\bullet$ is a morphism. Define $G_n : W^n \to Z^n$ by $G_n(x, x') = ((x, g_{n-1}(x')), (x', 0))$. Then $G : W^\bullet \to Z^\bullet$ is a morphism. Notice that $F \circ G$ is the identity map on $W^\bullet$. Hence, the induced maps on cohomology groups $H^n(W^\bullet) \xrightarrow{G^*} H^n(Z^\bullet) \xrightarrow{F^*} H^n(W^\bullet)$ imply that $F^* \circ G^* : H^n(W^\bullet) \to H^n(W^\bullet)$ is the identity map.

Let $s_n : Z^n \to Z^{n-1}$ be the map defined by

$$s(((x, y), (x', y'))) = ((0, (-1)^n y'), (0, 0)).$$

Then in the diagram

$$\begin{array}{ccccccc}
\cdots \to & Z^{n-1} & \xrightarrow{d_Z^{n-1}} & Z^n & \xrightarrow{d_Z^n} & Z^{n+1} & \to \cdots \\
& \text{id} \Big\downarrow \Big\downarrow G \circ F & \swarrow s_n & \text{id} \Big\downarrow \Big\downarrow G \circ F & \swarrow s_{n+1} & \text{id} \Big\downarrow \Big\downarrow G \circ F & \\
\cdots \to & Z^{n-1} & \longrightarrow & Z^n & \longrightarrow & Z^{n+1} & \to \cdots
\end{array}$$

$s_{n+1} \circ d_Z^n + d_Z^{n-1} \circ s_n = (G \circ F)_n - \mathrm{id}_{Z^n}$ holds. Therefore one has $G^* \circ F^* = \mathrm{id}_Z^* : H^n(Z^\bullet) \to H^n(Z^\bullet)$. Consequently, one obtains

$$H^k(X^\bullet \to Y^\bullet) = H^k(Z^\bullet) \cong H^k(W^\bullet) = H^k(A^\bullet \to B^\bullet).$$

A few lemmas are needed to define a generalized relative cohomology of sheaves.

Lemma 1. *Let X and Y be topological spaces, and let $f : X \to Y$ be a continuous map. Suppose that a sheaf $\mathscr{F}$ over Y is given; then there exists a canonical morphism $\mathscr{F} \to f_* f^{-1} \mathscr{F}$.*

Proof. Recall that the sheaf $f^{-1}\mathscr{F}$ is associated to the presheaf $\varinjlim_{U \supset f(V)} \mathscr{F}(U)$ for an open set V in X. Then, for an open set U of Y, there are induced maps

$$\mathscr{F}(U) \to \varinjlim_{U' \supset f(f^{-1}(U))} \mathscr{F}(U') \to (f^{-1}\mathscr{F})(f^{-1}(U)) = f_* f^{-1} \mathscr{F}(U),$$

which define the morphism $\mathscr{F} \to f_* f^{-1} \mathscr{F}$.

Lemma 2. *Let $\mathscr{G}$ and $\mathscr{F}$ be sheaves over topological spaces X and Y, respectively, and let $f : X \to Y$ be a continuous map. Then there is a natural bijection between the morphisms $\mathscr{F} \to f_* \mathscr{G}$ and the morphisms $f^{-1}\mathscr{F} \to \mathscr{G}$.*

Proof. Suppose that a morphism $f^{-1}\mathscr{F} \to \mathscr{G}$ is given. Then one has the morphism $f_* f^{-1} \mathscr{F} \to f_* \mathscr{G}$. Lemma 1 implies that a morphism $\mathscr{F} \to f_* \mathscr{G}$

exists. Conversely, assume that a morphism $\mathscr{F} \to f_*\mathscr{G}$ is given. That is, for an open set U in Y, the map $\mathscr{F}(U) \to \mathscr{G}(f^{-1}(U))$ is given. For an open set V in X with the property $f(V) \subset U$, the homomorphisms $\mathscr{F}(U) \to \mathscr{G}(f^{-1}(U)) \to \mathscr{G}(V)$ are induced, where we note that $f(V) \subset U$ implies $V \subset f^{-1}(U)$. Consequently, one obtains the homomorphism

$$\varinjlim_{f(V) \subset U} \mathscr{F}(U) \to \mathscr{G}(V) \qquad \text{for } V \text{ in } X.$$

Therefore, the morphism $f^{-1}\mathscr{F} \to \mathscr{G}$ exists.

Lemma 3. *Let $\mathscr{G}$ and $\mathscr{F}$ be sheaves over topological spaces X and Y respectively, let $f: X \to Y$ be a continuous map, and let $\rho: \mathscr{F} \to f_*\mathscr{G}$ be a morphism. For a flabby resolution of $\mathscr{F}$, $0 \to \mathscr{F} \to \mathscr{L}^\bullet$, there exist a flabby resolution of $\mathscr{G}$, $0 \to \mathscr{G} \to \mathscr{M}^\bullet$, and a morphism $\rho_k: \mathscr{L}^k \to f_*\mathscr{M}^k$ for each $k \geqq 0$, such that the diagram*

$$\begin{array}{ccccccccc} 0 \to & \mathscr{F} & \longrightarrow & \mathscr{L}^0 & \longrightarrow & \mathscr{L}^1 & \longrightarrow & \mathscr{L}^2 & \longrightarrow \cdots \\ & \downarrow \rho & & \downarrow \rho_0 & & \downarrow \rho_1 & & \downarrow \rho_2 & \\ 0 \to & f_*\mathscr{G} & \to & f_*\mathscr{M}^0 & \to & f_*\mathscr{M}^1 & \to & f_*\mathscr{M}^2 & \to \cdots \end{array}$$

commutes.

Proof. For a flabby resolution $0 \to \mathscr{F} \to \mathscr{L}^0 \to \mathscr{L}^1 \to \cdots$ of $\mathscr{F}$, the sequence $0 \to f^{-1}\mathscr{F} \to f^{-1}\mathscr{L}^0 \to f^{-1}\mathscr{L}^1 \to \cdots$ is exact. Then, by Lemma 2, one has a morphism $f^{-1}\mathscr{F} \to \mathscr{G}$. In the proof of Theorem 1.1.1, Lemma 1 implies that there exists a flabby resolution of $\mathscr{G}$, $0 \to \mathscr{G} \to \mathscr{M}^\bullet$, making the diagram

$$\begin{array}{ccccccc} 0 \to & f^{-1}\mathscr{F} & \to & f^{-1}\mathscr{L}^0 & \to & f^{-1}\mathscr{L}^1 & \to \cdots \\ & \downarrow & & \downarrow & & \downarrow & \\ 0 \longrightarrow & \mathscr{G} & \longrightarrow & \mathscr{M}^0 & \longrightarrow & \mathscr{M}^1 & \longrightarrow \cdots \end{array}$$

commutative. Hence one obtains the commutative diagram

$$\begin{array}{ccccccc} 0 \to & f_*f^{-1}\mathscr{F} & \to & f_*f^{-1}\mathscr{L}^0 & \to & f_*f^{-1}\mathscr{L}^1 & \to \cdots \\ & \downarrow & & \downarrow & & \downarrow & \\ 0 \longrightarrow & f_*\mathscr{G} & \longrightarrow & f_*\mathscr{M}^0 & \longrightarrow & f_*\mathscr{M}^1 & \longrightarrow \cdots. \end{array}$$

Then, one can complete the proof by the use of Lemma 1.

Definition 2.3.2. *Let the notations be the same as in Lemma 3. Abbreviate "$\rho: \mathscr{F} \to f_*\mathscr{G}$" as "$\mathscr{F} \xrightarrow{\rho} \mathscr{G}$." Then define*

$$H^k(X \xrightarrow{f} Y, \mathscr{G} \xleftarrow{\rho} \mathscr{F}) = H^k(\Gamma(Y, \mathscr{L}^\bullet) \to \Gamma(X, \mathscr{M}^\bullet))$$

and

$$\mathscr{D}\!ist^k_f(\mathscr{F} \xrightarrow{\rho} \mathscr{G}) = \mathscr{H}^k(\mathscr{L}^\bullet \to f_*\mathscr{M}^\bullet).$$

In particular, if $\mathscr{G} = f^{-1}\mathscr{F}$ holds, we define

$$H^k(X \to Y, \mathscr{F}) = H^k(X \to Y, f^{-1}\mathscr{F} \leftarrow \mathscr{F})$$

and

$$\mathscr{D}ist_f^k(\mathscr{F}) = \mathscr{D}ist_f^k(\mathscr{F} \to f^{-1}\mathscr{F}).$$

The continuous map f is said to be purely r-codimensional with respect to $\mathscr{F}$ if $\mathscr{D}ist_f^k(\mathscr{F}) = 0$ for $k \neq r$.

Remark. The sheaf $\mathscr{D}ist_f^n(\mathscr{F} \to \mathscr{G})$ over Y is associated to the presheaf $H^n(f^{-1}(U) \to U, \mathscr{G} \overset{\rho}{\leftarrow} \mathscr{F})$ for an open set U in Y.

The cohomology groups defined above are generalizations of $H_Z^k(X, \mathscr{F})$ and $\mathscr{H}_Z^k(\mathscr{F})$; i.e. we have:

Proposition 2.3.2. *Suppose that X is an open set in Y. The imbedding: $X \hookrightarrow Y$ is denoted by f. Then, one has the isomorphisms*

(1) $H^n(X \to Y, \mathscr{F}) \cong H_{Y-X}^n(Y, \mathscr{F})$ *and*

(2) $\mathscr{D}ist_f^n(\mathscr{F}) = \mathscr{H}_{Y-X}^n(\mathscr{F})$, *for* $n \geqq 0$.

We need a lemma.

Lemma. *If $0 \to E^\bullet \overset{e}{\to} A^\bullet \overset{f}{\to} B^\bullet \to 0$ is an exact sequence, then $H^n(E^\bullet) \cong H^n(A^\bullet \overset{f}{\to} B^\bullet)$ holds for $k \geqq 0$.*

Proof. Define a homomorphism $g_n: E^n \to A^n \oplus B^{n-1}$ by $g_n(x) = (e_n(x), 0)$. Then $g: E^\bullet \to C^\circ \underset{\text{def}}{=} C(A^\bullet \overset{f}{\to} B^\bullet)$ is a morphism. Therefore, from the exact sequence

$$0 \to B^\bullet[-1] \overset{h}{\to} C^\bullet \overset{k}{\to} A^\bullet \to 0,$$

one obtains the commutative diagram

$$\begin{array}{ccccccccccc}
\cdots \to & H^{n-1}(A^\bullet) & \xrightarrow{f^*} & H^{n-1}(B^\bullet) & \longrightarrow & H^n(E^\bullet) & \xrightarrow{e^*} & H^n(A^\bullet) & \xrightarrow{f^*} & H^n(B^\bullet) & \to \cdots \\
& \| & & \wr\downarrow (-1)^{n-1}\mathrm{id} & & \downarrow g^* & & \| & & \wr\downarrow (-1)^n\mathrm{id} & \\
\cdots \to & H^{n-1}(A^\bullet) & \longrightarrow & H^{n-1}(B^\bullet) & \xrightarrow{-h^*} & H^n(C^\bullet) & \xrightarrow{k^*} & H^n(A^\bullet) & \longrightarrow & H^n(B^\bullet) & \to \cdots .
\end{array}$$

One concludes $H^n(E^\bullet) \cong H^n(C^\bullet) = H^n(A^\bullet \overset{f}{\to} B^\circ)$ by Five Lemma.

Proof of Proposition 2.3.2. Let $0 \to \mathscr{F} \to \mathscr{L}^0 \to \mathscr{L}^1 \to \cdots$ be a flabby resolution of $\mathscr{F}$. Then $0 \to f^{-1}\mathscr{F} \to f^{-1}\mathscr{L}^0 \to f^{-1}\mathscr{L}^1 \to \cdots$ is a flabby resolution of $f^{-1}\mathscr{F}$, since $f^{-1}\mathscr{L}^k = \mathscr{L}^k|_X$ is flabby. Notice that, since $\mathscr{L}^k$ is flabby for each $k \geqq 0$, the sequence

$$0 \to \Gamma_{Y-X}(Y, \mathscr{L}^\bullet) \to \Gamma(Y, \mathscr{L}^\circ) \to \Gamma(X, \mathscr{L}^\bullet|_X) \to 0$$

is exact. The above lemma implies the isomorphism

$$H^n(X \to Y, \mathscr{F}) = H^n(\Gamma(Y, \mathscr{L}^\bullet) \to \Gamma(X, \mathscr{L}^\bullet)) \cong H^n(\Gamma_{Y-X}(Y, \mathscr{L}^\bullet)) \\ = H^n_{Y-X}(Y, \mathscr{F}).$$

In order to prove (2), first note that one has the exact sequence

$$0 \to \Gamma_{Y-X}(\mathscr{L}^\bullet) \to \mathscr{L}^\bullet \to f_* f^{-1}\mathscr{L}^\bullet \to 0.$$

Hence, by the Lemma above, one obtains

$$\mathscr{D}ist^n_f(\mathscr{F}) = \mathscr{H}^n(\mathscr{L}^\bullet \to f_* f^{-1}\mathscr{L}^\bullet) \cong \mathscr{H}^n(\Gamma_{Y-X}(\mathscr{L}^\bullet)) = \mathscr{H}^n_{Y-X}(\mathscr{F}).$$

Proposition 2.3.3.

(1) *Let $\mathscr{G}$ and $\mathscr{F}$ be sheaves over topological spaces X and Y, respectively, and let $f: X \to Y$ be a continuous map. If a morphism $\rho: \mathscr{F} \to f_*\mathscr{G}$ is given, then there is induced the long exact sequence*

(i) $\cdots \to H^k(X \to Y, \mathscr{G} \leftarrow \mathscr{F}) \to H^k(Y, \mathscr{F}) \to H^k(X, \mathscr{G}) \to \cdots$

and, in particular, one has the following:

$$\cdots \to H^k(X \to Y, \mathscr{F}) \to H^k(Y, \mathscr{F}) \to H^k(X, f^{-1}\mathscr{F}) \to \cdots.$$

(ii) *The sequence*

$$0 \to \mathscr{D}ist^0_f(\mathscr{F} \to \mathscr{G}) \to \mathscr{F} \xrightarrow{\rho} f_*\mathscr{G} \to \mathscr{D}ist^1_f(\mathscr{F} \to \mathscr{G}) \to 0$$

is exact, and for $k \geqq 2$, $\mathscr{D}ist^k_f(\mathscr{F} \to \mathscr{G}) = R^{k-1} f_*\mathscr{G}$ holds. In particular, the sequence

$$0 \to \mathscr{D}ist^0_f(\mathscr{F}) \to \mathscr{F} \to f_* f^{-1}\mathscr{F} \to \mathscr{D}ist^1_f(\mathscr{F}) \to 0$$

is exact, and $\mathscr{D}ist^k_f(\mathscr{F}) = R^{k-1}f_*(f^{-1}\mathscr{F})$ for $k \geqq 2$.

(2) *Let $\mathscr{G}$, $\mathscr{F}$, and $\mathscr{H}$ be sheaves over topological spaces X, Y, and Z respectively; and let $f: X \to Y$ and $g: Y \to Z$ be continuous maps. Suppose that morphisms $\mathscr{F} \to f_*\mathscr{G}$ and $\mathscr{H} \to g_*\mathscr{F}$ are given. Then there are induced long exact sequences*

$$\cdots \to H^k(Y \to Z, \mathscr{F} \leftarrow \mathscr{H}) \to H^k(X \to Z, \mathscr{G} \leftarrow \mathscr{H}) \\ \to H^k(X \to Y, \mathscr{G} \leftarrow \mathscr{F}) \to \cdots,$$

particularly

$$\cdots \to H^k(Y \to Z, \mathscr{H}) \to H^k(X \to Z, \mathscr{H}) \to H^k(X \to Y, g^{-1}\mathscr{H}) \to \cdots.$$

Proof. Definitions and Proposition 2.3.1 imply these claims plainly.

We apply (2) of Proposition 2.3.3 to the case $(X - M) \overset{\iota}{\hookrightarrow} \widetilde{{}^M X} \xrightarrow{\tau} X$ and the sheaf $\mathcal{O}_X$ over X. We have the following exact sequence:

$$\cdots \to H^k(\widetilde{{}^M X} \to X, \mathcal{O}_X) \to H^k((X - M) \hookrightarrow X, \mathcal{O}_X) \\ \to H^k((X - M) \hookrightarrow \widetilde{{}^M X}, \tau^{-1}\mathcal{O}_X) \to \cdots.$$

We will compute each term. From Proposition 2.3.2 we have

$$H^k((X - M) \hookrightarrow X, \mathcal{O}_X) = H^k_M(X, \mathcal{O}_X).$$

Recall that M is purely n-codimensional with respect to $\mathcal{O}_X$. Therefore, by Proposition 1.1.5,

$$H^k_M(X, \mathcal{O}_X) = 0 \qquad \text{holds for } k < n.$$

On the other hand, flabby dim $\mathcal{O}_X \leqq n$ (Theorem 1.1.4) and Theorem 1.1.3 imply that for $k > n$ we have

$$H^k_M(X, \mathcal{O}_X) = 0.$$

Consequently, we obtain

$$H^k((X - M) \hookrightarrow X, \mathcal{O}_X) = 0 \qquad \text{for } k \neq n.$$

Similarly, using Propositions 2.1.1 and 1.1.6, we have the following:

$$H^k((X - M) \hookrightarrow \widetilde{{}^M X}, \tau^{-1}\mathcal{O}_X) = H^k_{\sqrt{-1}SM}(\widetilde{{}^M X}, \tau^{-1}\mathcal{O}_X) = H^{k-1}(\sqrt{-1}SM, \mathcal{Q}).$$

We will prove a few propositions in order to compute $H^k(\widetilde{{}^M X} \to X, \mathcal{O}_X)$.

Proposition 2.3.4. *Let X and Y be topological spaces, let $f: X \to Y$ be a continuous map, and let $\mathcal{F}$ be a sheaf over Y. If the continuous map f is purely l-codimensional with respect to $\mathcal{F}$, then for an arbitrary integer $k \geqq 0$*

$$H^k(X \to Y, \mathcal{F}) = H^{k-l}(Y, \mathcal{D}ist^l_f(\mathcal{F}))$$

holds.

Proof. Let $0 \to \mathcal{F} \to \mathcal{L}^\bullet$ and $0 \to f^{-1}\mathcal{F} \to \mathcal{M}^\bullet$ be flabby resolutions of $\mathcal{F}$ and $f^{-1}\mathcal{F}$, and let $\mathcal{N}^\bullet = C(\mathcal{L}^\bullet \to f_*\mathcal{M}^\bullet)$. Then $\mathcal{D}ist^k_f(\mathcal{F}) = \mathcal{H}^k(\mathcal{N}^\bullet)$ holds. Note that $\mathcal{N}^\bullet$ is a cochain complex of flabby sheaves. By the hypothesis one has the exact sequences

$$0 \to \mathcal{N}^0 \to \mathcal{N}^1 \to \cdots \to \mathcal{N}^{l-1} \to \mathcal{I} \to 0,$$
$$0 \to \mathcal{I} \to \mathcal{Z} \to \mathcal{D}ist^l_f(\mathcal{F}) \to 0, \quad \text{and}$$
$$0 \to \mathcal{Z} \to \mathcal{N}^l \to \mathcal{N}^{l+1} \to \cdots.$$

One can complete the proof in a similar manner to the proofs of Proposition 1.1.5 or Proposition 1.1.6.

Proposition 2.3.5. *Let X and Y be topological spaces. Suppose that all the open sets in X or in Y are paracompact and Hausdorff. Let $\mathcal{F}$ be a sheaf over Y. If a continuous map $f: X \to Y$ is a closed map, then for an arbitrary $y \in Y$, there exists an isomorphism*

$$\mathcal{D}ist^k_f(\mathcal{F})_y = H^k(f^{-1}(y) \to \{y\}, \mathcal{F}_y).$$

Remark. The above proposition also holds under the assumption that X and Y are Hausdorff spaces and $f: X \to Y$ is a proper map.

Proof. Since f is a closed map, Proposition 2.1.4 can be applied. One has

$$\varinjlim_{y \in U} H^k(U, \mathscr{F}) = H^k(\{y\}, \mathscr{F}_y) \quad \text{and}$$

$$\varinjlim_{y \in U} H^k(f^{-1}(U), f^{-1}(\mathscr{F})) = H^k(f^{-1}(y), \mathscr{F}_y).$$

Take the direct limit of the diagram

$$\begin{array}{ccccccc}
\cdots \longrightarrow H^{k-1}(U, \mathscr{F}) & \longrightarrow & H^{k-1}(f^{-1}(U), f^{-1}\mathscr{F}) & \longrightarrow & H^k(f^{-1}(U) \to U, \mathscr{F}) \to \\
\downarrow & & \downarrow & & \downarrow \\
\cdots \to H^{k-1}(\{y\}, \mathscr{F}_y) & \longrightarrow & H^{k-1}(f^{-1}(y), \mathscr{F}_y) & \longrightarrow & H^k(f^{-1}(y) \to \{y\}, \mathscr{F}) \to
\end{array}$$

$$\begin{array}{ccc}
H^k(U, \mathscr{F}) & \longrightarrow & H^k(f^{-1}(U), f^{-1}\mathscr{F}) \to \cdots \\
\downarrow & & \downarrow \\
H^k(\{y\}, \mathscr{F}_y) & \longrightarrow & H^k(f^{-1}(y), \mathscr{F}_y) \longrightarrow \cdots .
\end{array}$$

Then one obtains, by Five Lemma,

$$\mathscr{D}ist_f^k(\mathscr{F})_y = \varinjlim_{y \in U} H^k(f^{-1}(U) \to U, \mathscr{F}) = H^k(f^{-1}(y) \to \{y\}, \mathscr{F}).$$

Proposition 2.3.6. *Let G be a constant sheaf on the n-sphere S^n. Then the following* (1), (2), *and* (3) *hold.*

(1) $$H^k(S^0, G) = \begin{cases} G \oplus G & \text{for } k = 0 \\ 0 & \text{for } k \neq 0 \end{cases}$$

and for $n > 0$,

$$H^k(S^n, G) = \begin{cases} G & \text{for } k = 0, n \\ 0 & \text{for } k \neq 0, n. \end{cases}$$

(2) $$H^k(S^n \to \{x_0\}, G) = \begin{cases} G & \text{for } k = n+1 \\ 0 & \text{for } k \neq n+1. \end{cases}$$

(3) *For an arbitrary k*

$$H^k(\{y_0\} \to \{x_0\}, G) = 0$$

holds.

Proof. Note that S^0 consists of two points. Since any sheaf on two points is a flabby sheaf by definition, the first part of (1) is immediate. Next suppose $n \geqq 1$. Let D_1^n and D_2^n be closed hemispheres of S^n such that $D_1^n \cap D_2^n = S^{n-1}$ holds. Then $H^k(D_i^n, G) = 0$ holds for $k > 0$ since D_i ($i = 1, 2$) is contractible. If $\mathscr{L}$ is a flabby sheaf, the sequence

$$0 \to \Gamma(S^n, \mathscr{L}) \xrightarrow{\varphi} \Gamma(D_1^n, \mathscr{L}) \oplus \Gamma(D_2^n, \mathscr{L}) \xrightarrow{\psi} \Gamma(S^{n-1}, \mathscr{L}) \to 0$$

is exact, where $\varphi(s) = (s|_{D_1^n}, s|_{D_2^n})$ and $\psi(s_1, s_2) = s_1|_{S^{n-1}} - s_2|_{S^{n-1}}$ (see Proposition 2.1.4). Therefore, by considering a flabby resolution of the constant sheaf G, one obtains the long exact sequence

$$\cdots \to H^k(S^n, G) \to H^k(D_1^n, G) \oplus H^k(D_2^n, G) \to H^k(S^{n-1}, G) \to \cdots.$$

In the case where $n = 1$, define $i: G \to G \oplus G$ by $i(x) = (x, x)$, and define $j: G \oplus G \to G \oplus G$ by $j(x, y) = (x - y, x - y)$. Then one has the exact sequence

$$0 \to G \xrightarrow{i} G \oplus G \xrightarrow{j} G \oplus G \to H^1(S^1, G) \to 0.$$

When $k \geqq 2$, $0 \to H^k(S^1, G) \to 0$ is exact. Hence the assertion for $n = 1$ follows. Next, in the case where $n \geqq 2$, define a map $j': G \oplus G \to G$ by $j'(x, y) = x - y$. For $k = 1$, one has the exact sequence

$$0 \to G \xrightarrow{i} G \oplus G \xrightarrow{j'} G \to H^1(S^n, G) \to 0.$$

For $k \geqq 2$ the sequence

$$0 \to H^{k-1}(S^{n-1}, G) \to H^k(S^n, G) \to 0$$

is exact, from which the proof can be completed inductively.

To prove (2), first notice that the sequence

$$\cdots \to H^k(S^n \to \{x_0\}, G) \to H^k(\{x_0\}, G) \to H^k(S^n, G) \to \cdots$$

is exact, and notice the fact that $H^k(\{x_0\}, G) = 0$ for $k \neq 0$. Then (2) follows immediately from (1).

Lastly we will prove (3). Note that (3) holds when $k \geqq 2$ from the exact sequence

$$\cdots \to H^k(\{y_0\} \to \{x_0\}, G) \to H^k(\{x_0\}, G) \to H^k(\{y_0\}, G) \to \cdots.$$

For $k = 0$ and 1, one must only observe the exactness of the following sequence to complete the proof:

$$0 \to H^0(\{y_0\} \to \{x_0\}, G) \to G \rightrightarrows G \to H^1(\{y_0\} \to \{x_0\}, G) \to 0.$$

We will state a theorem from Grauert (though we will not give a proof) that is fundamental to our subsequent discussion.

Theorem 2.3.1 (Grauert [1]). *Let $\mathscr{A}$ be sheaf of real analytic functions on a real analytic manifold M, and let ω be the orientation sheaf. Then*

$$H^k(M, \mathscr{A} \otimes \omega) = 0$$

holds for $k \neq 0$.

This theorem gives us the following proposition.

Proposition 2.3.7.

$$H^k(\widetilde{^M X} \xrightarrow{\tau} X, \mathcal{O}_X) = \begin{cases} 0 & \text{for } k \neq n \\ (\mathcal{A} \otimes \omega)(M) & \text{for } k = n \end{cases}$$

holds.

Proof. Since τ is a proper map, one has the isomorphism

$$\mathcal{D}ist_\tau^k(\mathcal{O}_X)_x = H^k(\tau^{-1}(x) \to \{x\}, \mathcal{O}_{X,x})$$

from Proposition 2.3.5.

Let $x \in X - M$; then $\tau^{-1}(x) \cong \{x\}$. Hence one has $\mathcal{D}ist_\tau^k(\mathcal{O}_{X,x}) = 0$ by (3) of Proposition 2.3.6. If $x \in M$, then $\tau^{-1}(x) = S^{n-1}$. One obtains $\mathcal{D}ist_\tau^k(\mathcal{O}_{X,x}) = 0$ for $k \neq n$ and $\mathcal{D}ist_\tau^n(\mathcal{O}_{X,x}) = \mathcal{O}_{X,x}$ from (2) of Proposition 2.3.6. Note that the isomorphism $H^n(S^{n-1} \to \{x\}, G) \cong G$ depends upon the orientation. Therefore, $\mathcal{D}ist_\tau^k(\mathcal{O}_X) = 0$ for $k \neq n$ and $\mathcal{D}ist_\tau^n(\mathcal{O}_X) = \mathcal{A} \otimes \omega$ hold. Proposition 2.3.4 provides

$$H^k(\widetilde{^M X} \xrightarrow{\tau} X, \mathcal{O}_X) = H^{k-n}(X, \mathcal{A} \otimes \omega) = H^{k-n}(M, \mathcal{A} \otimes \omega).$$

Hence, the above theorem from Grauert (Theorem 2.3.1) now completes the proof.

We will give a summary of what we have found. In the exact sequence

$$\cdots \to H^k(\widetilde{^M X} \to X, \mathcal{O}_X) \to H^k((X - M) \hookrightarrow X, \mathcal{O}_X)$$
$$\to H^k((X - M) \hookrightarrow \widetilde{^M X}, \tau^{-1}\mathcal{O}_X) \to \cdots,$$

we have obtained

$$H^k(\widetilde{^M X} \xrightarrow{\tau} X, \mathcal{O}_X) = \begin{cases} 0 & \text{for } k \neq n \\ (\mathcal{A} \otimes \omega)(M) & \text{for } k = n \end{cases}$$

$$H^k((X - M) \hookrightarrow X, \mathcal{O}_X) = \begin{cases} 0 & \text{for } k \neq n \\ H^n_M(X, \mathcal{O}_X) & \text{for } k = n \end{cases}$$

and

$$H^k((X - M) \hookrightarrow \widetilde{^M X}, \tau^{-1}\mathcal{O}_X) = H^{k-1}(\sqrt{-1}SM, \mathcal{Q}).$$

Consequently we have the following exact sequence:

$$0 \to H^{n-2}(\sqrt{-1}SM, \mathcal{Q}) \to (\mathcal{A} \otimes \omega)(M) \to H^n_M(X, \mathcal{O}_X)$$
$$\to H^{n-1}(\sqrt{-1}SM, \mathcal{Q}) \to 0 \tag{2.3.1}$$

and

$$H^k(\sqrt{-1}SM, \mathcal{Q}) = 0 \qquad \text{for } k \neq n-1, n-2. \tag{2.3.2}$$

Note that (2.3.1) and (2.3.2) still hold when M is replaced by an arbitrary open subset of M, and note that we have $H^k(\sqrt{-1}SM, \mathcal{Q}) = H^k(\tau^{-1}(M), \mathcal{Q})$.

Therefore, we have the localized version of (2.3.1) and (2.3.2) as follows. The sequence

$$0 \to R^{n-2}\tau_*\mathscr{Q} \to \mathscr{A} \otimes \omega \to \mathscr{H}^n_M(\mathscr{O}_X) \to R^{n-1}\tau_*\mathscr{Q} \to 0 \qquad (2.3.3)$$

is exact, and we have

$$R^k\tau_*\mathscr{Q} = 0 \qquad \text{for } k \neq n-1, n-2. \qquad (2.3.4)$$

Since $\omega \underset{\mathbf{Z}}{\otimes} \omega = \mathbf{Z}_M$, we also obtain the exact sequence

$$0 \to R^{n-2}\tau_*\mathscr{Q} \otimes \omega \to \mathscr{A} \to \mathscr{B} \to R^{n-1}\tau_*\mathscr{Q} \otimes \omega \to 0. \qquad (2.3.5)$$

We will compute the first and the last terms of the exact sequence (2.3.5).

Proposition 2.3.8.

$$R^k\tau_*\mathscr{Q} = \begin{cases} 0 & \text{for } k \neq n-1 \\ \pi_*\mathscr{C}^a \otimes \omega & \text{for } k = n-1 \end{cases}$$

holds.

We need a lemma to prove this proposition.

Lemma. *Let X, Y, and Z be topological spaces, and let $f: X \to Y$ and $g: Y \to Z$ be continuous maps. Suppose that $\mathscr{F}$ is a sheaf over X and that the continuous map f is purely l-dimensional with respect to $\mathscr{F}$. Then, for any integer k,*

$$R^k(g \circ f)_*\mathscr{F} = R^{k-l}g_*(R^lf_*\mathscr{F})$$

holds.

Proof. By Lemma 2 in the proof of Proposition 2.2.1, $H^k(X, \mathscr{F}) = H^{k-l}(Y, R^lf_*\mathscr{F})$ holds. This implies that for an open set U of Z there exists the isomorphism $H^k(f^{-1}(g^{-1}(U)), \mathscr{F}) = H^{k-l}(g^{-1}(U), R^lf_*\mathscr{F})$. Notice that the sheaf $R^k(g \circ f)_*\mathscr{F}$ over Z is the sheaf associated to the presheaf $H^k(f^{-1}(g^{-1}(U)), \mathscr{F})$ and that $R^{k-l}g_*(R^lf_*\mathscr{F})$ is the sheaf associated to $H^{k-l}(g^{-1}(U), R^lf_*(\mathscr{F}))$.

Note that $\pi^{-1}(x_0 + \sqrt{-1}v0)$, $x_0 + \sqrt{-1}v0 \in \sqrt{-1}SM$, is contractible. Hence we have

$$R^k\pi_*(\pi^{-1}\mathscr{Q})_{x+\sqrt{-1}v0} = H^k(\pi^{-1}(x + \sqrt{-1}v0), \mathscr{Q}_{x+\sqrt{-1}v0})$$
$$= \begin{cases} 0 & \text{for } k \neq 0 \\ \mathscr{Q}_{x+\sqrt{-1}v0} & \text{for } k = 0 \end{cases};$$

i.e.

$$R^k\pi_*(\pi^{-1}\mathscr{Q}) = \begin{cases} 0 & \text{for } k \neq 0 \\ \mathscr{Q} & \text{for } k = 0 \end{cases}.$$

By the above lemma, $R^k\tau_*\mathcal{Q} = R^k(\tau \circ \pi)_*(\pi^{-1}\mathcal{Q}) = R^k(\pi \circ \tau)_*(\pi^{-1}\mathcal{Q})$. On the other hand, Proposition 2.1.2′ implies

$$R^k\tau_*(\pi^{-1}\mathcal{Q}) = \begin{cases} \mathcal{C}^a \otimes \omega & \text{for } k = n-1 \\ 0 & \text{for } k \neq n-1 \end{cases}.$$

Again by the preceding lemma, one obtains

$$R^k(\pi \circ \tau)_*(\pi^{-1}\mathcal{Q}) = R^{k-n+1}\pi_*(\mathcal{C}^a \otimes \omega).$$

Hence, $R^k\tau_*\mathcal{Q} = (R^{k-n+1}\pi_*\mathcal{C}^a) \otimes \omega$. For $k = n-2$ and $k = n-1$, $R^{n-2}\tau_*\mathcal{Q} = 0$ and $R^{n-1}\tau_*\mathcal{Q} = \pi_*\mathcal{C}^a \otimes \omega$ hold respectively.

Hence we assert the following theorem.

Theorem 2.3.2.

(1) *There exists the exact squence*

$$0 \to \mathcal{A} \to \mathcal{B} \xrightarrow{\text{sp}} \pi_*\mathcal{C} \to 0$$

on M.

(2) *For $k \neq 0$, $R^k\pi_*\mathcal{C} = 0$ holds.*

(3) *The sequence*

$$0 \to \mathcal{A}(M) \to \mathcal{B}(M) \xrightarrow{\text{sp}} \mathcal{C}(\sqrt{-1}S^*M) \to 0$$

is exact.

Proof. Note $\pi_*\mathcal{C}^a = \pi_*\mathcal{C}$. Then the exact sequence (2.3.5) and Proposition 2.3.8 imply the short exact sequence in (1). The assertion (2) has been shown in the proof of Proposition 2.3.8.

There is induced the exact sequence

$$0 \to \mathcal{A}(M) \to \mathcal{B}(M) \to \mathcal{C}(\sqrt{-1}S^*M) \to H^1(M, \mathcal{A})$$

from the short exact sequence in (1). Then Grauert's theorem (Theorem 2.3.1) completes the proof of (3).

Definition 2.3.3. *For $u \in \mathcal{B}(M)$, $\text{sp}(u) \in \pi_*\mathcal{C}(M) = \mathcal{C}(\sqrt{-1}S^*M)$ is said to be the spectrum of u. The support of* sp(u), *denoted by* S.S. u, *is called the singularity spectrum of u.*

Corollary. *For $u \in \mathcal{B}(M)$* S.S. $u = \varnothing$ *holds if and only if u is a real analytic function on M.*

Proof. This is plain from the exact sequence (3) of Theorem 2.3.2.

Note. The terminology "singular spectrum" was coined by Boutet de Monvel. "Spectrum" originally meant the light decomposed according to the frequency or, more mathematically speaking, the support of the Fourier transform of a function. Hence this terminology gives picture of what it means (cf. Example 2.4.6 in the subsequent section). One problem with this terminology is that it becomes confounded with the existing terminology in spectral analysis of linear operators; and, in this respect, the terminology "singularity spectrum" proposed by Komatsu is preferable. Although some other terminologies are also proposed, here we use the terminology "singularity spectrum," which seems to be the most euphonious, self-explanatory, and commonly used.

We note that, for a distribution u, Hörmander [2] introduced a notion similar to S.S. u, in order to analyze the singularity structure of solutions of linear differential equations, and named it the (analytic) wave-front set of u. Bros and Iagolnitzer (Iagolnitzer [1] and the references cited therein) also introduced a similar notion "essential support" of a distribution u, starting from some physical motivation. When u is a distribution—as Bros and Iagolnitzer; Bony and Schapira; Kataoka; Nishiwada; and Hill showed (1975–1976) independently—these three concepts, i.e. the singularity spectrum, the analytic wave-front set, and the essential support, coincide. As we do not need this result in this book, we will not discuss it any further, but refer the reader to Bony [2] and references cited there.

We will construct a morphism $b:\tilde{\mathscr{A}} \to \tau^{-1}\mathscr{B}$ such that $0 \to \tilde{\mathscr{A}} \xrightarrow{b} \tau^{-1}\mathscr{B} \to \pi_*\tau^{-1}\mathscr{C} \to 0$ is an exact sequence on $\sqrt{-1}SM$.

Proposition 2.3.9. *Let π be the natural projection from $DM \underset{\sqrt{-1}S^*M}{\times} DM$ to $\sqrt{-1}SM \underset{M}{\times} \sqrt{-1}SM$, and let $\mathscr{F}$ be a sheaf on $\sqrt{-1}SM \underset{M}{\times} \sqrt{-1}SM$. Then*

$$\mathscr{D}\!\mathit{ist}^k_\pi(\mathscr{F}) = \begin{cases} 0 & \text{for } k \neq n-1 \\ \mathscr{F}|_{\Delta^a_{\sqrt{-1}SM}} \otimes \omega & \text{for } k = n-1 \end{cases}$$

holds, where the antidiagonal set $\Delta^a_{\sqrt{-1}SM} = \{(x + \sqrt{-1}v0, x - \sqrt{-1}v0) \in \sqrt{-1}SM \underset{M}{\times} \sqrt{-1}SM\}$.

Proof. Since π is proper, for $x(v_1, v_2) = (x + \sqrt{-1}v_1 0, x + \sqrt{-1}v_2 0) \in \sqrt{-1}SM \underset{M}{\times} \sqrt{-1}SM$ one has

$$\mathscr{D}\!\mathit{ist}^k_\pi(\mathscr{F})_{x(v_1,v_2)} = H^k(\pi^{-1}(x(v_1, v_2)) \to \{x(v_1, v_2)\}, \mathscr{F}_{x(v_1,v_2)})$$

by Proposition 2.3.5. On the other hand,

$$\pi^{-1}(x(v_1, v_2)) = \{\xi \in S^{n-1} | \langle v_1, \xi\rangle \leqq 0 \text{ and } \langle v_2, \xi\rangle \leqq 0\}.$$

Then, for $v_1 \neq -v_2$, $\pi^{-1}(x(v_1, v_2))$ is contractible to a point (see Figure 2.3.1(a)). Therefore one obtains

$$\mathscr{D}\!\mathit{ist}^k_\pi(\mathscr{F})_{x(v_1,v_2)} = 0 \quad \text{for } v_1 \neq -v_2.$$

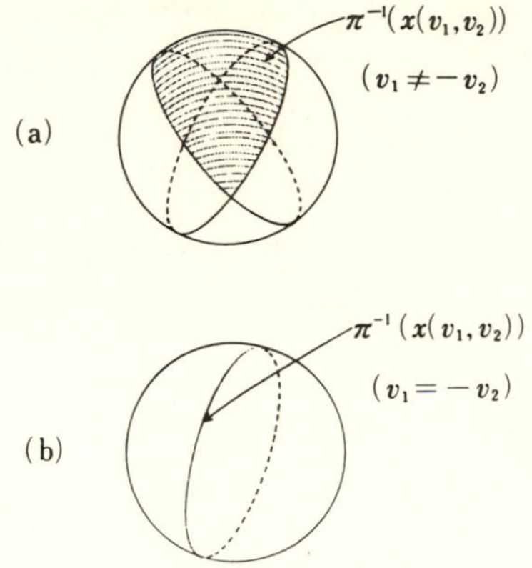

Figure 2.3.1

If $v_1 = -v_2$, one has $\pi^{-1}(x(v_1, v_2)) = S^{n-2}$ (see Figure 2.3.1(b)). Then

$$\mathscr{D}ist_{\pi}^{k}(\mathscr{F})_{x(v_1,v_2)} = \begin{cases} 0 & \text{for } k \neq n-1 \\ \mathscr{F}_{x(v_1,v_2)} \otimes \omega & \text{for } k = n-1 \end{cases}$$

by (2) of Proposition 2.3.6.

Proposition 2.3.10. *Let X, Y, and Z be topological spaces; let $f: X \to Y$ and $g: Y \to Z$ be continuous maps; and let $\mathscr{F}$ be a sheaf over Y. If f is purely l-codimensional with respect to $\mathscr{F}$, then there is induced the long exact sequence*

$$\cdots \to R^{k-l}g_*\mathscr{D}ist_f^l(\mathscr{F}) \to R^k g_*\mathscr{F} \to R^k(g \circ f)_*(f^{-1}\mathscr{F}) \to R^{k-l+1}g_*\mathscr{D}ist_f^l(\mathscr{F}) \to \cdots.$$

Proof. Let $0 \to \mathscr{F} \to \mathscr{L}^\bullet$ and $0 \to f^{-1}\mathscr{F} \to \mathscr{M}^\bullet$ be flabby resolutions of $\mathscr{F}$ and $f^{-1}\mathscr{F}$ respectively. Let $\mathscr{N}^\bullet = C(\mathscr{L}^\bullet \to f_*\mathscr{M}^\bullet)$. Then one has the following:

$$\begin{cases} \mathscr{D}ist_f^k(\mathscr{F}) = \mathscr{H}^k(\mathscr{N}^\bullet) \\ R^k g_*\mathscr{F} = \mathscr{H}^k(g_*\mathscr{L}^\bullet) \\ R^k(g \circ f)_*(f^{-1}\mathscr{F}) = \mathscr{H}^k((g \circ f)_*\mathscr{M}^\bullet). \end{cases}$$

The exact sequence $0 \to f_*\mathscr{M}^{k-1} \to \mathscr{N}^k \to \mathscr{L}^k \to 0$ induces the exact sequence $0 \to g_*f_*\mathscr{M}^{k-1} \to g_*\mathscr{N}^k \to g_*\mathscr{L}^k \to 0$, since $f_*\mathscr{M}^{k-1}$ is flabby. Hence one obtains the long exact sequence

$$\cdots \to \mathscr{H}^k(g_*\mathscr{N}^\bullet) \to R^k g_*\mathscr{F} \to R^k(g \circ f)_*(f^{-1}\mathscr{F}) \to \mathscr{H}^{k+1}(g_*\mathscr{N}^\bullet) \to \cdots,$$

where one notes $g_* f_* \mathscr{M}^{k-1} = (g \circ f)_* \mathscr{M}^{k-1}$. One needs to prove the isomorphism

$$\mathscr{H}^k(g_* \mathscr{N}^\bullet) = R^{k-1} g_* \mathscr{D}ist^l_f(\mathscr{F})$$

under the hypothesis $\mathscr{D}ist^k_f(\mathscr{F}) = 0$ for $k \neq l$.

Since $\mathscr{H}^k(\mathscr{N}^\bullet) = 0$ for $k \neq l$, there exists an exact sequence

$$0 \to \mathscr{N}^0 \to \mathscr{N}^1 \to \cdots \to \mathscr{N}^{l-1} \to \mathscr{N}^l \to \mathscr{Z} \to 0,$$

where $\mathscr{Z} = \mathscr{N}^l/\mathrm{Im}(\mathscr{N}^{l-1} \to \mathscr{N}^l)$. Notice that $\mathscr{N}^i$ and $\mathscr{Z}$ are flabby. Therefore, the following sequence is also exact:

$$0 \to g_*\mathscr{N}^0 \to g_*\mathscr{N}^1 \to \cdots \to g_*\mathscr{N}^l \to g_*\mathscr{Z} \to 0.$$

This implies that $\mathscr{H}^k(g_*\mathscr{N}^\bullet) = 0 = R^{k-l} g_* \mathscr{D}ist^l_f(\mathscr{F})$ holds for $k < l$. Next, notice that

$$0 \to \mathscr{H}^l(\mathscr{N}^\bullet) \to \mathscr{Z} \to \mathscr{N}^{l+1} \to \cdots$$

gives a flabby resolution of $\mathscr{H}^l(\mathscr{N}^\bullet)$. For $k \geqq l + 2$ one has

$$R^{k-l} g_* \mathscr{H}^l(\mathscr{N}^\bullet) = \mathscr{H}^k(g_*\mathscr{N}^\bullet).$$

For $k = l + 1$ $\mathrm{Im}(g_*\mathscr{Z} \to g_*\mathscr{N}^{l+1}) = \mathrm{Im}(g_*\mathscr{N}^l \to g_*\mathscr{N}^{l+1})$ holds. Hence this case can be treated as the above. In the case where $k = l$, from the exact sequence

$$0 \to g_*\mathscr{H}^l(\mathscr{N}^\bullet) \to g_*\mathscr{Z} \to g_*\mathscr{N}^{l+1},$$

one obtains

$$\begin{aligned} R^0 g_* \mathscr{D}ist^l_f(\mathscr{F}) \cong g_*\mathscr{H}^l(\mathscr{N}^\bullet) &= \mathrm{Ker}(g_*\mathscr{Z} \to g_*\mathscr{N}^{l+1}) \\ &= \frac{\mathrm{Ker}(g_*\mathscr{N}^l \to g_*\mathscr{N}^{l+1})}{\mathrm{Im}(g_*\mathscr{N}^{l-1} \to g_*\mathscr{N}^l)} = \mathscr{H}^l(g_*\mathscr{N}^\bullet). \end{aligned}$$

Proposition 2.3.11. *Let X, Y, and Y' be topological spaces, and let $\mathscr{F}$ be a sheaf over X. Suppose that $f: X \to Y$, $g: Y' \to Y$, $f': X' = X \underset{Y}{\times} Y' \to Y'$, and $g': X' \to X$ are continuous maps such that the diagram*

$$\begin{array}{ccc} X & \xleftarrow{\ g'\ } & X' = X \underset{Y}{\times} Y' \\ \downarrow{\scriptstyle f} & & \downarrow{\scriptstyle f'} \\ Y & \xleftarrow{\ g\ } & Y' \end{array}$$

is commutative. Further assume that f and *f' are closed maps, and that all the open subsets of X or X' are paracompact. Then one has*

$$R^k f'_* g'^{-1} \mathscr{F} = g^{-1} R^k f_* \mathscr{F}.$$

Remark. In order to claim the isomorphism above, one may assume that four of those topological spaces are Hausdorff and that f is proper (hence f' is proper).

Proof. The corollary of Proposition 2.1.4 implies

$$(R^k f'_* g'^{-1}\mathscr{F})_{y'} = H^k(f'^{-1}(y'), g'^{-1}\mathscr{F}|_{f'^{-1}(y')}).$$

Let $y = g(y')$. Since g' is a bijection from $f'^{-1}(y')$ onto $f^{-1}(y)$, one obtains

$$\begin{aligned} H^k(f'^{-1}(y'), g'^{-1}\mathscr{F}|_{f'^{-1}(y')}) &= H^k(f^{-1}(y), \mathscr{F}|_{f^{-1}(y)}) = (R^k f_* \mathscr{F})_y \\ &= (g^{-1}R^k f_* \mathscr{F})_{y'} \end{aligned}$$

Proposition 2.3.12. *Consider the following commutative diagram:*

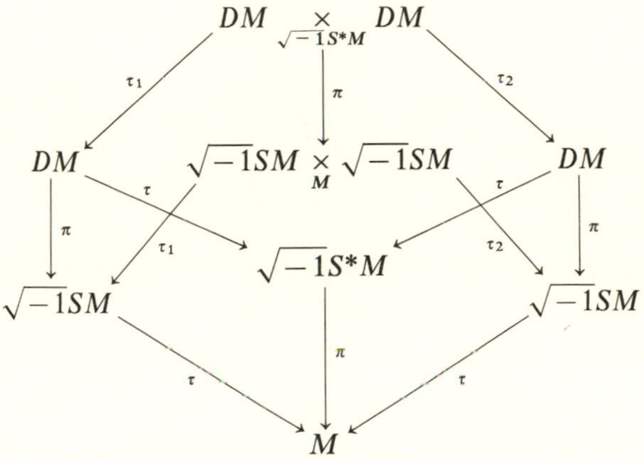

where τ and π are natural maps $\tau: \sqrt{-1}SM \to M$ and $\pi: \sqrt{-1}S^*M \to M$, respectively, and where τ_1 and τ_2 are projections on the first and the second components respectively. Let $\mathscr{F}$ be a sheaf on $\sqrt{-1}SM$. Then

(1) $$\begin{aligned} 0 &\to \tau^{-1}R^{n-2}\tau_*\mathscr{F} \to R^{n-2}(\pi \circ \tau_2)_*(\tau_1^{-1}\pi^{-1}\mathscr{F}) \\ &\to \mathscr{F}^a \otimes \omega \to \tau^{-1}R^{n-1}\tau_*\mathscr{F} \to R^{n-1}(\pi \circ \tau_2)_*(\tau_1^{-1}\pi^{-1}\mathscr{F}) \to 0 \end{aligned}$$

is an exact sequence.

(2) $$\tau^{-1}R^k\tau_*\mathscr{F} = R^k(\pi \circ \tau_2)_*(\tau_1^{-1}\pi^{-1}\mathscr{F})$$

holds for $k \neq n-1, n-2$.

Proof. For the sheaf $\tau_1^{-1}\mathscr{F}$ on $\sqrt{-1}SM \underset{M}{\times} \sqrt{-1}SM$, apply Propositions 2.3.9 and 2.3.10 to $DM \underset{\sqrt{-1}S^*M}{\times} DM \xrightarrow{\pi} \sqrt{-1}SM \underset{M}{\times} \sqrt{-1}SM \xrightarrow{\tau_2} \sqrt{-1}SM$. Then one obtains the exact sequence

$$\begin{aligned} \cdots &\to R^{k-n+1}\tau_{2*}(\tau_1^{-1}\mathscr{F}|_{\Delta^a_{\sqrt{-1}SM}} \otimes \omega) \to R^k\tau_{2*}\tau^{-1}\mathscr{F} \\ &\to R^k(\tau_2 \circ \pi)_*(\pi^{-1}\tau_1^{-1}\mathscr{F}) \to \cdots. \end{aligned}$$

We will compute each term in this sequence. Notice that $\tau_i|_{\Delta^a_{\sqrt{-1}SM}}$, $i = 1$ and 2, are homeomorphisms onto $\sqrt{-1}SM$ and that $\tau_2 \circ \tau_1^{-1}$ is the antipodal mapping $a: \sqrt{-1}SM \to \sqrt{-1}SM$, defined by $a(x + \sqrt{-1}v0) = x - \sqrt{-1}v0$. Therefore one finds

$$R^{k-n+1}\tau_{2*}(\tau_1^{-1}\mathscr{F}|_{\Delta^a_{\sqrt{-1}SM}} \otimes \omega) = R^{k-n+1}\tau_{2*}(\tau_1^{-1}\mathscr{F}|_{\Delta^a_{\sqrt{-1}SM}}) \otimes \omega$$
$$= \begin{cases} 0 & \text{for } k \neq n-1 \\ \mathscr{F}^a \otimes \omega & \text{for } k = n-1, \end{cases}$$

where $\mathscr{F}^a \underset{\text{def}}{=} a_*\mathscr{F} = a^{-1}\mathscr{F}$.

From the commutative diagram

$$\begin{array}{ccc} \sqrt{-1}SM & \xleftarrow{\tau_1} & \sqrt{-1}SM \underset{M}{\times} \sqrt{-1}SM \\ \downarrow{\scriptstyle \tau} & & \downarrow{\scriptstyle \tau_2} \\ M & \xleftarrow{\tau} & \sqrt{-1}SM, \end{array}$$

one obtains

$$R^k\tau_{2*}\tau_1^{-1}\mathscr{F} = \tau^{-1}R^k\tau_*\mathscr{F}$$

by Proposition 2.3.11. The commutativity of the diagram in Proposition 2.3.12 implies

$$R^k(\tau_2 \circ \pi)_*(\pi^{-1}\tau_1^{-1}\mathscr{F}) = R^k(\pi \circ \tau_2)_*(\tau_1^{-1}\pi^{-1}\mathscr{F}).$$

Hence these complete the proof of (2).

Proposition 2.3.13.

(1) *The sequence*

$$0 \to \mathscr{Q} \to \tau^{-1}\pi_*\mathscr{C} \to \pi_*\tau^{-1}\mathscr{C} \to 0$$

is exact on $\sqrt{-1}SM$.

(2) $$R^k\pi_*\tau^{-1}\mathscr{C} = 0 \qquad \text{for } k \neq 0.$$

Proof. Replace $\mathscr{F}$ in Proposition 2.3.12 by $\mathscr{Q} = \mathscr{H}^1_{\sqrt{-1}SM}(\tau^{-1}\mathscr{O}_X))$. First compute $R^k(\pi \circ \tau_2)_*(\tau_1^{-1}\pi^{-1}\mathscr{Q})$. Consider the commutative diagram

$$\begin{array}{ccc} DM & \xleftarrow{\tau_1} & DM \underset{\sqrt{-1}S^*M}{\times} DM \\ \downarrow{\scriptstyle \tau} & & \downarrow{\scriptstyle \tau_2} \\ \sqrt{-1}S^*M & \xleftarrow{\tau} & DM \end{array}$$

Then, for the sheaf $\pi^{-1}\mathscr{Q}$ on DM, Propositions 2.3.11 and 2.1.2′ give

$$R^k\tau_{2*}(\tau_1^{-1}\pi^{-1}\mathscr{Q}) = \tau^{-1}R^k\tau_*\pi^{-1}\mathscr{Q} = \begin{cases} 0 & \text{for } k \neq n-1 \\ \tau^{-1}\mathscr{C}^a \otimes \omega & \text{for } k = n-1. \end{cases}$$

By the lemma in the proof of Proposition 2.3.8,

$$\begin{aligned} R^k(\pi \circ \tau_2)_*(\tau_1^{-1}\pi^{-1}\mathscr{Q}) &= R^{k-n+1}\pi_* R^{n-1}\tau_{2*}(\tau_1^{-1}\pi^{-1}\mathscr{Q}) \\ &= R^{k-n+1}\pi_*\tau^{-1}\mathscr{C}^a \otimes \omega. \end{aligned}$$

In particular, one has

$$R^{n-2}(\pi \circ \tau_2)_*(\tau_1^{-1}\pi^{-1}\mathscr{Q}) = 0$$

and

$$R^{n-1}(\pi \circ \tau_2)_*(\tau_1^{-1}\pi^{-1}\mathscr{Q}) = \pi_*\tau^{-1}\mathscr{C}^a \otimes \omega.$$

On the other hand, from Proposition 2.3.8,

$$\tau^{-1}R^k\tau_*\mathscr{Q} = \begin{cases} 0 & \text{for } k \neq n-1 \\ \tau^{-1}\pi_*\mathscr{C}^a \otimes \omega & \text{for } k = n-1 \end{cases}.$$

Hence, (1) of Proposition 2.3.12 implies that

$$0 \to \mathscr{Q}^a \otimes \omega \to \tau^{-1}\pi_*\mathscr{C}^a \otimes \omega \to \pi_*\tau^{-1}\mathscr{C}^a \otimes \omega \to 0$$

is exact. Since $\omega \otimes \omega = \mathbf{Z}_M$, (1) of Proposition 2.3.13 is proved.

For $k \neq n-1$, $n-2$, one has the following from (2) of Proposition 2.3.12:

$$\tau^{-1}R^k\tau_*\mathscr{Q} = R^k(\pi \circ \tau_2)_*(\tau_1^{-1}\pi^{-1}\mathscr{Q}) = R^{k-n+1}\pi_*\tau^{-1}\mathscr{C}^a \otimes \omega.$$

Then, by Proposition 2.3.8, one obtains (2).

Proposition 2.3.14. *There is a canonical morphism* $b: \tilde{\mathscr{A}} \to \tau^{-1}\mathscr{B}$ *over* $\sqrt{-1}SM$.

Proof. From the exact sequence (1) of Proposition 2.3.12, there exists the morphism $\tilde{\mathscr{A}}^a \otimes \omega \to \tau^{-1}R^{n-1}\tau_*\tilde{\mathscr{A}}$ for $\mathscr{F} = \tilde{\mathscr{A}}$. Notations being the same as in Definition 2.1.2, recall the diagram

$$\begin{array}{ccc} \widetilde{{}^M X} & \xleftarrow{\ \tilde{\epsilon}\ } & \widetilde{{}^M X} - \sqrt{-1}SM \\ \big\downarrow{\scriptstyle \tau} & & \big\| \\ X & \xleftarrow{\ \epsilon\ } & X - M \end{array}$$

and recall the fact that $R^k\tilde{\epsilon}_*(\mathscr{O}_X|_{X-M}) = 0$ for $k \neq 0$. Then one obtains

$$\begin{aligned} R^{n-1}\tau_*\tilde{\mathscr{A}} = R^{n-1}\tau_*(\tilde{\epsilon}_*\mathscr{O}_X|_{X-M}) &= R^{n-1}(\tau \circ \tilde{\epsilon})_*(\mathscr{O}_X|_{X-M}) \\ &= R^{n-1}\epsilon_*(\mathscr{O}_X|_{X-M}). \end{aligned}$$

There is induced a canonical morphism $R^{n-1}\epsilon_*(\mathscr{O}_X|_{X-M}) \to \mathscr{H}^n_M(\mathscr{O}_X)$ from taking the direct limit of the sequence

$$\begin{aligned} \cdots \to H^{n-1}(U, \mathscr{O}_X) \to H^{n-1}(U - M, \mathscr{O}_X) \to H^n_{U \cap M}(U, \mathscr{O}_X) \\ \to H^n(U, \mathscr{O}_X) \to \cdots; \end{aligned}$$

furthermore, this morphism is an isomorphism for $n > 1$. Since $\tau^{-1}R^{n-1}\tau_*\tilde{\mathscr{A}} = \tau^{-1}R^{n-1}\epsilon_*(\mathcal{O}_X|_{X-M})$, one obtains the canonical morphism: $\tilde{\mathscr{A}}^a \otimes \omega \to \tau^{-1}\mathscr{H}^n_M(\mathcal{O}_X)$ by composing the above-constructed morphisms. Consequently, the functor ${}^a \otimes \omega$ induces

$$b:\tilde{\mathscr{A}} \to \tau^{-1}\mathscr{B}.$$

Remark. Proposition 2.3.14 asserts that a boundary value of a holomorphic function defines a hyperfunction. The terminology, "a boundary value of a holomorphic function," always has the connotation as that in Proposition 2.3.14. Note that, in spite of the wording, the (pointwise) "value" does not necessarily exist.

Theorem 2.3.3. *The sequence on $\sqrt{-1}SM$*

$$0 \to \tilde{\mathscr{A}} \xrightarrow{b} \tau^{-1}\mathscr{B} \to \pi_*\tau^{-1}\mathscr{C} \to 0$$

is exact.

Proof. Consider the following commutative diagram:

$$\begin{array}{ccccccccc}
 & & 0 & & 0 & & 0 & & \\
 & & \downarrow & & \downarrow & & \downarrow & & \\
0 & \longrightarrow & \tau^{-1}\mathscr{A} & = & \tau^{-1}\mathscr{A} & \longrightarrow & 0 & \longrightarrow & 0 \\
 & & \downarrow & & \downarrow & & \downarrow & & \\
0 & \longrightarrow & \tilde{\mathscr{A}} & \xrightarrow{b} & \tau^{-1}\mathscr{B} & \longrightarrow & \pi_*\tau^{-1}\mathscr{C} & \longrightarrow & 0 \\
 & & \downarrow & & \downarrow & & \downarrow & & \\
0 & \longrightarrow & \mathscr{Q} & \longrightarrow & \tau^{-1}\pi_*\mathscr{C} & \longrightarrow & \pi_*\tau^{-1}\mathscr{C} & \longrightarrow & 0 \\
 & & \downarrow & & \downarrow & & \downarrow & & \\
 & & 0 & & 0 & & 0 & &
\end{array}$$

In this diagram, all the vertical sequences and the first and the third horizontal sequences are exact as the result of Proposition 2.1.2, Theorem 2.3.2, and Proposition 2.3.13. Notice that the third vertical exact sequence makes the second horizontal sequence a cochain complex. Now, Nine Lemma implies that the second horizontal sequence, $0 \to \tilde{\mathscr{A}} \xrightarrow{b} \tau^{-1}\mathscr{B} \to \pi_*\tau^{-1}\mathscr{C} \to 0$, is exact.

We shall investigate the interplay of the singularity spectrum of a hyperfunction and the boundary value of the hyperfunction (Theorem 2.3.4).

Definition 2.3.4. *A subset Z of $\sqrt{-1}SM$ is said to be convex if each fibre $\tau^{-1}(x) \cap Z$ of $\tau:\sqrt{-1}SM \to M$ is convex.*

Note. A proper subset A of S^{n-1} is said to be convex if the inverse image of A under the projection $\tilde{\omega}:\mathbf{R}^n - \{0\} \to S^{n-1} \cong (\mathbf{R}^n - \{0\})/\mathbf{R}^{\times}_{+}$ is convex. Note that S^{n-1} is itself convex by definition.

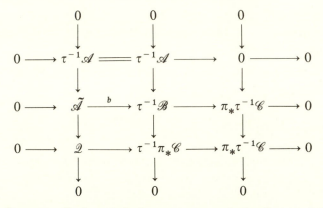

Similarly, a subset Z of $\sqrt{-1}SM$ is said to be *properly convex* if each fibre $\tau^{-1}(x) \cap Z$ is properly convex, where a convex subset A of S^{n-1} is called properly convex if the inverse image of A under the map $\varpi: \mathbf{R}^n - \{0\} \to S^{n-1}$ does not contain a line, $\cong \mathbf{R}$. A properly convex set is sometimes called a convex set without containing a line.

We similarly define a (properly) convex subset of $\sqrt{-1}S^*M$ as we defined the (properly) convex subset of $\sqrt{-1}SM$. The smallest convex set containing Z (the intersection of convex sets containing Z) is called the *convex hull* of Z.

Definition 2.3.5. *The polar set Z° of a subset $Z \subset \sqrt{-1}SM$ is defined as $Z^\circ = \{(x, \sqrt{-1}\xi_x\infty) \in \sqrt{-1}S^*M | \langle \xi_x, v_x \rangle > 0$ for an arbitary point $x + \sqrt{-1}v_x 0 \in Z\}$. The polar set of a subset of $\sqrt{-1}S^*M$ is defined similarly.*

Remark. The polar set Z° is always convex for any Z. Notice also that the correspondence between open convex sets U in $\sqrt{-1}SM$ and polar sets U° in $\sqrt{-1}S^*M$ is injective. Furthermore, any closed convex set Z can be expressed as the polar set U° of an open convex set U in $\sqrt{-1}SM$. Note that $U^{\circ\circ}$ is the convex hull of U.

The polar set of a convex subset A of $\sqrt{-1}SM$ (or of $\sqrt{-1}S^*M$) is non-empty if and only if A is properly convex.

Theorem 2.3.4. *Let U be an open set in $\sqrt{-1}SM$ and let $U \cap \tau^{-1}(x)$ be a non-empty connected set for each $x \in M$. Then*

(1) *the restriction map $\Gamma(V, \tilde{\mathscr{A}}) \to \Gamma(U, \tilde{\mathscr{A}})$ is bijective, where $V = U^{\circ\circ}$, and*

(2) *the sequence $0 \to \tilde{\mathscr{A}}(U) \xrightarrow{b} \mathscr{B}(M) \xrightarrow{\text{sp}} \mathscr{C}(\sqrt{-1}S^*M - U^\circ)$ is exact. In other words, for $\varphi \in \Gamma(U, \tilde{\mathscr{A}})$ one has* S.S. $b(\varphi) \subset U^\circ$*; and, if* S.S. $u \subset U^\circ$ *for $u \in \mathscr{B}(M)$, then $u = b(\varphi)$ for a unique $\varphi \in \Gamma(U, \tilde{\mathscr{A}})$.*

Proof. Since the sequence $0 \to \tilde{\mathscr{A}} \to \tau^{-1}\mathscr{B} \to \pi_*\tau^{-1}\mathscr{C} \to 0$ is exact, from Theorem 2.3.3, the following commutative diagram is obtained:

$$\begin{array}{ccccccc} 0 & \longrightarrow & \tilde{\mathscr{A}}(V) & \longrightarrow & \tau^{-1}\mathscr{B}(V) & \longrightarrow & \pi_*\tau^{-1}\mathscr{C}(V) \\ & & \downarrow & & \downarrow & & \downarrow \\ 0 & \longrightarrow & \tilde{\mathscr{A}}(U) & \longrightarrow & \tau^{-1}\mathscr{B}(U) & \longrightarrow & \pi_*\tau^{-1}\mathscr{C}(U) \end{array}$$

By the assumption, $\tau|_U$ and $\tau|_V$ are surjective open maps and the fibres are connected. Then one has

$$\tau^{-1}\mathscr{B}(V) = \tau^{-1}\mathscr{B}(U) = \mathscr{B}(M)$$

by (2) of Proposition 2.1.3. Hence the above diagram becomes

$$\begin{array}{ccccccc} 0 & \longrightarrow & \tilde{\mathscr{A}}(V) & \longrightarrow & \mathscr{B}(M) & \longrightarrow & \pi_*\tau^{-1}\mathscr{C}(V) \\ & & \downarrow & & \| & & \downarrow \\ 0 & \longrightarrow & \tilde{\mathscr{A}}(U) & \longrightarrow & \mathscr{B}(M) & \longrightarrow & \pi_*\tau^{-1}\mathscr{C}(U) \end{array}$$

This implies that the restriction $\tilde{\mathscr{A}}(V) \to \tilde{\mathscr{A}}(U)$ is a monomorphism. In order to show this restriction map is also onto, it is sufficient to prove that $\pi_*\tau^{-1}\mathscr{C}(V) \to \pi_*\tau^{-1}\mathscr{C}(U)$ is injective. Since the map

$$\tau:\pi^{-1}V \to \tau\pi^{-1}V = \sqrt{-1}S^*M - V^\circ = \sqrt{-1}S^*M - U^\circ$$

is surjective and open, and since each fibre is connected, (2) of Proposition 2.1.3 implies

$$\pi_*\tau^{-1}\mathscr{C}(V) = (\tau^{-1}\mathscr{C})(\pi^{-1}V) = \mathscr{C}(\tau\pi^{-1}V) = \mathscr{C}(\sqrt{-1}S^*M - U^\circ).$$

On the other hand, $\tau:\pi^{-1}U \to (\sqrt{-1}S^*M - U^\circ)$ is an epimorphism. Then it is plain that

$$\mathscr{C}(\sqrt{-1}S^*M - U^\circ) \to (\tau^{-1}\mathscr{C})(\pi^{-1}U) = \pi_*\tau^{-1}\mathscr{C}(U)$$

is a monomorphism. This completes the proof of (1), and also yields the exact sequence

$$0 \to \tilde{\mathscr{A}}(U) \to \mathscr{B}(M) \to \mathscr{C}(\sqrt{-1}S^*M - U^\circ).$$

Remark 1. Since the map $\tau:\sqrt{-1}SM \to M$ is smooth (the induced map by τ on the tangent space is surjective), locally τ is a projection. Consequently, it is an open map. Therefore, if each fibre is connected (including the case of the empty set), then statement (2) of Theorem 2.3.4 can be rephrased as the sequence

$$0 \to \tilde{\mathscr{A}}(U) \xrightarrow{b} \mathscr{B}(\tau(U)) \xrightarrow{\text{sp}} \mathscr{C}(\sqrt{-1}S^*M - U^\circ)$$

is exact.

Remark 2. We mention the following fact without proof: for an arbitrary open convex set V in $\sqrt{-1}SM$,

$$H^k(V, \tilde{\mathscr{A}}) = 0 \qquad \text{for } k \neq 0.$$

Hence, this fact and (1) of Theorem 2.3.4 indicate that the convex sets in $\sqrt{-1}SM$ play the role of Stein manifolds for the theory of several complex variables.

We need the following definition to rewrite Theorem 2.3.4 into a more applicable form.

Definition 2.3.6. *Let D be an open set in X—M. The open set D is said to be a conoidal neighborhood of $x_0 + \sqrt{-1}v0$ (of $U \subset \sqrt{-1}SM$) if $D \cup \sqrt{-1}SM$ is a neighborhood of $x_0 + \sqrt{-1}v0$ (of U). Denote the boundary value of $\varphi \in \mathscr{O}_X(D)$ by $b_D(\varphi)$. We denote the hyperfunction corresponding to $\varphi \in \tilde{\mathscr{A}}(U)$ by $b_U(\varphi)$ provided that each fibre of U is connected. We also write $b(\varphi; D)$ and $b(\varphi; U)$ instead of $b_D(\varphi)$ and $b_U(\varphi)$ respectively, or even $b(\varphi)$ when there is no fear of confusion.*

Note that

$$\tilde{\mathscr{A}}(U) = \varinjlim_{\substack{D \text{ runs through} \\ \text{the set of conoidal} \\ \text{neighborhoods of } U}} \mathcal{O}_X(D).$$

This and Definition 2.3.6 imply the following theorem.

Theorem 2.3.5. *Let M be a real analytic manifold and X be a complexification of M. Let D be an open set in $X—M$ and U be an open set in $\sqrt{-1}SM$ such that each fibre is connected. If D is a conoidal neighborhood of U, then the boundary value of $f(z) \in \mathcal{O}_X(D)$ determines a hyperfunction $f(x) \in \mathscr{B}_M(\tau(U))$ uniquely and such that the singularity spectrum* S.S. $f(x)$ *is contained in the polar set U°. Conversely, if the singularity spectrum* S.S. $u(x)$ *of a hyperfunction $u(x)$ on M is contained in a closed convex set Z in $\sqrt{-1}S^*M$, then there exists a conoidal neighborhood D of the polar set Z° such that $u(x) = b_D(f(z))$ for some $f(z) \in \mathcal{O}_X(D)$.*

It is clear that Theorem 2.4.4 implies Theorem 2.3.5. Notice that the larger U is, the smaller the polar set U° becomes. Therefore, Theorem 2.3.5 says that if a hyperfunction is defined by a boundary value from the larger U, then the singularity spectrum is smaller. In particular, if the convex hull of U is the whole $\sqrt{-1}SM$, then the corresponding hyperfunction is a real analytic function, provided $n \neq 1$ (if $n = 1$, fibres cannot be connected). In that case, we have $U^\circ = (\sqrt{-1}SM)^\circ = \varnothing$ which implies S.S. $u = \varnothing$.

Since the sequence

$$\pi^{-1}\mathscr{B} \xrightarrow{\text{sp}} \mathscr{C} \to 0$$

is exact, from Proposition 2.1.6, we have the exact sequence

$$0 \to \mathscr{A}^* \to \pi^{-1}\mathscr{B} \xrightarrow{\text{sp}} \mathscr{C} \to 0,$$

where the sheaf $\mathscr{A}^*$ on $\sqrt{-1}S^*M$ is defined as $\mathscr{A}^* = \operatorname{Ker}(\pi^{-1}\mathscr{B} \to \mathscr{C})$. Then $u \in \mathscr{A}^*_{(x,\sqrt{-1}\xi\infty)}$ can be expressed as

$$u = \sum_j b(\varphi_j),$$

where $\varphi_j \in \Gamma(U_j, \tilde{\mathscr{A}})$ and $(x, \sqrt{-1}\xi\infty) \notin U_j^\circ$. In this case, u is said to be *micro-analytic* at $(x, \sqrt{-1}\xi\infty)$.

§4. Examples

In this section we will denote the imaginary unit $\sqrt{-1}$ by i. For $\varphi(z) \in \tilde{\mathscr{A}}_{x+iv0}$ we denote $b(\varphi(z))$ by $\varphi(x + iv0)$.

Example 2.4.1.

$$\text{S.S.}(x+i0)^\lambda = \begin{cases} \{(0, i\,dx\infty)\} & \text{for } \lambda \neq 0, 1, \ldots \\ \varnothing & \text{for } \lambda = 0, 1, 2, \ldots \end{cases}$$

holds.

Since $(x + i0)^\lambda$ is real-analytic at each $x \neq 0$, we will focus on the fibre at $x = 0$. On the other hand, if $\lambda = 0, 1, 2, \ldots$, then $(x + i0)^\lambda$ is real-analytic at $x = 0$ as well. Therefore $\text{S.S.}(x + i0)^\lambda = \varnothing$ for $\lambda = 0, 1, 2, \ldots$. Notice that $(x + i0)^\lambda$ is not real-analytic at $x = 0$ if λ is neither a positive integer nor zero. Hence, the singularity spectrum is not an empty set and is contained in $\{(0, i\,dx\infty), (0, -i\,dx\infty)\}$. But, by Theorem 2.3.4, $(0, -i\,dx\infty) \notin \text{S.S.}(x + i0)^\lambda$ implies $\text{S.S.}(x + i0)^\lambda = \{(0, i\,dx\infty)\}$.

When $(x + i0)^\lambda$ is regarded as a microfunction, i.e. if we consider $\text{sp}(x + i0)^\lambda$, it has zeros of order one at $\lambda = 0, 1, 2, \ldots$, (see §8, Chapter III, for hyperfunctions [microfunctions] with holomorphic parameters). Since x^n is zero as a microfunction,

$$\lim_{\lambda\to n} \frac{(x+i0)^\lambda}{\lambda - n} = \lim_{\lambda\to n} \frac{(x+i0)^\lambda - x^n}{\lambda - n} = \frac{d}{d\lambda}(x+i0)^\lambda|_{\lambda=n}$$
$$= (x+i0)^\lambda \log(x+i0)|_{\lambda=n}$$
$$= (x+i0)^n \log(x+i0) = x^n \log(x+i0).$$

This is the boundary value of $z^n \log z$ from the upper half-plane, and it cannot be analytically continued beyond $z = 0$. Therefore, $x^n \log(x + i0)$ is not zero as a microfunction. That is, $(x + i0)^\lambda$ has zeros of order one at $\lambda = n$, non-negative integers. On the other hand, the gamma function $\Gamma(-\lambda)$ is never zero and has poles of order one at $\lambda = n$, non-negative integers. Therefore, for any λ, $\Gamma(-\lambda)(x + i0)^\lambda$ is not zero as a microfunction. Consequently,

$$\text{S.S. } \Gamma(-\lambda)(x+i0)^\lambda = \{(0, i\,dx\infty)\}$$

holds for an arbitrary λ. The value of $\Gamma(-\lambda)(x + i0)^\lambda$ at $\lambda = n \in \mathbf{Z}_+$, non-negative integers, can be computed as follows (note that, since $\Gamma(-\lambda)(x + i0)^\lambda$ is holomorphic in λ, the restriction $\lambda = n$ makes sense):

$$\Gamma(-\lambda)(x+i0)^\lambda|_{\lambda=n} = \left.\frac{(x+i0)^\lambda}{\lambda - n}\right|_{\lambda=n} (\lambda - n)\Gamma(-\lambda)|_{\lambda=n}$$
$$= \frac{(-1)^{n-1}}{n!} x^n \log(x+i0).$$

Example 2.4.2. Define

$$x_+^\lambda = \begin{cases} x^\lambda & \text{for } x > 0 \\ 0 & \text{for } x < 0, \end{cases} \tag{2.4.1}$$

where Re $\lambda > 0$. Then this is a well-defined continuous function. Therefore it is well defined as a hyperfunction, since distributions are hyperfunctions. For a proof, see the lecture notes by Komatsu mentioned in the Introduction.) We will express x_+^λ explicitly as boundary values of homomorphic functions. When λ is not an integer, define

$$x_+^\lambda = \frac{1}{e^{-\pi i\lambda} - e^{\pi i\lambda}} \{e^{-\pi i\lambda}(x + i0)^\lambda - e^{\pi i\lambda}(x - i0)^\lambda\}$$

$$= \frac{1}{-2i \sin \pi\lambda} \{e^{-\pi i\lambda}(x + i0)^\lambda - e^{\pi i\lambda}(x - i0)^\lambda\}. \tag{2.4.2}$$

Note that the right-hand side is well defined provided $\sin \pi\lambda \neq 0$ by Example 2.4.1. If Re $\lambda > 0$, the hyperfunction defined as (2.4.2) coincides with the continuous function defined in (2.4.1).

Proof. If $x > 0$, then $(x + i0)^\lambda = (x - i0)^\lambda = x^\lambda$, which is the case in (2.4.1). If $x < 0$, then $(x + i0)^\lambda = |x|^\lambda e^{\pi i\lambda}$ and $(x - i0)^\lambda = |x|^\lambda e^{-\pi i\lambda}$. This implies $e^{-\pi i\lambda}(x + i0) - e^{\pi i\lambda}(x - i0) = |x|^\lambda - |x|^\lambda = 0$. Therefore we can define the hyperfunction x_+^λ by (2.4.2). The hyperfunction defined as (2.4.2) has the following properties:

(1) for $\lambda = 0, 1, 2, \ldots$, it is well defined

and

(2) it has a pole of order one at $\lambda = -1, -2, \ldots$.

Suppose $\lambda = n \geqq 0$; then one has

$$e^{-\pi in}(x + i0)^n - e^{\pi in}(x - i0)^n = (-1)^n(x^n - x^n) = 0.$$

On the other hand, $1/(\sin \pi\lambda)$ has only a pole of order one at $\lambda = n$; i.e. (2.4.2) is well defined at $\lambda = n$. (Strictly speaking, we need the theory found in §6 of Chapter III to justify this claim. However, the following calculation guarantees it here.) Let $\lambda = -n < 0$; then

$$(e^{-\pi i\lambda}(x + i0)^\lambda - e^{\pi i\lambda}(x - i0)^\lambda)|_{\lambda = -n} = (-1)^n \left\{\frac{1}{(x + i0)^n} - \frac{1}{(x - i0)^n}\right\} \neq 0,$$

which implies that x_+^λ has a pole of order one.

For each $\lambda = 0, 1, \ldots$, as we have seen, x_+^λ defines a hyperfunction. The hyperfunction corresponding to $\lambda = 0$ is called the Heaviside function, denoted by $Y(x)$; i.e.

$$Y(x) = x_+^\lambda|_{\lambda=0} = \begin{cases} 1 & \text{for } x > 0 \\ 0 & \text{for } x < 0. \end{cases}$$

(The symbol $\theta(x)$ is also used to denote the Heaviside function.)

Furthermore,

$$Y(x) = \frac{-2\pi i + \log(x + i0) - \log(x - i0)}{-2\pi i},$$

where the branch of log is taken as $\log(1 + i0) = \log(1 - i0) = 0$. We have

$$x_+^\lambda = \frac{e^{-\pi i\lambda}(x + i0)^\lambda - e^{\pi i\lambda}(x - i0)^\lambda}{-2i \sin \pi\lambda},$$

and

$$\left.\frac{e^{-\pi i\lambda}[(x + i0)^\lambda - 1]}{-2i \sin \pi\lambda}\right|_{\lambda=0} = \left.\frac{e^{-\pi i\lambda}\lambda}{-2i \sin \pi\lambda} \cdot \frac{[(x + i0)^\lambda - 1]}{\lambda}\right|_{\lambda=0}$$

$$= \frac{1}{-2\pi i} \log(x + i0).$$

The other term can be rewritten similarly. Therefore, as a hyperfunction,

$$x_+^\lambda|_{\lambda=0} = \frac{-2\pi i + \log(x + i0) - \log(x - i0)}{-2\pi i}.$$

Similarly, we obtain

$$Y(x) = \frac{\log(-x + i0) - \log(-x - i0)}{2\pi i}.$$

If $x > 0$, then $\log(-x + i0) = \log x + \pi i$ and $\log(-x - i0) = \log x - \pi i$; i.e. the right-hand side of the above equation is 1. If $x < 0$, $\log(-x + i0) = \log(-x - i0) = \log(-x)$. Then the right-hand side is zero. Note also that

$$x_+^\lambda|_{\lambda=n} = x^n \cdot x_+^{\lambda-n}|_{\lambda=n} = x^n Y(x).$$

First we consider $x_+^\lambda/(\Gamma(1 + \lambda))$ in order to study the structure of x_+^λ for $\lambda = -1, -2, \ldots$. By the functional equation $\Gamma(1 + \lambda)\Gamma(-\lambda) = \pi/(-\sin \pi\lambda)$,

$$\left.\frac{x_+^\lambda}{\Gamma(1 + \lambda)}\right|_{\lambda=-n} = \left.\frac{e^{-\pi i\lambda}(x + i0)^\lambda - e^{\pi i\lambda}(x - i0)^\lambda}{-2i \sin \pi\lambda \cdot \Gamma(1 + \lambda)}\right|_{\lambda=-n}$$

$$= \frac{\Gamma(-\lambda)}{2\pi i} \{e^{-\pi i\lambda}(x + i0)^\lambda - e^{\pi i\lambda}(x - i0)^\lambda\}|_{\lambda=-n}$$

$$= \frac{(-1)^n(n - 1)!}{2\pi i} \left\{\frac{1}{(x + i0)^n} - \frac{1}{(x - i0)^n}\right\} = D_x^{n-1}\delta(x),$$

where the hyperfunction $\delta(x)$ is defined as $\delta(x) = (1/2\pi i)(1/(x - i0) - 1/(x + i0))$ and is called the Dirac δ-function.

Proof. Since $D_x x_+^\lambda = \lambda x_+^{\lambda-1}$,

$$D_x \frac{x_+^\lambda}{\Gamma(1+\lambda)} = \frac{x_+^{\lambda-1}}{\Gamma(\lambda)}.$$

On the other hand, we have

$$\delta(x) = \frac{x_+^\lambda}{\Gamma(1+\lambda)}\bigg|_{\lambda=-1}$$

by definition. Therefore,

$$D_x\delta(x) = \frac{x_+^{\lambda-1}}{\Gamma(\lambda)}\bigg|_{\lambda=-1} = \frac{x_+^\lambda}{\Gamma(1+\lambda)}\bigg|_{\lambda=-2}.$$

Repeating the above, we obtain

$$D_x^{n-1}\delta(x) = \frac{x_+^\lambda}{\Gamma(1+\lambda)}\bigg|_{\lambda=-n}.$$

Remark 1. We have supp $\delta(x) = \{x \in \mathbf{R} \mid x = 0\}$. That is, the δ-function is not a function in Dirichlet's sense (i.e. for each x there is a corresponding value). Dirac's introduction of δ-function in quantum mechanics, where the δ-function was used very effectively, gave an impetus to the theory of distributions and led to hyperfunction theory.

Remark 2. Let $\mathscr{D}$ be the sheaf of differential operators with coefficients in holomorphic functions. Then $P \in \mathscr{D}$ defines a homomorphism: $\mathcal{O}_X \to \mathcal{O}_X$. Therefore there is induced a homomorphism

$$\mathscr{B}_M = \mathscr{H}_M^n(\mathcal{O}_X) \otimes \omega_M \xrightarrow{P} \mathscr{H}_M^n(\mathcal{O}_X) \otimes \omega_M.$$

In particular, a holomorphic function acts on $\mathscr{B}$ by the multiplication; e.g. $x\delta(x) = 0$. In general, for $u = \sum_j b(\varphi_j) \in \mathscr{B}$, $Pu = \sum_j b(P\varphi_j)$.

We summarize what we have discussed as the following definition.

Definition 2.4.1.

(1) $x_+^\lambda = \dfrac{1}{-2i \sin \pi\lambda}\{e^{-\pi i\lambda}(x+i0)^\lambda - e^{\pi i\lambda}(x-i0)^\lambda\}$.

(2) *The Heaviside function* $Y(x) = x_+^\lambda|_{\lambda=0} = \dfrac{\log(-x+i0) - \log(-x-i0)}{2\pi i}$.

(3) *The Dirac δ-function of one variable* $\delta(x) = \dfrac{1}{2\pi i}\left(\dfrac{1}{x-i0} - \dfrac{1}{x+i0}\right)$.

(4) *The Dirac δ-function of several variables*

$$\delta(x) = \delta(x_1, \ldots, x_n) = \delta(x_1) \cdots \delta(x_n)$$
$$= \frac{1}{(-2\pi i)^n} \sum_{\epsilon_1, \ldots, \epsilon_n = \pm 1} \frac{\epsilon_1 \cdots \epsilon_n}{(x_1 + i\epsilon_1 0) \cdots (x_n + i\epsilon_n 0)},$$

where $1/((x_1 + i\epsilon_1 0) \cdots (x_n + i\epsilon_n 0))$ *is the boundary value of* $1/(z_1 \cdots z_n)$ *from* $\epsilon_i \operatorname{Im} z_i > 0$, $1 \leqq i \leqq n$.

Remark. For a real analytic function $a(x)$, we have $a(x)\delta(x) = a(0)\delta(x)$. That is, $x_1\delta(x) = \cdots = x_n\delta(x) = 0$ holds. *Proof*:

$$x_1\delta(x) = \frac{1}{(-2\pi i)^n} \sum_{\epsilon_1, \ldots, \epsilon_n = \pm 1} \frac{\epsilon_1 \cdots \epsilon_n}{(x_2 + i\epsilon_2 0) \cdots (x_n + i\epsilon_n 0)}$$
$$= \frac{1}{(2\pi i)^n} \left[\sum_{\epsilon_2, \ldots, \epsilon_n = \pm 1} \frac{\epsilon_2 \cdots \epsilon_n}{(x_2 + i\epsilon_2 0) \cdots (x_n + i\epsilon_n 0)} \right.$$
$$\left. - \sum_{\epsilon_2, \ldots, \epsilon_n = \pm 1} \frac{\epsilon_2 \cdots \epsilon_n}{(x_2 + i\epsilon_2 0) \cdots (x_n + i\epsilon_n 0)} \right] = 0.$$

Proposition 2.4.1. *Let $\delta(x)$ be the δ-function on $\mathbf{R}^n$, and let $GL(n, \mathbf{R})$ be the general linear group of degree n over $\mathbf{R}$ acting on $\mathbf{R}^n$. Then we have the following*:

(1) $\delta(gx) = \dfrac{1}{|\det g|} \delta(x)$, *where* $x \in \mathbf{R}^n$ *and* $g \in GL(n, \mathbf{R})$.

(2) S.S. $\delta(x) = \pi^{-1}(0) = \{(0, \sqrt{-1}\langle \xi, dx \rangle \infty) \mid \xi \in \mathbf{R}^n - \{0\}\}$, *where we often use a representative* $\xi \in \mathbf{R}^n - \{0\}$ *for a point in* S^{n-1}.

(3) $$\delta(x) = \frac{1}{(2\pi i)^n} \sum_{k=0}^{n} \frac{|\xi_0 \wedge \cdots \wedge \xi_{k-1} \wedge \xi_{k+1} \wedge \cdots \wedge \xi_n|}{(\langle x, \xi_0 \rangle + \sqrt{-1}0) \cdots (\langle x, \xi_{k-1} \rangle + \sqrt{-1}0)}$$
$$\times \frac{1}{(\langle x, \xi_{k+1} \rangle + \sqrt{-1}0) \cdots (\langle x, \xi_n \rangle + \sqrt{-1}0)},$$

where $\xi_0 + \cdots + \xi_n = 0$ *such that any n vectors of those* $\xi_0, \ldots, \xi_n$ *are linearly independent, and* $|\xi_1 \wedge \cdots \wedge \xi_n| \underset{\text{def}}{=} |\det(\xi_1, \ldots, \xi_n)|$.

Remark 1. From (1) above, $\delta(x)|dx_1 \wedge \cdots \wedge dx_n|$ is invariant under the action of $GL(n, \mathbf{R})$. Furthermore, if F is a real analytic map from a neighborhood of 0 to a neighborhood of 0 such that $F(0) = 0$ and $dF(0)$ is nondegenerate, one has

$$\delta(F(x)) = \frac{1}{|\det dF(0)|} \delta(x).$$

Generally, for a real analytic isomorphism $f:M_1 \to M_2$ there is induced a morphism $\mathscr{B}_{M_2} \to \mathscr{B}_{M_1}$ as follows: for $u(x_2) = \sum_j b(\varphi_j) \in \mathscr{B}_{M_2}$ there corresponds $u(f(x_1)) = \sum_j b(\varphi_j(f(x_1)) \in \mathscr{B}_{M_1}$. See Theorem 3.1.7.

Remark 2. Since $\delta(x)$ is a hyperfunction of n variables, the general theory in Chapter I guarantees that it can be expressed as a sum of boundary values of $(n + 1)$ holomorphic functions. The above Proposition 2.4.1 (3) realizes this statement concretely.

Proof of Proposition 2.4.1. Since $GL(n, \mathbf{R})$ is generated by diagonal matrices and matrices of the type

$$\begin{pmatrix} 1 & & & \\ & \ddots & \lambda & \\ & & \ddots & \\ & & & 1 \end{pmatrix},$$

it is sufficient to prove (1) for these matrices. First one has

$$\delta(a_1x_1, \ldots, a_nx_n) = \frac{1}{(-2\pi i)^n} \sum_{\epsilon_1, \ldots, \epsilon_n = \pm 1} \frac{\epsilon_1 \cdots \epsilon_n}{(a_1x_1 + i\epsilon_1 0) \cdots (a_nx_n + i\epsilon_n 0)},$$

where $(\epsilon_1 \cdots \epsilon_n)/((a_1x_1 + i\epsilon_1 0) \cdots (a_nx_n + i\epsilon_n 0))$ is the boundary value of the holomorphic function $1/((a_1z_1) \cdots (a_nz_n))$ restricted to $\{\epsilon_i \operatorname{Im} a_iz_i > 0, 1 \leqq i \leqq n\}$, i.e. the boundary value from $(\epsilon_i \operatorname{sgn} a_i) \operatorname{Im} z_i > 0$, $1 \leqq i \leqq n$, of $(1/(a_1 \cdots a_n)) \cdot (1/(z_1 \cdots z_n))$. Then

$$\frac{1}{(a_1x_1 + i\epsilon_1 0) \cdots (a_nx_n + i\epsilon_n 0)} = \frac{1}{a_1 \cdots a_n} \cdot \frac{1}{(x_1 + i\epsilon_1 \operatorname{sgn} a_1 0) \cdots (x_n + i\epsilon_n \operatorname{sgn} a_n 0)}.$$

Let $\epsilon_i' = \epsilon_i \operatorname{sgn} a_i$. Consequently, one obtains

$$\begin{aligned} \delta(a_1x_1, \ldots, a_nx_n) &= \frac{1}{(-2\pi i)^n} \cdot \frac{(\operatorname{sgn} a_1) \cdots (\operatorname{sgn} a_n)}{a_1 \cdots a_n} \sum_{\epsilon_1', \ldots, \epsilon_n' = \pm 1} \frac{\epsilon_1' \cdots \epsilon_n'}{(x_1 + i\epsilon_1' 0) \cdots (x_n + i\epsilon_n' 0)} \\ &= \left| \frac{1}{a_1 \cdots a_n} \right| \delta(x). \end{aligned}$$

Next, when

$$g = \begin{pmatrix} 1 & 1 & & 0 \\ & 1 & & \\ & & \ddots & \\ 0 & & & 1 \end{pmatrix},$$

one has

$$\delta(g(x)) = \delta(x_1 + x_2, x_2, \ldots, x_n)$$
$$= \frac{1}{(-2\pi i)^n} \sum_{\epsilon_1, \ldots, \epsilon_n = \pm 1} \frac{\epsilon_1 \cdots \epsilon_n}{(x_1 + x_2 + i\epsilon_1 0)(x_2 + i\epsilon_2 0) \cdots (x_n + i\epsilon_n 0)}.$$

Then

$$\delta(x_1 + x_2, x_2, \ldots, x_n) - \delta(x_1, \ldots, x_n)$$
$$= \frac{1}{(-2\pi i)^n} \sum_{\epsilon_1, \ldots, \epsilon_n = \pm 1} \epsilon_1 \cdots \epsilon_n$$
$$\times \left\{ \frac{1}{(x_1 + x_2 + i\epsilon_1 0)(x_2 + i\epsilon_2 0) \cdots (x_n + i\epsilon_n 0)} - \frac{1}{(x_1 + i\epsilon_1 0) \cdots (x_n + i\epsilon_n 0)} \right\}.$$

where, for each $\epsilon_1, \ldots, \epsilon_n$, $1/((x_1 + x_2 + i\epsilon_1 0)(x_2 + i\epsilon_2 0) \cdots (x_n + i\epsilon_n 0))$ is the boundary value from $\{\epsilon_1 \operatorname{Im}(z_1 + z_2) > 0$ and $\epsilon_i \operatorname{Im} z_i > 0$ for $i = 2, \ldots, n\}$, and $1/((x_1 + i\epsilon_1 0) \cdots (x_n + i\epsilon_n 0))$ is the boundary value from $\{\epsilon_i \operatorname{Im} z_i > 0$ if $i = 1, 2, \ldots, n\}$. Therefore, the difference of the above boundary values is the boundary value of

$$\frac{1}{(z_1 + z_2)z_2 \cdots z_n} - \frac{1}{z_1 \cdots z_n} = \frac{-1}{(z_1 + z_2)z_2 \cdots z_n}$$

from their intersection $\{\epsilon_1 \operatorname{Im}(z_1 + z_2) > 0$ and $\epsilon_i \operatorname{Im} z_i > 0$ for $i = 1, 2, \ldots, n\}$. If $-1/((x_1 + x_2 + i\epsilon_1 0)(x_1 + i\epsilon_1 0)(x_3 + i\epsilon_3 0) \cdots (x_n + i\epsilon_n 0))$ denotes the boundary value from $\{\epsilon_1 \operatorname{Im}(z_1 + z_2) > 0$, $\epsilon_1 \operatorname{Im} z_1 > 0$ and $\epsilon_i \operatorname{Im} z_i > 0$ for $i = 3, 4, \ldots, n\}$, we obtain

$$\delta(x_1 + x_2, x_2, \ldots, x_n) - \delta(x_1, \ldots, x_n)$$
$$= \frac{1}{(-2\pi i)^n} \sum_{\epsilon_1, \ldots, \epsilon_n = \pm 1} \frac{\epsilon_1 \cdots \epsilon_n}{(x_1 + x_2 + i\epsilon_1 0)(x_1 + i\epsilon_1 0)(x_3 + i\epsilon_3 0) \cdots (x_n + i\epsilon_n 0)}$$
$$= 0$$

The general case where

$$g = \begin{pmatrix} 1 & & \lambda \\ & \ddots & \\ & & 1 \end{pmatrix}$$

can be done similarly.

We will prove (2). Since $x_1\delta(x) = \cdots = x_n\delta(x) = 0$, $\operatorname{supp} \delta(x) = \{0\}$. Therefore, since $\delta(x)$ cannot be analytic,

$$\varnothing \neq \text{S.S. } \delta(x) \subset \pi^{-1}(0) = \{(0, i\langle \xi, dx\rangle\infty) \mid \xi \in \mathbf{R}^n - \{0\}\}.$$

Hence there exists $\xi \in \mathbf{R}^n - \{0\}$ so that $(0, i\langle\xi, dx\rangle\infty) \in \text{S.S.}\delta(x)$. Then, for this $\xi \in \mathbf{R}^n - \{0\}$, $(0, i\langle\xi, d(gx)\rangle\infty) = (0, i\langle{}^t g\xi, dx\rangle\infty)$ is contained in S.S. $\delta(gx)$. Since (1) implies S.S. $\delta(gx) =$ S.S. $\delta(x)$, $(0, i\langle{}^t g\xi, dx\rangle\infty)$ is contained in S.S. $\delta(x)$. On the other hand, $GL(n, \mathbf{R})$ acts transitively on $\mathbf{R}^n - \{0\}$. Hence we obtain S.S. $\delta(x) = \pi^{-1}(0)$.

Lastly we will prove (3) for the case $n = 2$, and we leave it to the reader to prove the general case. For the case $n = 2$, one can let $\xi_0 = (1, 0)$, $\xi_1 = (0, 1)$ and $\xi_2 = (-1, -1)$ without loss of generality. Then the right-hand side of (3) becomes

$$\frac{1}{(-2\pi i)^2}\left\{\frac{1}{(x_2 + i0)(-x_1 - x_2 + i0)} + \frac{1}{(x_1 + i0)(-x_1 - x_2 + i0)} + \frac{1}{(x_1 + i0)(x_2 + i0)}\right\}.$$

By definition,

$$\delta(x) = \frac{1}{(-2\pi i)^2}\left\{\frac{1}{(x_1 + i0)(x_2 + i0)} - \frac{1}{(x_1 + i0)(x_2 - i0)} - \frac{1}{(x - i0)(x_2 + i0)} + \frac{1}{(x_1 - i0)(x_2 - i0)}\right\}.$$

Therefore,

$$\begin{aligned}
&(-2\pi i)^2(\text{the right-hand side of } (3) - \delta(x)) \\
&= \left\{\frac{1}{(x_1 - i0)(x_2 + i0)} - \frac{1}{(x_1 + x_2 - i0)(x_2 + i0)}\right\} \\
&\quad + \left\{\frac{1}{(x_1 + i0)(x_2 - i0)} - \frac{1}{(x_1 + x_2 - i0)(x_1 + i0)}\right\} - \frac{1}{(x_1 - i0)(x_2 - i0)} \\
&= \frac{1}{(x_1 - i0)(x_1 + x_2 - i0)} + \frac{1}{(x_2 - i0)(x_1 + x_2 - i0)} - \frac{1}{(x_1 - i0)(x_2 - i0)} \\
&= 0.
\end{aligned}$$

Remark. One should keep in mind that the summation of two holomorphic functions can be performed only on the intersection of the domain of definition of each function. Although this is an obvious fact, forgetting it can lead to careless mistakes in such calculations as the one above.

The following proposition is often useful to consider explicit examples.

Proposition 2.4.2. *Let M be a real analytic manifold, and let X be a complexification of M. Let $f(x)$ be a real-valued real analytic function on M. Suppose that $df(x_0) \neq 0$ for $x_0 \in M$ and $\langle v_0, df(x_0)\rangle > 0$ for $v_0 \in \sqrt{-1}SM$; then $\{z \in X \mid \operatorname{Im} f(z) > 0\} \cup \sqrt{-1}SM$ is a neighborhood of $x_0 + iv_0 0$.*

Proof. It is sufficient to prove that $\operatorname{Im} f(z) > 0$ holds for $z = x + itv$ such that $|x - x_0| \ll 1$, $|v - v_0| \ll 1$, and $0 < t \ll 1$. Since $f(x + itv) = f(x) + it\langle v, df(x)\rangle + O(t^2)$, one obtains $\operatorname{Im} f(x + itv) = t\langle v, df(x)\rangle + O(t^2) > 0$ for a sufficiently small t.

As an example relating to Proposition 2.4.2, we consider the following.

Example 2.4.3. Let $u(x) = ((x_1 + i0)^2 - x_2^2 - \cdots - x_n^2)^\lambda$, and let $z_j = x_j + \sqrt{-1}y_j$. First we will show that $y_1^2 - y_2^2 - \cdots - y_n^2 > 0$ implies $z_1^2 - z_2^2 - \cdots - z_n^2 \neq 0$. Since $z_1^2 - z_2^2 - \cdots - z_n^2 = [(x_1^2 - x_2^2 - \cdots - x_n^2) - (y_1^2 - y_2^2 - \cdots - y_n^2)] + 2i(x_1y_1 - x_2y_2 - \cdots - x_ny_n)$, if $y_1^2 - y_2^2 - \cdots - y_n^2 > 0$ and $z_1^2 - z_2^2 - \cdots - z_n^2 = 0$ hold, then $(x_1^2 - x_2^2 - \cdots - x_n^2) > 0$ and $(x_1y_1 - x_2y_2 - \cdots - x_ny_n) = 0$ must hold. This would imply $|x_1y_1| = |x_2y_2 + \cdots + x_ny_n| \leqq \sqrt{x_2^2 + \cdots + x_n^2} \cdot \sqrt{y_2^2 + \cdots + y_n^2} < |x_1| \cdot |y_1|$. Therefore we have obtained $z_1^2 - z_2^2 - \cdots - z_n^2 \neq 0$. Let $D = \{(z_1, \ldots, z_n) \in \mathbf{C}^n | y_1 > 0 \text{ and } y_1^2 - y_2^2 - \cdots - y_n^2 > 0, \text{ where } z_j = x_j + \sqrt{-1}y_j\}$, and let $U = \{x + iv0 \in \sqrt{-1}S\mathbf{R}^n | v_1 > 0 \text{ and } v_1^2 - v_2^2 - \cdots - v_n^2 > 0, \text{ where } v = (v_1, \ldots, v_n)\}$. First note $\sqrt{-1}S\mathbf{R}^n \cong \mathbf{R}^n \times \sqrt{-1}S^{n-1}$. Then $D \cup \mathbf{R}^n \times \sqrt{-1}S^{n-1}$ is a neighborhood of U. Since $u(z) = (z_1^2 - z_2^2 - \cdots - z_n^2)^\lambda \in \mathcal{O}_{\mathbf{C}^n}(D)$, as we saw above, $u(x)$ is a well-defined hyperfunction on $\mathbf{R}^n$ by Theorem 2.3.5, and its singularity spectrum S.S. u is contained in U°. That is, S.S. $u \subset \{(x, i\langle \xi, dx\rangle\infty) | \xi_1^2 - \xi_2^2 - \cdots - \xi_n^2 \geqq 0 \text{ and } \xi_1 > 0\}$. On the other hand, if $(x_1^2 - x_2^2 - \cdots - x_n^2) \neq 0$ holds, then u is real-analytic, which implies S.S. $u \subset \{x_1^2 - x_2^2 - \cdots - x_n^2 = 0\}$. Suppose $x_1^2 - x_2^2 - \cdots - x_n^2 = 0$; then $x = (x_1, \ldots, x_n) \neq 0$ holds if and only if $x_1 \neq 0$. Therefore, we must consider the cases $x_1 > 0$, $x_1 = 0$, and $x_1 < 0$.

If $x_1 > 0$, $x_1^2 - x_2^2 - \cdots - x_n^2 = 0$, $y_1^2 - y_2^2 - \cdots - y_n^2 > 0$, and $y_1 > 0$, then we have $\operatorname{Im}(z_1^2 - z_2^2 - \cdots - z_n^2) > 0$. Hence, if we let $f(z) = z_1^2 - z_2^2 - \cdots - z_n^2$, then $u(z) = f(z)^\lambda$ can be continued in a neighborhood of $x_1 > 0$ to the domain $\{z | \operatorname{Im} f(z) > 0\}$, containing D. Then Proposition 2.4.2 implies that $\{z | \operatorname{Im} f(z) > 0\}$ is a conoidal neighborhood of $U = \{x + iv0 \in \sqrt{-1}S\mathbf{R}^n | x_1 > 0 \text{ and } \langle v, df(x)\rangle > 0\}$. Hence the singularity spectrum of the hyperfunction determined by the boundary values of $u(z)$ is contained in $U^\circ = \{(x, idf(x)\infty) | x_1 > 0\}$. That is, if $x_1 > 0$, then we have S.S. $u \subset \{(x, i(x_1, -x_2, \ldots, -x_n)\infty | x_1^2 - x_2^2 - \cdots - x_n^2 = 0\}$. Similarly, if $x_1 < 0$, then S.S. $u \subset \{(x, -i(x_1, -x_2, \ldots, -x_n)\infty | x_1^2 - x_2^2 - \cdots - x_n^2 = 0\}$. Consequently, we obtain

$$\text{S.S. } u \subset \left\{x, i\left(1, -\frac{x_2}{x_1}, \ldots, -\frac{x_n}{x_1}\right)\infty \,\middle|\, x \neq 0 \text{ and } x_1^2 - x_2^2 - \cdots - x_n^2 = 0\right\}$$
$$\cup \{(0, i\xi\infty) | \xi_1^2 - \xi_2^2 - \cdots - \xi_n^2 \geqq 0 \text{ and } \xi_1 > 0\}.$$

Note that Proposition 2.4.2 does not apply to this example at the origin as $df(0) = 0$. Hence, a priori, there was no guarantee that $\{z | \operatorname{Im} f(z) > 0\}$

determines a conoidal neighborhood, although it was verified to be the case for this example by a concrete calculation. Incidentally, let us calculate the value of $u = f^\lambda$ for $f \neq 0$. This computation will be useful in applications. If $x_1 > 0$, by taking the branch such that $u(1, 0, \ldots, 0) = (f(1, 0, \ldots, 0))^\lambda = 1$,

$$u(x) = (x_1^2 - x_2^2 - \cdots - x_n^2 + i0)^\lambda$$
$$= \begin{cases} (x_1^2 - x_2^2 - \cdots - x_n^2)^\lambda, & x_1 > \sqrt{x_2^2 + \cdots + x_n^2} \\ e^{\pi i\lambda}(x_2^2 + \cdots + x_n^2 - x_1^2)^\lambda, & x_1^2 < x_2^2 + \cdots + x_n^2 \end{cases},$$

where $(x_1^2 - x_2^2 - \cdots - x_n^2 + i0)^\lambda$ denotes the boundary value of $u(z) = (z_1^2 - z_2^2 - \cdots - z_n^2)^\lambda$ from $\operatorname{Im} f(z) = \operatorname{Im}(z_1^2 - z_2^2 - \cdots - z_n^2) > 0$. Similarly, for the case $x_1 < 0$,

$$u(x) = e^{\pi i\lambda}(x_2^2 + \cdots + x_n^2 - x_1^2 + i0)^\lambda$$
$$= \begin{cases} e^{\pi i\lambda}(x_1^2 - x_2^2 - \cdots - x_n^2)^\lambda, & x_1^2 < x_2^2 + \cdots + x_n^2 \\ e^{2\pi i\lambda}(x_2^2 + \cdots + x_n^2 - x_1^2)^\lambda, & x_1 < -\sqrt{x_2^2 + \cdots + x_n^2} \end{cases}$$

holds.

Let us next introduce the notion of positive type. This notion is quite important for the theory of partial differential equations and other applications. See §7 of Chapter III.

Definition 2.4.2. *Let M be a real analytic manifold, and let X be a complexification of M. A function $f(x)$ on X is said to be of positive type if* $\operatorname{Re} f(x) = 0$ *for $x \in M$ implies* $\operatorname{Im} f(x) \geqq 0$.

Proposition 2.4.3. *Let M be a real analytic manifold, and let X be a complexification of M. Let $f(x)$ be a complex-valued real analytic function of positive type such that for $x_0 \in M$ $f(x_0) = 0$ and $df(x_0)$ is a non-zero real vector. Then $\{f^{-1}(D_\epsilon) - M\} \cup \sqrt{-1}SM$ is a neighborhood of $x_0 + iv_0 0$, where $\langle v_0, df(x_0)\rangle > 0$; ϵ is an arbitrary positive real number; and $D_\epsilon = \{\tau \in \mathbf{C} \mid \operatorname{Im} \tau + \epsilon|\operatorname{Re} \tau| > 0\}$.*

Note. Recall the Weierstrass preparation theorem: let $f(z_1, z_2, \ldots, z_n, w)$ be a holomorphic function in a neighborhood of the origin such that $f(0) = 0$ and $f(0, w) = a_s w^s + a_{s+1} w^{s+1} + \cdots$, where $a_s \neq 0$. Then there exist holomorphic functions h and g_k, $k = 1, 2, \ldots, s$, such that $g_k(0) = 0$ and $h(0) \neq 0$, satisfying the equation

$$f = h(z_1, \ldots, z_n, w)\{w^s + g_1(z_1, \ldots, z_n)w^{s-1} + \cdots + g_s(z_1, \ldots, z_n)\}.$$

See, for example, Hitotumatu [1] for the proof.

Proof of Proposition 2.4.3. One may assume $x_0 = 0$, since the assertion is invariant under coordinate transformations. By the assumption,

$d \operatorname{Re} f(x) \neq 0$ in a neighborhood of the origin. Therefore one may let $\operatorname{Re} f(x) = x_1$. Then, by the Weierstrass preparation theorem, $f(x) = h(x)(x_1 - g(x'))$ holds for some h and g such that $g(0) = 0$ and $h(0) \neq 0$, where $x' = (x_2, x_3, \ldots, x_n)$.

Notice $dx_1(0) = df(0) = h(0)(dx_1(0) - dg(x'))$. From this, one obtains $h(0) = 1$ and $dg(0) = 0$. Since $x_1 = \operatorname{Re} f = x_1 \operatorname{Re} h - \operatorname{Re}(gh)$, $\operatorname{Re} f = 0$ implies $\operatorname{Re}(gh) = 0$. Furthermore, since $\operatorname{Im} f = x_1 \operatorname{Im} h - \operatorname{Im}(gh)$, $\operatorname{Re} f = 0$ implies $\operatorname{Im} f = -\operatorname{Im}(gh)$. Therefore, since f is of positive type, one has $h(0, x')g(x') = -\sqrt{-1}\varphi(x')$, where $\varphi(0) = 0$ and $\varphi(x') \geqq 0$.

We will prove this proposition for $\tilde{f}(x) = x_1 - g(x')$. Since $f(x + iy) = h(x + iy)\tilde{f}(x + iy)$ and $h(0) = 1$, the proof for $\tilde{f}(x)$ would imply that Proposition 2.4.3 holds sufficiently near the origin for f.

After a suitable coordinate transformation, one can assume $v_0 = \partial/\partial x_1$. In fact, for $v_0 = \partial/\partial x_1 + a_2(\partial/\partial x_2) + \cdots + a_n(\partial/\partial x_n)$, let the coordinate transformation be $x'_1 = x_1, x'_2 = x_2 - a_2 x_1, \ldots, x'_n = x_n - a_n x_1$. Then $v_0(x'_1) = 1$ and $v_0(x'_j) = 0$ for $j \neq 1$ hold. That is, with the new coordinates $v_0 = \partial/\partial x'_1$.

Then one obtains

$$\begin{aligned}
\operatorname{Im} \tilde{f}(x + iy) &+ \epsilon|\operatorname{Re} \tilde{f}(x + iy)| \\
&= y_1 - \operatorname{Im} g(x' + iy') + \epsilon|x_1 - \operatorname{Re} g(x' + iy')| \\
&\geqq y_1 - \operatorname{Im} g(x' + iy') + \epsilon\{|x_1| - |\operatorname{Re} g(x' + iy')|\} \\
&\geqq y_1 - \{\operatorname{Im} g(x' + iy') + \epsilon|\operatorname{Re} g(x' + iy')|\}.
\end{aligned}$$

Notice that, since $v_0 = \partial/\partial x_1$, $|v - v_0| \ll 1$ can be rephrased as $0 < |y'|/y_1 \ll 1$.

One has $g(x') = -i\varphi(x)\psi(x')$, where $\psi(x') = h(0, x')^{-1}$ and $\psi(0) = 1$, and $\varphi(x' + iy') = \varphi(x') + i\langle y', d_{x'}, \varphi(x')\rangle + O(|y'|^2) = \varphi(x') + O(|y'|)$. Therefore,

$$\begin{aligned}
-(\operatorname{Im} g(x' + iy') &+ \epsilon|\operatorname{Re} g(x' + iy')|) \\
&= O(|y'|) + (\operatorname{Im}(i\varphi(x')\psi(x' + iy')) - \epsilon|\operatorname{Re}(i\varphi(x')\psi(x' + iy'))|.
\end{aligned}$$

If, since $\varphi(x') \geqq 0$ and $\psi(0) = 1$, one lets $x' + iy'$ be sufficiently near 0, then for an arbitrary $\epsilon > 0$ one obtains

$$\operatorname{Im}(i\varphi(x')\psi(x' + iy')) - \epsilon|\operatorname{Re}(i\varphi(x')\psi(x' + iy'))| > 0.$$

Consequently, $\operatorname{Im} \tilde{f}(x + iy) + \epsilon|\operatorname{Re} \tilde{f}(x + iy)| \geqq y_1 + O(|y'|)$. If $0 < |y'|/y_1 \ll 1$, then $y_1 + O(|y'|) > 0$. Hence, one finally obtains $\operatorname{Im} f(x + itv) + \epsilon|\operatorname{Re} f(x + itv)| > 0$ for $|x| \ll 1$, $|v - v_0| \ll 1$, and $0 < t \ll 1$.

We will consider the hyperfunction $(x_1 + i(x_2^2 + \cdots + x_n^2) + i0)^\lambda$, where λ is neither zero nor a positive integer, as an application of Proposition 2.4.3. This hyperfunction has its singularity spectrum equaling one point, which is quite important not only from a theoretical point of view but also for applications. Theoretically, this fact and the flabbiness of the sheaf of microfunctions (see §7, Chapter III) are opposite sides of the same coin. As for applications, it is related to the (non-)solvability of linear differential equations; see Theorem 4.3.8.

Example 2.4.4. Let $f(x) = x_1 + i(x_2^2 + \cdots + x_n^2)$. Then $f(x)$ is of positive type on $\mathbf{R}^n$ such that $f(0) = 0$ and $df(0) = (1, 0, \ldots, 0)$. Therefore Proposition 2.4.3 can be applied to the function $f(x)$. That is, $\{z \in \mathbf{C}^n - \mathbf{R}^n | \operatorname{Im} f(x) + \epsilon|\operatorname{Re} f(z)| > 0\}$ is a conoidal neighborhood of $\{0 + iv0 | \langle v, idf(0)\rangle > 0\}$. Hence the singularity spectrum of $(f(x) + i0)^\lambda$ is contained in $\{(x, i\xi\infty) | \xi = (1, 0, \ldots, 0)\}$. On the other hand, $f(x) = 0$ if and only if $x = 0$; i.e. $(f(x) + i0)^\lambda$ is real-analytic everywhere but at the origin. These imply

$$\text{S.S.}(f(x) + i0)^\lambda \subset \{(x, i\xi\infty) | x = 0 \text{ and } \xi = (1, 0, \ldots, 0)\}.$$

For $\lambda \neq 0, 1, 2, \ldots, (f(x) + i0)^\lambda$ is not real-analytic at the origin. One can now conclude

$$\text{S.S.}(f(x) + i0)^\lambda = \{(x, i\xi0) | x = 0 \text{ and } \xi = (1, 0, \ldots, 0)\}.$$

Exercise. Sketch the domain $\{(z_1, z_2) \in \mathbf{C}^2 | \operatorname{Im}(z_1 + iz_2^2) > 0\}$. Then notice that one cannot treat $(x_1 + ix_2^2 - i0)^\lambda$ in the natural way, as above, by considering the boundary value of a holomorphic function.

We will consider Fourier series, which provide explicit examples of hyperfunctions.

Proposition 2.4.4. *For* $\alpha = (\alpha_1, \ldots, \alpha_n) \in \mathbf{Z}^n$ *let* $u(x) = \sum_{\alpha \in Z^n} a_\alpha e^{2\pi i\langle x,\alpha\rangle}$, *where* $x \in \mathbf{R}^n$. *Suppose that for an arbitrary* $\epsilon > 0$ *there exists a constant* C_ϵ *such that* $|a_\alpha| \leqq C_\epsilon e^{\epsilon|\alpha|}$ *holds for each* α, *where* $|\alpha| = \sum_{j=1}^n |\alpha_j|$. *Then* $u(x)$ *is a well-defined hyperfunction.*

Remark. If $|a_\alpha| \leqq C|\alpha|^N$ for some $N > 0$ and $C > 0$, then $u(x)$ is a distribution.

Proof. Let G_j, $j = 1, 2, \ldots, N$, be a closed convex cone containing no lines in $\mathbf{R}^n$ or having $a \in \mathbf{R}^n$ as its vertex, and let Γ_j satisfy $\Gamma_j \subset G_j$ and $\mathbf{Z}^n = \bigcup_{j=1}^N \Gamma_j$. Then write $u(z)$ as $u(z) = \sum_{j=1}^N \sum_{\alpha \in \Gamma_j} a_\alpha e^{2\pi i\langle z,\alpha\rangle}$. We will first show that $\sum_{\alpha \in \Gamma_j} a_\alpha e^{2\pi i\langle Z,\alpha\rangle}$, $z \in \mathbf{R}^n \times \sqrt{-1}G_j^\circ$, is holomorphic for each j, where G_j° is the polar set of G_j.

Lemma. *Let* G *be a closed cone without containing a line, and let* U *be an open cone such that* $\bar{U} \subset G^\circ \cup \{0\}$ *holds. Then there exists* $\delta > 0$ *such that*

$$|\langle x, y\rangle| \geqq \delta|x|\,|y| \qquad \text{for } x \in U \text{ and } y \in G.$$

Proof. Let $S = \{x \in \mathbf{R}^n \mid |x| = 1\}$. The map $(S \cap \bar{U}) \times (S \cap G) \to \mathbf{R}$, defined by $(x, y) \mapsto \langle x, y \rangle$, has the compact image. Therefore the image is closed, and $\langle x, y \rangle > 0$ by the definition of the polar set. Hence one obtains $|\langle x, y \rangle| \geqq \delta |x| \, |y|$, since U and G are cones.

Proof of Proposition 2.4.4. Let U be an arbitrary open cone such that $U \subset G_j^\circ$. By the lemma, there exists $\delta > 0$ such that $|\langle \operatorname{Im} z, \alpha \rangle| \geqq \delta |\operatorname{Im} z| \, |\alpha|$ for $\operatorname{Im} z \in U$. Therefore one has

$$\left| \sum_{\alpha \in \Gamma_j} a_\alpha e^{2\pi i \langle z, \alpha \rangle} \right| \leqq \sum_{\alpha \in \Gamma_j} C_\epsilon e^{\epsilon|\alpha|} e^{-2\pi \langle \operatorname{Im} z, \alpha \rangle} \leqq C_\epsilon \sum_{\alpha \in \Gamma_j} e^{\epsilon |\alpha| - 2\pi\delta |\operatorname{Im} z| \, |\alpha|}.$$

Note that for a sufficiently small $\epsilon > 0$ one has $e^{\epsilon|\alpha| - 2\pi\delta|\operatorname{Im} z|} < 1$. Hence $\sum_{\alpha \in \Gamma_j} a_\alpha e^{2\pi i \langle z, \alpha \rangle}$ is uniformly convergent on $\mathbf{R}^n \times \sqrt{-1}U$; i.e. it is a holomorphic function. There is defined a hyperfunction

$$u_j(x) = \sum_{\alpha \in \Gamma_j} a_\alpha e^{2\pi i \langle x, \alpha \rangle}$$

as the boundary value. Then $u(x) = \sum_{j=1}^{N} u_j(x)$ is a hyperfunction. The proof of Proposition 2.4.5 shows that the above definition is independent of the choice of partition of $\mathbf{Z}^n$.

Example 2.4.5 (*Poisson's Summation Formula*). Note that

$$\sum_{n=-\infty}^{\infty} e^{2\pi i n x} = \sum_{n=-\infty}^{\infty} \delta(x - n)$$

holds.

Proof. Write $u(x) = \sum_{n=0}^{\infty} e^{2\pi i n x} + \sum_{n=1}^{\infty} e^{-2\pi i n x}$.

Since

$$\sum_{n=0}^{\infty} e^{2\pi i n z} = \frac{1}{1 - e^{2\pi i z}} \qquad \text{for } \operatorname{Im} z > 0,$$

it is holomorphic. Therefore the first term equals $1/(1 - e^{2\pi i (x + i0)})$. If $\operatorname{Im} z < 0$, then

$$\sum_{n=1}^{\infty} e^{-2\pi i n z} = \frac{e^{-2\pi i z}}{1 - e^{-2\pi i z}} = -\frac{1}{1 - e^{2\pi i z}}.$$

The second term equals $-1/(1 - e^{2\pi i (x - i0)})$. Consequently, one has

$$u(x) = \frac{1}{1 - e^{2\pi i (x + i0)}} - \frac{1}{1 - e^{2\pi i (x - i0)}}.$$

Notice that in a complex neighborhood of $(-1, 1) \subset \mathbf{R}$,

$$\frac{1}{1 - e^{2\pi iz}} = \frac{1}{-2\pi iz} + (\text{a holomorphic function})$$

holds. Therefore, on $(-1, 1)$

$$u(x) = \frac{1}{-2\pi i}\left(\frac{1}{x + i0} - \frac{1}{x - i0}\right) = \delta(x).$$

On the other hand, $u(x + n) = u(x)$ holds for $n \in \mathbf{Z}$. One now concludes $u(x) = \sum_{n=-\infty}^{\infty} \delta(x - n)$.

We will consider a simple case from the theory of Fourier transformation, i.e. changing $\alpha \in \mathbf{Z}^n$ to the continuum $y \in \mathbf{R}^n$.

Proposition 2.4.5. *Let $f(y)$ be a locally Lebesgue integrable function on $\mathbf{R}^n$ such that for an arbitrary $\epsilon > 0$ there exists C_ϵ with the property $|f(y)| \leqq C_\epsilon e^{\epsilon|y|}$ almost everywhere. Then the Fourier transform of $f(y)$, $u(x) = \int_{\mathbf{R}^n} f(y)e^{2\pi i\langle x,y\rangle}\, dy$, is a well-defined hyperfunction.*

Proof. Let $\mathbf{R}^n = \Gamma_1 \cup \cdots \cup \Gamma_N$, where each Γ_j is a closed convex cone which contains no lines, such that for $j \neq k$ the measure of $\Gamma_j \cap \Gamma_k$ is zero. Then it is plain that Γ_i° is a non-empty open set. Furthermore, $\Gamma_j^\circ = \{x \mid \varphi_{\Gamma_j}(x) > 0\}$ holds, where $\varphi_{\Gamma_j}(x) = \inf_{\substack{y \in \Gamma_i \\ |y| = 1}} \langle x, y\rangle$. We now show that $u_{\Gamma_i}(z) \underset{\text{def}}{=} \int_{\Gamma_i} f(y)e^{2\pi i\langle z,y\rangle}\, dy$ is holomorphic in z for $\operatorname{Im} z \in \Gamma_i^\circ$. Since $\operatorname{Im}\langle z, y\rangle = \langle \operatorname{Im} z, y\rangle > 0$ holds for $\operatorname{Im} z \in \Gamma_i^\circ$ and $y \in \Gamma_i$, we have the inequality $\operatorname{Im}\langle z, y\rangle \geqq |y|\varphi_{\Gamma_i}(\operatorname{Im} z)$. Therefore we get

$$|e^{2\pi i\langle z,y\rangle}| = e^{-2\pi \operatorname{Im}\langle z,y\rangle} \leqq \exp(-2\pi|y|\varphi_{\Gamma_i}(\operatorname{Im} z)) \qquad \text{for } y \in \Gamma_i.$$

Consequently,

$$|u_{\Gamma_i}(z)| \leqq \int_{\Gamma_i} |f(y)|\,|e^{2\pi i\langle z,y\rangle}|\, dy \leqq \int_{\Gamma_i} C_\epsilon \exp(-|y|2\pi\varphi_{\Gamma_i}(\operatorname{Im} z) - \epsilon))\, dy.$$

Let $\alpha = 2\pi\varphi_{\Gamma_i}(\operatorname{Im} z) - \epsilon > 0$ for a sufficiently small ϵ. Then

$$|u_{\Gamma_i}(z)| \leqq C_\epsilon \int_{\mathbf{R}^n} e^{-\alpha|y|}\, dy = C_\epsilon\sigma_{n-1}\alpha^{-n}\Gamma(n) < +\infty,$$

from which we conclude that $u_{\Gamma_i}(z)$, $\operatorname{Im} z \in \Gamma_i$, is a holomorphic function in z, where σ_{n-1} denotes the area of sphere S^{n-1}.

Hence we can now consider the hyperfunction

$$u(x) = \sum_{i=1}^{N} u_{\Gamma_i}(x + \sqrt{-1}\Gamma_i^\circ 0)$$

induced from the above $u_{\Gamma_i} \in \mathcal{O}(\mathbf{R}^n \times \sqrt{-1}\Gamma_i^\circ)$, where $u_{\Gamma_i}(x + \sqrt{-1}\Gamma_i^\circ 0)$ denotes the boundary value from $\mathbf{R}^n \times \sqrt{-1}\Gamma_i$. Next we will prove that $u(x)$ is independently determined from the choice of partition $\mathbf{R}^n = \Gamma_1 \cup \cdots \cup \Gamma_N$. Let $\mathbf{R}^n = \Gamma_1 \cup \cdots \cup \Gamma_N = \Gamma'_1 \cup \cdots \cup \Gamma'_M$ be two partitions of $\mathbf{R}^n$. Then we can have the partition $\mathbf{R}^n = \Gamma''_1 \cup \cdots \cup \Gamma''_L$, where Γ''_k, $1 \leqq k \leqq L$, chosen from among non-zero measure sets $\Gamma''_{ij} = \Gamma_i \cap \Gamma'_j$, $1 \leqq i \leqq N$ and $1 \leqq j \leqq M$. Thus one can assume that one partition, $\Gamma'_1 \cup \cdots \cup \Gamma'_M$, is a refinement of the other, $\Gamma_1 \cup \cdots \cup \Gamma_N$. For each i, let $\Gamma_i = \Gamma'_{i_1} \cup \cdots \cup \Gamma'_{i_k}$. Abbreviate $u_{\Gamma_i}(z)$ and $u_{\Gamma'_j}(z)$ as $u_i(z)$ and $u'_j(z)$ respectively. Note that $\Gamma_i \supset \Gamma'_{i_j}$, $1 \leqq j \leqq k$, implies $\Gamma_i^\circ \subset \Gamma'^\circ_{i_1} \cap \cdots \cap \Gamma'^\circ_{i_k}$. Therefore, $\operatorname{Im} z \in \Gamma_i^\circ$ implies $\operatorname{Im} z \in \Gamma'^\circ_{i_j}$. That is, if $u_i(z)$ can be defined, then $u'_{i_j}(z)$ exists. Since $\Gamma_i = \Gamma'_{i_1} \cup \cdots \cup \Gamma'_{i_k}$, one has $u_i(z) = u'_{i_1}(z) + \cdots + u'_{i_k}(z)$. Hence $u_i(x + \sqrt{-1}\Gamma_i 0) = u'_{i_1}(x + \sqrt{-1}\Gamma_i 0) + \cdots + u'_{i_k}(x + \sqrt{-1}\Gamma_i 0)$ holds. We conclude that $u(x)$ is well defined.

Example 2.4.6. The function $f(y) \equiv 1$ clearly satisfies the condition of Proposition 2.4.5. Therefore $u(x) = \int_{\mathbf{R}} e^{2\pi i x y}\, dy$ is a hyperfunction. This hyperfunction is the δ-function; i.e.

$$\delta(x) = \int_{\mathbf{R}} e^{2\pi i x y}\, dy.$$

Proof. Let $\operatorname{Im} z > 0$. Then

$$\int_0^\infty e^{2\pi i z y}\, dy = \frac{e^{2\pi i z y}}{2\pi i z}\bigg|_0^\infty = -\frac{1}{2\pi i z}.$$

Hence one obtains

$$\int_0^\infty e^{2\pi i x y}\, dy = -\frac{1}{2\pi i(x + i0)}.$$

Similarly, one has

$$\int_{-\infty}^0 e^{2\pi i x y}\, dy = \frac{1}{2\pi i(x - i0)}.$$

Consequently,

$$\int_{\mathbf{R}} e^{2\pi i x y}\, dy = \int_0^\infty e^{2\pi i x y}\, dy + \int_{-\infty}^0 e^{2\pi i x y}\, dy = \delta(x).$$

Remark. We quote the following fact from the theory of Fourier transformations. Let D^n denote the compactification of $\mathbf{R}^n$ by adding S^{n-1} at infinity. There exists a flabby sheaf $\mathcal{Q}$ (this sheaf has nothing to do with the sheaf $\mathcal{Q}$ in Definition 2.1.3) such that $\mathcal{Q}|_{\mathbf{R}^n} = \mathcal{B}_{\mathbf{R}^n}$, and $\mathcal{Q}(D^n)$ and $\mathcal{Q}(D^{n*})$ correspond, via the Fourier transformation and the inverse Fourier transformation respectively, where D^{n*} denotes the dual of D^n and where it can be identified with D^n. See Kawai [1].

Let us now examine the singularity spectrum of the hyperfunction $u(x) = \int_{\mathbf{R}^n} f(y)e^{2\pi i\langle x,y\rangle}\,dy$, supposing some additional information on supp f.

Definition 2.4.3. *For a subset G of $\mathbf{R}^n$, define $G\infty = \{y \in \mathbf{R}^n - \{0\} \mid$ for an arbitrary $N > 0$ and $\epsilon > 0$, $G \cap \{y' \mid |y'| \geqq N$ and $|y'/|y'| - y/|y|| < \epsilon\} \neq \varnothing\}$.*

For example, let $G_1 = \{(x, y) \mid xy \geqq 1, x > 0\}$, and let $G_2 = \{(x, y) \mid y \geqq x^2\}$ in $\mathbf{R}^2$; then $G_1\infty = \{(x, y) \mid x \geqq 1$ and $y \geqq 1\}$, and $G_2\infty = \{(x, y) \mid x = 0\}$. Generally, $G\infty = \{y \in \mathbf{R}^n - \{0\} \mid$ the intersection of G and an arbitrary open cone containing y is never relatively compact$\}$.

Lemma. *Let G be a subset of $\mathbf{R}^n$. Let $\mathscr{G}$ be the collection of $\{G' \subset \mathbf{R}^n \mid G'$ is a closed cone such that $G' \supset G + a$ for some $a\}$. Then $G\infty = \bigcap_{G' \in \mathscr{G}} G'$ holds.*

Proof. Let $y \in G\infty$, and let U be an open cone containing y. Then one can find a closed cone T and an open cone V so that $y \in V \subset T \subset U$. Therefore, since $T \cap G$ is not relatively compact, $T \cap (G' - a)$ is not relatively compact where $a \in \mathbf{R}^n$, and G' is an arbitrary closed cone such that $G' \supset G + a$. But in general, for closed cones G_1 and G_2 and $a \in \mathbf{R}^n$, $(G + a) \cap G_2$ is relatively compact if and only if $G_1 \cap G_2 = \varnothing$. So one has $T \cap G' \neq \varnothing$ in this case. This implies $U \cap G' \neq \varnothing$. Hence $y \in G'$. This is because, if $y \notin G'$, then there exists an open set Ω such that $y \in \Omega$ and $\Omega \cap G' = \varnothing$ since G' is closed. Let $U = \mathbf{R}^+ \cdot \Omega = \{\lambda x \mid \lambda \in \mathbf{R}^+$ and $x \in \Omega\}$. Then U is an open cone containing y with the property $U \cap G' = \varnothing$. Therefore, one concludes $G\infty \subset \bigcap_{G' \in \mathscr{G}} G'$. Next suppose $y \in \bigcap_{G' \in \mathscr{G}} G'$ and $y \notin G\infty$. Then one can find an open cone U containing y such that $U \cap G$ is contained in a compact set K. Notice that there exists $a \in U$ such that $(U + a) \cap K = \varnothing$ (see Figure 2.4.1). Since one can assume that U is convex, $U + U \subset U$ holds. $(U + a) \cap G \subset U \cap G \subset K$ implies $(U + a) \cap G \subset (U + a) \cap K = \varnothing$. That is, $U \cap (G - a) = \varnothing$. Then $G' = \mathbf{R}^n - U$ is a closed cone such that $G - a \subset G'$ and $y \notin G'$, which is a contradiction. Consequently, $y \in G'$.

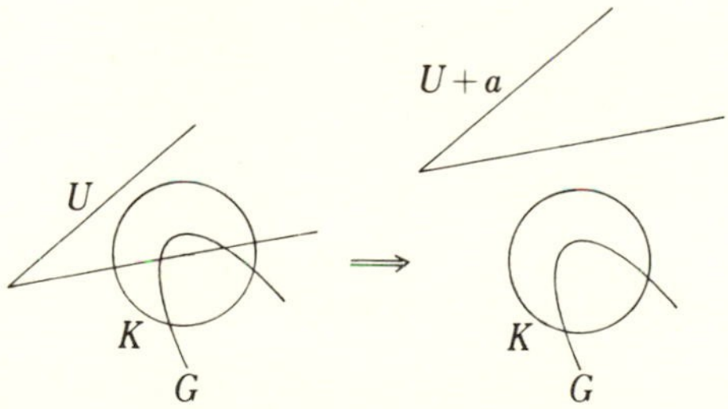

Figure 2.4.1

Proposition 2.4.6. *Keeping the assumptions on $f(x)$ the same as in Proposition 2.4.5, let $G = \operatorname{supp} f$. Then the singularity spectrum of the hyperfunction $u(x) = \int_{\mathbf{R}^n} f(y)e^{2\pi i\langle x,y\rangle}\,dy$ is contained in $\{(x, \sqrt{-1}\xi\infty) \mid \xi \in G\infty\}$.*

Note. The notion of essential support developed by Bros and Iagolnitzer (see the remark following Definition 2.3.3) results from Proposition 2.4.6 being carried to its limit. Probably this is the reason they chose the name "essential support" for their theory. Physically speaking, Proposition 2.4.6 is related to microcausality, but it is macrocausality that has immediate relevance for singularity spectrum theory. With respect to macrocausality, it is quite interesting that physicists were quite independently developing a prototype of the macrocausality notion of the singularity spectrum at around the same time that microfunction theory appeared. See Iagolnitzer and Stapp [1] and the references cited therein.

Proof of Proposition 2.4.6. Let $\xi \notin G\infty$. Then there exists a closed cone G' such that $\xi \notin G'$ and $G + a \subset G'$ for some $a \in \mathbf{R}^n$, from the above lemma. Let G_j be a closed convex cone without containing a line for each $j = 1, \ldots, N$ such that $G' \subset \bigcup_{j=1}^{N} G_j$ and $\xi \notin \bigcup_{j=1}^{N} G_j$. Proposition 2.4.5 implies that $u_j(z) = \int_{G_j - a} f(y)e^{2\pi i\langle z,y\rangle}\,dy$ is a holomorphic function for $z \in \mathbf{R}^n \times \sqrt{-1}G_j^{\circ}$. Hence $\text{S.S. } u_j(x + \sqrt{-1}G_j^{\circ}0) \subset \sqrt{-1}G_j$. Since $u(x) = \sum_{j=1}^{N} u_j(x + \sqrt{-1}G_j^{\circ}0)$, one has $\text{S.S. } u(x) \subset \bigcup_{j=1}^{N} \sqrt{-1}G_j$. Then $(x, \sqrt{-1}\xi\infty) \notin \text{S.S. } u$ holds, since $\xi \notin G_j$.

Remark. Note that $\int_{G_j - a} f(y)e^{2\pi i\langle x,y\rangle}\,dy = e^{-\langle x,a\rangle} \int_{G_j} f(y - a)e^{2\pi i\langle x,y\rangle}\,dy$. Therefore the shifting of G_j by a, $G_j - a$, does not affect the convergence.

Example 2.4.7. As an example of Proposition 2.4.6, we consider the hyperfunction $\int_{\mathbf{R}} y_+^{\lambda} e^{2\pi ixy}\,dy$, $\operatorname{Re}\lambda > 0$. Note that we have $G = G\infty = \overline{\mathbf{R}^+}$. By carrying out the integration, we have

$$\int y_+^{\lambda} e^{2\pi i\langle x,y\rangle}\,dy = \int_0^{\infty} y^{\lambda}e^{2\pi ixy}\,dy = \int_0^{\infty} \left(\frac{t}{-2\pi ix}\right)^{\lambda} e^{-t}\,\frac{dt}{(-2\pi ix)}$$

$$= (-2\pi i(x + i0))^{-(\lambda+1)}\Gamma(\lambda + 1) = \frac{\Gamma(\lambda + 1)e^{(\pi/2)i(1+\lambda)}}{(2\pi)^{1+\lambda}(x + i0)^{1+\lambda}}.$$

Then the singularity spectrum is contained in $(0, idx\infty)$.

Note. By the reasoning given in Chapter III, §8, we can verify that the above equality holds for any $\lambda \neq -1, -2, \ldots$.

To conclude this section, we will consider an example which has a different flavor from the previous ones. The equation (2.4.4), below, is important in application.

Example 2.4.8. We will define the hyperfunction $1/(x+\sqrt{-1}y)$ by boundary values. Let

$$\Omega_{\epsilon_1,\epsilon_2} = \{(z, w) \in \mathbf{C}^2 \,|\, \epsilon_1 \operatorname{Im} z > 0 \text{ and } \epsilon_2 \operatorname{Im} w > 0\},$$

where $\epsilon_1, \epsilon_2 = \pm$. Let $\log z$ denote the logarithm function defined on $\mathbf{C} - \{x \in \mathbf{R} \,|\, x \leqq 0\}$ with the prescription of the branch $\log z|_{z=1} = 0$. Then $\log z - \log w + \pi i/2$ has a zero on $z + iw = 0$ in $\Omega_{+,+}$. Therefore $\varphi_{+,+}(z, w) = (\log z - \log w + \pi i/2)/(z + \sqrt{-1}w)$ is holomorphic in $\Omega_{+,+}$. Similarly, we obtain that the function $\varphi_{\epsilon_1,\epsilon_2}$, where

$$\varphi_{-,+} = \left(\log z - \log w + \frac{\pi i}{2}\right)\Big/(z + \sqrt{-1}w),$$

$$\varphi_{+,-} = (\log z - \log w - \tfrac{3}{2}\pi i)/(z + \sqrt{-1}w),$$

and

$$\varphi_{-,-} = \left(\log z - \log w + \frac{\pi i}{2}\right)\Big/(z + \sqrt{-1}w),$$

is holomorphic in $\Omega_{\epsilon_1,\epsilon_2}$. Define

$$u(x, y) = \frac{1}{2\pi i}(b_{\Omega_{+,+}}(\varphi_{+,+}) - b_{\Omega_{+,-}}(\varphi_{+,-}) - b_{\Omega_{-,+}}(\varphi_{-,+}) + b_{\Omega_{-,-}}(\varphi_{-,-})),$$

where $b_{\Omega_{\epsilon_1,\epsilon_2}}(\varphi_{\epsilon_1,\epsilon_2})$ denotes the boundary value of $\varphi_{\epsilon_1,\epsilon_2}$ from $\Omega_{\epsilon_1,\epsilon_2}$. Then

$$\begin{aligned}
2\pi\sqrt{-1}(x+\sqrt{-1}y)u &= b_{\Omega_{+,+}}(\log z - \log w + \pi\sqrt{-1}/2) \\
&\quad - b_{\Omega_{+,-}}(\log z - \log w - 3\pi\sqrt{-1}/2) \\
&\quad - b_{\Omega_{-,+}}(\log z - \log w + \pi\sqrt{-1}/2) \\
&\quad + b_{\Omega_{-,-}}(\log z - \log w + \pi\sqrt{-1}/2) \\
&= \{\log(x+\sqrt{-1}0) - \log(y+\sqrt{-1}0) + \pi\sqrt{-1}/2\} \\
&\quad - \{\log(x+\sqrt{-1}0) - \log(y-\sqrt{-1}0) - 3\pi\sqrt{-1}/2\} \\
&\quad - \{\log(x-\sqrt{-1}0) - \log(y+\sqrt{-1}0) + \pi\sqrt{-1}/2\} \\
&\quad + \{\log(x-\sqrt{-1}0) - \log(y-\sqrt{-1}0) + \pi\sqrt{-1}/2\} \\
&= 2\pi\sqrt{-1}.
\end{aligned}$$

That is,

$$(x + \sqrt{-1}y)u = 1 \tag{2.4.3}$$

holds. In particular, $u = 1/(x + \sqrt{-1}y)$ for $(x, y) \neq (0, 0)$. Next we will prove the following equation which is important for applications:

$$\frac{1}{2}\left(\frac{\partial}{\partial x} + \sqrt{-1}\frac{\partial}{\partial y}\right)u = \pi\delta(x, y). \tag{2.4.4}$$

First notice that we have

$$\left(\frac{\partial}{\partial z}+\sqrt{-1}\,\frac{\partial}{\partial w}\right)\left(\frac{1}{z+\sqrt{-1}w}(\log z-\log w+c)\right)$$
$$=\frac{1}{z+\sqrt{-1}w}\left(\frac{1}{z}-\frac{\sqrt{-1}}{w}\right)=-\frac{\sqrt{-1}}{zw}\qquad\text{for } c\in\mathbf{C}.$$

From this we obtain the following:

$$\left(\frac{\partial}{\partial x}+\sqrt{-1}\,\frac{\partial}{\partial y}\right)u$$
$$=\frac{1}{2\pi\sqrt{-1}}\left\{\frac{-\sqrt{-1}}{(x+\sqrt{-1}0)(y+\sqrt{-1}0)}+\frac{\sqrt{-1}}{(x+\sqrt{-1}0)(y-\sqrt{-1}0)}\right.$$
$$\left.+\frac{\sqrt{-1}}{(x-\sqrt{-1}0)(y+\sqrt{-1}0)}-\frac{\sqrt{-1}}{(x-\sqrt{-1}0)(y-\sqrt{-1}0)}\right.$$
$$=(-\sqrt{-1})(2\pi\sqrt{-1})\delta(x,y).$$

Thus, we have verified (2.4.4). We will also give some other differential equations which u satisfies. On the other hand,

$$\left(z\frac{\partial}{\partial z}+w\frac{\partial}{\partial w}+1\right)\varphi_{\epsilon_1,\epsilon_2}=0$$

holds. Hence, we have

$$\left(x\frac{\partial}{\partial x}+y\frac{\partial}{\partial y}+1\right)u=0. \tag{2.4.5}$$

We also obtain

$$\left(y\frac{\partial}{\partial x}-x\frac{\partial}{\partial y}-\sqrt{-1}\right)u=0. \tag{2.4.6}$$

This is because equation (2.4.6) plus equation (2.4.5) $\cdot\sqrt{-1}$ implies

$$\left(y\frac{\partial}{\partial x}-x\frac{\partial}{\partial y}-\sqrt{-1}\right)u=\left((y+\sqrt{-1}x)\frac{\partial}{\partial x}-(x-\sqrt{-1}y)\frac{\partial}{\partial y}\right)u$$
$$=(y+\sqrt{-1}x)\left(\frac{\partial}{\partial x}+\sqrt{-1}\,\frac{\partial}{\partial y}\right)u$$
$$=2\pi(y+\sqrt{-1}x)\delta(x,y)=0.$$

CHAPTER III

Fundamental Operations

§1. Product, Restriction, and Substitution

We shall devote this section to definitions of the product, restriction, and substitution of hyperfunctions and microfunctions. We will treat Proposition 3.1.1, which follows, as an axiom for the succeeding argument. Note that in SKK [1] these operations are discussed first and that, as a result, the statement of Proposition 3.1.1 (and even the flabbiness of the microfunction sheaf, see §7) is proved. Our treatment in this book is intended to be a more elementary exposition of the material.

Proposition 3.1.1. *Let* $V_j, j = 1, 2, \ldots, N < \infty$, *be an open set in* $\sqrt{-1}S^*M$, *and let* $V = \bigcup_j V_j$.

(1) *If* S.S. $f \subset V$ *holds for* $f \in \mathscr{B}_M(M)$, *then for each* $x \in M$ *there exists* $f_j \in \mathscr{B}_M(W)$, *for some neighborhood* W *of* x, *such that* S.S. $f_j \subset V_j \cap \pi^{-1}(W)$ *and* $f = \sum_j f_j$ *on* W.

(2) *Let* $f_j \in \mathscr{B}_M(M)$ *such that* S.S. $f_j \subset V_j$ *and* $\sum f_j = 0$. *Then, for any* x *in* M *there exists* $f_{jk} \in \mathscr{B}_M(W)$, *for some neighborhood* W *of* x, *such that* $f_j = \sum_k f_{jk}$, $f_{jk} = -f_{kj}$ *and* S.S. $f_{jk} \subset V_j \cap V_k \cap \pi^{-1}(W)$.

Proposition 3.1.1 can be rephrased as follows.

Proposition 3.1.1′. *Let* $V_j \subset \sqrt{-1}S^*M$, $j = 1, 2, \ldots, N < \infty$, *be an open set and let* $V = \bigcup_j V_j$. *Then for each* $p \in \sqrt{-1}S^*M$ *there is a neighborhood* W *with the following properties*:

(1)′ *If* $u \in \mathscr{C}_M(W)$ *satisfies* supp $u \subset V \cap W$, *then* u *can be expressed as* $u = \sum_j u_j$ *for some* $u_j \in \mathscr{C}_M(W)$ *such that* supp $u_j \subset V_j \cap W$.

(2)′ *If* $u_j \in \mathscr{C}_M(W)$ *satisfies* supp $u_j \subset V_j$ *and* $\sum_j u_j = 0$, *then* $u_j = \sum_k u_{jk}$ *for some* $u_{jk} \in \mathscr{C}_M(W)$ *such that* $u_{jk} = -u_{kj}$ *and* supp $u_{jk} \subset V_j \cap V_k \cap W$.

We will define a combined concept of the support of a hyperfunction and that of a microfunction to obtain more systematic results. One may understand why we introduce this concept if one tries to rephrase the following Theorem 3.1.1 without using S.S., defined in Definition 3.1.1. One may also find it more convenient to work with T^*M rather than S^*M when one studies systems of differential equations. See also Definition 3.8.1, where the sheaf $\hat{\mathscr{C}}$, a combined notion of $\mathscr{B}$ and $\mathscr{C}$, is introduced such that $\widehat{\text{S.S.}}$ can be considered as the support of a section of $\hat{\mathscr{C}}$.

Definition 3.1.1. *Let u be a hyperfunction on a real analytic manifold M. Then define*

$$\widehat{\text{S.S.}}\, u \underset{\text{def}}{=} \{(x, \sqrt{-1}\xi) \in \sqrt{-1}T^*M \mid x \in \operatorname{supp} u \text{ and } \xi = 0\}$$
$$\cup \{(x, \sqrt{-1}\xi) \in \sqrt{-1}T^*M \mid \xi \neq 0 \text{ and } (x, \sqrt{-1}\xi\infty) \in \text{S.S. } u\}.$$

Remark. $\widehat{\text{S.S.}}\, u$ is closed in T^*M and is invariant under the transform

$$(x, \sqrt{-1}\xi) \mapsto (x, \sqrt{-1}c\xi) \qquad \text{for } c \geqq 0.$$

The reader may find it profitable to read first the examples at the end of this section in order to understand the following statements on products, restrictions, and substitutions of hyperfunctions or microfunctions.

Theorem 3.1.1. *Let M_1 and M_2 be real analytic manifolds, and let $M = M_1 \times M_2$. For hyperfunctions $u_1 = u_1(x_1)$ and $u_2 = u_2(x_2)$, on M_1 and M_2 respectively, one can define canonically the product $u = u(x_1, x_2) \underset{\text{def}}{=} u_1(x_1)u_2(x_2)$ so that*

$$\widehat{\text{S.S.}}\, u \subset \widehat{\text{S.S.}}\, u_1 \times \widehat{\text{S.S.}}\, u_2,$$

where we regard $\sqrt{-1}T^*M = \sqrt{-1}T^*M_1 \times \sqrt{-1}T^*M_2$.

Proof. Let $\sqrt{-1}S^*M_\nu = \bigcup_{j \in I_\nu} V_j^\nu$, $\nu = 1, 2$, where V_j^ν is an open and convex set such that $\tau(V_j^{\nu\circ}) = M_\nu$ and I_ν is a finite set. Then Proposition 3.1.1 can be applied to this situation since the question is local in nature. There exist a neighborhood W_ν and a hyperfunction u_j^ν on W_ν such that $u_\nu = \sum_j u_j^\nu$ and S.S. $u_j^\nu \subset V_j^\nu \cap \pi^{-1}(W_\nu)$. Let $W = W_1 \times W_2$. We will denote $V_j^\nu \cap \pi^{-1}(W_\nu)$ simply by V_j^ν. Let Z_j^ν be a closed convex set such that S.S. $u_j^\nu \subset Z_j^\nu \subset V_j^\nu$. Then $Z_j^{\nu\circ} \supset V_j^{\nu\circ}$ and $Z_j^{\nu\circ}$ is an open set. By Theorem 2.3.5, there is a conoidal neighborhood D_j^ν of $Z_j^{\nu\circ}$ and a holomorphic function $f_j^\nu \in \mathcal{O}(D_j^\nu)$ such that $u_j^\nu = b(f_j^\nu; D_j^\nu)$. Then $f_{j_1}^1(z_1)f_{j_2}^2(z_2) \in \mathcal{O}(D_{j_1}^1 \times D_{j_2}^2)$ holds. Since $D_{j_\nu}^\nu$ is a conoidal neighborhood of $Z_j^{\nu\circ}$, $D_{j_1}^1 \times D_{j_2}^2$ is a conoidal neighborhood of $Z_j^{1\circ} \hat{\times} Z_j^{2\circ} \underset{\text{def}}{=} \{(x_1, x_2) + \sqrt{-1}(\xi_1, \xi_2)0 \mid \xi_1, \xi_2 \neq 0$ and $(x_\nu, \sqrt{-1}\xi_\nu 0) \in Z_j^{\nu\circ}$ for $\nu = 1, 2\}$. Notice that $Z_j^{1\circ} \hat{\times} Z_j^{2\circ}$ is an open convex set such that $\tau(Z_j^{1\circ} \hat{\times} Z_j^{2\circ}) = M$. Hence $b(f_{j_1}^1(z_1)f_{j_2}^2(z_2); D_{j_1}^1 \times D_{j_2}^2)$ is a hyperfunction on M. Define, then, the product of $u_1(x_1)$ and $u_2(x_2)$ as $u_1(x_1)u_2(x_2) = \sum_{j_1 \in I_1} \sum_{j_2 \in I_2} b(f_{j_1}^1(z_1)f_{j_2}^2(z_2); D_{j_1}^1 \times D_{j_2}^2)$. Next we will show that this is well defined. Since one can find a refinement of the covering $\{V_j^\nu\}$, the assertion is independent of the choice of coverings. We will confirm that the assertion does not depend upon the choice of f_j^ν. Suppose $u_\nu = \sum_j b(f_j^\nu; D_j^\nu) = \sum_j b(\tilde{f}_j^\nu; D_j^\nu)$ for $\nu = 1, 2$. If one lets $g_j^\nu = \tilde{f}_j^\nu - f_j^\nu$, one obtains $\sum_j b(g_j^\nu; D_j^\nu) = 0$ and S.S. $g_j^\nu \subset V_j^\nu$. By Proposition 3.1.1, $b(g_j^\nu; D_j^\nu) = \sum_k w_{jk}^\nu$ holds for $w_{jk}^\nu \in \mathscr{B}(W_\nu)$ such that $w_{jk}^\nu = -w_{kj}^\nu$ and S.S. $w_{jk}^\nu \subset V_j^\nu \cap V_k^\nu$. Therefore one can find D_{jk}^ν and $g_{jk}^\nu \in \mathscr{O}(D_{jk}^\nu)$, where $D_{jk}^\nu \supset D_j^\nu \cup D_k^\nu$ and D_{jk}^ν is a conoidal neighborhood of $Z_j^{\nu\circ} \cup Z_k^{\nu\circ}$, so that $w_{jk}^\nu = b(g_{jk}^\nu; D_{jk}^\nu)$.

It is sufficient to prove the case $\tilde{f}_j^2 = f_j^2$. Then,

$$\begin{aligned}
\sum_{j_1,j_2} b(\tilde{f}_{j_1}^1(z_1)f_{j_2}^2(z_2)) &- \sum_{j_1,j_2} b(f_{j_1}^1(z_1)f_{j_2}^2(z_2)) \\
&= \sum_{j_1,j_2} b(g_{j_1}^1(z_1)f_{j_2}^2(z_2); D_{j_1}^1 \times D_{j_2}^2) \\
&= \sum_{j_1,j_2,k_1} b(g_{j_1k_1}^1(z_1)f_{j_2}^2(z_2); D_{j_1}^1 \times D_{j_2}^2) \\
&= \sum_{j_1,j_2,k_1} b(g_{j_1k_1}^1(z_1)f_{j_2}^2(z_2); D_{j_1k_1}^1 \times D_{j_2}^2) \\
&= \sum_{j_1,k_1,j_2} b(g_{j_1k_1}^1(z_1)f_{j_2}^2(z_2) + g_{k_1j_1}^1(z_1)f_{j_2}^2(z_2); D_{j_1k_1}^1 \times D_{j_2}^2) = 0,
\end{aligned}$$

i.e. independent of f_j^ν. We will prove $\widehat{\text{S.S.}}\, u \subset \widehat{\text{S.S.}}\, u_1 \times \widehat{\text{S.S.}}\, u_2$ next. Let S.S. $u_\nu \subset Z_\nu = \bigcup_j Z_{j_\nu}^\nu$, where $Z_{j_\nu}^\nu$ is a closed convex set for $\nu = 1, 2$. Then $D_{j_1}^1 \times D_{j_2}^2$ is a conoidal neighborhood of $Z_{j_1}^{1\circ} \hat{\times} Z_{j_2}^{2\circ}$. Therefore, one has S.S. $(b(f_{j_1}^1(z_1)f_{j_2}^2(z_2); D_{j_1}^1 \times D_{j_2}^2)) \subset (Z_{j_1}^{1\circ} \hat{\times} Z_{j_2}^{2\circ})^\circ$. We will show

$$(Z_{j_1}^{1\circ} \hat{\times} Z_{j_2}^{2\circ})^\circ \subset Z_{j_1}^1 \hat{\times} Z_{j_2}^2 \cup W_1 \times Z_{j_2}^2 \cup Z_{j_1}^1 \times W_2.$$

Let $((x_1, x_2), \sqrt{-1}(\xi_1, \xi_2)\infty) \in (Z_{j_1}^{1\circ} \hat{\times} Z_{j_2}^{2\circ})$. Then $\langle \xi_1, v_1\rangle + \langle \xi_2, v_2\rangle > 0$ holds for $x_1 + \sqrt{-1}v_1 0 \in Z_{j_1}^{1\circ}$ and $x_2 + \sqrt{-1}v_2 0 \in Z_{j_2}^{2\circ}$. This inequality holds for any tv_1 and sv_2, $s > 0$ and $t > 0$, which implies $\langle \xi_1, v_1\rangle \geqq 0$ and $\langle \xi_2, v_2\rangle \geqq 0$. Note that if $\xi_1 \neq 0$, then $\langle \xi_1, v_1\rangle > 0$ (and similarly for ξ_2). In fact, since $Z_{j_1}^\circ$ is an open set, if $\langle \xi_1, v_1\rangle = 0$, then $v_1 - \epsilon v_1'$ is contained in $Z_{j_1}^\circ$ for a sufficiently small $\epsilon > 0$ and v_1' such that $\langle \xi_1, v_1'\rangle > 0$. This implies $\langle \xi_1, v_1 - \epsilon v_1'\rangle = -\epsilon\langle \xi_1, v_1'\rangle < 0$. Hence for $\xi_1 \neq 0$, $\xi_2 \neq 0$, and an arbitrary $v_\nu \in Z_{j_\nu}^{\nu\circ}$, one has $\langle \xi_\nu, v_\nu\rangle > 0$ for $\nu = 1, 2$; i.e. $(x_\nu, \xi_\nu) \in Z_{j_\nu}^\nu$ for $\nu = 1, 2$. Consequently, one obtains $((x_1, x_2), \sqrt{-1}(\xi_1, \xi_2)\infty) \in Z_{j_1}^1 \hat{\times} Z_{j_2}^2$. If $\xi_1 = 0$, then $\langle \xi_2, v_2\rangle > 0$ holds for an arbitrary $v_2 \in Z_{j_2}^\circ$. Thus, $((x_1, x_2), \sqrt{-1}(0, \xi_2)\infty) \in M_1 \times Z_{j_2}^\circ$ holds (similarly for the case $\xi_2 = 0$). Therefore, one has S.S.$(b(f_{j_1}^1(z_1)f_{j_2}^2(z_2))) \subset Z_{j_1}^1 \hat{\times} Z_{j_2}^2 \cup W_1 \times Z_{j_2}^2 \cup Z_{j_1}^1 \times W_2$; i.e. S.S. $u \subset Z_1 \hat{\times} Z_2 \cup W_1 \times Z_2 \supset Z_1 \times W_2$ for an arbitrary point on $M_1 \times M_2$. Hence, S.S. $u \subset Z_1 \hat{\times} Z_2 \cup M_1 \times Z_2 \cup Z_1 \times M_2$

is true for any closed convex sets Z_ν with the property S.S. $u_\nu \subset Z_\nu$, $\nu = 1, 2$. Finally, one obtains S.S. $u \subset$ S.S. $u_1 \hat{\times}$ S.S. $u_2 \cup M_1 \times$ S.S. $u_2 \cup$ S.S. $u_1 \times M_2$. From this we will conclude $\widehat{\text{S.S.}}\, u \subset \widehat{\text{S.S.}}\, u_1 \times \widehat{\text{S.S.}}\, u_2$. Let $((x_1, x_2), \sqrt{-1}(\xi_1, \xi_2)) \in \widehat{\text{S.S.}}\, u$, $\xi_1 \neq 0$, and $\xi_2 \neq 0$. From the above result, one has $((x_1, x_2), \sqrt{-1}(\xi_1, \xi_2)\infty) \in$ S.S. $u_1 \hat{\times}$ S.S. u_2. Since $((x_1, x_2), \sqrt{-1}(0, \xi_2)\infty) \in M_1 \times$ S.S. u_2 for $\xi_1 = 0$ and $\xi_2 \neq 0$, one must show $x_1 \in \operatorname{supp} u_1$. But if $x_1 \notin \operatorname{supp} u_1$, then $(x_1, x_2) \notin \operatorname{supp} u$ (whose proof is equivalent to the one for independence of the choice of f_j^ν, as before); and similarly, where $\xi_1 \neq 0$ and $\xi_2 = 0$, we reach a contradiction. If $\xi_1 = \xi_2 = 0$, then $(x_1, x_2) \in \operatorname{supp} u$ implies $x_1 \in \operatorname{supp} u_1$ and $x_2 \in \operatorname{supp} u_2$.

Definition 3.1.2. *Let M_1 and M_2 be real analytic manifolds, and let $M = M_1 \times M_2$. Let $(\sqrt{-1}S^*M)' \underset{\text{def}}{=} \sqrt{-1}S^*M - \sqrt{-1}S^*M_1 \times M_2 - M_1 \times \sqrt{-1}S^*M_2$, and define $p_1 : (\sqrt{-1}S^*M)' \to \sqrt{-1}S^*M_1$ and $p_2 : (\sqrt{-1}S^*M)' \to \sqrt{-1}S^*M_2$ by $p_1((x_1, x_2), \sqrt{-1}(\xi_1, \xi_2)\infty)) = (x_1, \sqrt{-1}\xi_1\infty)$ and $p_2((x_1, x_2), \sqrt{-1}(\xi_1, \xi_2)\infty)) = (x_2, \sqrt{-1}\xi_2\infty)$ respectively. For an open set Ω_ν in $\sqrt{-1}S^*M_\nu$, $\nu = 1, 2$, let $\Omega_1 \hat{\times} \Omega_2 \underset{\text{def}}{=} \{((x_1, x_2), \sqrt{-1}(\xi_1, \xi_2)\infty) \mid \xi_1$ and $\xi_2 \neq 0$ and $(x_\nu, \sqrt{-1}\xi_\nu\infty) \in \Omega_\nu$ for $\nu = 1, 2\}$.*

Remark. We have $\Omega_1 \hat{\times} \Omega_2 = p_1^{-1}\Omega_1 \cap p_2^{-1}\Omega_2$.

Theorem 3.1.2. *There exists a canonical sheaf homomorphism*

$$p_1^{-1}\mathscr{C}_{M_1} \times p_2^{-1}\mathscr{C}_{M_2} \to \mathscr{C}_M|_{(\sqrt{-1}S^*M)'}.$$

Remark. We obtain the product $u_1(x_1)u_2(x_2) \in \mathscr{C}_M(\Omega_1 \hat{\times} \Omega_2)$ of $u_1(x_1) \in \mathscr{C}_{M_1}(\Omega_1)$ and $u_2(x_2) \in \mathscr{C}_{M_2}(\Omega_2)$ by the morphism in Theorem 3.1.2. The construction is as follows. By the sheaf homomorphism $\mathscr{C}_{M_\nu} \to p_{\nu *}p_\nu^{-1}\mathscr{C}_{M_\nu}$, $\nu = 1, 2$, for $u_\nu(x_\nu) \in \mathscr{C}_{M_\nu}(\Omega_\nu)$ we have $u'_\nu(x_\nu) \in p_\nu^{-1}\mathscr{C}_{M_\nu}(p_\nu^{-1}\Omega_\nu)$. Then the restrictions of $u_1(x_1)$ and $u_2(x_2)$ give $(u'_1(x_1), u'_2(x_2)) \in (p_1^{-1}\mathscr{C}_{M_1} \times p_2^{-1}\mathscr{C}_{M_2})(\Omega_1 \hat{\times} \Omega_2)$. Therefore, the image of the morphism in the above theorem is $u_1(x_1)u_2(x_2) \in \mathscr{C}_M(\Omega_1 \hat{\times} \Omega_2)$.

Proof of Theorem 3.1.2. For each $\nu = 1, 2$, let $(x_\nu, \sqrt{-1}\xi_\nu\infty) \in \sqrt{-1}S^*M_\nu$ and let $u_\nu \in \mathscr{C}_{M_\nu,(x_\nu,\sqrt{-1}\xi_\nu\infty)}$. From Proposition 2.1.6, $\pi^{-1}\mathscr{B}_{M_\nu} \xrightarrow{\text{sp}} \mathscr{C}_{M_\nu} \to 0$ is exact. This implies that there are $f_1 \in \mathscr{B}_{M_1,x_1}$ and $f_2 \in \mathscr{B}_{M_2,x_2}$ such that $u_1 = \operatorname{sp}(f_1)$ and $u_2 = \operatorname{sp}(f_2)$. Define, then, $u_1(x_1)u_2(x_2) \underset{\text{def}}{=} \operatorname{sp}(f_1 f_2) \in \mathscr{C}_{M,((x_1,x_2),\sqrt{-1}(\xi_1,\xi_2)\infty)}$. We will show next that this definition is independent of the choice of f_1 and f_2. Suppose $u_1 = \operatorname{sp}(f_1) = \operatorname{sp}(\tilde{f}_1)$ and $u_2 = \operatorname{sp}(f_2) = \operatorname{sp}(\tilde{f}_2)$. Then one has $f_1(x_1)f_2(x_2) - \tilde{f}_1(x_1)\tilde{f}_2(x_2) = f_1(x_1)(f_2(x_2) - \tilde{f}_2(x_2)) + (f_1(x_1) - \tilde{f}_1(x_1))\tilde{f}_2(x_2)$. Since $\operatorname{sp}(f_2 - \tilde{f}_2) = u_2 - u_2 = 0$, one obtains $(x_2, \sqrt{-1}\xi_2) \notin \widehat{\text{S.S.}}(f_2(x_2) - \tilde{f}_2(x_2))$. On the other hand, $\widehat{\text{S.S.}}\, f_1(f_2 - \tilde{f}_2) \subset \widehat{\text{S.S.}}\, f_1 \times \widehat{\text{S.S.}}(f_2 - \tilde{f}_2)$ holds by Theorem 3.1.1. Hence

$((x_1, x_2), \sqrt{-1}(\xi_1, \xi_2)\infty) \notin \widehat{\text{S.S.}}\, f_1(f_2 - \tilde{f}_2)$ holds; i.e. $\text{sp}\, f_1(f_2 - \tilde{f}_2) = 0$ at $((x_1, x_2), \sqrt{-1}(\xi_1, \xi_2)\infty)$. In a similar manner, one obtains $\text{sp}(f_1(x_1) - \tilde{f}_1(x_1))\tilde{f}_2(x_2) = 0$ at $((x_1, x_2), \sqrt{-1}(\xi_1, \xi_2)\infty)$. Consequently, $\text{sp}\, f_1 f_2 = \text{sp}\, \tilde{f}_1 \tilde{f}_2$ holds.

Remark. We proved Theorem 3.1.2 locally at each stalk. In order to prove the theorem globally, notice first that $\mathscr{C}_{M_1}$ is the sheaf associated to the presheaf $\mathscr{B}_{M_1}(\pi(U))/\mathscr{A}^*_{M_1}(U)$ for an open set U. Then one can complete the proof by constructing the map: $(\mathscr{B}_{M_1}(\pi(U_1))/\mathscr{A}^*_{M_1}(U_1) \times (\mathscr{B}_{M_2}(\pi(U_2))/\mathscr{A}^*_{M_2}(U_2)) \to \mathscr{C}_{M_1 \times M_2}(p_1^{-1}U_1 \cap p_2^{-1}U_2)$ via the spectrum map sp.

Definition 3.1.3. *Let N and M be real analytic manifolds, and let $f: N \to M$ be a real analytic map. For $y \in N$ and $\xi \in T^*_{f(y)}M$, define a map $\hat{\rho}: N \underset{M}{\times} T^*M \to T^*N$ by $\hat{\rho}(y, \xi) = (y, f^*(\xi))$. The kernel of $\hat{\rho}$ is said to be the conormal bundle with supports in N, denoted by $T^*_N M$.*

We denote $(T^*_N M - N)/\mathbf{R}^\times_+$ by $S^*_N M$, regarding $N = \{(y, \xi) \in N \underset{M}{\times} T^*M \mid y \in N$ and $\xi = 0\} \subset T^*_N M$. Notice that $\sqrt{-1}S^*_N M$ is a closed set in $N \underset{M}{\times} \sqrt{-1}S^*M$. Let the maps $\rho = \rho_f: (N \underset{M}{\times} \sqrt{-1}S^*M - \sqrt{-1}S^*_N M) \to \sqrt{-1}S^*N$ and $\varpi = \varpi_f: (N \underset{M}{\times} \sqrt{-1}S^*M - \sqrt{-1}S^*_N M) \to \sqrt{-1}S^*M$ be as follows:

$$\rho((y, \sqrt{-1}\xi\infty)) = (y, \sqrt{-1}f^*(\xi)\infty)$$
$$\varpi((y, \sqrt{-1}\xi\infty)) = (f(y), \sqrt{-1}\xi\infty).$$

Note. In the case when f is an embedding, the above definition of a conormal bundle agrees with the one in Definition 2.1.1.

Remark. The important maps ρ and ϖ, above, will be frequently used hereafter in this book. When N is a submainfold of M, the map ρ is an epimorphism and ϖ is a monomorphism. (Compare with Definition 2.1.1.)

Theorem 3.1.3 (Restriction of a Hperfunction). *Let N be a submanifold of M, and ι denotes the embedding $N \hookrightarrow M$. Let u be a hyperfunction on M, $u \in \mathscr{B}_M(M)$, such that* S.S. $u \cap \sqrt{-1}S^*_N M = \varnothing$. *Then one can define the restriction of u to N, $u|_N \in \mathscr{B}_N$, such that*

$$\widehat{\text{S.S.}}(u|_N) \subset \hat{\rho}(N \underset{M}{\times} \sqrt{-1}T^*M \cap \widehat{\text{S.S.}}\, u)$$

$$\text{S.S.}(u|_N) \subset \rho(N \underset{M}{\times} \sqrt{-1}S^*M \cap \text{S.S.}\, u).$$

Proof. Let X and Y be complexifications of M and N, respectively, and let Y be a submanifold of X. Let each U_j be an open cone containing no lines such that S.S. $u \subset \bigcup_j U_j$ and $U_j \cap \sqrt{-1}S^*_N M = \varnothing$. Then one has $[\rho(U_j \cap N \underset{M}{\times} \sqrt{-1}S^*M)]^\circ = U_j^\circ \cap \sqrt{-1}SN$, since the left-hand side of this equality is $\{x + \sqrt{-1}v0 \in \sqrt{-1}SN \mid \langle v, \iota^*(\xi)\rangle > 0$ for an arbitrary $(x, \sqrt{-1}\xi\infty) \in U_j \cap N \underset{M}{\times} \sqrt{-1}S^*M\} = \{x + \sqrt{-1}v0 \in \sqrt{-1}SN \mid \langle v, \xi\rangle > 0$ for an arbitrary $(x, \sqrt{-1}\xi\infty) \in U_j\}$, which is the right-hand side. Here we used the fact $\langle v, \iota^*(\xi)\rangle = \langle \iota_*(v), \xi\rangle = \langle v, \xi\rangle$. Furthermore, one has $\tau(U_j^\circ \cap \sqrt{-1}SN) = N$. Here is a proof for this. If there exists x in N such that $x \notin \tau(U_j^\circ \cap \sqrt{-1}SN)$, then $\rho(U_j \cap N \underset{M}{\times} \sqrt{-1}S^*M)_x \cap (\sqrt{-1}S^*M)_x = (\sqrt{-1}S^*N)_x$, since $\rho(U_j \cap N \underset{M}{\times} \sqrt{-1}S^*M)^\circ \cap (\sqrt{-1}SN)_x = \varnothing$ and $\rho(U_j \cap N \underset{M}{\times} \sqrt{-1}S^*M)$ is convex. Therefore, there can be found ξ and ξ' in $\sqrt{-1}T_x^*M - \{0\}$ such that $\iota^*(\xi) \neq 0$ and $\iota^*(\xi') = 0$, satisfying $(x, \sqrt{-1}\xi\infty) \in U_j$ and $(x, -\sqrt{-1}(\xi - 2\xi')\infty) \in U_j$. Note that $\frac{1}{2}\xi + \frac{1}{2}(-\xi + 2\xi') = \xi'$, since U_j is convex, implies $(x, \sqrt{-1}\xi'\infty) \in U_j$. But this contradicts $U_j \cap \sqrt{-1}S_N^*M = \varnothing$, since $(x, \sqrt{-1}\xi'\infty) \in S_N^*M$. Conversely, let $(x, \sqrt{-1}\xi'\infty) \in U_j \cap \sqrt{-1}S_N^*M$. For an arbitrary $\xi \in T_x^*M$, one has $(x, \sqrt{-1}(\epsilon\xi + \xi')\infty) \in U_j$ for a sufficiently small $\epsilon > 0$. Then $\rho(U_j \cap N \underset{M}{\times} \sqrt{-1}S^*M) \cap (\sqrt{-1}S^*M)_x = (\sqrt{-1}S^*N)_x$ holds. Hence, the polar set is an empty set; i.e. $U_j^\circ \cap (\sqrt{-1}SN)_x = \varnothing$.

Lemma. *If D is a conoidal neighborhood of $U \subset \sqrt{-1}SM$, then for a complex neighborhood Y of N, $D \cap Y$ is a conoidal neighborhood of $U \cap \sqrt{-1}SN$.*

Proof. Since the assertion is of a local nature, one may assume $N = \{x_1 = \cdots = x_r = 0\} \subset M \subset \mathbf{R}^n$. If D is a conoidal neighborhood of $0 + \sqrt{-1}(0, v_{r+1}, \ldots, v_n)0 \in \sqrt{-1}SM$, then $x + \sqrt{-1}tv \in D$ for $|x| \ll 1$, $0 < t \ll 1$ and $|v - v_0| \ll 1$. That is, for $|(x_{r+1}, \ldots, x_n)| \ll 1$, $0 < t \ll 1$ and $|(v'_{r+1}, \ldots, v'_n) - (v_{r+1}, \ldots, v_n)| \ll 1$, one obtains $(0, x_{r+1}, \ldots, x_n) + \sqrt{-1}(0, v'_{r+1}, \ldots, v'_n)t \in D \cap Y$, completing the proof of the lemma.

Let $u = \sum_j b_{D_j}(f_j)$ for $u \in \mathscr{B}_M(M)$, where D_j is a conoidal neighborhood of U_j° and $f_j \in \mathcal{O}_X(D_j)$. The above lemma implies that $D_j \cap Y$ is a conoidal neighborhood of $U_j^\circ \cap \sqrt{-1}SN$ and $\tau(U_j^\circ \cap \sqrt{-1}SN) = N$. Since $\sum_j b_{D_j \cap Y}(f_j|_Y)$ defines a hyperfunction on N, one can now let $u|_N = \sum_j b_{D_j \cap Y}(f_j|_Y)$. On the other hand, one has $U_j^\circ \cap \sqrt{-1}SN = \rho(U_j \cap N \underset{M}{\times} \sqrt{-1}S^*M)^\circ$. Then $\mathrm{S.S.}(u|_N) \subset \bigcup_j \rho(U_j \cap N \underset{M}{\times} \sqrt{-1}S^*M)$ holds by Theorem 2.3.4. This implies $\mathrm{S.S.}(u|_N) \subset \rho(N \underset{M}{\times} \sqrt{-1}S^*M \cap \mathrm{S.S.}\, u)$. Therefore, since $\mathrm{supp}(u|_N) \subset N \cap \mathrm{supp}\, u$, $\widehat{\mathrm{S.S.}}(u|_N) \subset \hat{\rho}(N \underset{M}{\times} \sqrt{-1}S^*M \cap \widehat{\mathrm{S.S.}}\, u)$.

Definition 3.1.4. *Let X and Y be topological spaces, let $f: X \to Y$ be a continuous map, and let $\mathscr{F}$ be a sheaf over X. Then we denote $\{s \in \Gamma(X, \mathscr{F}) \mid f|_{\text{supp}\, s}$ is a proper map$\}$ by $\Gamma_{f-pr}(X, \mathscr{F})$. Let $f_!(\mathscr{F})$ be the sheaf over Y associated to the presheaf $\Gamma_{f-pr}(f^{-1}(U), \mathscr{F})$, where U is an open set in Y. The sheaf $R^k f_!(\mathscr{F})$ over Y denotes the sheaf associated to the presheaf $H^k(\Gamma_{f-pr}(f^{-1}(U), \mathscr{L}^\bullet))$, where $0 \to \mathscr{F} \to \mathscr{L}^\bullet$ is a flabby resolution of $\mathscr{F}$ and U is an open set in Y.*

Remark. $R^0 f_!(\mathscr{F}) = f_!(\mathscr{F})$ holds and $R^k f_!(\mathscr{F}) = R^k f_*(\mathscr{F})$ if f itself is a proper map.

Theorem 3.1.4 (Restriction of a Microfunction). *Let N be a submanifold of M; then there exists a sheaf homomorphism*

$$\rho_! \varpi^{-1} \mathscr{C}_M \to \mathscr{C}_N.$$

Proof. Let U be an open set in $\sqrt{-1}S^*N$. For W $(\subset N)$, a neighborhood of $\pi_N(U)$, we let

$$G_1(W, U) = \{s \in \mathscr{B}_M(W) \mid \text{S.S. } s \cap \sqrt{-1}S_N^*M = \varnothing\}$$

$$G_2(W, U) = \{s \in \mathscr{B}_W(W) \mid \text{S.S. } s \cap \rho^{-1}(U) = \varnothing, \text{S.S. } s \cap \sqrt{-1}S_N^*M = \varnothing\}$$

and let

$$G(W, U) = G_1(W, U)/G_2(W, U).$$

Next we will demonstrate that the sheaf associated to the presheaf $\varinjlim_{W \supset \pi_N(U)} G(W, U)$, for an open set U in $\sqrt{-1}S^*N$, is the sheaf $\rho_! \varpi^{-1} \mathscr{C}_M$. First we will show that a map from $G(W, U)$ to $\Gamma(U, \rho_! \varpi^{-1} \mathscr{C}_M)$ can be defined by the correspondence $s \in \mathscr{B}_M(W)$ to $\text{sp}(s)$. Since $\text{sp}(s) \in \mathscr{C}_M(\pi_M^{-1}(W))$, one may regard $\text{sp}(s) \in \Gamma(\rho^{-1}(U), \varpi^{-1}\mathscr{C}_M)$. Now one needs to show that $\rho|_{\text{supp sp}(s) \cap \rho^{-1}(U)}$ is a proper map. Consider the commutative diagrams (3.3.1) and (3.3.2) for $x \in W$.

$$\begin{array}{ccc} N \underset{M}{\times} \pi_M^{-1}(W) & \longrightarrow & (\sqrt{-1}S^*M)_x \\ \big\downarrow & & \big\downarrow \text{proper} \\ W & \longrightarrow & \{x\} \end{array} \tag{3.3.1}$$

$$\begin{array}{ccc} & \text{supp sp}(s) \cap (N \underset{M}{\times} \pi_M^{-1}(W) - \sqrt{-1}S_N^*M) & \\ \swarrow \rho & & \big\downarrow \pi_M \\ \pi_N^{-1}(W) & \searrow & \\ & & W \end{array} \tag{3.3.2}$$

Since the map: $N \underset{M}{\times} \pi_M^{-1}(W) \to W \cap N$ is proper and since $\operatorname{supp}\operatorname{sp}(s) \cap N \underset{M}{\times} \pi_M^{-1}(W)$ is closed in $N \underset{M}{\times} \pi_M^{-1}(W)$, the map:

$$\operatorname{supp}\operatorname{sp}(s) \cap N \underset{M}{\times} \pi_M^{-1}(W) \to W$$

is proper. Therefore $\operatorname{supp}\operatorname{sp}(s) \cap (N \underset{M}{\times} \pi_M^{-1}(W) - \sqrt{-1}S_N^*M) \to W$ is a proper map, since $\operatorname{supp}\operatorname{sp}(s) \cap \sqrt{-1}S_N^*M = \varnothing$. Consequently, one needs the following lemma to claim that ρ: $\operatorname{supp}\operatorname{sp}(s) \cap (N \underset{M}{\times} \pi_M^{-1}(W) - \sqrt{-1}S_N^*M) \to \pi_N^{-1}(W)$ is a proper map; see (3.3.2).

Lemma 1. *Let X, Y, and Z be topological spaces, and let $f: X \to Y$ and $g: Y \to Z$ be continuous maps. Assume that Y is a Hausdorff space. Then, if $g \circ f$ is a proper map, f is a proper map.*

Proof. In the commutative diagram

$$\begin{array}{ccccc}
X & \overset{\iota}{\hookrightarrow} & X \underset{Z}{\times} Y & \longrightarrow & X \\
\downarrow{\scriptstyle f} & & \downarrow{\scriptstyle \tilde{f}} & & \downarrow{\scriptstyle g \circ f} \\
Y & = & Z \underset{Z}{\times} Y & \longrightarrow & Z
\end{array}$$

ι is defined by $\iota(x) = (x, f(y))$, and $\tilde{f}$ is defined by $\tilde{f}(x, y) = ((g \circ f)(x), y) = (g(y), y)$. Then $\tilde{f}$ is a proper map, since the map $g \circ f$ is proper. Hence, it is sufficient to prove that $\iota(X)$ is closed in $X \underset{Z}{\times} Y$. Suppose $(x, y) \in X \underset{Z}{\times} Y$ does not belong to $\iota(X)$. Then $y \neq f(x)$ holds. Since Y is Hausdorff, for some open set U and V, where $U \cap V = \varnothing$, one has $f(x) \in U$ and $y \in V$. For such U and V, the neighborhood $f^{-1}(U) \underset{Z}{\times} V$ of (x, y) does not intersect with X; i.e. X is closed. Therefore, the map $f = \tilde{f}|_X$ is a proper map.

By this lemma, the restriction ρ on $\rho^{-1}(U) \cap \operatorname{supp}\operatorname{sp}(s)$ is also a proper map. Hence $\operatorname{sp}(s) \in \Gamma(U, \rho_! \varpi^{-1}\mathscr{C}_M)$. If S.S. $s \cap \rho^{-1}(U) = \varnothing$ for $s \in \mathscr{B}_M(W)$, then $\operatorname{sp}(s) = 0$ on $\rho^{-1}(U)$ holds. Therefore, the above correspondence is a well-defined map. Conversely, if $\operatorname{sp}(s) = 0$ on $\rho^{-1}(U)$, then $s = 0$ as an element of $G(W, U)$; i.e. the map is monomorphic. It remains to be shown that the map is epimorphic at each stalk. First notice that, for $p \in \sqrt{-1}S^*N$, one has $(\rho_! \varpi^{-1}\mathscr{C}_M)_p = \{s \in \Gamma(\rho^{-1}(p), \varpi^{-1}\mathscr{C}_M) \mid \operatorname{supp}\operatorname{sp}(s) \text{ is compact}\}$ as the corollary of Proposition 2.1.4. Let $x = \pi_N(p)$ in N. Then define a section $\tilde{s}$ of $\mathscr{C}_M$ on $F \underset{\text{def}}{=} \rho^{-1}(p) \cup (\sqrt{-1}S_N^*M)_x$ such that $\tilde{s}|_{\rho^{-1}(p)} = s$ and $\tilde{s}|_{\sqrt{-1}S_N^*M} = 0$; i.e. $\tilde{s} \in \Gamma(F, \mathscr{C}_M|_F)$. Since $\rho^{-1}(p)$ is closed in $(\sqrt{-1}S^*M)_x - (\sqrt{-1}S_N^*M)_x$, F is a closed set in $(\sqrt{-1}S^*M)_x = \pi_M^{-1}(x)$.

Lemma 2. *If F is a closed set in $\pi^{-1}(x)$, the restriction map*

$$\Gamma(\pi^{-1}(x), \mathscr{C}_M|_{\pi^{-1}(x)}) \to \Gamma(F, \mathscr{C}_M|_F)$$

is epimorphic.

By Lemma 2, and since the map $\mathscr{B}_x \to \Gamma(\pi^{-1}(x), \mathscr{C}_M|_{\pi^{-1}(x)})$ is an epimorphism, there exists $u \in \mathscr{B}_x$ such that u is mapped onto $\tilde{s} \in \Gamma(F, \mathscr{C}_M|_F)$. From the definition of $\tilde{s}$, $\text{sp}(u) = s$ on $\rho^{-1}(p)$ and S.S. $u \cap \sqrt{-1}s_N^*M = \varnothing$ hold. Hence the sheafication of the presheaf $\left\{U \mapsto \varinjlim_{W \supset \pi_N(U)} G(W, U)\right\}$ is $\rho_! \pi^{-1}\mathscr{C}_M$.

Then the correspondence $s \in \mathscr{B}_M(W)$, where S.S. $s \cap \sqrt{-1}S_N^*M = \varnothing$, to $\text{sp}(s|_N) \in \mathscr{C}_N$ gives a sheaf homomorphism $\rho_! \varpi^{-1}\mathscr{C}_M \to \mathscr{C}_N$.

Proof of Lemma 2. Since $\mathscr{B}_x \to \mathscr{C}_{M,p}$ is an epimorphism for $p \in F$, F has a covering $\{U_j\}$ such that for $s \in \Gamma(F, \mathscr{C}_M|_F)$ $\text{sp}(u_j)|_{U_j \cap F} = s$ for some $u_j \in \mathscr{B}_x$. One can let $F \subset \bigcup_{i=1}^{N} U_i$, since F is compact. We will prove by induction on N the case when $N = 1$ is trivial. Assume the assertion for the case of $N - 1$. Since $F - U_N \subset \bigcup_{i=1}^{N-1} U_i$, there exists $u \in \mathscr{B}_x$ such that $\text{sp}(u) = s$ in a neighborhood $\bar{V}$ of $F - U_N$. Then $F \subset U_N \cup V$; i.e. we need only prove for $N = 2$. If $N = 2$, then S.S. $(u_1 - u_2) \subset \complement\bar{U}_1 \cup \complement\bar{U}_2 \cup \complement F$ holds, where $\complement$ denotes the complement, so that $u_1 - u_2 = w_1 - w_2 + t$ where S.S. $w_1 \subset \complement\bar{U}_1$, S.S. $w_2 \subset \complement\bar{U}_2$, and S.S. $t \subset \complement F$. Then let $u \underset{\text{def}}{=} u_1 - w_1 = u_2 - w_2 + t$. One obtains

$$\text{sp}\, u|_{U_1 \cap F} = \text{sp}\, u_1|_{U_1 \cap F} = s|_{U_1 \cap F}$$

and

$$\text{sp}\, u|_{U_2 \cap F} = \text{sp}\, u_2|_{U_2 \cap F} = s|_{U_2 \cap F};$$

i.e. $\text{sp}(u)|_F = s$, completing the proof of Lemma 2. Therefore, we have proved Theorem 3.1.4.

We will consider the product of hyperfunctions on the same manifold.

Theorem 3.1.5 (Product of Hyperfunctions). *Let $u(x)$ and $v(x)$ be hyperfunctions on a real analytic manifold M such that* S.S. $u \cap (\text{S.S. } v)^a = \varnothing$. *Then the product $u(x)v(x) \in \mathscr{B}(M)$ exists with the following properties:* $\widehat{\text{S.S.}}\,(uv) \subset \{(x, \sqrt{-1}(\xi_1 + \xi_2)) | (x, \sqrt{-1}\xi_1) \in \widehat{\text{S.S.}}\, u$ and $(x, \sqrt{-1}\xi_2) \in \widehat{\text{S.S.}}\, v\}$. S.S. $(uv) \subset \{(x, \sqrt{-1}(\theta\xi_1 + (1-\theta)\xi_2)\infty) | (x, \sqrt{-1}\xi_1\infty) \in \text{S.S. } u, (x, \sqrt{-1}\xi_2\infty) \in \text{S.S. } v$ and $0 \leqq \theta \leqq 1\} \cup \text{S.S. } u \cup \text{S.S. } v$, *where $a{:}\sqrt{-1}S^*M \to \sqrt{-1}S^*M$ is the antipodal mapping* (i.e. $a(x, \sqrt{-1}\xi\infty) = (x, -\sqrt{-1}\xi\infty)$) *and where* $(\text{S.S. } v)^a$ *denotes the image of* S.S. v *under the antipodal mapping.*

Proof. First of all, the hyperfunction $u(x_1)v(x_2)$ on $M \times M$ can be defined from Theorem 3.1.1. Let the map $\iota: M \to M \times M$ be defined as $\iota(x) = (x, x) \in M \times M$ so that M may be regarded as a subset of $M \times M$. Then define $u(x)v(x) \underset{\text{def}}{=} u(x_1)v(x_2)|_M$. By Theorem 3.1.3, the restriction on M exists if $\text{S.S.}(u(x_1)v(x_2)) \cap \sqrt{-1}S_M^*(M \times M) = \emptyset$ holds, which will be proved next. Note that, by Theorem 3.1.1, one has

$$\widehat{\text{S.S.}}\ u(x_1)v(x_2) \subset \{((x_1, x_2), \sqrt{-1}(\xi_1, \xi_2)) \in \sqrt{-1}T^*(M \times M) | (x_1, \sqrt{-1}\xi_1) \in \widehat{\text{S.S.}}\ u \text{ and } (x_2, \sqrt{-1}\xi_2) \in \widehat{\text{S.S.}}\ v\}.$$

On the other hand, the map $\iota: M \to M \times M$ induces the map

$$\hat{\rho}: M \underset{M \times M}{\times} \sqrt{-1}T^*(M \times M) \to \sqrt{-1}T^*M$$

such that $\sqrt{-1}T_M^*(M \times M) = \text{Ker}\ \hat{\rho} = \{((x_1, x_2), \sqrt{-1}(\xi_1, \xi_2)) | x_1 = x_2$ and $\xi_1 + \xi_2 = 0\}$, where $\hat{\rho}((x, x), \sqrt{-1}(\xi_1, \xi_2)) = (x, \sqrt{-1}(\xi_1 + \xi_2))$. The hypothesis, $\text{S.S.}\ u \cap (\text{S.S.}\ v)^a = \emptyset$, implies $\widehat{\text{S.S.}}\ u(x_1)v(x_2) \cap \sqrt{-1}T_M^*(M \times M) = \emptyset$. Hence $\text{S.S.}\ u(x_1)v(x_2) \cap \sqrt{-1}S_N^*(M \times M) = \emptyset$ holds. Consequently, $u(x)v(x)$ is defined by Theorem 3.1.1.

Remark 1. Theorem 3.1.5 can be proved more directly as follows. Let u and v be hyperfunctions on M, and let $\text{S.S.}\ u \cap (\text{S.S.}\ v)^a = \emptyset$. Then one can find finitely many properly closed convex sets U_j and V_k so that

$$\text{S.S.}\ u \subset \bigcup_j U_j,$$

$$\text{S.S.}\ v \subset \bigcup_k V_k,$$

and

$$U_j \cap V_k^a = \emptyset.$$

Let D_j be a conoidal neighborhood of U_j°, and let D_k' be a conoidal neighborhood of V_k°. If $\tau(U_j^\circ \cap V_k^\circ) = M$, then $D_j \cap D_k'$ is a conoidal neighborhood of $U_j^\circ \cap V_k^\circ$ and $\varphi_j \cdot \varphi_k' \in \mathcal{O}(D_j \cap D_k')$, where $u = \sum_j b_{D_j}(\varphi_j)$ and $v = \sum_k b_{D_k'}(\varphi_k')$. Therefore, one can define $u \cdot v = \sum_{j,k} b_{D_j \cap D_k'}(\varphi_j \cdot \varphi_k')$ since $b_{D_j \cap D_k'}(\varphi_j \cdot \varphi_k')$ is a hyperfunction on M. If there exists $x \in M$ such that $U_j^\circ \cap U_k^\circ \cap (\sqrt{-1}SM)_x = \emptyset$, then one obtains $(U_j^\circ \cap V_k^\circ) \cap (\sqrt{-1}SM)_x = (U_j \cup V_k)^\circ \cap (\sqrt{-1}SM)_x = \emptyset$; i.e. there is ξ with the properties $(x, \sqrt{-1}\xi\infty) \in U_j \cup V_k$ and $(x, -\sqrt{-1}\xi\infty) \in U_j \cup V_k$. But since U_j and V_k are properly convex, $(x, \sqrt{-1}\xi\infty) \in U_j$ implies $(x, -\sqrt{-1}\xi\infty) \in V_k$; i.e. $(x, \sqrt{-1}\xi\infty) \in U_j \cap V_k^a = \emptyset$, a contradiction. Hence one concludes $\tau(U_j^\circ \cap V_k^\circ) = M$. Therefore, the product $u \cdot v$ exists.

Remark 2. Under the assumption in Theorem 3.1.5, we have the commutativity $u(x)v(x) = v(x)u(x)$, but the associativity $(u_1(x)u_2(x))u_3(x) = u_1(x)(u_2(x)u_3(x))$ does not hold in general even if both sides are well defined. For example, let $u_1(x) = 1/(x + i0)$, $u_2(x) = x$, and $u_3(x) = \delta(x)$; then $(u_1u_2)u_3 = \delta(x)$ and $u_1(u_2u_3) = 0$.

Let S.S. $u_j = Z_j$, $1 \leqq j \leqq 3$, and let $Z_1 \cap Z_2^a = \varnothing$. Then u_1u_2 is defined, and one can also define $Z_1 + Z_2 = \{(x, \sqrt{-1}(\xi_1 + \xi_2)\infty) | (x, \sqrt{-1}\xi_1\infty) \in Z_1$ and $(x, \sqrt{-1}\xi_2\infty) \in Z_2\}$. Furthermore, if $(Z_1 + Z_2)^a \cap Z_3 = \varnothing$ holds, then the associativity can hold. In order to see this, let $u_i = \sum_j b_{D_{ij}}(f_{ij})$, where $1 \leqq i \leqq 3$ as in Remark 1. Then $b_{D_{1i} \cap D_{2j} \cap D_{3k}}(f_{1i} \cdot f_{2j} \cdot f_{3k})$ determines a hyperfunction on M so that $(u_1u_2)u_3 = \sum_{i,j,k} b_{D_{1i} \cap D_{2j} \cap D_{3k}}(f_{1i} \cdot f_{2j} \cdot f_{3k}) = u_1(u_2 \cdot u_3)$ may hold.

Remark 3. Note that the product and the restriction defined above are stable under the action of differential operators. This is quite remarkable, as the definition of the product based upon conditions of "quantitative" regularity lacks this property. For example, let $u(x)$ and $v(x)$ be continuous functions on M. Then $u(x)v(x)$ as a continuous function (and therefore as a hyperfunction) can always be defined. However, since for a differential operator $P(D)$, $P(D)u(x)$ need not be a continuous function, the product of $P(D)u(x)$ and $v(x)$ may not exist.

Theorem 3.1.6 (Product of Microfunctions). *Let M be a real analytic manifold, and let Δ_M be the diagonal set of $M \times M$. Then define*

$$N = \Delta_M \underset{M \times M}{\times} (\sqrt{-1}S^*(M \times M)) - \Delta_M \underset{M \times M}{\times} (M \times \sqrt{-1}S^*M)$$
$$- \Delta_M \underset{M \times M}{\times} (\sqrt{-1}S_M^*M \times M) - \sqrt{-1}S_M^*(M \times M).$$

For a point $z = (x, x, \sqrt{-1}(\xi_1, \xi_2)\infty) \in N$, where $\xi_1 \neq 0$, $\xi_2 \neq 0$, and $\xi_1 + \xi_2 \neq 0$, one lets $p_1(z) = (x, \sqrt{-1}\xi_1\infty) \in \sqrt{-1}S^*M$, $p_2(z) = (x, \sqrt{-1}\xi_2\infty) \in \sqrt{-1}S^*M$, and $q(z) = (x, \sqrt{-1}(\xi_1 + \xi_2)\infty) \in \sqrt{-1}S^*M$. Then there exists a sheaf homomorphism

$$q_!(p_1^{-1}\mathscr{C}_M \times p_2^{-1}\mathscr{C}_M) \to \mathscr{C}_M.$$

Proof. Theorem 3.1.2 implies that $p_1^{-1}\mathscr{C}_M \times p_2^{-1}\mathscr{C}_M \to \mathscr{C}_{M \times M}|_N$ exists. On the other hand, one has a homomorphism $q_!(\mathscr{C}_{M \times M}|_N) \to \mathscr{C}_M$ by Theorem 3.1.4. Hence the composite of these homomorphisms gives us what we need.

Lastly, we will discuss substitutions for hyperfunctions and microfunctions. Let N and M be real analytic manifolds, and let $f: N \to M$ be a smooth real analytic map (i.e. $T_yN \overset{f^*}{\to} T_{f(y)}M$ is surjective for each $y \in N$). Then let X and Y be complexifications of M and N, respectively, and let $f: Y \to X$ be an extension of the smooth map $N \to M$. We shall prove that there can be defined a hyperfunction $u(f(x)) = f^*u \in \mathscr{B}_N(N)$ for a given $u \in \mathscr{B}_M(M)$.

Theorem 3.1.7

(1) (Substitution for Hyperfunction). *If a map $f: N \to M$ is smooth, there is induced a sheaf homomorphism*

$$f^*: f^{-1}\mathscr{B}_M \to \mathscr{B}_N.$$

Furthermore, define $\hat{\rho}: N \underset{M}{\times} (\sqrt{-1}T^*M) \to (\sqrt{-1}T^*N$ and $\varpi: N \underset{M}{\times} (\sqrt{-1}T^*M) \to \sqrt{-1}T^*M$ by $\hat{\rho}((y, \sqrt{-1}\xi)) = (y, \sqrt{-1}f^*(\xi))$ and $\varpi((y, \sqrt{-1}\xi)) = (f(y), \sqrt{-1}\xi)$. Then

$$\widehat{\text{S.S.}}(f^*u) = \hat{\rho}\hat{\varpi}^{-1}(\widehat{\text{S.S.}}\ u)$$

and

$$\text{S.S.}(f^*u) = \rho\varpi^{-1}(\text{S.S.}\ u)$$

hold (see Definition 3.1.3). *Note that, since f is smooth, ρ and $\hat{\rho}$ are monomorphic.*

(2) (Substitution for Microfunction). *There exists a homomorphism*

$$f^*: \varpi^{-1}\mathscr{C}_M \to \mathscr{H}^0_{N \underset{M}{\times} \sqrt{-1}S^*M}(\mathscr{C}_N).$$

Lemma. *Let $D \subset X - M$ be a conoidal neighborhood of $U \subset \sqrt{-1}SM$. Then, if f is smooth, $f^{-1}(D)$ is a conoidal neighborhood of $(f_*)^{-1}U = \{y + \sqrt{-1}v0 \in \sqrt{-1}SN \,|\, f(y) + \sqrt{-1}f_*v0 \in U\}$.*

Proof of Lemma. The question being local and f being smooth, one may assume $M = \mathbf{R}^n$ and $N = \mathbf{R}^n \times \mathbf{R}^l$. Furthermore, one may let $(x, y) + \sqrt{-1}v0 = 0 + \sqrt{-1}(\xi, \eta)0$. When D is a conoidal neighborhood of $0 + \sqrt{-1}\xi 0$, by definition $x + \sqrt{-1}t\xi' \in D$ for $|x| \ll 1$, $0 < t \ll 1$ and $|\xi' - \xi| \ll 1$. On the other hand, if $(x, y) + \sqrt{-1}t(\xi', \eta') \in \tilde{D}$ holds for some (ξ, η) where $|x| \ll 1$, $|y| \ll 1$, $|\xi' - \xi| \ll 1$, and $|\eta' - \eta| \ll 1$, then $\tilde{D}$ is a conoidal neighborhood of $0 + \sqrt{-1}(\xi, \eta)0$. So if one lets $\tilde{D} = f^{-1}(D)$, the assertion follows from the choice of the coordinate system.

Proof of Theorem 3.1.7. For $u \in \mathscr{B}(M)$, choose finitely many properly closed convex sets U_j so that S.S. $u \subset \bigcup_j U_j$. Let D_j be a conoidal neighborhood of $U_j^\circ \subset \sqrt{-1}SM$, and let $u = \sum_j b_{D_j}(\varphi_j)$. Then $b_{f^{-1}(D_j)}(\varphi_j \circ f)$ is a hyperfunction on N since $f^{-1}(D_j)$ is a conoidal neighborhood of

$((f_*)^{-1}U_j)^\circ$, where $f_*:TN \to N \underset{M}{\times} TM$ is the transposed mapping of $\rho:N \underset{M}{\times} T^*M \to T^*N$. Then define $f^*u = \sum_j b_{f^{-1}(D_j)}(\varphi_j \circ f)$.

Let us study the singularity spectrum of f^*u. We need to show $((f_*)^{-1}U_j)^\circ \subset \rho\varpi^{-1}U_j^\circ$ to claim $\text{S.S.}(f^*u) \subset \rho\varpi^{-1}(\text{S.S. }u)$. Let $W = \text{Ker}(f_*:TN \to N \underset{M}{\times} TM)$, and let $(x, \sqrt{-1}\xi\infty) \in ((f_*)^{-1}U_j)^\circ \subset \sqrt{-1}S^*N$. Since $\langle \xi, v\rangle > 0$ holds for an arbitrary v such that $f(x) + \sqrt{-1}f_*(v)0 \in U_j$, one has $\langle \xi, v + aw\rangle = \langle \xi, v\rangle + a\langle \xi, w\rangle > 0$ for any $w \in W$ and $a \in \mathbf{R}^\times$. Hence $\langle \xi, w\rangle = 0$; i.e. $\xi \in W^\perp = \rho(N \underset{M}{\times} T^*M)$. This implies $\xi = \rho(\xi')$ for some $\xi' \in T^*_{f(x)}M$. Therefore one obtains $\langle \xi, v\rangle = \langle \rho(\xi'), v\rangle = \langle \xi', f_*(v)\rangle > 0$ for any v with the condition $f(x) + \sqrt{-1}f_*(v)0 \in U_j$. Since f_* is surjective, one gets $\xi' \in \varpi^{-1}U_j^\circ$, i.e. $(x\sqrt{-1}\xi\infty) \in \rho\varpi^{-1}U_j^\circ$.

$\text{S.S.}(f^*u) \supset \rho\varpi^{-1}(\text{S.S. }u)$ will be shown next. One may assume that $N = M \times L$ and that f is a projection on M, owing to the local nature of the assertion. First note that $f^*u|_{M\times\{x_0\}} = u$ for $x_0 \in L$. This can be seen if one represents u as boundary values of a holomorphic function. Then, by Theorem 3.1.3, for $q_{x_0}:(M \times \{x_0\} \underset{M\times L}{\times} \sqrt{-1}T^*(M \times L) \to \sqrt{-1}T^*(M \times \{x_0\})$, one has $\widehat{\text{S.S.}}\, u \subset q_{x_0}(\widehat{\text{S.S.}}\, f^*u)$. By varying x_0 in L, one can consider $L \times \widehat{\text{S.S}}\, u \subset \widehat{\text{S.S.}}(f^*u)$. On the other hand, one may derive $\hat{\rho}\hat{\varpi}^{-1}\, \widehat{\text{S.S.}}\, u = L \times \widehat{\text{S.S.}}\, u$ from the definition. Consequently $\hat{\rho}\hat{\varpi}^{-1}\widehat{\text{S.S.}}\, u \subset \widehat{\text{S.S.}}(f^*u)$, completing (1) of this theorem.

As for (2), one has the epimorphisms

$$\mathscr{B}_{M,f(y)} \xrightarrow{\text{sp}} \mathscr{C}_{M,(f(y),\sqrt{-1}\xi\infty)}$$

and

$$\mathscr{B}_{N,y} \xrightarrow{\text{sp}} \mathscr{H}^0_{\sqrt{-1}S^*M \underset{M}{\times} N}(\mathscr{C}_N)_{(y,\sqrt{-1}f^*(\xi)\infty)},$$

and the substitution $f^*:\mathscr{B}_{M,f(y)} \to \mathscr{B}_{N,y}$. Hence, one has the following homomorphism:

$$\mathscr{C}_{M,(f(y),\sqrt{-1}\xi\infty)} \to \mathscr{H}^0_{\sqrt{-1}S^*M \underset{M}{\times} N}(\mathscr{C}_N)_{(y,\sqrt{-1}f^*(\xi)\infty)}.$$

One notices that the above homomorphism can be defined independently from the choice of the inverse map of sp.

Remark. Suppose that a real analytic map $f:N \to M$ is smooth and that $\mathscr{X}$ is a vector field in N. If $\mathscr{X}(\varphi \circ f) = 0$ holds for any function φ on M, i.e. $\mathscr{X}$ is tangent to each fibre of f, then we have $\mathscr{X}(f^*u) = 0$ for $u \in \mathscr{B}(M)$. Hence, for $u = \sum_j b(\varphi_j)$ we have

$$\mathscr{X}(f^*u) = \mathscr{X}\left(\sum_j b(\varphi_j \circ f)\right) = \sum_j b\left(\mathscr{X}(\varphi_j \circ f)\right) = 0.$$

We will give some examples of the various operations discussed in this section.

Example 3.1.1. Let $\delta(x)$ denote the δ-function of one variable, and let $f(x_1, \ldots, x_n)$ be a real-valued real analytic function defined in $U \subset \mathbf{R}^n$, satisfying the following condition.

$$\text{If } f(x) = 0, \text{ then } d_x f(x) \neq 0. \tag{3.1.1}$$

Then $f^*\delta(t)$ is well defined as an element of $\mathscr{B}_{\mathbf{R}^n}(U)$. Furthermore, we have the following inclusion:

$$\text{S.S. } f^*\delta(t) \subset \{(x, \sqrt{-1}\xi\infty) \in \sqrt{-1}S^*U \,|\, f(x) = 0 \text{ and } \xi = c \operatorname{grad}_x f(x), c \in \mathbf{R} - \{0\}\}. \tag{3.1.2}$$

In order to see (3.1.2), notice that $\delta(t) = 0$ for $t \neq 0$ and that the map $f: f^{-1}(V) \to V$ is smooth for a neighborhood V of the origin. Then, by Theorem 3.1.7, $f^*\delta(t)$ is well defined and (3.1.2) holds. We often denote $f^*\delta(t)$ by $\delta(f(x))$.

Example 3.1.2. Let $f(x)$ be as in Example 3.1.1. Then $f^*(1/(t + \sqrt{-1}0))$ is well defined, as above, and is sometimes denoted by $1/(f(x) + \sqrt{-1}0)$. As before, we have:

$$\text{S.S.}\left(\frac{1}{f(x) + \sqrt{-1}0}\right) \subset \{(x, \sqrt{-1}\xi\infty) \in \sqrt{-1}S^*U \,|\, f(x) = 0 \text{ and } \xi = c \operatorname{grad}_x f(x), c > 0\}. \tag{3.1.3}$$

Let W be a sufficiently small Stein neighborhood of U, and let $D = W - \{z \in W \,|\, \operatorname{Im} f(z) \leqq 0\}$. The boundary value $b_D(1/f(z))$ of $1/f(z)$ from the domain D is the above $1/(f(x) + \sqrt{-1}0)$ from the proof of Theorem 3.1.7; see also Theorem 2.3.4.

Example 3.1.3. Let U be an open set in $\mathbf{R}^n$, and let $H = \{x \in U \,|\, h(x) = 0\}$ be a non-singular hypersurface; i.e. h is a real-valued real analytic function such that $dh \neq 0$ on H. Let $f(x)$ be as in Example 3.1.1, with an additional condition:

$$\operatorname{grad}_x h \not\parallel \operatorname{grad}_x f \text{ on } H \cap \{f = 0\}.$$

Then the restriction of $\delta(f(x))$ to H, i.e. $\delta(f(x))|_H$, is well defined as an element of $\mathscr{B}_H(H)$ so that $(f|_H)^*\delta(t) \equiv \delta(f(x)|_H)$. Since the sequence

$$0 \to T_H^*U \to T^*U \underset{U}{\times} H \to T^*H \to 0 \tag{3.1.5}$$

is exact, a point on T^*H can be described as $(x, \xi) \in U \times \mathbf{R}^n \cong T^*U$

modulo $\{(x, \xi) | h(x) = 0 \text{ and } \xi = c \operatorname{grad}_x h(x), c \in \mathbf{R}\}$. Then the following (3.1.6) holds:

$$\text{S.S.}(\delta(f(x))|_H) \subset A/B, \tag{3.1.6}$$

where

$$A = \{(x, \sqrt{-1}\xi) \in U \times \sqrt{-1}\mathbf{R}^n | f(x) = h(x) = 0 \text{ and}$$
$$\xi = c_1 \operatorname{grad}_x f(x) + c_2 \operatorname{grad}_x h(x), (c_1, c_2) \in \mathbf{R}^2, c_1 \neq 0\}$$
$$B = \{(x, \sqrt{-1}\xi) \in U \times \sqrt{-1}\mathbf{R}^n | h(x) = 0 \text{ and } \xi = c \operatorname{grad}_x h(x), c \in \mathbf{R}\}.$$

For example, $\delta(x_1 - x_2)|_{\{x_2=0\}} = \delta(x_1)$.

Note. In Example 3.1.3, one may let H be a submanifold of $U \subset \mathbf{R}^n$, as well.

Example 3.1.4. Let $f_1(x)$ and $f_2(x)$ be real-valued real analytic functions satisfying the condition (3.1.1). Assume further that the following condition is satisfied:

On $\{f_1(x) = f_2(x) = 0\}$ we have $\operatorname{grad}_x f_1(x)$ and $\operatorname{grad}_x f_2(x)$ being linearly independent. (3.1.7)

Then $\delta(f_1(x))\delta(f_2(x))$ is well defined, and we have the inclusion

$$\text{S.S. } \delta(f_1)\delta(f_2)$$
$$\subset \{(x, \sqrt{-1}\xi\infty) \in \sqrt{-1}S^*U \,|\, f_1(x) = f_2(x) = 0 \text{ and}$$
$$\xi = c_1 \operatorname{grad}_x f_1 + c_2 \operatorname{grad}_x f_2, (c_1, c_2) \in \mathbf{R}^2 - \{0\}\}. \tag{3.1.8}$$

Example 3.1.5. Let us assume the condition (3.1.9), which is weaker than (3.1.7), on f_1 and f_2.

For arbitrary $\alpha_1 \geqq 0$ and $\alpha_2 \geqq 0$ such that $\alpha_1 + \alpha_2 \neq 0$,
$\alpha_1 \operatorname{grad}_x f_1(x) + \alpha_2 \operatorname{grad}_x f_2(x) \neq 0$ on $\{f_1(x) = f_2(x) = 0\}$. (3.1.9)

Then $1/(f_1(x) + \sqrt{-1}0) \cdot 1/(f_2(x) + \sqrt{-1}0)$ is well defined, and

$$\widehat{\text{S.S.}} \frac{1}{f_1(x) + \sqrt{-1}0} \cdot \frac{1}{f_2(x) + \sqrt{-1}0}$$
$$\subset \{(x, \sqrt{-1}\xi) \in \sqrt{-1}T^*U \,|\, \alpha_1 f_1 = 0, \alpha_2 f_2 = 0 \text{ and}$$
$$\xi = \alpha_1 \operatorname{grad}_x f_1(x) + \alpha_2 \operatorname{grad}_x f_2(x), (\alpha_1, \alpha_2) \in \mathbf{R}^2, \alpha_1, \alpha_2 \geqq 0\}. \tag{3.1.10}$$

For example, $1/(x_1 + \sqrt{-1}0) \cdot 1/(x_1 - x_2^2 + \sqrt{-1}0)$ and $1/(x_1 + \sqrt{-1}0))^2$ are well defined. On the contrary, $\delta(x_1)\delta(x_1 - x_2^2)$ and $\delta(x_1)^2$ are not well defined in the sense of Theorem 3.1.5.

§2. Integration

We shall describe the integrations of a hyperfunction and a microfunction. The next proposition states that "indefinite integrals" exist.

Proposition 3.2.1. *If* $M = \mathbf{R}^n$, *then one has the following* (a), (b), *and* (c):

(a) $D_1 \underset{\text{def}}{=} \partial/\partial x_1 : \mathscr{B}_M \to \mathscr{B}_M$ *and* $D_1 : \mathscr{C}_M \to \mathscr{C}_M$ *are epimorphisms.*

(b) *Let* U_1 *be a connected open set in* $\mathbf{R}$, *let* U_2 *be an open set in* $\mathbf{R}^{n-1}$, *and let* $u \in \mathscr{B}_M(U)$ *where* $U = U_1 \times U_2$. *If* $D_1 u = 0$, *there exists a unique* $v \in \mathscr{B}_{\mathbf{R}^{n-1}}(U_2)$ *such that* $u(x) = v(x')$, *where* $x' = (x_2, \ldots, x_n)$.

(c) $D_1 : \mathscr{C}_M \to \mathscr{C}_M$ *is an isomorphism over a neighborhood of* $(x, \sqrt{-1}\xi\infty)$, *where* $\xi = (\xi_1, \ldots, \xi_n)$, $\xi_1 \neq 0$. *Let* U_1 *be an open interval in* $\mathbf{R}$, *and let* Ω *be an open set in* $\sqrt{-1}S^*\mathbf{R}^{n-1}$. *Then, in a neighborhood of* $(x, \sqrt{-1}\xi\infty)$, $\xi_1 = 0$, *if* $D_1 u = 0$ *for* $u \in \mathscr{C}_{\mathbf{R}^n}(U_1 \times \Omega)$, *there exists* $v \in \mathscr{C}_{\mathbf{R}^{n-1}}(\Omega)$ *such that* $u = v(x')$.

Recall the following well-known lemma, with the interesting proof from Suzuki [1] applied to the convex case.

Lemma. *Define* $F : \mathbf{C}^n \to \mathbf{C}^{n-1}$ by $F(z_1, \ldots, z_n) = (z_2, \ldots, z_n)$. *Let* Ω *be a convex set in* $\mathbf{C}^n$, *and let* $\Omega_1 = F(\Omega)$. *Then,*

(1) $D_1 \mathscr{O}_{\mathbf{C}^n}(\Omega) = \mathscr{O}_{\mathbf{C}^n}(\Omega)$ *and*

(2) *if* $D_1 u = 0$ *for* $u \in \mathscr{O}_{\mathbf{C}^n}(\Omega)$, *there exists* $v \in \mathscr{O}_{\mathbf{C}^{n-1}}(\Omega_1)$ *such that* $u(z) = v(z')$, *where* $z' = F(z)$.

Proof. We will give a proof for (1), since (2) is clearly true. For $f \in \mathscr{O}_{\mathbf{C}^n}(\Omega)$ we will show that $D_1 u = f$ for some $u \in \mathscr{O}_{\mathbf{C}^n}(\Omega)$. Let open convex sets $\{U_j\}$ be a covering of Ω_1, where $\Omega \supset \{z_1 = \alpha_j\} \times U_j$ for $\alpha_j \in \mathbf{C}$. For $f(z) \in \mathscr{O}_{\mathbf{C}^n}(\Omega)$ define $u_j(z) = \int_{\alpha_j}^{z_1} f(z)\, dz_1$. Then $u_j(z)$ is a holomorphic function defined in $\tilde{U}_j \underset{\text{def}}{=} \Omega \cap F^{-1}(U_j)$. Since $D_1 u_j = f$ on $\tilde{U}_j$, one obtains $D_1(u_j - u_k) = 0$ on $\tilde{U}_j \cap \tilde{U}_k$. Then (2) implies $u_j - u_k \underset{\text{def}}{=} u_{jk} \in \mathscr{O}_{\mathbf{C}^{n-1}}(U_j \cap U_k)$, and $\{u_{jk}\}$ is plainly a cocycle. Since Ω_1 is convex, it is a Stein manifold. Hence $H^1(\Omega_1, \mathscr{O}) = 0$ holds. Further, U_j is also a Stein manifold, since it is convex. Therefore $\{U_j\}$ is a Leray covering. The Leray theorem (Theorem 1.2.1) implies that there exists $u'_j \in \mathscr{O}_{\mathbf{C}^{n-1}}(U_j)$ such that $u_{jk} = u'_j - u'_k$. Let $\tilde{u}_j(z) \underset{\text{def}}{=} u_j(z) - u'_j(F(z)) \in \mathscr{O}_{\mathbf{C}^n}(\tilde{U}_j)$. Then one has $D_1 \tilde{u}_j = f$, and $\tilde{u}_j = \tilde{u}_k$ holds in $\tilde{U}_j \cap \tilde{U}_k$. Therefore, one can let $u \in \mathscr{O}_{\mathbf{C}^n}(\Omega)$ so that $u|_{\tilde{U}_j} = \tilde{u}_j$ holds. Consequently, u is a solution for $D_1 u = f$.

Proof of Proposition 3.2.1. We will prove (a) as follows. Let U be an open convex set in $\mathbf{R}^n$, and let $u \in \mathscr{B}_M(U)$. Then u can be expressed as $u =$

$\sum_j b_{U\times\sqrt{-1}G_j}(\varphi_j)$, where G_j, $1 \leqq j \leqq N$, are properly chosen open convex cones in $\mathbf{R}^n$ and $\varphi_j \in \mathcal{O}_{\mathbf{C}^n}(U \times \sqrt{-1}G_j)$. The above lemma implies that there exists $\psi_j \in \mathcal{O}_{\mathbf{C}^n}(U \times \sqrt{-1}G_j)$ such that $D_1\psi_j = \varphi_j$. If one lets $v = \sum_j b(\psi_j)$, then $v \in \mathscr{B}_M(U)$ and $D_1 v = u$. Therefore, $D_1 : \mathscr{B}_M \to \mathscr{B}_M$ is epimorphic. It can be shown similarly that $D_1 : \mathscr{C}_M \to \mathscr{C}_M$ is an epimorphism.

Next we will prove (b). Let us recall a fact from §2 of Chapter I for this case of (b). Let $\xi_j \in \mathbf{R}^n$ for $0 \leqq j \leqq n$ such that the convex hull of ξ_j, $j = 0, 1, \ldots, n$, is a neighborhood of the origin; let $G_j = \{y \in \mathbf{R}^n | \langle y, \xi_k\rangle > 0 \text{ for } k \neq j\}$, and let $G_{jk} = \{y \in \mathbf{R}^n | \langle y, \xi_1\rangle > 0 \text{ for } l \neq j, k\}$. Then one has

$$\mathscr{B}(U) = \bigoplus_j \mathcal{O}_{\mathbf{C}^n}(U \times \sqrt{-1}G_j) \Big/ \bigoplus_{j,k}{}' \mathcal{O}_{\mathbf{C}^n}(U \times \sqrt{-1}G_{jk}),$$

where $\bigoplus'$ denotes the alternative sum. Hence $u \in \mathscr{B}(U)$ can be expressed as $u = \sum_j b(\varphi_j)$, $\varphi_j \in \mathcal{O}_{\mathbf{C}^n}(U \times \sqrt{-1}G_j)$; and if $\sum_j b(\varphi_j) = 0$, then $\varphi_j = \sum_{j,k} \varphi_{jk}$ for $\varphi_{jk} \in \mathcal{O}_{\mathbf{C}^n}(U \times \sqrt{-1}G_{jk})$ such that $\varphi_{jk} = -\varphi_{kj}$.

We will apply the above result in the following manner. Let U_2 be an open convex set in $\mathbf{R}^{n-1}$, and let $U \underset{\text{def}}{=} (a, b) \times U_2 \subset \mathbf{R}^1 \times \mathbf{R}^{n-1}$. For $u \in \mathscr{B}(U)$, one lets $u = \sum_j b(\varphi_j)$, $\varphi_j \in \mathcal{O}_{\mathbf{C}^n}(U \times \sqrt{-1}G_j)$. If $D_1 u = \sum_j b(D_1\varphi_j) = 0$, there exists $\psi_{jk} \in \mathcal{O}_{\mathbf{C}^n}(U \times \sqrt{-1}G_{jk})$, with the property $\psi_{jk} = -\psi_{kj}$, such that $D_1\varphi_j = \sum \psi_{jk}$. Since $U \times \sqrt{-1}G_{jk}$ is convex, $\psi_{jk} = D_1\varphi_{jk}$ for $\varphi_{jk} \in \mathcal{O}_{\mathbf{C}^n}(U \times \sqrt{-1}G_{jk})$ such that $\varphi_{jk} = -\varphi_{kj}$. (For $j = k$, let $\varphi_{jk} = 0$; for $j > k$, one can choose a solution φ_{jk} for $\psi_{jk} = D_1\varphi_{jk}$; and for $j < k$, one can define $\{\varphi_{jk}\}$ such that $\varphi_{jk} = -\varphi_{kj}$.) Then $D_1\varphi_j = \sum_k D_1\varphi_{jk}$ holds. Hence one obtains $D_1\left(\varphi_j - \sum_j \varphi_{jk}\right) = 0$. If one lets $\psi_j = \varphi_j - \sum_j \varphi_{jk}$, then ψ_j does not depend upon z_1. Hence $v = \sum_j b(\psi_j(z'))$, $z' = (z_2, \ldots, z_n)$, is a hyperfunction on $\mathbf{R}^{n-1}$. Since one has $\sum_{j,k} b(\varphi_{jk}) = 0$ on $\mathbf{R}^n$, then $v = \sum_j b(\psi_j) = u$.

We will give a proof of (c) last. In order to prove the first half of the assertion, it is sufficient to show that $(0, \sqrt{-1}\,dx_1\infty) \notin \text{S.S. } u$, provided that $(0, \sqrt{-1}\,dx_1\infty) \notin \text{S.S.}(D_1 u)$, i.e. $\text{sp}(D_1 u) = 0$ at $(0, \sqrt{-1}\,dx_1\infty)$. If $(0, \sqrt{-1}\,dx_1\infty) \notin \text{S.S.}(D_1 u)$ is true, then $D_1 u = \sum_j b(\psi_j)$, $\psi_j \in \mathcal{O}(U_\epsilon \times \sqrt{-1}\Gamma_j)$, where $U_\epsilon = \{x \,|\, |x| < \epsilon\}$ and Γ_j is an open convex cone such that $\Gamma_j \subset \{y \in \mathbf{R}^n | y_1 < 0\}$ (see Theorem 2.3.4). If $\Gamma_{j,\epsilon} = \Gamma_j \cap \{y \,|\, |y| < \epsilon\}$, since $U_\epsilon \times \sqrt{-1}\Gamma_{j,\epsilon}$ is convex, then $D_1\varphi_j = \psi_j$ holds for $\varphi_j \in \mathcal{O}_{\mathbf{C}^n}(U_\epsilon \times \sqrt{-1}\Gamma_{j,\epsilon})$ by the above lemma. Then let $v = \sum_j b(\varphi_j)$. One obtains $D_1 u = D_1 v$. Therefore, $D_1(u - v) = 0$. Now (b) implies $u - v = w(x')$, $x' = (x_2, \ldots, x_n)$; i.e. $(0, \sqrt{-1}\, dx_1 \infty) \notin \text{S.S.}(u - v)$. On the other hand, $(0, \sqrt{-1}\, dx_1 \infty) \notin \text{S.S. } v$ holds from the definition of v and from Theorem 2.3.5. Consequently, one obtains $(0, \sqrt{-1}\, dx_1 \infty) \notin \text{S.S. } u$. We will prove the latter half of (c). Let $u = \text{sp}(f)$ for $f \in \mathscr{B}(\mathbf{R}^n)$. Suppose $(x, \sqrt{-1}\xi\infty) \notin \text{S.S.}(D_1 f)$; then, by Theorem 2.3.5, one can express $D_1 f = \sum_j b(\psi_j)$, where $\psi_j \in \mathcal{O}_{\mathbf{C}^n}(U_\epsilon \times \sqrt{-1}\Gamma_{j,\epsilon})$ and $\Gamma_j \subset \{y | \langle y, \xi \rangle < 0\}$. Define $v = \sum_j b(\varphi_j)$ for $\varphi_j \in \mathcal{O}_{\mathbf{C}^n}(U_\epsilon \times \sqrt{-1}\Gamma_{j,\epsilon})$ such that $D_1 \varphi_j = \psi_j$. One has $(x, \sqrt{-1}\xi\infty) \notin \text{S.S. } v$. Since $D_1 f = D_1 v$, then $D_1 w = 0$ for $w = f - v$. Since in a neighborhood of $(x, \sqrt{-1}\xi\infty)$, $\xi_1 = 0$, $\text{sp}(v) = 0$ holds, one has $\text{sp}(u) = \text{sp}(w)$.

Let N be a real analytic manifold, and let $u(t, x)$ be a hyperfunction on $\mathbf{R} \times N$ such that $\text{supp } u(t, x) \subset \{(t, x) \in \mathbf{R} \times N | a < t < b\}$. Then there exists $v(t, x)$ such that $D_t v(t, x) = u(t, x)$ by Proposition 3.2.1. (In the case when N is a manifold, Grauert's theorem can be applied to prove the assertions in Proposition 3.2.1.) Then define the integration $\int u(t, x)\, dt = v(b, x) - v(a, x)$. Since $D_t v(t, x) = u(t, x) = 0$ in a neighborhood of a and b, we have $\text{S.S. } v \cap \sqrt{-1}S^*_{\{b\} \times N}(\mathbf{R} \times N) = \varnothing$ by the first half of the statement in (c) of Proposition 3.2.1. Notice that $v(t, x)|_{t=b} = v(b, x)$ is well defined by Theorem 3.1.3. Furthermore, we have $v(b', x) = v(b, x)$ for $b' > b$ and $v(a', x) = v(a, x)$ for $a' < a$ from (b) of Proposition 3.2.1. Therefore, $\int u(t, x)\, dt$ does not depend upon the choice of (a, b). Next, it can be seen also from the fact that $\int u(t, x)\, dt$ can be defined as $\tilde{v}(b, x) - \tilde{v}(a, x)$, where $\tilde{v}(t, x)$ is a hyperfunction satisfying $D_t \tilde{v}(t, x) = u(t, x)$ by Proposition 3.2.1(b). Hence $\int u(t, x)\, dt$ can be defined as in the above.

If a hyperfunction $u(t, x) = u(t_1, \ldots, t_m, x)$ on $\mathbf{R}^m \times \mathbf{N}$ satisfies $\text{supp}(u(t, x)) \subset K \times N$ for a relatively compact set K in $\mathbf{R}^m$, then by repeated use of the above argument one can define $\int u(t, x)\, dt = \int (\ldots \int (\ldots (\int u(t, x)\, dt_1)\, dt_2 \ldots)\, dt_m$. Let $u(t)$ be a hyperfunction of one variable, let $\text{supp } u(t) = K \subset (a, b)$, and let D be a complex neighborhood of K. One may assume that D contains $[a, b]$, and we denote the restriction of $\tau \in \mathbf{C}$ to $\mathbf{R}$ by $t = \tau|_{\mathbf{R}}$. Then $u(t)$ can be expressed as the boundary values of a holomorphic function $\varphi(\tau)$ in $D - K$. That is, $u(t) = \varphi_+(\tau) - \varphi_-(\tau)$ where $\varphi_\pm(\tau) = \lim_{\text{Im}\,\tau \to \pm 0} \varphi(\tau)$. Let $\psi_+(\tau) = \int_a^\tau \varphi(\tau)\, d\tau$ for $\text{Im}\,\tau > 0$; let $\psi_-(\tau) = \int_a^\tau \varphi(\tau)\, d\tau$ for $\text{Im}\,\tau < 0$; and let $v(t) = \psi_+(t + i0) - \psi_-(t - i0)$. Then $D_t v(t) = \varphi(t + i0) - \varphi(t - i0) = u(t)$ holds. Hence we obtain

$$\int u(t)\, dt = v(b) - v(a) = \underset{\gamma+}{\int_a^b} \varphi(\tau)\, d\tau - \underset{\gamma-}{\int_a^b} \varphi(\tau)\, d\tau = \oint_\gamma \varphi(\tau)\, d\tau,$$

where γ is a path around K as in Figure 3.2.1.

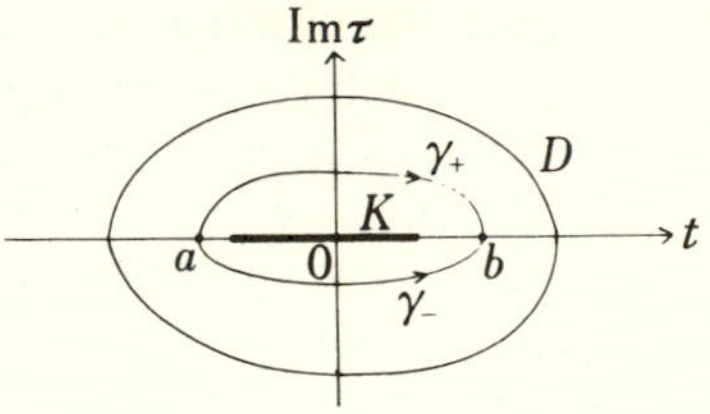

Figure 3.2.1

Example 3.2.1. $\int_{\mathbf{R}} \delta(t)\, dt = 1.$

Proof. Since

$$\delta(t) = \frac{-1}{2\pi i}\left(\frac{1}{t+i0} - \frac{1}{t-i0}\right),$$

the integration around the unit circle of $\varphi(\tau) = -1/2\pi i\tau$ gives

$$\oint_\gamma \frac{d\tau}{\tau} = \int_\pi^0 \frac{ie^{i\theta}}{e^{i\theta}}\, d\theta + \int_0^{-\pi} \frac{ie^{i\theta}}{e^{i\theta}}\, d\theta = [i\theta]_\pi^0 + [i\theta]_0^{-\pi} = -2\pi i.$$

Hence $\oint_\gamma \varphi(\tau)\, d\tau = 1$. One may also prove this more directly, as follows. The derivative of $Y(x) = (-2\pi i + \log(x+i0) - \log(x-i0))/-2\pi i$ with respect to x gives $(d/dx)/Y(x) = -(1/2\pi i)\{1/(x+i0) - 1/(x-i0)\} = \delta(x)$. Therefore

$$\int \delta(x)\, dx = Y(1) - Y(-1) = 1.$$

Note also that x_+ is a primitive function of $Y(x)$.

Example 3.2.2. $\int \delta^{(n)}(x)\, dx = 0$ for $n \geqq 1$, where $\delta^{(n)}(x) = (d^n/dx^n)\, \delta(x)$.

Proof. For $x \neq 0$ one has $\delta^{(n-1)}(x) = 0$ for $n \geqq 1$ and $(d/dx)\delta^{(n-1)}(x) = \delta^{(n)}(x)$.

Example 3.2.3. $\int_{-\infty}^1 x_+^\lambda\, dx = 1/(\lambda+1)$ for $\lambda \neq -1, -2, \ldots$.

Proof. Recall that x_+^λ is well defined as a hyperfunction for $\lambda \neq -1, -2, \ldots$. One also has

$$\frac{d}{dx}\left(\frac{1}{\lambda+1}\, x_+^{\lambda+1}\right) = x_+^\lambda.$$

The proof follows from this (see Example 2.4.2).

When a hyperfunction $u(t, x)$ on $\mathbf{R} \times N$ satisfies $\operatorname{supp} u(t, x) \subset \{(t, x) \in \mathbf{R} \times N \mid a < t < b\}$, we define, as before, the definite integral $\int u(t, x)\, dt$

as $v(b, x) - v(a, x)$ via a primitive function $v(t, x)$; i.e. $(\partial/\partial t)v(t, x) = u(t, v)$. For the case $t = (t_1, \ldots, t_m)$, we defined it by repeating the above m times. Thus we obtain a hyperfunction on N through the integration of a hyperfunction on $\mathbf{R} \times N$, satisfying a certain condition on the support. (If $N = \varnothing$, then we get a complex number.) One is naturally led to study the singularity spectrum of $\int u(t, x)\,dt$ when one is provided with the singularity spectrum of $u(t, x)$. We will give an extremely concise answer to this question. The next proposition can be regarded as a modern version of the classical stationary-phase method (see, for example, Lax [1]). It is a geometric interpretation of the smoothing effect of an integral operator. The treatment in Hörmander [3] is closer to the classical one.

Proposition 3.2.2. *Let $u(t, x)$ be a hyperfunction on $\mathbf{R} \times N$ such that* $\operatorname{supp} u(t, x) \subset \{(t, x) \in \mathbf{R} \times N \mid a < t < b\}$. *If, for* $(x_0, \sqrt{-1}\langle \xi_0, dx\rangle\infty) \in \sqrt{-1}S^*N$,

$$(t, v_0, \sqrt{-1}(0 \cdot dt + \langle \xi_0, dx\rangle)\infty) \notin \text{S.S. } u(t, x)$$

holds for an arbitrary $t \in \mathbf{R}$, then one has

$$(x_0, \sqrt{-1}\langle \xi_0, dx\rangle\infty) \notin \text{S.S.}\left(\int u(t, x)\,dt\right).$$

Remark. It is a quite important fact that such a small set in the singularity spectrum of $u(t, x)$ contributes to the singularity spectrum of $\int u(t, x)\,dt$.

Proof. Let f be the projection $\mathbf{R} \times N \to N$. Since $(\partial/\partial t)\operatorname{sp} v(t, x) = \operatorname{sp} u(t, x) = 0$ holds in a neighborhood of $(t, x_0, \sqrt{-1}\langle \xi_0, dx\rangle\infty)$, by the assumption, then (c) of Proposition 3.2.1 implies that $\operatorname{sp} v(t, x) = f^*w(x)$ for some $w(x) \in \mathscr{C}_{N,(x_0,\sqrt{-1}\xi_0\infty)}$. Consider N as a submanifold of $\mathbf{R} \times N$ by the map $\iota_b\colon x \mapsto (b, x) \in \mathbf{R} \times N$. Then $f \circ \iota_b = \mathrm{id}_N$ holds, where id_N is the identity map on N. Hence, one obtains $\operatorname{sp} v(b, x) = \operatorname{sp} v(t, x)|_{t=b} = w(x)$; and $\operatorname{sp} v(a, x) = w(x)$ similarly holds. Therefore $\operatorname{sp}(v(b, x) - v(a, x)) = w(x) - w(x) = 0$.

Next we will consider the integration of a microfunction. Let M be a real analytic manifold. By the embedding

$$\mathbf{R} \times \sqrt{-1}S^*M \hookrightarrow \sqrt{-1}S^*(\mathbf{R} \times M),$$

sending $(t, x, \sqrt{-1}\langle \xi, dx\rangle\infty)$ to $((t, x), \sqrt{-1}(0 \cdot dt + \langle \xi, dx\rangle)\infty)$, one can regard $\mathbf{R} \times \sqrt{-1}S^*M$ a subset of $\sqrt{-1}S^*(\mathbf{R} \times M)$. Therefore, $\mathbf{R} \times U$ is a subset of $\sqrt{-1}S^*(\mathbf{R} \times M)$ where U is an open subset of $\sqrt{-1}S^*M$. Let $u(t, x) \in \mathscr{C}_{\mathbf{R}\times M}(\mathbf{R} \times U)$ such that $\operatorname{supp} u(t, x) \subset \{(t, x) \mid a < t < b\}$. Rigorously speaking, $u(t, x)$ is a microfunction on an open subset containing $\mathbf{R} \times U$, since $\mathbf{R} \times U$ is not an open subset of $\sqrt{-1}S^*(\mathbf{R} \times M)$; i.e. $u(t, x) \in \mathscr{C}_{\mathbf{R}\times M}|_{\mathbf{R}\times\sqrt{-1}S^*M}$. Let $v(t, x)$ be a microfunction on $\mathbf{R} \times U$ such that

$(\partial/\partial t)v(t, x) = u(t, x)$. Then define $w(x) = \int u(t, x)\,dt = v(b, x) - v(a, x)$. Consider the diagram

$$\begin{array}{ccc} \sqrt{-1}S^*(\mathbf{R} \times M) \overset{\varpi}{\hookleftarrow} & \Lambda & \supset (\{b\} \times M) \underset{\mathbf{R}\times M}{\times} (\mathbf{R} \times \sqrt{-1}S^*M) \\ & \downarrow \rho & \swarrow \wr \\ & \sqrt{-1}S^*(\{\mathbf{b}\} \times M) & \end{array}$$

where $\Lambda = (\{b\} \times M) \underset{\mathbf{R}\times M}{\times} \sqrt{-1}S^*(\mathbf{R} \times M) - \sqrt{-1}S^*_{\{b\}\times M}(\mathbf{R} \times M)$. Since one has $\varpi^{-1} \operatorname{supp} u \subset (\{b\} \times M) \underset{\mathbf{R}\times M}{\times} (\mathbf{R} \times \sqrt{-1}S^*M)$, the map ρ: $\varpi^{-1} \operatorname{supp} u \to \sqrt{-1}S^*M$ is proper. Then Theorem 3.1.4 implies that $v(t, x)$ has the restriction to $t = b$ (or $t = a$). Therefore $v(b, x)$ and $v(a, x)$ exist. Notice, also, that the definition of $\int u(t, x)\,dt$ does not depend upon the choice of (a, b) and that of v as we showed for the hyperfunction case. Similarly, as before, we define $\int u(t_1, \ldots, t_m, x)\,dt_1 \ldots dt_m = \int(\ldots(\int(\int u(t_1, \ldots, t_m x)\,dt_1)\,dt_2)\ldots)\,dt_m$. Hence we obtain the following theorem.

Theorem 3.2.1

(I) (Integration of Hyperfunction). *Let M and N be real analytic manifolds, and let $f: M \times N \to N$ be the natural projection. If $f|_{\operatorname{supp} u}$ is a proper map for a hyperfunction $u(t, x)$ on $M \times N$, then the integration of $u(t, x)$ along the fibre*

$$v(x) = \int_{f^{-1}(x)} u(t, x)\,dt$$

can be defined. Furthermore, one has

$$\text{S.S. } v(x) \subset \pi(\text{S.S. } u \cap M \times \sqrt{-1}S^*N),$$

where π denotes the natural projection from $M \times \sqrt{-1}S^*N$ to $\sqrt{-1}S^*N$. That is, there exists a homomorphism

$$f_!(\mathscr{B}_{M\times N} \otimes v_M) \to \mathscr{B}_N,$$

where $v_M = \Omega^m_M \otimes \omega_M$, Ω^m_M is a sheaf of holomorphic differential forms of degree m on M of dimension m, and ω_M is the orientation sheaf on M.

(II) (Integration of Microfunction). *Let M, N and π be as in* (I), *and let U be an open subset of $\sqrt{-1}S^*N$. If, for $u(t, x) \in \mathscr{C}_{M\times N}(\pi^{-1}(U))$, $\pi|_{\operatorname{supp} u(t,x)}$ is a proper map, then the integration $v(x) = \int_{f^{-1}(x)} u(t, x)\,dt$ is well-defined as a microfunction. Therefore, there exists a homomorphism*

$$\pi_!(\mathscr{C}_{M\times N}|_{M\times\sqrt{-1}S^*N} \otimes v_M) \to \mathscr{C}_N.$$

Proof. We have proved the case when $M = \mathbf{R}^m$. In order to prove the general case, first cover M with coordinate neighborhoods. Then use Proposition 3.1.1 to define the integration in each coordinate neighborhood and paste those together.

Exercise. Find a sufficient condition so that the Fubini theorem $\int dy \int f(x, y)\, dx = \int dx \int f(x, y)\, dy$ may hold. Also find sufficient conditions so that $\int f(x)g(x, y)\, dy = f(x) \int g(x, y)\, dy$ and $\int f(x, y)\, dy)|_{x=a} = \int (f(x, y)|_{x=a})\, dy$ hold.

The following examples are important for applications.

For the δ-function of several variables, we will show $\int_{\mathbf{R}^n} \delta(x)\, dx = 1$. We have

$$\delta(x) = \frac{1}{(-2\pi\sqrt{-1})^n} \sum_{\epsilon_1, \ldots, \epsilon_n = \pm 1} \frac{\epsilon_1 \cdots \epsilon_k}{\prod_{k=1}^n (x_k + \sqrt{-1}\epsilon_k 0)}$$

$$= \prod_{k=1}^n \frac{1}{(-2\pi\sqrt{-1})} \left(\frac{1}{x_k + \sqrt{-1}0} - \frac{1}{x_k - \sqrt{-1}0} \right) = \delta(x_1)\delta(x_2) \cdots \delta(x_n).$$

Therefore, $\int \delta(x)\, dx = \prod_{i=1}^n (\int \delta(x_i)\, dx_i) = 1$ holds, as $\int \delta(x_i)\, dx_i = 1$ was shown before.

We will prove the plane-wave decomposition formula of the δ-function, which is not only of theoretical importance but is also important for applications. This plane-wave decomposition formula is essentially done in John [1] and [2], in which the fundamental solution for elliptic partial differential equations is elegantly constructed, based on this formula. Apparently the plane-wave decomposition formula was crucially important for Sato when he constructed the microfunction theory. We will interpret John's work in the microfunction theoretic framework; see Theorem 3.4.3. Leray [1] is an attempt to apply John's theory to partial differential equations of hyperbolic type. We will give our construction of the fundamental solution for hyperbolic differential equations in §6 of this chapter.

Define an $(n-1)$-form $\omega(\xi)$ on $\mathbf{R}^n$ as

$$\omega(\xi) = \sum_{i=1}^n (-1)^{i-1} \xi_i\, d\xi_1 \wedge \cdots \wedge d\xi_{i-1} \wedge d\xi_{i+1} \wedge \cdots \wedge d\xi_n.$$

Denote $\langle x, \xi \rangle = \sum_{i=1}^n x_i \xi_i$ for $x = (x_1, \ldots, x_n)$ and $\xi = (\xi_1, \ldots, \xi_n)$. Since $(t + \sqrt{-1}0)^{-n}$ is well-defined as a hyperfunction of one variable, we have the hyperfunction $(\langle x, \xi \rangle + \sqrt{-1}0)^{-n}$, where t is replaced by $\langle x, \xi \rangle$; see Example 3.1.2. Then the $(n-1)$-form $\omega(\xi)/(\langle x, \xi \rangle + \sqrt{-1}0)^n$ is invariant under the multiplication of ξ by a positive real number. Hence, it can be regarded as an $(n-1)$-form on $S^{n-1} = (\mathbf{R}^n - \{0\})/\mathbf{R}_+^\times$.

Lemma. *To be more precise, let N and M be real analytic manifolds of dimension n and $(n+1)$ respectively, let $f: M \to N$ be a smooth real analytic epimorphism, and let each fibre $f^{-1}(y)$, $y \in N$, be connected. Let v be a vector field on M, such that v is tangent to each fibre, and that v never vanishes on M. If an n-form η on M with hyperfunctions as coefficients satisfies $\iota_v\eta = 0$ and $L_v\eta = 0$, where ι_v denotes the interior product and L_v denotes the Lie derivative, then there exists an n-form, η_0 on N, with coefficients in hyperfunctions such that $\eta = f^*(\eta_0)$.*

Proof. Fix a local coordinate system, and let $f:(t, x) \mapsto x$. Then one can express $v = a(t, x)(\partial/\partial t)$, $\eta = \varphi_0\, dx_1 \wedge \cdots \wedge dx_n + dt \wedge \left(\sum_j g_j(t, x)\, dx_1 \wedge \cdots \wedge dx_{j-1} \wedge dx_{j+1} \wedge \cdots \wedge dx_n\right)$. One has $\iota_v\eta = a(t, x) \sum_j g_j(t, x)\, dx_1 \wedge \cdots \wedge dx_{j-1} \wedge dx_{j+1} \wedge \cdots \wedge dx_n = 0$ and $a(t, x) \neq 0$ by the hypothesis. Hence $\eta = \varphi_0\, dx_1 \wedge \cdots \wedge dx_n$. On the other hand, one has

$$L_v\eta = a(t, x)\frac{\partial\varphi_0}{\partial t}\, dx_1 \wedge \cdots \wedge dx_n = 0.$$

Therefore, $\partial\varphi_0/\partial t = 0$ holds. Let $\eta_0 = \varphi_0\, dx_1 \wedge \cdots \wedge dx_n$. Then η_0 is an n-form on N such that $\eta = f^*(\eta_0)$.

In the above, if one lets $v = \sum_{i=1}^{n} \xi_i(\partial/\partial\xi_i)$, one obtains

$$L_v \frac{\omega(\xi)}{(\langle x, \xi\rangle + i0)^n} = 0,$$

since $L_v\omega(\xi) = n\omega(\xi)$. Similarly,

$$\iota_v \frac{\omega(\xi)}{(\langle x, \xi\rangle + i0)^n} = 0$$

holds. Therefore, by the above lemma, $\omega(\xi)/(\langle x, \xi\rangle + i0)^n$ can be regarded as an $(n$-$1)$-form on S^{n-1}. Next we will prove the following celebrated result of John ([1] and [2]) in our framework.

Proposition 3.2.3 (Plane-Wave Decomposition of the δ-Function).

$$\delta(x) = \frac{(n-1)!}{(-2\pi\sqrt{-1})^n}\int_{S^{n-1}} \frac{\omega(\xi)}{(\langle x, \xi\rangle + \sqrt{-1}0)^n}$$

holds.

Proof. Since $\omega(\xi)$ is a volume element on S^{n-1}, for $n = 1$, in which case $S^0 = \{+1\} \cup \{-1\}$, one has

$$\frac{1}{-2\pi\sqrt{-1}} \int_{\{\pm 1\}} \frac{|\xi|}{(x\xi + \sqrt{-1}0)}$$

$$= \frac{1}{-2\pi\sqrt{-1}} \left\{ \frac{1}{x + \sqrt{-1}0} + \frac{1}{-x + \sqrt{-1}0} \right\}$$

$$= \frac{1}{-2\pi\sqrt{-1}} \left\{ \frac{1}{x + \sqrt{-1}0} - \frac{1}{x - \sqrt{-1}0} \right\} = \delta(x).$$

The general case will be proved, after the following lemma, by the method of integration along fibres.

Lemma (Feynman). *Let $a = (a_1, \ldots, a_n) \in \mathbf{C}^n$ such that $\langle a, \xi \rangle \neq 0$ holds for an arbitrary $\xi = (\xi_1, \ldots, \xi_n) \neq 0$, $\xi_i \geqq 0$ for $i = 1, 2, \ldots, n$. Then*

$$\frac{1}{a_1 \cdots a_n} = (n-1)! \int_{\xi_1 \geqq 0, \ldots, \xi_n \geqq 0} \frac{\omega(\xi)}{\langle a, \xi \rangle^n}$$

holds.

Proof. We will prove this lemma in the case $\operatorname{Im} a_j < 0$ for $1 \leqq j \leqq n$. The general case can be proved by the analytic continuation. Let $\operatorname{Im} a_k < 0$ and then

$$\frac{1}{a_k} = \sqrt{-1} \int_0^\infty e^{-\sqrt{-1}a_k x_k}\, dx_k$$

holds. Therefore one obtains

$$\frac{1}{a_1 \cdots a_n} = (\sqrt{-1})^n \int_{x_1 \geqq 0, \ldots, x_n \geqq 0} e^{-\sqrt{-1}\langle a, x \rangle}\, dx.$$

Let $x = t\xi$ for $t \in \mathbf{R}$ and $\xi \in S^{n-1}$. Then $dx = t^{n-1}\, dt\omega(\xi)$ holds. Consequently, one obtains

$$\frac{1}{a_1 \cdots a_n} = (\sqrt{-1})^n \int_{\substack{\xi_1 \geqq 0, \ldots, \xi_n \geqq 0 \\ \xi \in S^{n-1}}} \omega(\xi) \int_0^\infty t^{n-1} e^{-\sqrt{-1}\langle a, \xi \rangle t}\, dt$$

$$= (n-1)! \int_{\substack{\xi_1 \geqq 0, \ldots, \xi_n \geqq 0 \\ \xi \in S^{n-1}}} \frac{\omega(\xi)}{\langle a, \xi \rangle^n}.$$

Note. The above formula was used ingeniously by Feynman for the study of Feynman integrals (see §3 of this chapter).

We now return to the proof of Proposition 3.2.3. One has

$$\int \frac{\omega(\xi)}{(\langle x, \xi\rangle + \sqrt{-1}0)^n} = \sum_{\epsilon_1, \ldots, \epsilon_n = \pm 1} \int_{\epsilon_i\xi_i \geqq 0} \frac{\omega(\xi)}{(\langle x, \xi\rangle + \sqrt{-1}0)^n}.$$

Then, by the following Proposition 3.2.4, one can first integrate with x being complex and then take the boundary value. That is

$$\int_{\epsilon_i\xi_i \geqq 0} \frac{\omega(\xi)}{(\langle x, \xi\rangle + \sqrt{-1}0)^n} = b_{\{\mathrm{Im}\langle z,\xi\rangle > 0\}}\left(\int_{\epsilon_i\xi_i \geqq 0} \frac{\omega(\xi)}{\langle z, \xi\rangle^n}\right)$$

$$= b\left(\int_{\eta_1 \geqq 0, \ldots, \eta_n \geqq 0} \frac{\omega(\eta)}{\langle z, \xi\eta\rangle^n}\right).$$

The above lemma implies that

$$\frac{1}{(n-1)!} b\left(\frac{1}{(\epsilon_1 z_1) \cdots (\epsilon_n z_n)}\right)$$

$$= \frac{1}{(n-1)!} \cdot \frac{\epsilon_1 \cdots \epsilon_n}{(x_1 + \sqrt{-1}\epsilon_1 0) \cdots (x_n + \sqrt{-1}\epsilon_n 0)}.$$

Therefore one obtains

$$\int \frac{\omega(\xi)}{(\langle x, \xi\rangle + \sqrt{-1}0)^n}$$

$$= \frac{1}{(n-1)!} \sum_{\epsilon_1, \ldots, \epsilon_n = \pm 1} \frac{\epsilon_1 \cdots \epsilon_n}{(x_1 + \sqrt{-1}\epsilon_1 0) \cdots (x_n + \sqrt{-1}\epsilon_n 0)}$$

$$= \frac{(-2\pi\sqrt{-1})^n}{(n-1)!} \delta(x).$$

Proposition 3.2.4. *Let M and N be real analytic manifolds, and let $M^{\mathbf{C}}$ and $N^{\mathbf{C}}$ be complexifications of M and N respectively. Further assume that N is compact. Suppose that $N = \bigcup_j K_j$ and $v(N) = \sum_j v(K_j)$ where K_j is a closed subset of N and v denotes the volume. Let U_j be an open subset of $\sqrt{-1}SM$ such that $\tau(U_j) = M$, and let $D_j \subset M^{\mathbf{C}} \times N^{\mathbf{C}}$ be a conoidal neighborhood of $U_j \times K_j$, regarded as a subset of $\sqrt{-1}S(M \times N)$ via the map $\sqrt{-1}SM \times N \hookrightarrow \sqrt{-1}S(M \times N)$. If, in a neighborhood of $M \times K_j$, $u(x, t) = b_{D_j}(f_j(z, \tau))$ holds for some $f_j(z, \tau) \in \mathcal{O}(D_j)$, then there exists a conoidal neighborhood V_j of U_j such that $V_j \times K_j \subset D_j$ and*

$$\int u(x, t)\, dt = \sum_j b_{V_j}\left(\int_{K_j} f_j(z, t)\, dt\right)$$

holds.

Proof. It is sufficient to prove the case when $K_j = [a_1, c_1] \times \cdots \times [a_n, c_n] \subset \mathbf{R}^n$. Let $Y_{a,c}(t) = Y(c-t)Y(t-a)$ for $a, c \in \mathbf{R}$, $a < c$. Then one has

$$\int_{a_1}^{c_1} \cdots \int_{a_n}^{c_n} u(x, t)\, dt = \int u(x, t) Y_{a_1,c_1}(t_1) \cdots Y_{a_n,c_n}(t_n)\, dt.$$

In order to prove that the right-hand side is equal to $b(\int_{a_1}^{c_1} \cdots \int_{a_n}^{c_n} f_j(z, t)\, dt)$, it is enough to prove the case when $n = 1$; i.e.

$$\int_a^c u(x, t)\, dt = b\left(\int_a^c f(z, t)\, dt\right).$$

Let $v(x, t) = b(g(z, t))$ where $g(z, t) = \int_a^t f(z, t)\, dt$. Then it is plain that $(\partial/\partial t)v(x, t) = u(x, t)$ holds. Therefore, one has $b(\int_a^c f(z, t)\, dt) = v(x, c) - v(x, a)$. One now needs to show $\int_a^c u(x, t)\, dt = v(x, c) - v(x, a)$.

Lemma. *If hyperfunctions* $u(x, t)$ *and* $v(x, t)$ *satisfy* $(\partial/\partial t)v(x, t) = u(x, t)$, *then one has*

$$\frac{\partial}{\partial t} w(x, t) = u(x, t) Y_{a,c}(t)$$

where

$$w(x, t) = v(x, t) + Y(t - c)\{v(x, c) - v(x, t)\} + Y(a - t)\{v(x, a) - v(x, t)\}.$$

Proof. One can prove this lemma by direct computation using the next proposition.

We can complete the proof of Proposition 3.2.4 by the above lemma. For the w in the lemma above, one has $\int_a^c u(x, t)\, dt = \int u(x, t) Y_{a,c}(t)\, dt = \int (\partial/\partial t) w(x, t)\, dt = \int_{a'}^{c'} (\partial/\partial t) w(x, t)\, dt = w(x, c') - w(x, a')$, where $a' < a < c < c'$. Furthermore, $w(x, c') = v(x, c)$ and $w(x, a') = v(x, a)$ hold. Hence one obtains

$$\int_a^c u(x, t)\, dt = v(x, c) - v(x, a) = b\left(\int_a^c f(x, t)\, dt\right).$$

Proposition 3.2.5. *If a hyperfunction* $u(x, t)$ *satisfies the condition* $(x, 0; \pm\sqrt{-1}\, dt\infty) \notin \text{S.S. } u$, *then the restriction* $u(x, 0) = u(x, t)|_{t=0}$ *and the product* $u(x, t)\delta(t)$ *are defined and satisfy*

$$u(x, t)\delta(t) = u(x, 0)\delta(t).$$

Note. The restriction $u(x, 0)$ and $u(x, t)\delta(t)$ are well defined from Theorems 3.1.3 and 3.1.5. Hence, it is sufficient to prove $u(x, t)\delta(t) = u(x, 0)\delta(t)$.

Lemma. *If a hyperfunction* $u(x, t)$ *satisfies the condition*

$$(x, 0; \pm\sqrt{-1}\, dt\infty) \notin \text{S.S. } u(x, t),$$

then the following two statements are equivalent:

(1) $u(x, 0) = 0$.

(2) *There exists a hyperfunction* $v(x, t)$ *such that* $(x, 0; \pm\sqrt{-1}\,dt\infty) \notin$ S.S. $v(x, t)$ *and* $u(x, t) = tv(x, t)$.

Proof. It is plain that (2) implies (1). We will prove that (1) implies (2). Let $u = \sum_j b(f_j(z, \tau))$. Then $f_j(z, 0)$ is well defined by the assumption on S.S. $u(x, t)$. Since $u|_{t=0} = \sum_j b(f_j(z, 0)) = 0$ by (1), then $f_j(z, 0)$ is a coboundary; i.e. $f_j(z, 0) = \sum_k g_{jk}(z)$ where $g_{jk} = -g_{kj}$. Then let $h_j(z, \tau) = f_j(z, \tau) - \sum_k g_{jk}(z)$. One obtains $u = \sum_j b(f_j) = \sum_j b(f_j - \sum_k g_{jk}) = \sum_j b(h_j)$. Since $h_j(z, \tau)$ is holomorphic and $h_j(z, 0) = 0$, one can express $h_j(z, \tau) = \tau\varphi_j(z, \tau)$. Therefore, if one lets $v = \sum_j b(\varphi_j(z, \tau))$, then $u(x, t) = tv(x, t)$.

From this lemma, $u(x, t) - u(x, 0) = tv(x, t)$ holds. Then, $u(x, t)\delta(t) - u(x, 0)\delta(t) = (tv(x, t))\delta(t) = (v(x, t)t)\delta(t)$ is obtained. The associative law $(v(x, t)t)\delta(t) = u(x, t) \cdot (t\delta(t)) = 0$ holds by the assumption

$$(x, 0; \pm\sqrt{-1}\,dt\infty) \notin \text{S.S. } v(x, t).$$

The following proposition on the δ-function is important for applications.

Proposition 3.2.6. *Let f be a real analytic map from a neighborhood of the origin in* $\mathbf{R}$ *to a neighborhood of the origin in* $\mathbf{R}$ *such that* $f(0) = 0$ *and* $f'(0) \neq 0$. *Then one has*

$$\delta(f(x)) = \frac{1}{|f'(0)|}\,\delta(x)$$

in a neighborhood of the origin. Generally, if f is a real analytic map from a neighborhood of the origin in $\mathbf{R}^n$ *to a neighborhood of the origin in* $\mathbf{R}^n$ *such that* $f(0) = 0$ *and* $\det df(0) \neq 0$, *then*

$$\delta(f(x)) = \frac{1}{|\det df(0)|}\,\delta(x)$$

holds.

Proof. Here we will prove the case when $n = 1$, and it is recommended that the reader prove the case in several variables. First consider the case where $f(x) = x\varphi(x)$, $\varphi(x) > 0$. Then one has

$$\delta(f(x)) = -\frac{1}{2\pi i}\left(\frac{1}{x\varphi(x) + i0} - \frac{1}{x\varphi(x) - i0}\right) = \frac{1}{\varphi(x)}\,\delta(x) = \frac{1}{\varphi(0)}\,\delta(x).$$

Since $\varphi(0) = |f'(0)|$, then $\delta(f(x)) = (1/|f'(0)|)\delta(x)$ holds. If $\varphi(x) < 0$, Proposition 2.4.1 implies $\delta(f(x)) = \delta(-f(x))$. Then, use $-f(x)$ instead to complete the proof.

Remark. This proposition suggests that coordinate-transformationwise it is $\delta(x)\,dx$ rather than $\delta(x)$ which is defined intrinsically.

Example 3.2.4. $\delta(x^2 - 1) = \frac{1}{2}\delta(x - 1) + \frac{1}{2}\delta(x + 1)$ holds.

Let $X = x - 1$, and let $F(X) = X(2 + X)$ in a neighborhood of $x = 1$. Then we have

$$\delta(x^2 - 1) = \delta(F(X)) = \frac{1}{|F'(0)|}\,\delta(X) = \frac{1}{2}\,\delta(x - 1).$$

We also obtain $\delta(x^2 - 1) = \frac{1}{2}\delta(x + 1)$ in a neighborhood of $x = -1$. The next example can be proved in a similar manner.

Example 3.2.5. $\delta(\cos\theta) = \delta(\theta + \pi/2) + \delta(\theta - \pi/2)$ holds for $-\pi \leqq \theta \leqq \pi$.

We will give useful examples of integrals using the above obtained results.

Example 3.2.6. We have $\int_{\mathbf{R}} \delta(t - x^2)\,dx = t_+^{-1/2}$.

Proof. It is convenient to employ differential equations for this type of computation. We have $x\delta'(x) + \delta(x) = 0$ by differentiating $x\delta(x) = 0$. Hence we have $(t(\partial/\partial t) + \frac{1}{2} + \frac{1}{2}(\partial/\partial x)x)\delta(t - x^2) = t\delta'(t - x^2) + \frac{1}{2}\delta(t - x^2) + \frac{1}{2}\{\delta(t - x^2) + x(-2x)\delta'(t - x^2)\} = (t - x^2)\delta'(t - x^2) + \delta(t - x^2) = 0$. Let $u(t) = \int\delta(t - x^2)\,dx$. Then $(t(\partial/\partial t) + \frac{1}{2})u(t) = \int(t(\partial/\partial t) + \frac{1}{2})\delta(t - x^2)\,dx = \int(-\frac{1}{2}(\partial/\partial x)(x\delta(t - x^2)))\,dx = -\frac{1}{2}\{b\delta(t - x^2) - a\delta(t - a^2)\} = 0$, where, for t in the equations, a and b are chosen so that $t < a^2, b^2$. If $t < 0$, then clearly $u(t) = 0$ holds. Therefore, $u(t) = ct_+^{-1/2}$ for some constant c. Then $c = u(1) = \int\delta(1 - x^2)\,dx = \int(\frac{1}{2}\delta(x - 1) + \frac{1}{2}\delta(x + 1))\,dx = 1$ holds. Consequently, $u(t) = t_+^{-1/2}$.

Example 3.2.7. $\int_{-\pi}^{\pi} d\theta/(\cos\theta + \sqrt{-1}0) = -2\pi\sqrt{-1}$.

Proof. One can prove easily that $A = \int_{-\pi}^{\pi} d\theta/(\cos\theta + i0)$ exists and is finite. By transforming the variable θ to $\theta + \pi$, we have

$$A = \int_{-\pi}^{\pi} \frac{d\theta}{-\cos\theta + i0}.$$

Hence

$$2A = \int_{-\pi}^{\pi}\left(\frac{1}{\cos\theta + i0} + \frac{1}{-\cos\theta + i0}\right)d\theta = \int_{-\pi}^{\pi}(-2\pi i)\delta(\cos\theta)\,d\theta$$

$$= \int_{-\pi}^{\pi}(-2\pi i)\left(\delta\left(\theta + \frac{\pi}{2}\right) + \delta\left(\theta - \frac{\pi}{2}\right)\right)d\theta = -4\pi i$$

holds by the previous example. We conclude

$$A = \int_{-\pi}^{\pi} \frac{d\theta}{(\cos\theta + i0)} = -2\pi i.$$

Example 3.2.8. Let $q(x)$ be a non-degenerate quadratic form, and denote the signature by sgn q. Then we have

$$\int_{S^{n-1}} \frac{\omega(\xi)}{(q(\xi) + \sqrt{-1}0)^{n/2}} = \frac{2\pi^{n/2}}{\Gamma\left(\frac{n}{2}\right)} \cdot \frac{1}{\sqrt{\det(q + \sqrt{-1}0)}}$$

$$= \frac{2\pi^{n/2}}{\Gamma\left(\frac{n}{2}\right)} \cdot \frac{\exp\left(-\frac{n\sqrt{-1}}{4}(n - \operatorname{sgn} q)\right)}{\sqrt{|\det q|}}.$$

Remark. Recall that $q(x)$ is said to be a quadratic form on $\mathbf{R}^n$ if there exists a real symmetric matrix $A = (a_{ij})$ of order n such that $q(x) = \langle Ax, x\rangle = \sum_{i,j=1}^{n} a_{ij}x_ix_j$. If $\det A \neq 0$, then $q(x)$ is said to be non-degenerate and $\det q$ is defined by $\det A$. All the eigenvalues of A are real, and non-zero if q is non-degenerate. The difference between the number of positive eigenvalues and the number of negative eigenvalues is called the signature of q. (Sometimes the pair of the numbers of the positive and the negative eigenvalues is called the signature of q.) As in Proposition 3.2.3, we consider $\omega(\xi)/(q(\xi) + \sqrt{-1}0)^{n/2}$ as an $(n-1)$-form on S^{n-1}.

Proof. We first consider the case when $q(x)$ is a positive definite form, i.e. sgn $q = n$. Then $q = \langle Ax, x\rangle$ such that $A = {}^tTT$ with T being invertible. Hence one has the following:

$$\int \frac{\omega(\xi)}{(\langle A\xi, \xi\rangle + \sqrt{-1}0)^{n/2}} = \int \frac{\omega(\xi)}{(\langle T\xi, T\xi\rangle + \sqrt{-1}0)^{n/2}}$$

$$= \int \frac{(\det T)^{-1}\omega(\eta)}{(\langle \eta, \eta\rangle + \sqrt{-1}0)^{n/2}}$$

$$= \frac{1}{\sqrt{\det q}} \int_{\xi\in S^{n-1}} \frac{\omega(\xi)}{(\langle \xi, \xi\rangle)^{n/2}}$$

$$= \frac{1}{\sqrt{\det q}} \int_{\xi\in S^{n-1}} \omega(\xi).$$

Notice in the above that $\langle \xi, \xi\rangle = 1$ for $\xi \in S^{n-1}$ holds, and that one then obtains $(\langle \xi, \xi \rangle + \sqrt{-1}0)^{-n/2} = \langle \xi, \xi \rangle^{-n/2}$. Since

$$\int_{\xi \in S^{n-1}} \omega(\xi) = \frac{2\pi^{n/2}}{\Gamma\left(\dfrac{n}{2}\right)}$$

holds, consequently one obtains

$$\int_{S^{n-1}} \frac{\omega(\xi)}{(q(\xi) + \sqrt{-1}0)^{n/2}} = \frac{2\pi^{n/2}}{\Gamma\left(\dfrac{n}{2}\right)} \cdot \frac{1}{\sqrt{\det q}}$$

for $\operatorname{sgn} q = n$.

For the general case, we consider each entry of A to be a variable for $q(x) = \langle Ax, x \rangle$. Then $1/(\langle Ax, x \rangle + \sqrt{-1}0)^{n/2}$ can be considered as the boundary value from the domain where $\operatorname{Im} A$ is positive definite; more precisely, $\operatorname{Im}\langle A\xi, \xi \rangle \geqq 0$ regarding ξ as a complex number. If $\operatorname{Re} A$ is positive definite, one has

$$\int \frac{\omega(\xi)}{(q(\xi))^{n/2}} = \frac{2\pi^{n/2}}{\Gamma\left(\dfrac{n}{2}\right)} \cdot \frac{1}{\sqrt{\det q}}.$$

Therefore, one can continue this equation from the domain where both $\operatorname{Re} A$ and $\operatorname{Im} A$ are positive-definite to the one where $\operatorname{Im} A$ is positive-definite. Hence, one obtains the equation

$$\int \frac{\omega(\xi)}{(\langle A\xi, \xi \rangle + \sqrt{-1}0)^{n/2}} = \frac{2\pi^{n/2}}{\Gamma\left(\dfrac{n}{2}\right)} \cdot \frac{1}{\sqrt{\det(A + \sqrt{-1}0)}}$$

of hyperfunctions. We will compute $\sqrt{\det(A + \sqrt{-1}0)}$ next.

One may let

$$A = T \begin{bmatrix} \lambda_1 & & \\ & \ddots & \\ & & \lambda_n \end{bmatrix} {}^tT, \qquad \operatorname{Im} \lambda_i > 0 \qquad \text{for } 1 \leqq i \leqq n$$

for a positive definite $\operatorname{Im} A$. This implies

$$\sqrt{\det(A(\lambda) + \sqrt{-1}0)} = |\det T| \sqrt{\lambda_1 + \sqrt{-1}0} \cdots \sqrt{\lambda_n + \sqrt{-1}0}.$$

Taking the branch $\sqrt{1 + \sqrt{-1}0} = 1$, one has $\sqrt{-1 + \sqrt{-1}0} = \exp(\pi\sqrt{-1}/2)$. Since the number of negative eigenvalues is $(n - \operatorname{sgn} q)/2$, one obtains $\sqrt{\det(A + \sqrt{-1}0)} = \sqrt{|\det A|} \cdot \exp(\pi\sqrt{-1}(n - \operatorname{sgn} q)/4)$, completing the proof.

Example 3.2.9. $\int_{\mathbf{R}^n} (x_1^2 + \cdots + x_n^2 - t)_-^\lambda \, dx_1 \cdots dx_n = \dfrac{\pi^{n/2}\Gamma(\lambda+1)}{\Gamma(\lambda+n/2+1)} t_+^{\lambda+n/2}$.

Proof. As in Example 3.2.6, let $u(t)$ be the left-hand side of this equation. Then $(t(\partial/\partial t) - \lambda - n/2)u(t) = 0$ holds. It is also plain that $u(t) = 0$ for $t < 0$. Therefore, one can express $u(t) = ct_+^{\lambda+n/2}$. Then one obtains

$$c = u(1) = \int_{\mathbf{R}^n} (1 - x_1^2 - \cdots - x_n^2)_+^\lambda \, dx = \int (1 - r^2)_+^\lambda r^{n-1} \, dr\omega(\xi)$$

$$= \int_0^1 (1 - r^2)^\lambda r^{n-1} \, dr \int_{S^{n-1}} \omega(\xi) = \frac{\pi^{n/2}\Gamma(\lambda+1)}{\Gamma\left(\lambda + \dfrac{n}{2} + 1\right)}.$$

§3. Analyticity of Feynman Integrals

Before discussing the various aspects of the usefulness of microfunction theory for linear partial differential equations, we will study the analyticity of Feynman integrals as an application with a different flavor. Since the results in this section will not be used later on in this book, one may proceed to the next section. As we will explain, Feynman integrals are integrals of hyperfunctions corresponding to Feynman diagrams, which are quite important for the study of the quantum field theory. Even though we will go neither into the background details nor into the applications of algebraic analysis to theoretical physics, now being developed chiefly by Sato, we will give an elementary example here. Those interested are advised to read Eden et al. [1] and Nakanishi [1], and to see *Publ. RIMS, Kyoto Univ. 12*, suppl. (1977) for applications of hyperfunction theory to theoretical physics. Incidentally, it might be worth mentioning that the dispersion relation (see Vladimirov [1]) in quantum field theory had some influence on Sato when he constructed the hyperfunction theory, and that the edge-of-the-wedge theorem (see Morimoto [1]) was apparently within Sato's subconcious when he constructed the microfunction theory. It is also worthy of note that physicists independently found a notion pretty similar to that of microfunctions; see Iagolnitzer and Stapp [1]. It might be said that the naturalness of the theory of hyperfunctions is demonstrating itself.

Definition 3.3.1. *A Feynman diagram D consists of finitely many points, called vertices, $V_1, \ldots, V_{n'}$; finitely many one-dimensional segments, called internal lines, $L_1, \ldots, L_N$; and finitely many half-lines, called external lines, $L_1^e, \ldots, L_n^e$. Each pair of end points W_l^+ and W_l^- of L_l, $l = 1, 2, \ldots, N$, are* V_i and V_j *for some i and j, $1 \leqq i, j \leqq n'$. We also assume $W_l^+ \neq W_l^-$ in the following discussion. This vertex of L_r^e, $r = 1, \ldots, n$, coincides with some V_j, $j = 1, \ldots, n'$. Each external line L_r^e associates a real four-dimensional* *vector* $p_r = (p_{r,0}, p_{r,1}, p_{r,2}, p_{r,3})$, *and each internal line* L_l *associates a positive constant* $m_l^2 \gneqq 0$. *External and internal lines are oriented, and* $\rightarrowtail$ *denotes the orientation. Then the incidence number* $[j:l]$ *is defined as follows: if* V_j *is the initial point of* L_l, *then let* $[j:l] = -1$; *if* V_j *is the terminal point of* L_l, *then* $[j, l] = +1$; *and if the vertex* V_j *is neither the initial point nor the terminal point, then the incidence number* $[j:l] = 0$. *The incidence number* $[j, r]$ *for* V_j *and* L_r^e *is defined in a similar manner.*

An example of a Feynman diagram D is given in Figure 3.3.1, below. We have the incidence numbers as follows in this example: $[1:1] = [1:2] = [1:3] = -1$; $[2:1] = [2, 2] = [2:3] = +1$; $[3:1] = [3:3] = 0$; and so on.

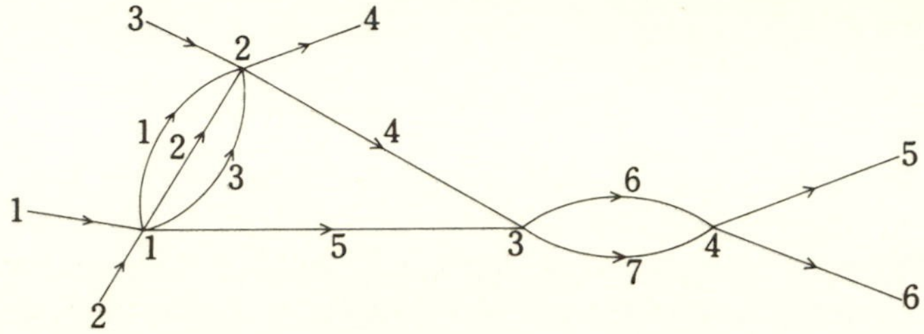

Figure 3.3.1

Remark 1. For the sake of simplicity, we will consider only Feynman diagrams that are connected.

Remark 2. There are some important cases where $m_l^2 = 0$; however, for simplicity, we restrict ourselves to the case $m_l^2 > 0$.

Remark 3. A Feynman diagram expresses (diagrammatically) the interaction of elementary particles, where m_l represents the mass of an elementary particle.

Definition 3.3.2. *Let D be a Feynman diagram. The Feynman integral* $F_D(p)$ *is formally defined by the integral*

$$F_D(p) = F_D(p_1, \ldots, p_n)$$

$$\underset{\text{def}}{=} \int \frac{\prod_{j=1}^{n'} \delta^4\left(\sum_{r=1}^{n} [j:r]p_r + \sum_{l=1}^{N} [j:l]k_l\right)}{\prod_{l=1}^{N} (k_l^2 - m_l^2 + \sqrt{-1}0)} \prod_{l=1}^{N} d^4k_l, \qquad (3.3.1)$$

where $k_l^2 = k_{l,0}^2 - \sum_{v=1}^{3} k_{l,v}^2$. *(The square of a four-dimensional vector is defined in this way in this section.)*

Note that the above definition is quite formal and that Feynman integrals are divergent integrals in general. The renormalization theory, or self-consistent subtraction theory, gives them a definite meaning consistently. For details, see Nakanishi [1]. We will consider the case where the integral exists, in the microfunction theoretic sense, following Sato [2].

We will assume that for any j there is r such that $[j:r] \neq 0$; i.e. each vertex has at least one external line attached. Then, by replacing $\sum_r [j:r]p_r$ by $-p_j$ in the definition of the integral, one can assume that there is a unique external line leaving each vertex, i.e. identifying j and r. An example of the Feynman diagram under consideration is shown in Figure 3.3.2. Then we have

$$\text{S.S. } \delta^4(p) = \{(p; \sqrt{-1}u\infty) \in \sqrt{-1}S^*\mathbf{R}^4 \mid p = 0, u \neq 0\}$$

$$\text{S.S.} \left(\frac{1}{k^2 - m^2 + \sqrt{-1}0}\right) = \{(k; \sqrt{-1}v\infty) \in \sqrt{-1}S^*\mathbf{R}^4 \mid k^2 = m^2, v = ck \ (c > 0)\}.$$

Note. Here the assumption $m^2 \neq 0$ is used. We also identify $\operatorname{grad}_k k^2$ with $2k$ via the Minkowsky metric. When one regards $2k_1$ as a cotangent vector, it may be proper to denote it by $\operatorname{grad}_{k_1} k_1^2$ in order to be rigorous. We will follow the notation that physicists use.

Therefore, Theorem 3.1.3 implies that the integrand of (3.3.1) is well defined, and the singularity spectrum is contained in the following set Λ:

$$\Lambda = \left\{(p, k; \sqrt{-1}(u, v)\infty \in \sqrt{-1}S^*\mathbf{R}^{4(n+N)} \,\middle|\, -\sum_{j=1}^{n} [j:l]u_j + \alpha_l k_l = v_l, \right.$$
$$\alpha_l(k_l^2 - m_l^2) = 0 \text{ for } \alpha_l \geqq 0, \ l = 1, \ldots, N \text{ and}$$
$$\left. -p_j + \sum_{l=1}^{N} [j:k]k_l = 0, \ j = 1, \ldots, n\right\}. \tag{3.3.2}$$

We wish to locate the singularity spectrum of $F_D(p)$ by applying Theorem 3.2.1 to the integral (3.3.1). But generally the condition of the theorem

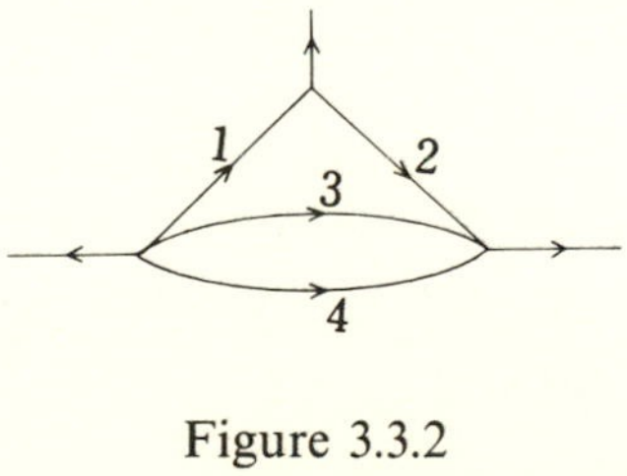

Figure 3.3.2

is not satisfied. Hence, we will look for a point where (3.3.1) determines a microfunction and where it is zero or not zero.

Let

$$\mathscr{L} \underset{\text{def}}{=} \{(p; \sqrt{-1}u\infty) \in \sqrt{-1}S^*\mathbf{R}^{4n} \mid \text{there exist } \alpha_l \text{ and } k_l,\ l = 1, \ldots, N$$

such that

$$\left.\begin{array}{llr} (3.3.3\text{a}) & p_j = \sum_{l=1}^{N} [j:l]k_l, & j = 1, \ldots, n. \\ (3.3.3\text{b}) & \sum_{j=1}^{n} [j:l]u_j = \alpha_l k_l, & l = 1, \ldots, N. \\ (3.3.3\text{c}) & \alpha_l(k_l^2 - m_l^2) = 0, & l = 1, \ldots, N. \\ (3.3.3\text{d}) & \alpha_l \geqq 0, & l = 1, \ldots, N. \end{array}\right\} \quad (3.3.3)$$

hold$\}$. Furthermore, let

$$\mathscr{L}_0 = \{(p; \sqrt{-1}u\infty) \in \sqrt{-1}S^*\mathbf{R}^{4n} \mid (p; \sqrt{-1}u\infty) \in \mathscr{L}$$
$$\text{and all the } \alpha_1 \text{ are strictly positive}\}.$$

For $(p; \sqrt{-1}u\infty) \in \mathscr{L}_0$, the condition of Theorem 3.2.1 (II) is satisfied in a neighborhood of this point $(p; \sqrt{-1}u\infty)$, since (α_l, k_l) in (3.3.3) uniquely determined (p, u). Notice also that if $(p; \sqrt{-1}u\infty) \notin \mathscr{L}$, then $F_D(p)$ as a microfunction is zero in a neighborhood of $(p; \sqrt{-1}u\infty)$. Therefore, we have the following theorem.

Theorem 3.3.1. *For the Feynman diagram considered in this section, $F_D(p)$ determines a microfunction outside $(\mathscr{L} - \mathscr{L}_0)$ and the support is within $\mathscr{L}_0$.*

Remark 1. Note that $\mathscr{L}$ is called the Landau-Nakanishi variety. Landau and Nakanishi independently found the notion (1959) in their study of the Feynman integral.

Remark 2. As an example, consider Figure 3.3.2. Let a four-dimensional vector u_j be assigned to each vertex in Figure 3.3.2. Then (3.3.3b) indicates that, by assigning a four-dimensional vector k_l to each internal line L, the sum $\alpha_1 k_1 + \alpha_2 k_2 - \alpha_3 k_3$ around a loop in D is zero, where α_1, α_2, and α_3 are properly chosen. Note that treatises in physics often use this form, rather than the condition (3.3.3b).

Remark 3. If $(p; u)$ is a solution to (3.3.3), then $(p; u + a)$ is a solution of (3.3.3) for $a \in \mathbf{R}^4$. This indicates diagrammatically that the diagram may be shifted to any place. In a phrase of analysis, $F_D(p)$, if it is well defined, takes the form $\delta^4\left(\sum_{j=1}^{n} p_j\right) f_D(p)$.

Remark 4. The reader may notice that the obstruction to applying Theorem 3.2.1 to $F_D(p)$ is the fact that the domain of integration is not finite. One of the primary objectives of the self-consistent subtraction theory is to deal with this divergence (ultraviolet divergence) originating from the points where k_l is large. Mathematically this is usually done by compactifying the domain of integration. But we will not go into that here. It is known that $F_D(p)$ is well defined as a hyperfunction under a proper self-consistent subtraction and whose singularity spectrum is contained in $\mathscr{L}$.

Exercise. Let $I_D(p)$ be the integral obtained by replacing $1\Big/\prod_{l=1}^{N}(k_l^2 - m_l^2 + \sqrt{-1}0)$ by $\delta(k_l^2 - m_l^2)Y(k_{l,0})$ in Definition 3.3.2. Consider how far one can carry on an argument similar to the one above. Note that $I_D(p)$ is also an important function in physics, called the phase space integral, and that, contrary to $F_D(p)$, the product of integrands is not well defined. See Kawai and Stapp [1] on these topics.

§4. Microlocal Operators and the Fundamental Theorem of Sato

Let M be a real analytic manifold, and let X be a complexification of M. Then let $\Delta_M = \{(x, y) \in M \times M \,|\, x = y\}$, and let

$$\Delta^a_{\sqrt{-1}S^*M} = \{(x, y; \sqrt{-1}(\xi, \eta)\infty \,|\, x = y \text{ and } \xi = -\eta\}.$$

If $K(x, y)$ is a hyperfunction on $M \times M$ with the support in Δ_M and S.S. $K(x, y) \subset \Delta^a_{\sqrt{-1}S^*M}$, then the integral operator $\mathscr{K}$ defined by

$$\mathscr{K}(u(y)) = \int K(x, y)u(y)\,dy$$

is a sheaf homomorphism from $\mathscr{B}_M$ to $\mathscr{B}_M$. Then such an integral operator is a natural generalization of a differential operator of finite order. In this section, we will treat the analogous case for $\mathscr{C}_M$. When one is interested in studying the case for $\mathscr{C}_M$, one may expect that the support of the above hyperfunction $K(x, y)$ need not be contained in Δ_M, judging from §1 and §2 of this chapter. For example, let $M = \mathbf{R}^1$. Then

$$u(y) \mapsto \frac{1}{-2\pi\sqrt{-1}} \int \frac{u(y)}{x - y + \sqrt{-1}0}\,dy$$

induces a sheaf homomorphism on the sheaf $\mathscr{C}_\mathbf{R}$ of microfunctions at $\{(x, \sqrt{-1}\,dx\infty)\}$; it is actually an identity map on microfunctions at $\{(x, \sqrt{-1}\,dx\infty)\}$, since

$$\delta(x) = \frac{1}{-2\pi\sqrt{-1}}\left(\frac{1}{x + \sqrt{-1}0} - \frac{1}{x - \sqrt{-1}0}\right).$$

Our intent is to define a class (as general as possible) of integral operators, based on the results in §1 and §2 of this chapter, inducing sheaf homomorphisms on $\mathscr{C}_M$. The class to be considered in this section is too general to enjoy algebraic properties. A desired class for algebraic consideration is the one of microdifferential operators defined in §1 of Chapter IV.

We begin with notations for the discussion which follows. Let M and N be real analytic manifolds, and let Δ_N be the diagonal set of N; i.e. $\Delta_N = \{(x, y) \in N \times N \mid x = y\}$. Define L_1, L_2, and L_3 as follows:

$$L_1 = \sqrt{-1}S^*(M \times N) - M \times \sqrt{-1}S^*N - \sqrt{-1}S^*M \times N,$$

$$\begin{aligned} L_2 = {} & (M \times \Delta_N) \underset{M \times N \times N}{\times} \sqrt{-1}S^*(M \times N \times N) \\ & - \sqrt{-1}S^*_{M \times \Delta_N}(M \times N \times N) - M \times N \times \sqrt{-1}S^*N \\ & - \sqrt{-1}S^*(M \times N) \times N, \end{aligned}$$

and

$$\begin{aligned} L_3 = {} & \sqrt{-1}S^*(M \times N \times N) - M \times N \times \sqrt{-1}S^*N \\ & - \sqrt{-1}S^*(M \times N) \times N. \end{aligned}$$

Consider the following diagrams:

$$\begin{array}{ccc} & \sqrt{-1}S^*(M \times N) - M \times \sqrt{-1}S^*N - \sqrt{-1}S^*M \times N & \\ \swarrow^{p_1} & & \searrow^{p_2^a} \\ \sqrt{-1}S^*M & & \sqrt{-1}S^*N \end{array}$$

Diagram 3.4.1

where $p_1((x, y; \sqrt{-1}(\xi, \eta)\infty)) = (x, \sqrt{-1}\xi\infty)$ and

$$p_2^a((x, y; \sqrt{-1}(\xi, \eta)\infty)) = (y, -\sqrt{-1}\eta\infty).$$

$$\begin{array}{ccccccc} & & & & & \overset{k}{\nearrow} & \sqrt{-1}S^*(M \times N \times N) \\ & & L_1 & \overset{\iota'}{\hookrightarrow} & L_2 \overset{j}{\hookrightarrow} L_3 & \xrightarrow{\beta_2} & \sqrt{-1}S^*N \\ & \swarrow^{\gamma_1'} & & & \searrow^{\gamma_1} & \underset{\beta_1}{\searrow} & \sqrt{-1}S^*(M \times N) \\ \sqrt{-1}S^*M \times N & & \overset{\iota}{\hookrightarrow} & & \sqrt{-1}S^*(M \times N) \cong \sqrt{-1}S^*(M \times \Delta_N) & & \\ \downarrow{\scriptstyle \gamma_2} & & & & & & \\ \sqrt{-1}S^*M & & & & & & \end{array}$$

Diagram 3.4.2

where ι, ι', j, and k are natural embeddings. That is, for example, $\iota'((x, y; \sqrt{-1}(\xi, \eta)\infty) = (x, y, y; \sqrt{-1}(\xi, \eta, -\eta)\infty)$, and $\gamma_1((x, y, y; \sqrt{-1}(\xi, \eta_1, \eta_2)\infty)) = (x, y; \sqrt{-1}(\xi, \eta_1 + \eta_2)\infty)$, $\gamma'_1 = \gamma_1|_{L_1}$; and β_1, β_2, and γ_2 are natural projections.

Under these notations, a main theorem can be phrased as follows.

Theorem 3.4.1. *Let M and N be real analytic manifolds, and let Z be a locally closed subset of $\sqrt{-1}S^*(M \times N)$ such that $Z \cap M \times \sqrt{-1}S^*N = \varnothing$ and $Z \cap \sqrt{-1}S^*M \times N = \varnothing$. Further assume that $p_1(Z)$ is a locally closed subset of $\sqrt{-1}S^*M$. We denote the sheaf of volume elements of N by v_N. Then define an integral operator $\mathscr{K}$ as*

$$\mathscr{K}(u) = \int_N K(x, y)u(y)\,dy$$

for $K(x, y)\,dy \in H^0_Z(\sqrt{-1}S^*(M \times N), \mathscr{C}_{M\times N} \otimes v_N)$, and $u \in \mathscr{C}_N$. Hence one obtains a sheaf homomorphism

$$\mathscr{K}: (p_1|_Z)_!(p_2^a|_Z)^{-1}\mathscr{C}_N \to \mathscr{C}_M.$$

Proof. We will consider the restriction of a section $w(x, y_1, y_2)$ of the sheaf $\mathscr{C}_{M\times N\times N}$ over $\sqrt{-1}S^*(M \times N \times N)$ to $\{y_1 = y_2\}$. This operation corresponds to giving a sheaf homomorphism $\gamma_{1!}(k \circ j)^{-1}\mathscr{C}_{M\times N\times N} \to \mathscr{C}_{M\times\Delta_N}$ over $\sqrt{-1}S^*(M \times \Delta_N)$. This induces a sheaf homomorphism

$$\iota^{-1}\gamma_{1!}(k \circ j)^{-1}\mathscr{C}_{M\times N\times N} \to \iota^{-1}\mathscr{C}_{M\times\Delta_N}$$

over $\sqrt{-1}S^*M \times N$. On the other hand, one can show that

$$\begin{aligned}\iota^{-1}\gamma_{1!}(k \circ j)^{-1}\mathscr{C}_{M\times N\times N} &= \gamma'_{1!}\iota'^{-1}(k \circ j)^{-1}\mathscr{C}_{M\times N\times N}\\ &= \gamma'_{1!}(k \circ j \circ \iota')^{-1}\mathscr{C}_{M\times N\times N}\end{aligned}$$

holds. One may identify $\mathscr{C}_{M\times\Delta_N} = \mathscr{C}_{M\times N}$. For the sake of clarity, N_1 and N_2 will be used in place of N. Consequently, one obtains the sheaf homomorphism

$$\gamma'_{1!}(k \circ j \circ \iota')^{-1}(\mathscr{C}_{M\times N_1\times N_2} \otimes v_{N_1}) \to \iota^{-1}\mathscr{C}_{M\times N} \otimes v_N$$

over $\sqrt{-1}S^*M \times N$. Therefore, there is induced a sheaf homomorphism

$$(\gamma_2 \circ \gamma'_1)_!(k \circ j \circ \iota')^{-1}(\mathscr{C}_{M\times N_1\times N_2} \otimes v_{N_1}) \to \gamma_{2!}\iota^{-1}(\mathscr{C}_{M\times N} \otimes v_N)$$

over $\sqrt{-1}S^*M$. This map is nothing but the map which assigns $w(x, y, y)\,dy$ for $w(x, y_1, y_2)\,dy_1$. On the other hand, the assignment $w(x, y, y,)\,dy$ to the integral $\int_N w(x, y, y)\,dy$ is a sheaf homomorphism $\gamma_{2!}\iota^{-1}(\mathscr{C}_{M\times N} \otimes v_N) \to \mathscr{C}_M$. Hence we have obtained a sheaf homomorphism

$$(\gamma_2 \circ \gamma'_1)_!(k \circ j \circ \iota')^{-1}(\mathscr{C}_{M\times N_1\times N_2} \otimes v_{N_1}) \to \mathscr{C}_M$$

over $\sqrt{-1}S^*M$.

Since we are interested in replacing $w(x, y, y_1, y_2)\,dy_1$ by $K(x, y_1)\cdot u(y_2)\,dy_1$, we first consider the product of K and u. The product corresponds to a sheaf homormophism over L_3,

$$\beta_1^{-1}(\mathscr{C}_{M\times N_1} \otimes v_{N_1}) \times \beta_2^{-1}\mathscr{C}_{N_2} \to k^{-1}\mathscr{C}_{M\times N_1\times N_2} \otimes v_{N_1}.$$

Therefore, for a given $K(x, y_1)\,dy_1 \in H^0_Z(M \times N_1, \mathscr{C}_{M\times N_1} \otimes v_{N_1})$ we have a sheaf homomorphism

$$\beta_2^{-1}\mathscr{C}_{N_2} \to \mathscr{H}^0_{\beta_1^{-1}(Z)}(k^{-1}\mathscr{C}_{M\times N_1\times N_2} \otimes v_{N_1}).$$

From these sheaf homomorphisms we obtain

$$\begin{aligned}(\beta_2 \circ j \circ \iota')^{-1}\mathscr{C}_{N_2} &= \iota'^{-1}j^{-1}\beta_2^{-1}\mathscr{C}_{N_2}\\ &\to \iota'^{-1}j^{-1}\mathscr{H}^0_{\beta_1^{-1}(Z)}(k^{-1}\mathscr{C}_{M\times N_1\times N_2} \otimes v_{N_1})\\ &\to \mathscr{H}^0_{\iota'^{-1}j^{-1}\beta_1^{-1}(Z)}(\iota'^{-1}j^{-1}k^{-1}(\mathscr{C}_{M\times N_1\times N_2} \otimes v_{N_1})\\ &= \mathscr{H}^0_{(\beta_1\circ j\circ\iota')^{-1}(Z)}((k \circ j \circ \iota')^{-1}(\mathscr{C}_{M\times N_1\times N_2} \otimes v_{N_1})).\end{aligned}$$

Note that $p_1 = \gamma_2 \circ \gamma_1'$, $\beta_1 \circ j \circ \iota' = \mathrm{id}$, and that $\beta_2 \circ j \circ \iota' = p_2^a$ hold. Consequently, one obtains sheaf homomorphisms

$$p_{1!}(k \circ j \circ i')^{-1}(\mathscr{C}_{M\times N_1\times N_2} \otimes v_{N_1}) \to \mathscr{C}_M$$

and

$$(p_2^a)^{-1}\mathscr{C}_{N_2} \to \mathscr{H}^0_Z(k \circ j \circ \iota')^{-1}(\mathscr{C}_{M\times N_1\times N_2} \otimes v_{N_1}).$$

By restricting the latter homomorphism to Z and operating $p_{1!}$ to it, one obtains, combined with the former homomorphism, a sheaf homomorphism

$$(p_1|_Z)_!(p_2^a|_Z)^{-1}\mathscr{C}_N \to \mathscr{C}_M$$

over $\sqrt{-1}S^*M$. This homomorphism is the one which, for a given $K(x, y)\,dy \in H^0_Z(\sqrt{-1}S^*(M \times N), \mathscr{C}_{M\times N} \otimes v_N)$, assigns $\int K(x, y)u(y)\,dy$ for $u \in \mathscr{C}_N$.

Corollary. *Let M be a real analytic manifold. Then, for*

$$K(x, y)\,dy \in \mathscr{H}^0_{\sqrt{-1}S^*_M(M\times M)}(\mathscr{C}_{M\times M} \otimes v_M)$$

the integral operator $\mathscr{K}(u) = \int K(x, y)u(y)\,dy$ defines a sheaf homomorphism from $\mathscr{C}_M$ to $\mathscr{C}_M$.

Proof. Suppose $M = N$ in Theorem 3.4.1; then one has a sheaf homomorphism $(p_1|_Z)_!(p_2^a|_Z)^{-1}\mathscr{C}_M \to \mathscr{C}_M$. Note that in this case $p_1|_Z$ and $p_2^a|_Z$ are isomorphisms; in particular, they are proper and $(p_1|_Z) \circ (p_2^a|_Z)^{-1}$ is an identity on $\sqrt{-1}S^*M$. Hence, $(p_1|_Z)_!(p_2^a|_Z)^{-1}\mathscr{C}_M = \mathscr{C}_M$.

Remark. We sometimes call $K(x, x')\,dx'$ the kernel function of a microlocal operator $\mathscr{K}$. When there is no fear of confusion, the kernel function and the corresponding microlocal operator are identified and denoted with the same notation.

Definition 3.4.1. *$\mathscr{H}^0_{\sqrt{-1}S^*_M(M\times M)}(\mathscr{C}_{M\times M}\otimes v_M)$ is said to be the sheaf of microlocal operators and denoted by $\mathscr{L}_M$.*

Then we have the following corollary from the corollary of Theorem 3.4.1 and the above definition.

Corollary.

(1) *$\mathscr{L}_M$ acts on $\mathscr{C}_M$ as a sheaf homomorphism.*

(2) *Let $K_1(x, y)\,dy$ and $K_2(x, y)\,dy$ be elements of $\mathscr{L}_M$. Let $\mathscr{K}_1$ and $\mathscr{K}_2$ be the integral operators determined by K_1 and K_2 respectively. Then $\mathscr{K}_1 \circ \mathscr{K}_2$ is the microlocal operator having $(\int K_1(x, y')K_2(y', y)\,dy')\,dy$ as the kernel function.*

(3) *$\mathscr{L}_M$ is a ring with identity $\delta(x - y)\,dy$, which acts on $\mathscr{C}_M$ as an identity map.*

Definition 3.4.2. *Let $P(x, D_x) = \sum_{|\alpha|\leqq m} a_\alpha(x)D_x^\alpha$ be a linear differential operator of order m. The principal symbol $\sigma(P)(x, \xi)$ is defined as $\sigma(P)(x, \xi) = \sum_{|\alpha|=m} a_\alpha(x)\xi^\alpha \in \mathscr{O}_{T^*M}$.*

Note. $D_x^\alpha = (\partial/\partial x_1)^{\alpha_1}\cdots(\partial/\partial x_n)^{\alpha_n}$ for $\alpha = (\alpha_1, \ldots, \alpha_n) \in (\mathbf{Z}^+ \sqcup \{0\})^n$, $|\alpha| = \alpha_1 + \cdots + \alpha_n$, and $\xi^\alpha = \xi_1^{\alpha_1}\cdots\xi_n^{\alpha_n}$.

Remark 1. Since we seldom deal with non-linear differential operators in this book, we will simply call them differential operators.

Remark 2. A differential operator $P(x, D_x)$ is clearly a microlocal operator with the kernel function $P(x, D_x)\delta(x - y)\,dy$.

Remark 3. Even though the lower-order terms of $P(x, D_x)$ are not invariant under a coordinate transformation, the principal symbol is invariant under a coordinate transformation.

The main aim of the rest of this section is to give a proof of Sato's fundamental theorem, which is the microlocalization of John's construction of a fundamental solution for an elliptic differential equation; see John [1] and [2]. First we will recall the Cauchy-Kovalevsky theorem, which is the most fundamental in the theory of differential equations. Note that the Cauchy-Kovalevsky theorem is valid for the case of non-linear differential equations.

Theorem 3.4.2 (Cauchy-Kovalevsky). *Let $\psi(x)$ be a real analytic function defined in a neighborhood of x_0 such that the hypersurface $\{x\,|\,\psi(x) = 0\}$ is non-characteristic at $x = x_0$ with respect to a differential operator $P(x, D_x)$;*

i.e. $\sigma(P)(x_0, d\psi(x_0)) \neq 0$. *Then, for real analytic functions f and* φ *defined in a neighborhood of* $x = x_0$, *there exists a unique real analytic function* $u(x)$ *such that*

$$Pu = f \quad \text{and} \quad u \equiv \varphi \bmod \psi^m.$$

Proof. See, for example, Oshima and Komatsu [1].

We will prove Sato's fundamental theorem as a consequence of the Cauchy-Kovalevsky theorem. This fundamental theorem from Sato is so important and exquisite that it might not be an exaggeration to say that it was this theorem which convinced all analysts of the importance and the profoundness of the theory of microfunctions. Nowadays, the theorem is so fundamental among specialists that proving the theorem does not occur to them; but Sato's theorem was really revolutionary in the theory of differential equations.

Theorem 3.4.3 (Sato). *A linear differential operator of finite order* $P(x, D_x)$ *is left- and right-invertible in the ring* $\mathscr{L}_M$ *over* $\{(x, \sqrt{-1}\xi\infty) \in \sqrt{-1}S^*M \mid \sigma(P)(x, \sqrt{-1}\xi) \neq 0\}$.

Remark. We will later (Corollary of Theorem 4.1.6) prove a stronger result which claims that the left and right inverse is a microdifferential operator.

Proof. Since the statement is local, we will consider it in a neighborhood of $x = 0$ of a local coordinate system. Let $\sigma(P)(0, \sqrt{-1}\xi_0) \neq 0$. By Theorem 3.4.2, there is a real analytic function $u(x, \xi, p)$ defined in a neighborhood of $(x, \xi, p) = (0, \xi_0, 0) \in S^*M \times \mathbf{R}$ so that

$$P(x, D_x)u(x, \xi, p) = 1 \quad \text{and} \quad u \equiv 0 \bmod(\langle x, \xi\rangle - p)^m.$$

Note that $\sigma(P)(0, \xi_0) = (-\sqrt{-1})^m\sigma(P)(0, \sqrt{-1}\xi_0) \neq 0$. We sometimes call such a solution $u(x, \xi, p)$ a unitary solution after Leray. The method of obtaining a fundamental solution by superposing unitary solutions is sometimes called the Duhamel principle. (Generally speaking, the term "Duhamel principle" refers to obtaining a solution v of inhomogeneous equation $Pv = f$ by integrating a family of solutions of homogeneous equation $Pu = 0$ with respect to a parameter, assuming suitable initial conditions on parameterized surfaces.) For the solution $u(x, \xi, p)$, let $v(x, \xi, p) = u(x, \xi, p)Y(\langle x, \xi\rangle - p)$, where $Y(t)$ is the Heaviside function.

We need two lemmas.

Lemma 1. *One has*

$$D_1^j(u(x)Y(x_1)) = (D_1^j u(x))Y(x_1) + \sum_{k=1}^{j} D_1^{j-k}u(0, x')\delta^{(k-1)}(x_1)$$

where $x' = (x_2, \ldots, x_n)$.

Proof. It is plainly true for $j = 0, 1$. We will prove it by induction on j. Suppose the equation holds for $j = j_0$. Then

$$D_1^{j_0+1}(u(x)Y(x_1)) = (D_1^{j_0+1}u(x))Y(x_1) + (D_1^{j_0}u(x))\delta(x_1) + \sum_{k=1}^{j_0} D_1^{j_0-k}u(0, x')\delta^{(k)}(x_1)$$

$$= (D_1^{j_0+1}u(x))Y(x_1) + \sum_{k=1}^{j_0+1} D_1^{j_0+1-k}u(0, x')\delta^{(k-1)}(x_1)$$

holds. Hence, the inductive assumption implies the above equation and completes the proof.

Lemma 2. *Let $\sigma(P)(x, d\psi(x)) \neq 0$. A real analytic function $u(x)$ such that $P(x, D)u(x) = 1$ and $u \equiv 0 \bmod \psi(x)^m$ satisfies*

$$P(x, D_x)(u(x)Y(\psi(x))) = Y(\psi(x)).$$

Proof. One may assume $\psi = x_1$ after a suitable coordinate transformation, since $d\psi \neq 0$. Let $P(x, D) = \sum_{j=0}^{m} A_j(x, D')D_1^j$, where $A_j(x, D') = \sum_{\alpha' \in \mathbf{Z}_+^{n-1}} a_{\alpha'}(x)D'^{\alpha'}$ and $D' = (D_2, \ldots, D_n)$. By Lemma 1, we obtain

$$P(x, D)(u(x)Y(x_1)) = \sum_{j=0}^{m} A_j(x, D')D_1^j(uY(x_1))$$

$$= \sum_{j=0}^{m} A_j(x, D')(D_1^j u(x))Y(x_1) + \sum_{j=0}^{m}\sum_{k=1}^{j} A_j(x, D')(D_1^{j-k}u(0, x'))\delta^{(k-1)}(x_1).$$

On the other hand, since $u \equiv 0 \bmod x_1^m$ and $j - k < m$, one concludes that $D_1^{j-k}u(0, x') = 0$. Consequently, we have

$$P(x, D)(u(x)Y(x_1)) = \sum_{j=0}^{m} A_j(x, D')(D_1^j u(x))Y(x_1) = (Pu)Y(x_1) = Y(x_1),$$

since $Pu = 1$. This completes the proof of Lemma 2.

By Lemma 2, $P(x, D)v(x, \xi, p) = Y(\langle x, \xi\rangle - p)$ holds, and the n-time differentiation with respect to p gives

$$P(x, D_x)(\partial/\partial p)^n v(x, \xi, p) = (-1)^n\delta^{(n-1)}(\langle x, \xi\rangle - p).$$

Let $w(x, \xi, p) = (\partial/\partial p)^n v(x, \xi, p)$. We obtain

$$P(x, D_x)w(x, \xi, p) = (-1)^n\delta^{(n-1)}(\langle x, \xi\rangle - p).$$

If one lets $p = \langle x', \xi \rangle$, then one has

$$P(x, D_x)w(x, \xi, \langle x', \xi \rangle) = (-1)^n \delta^{(n-1)}(\langle x - x', \xi \rangle).$$

We have the following lemma.

Lemma 3. *In a neighborhood of* $(0, \sqrt{-1}\xi_0\infty)$, $\delta(x)$ *as a microfunction, i.e.* $\operatorname{sp} \delta(x)$, *is equal to*

$$\frac{(-1)^{n-1}}{(-2\pi\sqrt{-1})^n} \int \delta^{(n-1)}(\langle x, \xi \rangle)\chi_U(\xi)\omega(\xi),$$

where U *is a neighborhood of* $\xi_0 \in S^{n-1}$ *and the function* $\chi_U(\xi)$ *takes values* 1 *in* U *and* 0 *outside* $\bar{U}$.

Proof. First note that

$$\delta^{(n-1)}(t) = \frac{(-1)^{n-1}(n-1)!}{(-2\pi\sqrt{-1})}\left(\frac{1}{(t + \sqrt{-1}0)^n} - \frac{1}{(t - \sqrt{-1}0)^n}\right).$$

Then Proposition 3.2.3 and the above expression imply the conclusion of Lemma 3. The details are left for the reader, as an exercise.

We now integrate $P(x, D)w(x, \xi, \langle x', \xi \rangle) = (-1)^n \delta^{(n-1)}(\langle x - x', \xi \rangle)$ with respect to ξ in a neighborhood of ξ_0. Then, by Lemma 3, in a neighborhood of $(0, 0; \sqrt{-1}\langle \xi_0, d(x - x') \rangle)$, one obtains

$$P(x, D_x) \int_U w(x, \xi, \langle x', \xi \rangle)\omega(\xi) = (-1)^n \int_U \delta^{(n-1)}(\langle x - x', \xi \rangle)\omega(\xi)$$
$$= (-2\pi\sqrt{-1})^{n-1}\delta(x - x').$$

Therefore, if one lets

$$E(x, x') = \frac{1}{(-2\pi\sqrt{-1})^{n-1}} \int_U w(x, \xi, \langle x', \xi \rangle)\omega(\xi),$$

then, in a neighborhood of $(0, 0; \sqrt{-1}\langle \xi_0, d(x - x') \rangle\infty)$, one obtains

$$P(x, D_x)E(x, x') = \delta(x - x')$$

as an equality of microfunctions. Hence, if one can show that the singularity spectrum of $E(x, x')\,dx$ in a sufficiently small neighborhood of $(0, 0; \sqrt{-1}\langle \xi_0, d(x - x') \rangle\infty)$ is contained in $\sqrt{-1}S^*_M(M \times M)$, then $E(x, x')\,dx'$ is an element of $\mathscr{L}_M$. Denoting $E(x, x')\,dx'$ by E, one has $PE = 1$; i.e. E is a right inverse of P in $\mathscr{L}_M$. In fact, we have $PE = (\int P(x, D_x)\delta(x - y)E(y, x')\,dy)\,dx' = P(x, D_x)E(x, x')\,dx' = \delta(x - x')\,dx'$. Next we will consider the singularity spectrum of $E(x, x')dx'$. By the definition of u, $u(x, \xi, p)$ is a real analytic function defined in a neighborhood of $(x, \xi, p) = (0, \xi_0, 0)$. Hence $v(x, \xi, p) = u(x, \xi, p)Y(\langle x, \xi \rangle - p)$ is a hyperfunction defined in a neighborhood of $(x, \xi, p) = (0, \xi_0, 0)$, and its singularity spectrum must be in $\{\langle x, \xi \rangle = p, \pm\sqrt{-1}d(\langle x, \xi \rangle - p)\infty\}$. Therefore, by Theorem 3.2.1, the singularity spectrum of $E(x, x') = \int_U w(x, \xi, \langle x', \xi \rangle)\omega(\xi)$, in a sufficiently small neighborhood of $(0, 0; \sqrt{-1}\langle \xi_0, d(x - x')\rangle\infty)$, is restricted to $\{x = x', \sqrt{-1}\langle \xi, d(x - x')\rangle\infty\}$. Consequently, $E(x, x')\,dx$ is an element of $\mathscr{L}_M$ in a neighborhood of $(0, \sqrt{-1}\langle \xi_0, dx\rangle\infty)$.

Next we will show that there exists a microlocal operator E' such that $E'P = 1$. We will fix a volume element $dx = |dx_1 \wedge \ldots \wedge dx_n|$. For the sake of smoothness in the following argument, we will introduce the notion of a conjugate operator.

Definition 3.4.3. *Let $K(x, x')\,dx'$ be the kernel function of a microlocal operator $\mathscr{K}$. Then the microlocal operator defined by the kernel function $K(x', x)dx'$ is said to be the conjugate operator (or adjoint operator) of $\mathscr{K}$, denoted with $\mathscr{K}^*$.*

Remark 1. If $\mathscr{K}$ is a microlocal operator defined in a neighborhood of $(x_0, \sqrt{-1}\langle \xi_0, dx\rangle\infty)$, then $\mathscr{K}^*$ is a microlocal operator defined in a neighborhood of $(x_0, -\sqrt{-1}\langle \xi_0, dx\rangle\infty)$. That is, the conjugate operation * induces a sheaf homomorphism $\mathscr{L}_M \to \mathscr{L}^a_M$, where a is the antipodal map: $(x, \sqrt{-1}\langle \xi, dx\rangle\infty) \mapsto (x, -\sqrt{-1}\langle \xi, dx\rangle\infty)$. Note that we have $\mathscr{L}^a_M = a^*\mathscr{L}_M(=a^{-1}\mathscr{L}_M)$ by definition.

Remark 2. The notion of a conjugate operator depends upon the choice of a volume element.

Lemma 4.

(1) *Let $\mathscr{K}$, $\mathscr{K}_1$, and $\mathscr{K}_2$ be microlocal operators defined in a neighborhood of $(x_0, \sqrt{-1}\langle \xi_0, dx\rangle\infty)$. Then $(\mathscr{K}_1\mathscr{K}_2)^* = \mathscr{K}_2^*\mathscr{K}_1^*$ and $(\mathscr{K}^*)^* = \mathscr{K}$ hold.*

(2) *For a linear differential operator $P(x, D_x) = \sum_{|\alpha| \leqq m} a_\alpha(x)D_x^\alpha$, the conjugate operator $P^*(x, D_x)$ is a linear differential operator given by $\sum_{|\alpha| \leqq m} (-1)^{|\alpha|}D_x^\alpha(a_\alpha(x)\cdot)$, where $a_\alpha(x)\cdot$ denotes the multiplication by $a_\alpha(x)$.*

Proof. Statement (1) is plain by the definition of a conjugate operator. Let us prove (2). First we will find the conjugate operator of the multiplication operator $a_\alpha(x)\cdot$. Since the corresponding kernel function of $a_\alpha(x)\cdot$ is $a(x)\delta(x' - x)$, the kernel function corresponding to the conjugate operator $(a(x)\cdot)^*$ of $a(x)\cdot$ is $a(x')\delta(x' - x)$ by the definition. Hence, $a(x')\delta(x' - x) = a(x)\delta(x - x')$ holds, which implies $(a(x)\cdot)^* = a(x)\cdot$. Next we will find the conjugate operator of $D_j = \partial/\partial x_j$. The kernel function corresponding to D_j is $(\partial/\partial x_j)\delta(x - x')$. Therefore, the kernel function corresponding to D_j^*

is, by definition, given by $(\partial/\partial x'_j)\delta(x'-x) = (\partial/\partial x'_j)\delta(x'_1 - x_1)\delta(x'_2 - x_2)\cdots \delta(x'_n - x_n) = -(\partial/\partial x_j)\delta(x_1 - x'_1)\delta(x_2 - x'_2)\cdots\delta(x_n - x'_n)$. Hence, one has

$$(D^\alpha_x)^* = (D_1^{\alpha_1}\cdots D_n^{\alpha_n})^* = (-D_n)^{\alpha_n}\cdots(-D_1)^{\alpha_1} = (-1)^{|\alpha|}D^\alpha_x.$$

Consequently, one obtains

$$P^*(x, D_x) = \sum_{|\alpha|\leqq m}(a_\alpha(x)D^\alpha)^* = \sum_{|\alpha|\leqq m}(D^\alpha)^*(a_\alpha(x)\cdot)^* = \sum_{|\alpha|\leqq m}(-1)^{|\alpha|}D(a_\alpha(x)\cdot).$$

Now we return to the proof of Theorem 3.4.3. Recall that if $\sigma(P)(x, \sqrt{-1}\xi) \neq 0$ holds, then there exists a microlocal operator E in a neighborhood of $(x, \sqrt{-1}\langle\xi, dx\rangle\infty)$ such that $PE = 1$. Let P^* be the conjugate operator of P. Then $\sigma(P^*)(x, -\sqrt{-1}\xi) = \sigma(P)(x, \sqrt{-1}\xi) \neq 0$ holds. Hence there exists a microlocal operator E' such that $P^*E' = 1$ holds in a neighborhood of $(x, -\sqrt{-1}\langle\xi, dx\rangle\infty)$. Then the conjugate operator $(E')^*$ of E' is a microlocal operator defined in a neighborhood of $(x, \sqrt{-1}\langle\xi, dx\rangle\infty)$. One has $(E')^*P = (E')^*P^{**} = (P^*E')^* = 1$; i.e. $(E')^*$ is a left inverse of P. Furthermore, $(E')^* = (E')^*PE = (E'^*P)E = E$ holds. Therefore, for $P(x, D_x)$ there exists the left and right inverse E in $\mathscr{L}_M$ such that $PE = EP = 1$.

Remark. The most crucial point in the proof of Theorem 3.4.3 is the construction of $E(x, x')$ such that $P(x, D_x)E(x, x') = \delta(x - x')$. Such a hyperfunction $E(x, x')$ is sometimes called an elementary solution or a fundamental solution for the differential operator $P(x, D_x)$. Since $f(x) = \int f(x')\delta(x - x')\,dx'$ holds for an arbitrary hyperfunction $f(x)$, then $u(x) = \int E(x, x')f(x')\,dx'$ is a solution of $P(x, D_x)u(x) = f(x)$, provided $\int E(x, x')f(x')dx'$ makes sense and the differentiation and integration commute. The construction of a fundamental solution is one of the most effective methods in the local theory of linear differential equations.

Definition 3.4.4. *A linear differential operator $P(x, D_x)$ is said to be an elliptic operator at x_0 if for an arbitrary $\xi \in \mathbf{R}^n - \{0\}$, $\sigma(P)(x_0, \sqrt{-1}\xi) \neq 0$ holds.*

Theorem 3.4.4.

(i) *If hyperfunctions $u(x)$ and $f(x)$ satisfy $P(x, D_x)u(x) = f(x)$, then one has the inclusion*

$$\begin{aligned}\text{S.S. } u \subset \{&(x, \sqrt{-1}\langle\xi, dx\rangle\infty)\\ &\in \sqrt{-1}S^*M \,|\, \sigma(P)(x, \sqrt{-1}\xi) = 0\} \cup \text{S.S. } f.\end{aligned}$$

Particularly, if $P(x, D_x)$ is elliptic at arbitrary point on M, and if f is a real analytic function, then $u(x)$ is also a real analytic function.

(ii) *If $P(x, D_x)$ is elliptic at x_0, then $P: \mathscr{B}_{x_0} \to \mathscr{B}_{x_0}$ is an epimorphism.*

Note. The latter half of (i), above, asserts that Weyl's lemma holds. Weyl's lemma, in connection to Hilbert's 19th problem (see Hilbert [1]), is one of the most interesting results for applications in harmonic integral theory and for the theory of partial differential equations. Weyl's lemma triggered the development of the general theory of partial differential equations.

Proof of (i): The equation $P(x, D_x)u(x) = f(x)$ implies that $P(x, D_x)\,\mathrm{sp}(u(x)) = \mathrm{sp}(f(x))$. By Theorem 3.4.3, if $\sigma(P)(x, \sqrt{-1}\xi) \neq 0$, then $EP = 1$ for a microlocal operator E. Therefore, $\mathrm{sp}(u) = (EP)\,\mathrm{sp}(u) = E\,\mathrm{sp}(f)$ holds. From this, one obtains

$$\mathrm{S.S.}\, u \subset \{(x, \sqrt{-1}\xi\infty) \in \sqrt{-1}S^*M \,|\, \sigma(P)(x, \sqrt{-1}\xi) = 0\} \cup \mathrm{S.S.}\, f.$$

Next we will prove (ii). Theorem 3.4.3 implies that $P(x, D_x)$ has the right inverse E in a neighborhood of each point $(x_0, \sqrt{-1}\xi\infty)$. Since a left inverse also exists, the right inverse is unique. Therefore, the right inverse operator of P exists in a neighborhood of $\pi^{-1}(x_0)$, globally in the fibre direction. Since $\mathscr{B} \xrightarrow{\mathrm{sp}} \pi_*\mathscr{C} \to 0$ is exact, there exists $u \in \mathscr{B}_{x_0}$ such that E $\mathrm{sp}(f) = \mathrm{sp}(u)$; i.e. $\mathrm{sp}(Pu - f) = 0$ holds for such a u in a neighborhood of $\pi^{-1}(x_0)$. Hence, by (i), $g \underset{\mathrm{def}}{=} Pu - f$ is a real analytic function in a neighborhood of x_0. On the other hand, there exists a real analytic function v in a neighborhood of x_0 such that $Pv = g$ holds by Theorem 3.4.2. If one lets u be $u - v$, then $Pu = f$ holds in a neighborhood of x_0.

Remark. From (i) of Theorem 3.4.4, a solution $u(x)$ of $P(x, D_x)u(x) = 0$ has the singularity spectrum in $\{(x, \sqrt{-1}\xi\infty) \in \sqrt{-1}S^*M \,|\, \sigma(P)(x, \sqrt{-1}\xi) = 0\}$. Therefore, the study of the structures of solutions for $Pu = 0$ on this variety is our main goal in the theory of linear differential equations. (In general, the set of zeros of $\sigma(P)(x, \xi) = 0$ is called the characteristic variety.) This central problem has long been recognized in the case of equations with constant coefficients (i.e. Ehrenpreis' fundamental principle; see Ehrenpreis [1]). But one had to wait until the advent of microfunction theory to consider the above problem for the case of variable coefficients. (Though some distinguished experts, for example Hörmander [1] and Mizohata [1], had shown the trend implicitly, their virtuoso performances were really too ingenious to be appreciated by everybody.)

§5. The Wave Equation

In this section we will treat the wave equation, the most elementary example of partial differential equations of hyperbolic type, and consider the initial value problem.

Let $P(x, D_x)$ be a linear differential operator of order m, and let $\varphi_\nu(x')$, $0 \leqq \nu \leqq m - 1$, be a hyperfunction, where $x' = (x_2, \ldots, x_n)$. The Cauchy problem, or the initial value problem, is to find a hyperfunction $u(x)$ such that

$$\begin{cases} P(x, D_x)u(x) = 0 \\ \left(\dfrac{\partial}{\partial x_1}\right)^\nu u|_{x_1=0} = \varphi_\nu(x'), \qquad 0 \leqq \nu \leqq m-1 \end{cases}$$

hold. Note that $\varphi_\nu(x')$, $0 \leqq \nu \leqq m-1$, are called the Cauchy data, or the initial values.

Note. For given hyperfunctions $f(x)$ and $\varphi_\nu(x')$, $\nu = 0, \ldots, m-1$, consider the problem to find $u(x)$ such that

$$\begin{cases} P(x, D_x)u(x) = f(x) \\ \left(\dfrac{\partial}{\partial x_1}\right)^\nu u|_{x_1=0} = \varphi_\nu(x'), \qquad 0 \leqq \nu \leqq m-1. \end{cases}$$

This problem, as well, is sometimes called the Cauchy problem. In view of the terminology used for overdetermined systems, we will consider the case $f(x) = 0$ in this book.

The next proposition explains why we take $\{\varphi_\nu(x')\}_{\nu=0}^{m-1}$ as initial values.

Proposition 3.5.1. *Let $P(x, D_x)$ be a differential operator of order m. Assume that the hyperplane $\{x \mid x_1 = 0\}$ is non-characteristic with respect to P; i.e. if $x_1 = 0$, then $\sigma(P)(x, dx_1) \neq 0$. Then, for a hyperfunction u(x) such that $P(x, D_x)u(x) = 0$, the restriction $D_1^\nu u(x)|_{x_1=0}$ is well defined for any $\nu \in \mathbf{Z}^+ \cup \{0\}$, and all the $D_1^\nu u(x)|_{x_1=0}$ are uniquely determined by $D_1^\nu u(x)|_{x_1=0}$, $\nu = 0, \ldots, m-1$.*

Proof. Since $\{x \mid x_1 = 0\}$ is non-characteristic with respect to P, one has, by Theorem 3.4.4, $(x, \pm\sqrt{-1}\, dx_1\infty) \notin \text{S.S. } u(x)$ if $x_1 = 0$. Therefore, if $x_1 = 0$, then $(x, \pm\sqrt{-1}\, dx_1\infty) \notin \text{S.S. } D_1^\nu u(x)$ for an arbitrary ν. Hence, by Theorem 3.1.3, $D_1^\nu u(x)|_{x_1=0}$ is well defined. Next we will prove the uniqueness in the latter half of this proposition. Let

$$P(x, D_x) = \sum_{|\alpha| \leqq m} a_\alpha(x) D_x^\alpha = \sum_{j=0}^{m} A_j(x, D') D_1^j \cdot$$

Then $A_m(x, D') = a_{(m,0,\ldots,0)}(x) \neq 0$ since $\sigma(P)(0, dx_1) = a_{(m,0,\ldots,0)}(0) \neq 0$. Then, $Pu = \sum_{j=0}^{m} A_j(x, D') D_1^j u = 0$ implies

$$a_{(m,0,\ldots,0)}(x) D_1^m u = -\sum_{j=0}^{m-1} A_j(x, D') D_1^j u.$$

Hence, $D_1^m u|_{x_1=0} = 1/a_{(m,0,\ldots,0)}(x') \sum_{j=0}^{m-1} A_j(0, x', D')D_1^j u(x)|_{x_1=0}$ holds. That is, $D_1^m u|_{x_1=0}$ is uniquely determined by $D_1^\nu u(x)|_{x_1=0}$, $0 \leqq \nu \leqq m-1$. Similarly, if $D_1^{m+l} u|_{x_1=0}$, $l \geqq 0$, are given, then

$$D_1^{m+l+1} u = -D_1^{l+1}\left(\frac{1}{a_{(m,0,\ldots,0)}(x')} \sum_{j=0}^{m-1} A_j(0, x', D')D_1^j u\right)$$

implies that $D_1^{m+l+1} u|_{x_1=0}$ is determined uniquely.

Remark. As in the first half of the proof, if a hyperfunction $u(x)$ satisfies $(x, \pm\sqrt{-1}\, dx_1 \infty) \notin \text{S.S. } u(x)$, then $D_1^\nu u(x)|_{x_1=0}$ is always well defined. Hyperfunctions with this property, i.e. $(x, \pm\sqrt{-1}\, dx_1 \infty) \notin \text{S.S. } u(x)$, are said (Sato [1]) to be hyperfunctions containing real holomorphic parameters. (Since Sato [1] appeared long before the microfunction theory, the notation S.S. was not used there, but the concept is equivalent to the one above.) The above notion is essentially different from that of hyperfunctions containing holomorphic parameters in §8 of this chapter. (Not every hyperfunction containing real holomorphic parameters is obtained from the restriction of a hyperfunction containing holomorphic parameters to the real domain. The latter class of hyperfunctions is much more restrictive than the first one.) Because of this, we will not use such terminology in this book. The psychological effect of this term led us to consider whether the unique continuation theorem might hold, with respect to the real holomorphic parameter, when we attempted to grasp Holmgren's uniqueness theorem (Theorem 3.5.1) from the microfunction point of view. This may suggest to the reader the nature of the atmosphere in which Sato worked during the early stages of microfunction theory, when things were foggy for everyone except Sato. The section on "Hyperfunctions Containing Real Holomorphic Parameters" in Sato [1] gave the impression that the notion was not completely exposed: one had to wait for the appearance of microfunction theory to see the full picture.

In this way, the Cauchy problem for hyperfunction solutions has been formulated. Contrary to the case of real analytic functions (Theorem 3.4.2), there is generally no guarantee that a hyperfunction solution exists for the Cauchy problem. In fact, if $P(x, D_x)$ is elliptic, then $u(x)$ and, therefore, $D_1^\nu u|_{x_1=0}$ are all real analytic functions (Theorem 3.4.4 (i)). Hence, the Cauchy data must be analytic functions, which creates the necessity of introducing the notion of hyperbolic equations in the following section. However, if a hyperfunction solution exists for the Cauchy problem, then it is unique. This is Holmgren's theorem (Theorem 3.5.1). We have the following general proposition in light of the microfunction theory.

Proposition 3.5.2. *Let $\varphi(x)$ be a real-valued real analytic function on a real analytic manifold M such that $\varphi(x_0) = 0$ and $d\varphi(x_0) \neq 0$ at $x_0 \in M$. If* *a hyperfunction* $u(x)$ *on* M *satisfies* supp $u \subset \{x \in M \mid \varphi(x) \geqq 0\}$, *and if either* $(x_0, \sqrt{-1}\, d\varphi(x_0)\infty)$ *or* $(x_0, -\sqrt{-1}\, d\varphi(x_0)\infty)$ *is not contained in* S.S. u, *then* $u = 0$ *in a neighborhood of* x_0.

Proof. Since $d\varphi(x_0) \neq 0$ holds, after a coordinate transformation in a neighborhood of x_0, one may assume $(t, x) \in M = \mathbf{R}_t \times \mathbf{R}_x^n$, $\varphi = \varphi(t, x) = t - x_1^2 - \cdots - x_n^2$, and $x_0 = (t, x) = (0, 0)$. Then one has supp $u \subset \{(t, x) \mid t \geqq x_1^2 + \cdots + x_n^2\}$. Let us assume $(0, 0; \sqrt{-1}\, dt\infty) \notin$ S.S. $u(t, x)$. For $\xi \in S^{n-1}$, define

$$v(t, \xi, p) = \frac{(n-1)!}{(-2\pi\sqrt{-1})^n} \int \frac{u(t, x)}{(p - \langle x, \xi \rangle + \sqrt{-1}0)^n}\, dx.$$

If $t < \epsilon$, then $u(t, x) = 0$ for $|x|^2 > \epsilon$. Thus the above integral makes sense. (We did in fact choose the coordinate transformation so that the integral might make sense; this transformation is called the Holmgren transformation, which Holmgren used in the proof of his celebrated uniqueness theorem.) Next, substitute $p = \langle x, \xi \rangle$ in $v(t, \xi, p)$, and integrate with respect to ξ to obtain

$$\begin{aligned}
&\int_{S^{n-1}} v(t, \xi, \langle x, \xi \rangle)\omega(\xi) \\
&\quad = \frac{(n-1)!}{(-2\pi\sqrt{-1})^n} \int_{S^{n-1}} \omega(\xi) \int \frac{u(t, x')}{(\langle x - x', \xi \rangle + \sqrt{-1}0)^n}\, dx' \\
&\quad = \int u(t, x')\, dx' \left(\frac{(n-1)!}{(-2\pi\sqrt{-1})^n} \int_{S^{n-1}} \frac{\omega(\xi)}{(\langle x - x', \xi \rangle + \sqrt{-1}0)^n} \right) \\
&\quad = \int u(t, x')\delta(x - x')\, dx' = u(t, x).
\end{aligned}$$

We will first consider the singularity spectrum of $v(t, \xi, p)$. Recall that $(0, 0; \sqrt{-1}\, dt\infty) \notin$ S.S. $u(t, x)$ is assumed. If $|t| < \epsilon$ (hence $|x^2| < \epsilon$) and $a > M|\eta|$ hold for a sufficiently small $\epsilon > 0$ and a sufficiently large $M > 0$, then one has

$$(t, x; \sqrt{-1}(a\, dt + \langle \eta, dx \rangle)\infty) \notin \text{S.S. } u(t, x).$$

Therefore,

$$\widehat{\text{S.S.}}\, u(t, x) \cap \{|t| < \epsilon\} \subset \{(t, x; \sqrt{-1}(a\, dt + \langle \eta, dx \rangle)) \mid a \leqq M|\eta|\}$$

holds. On the other hand, we have

$$\begin{aligned}
\widehat{\text{S.S.}}(p - \langle x, \xi \rangle + \sqrt{-1}0)^{-n} \subset \{(p, x, \xi; \sqrt{-1}\alpha d(p - \langle x, \xi \rangle)) \mid \alpha \geqq 0 \\
\text{and } \alpha(p - \langle x, \xi \rangle) = 0\}.
\end{aligned}$$

Hence, Theorem 3.1.5 implies

$$\widehat{\text{S.S.}}(u(t,x)(p-\langle x,\xi\rangle+\sqrt{-1}0)^{-n}\cap\{|t|<\epsilon\}$$
$$\subset\{(t,x,\xi,p,\sqrt{-1}(a\,dt+\langle\eta-\alpha\xi,dx\rangle-\alpha\langle x,d\xi\rangle+\alpha\,dp)|a\leqq M|\eta|,$$
$$\alpha\geqq 0\text{ and }\alpha(p-\langle x,\xi\rangle)=0\}.$$

Then by Theorem 3.2.1, one obtains

$$\widehat{\text{S.S.}}\,v(t,\xi,p)\cap\{|t|<\epsilon\}\subset\{(t,\xi,p;\sqrt{-1}(a\,dt+\alpha\,dp-\alpha\langle x,d\xi\rangle)|\,|x|^2\leqq\epsilon,$$
$$a\leqq M\alpha|\xi|,\ \alpha\geqq 0\text{ and }\alpha(p-\langle x,\xi\rangle)=0\}.$$

Since $a\,dt+\alpha\,dp-\alpha\langle x,d\xi\rangle=\alpha(dp-\langle x,d\xi\rangle+M|\xi|\,dt)+(\alpha M|\xi|-a)(-dt)$ holds, and since $\alpha M|\xi|-a\geqq 0$ by the assumption, consequently S.S. $v(t,\xi,p)\cap\{|t|<\epsilon\}$ is contained in the convex hull of $\{(t,\xi,p;\sqrt{-1}(dp-\langle\zeta,d\xi\rangle+M|\xi|\,dt)\infty|\,|\zeta|\leqq\sqrt{\epsilon}\}\cup\{(t,\xi,p;-\sqrt{-1}\,dt\infty)\}$. Note that this convex hull is properly convex. Therefore, from Theorem 2.3.5, there exists a holomorphic function $\psi(t,\xi,p)$ such that $v(t,\xi,p)=b(\psi(t,\xi,p))$. Note that $\psi(t,\xi,p)$ is holomorphic in t for $|t|<\epsilon$. For $t<0$, $u(t,x)=0$ holds; i.e. $v(t,\xi,p)=0$ holds. The exact sequence $0\to\tilde{\mathscr{A}}\xrightarrow{b}\tau^{-1}\mathscr{B}$ implies that one has $\psi(t,\xi,p)=0$ for $t<0$. By the uniqueness of analytic continuation, we have $\psi(t,\xi,p)\equiv 0$. Then $v(t,\xi,p)=0$ holds. Hence we obtain $u(t,x)=\int v(t,\xi,\langle x,\xi\rangle)\omega(\xi)=0$.

Theorem 3.5.1 (Holmgren). *Let $P(x,D_x)$ be a linear differential operator of order m such that $\sigma(P)(0,dx_1)\neq 0$ holds. If a hyperfunction $u(x)$ satisfies $Pu=0$ and $D_1^\nu u|_{x_1=0}=0$, $0\leqq\nu\leqq m-1$, in a neighborhood of the origin, then $u=0$ holds in a neighborhood of $x=0$. That is, a hyperfunction solution for the Cauchy problem, if one exists, is unique.*

Proof. Since $\sigma(P)(0,dx_1)\neq 0$ implies $\sigma(P)(0,\pm\sqrt{-1}\,dx_1)\neq 0$, one has $(0,\pm\sqrt{-1}\,dx_1)\notin$ S.S. u by Theorem 3.4.4. Then $v(x)=u(x)Y(x_1)$ is well defined. Furthermore,

$$P(x,D)v=(P(x,D)u)Y(x_1)+\sum_{j=0}^{m}\sum_{k=1}^{j}A_j(x,D')(D_1^{j-k}u(0,x'))\delta^{(k-1)}(x_1)=0$$

holds (see Lemma 2 in the proof of Theorem 3.4.3). Therefore, $(0,\pm\sqrt{-1}\,dx_1\infty)\notin$ S.S. v by Theorem 3.4.4. On the other hand, one has $\operatorname{supp}v\subset\{x|x_1\geqq 0\}$. Then, in a neighborhood of $x=0$, $v(x)=u(x)Y(x_1)=0$ holds by Proposition 3.5.2. Similarly, one obtains $u(x)Y(-x_1)=0$. Consequently, $u(x)=u(x)(Y(x_1)+Y(-x_1))=0$ holds.

Note. $(d/dx_1)(Y(x_1)+Y(-x_1))=\delta(x_1)-\delta(-x_1)=0$ holds, and $Y(x_1)+Y(-x_1)=1$ for $x_1\neq 0$. Hence, $Y(x_1)+Y(-x_1)=1$ identically holds.

Remark. We will give refined statements, without proofs, of Holmgren's Theorem.

(1) The Split Watermelon Theorem (Morimoto [2]): If a hyperfunction $u(x)$ satisfies $\operatorname{supp} u \subset \{x \mid x_1 \geqq 0\}$, then there exists a closed convex cone G in $\mathbf{R}^{n-1}$ such that $(\overline{\text{S.S. } u}) \cap \{x = 0\} = \{(0, \sqrt{-1}\langle \xi, dx \rangle) \mid \xi' = (\xi_2, \ldots, \xi_n) \in G\}$. ["Split watermelon" is the name of a Japanese game.] The reader is advised to draw a figure for himself to consider why this name is given to the theorem.

(2) Let u be a hyperfunction on a real analytic manifold M. Denote $G = \operatorname{supp} u \subset M$ and $T = \text{S.S. } u \subset \sqrt{-1}S^*M$. Let φ be a real-valued real analytic function such that $\varphi(x_0) = 0$ and $d\varphi(x_0) \neq 0$. Then Proposition 3.5.2 can be rephrased: "If $x_0 \in G \subset \{x \mid \varphi(x) \geqq 0\}$ holds, then $(x_0, \pm\sqrt{-1}\, d\varphi(x_0)\infty) \in T$ holds" under the above notation. Furthermore, the following assertion is true (Bony [1]).

Let $\mathscr{I} = \{C^\infty\text{-functions on } \sqrt{-1}T^*M - M, \text{ homogeneous in } \xi, \text{ which are zero on } T\}$. Then let $\tilde{\mathscr{I}}$ be the smallest ideal among those containing $\mathscr{I}$ and with the following property: if f and g belong to $\tilde{\mathscr{I}}$, then the Poisson bracket

$$\{f, g\} = \sum_{i=1}^{n} \left(\frac{\partial f}{\partial \xi_i} \frac{\partial g}{\partial x_i} - \frac{\partial g}{\partial \xi_i} \frac{\partial f}{\partial x_i} \right) \in \tilde{\mathscr{I}}.$$

Note that for the zero set $\tilde{T}$ of $\tilde{\mathscr{I}}$ we have $\tilde{T} \subset T$. Then, for the pair $(G, \tilde{T})$ the corresponding statement of Proposition 3.5.2 holds. That is, if $x_0 \in G \subset \{x \mid \varphi(x) \geqq 0\}$ holds, then one has $(x_0, \pm\sqrt{-1}\, d\varphi(x_0)\infty) \in \tilde{T}$.

Now we consider a differential operator $P = D_t^2 - D_{x_1}^2 - \cdots - D_{x_n}^2$ on $M = R_t \times R_x^n$, which is called the d'Alambertian (or wave operator). The equation $Pu = 0$ is said to be the wave equation. We will study the structure of solutions of the Cauchy problem, using the wave equation as an example, and consider the problems raised by this example. We abbreviate the Laplacian $D_{x_1}^2 + \cdots + D_{x_2}^2$ as Δ_x or Δ.

By direct computation, we obtain

$$P(x_1^2 + \cdots + x_n^2 - t^2)^\lambda = -4\lambda\left(\lambda + \frac{n-1}{2}\right)(x_1^2 + \cdots + x_n^2 - t^2)^{\lambda - 1}.$$

On the other hand, $\tilde{u}_\pm = (x_1^2 + \cdots + x_n^2 - (t \pm i0)^2)^\lambda$ is well defined as a hyperfunction on $\mathbf{R}_x^n \times \mathbf{R}_t$ by Example 2.4.3. Then we have

$$\begin{aligned} \text{S.S. } \tilde{u}_\pm \subset \{&((x, t); \sqrt{-1}(\pm x_1/t, \ldots, \pm x_n/t, \mp 1)\infty) \mid t \neq 0, \\ &t^2 = x_1^2 + \cdots + x_n^2\} \cup \{((0, 0), \sqrt{-1}(\xi_1, \ldots, \xi_n, \eta)\infty) \mid \\ &\eta^2 \geqq \xi_1^2 + \cdots + \xi_n^2 \text{ and } \pm\eta > 0\}, \end{aligned}$$

where the order of the signs $\pm$ is taken respectively. In particular, we will consider the case $\lambda = (1-n)/2$. Let $u_\pm = (x_1^2 + \cdots + x_n^2 - (t \pm i0)^2)^{(1-n)/2}$. Then, for the d'Alambertian P, we have $Pu_\pm = 0$. Hence, by Theorem 3.4.4 (1), (S.S. $u_\pm) \cap \{t = x = 0\} \subset \{(0, 0); \sqrt{-1}(\xi_1, \ldots, \xi_n, \eta)\infty \,|\, \eta^2 = \xi_1^2 + \cdots + \xi_n^2$ and $\pm\eta > 0\}$. (We will show later that actually the equality holds.) Therefore, $u_\pm(t, x)$ can be restricted to $t = 0$. Similarly, we can restrict $(\partial/\partial t)u_\pm(t, x)$ to $t = 0$. For $x \neq 0$, we have $u_+(0, x) - u_-(0, x) = 0$ and

$$(\partial/\partial t)u_\pm(t, x)|_{t=0} = \frac{1-n}{2} t(x^2 - (t + i0)^2)^{(1-n)/2-1}|_{t=0} = 0.$$

That is, the hyperfunctions $u_+(0, x) - u_-(0, x)$ and $(\partial/\partial t)u_\pm(0, x)$ on $\mathbf{R}^n$ have their supports only at the origin $x = 0$. The following fact is known about the structure of hyperfunctions on $\mathbf{R}^n$ whose supports are only at the origin.

Proposition 3.5.3. *Let $u(x)$ be an element of $H^0_{\{0\}}(\mathbf{R}^n, \mathscr{B}_{\mathbf{R}^n})$. Then $u(x)$ can be expressed as $\sum_{\alpha \in (\mathbf{Z}^+ \cup \{0\})^n} a_\alpha \delta^{(\alpha)}(x)$, where $\mathbf{Z}^+ = \{1, 2, \ldots\}$ and $\delta^{(\alpha)}(x) = (\partial/\partial x_1)^{\alpha_1} \cdots (\partial/\partial x_n)^{\alpha_n}\delta(x)$. The coefficients $a_\alpha \in \mathbf{C}$ satisfy the following estimate: for an arbitrary $\epsilon > 0$, there exists C_ϵ such that*

$$|a_\alpha| \leqq C_\epsilon \frac{\epsilon^{|\alpha|}}{|\alpha|!} \qquad \text{for any } \alpha.$$

Note that a_α can be determined from $u(x)$ as $a_\alpha = (-1)^{|\alpha|}/\alpha! \int x^\alpha u(x)\, dx$.

Proof. Let

$$v(\xi, p) = \frac{(n-1)!}{(-2\pi i)^n} \int \frac{u(x)}{(p - \langle x, \xi \rangle - \sqrt{-1}0)^n} dx$$

Then one obtains $u(x) = \int_{S^{n-1}} v(\xi, \langle x, \xi \rangle)\omega(\xi)$ in the same way as in the proof of Proposition 3.5.2. Define

$$v(\zeta, \tau) = \frac{(n-1)!}{(-2\pi\sqrt{-1})^n} \int \frac{u(x)}{(\tau - \langle x, \zeta \rangle)^n} dx \qquad \text{for } \zeta \in \mathbf{C}^n \text{ and } \tau \in \mathbf{C}.$$

Then v is a holomorphic function defined for $\tau \neq 0$. Note that $u(x) = 0$ for $x \neq 0$. By definition, one has $v(\xi, p) = v(\xi, p + \sqrt{-1}0)$. Since $v(\zeta, \tau)$ is homogeneous in (ζ, τ) of degree $(-n)$, one can write $v(\zeta, \tau) = \sum_{\alpha \in (\mathbf{Z}^+ \cup \{0\})^n} a_\alpha \zeta^\alpha \tau^{-n-|\alpha|}$. In particular for $\tau = 1$, one has $\sum_{\alpha \in (\mathbf{Z}^+ \cup \{0\})^n} a_\alpha \zeta^\alpha$, which is an entire function. Hence the Cauchy-Hadamard theorem for power series implies that for an arbitrary $\epsilon > 0$ there exists C_ϵ such that

$|a_\alpha| \leqq C_\alpha \epsilon^{|\alpha|}$ holds for any $\alpha \in (\mathbf{Z}^+ \cup \{0\})^n$. Since $v(\xi, p) = \sum_\alpha a_\alpha \xi^\alpha (p + \sqrt{-1}0)^{-n-|\alpha|}$, then

$$u(x) = \int_{S^{n-1}} \sum_\alpha a_\alpha \xi^\alpha (\langle x, \xi \rangle + \sqrt{-1}0)^{-n-|\alpha|} \omega(\xi)$$

holds. On the other hand, one has $D_1(\langle x, \xi \rangle)^{-n} = -n\xi_1(\langle x, \xi \rangle)^{-n-1}$. Generally, one has

$$D_x^\alpha(\langle x, \xi \rangle)^{-n} = (-n)(-n-1)\cdots(-n-|\alpha|+1)\xi^\alpha(\langle x, \xi \rangle)^{-n-|\alpha|}.$$

Therefore,

$$\begin{aligned} u(x) &= \int \sum_\alpha \frac{a_\alpha}{(-n)(-n-1)\cdots(-n-|\alpha|+1)} D_x^\alpha(\langle x, \xi \rangle + \sqrt{-1}0)^{-n}\omega(\xi) \\ &= \sum_\alpha C_\alpha D_x^\alpha \frac{(n-1)!}{(2\pi\sqrt{-1})^n} \int \frac{\omega(\xi)}{(\langle x, \xi \rangle + \sqrt{-1}0)^n}, \end{aligned}$$

where

$$C_\alpha = \frac{a_\alpha}{(-n)(-n-1)\cdots(-n-|\alpha|+1)} \cdot \frac{(-2\pi\sqrt{-1})^n}{(n-1)!}.$$

By Feynman's lemma in the proof of Proposition 3.2.3,

$$\frac{1}{(-2\pi\sqrt{-1})^n} \int_{\epsilon_j \xi_j \geqq 0} \frac{\omega(\xi)}{\langle x, \xi \rangle^n} = \frac{\operatorname{sgn}(\epsilon_1 \cdots \epsilon_n)}{(-2\pi\sqrt{-1})^n} \cdot \frac{1}{x_1 \cdots x_n}$$

holds, where $\epsilon_1, \ldots, \epsilon_n = \pm 1$. If one lets

$$f(z) = \frac{1}{(-2\pi\sqrt{-1})^n} \sum_\alpha c_\alpha D_z^\alpha \frac{1}{z_1 \cdots z_n},$$

then one obtains $u(x) = \sum_{\epsilon_1, \ldots, \epsilon_n = \pm 1} \operatorname{sgn}(\epsilon_1 \cdots \epsilon_n) f(x_1 + \sqrt{-1}\epsilon_1 0, \ldots, x_n + \sqrt{-1}\epsilon_n 0) = \sum c_\alpha \delta^{(\alpha)}(x)$. Recall $\delta(x) = 1/(-2\pi\sqrt{-1})^n \sum_{\epsilon_1, \ldots, \epsilon_n = \pm 1} (\epsilon_1 \cdots \epsilon_n)/((x_1 + \sqrt{-1}\epsilon_1 0) \cdots (x_n + \sqrt{-1}\epsilon_n 0))$. Conversely, if $u(x)$ is expressed as $\sum_{\epsilon_1, \ldots, \epsilon_n = \pm 1} \operatorname{sgn}(\epsilon_1 \cdots \epsilon_n) f(x_1 + \sqrt{-1}\epsilon_1 0, \ldots, x_n + \sqrt{-1}\epsilon_n 0)$, then $u(x) = 0$ for $x_1 \neq 0$ holds. This is because

$$\begin{aligned} f(x_1 + \sqrt{-1}0, x_2 + \sqrt{-1}\epsilon_2 0, \ldots, x_n + \sqrt{-1}\epsilon_n 0) = f(x_1 - \sqrt{-1}0, x_2 \\ + \sqrt{-1}\epsilon_2 0, \ldots, x_n + \sqrt{-1}\epsilon_n 0) \end{aligned}$$

holds for $x_1 \neq 0$. Repeated arguments for $x_2, \ldots, x_n$ imply that $u(x) = 0$ for $x \neq 0$; i.e. $u(x)$ has its support only at the origin.

Lastly, we will find a formula for C_α when a hyperfunction having the support only at the origin $u(x)$ is given. For each j, $1 \leqq j \leqq n$, let γ_j be a closed path in the z_j-plane around the origin, oriented counter clockwise. Then, for a hyperfunction $u(x)$ having support only at the origin, one has

$$\int x^\alpha u(x)\,dx = (-1)^n \oint_{\gamma_1} \cdots \oint_{\gamma_n} z^\alpha f(z)\,dz.$$

By definition,

$$f(z) = \frac{1}{(-2\pi\sqrt{-1})^n} \sum_\alpha (-1)^{|\alpha|} \alpha! C_\alpha \frac{1}{z_1^{1+\alpha_1} \cdots z_n^{1+\alpha_n}}$$

holds. Hence, the coefficient of $1/(z_1 \cdots z_n)$ in $z^\alpha f(z)$ is given by $(1/(-2\pi i)^n)(-1)^{|\alpha|}\alpha! C_\alpha$. On the other hand, we have

$$\oint_{\gamma_i} \frac{dz_i}{z_i^n} = \begin{cases} 2\pi i & \text{for } n = 1 \\ 0 & \text{for } n \neq 1. \end{cases}$$

Therefore, $\int x^\alpha u(x)\,dx = (-1)^{|\alpha|}\alpha! C_\alpha$ holds; i.e. $C_\alpha = ((-1)^{|\alpha|}/\alpha!) \int x^\alpha u(x)\,dx$.

Corollary. *Let $u(x)$ be a hyperfunction on $\mathbf{R}^n$ whose support is only at the origin and such that the homogeneous degree is λ; i.e. Euler's formula $\sum_{i=1}^{n} (x_i D_i - \lambda)u = 0$ holds. Then the following two statements are true:*

(1) *If $\lambda \neq -n, -n-1, -n-2, \ldots$, then $u(x) = 0$.*

(2) *If $\lambda = -n - m$, $m \geqq 0$, then $u(x) = \sum_{|\alpha| = m} a_\alpha \delta^{(\alpha)}(x)$.*

Proof. From Proposition 3.5.3, $u(x)$ can be expressed as $u(x) = \sum_\alpha a_\alpha \delta^{(\alpha)}(x)$. Since $\left(\sum_{i=1}^{n} x_i D_i\right) \delta^{(\alpha)}(x) = -(n + |\alpha|)\delta^{(\alpha)}(x)$ holds, one has

$$\begin{aligned} 0 = \left(\sum_{i=1}^{n} x_i D_i - \lambda\right) u &= \sum_\alpha a_\alpha \left(\sum_{i=1}^{n} x_i D_i - \lambda\right) \delta^{(\alpha)}(x) \\ &= -\sum_\alpha (n + |\alpha| - \lambda) a_\alpha \delta^{(\alpha)}(x). \end{aligned}$$

Hence, $(n + |\alpha| - \lambda)a_\alpha = 0$ holds for any α; i.e. $a_\alpha = 0$ holds for $\lambda \neq n + |\alpha|$, which completes the proof.

We will now return to the wave equation. First we will consider $u_+(0, x) - u_-(0, x)$. Recall that $u_\pm(t, x) = (x_1^2 + \cdots + x_n^2 - (t \pm \sqrt{-1}0)^2)^{(1-n)/2}$; then we have $\left(\sum_{i=1}^{n} x_i D_i + n - 1\right) u_\pm(0, x) = 0$. That is, $u_+(0, x) - u_-(0, x)$ is a hyperfunction of homogeneous degree $(-n + 1)$, whose support is only at the origin. Then the above corollary implies $u_+(0, x) = u_-(0, x)$.

Next, we will study the structure of $((\partial/\partial t)u_\pm(0, x))$. Note first that

$$0 = D_t\left(tD_t + \sum_{i=1}^{n} x_i D_i + (n-1)\right)u_\pm(t, x)$$

$$= \left(tD_t + \sum_{i=1}^{n} x_i D_i + n\right)D_t u_\pm(t, x)$$

holds. Hence the restriction $t = 0$ provides $\sum_{i=1}^{n} (x_i D_i + n)D_t u_\pm(0, x) = 0$, which implies that $D_t u_\pm(0, x)$ is a homogeneous hyperfunction of degree $(-n)$. Again by the above corollary, $D_t u_\pm(0, x) = C_\pm \delta(x)$ for some constant $C_\pm$. $C_\pm$ can be computed as follows: $C_\pm = \int C_\pm \delta(x)\, dx = \int D_t u_\pm(0, x)\, dx$ holds, and the origin is the only support of $D_t u_\pm(0, x)$. Therefore, we have $C_\pm = \int_{|x| \leqq a} D_t u_\pm(0, x)\, dx$ for $a > 0$. On the other hand, we know

$$\text{S.S.}(D_t u_\pm(t, x)) \subset \{((t, x), \sqrt{-1}(\tau\, dt + \langle \xi, dx \rangle)\infty) | \tau^2 = |\xi|^2\}.$$

Hence one can define the product $D_t u_\pm(t, x) Y(a - |x|)$. Then, the integral

$$\psi_\pm(t) = \int_{|x| \leqq a} D_t u_\pm(t, x)\, dx = \int D_t u_\pm(t, x) \cdot Y(a - |x|)\, dx$$

$$= D_t \int u_\pm(t, x) Y(a - |x|)\, dx$$

is also well defined. (Notice that the integrand as a function of x has the compact support.)

In order to compute $\psi_\pm$, we will first compute $\int_{|x| \leqq a} (x^2 - (t + i0)^2)^\lambda\, dx$. This integral is the boundary value of the integral $\int_{|x| \leqq a} (x^2 - \tau^2)^\lambda\, dx$ from $\operatorname{Im} \tau > 0$, by Proposition 3.2.4; i.e. $\int_{|x| \leqq a} (x^2 - (t + i0)^2)^\lambda\, dx = b(\int_{|x| \leqq a} (x^2 - \tau^2)^\lambda\, dx)$. With the coordinate transformation $x = s\xi$ and $|\xi| = 1$, one has

$$\int_{|x| \leqq a} (x^2 - \tau^2)^\lambda\, dx = c \int_0^a (s - \tau^2)^\lambda s^{n-1}\, ds,$$

where c = the surface area of n-dimensional sphere $= 2\pi^{n/2}/\Gamma(n/2)$. Therefore, we obtain

$$\psi(t) = c \frac{d}{dt} \int_0^a (s^2 - (t + i0)^2)^\lambda s^{n-1}\, ds$$

$$= -2\lambda c t \int_0^a (s^2 - (t + i0)^2)^{\lambda - 1} s^{n-1}\, ds.$$

Let $t = \sqrt{-1}t'$ $(t' > 0)$. Then

$$\psi(\sqrt{-1}t') = -2\lambda c\sqrt{-1}t' \int_0^a (s^2 + t'^2)^{\lambda-1} s^{n-1}\, ds$$
$$= -2\lambda c\sqrt{-1}t' \int_0^{a/t'} t'^{2\lambda+n-2}(s^2+1)^{\lambda-1}s^{n-1}\, ds$$

holds. In particular, for $\lambda = (1-n)/2$ we have

$$\psi(\sqrt{-1}t') = (n-1)c\sqrt{-1}\int_0^{a/t'} (s^2+1)^{-(n+1)/2}s^{n-1}\, ds.$$

Consequently, we obtain

$$C_+ = \lim_{t'\to 0} \psi(\sqrt{-1}t') = (n-1)c\sqrt{-1}\int_0^{\infty} (s^2+1)^{-(n+1)/2}s^{n-1}\, ds$$
$$= (n-1)c\sqrt{-1}\,\frac{\sqrt{\pi}\,\Gamma\left(\frac{n}{2}\right)}{2\Gamma\left(\frac{1+n}{2}\right)} = \frac{2\pi^{(n+1)/2}}{\Gamma\left(\frac{n-1}{2}\right)}\cdot\sqrt{-1}.$$

By a similar computation for C_-, we obtain the final form

$$D_t u_\pm(0, x) = \pm\frac{2\pi^{(n+1)/2}\sqrt{-1}}{\Gamma\left(\frac{n-1}{2}\right)}\,\delta(x).$$

Note also from the above that we have

$$(\text{S.S. } u_\pm) \cap \{t = x = 0\} = \{(0, 0; \sqrt{-1}(\xi_1, \ldots, \xi_n, \eta)\infty) \,|\, \eta^2 = |\xi|^2, \pm\eta > 0\},$$

where the order of the double sign is taken respectively.

We will summarize what we have obtained as follows. Let $P = D_t^2 - \Delta$, and let

$$u_\pm = (x_1^2 + \cdots + x_n^2 - (t \pm \sqrt{-1}0)^2)^{(1-n)/2}.$$

Then $Pu_\pm = 0$ holds, and we also have

$$u_+(0, x) = u_-(0, x) \quad \text{and} \quad D_t u_\pm(0, x) = \pm\frac{2\pi^{(n+1)/2}\sqrt{-1}}{\Gamma\left(\frac{n-1}{2}\right)}\,\delta(x).$$

Define a hyperfunction $K_1(t, x)$ as

$$K_1(t, x) = \frac{\Gamma\left(\frac{n-1}{2}\right)}{4\sqrt{-1}\,\pi^{(n+1)/2}}\{(x_1^2 + \cdots + x_n^2 - (t + \sqrt{-1}0)^2)^{(1-n)/2}$$
$$- (x_1^2 + \cdots + x_n^2 - (t - \sqrt{-1}0)^2)^{(1-n)/2}\}.$$

Then we have

$$\begin{cases} (D_t^2 - \Delta)K_1(t, x) = 0 \\ K_1(0, x) = 0 \\ \dfrac{\partial K_1}{\partial t}(0, x) = \delta(x). \end{cases}$$

A similar computation to Example 2.4.3 provides $\operatorname{supp} K_1 \subset \{(t, x) \mid |x|^2 \leqq t^2\}$.

For this hyperfunction $K_1(t, x)$ we have the following theorem.

Theorem 3.5.2. *Let $K_0(t, x) = D_t K_1(t, x)$.*

(1) *If $(D_t^2 - \Delta)u(t, x) = 0$ holds on $\mathbf{R}^{n+1}$, then we have*

$$u(t, x) = \int K_0(t, x - x')u(0, x')\,dx' + \int K_1(t, x - x')\frac{\partial u}{\partial t}(0, x')\,dx'.$$

(2) *Conversely, for arbitrary hyperfunctions $\varphi_0(x)$ and $\varphi_1(x)$ on $\mathbf{R}^n$, one lets*

$$u(t, x) = \int K_0(t, x - x')\varphi_0(x')\,dx' + \int K_1(t, x - x')\varphi_1(x')\,dx'.$$

Then $u(t, x)$ satisfies

$$\begin{cases} (D_t^2 - \Delta)u(t, x) = 0 \\ u(0, x) = \varphi_0(x) \\ \dfrac{\partial u}{\partial t}(0, x) = \varphi_1(x). \end{cases}$$

That is, the Cauchy problem for the wave equation can be solved uniquely.

Proof. We will prove (2) first. As we noted before, we have

$$\begin{cases} (D_t^2 - \Delta)K_1 = 0 \\ K_1(0, x) = 0 \\ \dfrac{\partial K_1}{\partial t}(0, x) = \delta(x). \end{cases}$$

Hence, $(D_t^2 - \Delta)K_0 = D_t(D_t^2 - \Delta)K_1 = 0$ and $K_0(0, x) = \delta(x)$ hold. Furthermore, we also have $(\partial K_0/\partial t)(0, x) = 0$. This is because $D_t K_0 = D_t^2 K_1 = \Delta K_1$ implies $D_t K_0|_{t=0} = \Delta K_1|_{t=0} = \Delta(K_1|_{t=0}) = 0$. Notice that the support of K_1 (therefore of K_0) is restricted to $\{|x|^2 \leq t^2\}$. Then the integral defining $u(t, x)$ is well defined and, furthermore, we obtain the following:

$$(D_t^2 - \Delta)u(t, x) = \int [(D_t^2 - \Delta_x)K_0](t, x - x')\varphi_0(x')\, dx'$$
$$+ \int [(D_t^2 - \Delta_x)K_1](t, x - x')\varphi_1(x')\, dx' = 0,$$
$$u(0, x) = \int K_0(0, x - x')\varphi_0(x')\, dx' + \int K_1(0, x - x')\varphi_1(x')\, dx'$$
$$= \int \delta(x - x')\varphi_0(x')\, dx' = \varphi_0(x)$$

and

$$\frac{\partial u}{\partial t}(0, x') = \int \frac{\partial K_0}{\partial t}(0, x - x')\varphi_0(x')\, dx' + \frac{\partial K_1}{\partial t}(0, x - x')\varphi_1(x')\, dx'$$
$$= \int \delta(x - x')\varphi_1(x')\, dx' = \varphi_1(x),$$

completing the proof for (2).

In order to prove (1), note that one has the following:

$$\int_0^t \left(\int D_{t'}(D_t K_1(t - t', x - x')u(t', x'))\, dx' \right) dt'$$
$$= \int D_t K_1(0, x - x')u(t, x')\, dx' - \int D_t K_1(t, x - x')u(0, x')\, dx'$$
$$= u(t, x) - \int D_t K_1(t, x - x')u(0, x')\, dx'$$
$$= u(t, x) - \int K_0(t, x - x')u(0, x')\, dx'.$$

We also have

$$\int_0^t \left(\int D_{t'}[K_1(t - t', x - x')D_{t'}u(t', x')]\, dx' \right) dt'$$
$$= \int K_1(0, x - x')D_t u(t, x')\, dx' - \int K_1(t, x - x')(D_t u)(0, x')\, dx'$$
$$= -\int K_1(t, x - x')D_t u(0, x')\, dx'.$$

From the above, then, we obtain the following:

$$u(t, x) - \int K_0(t, x - x')u(0, x')\, dx' - \int K_1(t, x - x')(D_t u)(0, x')\, dx'$$
$$= \int_0^t \left(\int D_{t'}[K_1(t - t', x - x')D_{t'}u(t', x') + D_t K_1(t - t', x - x')u(t', x')]\, dx' \right) dt'$$
$$= \int_0^t \left(\int [K_1(t - t', x - x')D_{t'}^2 u(t', x') - (D_{t'}^2 K_1(t - t', x - x')u(t', x'))]\, dx' \right) dt'$$
$$= \int_0^t \left(\int [K_1(t - t', x - x')\Delta_{x'} u(t', x') - (\Delta_{x'} K_1)(t - t', x - x')u(t', x')]\, dx' \right) dt'.$$

Let $\Omega_{t'}$ be a domain with smooth boundary $\partial\Omega_{t'}$ in $\mathbf{R}^n_{x'}$ such that

$$\{x' \in \mathbf{R}^n \mid (t-t')^2 \leqq |x-x'|^2\} \subset \Omega_{t'}.$$

From the condition on the support of K_1, the Stokes theorem can be applied to the above. Then one obtains that the above integral is

$$\int_0^\infty \left[K_1(t-t', x-x') \sum_i \frac{\partial}{\partial x_i} u(t', u') \right.$$
$$\left. - \sum_i \left(\frac{\partial}{\partial x_i} K_1 \right)(t-t', x-x') u(t', x')\Big|_{\partial\Omega_{t'}} \right] dt' = 0,$$

which completes the proof of (1).

Remark. The pair of hyperfunctions $(K_0(t, x), K_1(t, x))$, in Theorem 3.5.2, is sometimes called a fundamental solution for the Cauchy problem.

Exercise. In Theorem 3.5.2, we consider the problem in $\mathbf{R}^{n+1}$. If $\varphi_0(x)$ and $\varphi_1(x)$ are given in $\Omega \subset \mathbf{R}^n$, then consider the domain where the hyperfunction $u(t, x)$ defined as

$$u(t, x) = \int K_0(t, x-x')\varphi_0(x')\, dx' + \int K_1(t, x-x')\varphi_1(x')\, dx'$$

is well defined. Generally, a domain in $\{(t, x) \mid t = 0\}$ which determines the solution near (t, x) is said to be a domain of dependence of u at (t, x). Conversely, when $\varphi_0(x)$ and $\varphi_1(x)$ vary in a neighborhood of x_0 in $\{(t, x) \mid t = 0\}$, a domain on which the solution $u(t, z)$ is affected is said to be a domain of influence.

One can show that the support of $K_0(t, x)$ in Theorem 3.5.2 is contained in $\{(t, x) \mid t^2 = |x|^2\}$ for any odd integer strictly greater than 1; otherwise by direct computation, it can be shown that the support is the entirety of $\{(t, x) \mid t^2 \leqq |x|^2\}$. (When this implication holds, one says that Huyghens' principle holds for an odd integer $n \geqq 3$, or that diffusion of waves does not occur.) This property—that the support of K_0 varies sensitively for the dimension n, which interested many analysts,—was crystallized as the lacuna theory in Petrowsky [1]. However, as Atiyah, Bott, and Gårding ([1] and [2]) indicated, the study of supports seems to be too specialized a topic in the whole structure of the theory of linear partial differential equations (in fact, Huyghens' principle holds as an extremely exceptional case). In the theory of microfunctions, to one's astonishment, the Huygens principle for microfunctions (sometimes called Huyghen's principle in the wider sense) holds quite generally. We will return to this topic in the next section. In this section, we will discuss how a microfunction solution of the wave equation propagates. First, we will recall the definition of a bicharacteristic strip in the classical theory of partial differential equations.

Definition 3.5.1. *Let $p_m(x, \xi)$ be the principal symbol $\sigma(P)$* of a linear differential operator *$P(x, D_x)$. An integral curve $(x(t), \xi(t))$ of*

$$\frac{dx_1}{\dfrac{\partial p_m}{\partial \xi_1}} = \cdots = \frac{dx_n}{\dfrac{\partial p_m}{\partial \xi_n}} = \frac{d\xi_1}{-\dfrac{\partial p_m}{\partial x_1}} = \cdots = \frac{d\xi_n}{-\dfrac{\partial p_m}{\partial x_n}}$$

with the property $p_m(x(t), \xi(t)) = 0$ is said to be a bicharacteristic strip of the equation $P(x, D_x)u(x) = 0$. The image $\{x(t)\}$ of the projection of a bicharacteristic strip onto the base space is called the bicharacteristic curve.

We obtain the following theorem, in terms of a bicharacteristic strip on the structure of microfunction solutions for the wave equation $(D_t^2 - \Delta_x)u(t, x) = 0$.

Theorem 3.5.3. *Let Ω be an open subset of $\sqrt{-1}S^*\mathbf{R}^{n+1}$. For a microfunction solution u and an arbitrary bicharacteristic strip b,* supp *$u \cap b$ is open and closed in $b \cap \Omega$; i.e.* supp *$u \cap b$ is a union of connected subsets of $b \cap \Omega$.*

Remark. This theorem implies that the microfunction solution of the wave equation propagates only along a bicharacteristic strip. On the other hand, the characteristic manifold $\{(t, x; \sqrt{-1}(\tau, \xi)\infty) \in \sqrt{-1}S^*\mathbf{R}^{n+1} | \tau^2 = \xi^2\}$ of the wave equation is covered by an n-parameter family of bicharacteristic strips. Hence, the structure problem, mentioned in the last section, for microfunction solutions of the wave equation is solved.

Proof of Theorem 3.5.3. Let $K(t, x)$ be a solution of

$$\begin{cases} (D_t^2 - \Delta_x)K(t, x) = 0 \\ K(0, x) = 0 \\ \dfrac{\partial K}{\partial t}(0, x) = \delta(x), \end{cases}$$

and let $E(t, x) = K(t, x)Y(t)$, where $Y(t)$ is the Heaviside function. First we need the following lemma.

Lemma.

(i) $(D_t^2 - \Delta_x)E(t, x) = \delta(t, x)$.

(ii) supp $E \subset \{(t, x) \in \mathbf{R}^{n+1} | |x| \leqq t\}$.

(iii) S.S. $E \subset \{(0, 0; \sqrt{-1}(\tau, \xi)\infty) \in \sqrt{-1}S^*\mathbf{R}^{n+1} | (\tau, \xi) \neq 0\}$
$\cup \{(t, x; \sqrt{-1}(\tau, \xi)\infty) \in \sqrt{-1}S^*\mathbf{R}^{n+1} | \tau^2 = \xi^2 \neq 0, t \geqq 0, x = -(\xi/\tau)t\}$.

Proof. We will prove (i) first. Since one has

$$D_tE(t, x) = (D_t(K(t, x))Y(t) + K(t, x)\delta(t) = (D_tK)Y(t) + K(0, x)\delta(t)$$
$$= (D_tK)Y(t),$$

then

$$D_t^2E(t, x) = (D_t^2K)Y(t) + D_tK(0, x)\delta(t) = (D_t^2K)Y(t) + \delta(t, x)$$

holds. On the other hand, $\Delta_xE(t, x) = (\Delta_xK)Y(t)$; hence one obtains

$$(D_t^2 - \Delta)E(t, x) = ((D_t^2 - \Delta)K)Y(t) + \delta(t, x) = \delta(t, x).$$

Statement (ii) follows from the facts $\operatorname{supp} K \subset \{(t, x) \in \mathbf{R}^{n+1} \mid |x|^2 \leqq t^2\}$ and $\operatorname{supp} Y(t) \subset \{(t, x) \in \mathbf{R}^{n+1} \mid t \geqq 0\}$. We will prove (iii). One has S.S. $E. \cap \{t \leqq 0\} \subset \{t = x = 0\}$ by (ii). Since $Y(t) = 1$ for $t > 0$, one has

$$E(t, x) = K(t, x) = \frac{\Gamma\left(\frac{n-1}{2}\right)}{4\sqrt{-1}\pi^{(n+1)/2}}\{(x_1^2 + \cdots + x_n^2 - (t + \sqrt{-1}0)^2)^{(1-n)/2}$$
$$- (x_1^2 + \cdots + x_n^2 - (t - \sqrt{-1}0)^2)^{(1-n)/2}\}.$$

By Example 2.4.3, as the boundary value from $\operatorname{Im} t > 0$ and $\operatorname{Im}(x_1^2 + \cdots + x_n^2 - t^2) < 0$, then $(x_1^2 + \cdots + x_n^2 - (t \pm \sqrt{-1}0)^2)^\lambda = (x_1^2 + \cdots + x_n^2 - t^2 \pm \sqrt{-1}0)^\lambda$ holds. Therefore, the singularity spectrum is contained in $\{(x, t; (\xi, \tau)\infty) \mid x_1^2 + \cdots + x_n^2 - t^2 = 0, \quad (\xi, \tau) = (2x, 2t)\}$. Hence, there exists a positive number $c > 0$ such that $\xi = cx$ and $\tau = -ct$. Then $x = -(\xi/\tau)t$ holds. On the other hand, one has $(D_t^2 - \Delta_x)E = 0$ $(t > 0)$. Hence, $\tau^2 = \xi^2 \neq 0$ holds by Theorem 3.4.4. This completes the proof.

Now we will prove Theorem 3.5.3, using the above lemma. It is sufficient to prove the following statement:

Let $u(t, x)$ be a microfunction defined in a neighborhood of $(0, 0; \sqrt{-1}(\tau_0, \xi_0)\infty)$, where $\tau_0^2 = \xi_0^2$, such that $(D_t^2 - \Delta)u(t, x) = 0$ holds. If $u(t, x)$ is 0 at $(0, 0; \sqrt{-1}(\tau_0, \xi_0)\infty)$, then $u(t, x)$ is 0 at $(t, -(t/\tau_0)\xi_0, \sqrt{-1}(\tau_0, \xi_0)\infty)$.

Define a microfunction $\tilde{u}$ as $\tilde{u} = u(t, x)Y(t)$. Then it is sufficient to prove $\tilde{u} = 0$. Notice that $(D_t^2 - \Delta_x)\tilde{u}(t, x) = 0$ clearly holds in a neighborhood of $\{(t, x; \sqrt{-1}(\tau, \xi)\infty) \mid \tau = \tau_0 \text{ and } \xi = \xi_0\}$. The above lemma (i) implies

$$\tilde{u}(t, x) = \int \delta(t - t', x - x')\tilde{u}(t', x')\, dt'\, dx'$$
$$= \int [(D_{t'}^2 - \Delta_{x'})E(t - t', x - x')]\tilde{u}(t', x')\, dt'\, dx'.$$

Through integration by parts and (ii) and (iii) of our lemma,

$$\tilde{u}(t, x) = \int E(t - t', x - x')(D_{t'}^2 - \Delta_{x'})\tilde{u}(t', x'))\, dt'\, dx' = 0$$

holds in a neighborhood of $\{(t, x; \sqrt{-1}(\tau, \xi)\infty) \in \sqrt{-1}S^*\mathbf{R}^{n+1} \mid \tau = \tau_0$ and $\xi = \xi_0\}$. Details are left as an exercise for the reader.

§6. Fundamental Solutions for Regularly Hyperbolic Operators

We constructed the inverse in $\mathscr{L}_M$ of a linear differential operator $P(x, D_x)$ under the provision that $\sigma(P)(x, \sqrt{-1}\xi) \neq 0$. In the preceding section, we discussed the solvability of an initial value problem relating to Holmgren's theorem. There we treated the wave operator as an example when one can explicitly construct a fundamental solution for the initial value problem. One can ask, then, how to locate wave operators among non-elliptic operators. In this section, we will show that an operator called a regularly hyperbolic operator has properties similar to those of the wave operator. The material in this section is connected with that in §3 of Chapter IV.

We begin with the definition of a regularly hyperbolic operator.

Definition 3.6.1. *A linear differential operator $P(x, D_x)$ is said to be hyperbolic at the origin in the direction* $(1, 0, \ldots, 0)$ *if*

$$\sigma(P)(x, \sqrt{-1}\xi) = a(x) \prod_{l=1}^{m} (\xi_1 - \lambda_l(x, \xi'))$$

such that $a(0) \neq 0$ and $\lambda_l(x, \xi')$ is real for $(x, \xi') \in \mathbf{R}^n \times \mathbf{R}^{n-1}$. A hyperbolic operator is said to be regularly hyperbolic when $\lambda_l(x, \xi') \neq \lambda_k(x, \xi')$ holds for $l \neq k$ and $\xi' \neq 0$.

Remark. The above definition is the most natural one for the study of hyperfunction solutions; for distribution solutions, however, the notion of hyperbolic operators is usually defined in a more narrow sense. The facts which led to the above definition are, first of all, that it is such an operator, as it has a fundamental solution for the initial value problem (see Gårding [1]), and secondly that in the framework of hyperfunction theory there exists a fundamental solution of the initial value problem for the operator defined as above (Bony and Schapira [1]). We will not treat this problem in this book, since it is rather technical. The micro-local version of a hyperbolic operator, called a micro-hyperbolic operator (Kashiwara and Kawai [3]), is theoretically more important. But it will not be considered since it is a bit beyond this book's scope.

For a regularly hyperbolic operator, we will locally construct a fundamental solution for the initial value problem and study its singularity spectrum. The following, very fundamental theorem of Hamada will be used most effectively to carry out our plan. Basically we will follow Kawai [2] for our treatment of this topic.

Theorem (Hamada). *Let* $P(z, D_z) = P(z_1, \ldots, z_n, \partial/\partial z_1, \ldots, \partial/\partial z_n)$ *be a linear differential operator with holomorphic coefficients defined in a neighborhood of* $z = 0 \in \mathbf{C}^n$. *Regarding* $\sigma_m(P)(z, \zeta_1, \zeta') = 0$, $\zeta' = (\zeta_2, \ldots, \zeta_n)$, *as an equation of* ζ_1, *assume that all the roots* $\lambda_1(z, \zeta'), \ldots, \lambda_m(z, \zeta')$ *are distinct in a neighborhood of* $(z_1, z'; \zeta') = (0, z'_0; \zeta'_0)$, $z' = (z_2, \ldots, z_n)$. *Then the following initial value problem* (3.6.2) *has a solution* $u(z, w', \zeta')$, *as in* (3.6.3). *We let* $\varphi_k(z, w', \zeta')$ *be a solution of the initial value problem*

$$\left.\begin{aligned} &\frac{\partial \varphi_k}{\partial z_1} - \lambda_k(z, \operatorname{grad}_{z'} \varphi_k) = 0 \\ &\varphi_k(z, w', \zeta')|_{z_1=0} = \langle z' - w', \zeta' \rangle \end{aligned}\right\} \tag{3.6.1}$$

$$\begin{cases} P(z, D_z)u(z, w', \zeta) = 0 \\ \dfrac{\partial^j}{\partial z_1^j} u(z, w', \zeta')|_{z_1=0} = \dfrac{c_j}{\langle z' - w', \zeta' \rangle^{l_j}} \qquad \text{for } j = 0, \ldots, m-1. \end{cases} \tag{3.6.2}$$

where c_j *is a constant and* l_j *is a positive integer.*

$$u = \sum_{k=1}^{m} u_k, \tag{3.6.3}$$

where

$$u_k(z, w', \zeta') = \sum_{j \geqq j_k} a_{j_k}(z, w', \zeta')\varphi_k(z, w', \zeta')^j + b_k(z, w', \zeta') \log \varphi_k(z, w', \zeta')$$

and where a_{j_k} *and* b_k *are holomorphic in a neighborhood of* $(0, z'_0, \zeta'_0)$.

Remark 1. Since $\log \varphi_k$ is many-valued in (3.6.2), perhaps one should consider (3.6.3) over a covering space to be precise. In what follows, we choose a branch of $\log \varphi_k$ so that the equation (3.6.3) may have a clear meaning.

Remark 2. It is known that one can have u_k such that $P(z, D_z)u_k = 0$.

Remark 3. It is the classical Hamilton-Jacobi theory which says that there exists a solution φ_k to the equation (3.6.1). Consult Yosida [1], pt. 2, chap. 4, for details. Notice, also, that $\operatorname{grad}_z \varphi_k(z, w', \zeta')|_{z=z(s)} = \zeta(s)$ holds on the bicharacteristic strip $(z(s), \zeta(s))_{|s|<<1}$ of $\zeta_1 - \lambda_k(z, \zeta')$. The Hamilton-Jacobi theory implies that the solution to (3.6.1) is a real-valued real analytic function, under the assumption that $\lambda_k(x, \xi')$ is real for a real (x, ξ'). From the uniqueness of a solution to the initial value problem (3.6.1), φ_k is homogeneous in ζ' of degree one since λ_k is homogeneous in ζ' of degree one (i.e. $\sum_{j=2}^{n} \zeta_j(\partial/\partial\zeta_j)\lambda_k(z, \zeta') = \lambda_k(z, \zeta')$). This is because $\varphi_k(z, w', c\zeta')/c$ $(c > 0)$ and $\varphi_k(z, w', \zeta')$ are both solutions to (3.6.1).

We will not prove the above theorem here. It is achieved by finding a formal solution of the above form and then proving its convergence.

Theorem 3.6.2. *Let a linear differential operator* $P(x, D_x)$ *be regularly hyperbolic at the origin in the direction of* $(1, 0, \ldots, 0)$. *Then there exists a solution* $E(x, y')$ *in a neighborhood of* $(x, y') = (0, 0)$, $y' = (y_2, \ldots, y_n)$ *to the initial value problem* (3.6.4), *and it is unique*:

$$\left.\begin{aligned} &P(x, D_x)E(x, y') = 0 \\ &\frac{\partial^j}{\partial x_1^j} E(x, y')|_{x_1=0} = \delta_{j,m-1}\delta(x' - y') \qquad \text{for } j = 0, \ldots, m-1 \end{aligned}\right\} \tag{3.6.4}$$

Proof. Considering the plane-wave decomposition of the δ-function in Proposition 3.2.3, we will solve the following initial value problem:

$$\left.\begin{aligned} &P(x, D_x)F(x, y', \xi') = 0 \\ &\frac{\partial^j}{\partial x_1^j} F(x, y', \xi')|_{x_1=0} = \frac{(n-2)\delta_{j,m-1}}{(2\pi\sqrt{-1})^{n-1}(\langle x' - y', \xi'\rangle + \sqrt{-1}0)^{n-1}}, \\ &\qquad\qquad j = 0, \ldots, m-1. \end{aligned}\right\} \tag{3.65}$$

Since $P(x, D_x)$ can be extended to an operator $P(z, D_z)$, defined in a neighborhood of the origin in $\mathbf{C}^n$, one can apply Theorem 3.6.1 to the case where

$$c_j = \frac{(n-2)!\delta_{j,m-1}}{(2\pi\sqrt{-1})^{n-1}} \quad \text{and} \quad l_j = n - 1.$$

As in Remark 3, $\varphi_k(z, w', \xi')$ is real if (z, w', ξ') is real and $\operatorname{grad}_z \varphi_k \neq 0$ holds for the points $\varphi_k = 0$. Hence, by Theorem 2.3.4 (see also Example 3.1.2), a single-valued holomorphic function in

$$\Omega_k = \{(z, w', \zeta') \in \omega \,|\, \operatorname{Im} \varphi_k(z, w', \zeta') > 0\},$$

where ω is a sufficiently small neighborhood of $(0, 0, \xi_0')$ for $\xi_0' \in \mathbf{R}^{n-1} - \{0\}$, defines a hyperfunction. On the other hand, φ_k^j and $\log \varphi_k$ are single-valued holomorphic functions in Ω_k. Therefore, $u_k(z, w', \zeta')$, which exists by Theorem 3.6.1, is a single-valued holomorphic function in Ω_k for a sufficiently small neighborhood ω. Since $P(z, D_z)u_k = 0$ holds from Remark 2 above, the hyperfunction $F_k(x, y', \xi')$, induced by u_k, satisfies $P(x, D_x)F_k(x, y', \xi') = 0$. Furthermore, one has $\varphi_k(z, w', \zeta')|_{z_1=0} = \langle z' - w', \zeta'\rangle$. Hence the restriction

$$\frac{\partial^j}{\partial z_1^j}\left(\sum_{k=1}^m u_k(z, w', \zeta')\right)\Bigg|_{z_1=0}$$

is a well-defined, single-valued holomorphic function in $\Omega_k \cap \{z_1 = 0\}$ and is equal to

$$\frac{(n-2)!\delta_{j,m-1}}{(2\pi\sqrt{-1})^{n-1}(\langle z' - w', \zeta' \rangle)^{n-1}}.$$

That is, one obtains

$$\frac{\partial^j}{\partial x_1^j}\left(\sum_{k=1}^m F_k(x, y', \xi')\right)\bigg|_{x_1=0} = \frac{(n-2)!\delta_{j,m-1}}{(2\pi\sqrt{-1})^{n-1}(\langle x' - y', \xi' \rangle + \sqrt{-1}0)^{n-1}}.$$

Lastly, we will construct a solution $E(x, y')$ by integrating $F_k(x, y', \xi')$ with respect to ξ' over $S_{\xi'}^{n-1}$. As we noted previously, $\varphi_k(z, w', \zeta')$ is homogeneous in ζ' of degree one. Hence, by Euler's identity, one has

$$\left(\sum_{j=2}^n \zeta \frac{\partial}{\partial \zeta_j} - j\right)\varphi_k^j = 0 \quad \text{and} \quad \sum_{j=2}^n \zeta_j \frac{\partial}{\partial \zeta_j} \log \varphi_k = 1.$$

Therefore, their boundary values satisfy the same equations. Then $F_k(x, y', \xi')|_{|\xi'|=1}$ is well defined from Proposition 3.5.1. Applying Theorem 3.2.1 to this case,

$$E_k(x, y') = \int_{\xi' \in S^{n-1}} F_k(x, y', \xi')\omega(\xi')$$

is well defined. Let $E(x, y') = \sum_{k=1}^m E_k(x, y')$. Then $E(x, y')$ satisfies the required equations. The uniqueness of $E(x, y')$ is obtained from Holmgren's theorem (Theorem 3.5.1).

Next we will study where the singularity spectrum of $E(x, y')$ is located. It will be self-evident how useful the microfunction theory is to the study of the singularity of a hyperfunction.

Theorem 3.6.3. *If* $(x, y'; \sqrt{-1}(\eta, \theta')\infty)$ *is in* S.S. $E(x, y')$, *then* $\theta' = -\xi'$, *and* (x, η) *is on the bicharacteristic strip of* $\eta_1 - \lambda_k(x, \eta')$ *going through* $(0, y'; \lambda_k(0, y', \xi'), \xi')$ *for some* $\xi' \in \mathbf{R}^{n-1} - \{0\}$.

Remark. Since we assume $\lambda_k(x, \eta') \neq \lambda_l(x, \eta')$ for $k \neq l$, the union of bicharacteristic strips of $\eta_1 - \lambda_k(x, \eta')$ is the one of P.

Corollary 1. *The hyperfunction* $E(x, y')$ *is real-analytic outside the set of bicharacteristic curves through* y' (*i.e. the image of the projection of a bicharacteristic strip on the base space*).

Remark. Such a set in Corollary 1 is called a characteristic conoid of $P(x, D_x)$ with its vertex at y'.

Corollary 2. *The support of* $E(x, y')$ *is contained in a cone with* y' *as its vertex*; *i.e. for the operator* P, *it has a finite propagation.*

Proof. Since P is regularly hyperbolic, a characteristic conoid with its vertex at y' is contained in some cone, provided $|x_1| \ll 1$ and $|x' - y'| \ll 1$. On the other hand, Holmgren's theorem implies that $E(x, y')$ is 0 in a neighborhood of $\{x | x_1 = 0 \text{ and } x' \neq y'\}$. Hence, by the uniqueness of the continuation of an analytic function, the assertion of Corollary 2 follows, since $E(x, y')$ is real-analytic outside the characteristic conoid.

Proof of Theorem 3.6.3. Denote a cotangent vector at $(x, y', \xi') \in \mathbf{R}^n \times \mathbf{R}^{n-1} \times (\mathbf{R}^{n-1} - \{0\})$ by (η, θ', τ'), where $\eta \in \mathbf{R}^n$ and $\theta', \tau' \in \mathbf{R}^{n-1}$. By the definition of Ω_k and by Theorem 2.3.4 (see, also, Example 3.1.2), one has

$$\begin{aligned}\text{S.S. } F_k(x, y', \xi') \subset \{(x, y', \xi'; \sqrt{-1}(\eta, \theta', \tau')\infty) | \varphi_k(x, y', \xi') = 0 \text{ and} \\ (\eta, \theta', \tau') = \alpha \operatorname{grad}_{(x,y',\xi')}\varphi_k(x, y', \xi'), (\alpha > 0)\}.\end{aligned} \tag{3.6.6}$$

Furthermore, by Euler's identities, satisfied by $\varphi_k(z, w', \zeta')^j$ and $\log \varphi_k(z, w', \zeta')$ in the proof of Theorem 3.6.2, and with Theorem 3.4.4, one obtains

$$\begin{aligned}\text{S.S. } F_k(x, y', \xi') \subset \Big\{(x, y', \xi'; \sqrt{-1}(\eta, \theta', \tau') | \varphi_k(x, y', \xi') = 0 \text{ and} \\ (\eta, \theta', \tau') = \alpha \operatorname{grad}_{(x,y',\xi')} \varphi_k(x, y', \xi'), (\alpha > 0) \\ \text{and } \sum_{j=2}^{n} \xi_j \tau_j = 0\Big\}.\end{aligned} \tag{3.6.7}$$

Therefore, Theorem 3.1.4 implies

$$\begin{aligned}\text{S.S.}(F_k(x, y', \xi')|_{S^{n-1}_{\xi'}}) \subset \Big\{(x, y', \xi'; \sqrt{-1}(\eta, \theta', \tau')\infty) | \varphi_k(x, y', \xi') = 0, \\ (\eta, \theta', \tau' + c\xi') = \alpha \operatorname{grad}_{(x,y',\xi')} \varphi_k(x, y', \xi'), \\ (\alpha > 0 \text{ and } c \in \mathbf{R}) \text{ and } \sum_{j=2}^{n} \xi_j(\tau_j + c\xi_j) = 0\Big\}.\end{aligned} \tag{3.6.8}$$

Generally, one has, for a submanifold M of $\mathbf{R}^n$, the exact sequence

$$0 \to T^*_M\mathbf{R}^n \to T^*\mathbf{R}^n \underset{\mathbf{R}^n}{\times} M \to T^*M \to 0. \tag{3.6.9}$$

In the equation (3.6.8), we used the fact that a point on T^*M is identified with $(x, \xi) \in T^*\mathbf{R}^n \cong \mathbf{R}^n \times \mathbf{R}^n$ modulo elements of $T^*_M\mathbf{R}^n$ (see Example 3.1.3). That is, τ' is actually τ' mod $\operatorname{grad}_{\xi'} \xi'^2$ or τ' mod ξ'. Therefore, one obtains, by Theorem 3.2.1,

$$\text{S.S. } \int_{S^{n-1}_{\xi'}} F_k(x, y', \xi')\omega(\xi') \subset \Lambda_k, \tag{3.6.10}$$

where $\Lambda_k = \{(x, y'; \sqrt{-1}(\eta, \theta')\infty) \mid \varphi_k(x, y', \xi') = 0$ for some $\xi' \in \mathbf{R}^n - \{0\}$, $(\eta, \theta') = \alpha\, \mathrm{grad}_{(x,y',\xi')}\, \varphi_k(x, y', \xi')$, $\alpha > 0$ and $\alpha\, \mathrm{grad}_{\xi'}\, \varphi_k(x, y', \xi') = 0\}$.

Note. By the above agreement, the last equation in Λ_k ought to be phrased as "$\alpha\, \mathrm{grad}_{\xi'}\, \varphi_k(x, y', \xi') + c'\xi' = 0$ for some $c' \in \mathbf{R}$." But the homogeneity of φ_k in ξ' implies that $0 = \langle \xi', \mathrm{grad}_{\xi'}\, \varphi_k(x, y', \xi') + c\xi' \rangle = \varphi_k(x, y', \xi') + c'\xi'^2 = c'\xi'^2 = c'$. If one pays close attention to the construction of F_k in Hamada [1], one notices, without going through the somewhat lengthy argument above, that $F_k(x, y', \xi')$ as a microfunction is homogeneous in ξ' of degree $n - 1$. Hence, one can regard $F_k\omega(\xi')$ as being defined on $S^{n-1}_{\xi'}$. In that case, one may take the normalized (ξ', τ') on $T^*S^{n-1}_{\xi'}$ so that $\langle \xi', \tau' \rangle = 0$ holds (Exercise: explain why). Then $\varphi_k(x, y', \xi') = 0$ implies $\langle \xi', \mathrm{grad}_{\xi'}\, \varphi_k(x, y', \xi') \rangle = 0$. Thus, one can obtain (3.6.10) directly from (3.6.6). We often use this kind of computational skill on homogeneous functions defined on a sphere (or on a projective space).

We have obtained an estimate on S.S. $E_k(x, y')$. Next, we shall consider the geometric meaning of Λ_k in (3.6.10). The next lemma gives its meaning.

Lemma. *On a bicharacteristic strip of* $\eta_1 - \lambda_k(x, \eta')$, $\mathrm{grad}_{\xi'}\, \varphi_k(x, y', \xi')$ *and* $\mathrm{grad}_{y'}\, \varphi_k(x, y', \xi')$ *are constant. In particular, one has* $\mathrm{grad}_{y'}\, \varphi_k = -\xi'$.

Proof. Let $(x(s), \eta(s))_{|s| << 1}$ be the bicharacteristic strip under consideration. Then the following (3.6.11) holds:

$$\begin{aligned}
\frac{d}{ds}&\left(\frac{\partial}{\partial \xi_j}\varphi_k(x(s), y', \xi')\right) \\
&= \sum_{l=1}^{n} \frac{\partial^2 \varphi_k}{\partial x_l\, \partial \xi_j} \frac{dx_l}{ds} \\
&= \frac{\partial}{\partial \xi_j}(\lambda_k(x, \mathrm{grad}_{x'}\, \varphi_k(x, y', \xi')))\Big|_{x = x(s)} \\
&\quad + \sum_{l=2}^{n} \left(\frac{\partial^2 \varphi_k}{\partial x_l\, \partial \xi_j}\left(\left(-\frac{\partial \lambda_k(x, \eta')}{\partial \eta_l}\right)\Big|_{\eta' = \mathrm{grad}_{x'}\varphi}\right)\right)\Big|_{x = x(s)} \\
&= \sum_{l=2}^{n} \left(\frac{\partial^2 \varphi_k}{\partial x_l\, \partial \xi_j}\left(\frac{\partial \lambda_k}{\partial \eta_l}\Big|_{\eta' = \mathrm{grad}_{x'}\varphi}\right)\right)\Big|_{x = x(s)} \\
&\quad - \sum_{l=2}^{n} \left(\frac{\partial^2 \varphi_k}{\partial x_l\, \partial \xi_j}\left(\frac{\partial \lambda_k}{\partial \eta_l}\Big|_{\eta' = \mathrm{grad}_{x'}\varphi}\right)\right)\Big|_{x = x(s)} \\
&= 0. \qquad (3.6.11)
\end{aligned}$$

Hence, $\mathrm{grad}_{\xi'}\, \varphi_k(x, y', \xi')$ is constant on the bicharacteristic strip, and similarly for the second assertion.

We return to the proof of Theorem 3.6.3. Let $(x, y'; \sqrt{-1}(\eta, \theta')\infty) \in \Lambda_k$. Take a bicharacteristic strip $(x(s), \eta(s))$ of $\eta_1 - \lambda_k(x, \eta')$ such that $(x(0), \eta(0)) = (x, \eta)$. Then p^* denotes the point of intersection with $\{x_1 = 0\}$. By the above lemma, $\mathrm{grad}_{\xi'} \varphi_k = 0$ holds at p^*. The equation (3.6.1) for φ_k implies $\mathrm{grad}_{\xi'} \varphi_k = x' - y'$ for $x_1 = 0$. Hence, one has $x' = y'$ at p^*. Again by (3.6.1), one has $\mathrm{grad}_x \varphi_k = (\lambda_k(0, y', \xi'), \xi')$ at p^*. From Remark 3 on the construction of φ_k, one concludes that (x, η) is on the bicharacteristic strip of $\eta_1 - \lambda_k(x, \eta')$ passing through $(0, y', \lambda_k(0, y, \xi'), \xi')$. The above lemma implies that $\theta' = -\xi'$.

Theorem 3.6.4. *Let a differential operator $P(x, D_x)$ be regularly hyperbolic in the direction of* $(1, 0, \ldots, 0)$ *at the origin. Then the following initial value problem* (3.6.12) *has a unique solution in a neighborhood of the origin, and the singularity spectrum of the solution $u(x)$ is contained in the set of bicharacteristic strips of $P(x, D_x)$ initiating from the singularity spectrums of the initial conditions $\mu_j(x')$:*

$$\left.\begin{aligned} &P(x, D_x)u(x) = 0 \\ &\frac{\partial^j}{\partial x_1^j} u(x)\Big|_{x_1=0} = \mu_j(x'), \qquad u = 0, 1, \ldots, m-1. \end{aligned}\right\} \tag{3.6.12}$$

Proof. By the same argument for Theorem 3.6.3, one can construct $E_j(x, y')$, satisfying

$$\begin{cases} P(x, D_x)E_j(x, y') = 0 \\ \dfrac{\partial^l}{\partial x_1^l} E_j(x, y')\Big|_{x_1=0} = \delta_{l,j}\delta(x' - y'), \qquad (0 \leqq l, j \leqq m-1) \end{cases}$$

and having the same singularity spectrum as $E(x, y')$ in Theorem 3.6.3. By defining $u(x)$ as

$$u(x) = \sum_{j=0}^{m-1} \int E_j(x, y')\mu_j(y')\, dy, \tag{3.6.13}$$

one can show the existence of a solution. The details are left to the reader. (One ought to check carefully that the integral is well defined. Then notice that the domain of the integration is compact. One can use the flabbiness of the hyperfunction sheaf to cut off $\mu_j(y')$ and then show that the operation does not affect the solution $u(x)$ in a sufficiently small neighborhood of the origin. See the problem below.) The uniqueness is nothing but Holmgren's theorem (Theorem 3.5.1). The statement on S.S. $u(x)$ is left to be proved. This should be done during the process of showing the well-definedness of the integral in (3.6.13). The reader is expected to prove it. One should then be careful to check that there is no contribution from the real analytic region of $E(x, y')$ or $\mu(y')$.

Problem. For what kind of domain does the uniqueness of a solution of (3.6.12) hold?

As in Theorems 3.6.3 and 3.6.4, bicharacteristic strips play an important role. One might think that the importance of bicharacteristic strips as carriers of singularities would have been recognized implicitly since the publication of Courant and Hilbert [1]. However, microfunction theory was needed to make possible the clear formulation given here. For example, consider the hyperfunction in Example 2.4.4 as a Cauchy datum. Then one can find a solution u, for a regularly hyperbolic operator P, such that $Pu = 0$ and S.S. u is contained in one bicharacteristic strip (see Kawai [2]). From this, for $P = \partial^2/\partial t^2 - \partial^2/\partial x_1^2 - \partial^2/\partial x_2^2$, we can easily find a solution u of $Pu = 0$ so that u is not analytic outside the cylinder $Z = \{(t, x_1, x_2) \in \mathbf{R}^3 | x_1^2 + x_2^2 \leqq 1\}$ and analytic in the interior of Z. (Hint: For each point on the boundary of Z, there is a bicharacteristic curve going through the point and tangent to Z.) The existence of such a solution, however, seems to have puzzled even such an expert as John in 1960. On finding such a solution, John says: "What is remarkable is that this cylinder is not a characteristic surface for the differential equation. Apparently not *all* types of singularities propagate along characteristic surfaces" (John [3], p. 574). For those of us who have succeeded in grasping the notion of bicharacteristic strips as building blocks for the singularities of a solution, the above fact is a natural consequence. This is a very good example for showing just how profoundly the theory of linear differential equations has advanced over the past twenty years. Furthermore, the notion of a bicharacteristic strip has its origins in contact geometry, and S^*M is the most typical contact manifold (for example, the classical Darboux theorem states that an arbitrary contact manifold is locally isomorphic to an open subset of S^*M as contact manifolds). Hence, it is natural that one should visualize the interplay between the theory of linear differential equations and contact geometry; note, also, that the theory of first order (non-linear) partial differential equations and contact geometry are two sides of the same coin. This visualization will be realized in the next chapter. There we reach a more sophisticated level, where we study not only the structure of solutions but also the structure of equations.

§7. The Flabbiness of the Sheaf of Microfunctions

The sheaf of hyperfunctions has a remarkable property of being a flabby sheaf, which provided a great clue to its usefulness in analysis. As we will show below, it is easy to prove that $\mathscr{B}_M/\mathscr{A}_M$ is a flabby sheaf. The next question is whether the sheaf of microfunctions is flabby or not. Since $\pi_*\mathscr{C}_M \cong \mathscr{B}_M/\mathscr{A}_M$ holds, the flabbiness of $\mathscr{C}_M$ would imply the flabbiness of $\mathscr{B}_M/\mathscr{A}_M$. But the converse does not follow immediately. We will give an affirmative answer to the question.

Theorem 3.7.1. *The sheaf of microfunctions is a flabby sheaf.*

Proof. Let $N = \sqrt{-1}S^*M$, and let $\pi: N \to M$. $\mathscr{B}_N$ and $\mathscr{A}_N$ denote the sheaf of hyperfunctions on N and the sheaf of real analytic functions on N respectively.

Lemma 1. *$\mathscr{B}_N/\mathscr{A}_N$ is a flabby sheaf.*

Proof. By Grauert's theorem (Theorem 1.2.3) one has $H^1(\Omega, \mathscr{A}_N) = 0$ for an (oriented) open subset Ω in N. Hence one has the canonical exact sequence

$$0 \to \mathscr{A}_N \to \mathscr{B}_N \to \mathscr{B}_N/\mathscr{A}_N \to 0.$$

From this, the sequence

$$\Gamma(\Omega, \mathscr{B}_N) \to \Gamma(\Omega, \mathscr{B}_N/\mathscr{A}_N) \to H^1(\Omega, \mathscr{A}_N) = 0$$

is exact. On the other hand, $\Gamma(N, \mathscr{B}_N) \to \Gamma(\Omega, \mathscr{B}_N) \to 0$ is also an exact sequence since $\mathscr{B}_N$ is flabby. From the commutative diagram

$$\begin{array}{ccccc}
\Gamma(N, \mathscr{B}_N) & \longrightarrow & \Gamma(N, \mathscr{B}_N/\mathscr{A}_N) & & \\
\downarrow & & \downarrow & & \\
\Gamma(\Omega, \mathscr{B}_N) & \longrightarrow & \Gamma(\Omega, \mathscr{B}_N/\mathscr{A}_N) & \longrightarrow & 0 \\
\downarrow & & & & \\
0 & & & &
\end{array}$$

one concludes that $\Gamma(N, \mathscr{B}_N/\mathscr{A}_N) \to \Gamma(\Omega, \mathscr{B}_N/\mathscr{A}_N)$ is an epimorphism. Hence, the sheaf $\mathscr{B}_N/\mathscr{A}_N$ is flabby.

We plan to derive the flabbiness of $\mathscr{C}_M$ from that of $\mathscr{B}_N/\mathscr{A}_N$. Suppose that there exist sheaf homomorphisms $\Phi: \mathscr{C}_M \to \mathscr{B}_N/\mathscr{A}_N$ and $\Psi: \mathscr{B}_N/\mathscr{A}_N \to \mathscr{C}_M$ such that $\Psi \circ \Phi = 1$ holds. Then $\Phi(u) \in \Gamma(\Omega, \mathscr{B}_N/\mathscr{A}_N)$ for $u \in \mathscr{C}_M(\Omega)$. By Lemma 1, there exists $s \in \Gamma(N, \mathscr{B}_N/\mathscr{A}_N)$ such that $s|_\Omega = \Phi(u)$ holds. Then one has $\Psi(s) \in \Gamma(N, \mathscr{C}_M)$ and $\Psi(s)|_\Omega = \Psi \circ \Phi(u) = u$; i.e. $\mathscr{C}_M$ is flabby. Therefore, we must construct Φ and Ψ such that $\Psi \circ \Phi = 1$. Flabbiness is a local property (see Theorem 1.1.3). Hence it is sufficient to prove for the case where $M = \mathbf{R}^n$ and $N = \sqrt{-1}S^*M \cong \mathbf{R}^n \times S^{n-1}$. Let us recall Theorem 3.4.1. Let $K(x, \xi) \in \mathscr{C}_N$ such that $\operatorname{supp} K(x, \xi) \subset \{(x, \xi; \sqrt{-1}\langle \xi, dx \rangle \infty) \in \sqrt{-1}S^*N \mid x = 0\}$. Then one has $\operatorname{supp} K(x - y, \xi)\, dy \subset Z = \{(x, \xi, y; \sqrt{-1}\langle \xi, d(x - y) \rangle \infty) \mid x = y\}$ for $K(x - y, \xi)\, dy \in \mathscr{C}_{N \times M} \otimes v_M$. From the definitions, for $z = (x, \xi, y; \sqrt{-1}\langle \xi, d(x - y) \rangle \infty) \in Z$, one has $p_1(z) = (x, \xi, \sqrt{-1}\langle \xi, dx \rangle \infty) \in \sqrt{-1}S^*N$ and $p_2^a(z) = (y, \sqrt{-1}\langle \xi, dy \rangle \infty) \in \sqrt{-1}S^*M$. Define a subset G of $\sqrt{-1}S^*N$ by $G = \{(x, \xi; \sqrt{-1}\langle \xi, dx \rangle \infty)\}$. Then $(p_1|_Z)_!(p_2^a|_Z)^{-1}\mathscr{C}_M = (\pi_N|_G)^{-1}\mathscr{C}_M$ holds.

Hence, by Theorem 3.4.1, there exists a sheaf homomorphism

$$(\pi_N|_G)^{-1}\mathscr{C}_M \to \mathscr{H}^0_G(\mathscr{C}_N).$$

That is, one obtains a sheaf homomorphism

$$\mathscr{C}_M \to (\pi_N|_G)_*\mathscr{H}^0_G(\mathscr{C}_N) = \pi_{N*}(\mathscr{H}^0_G(\mathscr{C}_N)) \subset \pi_{N*}(\mathscr{C}_N) = \mathscr{B}_N/\mathscr{A}_N.$$

We denote this homomorphism with Φ, i.e.

$$\Phi : u(x) \mapsto \int K(x - y, \xi)u(y)\, dy.$$

Next, let $T(x, \xi) \in \mathscr{C}_N$ satisfy

$$\operatorname{supp} T(x, \xi) \subset \{(x, \xi; \sqrt{-1}\langle \xi, dx\rangle\infty) | x = 0\}.$$

Then one has $\operatorname{supp} T(x - y, \xi) \subset \{(x, y, \xi; \sqrt{-1}\langle \xi, d(x - y)\rangle\infty) | x = y\} = Z$, and also $p_1(z) = (x; \sqrt{-1}\langle \xi, dx\rangle\infty) \in N = \sqrt{-1}S^*M$, $p^a_2(z) = (y, \xi; \sqrt{-1}\langle \xi, dy\rangle\infty) \in \sqrt{-1}S^*N$ for $z = (x, y, \xi; \sqrt{1}\langle \xi, d(x - y)\rangle\infty)$. Theorem 3.4.1 implies that there exists a sheaf homomorphism

$$(p_1|_Z)_!(p^a_2|_Z)^{-1}\mathscr{C}_N \to \mathscr{C}_M.$$

Since $(p_1|_Z)_!(p^a_2|_Z)^{-1}\mathscr{C}_N = \pi_{N*}(\mathscr{C}_N|_G) \subset \pi_{N*}(\mathscr{C}_N) = \mathscr{B}_N/\mathscr{A}_N$, consequently a sheaf homomorphism $\Psi : \mathscr{B}_N/\mathscr{A}_N \to \mathscr{C}_M$ has been obtained. That is,

$$\Psi : u(x, \xi) \mapsto \int T(x - y, \xi)u(y, \xi)\, dy\omega(\xi).$$

We will compute $\Psi \circ \Phi$ next:

$$\begin{aligned}\Psi \circ \Phi : u(x) \mapsto &\int T(x - x', \xi)\, dx'\omega(\xi) \int K(x' - y, \xi)u(y)\, dy\\ &= \int u(y)\, dy \int T(x - x', \xi)K(x' - y, \xi)\, dx'\omega(\xi)\\ &= \int u(y)\, dy \int T(x - y - x', \xi)K(x', \xi)\, dx'\omega(\xi).\end{aligned}$$

Therefore, if one can choose T and K such that

$$\int T(x - y - x', \xi)K(x', \xi)\, dx'\omega(\xi) = \delta(x - y),$$

then one obtains $\Psi \circ \Phi = 1$. Hence our proof rests upon the construction of $K(x, \xi)$ and $T(x, \xi)$ in $\mathscr{C}_N$ such that

$$\operatorname{supp} K(x, \xi) \quad \text{and} \quad \operatorname{supp} T(x, \xi) \subset \{(x, \xi; \sqrt{-1}\langle \xi, dx\rangle\infty) | x = 0\} \tag{3.7.1}$$

and

$$\int T(x - y, \xi)K(y, \xi)\, dy\omega(\xi) = \delta(x) \tag{3.7.2}$$

hold. In order to construct K and T, we need several lemmas in connection with the plane-wave decomposition of the δ-function. The following Lemma 2 is of particular interest, as it stands and has theoretical importance.

Lemma 2. *Let $\varphi_j(x, \xi)$ be a real analytic function defined in $\mathbf{R}^n \times \mathbf{R}^n$ for $j = 1, 2, \ldots, n$. Assume that* $\Phi(x, \xi) \underset{\text{def}}{=} (\varphi_1(x, \xi),\ \varphi_2(x, \xi), \ldots, \varphi_n(x, \xi))$

is homogeneous in ξ of degree one such that $\Phi(0, \xi) = \xi$. Furthermore, we assume that $\langle x, \Phi(x, \xi)\rangle$ is of positive type (see Definition 2.4.2). Then we have

$$\delta(x) = \frac{(n-1)!}{(-2\pi\sqrt{-1})^n} \int \frac{J(x, \xi)}{(\langle x, \Phi(x, \xi)\rangle + \sqrt{-1}0)^n}\, \omega(\xi),$$

where $J(x, \xi) = \det\,(\partial\Phi(x, \xi)/\partial\xi) = \det\,(\partial\varphi_i(x, \xi)/\partial\xi_j)_{1 \leqq i,j \leqq n}$.

Remark. The above equation is obtained formally from the plane-wave decomposition by the change of variables $\zeta = \Phi(x, \xi)$,

$$\delta(x) = \frac{(n-1)!}{(-2\pi\sqrt{-1})^n} \int \frac{\omega(\zeta)}{(\langle x, \xi\rangle + \sqrt{-1}0)^n}.$$

Since $\Phi(x, \xi)$ is not assumed to be real-valued, we need a proof.

Proof. First note that

$$F(x) = \int \frac{J(x, \xi)}{(\langle x, \Phi(x, \xi)\rangle + \sqrt{-1}0)^n}\, \omega(\xi)$$

is well defined as a hyperfunction since $\langle x, \Phi(x, \xi)\rangle$ is of positive type. Define a multivalued function Φ with

$$\Phi(z) = \int \frac{J(x, \xi)}{\langle z, \Phi(z, \xi)\rangle^n}\, \omega(\xi) \qquad \text{for } z \in \mathbf{C}^n.$$

If one can prove $z_j\Phi(z) = 0$ for $j = 1, \ldots, n$, then $J(x, \xi)/(\langle x, \Phi(x, \xi)\rangle + \sqrt{-1}0)^n$ does make sense by taking the boundary values of $J(z, \xi)/\langle z, \Phi(z, \xi)\rangle^n$ from $\operatorname{Im}\langle z, \Phi(z, \xi)\rangle > 0$, depending on ξ. Then $x_jF(x) = 0$ holds for $j = 1, \ldots, n$; i.e. the support of $F(x)$ is only at the origin. We will prove $z_j\Phi(z) = 0$ for $j = 1, \ldots, n$ first. We will denote the exterior differential operator and the interior product by d and ι respectively. Then, by direct computation,

$$d(f\iota_{\partial/\partial\zeta_j}\omega(\zeta)) = \frac{\partial f}{\partial\zeta_j}\, \omega(\zeta) + (-1)^j\left((n-1)f + \sum_k \zeta_k \frac{\partial f}{\partial\zeta_k}\right) d\zeta_1 \wedge \cdots \wedge d\zeta_{j-1} \wedge d\zeta_{j+1} \wedge \cdots \wedge d\zeta_n$$

(see Matsushima [1]). In particular, if $(1 - n)$ is the homogeneous degree of f, then $\partial f/\partial\zeta_j\omega(\zeta) = d(f\iota_{\partial/\partial\zeta_j}\omega(\zeta))$. From this equation, one obtains

$$z_jF(z) = \int \frac{z_jJ(z, \xi)}{\langle z, \Phi\rangle^n}\, \omega(\xi) = \int \frac{z_j}{\langle z, \zeta\rangle^n}\, \omega(\zeta) = \int \frac{\partial}{\partial\zeta_j}\left(\frac{1}{(1-n)\langle z, \zeta\rangle^{n-1}}\right) \omega(\zeta)$$
$$= \int d\left(\frac{1}{(1-n)\langle z, \zeta\rangle^{n-1}}\, \iota_{\partial/\partial\zeta_j}\omega(\zeta)\right) = 0.$$

Therefore, $F(x)$ is a hyperfunction whose support is only at the origin. On the other hand,

$$\left(\sum_{i=1}^{n} x_i D_i\right) F = \sum_{i=1}^{n} (D_i x_i - 1) F = -nF$$

holds. Hence, the homogeneous degree of F is $-n$. From the corollary of Proposition 3.5.3, one concludes $F(x) = c\delta(x)$. Next, we shall determine the constant c. Since $\operatorname{sp} F = c \operatorname{sp} \delta$, it is sufficient to consider a neighborhood of $(0, \sqrt{-1}d(x_1 + \ldots + x_n)\infty)$ for the computation of c. We let:

$$G_z = \{\zeta = \Phi(tz, \xi) \mid 0 \leqq t \leqq 1, \xi = (\xi_1, \ldots, \xi_n), \xi_1 + \ldots + \xi_n = 1, \xi_i \geqq 0 \text{ for } 1 \leqq i \leqq n\};$$

$$\Gamma_z = \{\zeta = \Phi(z, \xi) \mid \xi_1 + \ldots + \xi_n = 1, \xi_i \geqq 0 \text{ for } 1 \leqq i \leqq n\};$$

$$\Gamma_0 = \{\xi \mid \xi_1 + \ldots + \xi_n = 1, \xi_i \geqq 0 \text{ for } 1 \leqq i \leqq n\};$$

$$G_j = \{\zeta = \Phi(tz, \xi) \mid 0 \leqq t \leqq 1, \xi_1 + \ldots + \xi_n = 1, \xi_j = 0, \text{ and for an arbitrary } i (\neq j)\ \xi_i \geqq 0\}, j = 1, 2, \ldots, n.$$

Then the boundary ∂G_z of G_z is clearly $\Gamma_z \cup \Gamma_0 \cup G_1 \cup \ldots \cup G_n$. Stokes' theorem implies

$$\int_{\partial G_z} \frac{\omega(\zeta)}{\langle z, \zeta \rangle^n} = \int_{G_z} d \frac{\omega(\zeta)}{\langle z, \zeta \rangle^n} = 0,$$

since n-forms are 0 on the $(n-1)$-dimensional space. By taking the orientation into consideration, one has,

$$\int_{\Gamma_z} \frac{\omega(\zeta)}{\langle z, \zeta \rangle^n} = \int_{\Gamma_0} \frac{\omega(\zeta)}{\langle z, \zeta \rangle^n} + \sum_{j=1}^{n} \int_{G_j} \frac{\omega(\zeta)}{\langle z, \zeta \rangle^n}.$$

Therefore, in a neighborhood of $(0, \sqrt{-1}\, d(x_1 + \cdots + x_n)\infty)$, one gets

$$F(x) = b\left(\int_{\xi_i \geqq 0,\, 1 \leqq i \leqq n} \frac{J(z, \xi)}{\langle z, \Phi(z, \xi) \rangle^n} \omega(\xi)\right) = b\left(\int_{\Gamma_z} \frac{\omega(\xi)}{\langle z, \zeta \rangle^n}\right)$$

$$= b\left(\int_{\Gamma_0} \frac{\omega(\zeta)}{\langle z, \zeta \rangle^n} + \sum_{j=1}^{n} b\left(\int_{G_j} \frac{\omega(\zeta)}{\langle z, \zeta \rangle^n}\right)\right).$$

Then,

$$b\left(\int_{\Gamma_0} \frac{\omega(\zeta)}{\langle z, \zeta \rangle^n}\right) = \int \frac{\omega(\xi)}{(\langle x, \xi \rangle + \sqrt{-1}0)^n} = \frac{(-2\pi\sqrt{-1})^n}{(n-1)!} \delta(x)$$

holds. Therefore, if one can prove

$$b\left(\int_{G_j} \frac{\omega(\zeta)}{\langle z, \zeta \rangle^n}\right) = 0 \qquad \text{for all } j$$

as a microfunction, then one obtains

$$F(x) = \frac{(-2\pi\sqrt{-1})^n}{(n-1)!}\,\delta(x),$$

and

$$\delta(x) = \frac{(n-1)!}{(-2\pi\sqrt{-1})^n}\int \frac{J(x,\xi)}{\langle x, \Phi\rangle^n}\,\omega(\xi).$$

If a real analytic function $\varphi(x)$ is of positive type such that $\varphi(0) = 0$ and $d\varphi(0) = \xi$ hold, then by Proposition 2.4.3 one has $\varphi(z) \neq 0$ for a sufficiently small z such that $\langle \operatorname{Im} z, \xi\rangle > \epsilon|\operatorname{Im} z|$. Let us apply this to $\langle z, \Phi(z,\xi)\rangle = \langle z, \zeta\rangle$. Then one has $\langle z, \zeta\rangle \neq 0$ in G_j if $\sum_{k\neq j} (\operatorname{Im} z_k)\xi_k = \sum_k (\operatorname{Im} z_k)\xi_k > \epsilon|\operatorname{Im} z|$ holds. Let $D_j = \{z \mid \operatorname{Im} z_k > \epsilon|\operatorname{Im} z|$ for an arbitrary $k \neq j\}$. Hence, in particular, for $z \in D_j$ and $\zeta \in G_j$ one obtains $\langle z, \zeta\rangle \neq 0$. Therefore, $\int_{G_j} \omega(\zeta)/\langle z, \zeta\rangle^n$ is holomorphic on D_j. Consequently,

$$b\left(\int_{G_j} \frac{\omega(\zeta)}{\langle z, \zeta\rangle^n}\right) = b_{D_j}\left(\int_{G_j} \frac{\omega(\zeta)}{\langle z, \zeta\rangle^n}\right)$$

holds. On the other hand, D_j is a conoidal neighborhood of $U_j = \{(0 + \sqrt{-1}(v_1, \ldots, v_n)0) \mid v_k > \epsilon\sqrt{v_1^2 + \cdots + v_n^2}$ for any $k \neq j\}$. By Theorem 2.3.5, one can conclude that the singularity spectrum of $b_{D_j}(\int_{G_j} \omega(\zeta)/\langle z, \zeta\rangle^n)$ is contained in U_j°. But, by definition,

$$U_j^\circ \cap \{x = 0\} = \{(0, \sqrt{-1}(\xi_1, \ldots, \xi_n)\infty \,\Big|\, \sum_{i=1}^n \xi_i v_i > 0$$

for $(v_1, \ldots, v_n)$, such that $v_k > \epsilon\sqrt{v_1^2 + \cdots + v_n^2}$ for any $k \neq j\}$. Hence, $(0, \sqrt{-1}(1, \ldots, 1)\infty) \notin U_j^\circ$. In fact, for an arbitrary $k \neq j$, one can clearly find $(v_1, \ldots, v_n) \in \mathbf{R}^n$ such that $v_k > \epsilon\sqrt{v_1^2 + \cdots + v_n^2}$ and $\sum_{i=1}^n v_i < 0$ hold for a sufficiently small ϵ. Therefore, one obtains $b(\int_{Gj} \omega(\zeta)/\langle z, \zeta\rangle^n) = 0$ in a neighborhood of $(0, \sqrt{-1}(dx_1 + \cdots + dx_n))$.

Lemma 3. *Let* $\alpha, \beta \in \mathbf{C}$ *such that* $\operatorname{Re}\alpha = \operatorname{Re}\beta = 0$ *and* $\operatorname{Im}\alpha > 0$ *hold. Then one has*

$$\delta(x) = \frac{(n-1)!}{(-2\pi\sqrt{-1})^n}\int_{|\xi|=1}$$

$$\cdot\frac{(1-\beta\langle x,\xi\rangle)^{n-1}(1+(\alpha-\beta)\langle x,\xi\rangle+\alpha\beta(1-\beta\langle x,\xi\rangle))^{n-2}(|x|^2-\langle x,\xi\rangle^2)}{(\langle x,\xi\rangle+\alpha|x|^2-\beta\langle x,\xi\rangle^2+\sqrt{-1}0)^n}\cdot\omega(\xi).$$

In particular, for $\alpha = \beta$,

$$\delta(x) = \frac{(n-1)!}{(-2\pi\sqrt{-1})^n} \cdot \int_{|\xi|=1} \frac{(1-\alpha\langle x, \xi\rangle)^{n-1} + \alpha^2(1-\alpha\langle x, \xi\rangle)^{n-2}(|x|^2 - \langle x, \xi\rangle^2)}{(\langle x, \xi\rangle + \alpha(|x|^2 - \langle x, \xi\rangle^2) + \sqrt{-1}0)^n}\, \omega(\xi)$$

holds.

Note. The right-hand sides of the above equations are not homogeneous in ξ. Hence, our computation has to be done for $|\xi| = 1$.

Proof. Let $\Phi(x, \xi) = \xi + \alpha|\xi|x - (\beta\langle x, \xi\rangle/|\xi|)\xi$. Then Φ satisfies the conditions for Lemma 2. First we will compute $J(x, \xi) = \det(\partial\Phi(x, \xi)/\partial\xi)$. If one lets $\Phi(x, \xi) = (\varphi_1, \ldots, \varphi_n)$, then $\varphi_j(x, \xi) = \xi_j + \alpha|\xi|x_j - (\beta\langle x, \xi\rangle/|\xi|)\cdot \xi_j$. Hence, $J(x, \xi)$ can be computed from $d\varphi_1 \wedge \cdots \wedge d\varphi_n = J(x, \xi)d\xi_1 \wedge \cdots \wedge d\xi_n$. One has $d\varphi_i = (1 - \beta\langle x, \xi\rangle)\, d\xi_i + (\alpha x_i + \beta\langle x, \xi\rangle\xi_i)d|\xi| - \beta\xi_i\langle x, d\xi\rangle$ and $d|\xi| = \sum_{i=1}^{n} \xi_i\, d\xi_i$ since $|\xi| = 1$. Then, one obtains

$$\begin{aligned}
&d\varphi_1 \wedge \cdots \wedge d\varphi_n \\
&\quad = (1 - \beta\langle x, \xi\rangle)^n\, d\xi_1 \wedge \cdots \wedge d\xi_n + (1 - \beta\langle x, \xi\rangle)^{n-1} \\
&\qquad \cdot \left(\sum_{i=1}^{n} (\alpha x_i + \beta\langle x, \xi\rangle\xi_i)\, d\xi_1 \wedge \cdots \wedge \overset{i}{\overbrace{d|\xi|}} \wedge \cdots \wedge d\xi_n\right) \\
&\qquad + (1 - \beta\langle x, \xi\rangle)^{n-1} \\
&\qquad \cdot \left(\sum_{i=1}^{n} (-\beta\xi_i)\, d\xi_1 \wedge \cdots \wedge \overset{i}{\overbrace{\langle x, d\xi\rangle}} \wedge \cdots \wedge d\xi_n\right) \\
&\qquad + (1 - \beta\langle x, \xi\rangle)^{n-2} \cdot \\
&\qquad \cdot \sum_{i \neq j} (\alpha x_i + \beta\langle x, \xi\rangle\xi_i)(-\beta\xi_j)\, d\xi_1 \wedge \cdots \wedge \overset{i}{\overbrace{d|\xi|}} \wedge \cdots \wedge \overset{j}{\overbrace{\langle x, d\xi\rangle}} \wedge \cdots \wedge d\xi_n \\
&\quad = [(1 - \beta\langle x, \xi\rangle)^n + (1 - \beta\langle x, \xi\rangle)^{n-1}(\alpha + \beta)\langle x, \xi\rangle) \\
&\qquad - (1 - \beta\langle x, \xi\rangle)^{n-1}\beta\langle x, \xi\rangle + \alpha\beta(|x|^2 \\
&\qquad - \langle x, \xi\rangle^2)(1 - \beta\langle x, \xi\rangle)^{n-2}]\, d\xi_1 \wedge \cdots \wedge d\xi_n \\
&\quad = [(1 - \beta\langle x, \xi\rangle)^{n-1}(1 + (\alpha - \beta)\langle x, \xi\rangle) \\
&\qquad + \alpha\beta(1 - \beta\langle x, \xi\rangle)^{n-2}(|x|^2 - \langle x, \xi\rangle^2)]\, d\xi_1 \wedge \cdots \wedge \xi_n.
\end{aligned}$$

Note. The notation $\overset{i}{\overbrace{d|\xi|}}$ means that $d\xi_i$ is replaced by $d|\xi|$.

Lemma 4. *For* $\alpha \in \mathbf{C}$, $\operatorname{Im} \alpha > 0$, *one can express*

$$\delta(x) = \sum_{j,k} c_{jk} \int_{|\xi|=1} \frac{(|x|^2 - \langle x, \xi\rangle^2)^j}{(\langle x, \xi\rangle + \alpha(|x|^2 - \langle x, \xi\rangle^2) + \sqrt{-1}0)^{n-k}}\, \omega(\xi)$$

for some finitely many $c_{jk} \in \mathbf{C}$.

Proof. Let $t = \langle x, \xi\rangle + \alpha(|x|^2 - \langle x, \xi\rangle^2)$, and let $y = |x|^2 - \langle x, \xi\rangle^2$. Then one can verify easily that there exists $c_{jk} \in \mathbf{C}$ such that

$$\frac{(n-1)!}{(-2\pi\sqrt{-1})^n}\frac{(1-\alpha\langle x, \xi\rangle)^{n-1} + \alpha^2(1-\alpha\langle x, \xi\rangle)^{n-2}(|x|^2 - \langle x, \xi\rangle^2)}{(\langle x, \xi\rangle + \alpha(|x|^2 - \langle x, \xi\rangle^2))^n}$$
$$= \frac{(n-1)!}{(-2\pi\sqrt{-1})^n}\frac{(1-\alpha t + \alpha^2 y)^{n-1} + \alpha^2 y(1-\alpha t + \alpha^2 y^{n-2})}{t^n} = \sum_{j,k} c_{jk}\frac{y^j}{t^{n-k}}$$

By Lemma 3, one obtains

$$\delta(x) = \sum_{j,k} c_{jk}\int_{|\xi|=1}\frac{y^j}{(t+\sqrt{-1}0)^{n-k}}\,\omega(\xi)$$
$$= \sum_{j,k} c_{jk}\int_{|\xi|=1}\frac{(|x|^2 - \langle x, \xi\rangle^2)^j}{(\langle x, \xi\rangle + \alpha(|x|^2 - \langle x, \xi^2) + \sqrt{-1}0)^{n-k}}\,\omega(\xi).$$

Lemma 5. *The following three equations hold:*

(1) $$\int (t+\sqrt{-1}0)^{\alpha}(x-t+\sqrt{-1}0)^{\beta}\,dt = 2\pi\sqrt{-1}\,\frac{\Gamma(-\alpha-\beta-1)}{\Gamma(-\alpha)\Gamma(-\beta)}(x+\sqrt{-1}0)^{\alpha+\beta+1}.$$

(2) $$\int (x_1^2 + \cdots + x_n^2 + c)^{\lambda}\,dx_1\cdots dx_n = \pi^{n/2}\frac{\Gamma\left(-\lambda-\dfrac{n}{2}\right)}{\Gamma(-\lambda)}c^{\lambda+n/2},\ c > 0.$$

(3) $$\int (\langle x-y, \xi\rangle + \alpha(|x-y|^2 - \langle x-y, \xi\rangle^2))^{\lambda}(\langle y, \xi\rangle + \beta(|y|^2 - \langle y, \xi\rangle^2))^{\mu}\,dy$$
$$= -2\pi^{(n+1)/2}\sqrt{-1}\,\frac{\Gamma\left(-\lambda-\mu-\dfrac{n+1}{2}\right)}{\Gamma(-\lambda)\Gamma(-\mu)}(\alpha+\beta)^{-(n-1)/2}\cdot\left(\langle x, \xi\rangle + \frac{\alpha\beta}{\alpha+\beta}(|x|^2 - \langle x, \xi\rangle^2)\right)^{\lambda+\mu+(n+1)/2},$$

where Im α, Im $\beta > 0$.

Proof. First note that $u(x) = \int (t+\sqrt{-1}0)^{\alpha}(x-t+\sqrt{-1}0)^{\beta}\,dt$ is well defined as the integration of a microfunction. Then

$$\operatorname{supp} u(x) = \{0, \sqrt{-1}\,dx\infty\}$$

holds. On the other hand, it is clear that $u(cx) = c^{\alpha+\beta+1}u(x)$ holds. Therefore, one obtains $u(x) = c(x+\sqrt{-1}0)^{\alpha+\beta+1}$. Next we will compute c. By

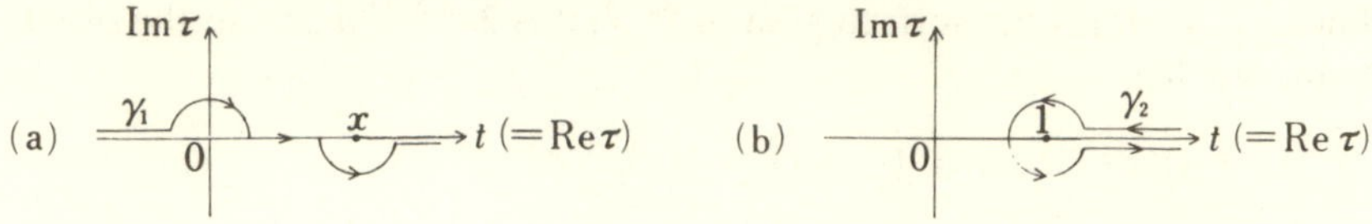

Figure 3.5.1

the definition of integration of a microfunction, one has

$$u(x) = \int_{\gamma_1} \tau^\alpha (x - \tau)^\beta \, d\tau,$$

where γ_1 is as in Figure 3.5.1(a). As an integration of a microfunction, the integral using γ_2 in Figure 3.5.1(b),

$$\int_{\gamma_2} \tau^\alpha (x - \tau)^\beta \, d\tau,$$

is equal to the above integral in the term of γ_1. Hence one obtains

$$\begin{aligned} c &= \int (t + \sqrt{-1}0)^\alpha (1 - t + \sqrt{-1}0)^\beta \, dt = \int_{\gamma_2} \tau^\alpha (1 - \tau)^\beta \, dt \\ &= \int_1^\infty t^\alpha [(t-1)^\beta e^{\pi\sqrt{1}\beta} - (t-1)^\beta e^{-\pi\sqrt{1}\beta}] \, dt \\ &= 2\sqrt{-1} \sin \pi\beta \, B(\beta + 1, -\alpha - \beta - 1) \\ &= 2\sqrt{-1} \sin \pi\beta \frac{\Gamma(\beta + 1)\Gamma(-\alpha - \beta - 1)}{\Gamma(-\alpha)} \\ &= -2\pi\sqrt{-1} \frac{\Gamma(-\alpha - \beta - 1)}{\Gamma(-\alpha)\Gamma(-\beta)}. \end{aligned}$$

Note. Here, $B(p, q)$ is the beta function defined as

$$\int_0^1 x^{p-1}(1 - x)^{q-1} \, dx \left(= \frac{\Gamma(p)\Gamma(q)}{\Gamma(p + q)} \right).$$

Even though the above computation is valid, with the proviso that $\int_{\gamma_2} \tau^\alpha (1 - \tau)^\beta \, d\tau$ converges absolutely, the result is still correct for any (α, β) by virtue of the method of analytic continuation. This note will apply to the computations below.

We will prove (2). By the transformation of variables $x_i = r\xi_i$, $(\xi_1^2 + \cdots + \xi_n^2 = 1)$, the left-hand side of (2) equals

$$\int_0^\infty (r^2 + c)^\lambda r^{n-1} \, dr \int_{|\xi| = 1} \omega(\xi).$$

Since $\int_{|\xi|=1} \omega(\xi)$ is the surface area of S^{n-1}, it is $2\pi^{n/2}/\Gamma(n/2)$. On the other hand, one has

$$\begin{aligned}\int_0^\infty (x+1)^\lambda x^{n/2-1}\,dx &= \int_1^\infty x^\lambda (x-1)^{n/2-1}\,dx \\ &= \int_1^0 \frac{1}{x^\lambda}\cdot\left(\frac{1}{x}-1\right)^{n/2-1}\left(-\frac{dx}{x^2}\right) \\ &= \int_0^1 x^{-\lambda-n/2-1}(1-x)^{n/2-1}\,dx = B\left(-\lambda-\frac{n}{2},\frac{n}{2}\right) \\ &= \frac{\Gamma\left(-\lambda-\dfrac{n}{2}\right)\Gamma\left(\dfrac{n}{2}\right)}{\Gamma(-\lambda)}.\end{aligned}$$

Since one has

$$\int_0^\infty (r^2+c)^\lambda r^{n-1}\,dr = \tfrac{1}{2}c^{\lambda+n/2}\int_0^\infty (x+1)^\lambda x^{(n-1)/2}\,dx,$$

the left-hand side of (2) is equal to

$$\frac{2\pi^{n/2}}{\Gamma\left(\dfrac{n}{2}\right)}\cdot\frac{c^{\lambda+n/2}}{2}\cdot\frac{\Gamma\left(-\lambda-\dfrac{n}{2}\right)\Gamma\left(\dfrac{n}{2}\right)}{\Gamma(-\lambda)} = \pi^{n/2}\frac{\Gamma\left(-\lambda-\dfrac{n}{2}\right)}{\Gamma(-\lambda)}c^{\lambda+n/2}.$$

Finally, we will prove (3). Since both sides of the equation in (3) are invariant with respect to ξ under the action of the rotation group, it is sufficient to prove the case where $\xi = (1, 0, \ldots, 0)$; i.e.

$$\begin{aligned}\int (x_1 - y_1 + \alpha((x_2-y_2)^2 + \cdots + (x_n-y_n)^2))^\lambda (y_1 + \beta(y_2^2+\cdots+y_n^2))^\mu\,dy \\ = -2\pi^{(n+1)/2}\sqrt{-1}\,\frac{\Gamma\left(-\lambda-\mu-\dfrac{n+1}{2}\right)}{\Gamma(-\lambda)\Gamma(-\mu)}(\alpha+\beta)^{-(n-1)/2} \\ \cdot\left(x_1 + \frac{\alpha\beta}{\alpha+\beta}(x_2^2+\cdots+x_n^2)^{\lambda+\mu+(n+1)/2}\right).\end{aligned}\tag{3.7.3}$$

By virtue of (1) for the integration with respect to y_1, the left-hand side of (3.7.3), letting $y_i - \alpha x_i/(\alpha+\beta)$ be y_i, is equal to

$$-2\pi\sqrt{-1}\,\frac{\Gamma(-\lambda-\mu-1)}{\Gamma(-\lambda)\Gamma(-\mu)}\cdot\int\left(x_1 + \frac{\alpha\beta}{\alpha+\beta}(x_2^2+\cdots+x_n^2) + (\alpha+\beta)(y_2^2+\cdots+y_n^2)\right)^{\lambda+\mu+1} dy_2\cdots dy_n.$$

Then, by (2), this equation can be written as

$$-2\pi\sqrt{-1}\,\frac{\Gamma(-\lambda-\mu-1)}{\Gamma(-\lambda)\Gamma(-\mu)}(\alpha+\beta)^{\lambda+\mu+1}\,\pi^{(n-1)/2}\,\frac{\Gamma\left(-\lambda-\mu-1-\frac{n-1}{2}\right)}{\Gamma(-\lambda-\mu-1)}$$
$$\cdot\left\{\frac{x_1}{\alpha+\beta}+\frac{\alpha\beta}{(\alpha+\beta)^2}(x_2^2+\cdots+x_n^2)\right\}^{\lambda+\mu+1+(n-1)/2},$$

which is the right-hand side of (3).

We have finished the necessary preparation. Now we return to the proof of Theorem 3.7.1. Let

$$L_{\lambda,\mu}(\alpha, x, \xi) = \frac{\Gamma(-\lambda)\Gamma(-\mu)}{2\pi^{(n+1)/2}\sqrt{-1}\,\Gamma\left(-\lambda-\mu-\frac{n+1}{2}\right)}$$
$$\cdot(\alpha+\sqrt{-1})^{(n-1)/2}(\langle x,\xi\rangle+\alpha(|x|^2-\langle x,\xi\rangle^2))^{\lambda},$$

and let

$$K_\mu(x,\xi) = (\langle x,\xi\rangle+\sqrt{-1}(|x|^2-\langle x,\xi\rangle^2))^{\mu}.$$

Then Lemma 5, (3) implies

$$\int L_{\lambda,\mu}(\alpha, x-y, \xi)K_\mu(y,\xi)\,dy$$
$$= -\left(\langle x,\xi\rangle+\frac{\alpha\sqrt{-1}}{\alpha+\sqrt{-1}}(|x|^2-\langle x,\xi\rangle^2)\right)^{\lambda+\mu+(n+1)/2}.$$

Hence one obtains

$$\left(-(\alpha+\sqrt{-1})^2\frac{d}{d\alpha}\right)^j\int L_{\lambda,\mu}(\alpha, x-y, \xi)K_\mu(y,\xi)\,dy$$
$$= -(|x|^2-\langle x,\xi\rangle^2)^j\left(\langle x,\xi\rangle+\frac{\alpha\sqrt{-1}}{\alpha+\sqrt{-1}}(|x|^2-\langle x,\xi\rangle^2)\right)^{\lambda+\mu+(n+1)/2-j}.$$

On the other hand, by Lemma 4, one has,

$$\delta(x) = \sum_{j,k} c_{j,k}\int_{|\xi|=1}\frac{(|x|^2-\langle x,\xi\rangle^2)^j}{\langle x,\xi\rangle+\frac{\sqrt{-1}}{2}(|x|^2-\langle x,\xi\rangle^2+\sqrt{-1}0)^{n-k}}\,\omega(\xi)$$

for some $c_{j,k}$. Then let

$$T_\mu(x,\xi) = -\sum_{j,k} c_{j,k}\left(-(\alpha+\sqrt{-1})^2\frac{d}{d\alpha}\right)^j L_{j+k-\mu-(3/2)n-1,\mu}(\alpha, x, \xi).$$

One obtains

$$\int\left(\int T_\mu(x-y,\xi)K_\mu(y,\xi)\,dy\right)\omega(\xi)=\delta(x).$$

By Proposition 2.4.3 and by definition, S.S. $T_\mu(x,\xi)$ and S.S. $K_\mu(x,\xi)$ are contained in $\{(x,\xi;\sqrt{-1}d\langle x,\xi\rangle\infty)\in\sqrt{-1}S^*N|x=0\}=\{(x,\xi;\sqrt{-1}\langle\xi, dx\rangle\infty|x=0\}$. Fixing μ, let $T(x,\xi)=T_\mu(x,\xi)$, and let $K(x,\xi)=K_\mu(x,\xi)$. Then define sheaf homomorphisms $\Phi:\mathscr{C}_M\to\mathscr{B}_N/\mathscr{A}_N$ and $\Psi:\mathscr{B}_N/\mathscr{A}_N\to\mathscr{C}_M$ as

$$\Phi:u(x)\mapsto\int K(x-y,\xi)u(y)\,dy$$

and

$$\Psi:u(x,\xi)\mapsto\int T(x-y,\xi)u(y,\xi)\,dy\omega(\xi).$$

The sheaf homomorphisms Φ and Ψ satisfy the conditions required to claim the flabbiness of the sheaf $\mathscr{C}_M$ of microfunctions.

§8. Appendix

We will supply a few important motions concerning microfunction theory. First a sheaf $\hat{\mathscr{C}}_M$ can be introduced in connection with Definition 3.1.1.

Definition 3.8.1. *Define a sheaf* $\hat{\mathscr{C}}_M$ *on* $\sqrt{-1}T^*M(=T^*_MX)$ *such that the stalk of* $\hat{\mathscr{C}}_M$ *at* $(x,\sqrt{-1}\xi)$ *is given by*

$$\hat{\mathscr{C}}_{M,(x,\sqrt{-1}\xi)}=\begin{cases}\mathscr{C}_{M,(x,\sqrt{-1}\xi\infty)} & \text{for } \xi\neq 0\\ \mathscr{B}_{M,x} & \text{for } \xi=0.\end{cases}$$

Remark 1. The notion $\widehat{\text{S.S.}}\,u$ in Definition 3.1.1 can be interpreted as the support of a section of $\hat{\mathscr{C}}_M$.

Remark 2. By virtue of the sheaf $\hat{\mathscr{C}}_M$, one can conveniently treat the hyperfunction sheaf and the microfunction sheaf simultaneously. The combined version of Theorems 3.1.5 and 3.1.6 is the following theorem.

Theorem 3.8.1. *Define* $p_1:\sqrt{-1}T^*(M\times M)\underset{M\times M}{\times}M\to\sqrt{-1}T^*M$ *by*

$$p_1((x,\sqrt{-1}\xi_1),(x,\sqrt{-1}\xi_2))=(x,\sqrt{-1}\xi_1);$$

$p_2:\sqrt{-1}T^*(M\times M)\underset{M\times M}{\times}M\to T^*M$ *by*

$$p_2((x,\sqrt{-1}\xi_1),(x,\sqrt{-1}\xi_2))=(x,\sqrt{-1}\xi_2);$$

and $q:\sqrt{-1}T^*(M\times M)\underset{M\times M}{\times}M\to\sqrt{-1}T^*M$ *by*

$$q((x,\sqrt{-1}\xi_1),(x,\sqrt{-1}\xi_2))=(x,\sqrt{-1}(\xi_1+\xi_2)).$$

Then there exists a sheaf homomorphism

$$q_!(p_1^{-1}\hat{\mathscr{C}}_M \times p_2^{-1}\hat{\mathscr{C}}_M) \to \hat{\mathscr{C}}_M.$$

Exercise. Rephrase Theorem 3.2.1 in terms of $\hat{\mathscr{C}}_M$.

The next proposition gives us a link between $\hat{\mathscr{C}}_M$ and $\hat{\mathscr{C}}_{M\times\mathbf{R}}$, which is of theoretical importance.

Proposition 3.8.1. *Let M be a real analytic manifold. Define $s:\mathscr{B}_M \to \mathscr{B}_{M\times\mathbf{R}}$ by $s(u(x)) = u(x)\delta(t)$. Then define a map $s:\hat{\mathscr{C}}_M \to \hat{\mathscr{C}}_{M\times\mathbf{R}}$ in the same manner. The following sequences are exact:*

$$0 \to \mathscr{B}_M \xrightarrow{s} \mathscr{B}_{M\times\mathbf{R}} \xrightarrow{t\cdot} \mathscr{B}_{M\times\mathbf{R}} \to 0. \tag{3.8.1}$$

$$0 \to \hat{\mathscr{C}}_M \xrightarrow{s} \hat{\mathscr{C}}_{M\times\mathbf{R}} \xrightarrow{t\cdot} \hat{\mathscr{C}}_{M\times\mathbf{R}} \to 0. \tag{3.8.2}$$

Here, $t\cdot$, $t \in \mathbf{R}$, means the multiplication by t.

Proof. First we will prove the exactness of (3.8.1).

(i) We will prove that s is monomorphic. Consider the map $u(x)\delta(x) \mapsto \int u(x)\delta(t)\,dt = u(x)$. Then this map is a left inverse of s. Hence s is monomorphic.

(ii) We will prove that $t\cdot$ is epimorphic. Since the question is local in nature, we may assume $M \subset \mathbf{R}^n$ without loss of generality. Then, for $u(x, t) \in \mathscr{B}_{M\times\mathbf{R}}$, there exist finitely many open convex cones Γ_j in $\mathbf{R}^n \times \mathbf{R}$, such that $u = \sum_j b(\varphi_j)$ for $\varphi_j \in \mathscr{O}(M \times \mathbf{R}) \times \sqrt{-1}\Gamma_j \cap \Omega)$, where Ω is a complex neighborhood of $M \times \mathbf{R}$. If necessary, one can arrange Γ_j so that $\Gamma_j \cap (\mathbf{R}^n \times \{0\}) = \varnothing$ holds. Then $\psi_j = \varphi_j/t$ is a holomorphic function in $(M \times \mathbf{R}) \times \sqrt{-1}\Gamma_j) \cap \Omega$. Hence $v(x, t) \in \mathscr{B}_{M\times\mathbf{R}}$ can be defined by $\sum_j b(\psi_j)$. Clearly, $u(x, t) = tv(x, t)$ holds; i.e. $t\cdot$ is an epimorphism.

(iii) We will prove Ker $(t\cdot)$ = Im s. Since Im $s \subset$ Ker $(t\cdot)$ is plainly true, we will give a proof for Ker $(t\cdot) \subset$ Im s. As shown in §2 of Chapter I, an arbitrary hyperfunction $u(x, t) \in \mathscr{B}_{M\times\mathbf{R}}$ can be expressed as $\sum_\epsilon f_\epsilon(x_1 + \sqrt{-1}\epsilon_1 0, \ldots, x_n + \sqrt{-1}\epsilon_n 0, t + \sqrt{-1}\epsilon_{n+1}0)$, where $\epsilon = (\epsilon_1, \ldots, \epsilon_n, \epsilon_{n+1}) \in \{-1, 1\}^{n+1}$. Let $u(x, t) \in$ Ker $(t\cdot)$. For the sake of simplicity, rewrite $t = x_{n+1}$. Then $x_{n+1}u(x, x_{n+1}) = 0$ holds. Hence, by using a holomorphic function $\varphi_{\hat{\epsilon}_k}(x, x_{n+1}) \underset{\text{def}}{=} \varphi_{\epsilon_1,\ldots,\epsilon_{k-1},\epsilon_{k+1},\ldots,\epsilon_{n+1}}(x, x_{n+1})$ on $\Omega_k \underset{\text{def}}{=} \{(x, x_{n+1}) \in \Omega \mid \epsilon_j \operatorname{Im} x_j > 0$ for any $j \neq k\}$, one can express $x_{n+1}u(x, x_{n+1}) = \sum_{k=1}^{n+1} (-1)^{k+1} \cdot \varphi_{\hat{\epsilon}_k}(x, x_{n+1})$. Since $x_{n+1} \neq 0$ holds where $\varphi_{\hat{\epsilon}_k}$ is defined for $k \leqq n$, then $\varphi_{\hat{\epsilon}_k} \underset{\text{def}}{=} \psi_{\hat{\epsilon}_k}/x_{n+1}$ is holomorphic in Ω_k. If $k = n + 1$, then construct holomorphic functions $\psi_{\hat{\epsilon}_{n+1}}$ and $\xi_{\epsilon_1,\ldots,\epsilon_n}$ such that one has $\varphi_{\hat{\epsilon}_{n+1}} = x_{n+1}\psi_{\hat{\epsilon}_{n+1}}(x, x_{n+1}) + \xi_{\epsilon_1,\ldots,\epsilon_n}(x)$. (Note that it is not trivial that $\psi_{\hat{\epsilon}_{n+1}}$ and $\xi_{\epsilon_1,\ldots,\epsilon_n}$ exist with the above property, since one needs to construct them globally. The reader is expected to verify this point.) Then one obtains

$$\begin{aligned} u(x, x_{n+1}) &= \sum_{\epsilon} f_\epsilon = \sum_{\epsilon}\left(f_\epsilon + \sum_{k=1}^{n+1}(-1)^k\epsilon_k\psi_{\epsilon_k}\right) \\ &= \sum_{\epsilon}\frac{(-1)^{n+1}\epsilon_{n+1}}{x_{n+1}}\xi_{\epsilon_1,\ldots,\epsilon_n}(x) \\ &= (-1)^{n+1}\left(\frac{1}{x_{n+1}+\sqrt{-1}0}\right. \\ &\quad \left.-\frac{1}{x_{n+1}-\sqrt{-1}0}\right)\sum_{\epsilon_1,\ldots,\epsilon_n=\pm 1}\xi_{\epsilon_1,\ldots,\epsilon_n}(x). \end{aligned}$$

Then, if one lets $v(x) = (-1)^{n+1}2\pi\sqrt{-1}\sum_{\epsilon_1,\ldots,\epsilon_n=\pm 1}\xi_{\epsilon_1,\ldots,\epsilon_n}(x)$, then $u(x, x_{n+1}) = v(x)\delta(x_{n+1})$ holds, i.e. $u(x, x_{n+1}) \in \operatorname{Im} s$.

Next we will prove (3.8.2).

(i) We begin with the proof that $t\cdot$ is an epimorphism. Since the sheaf of microfunctions is a flabby sheaf by Theorem 3.7.1, then Theorem 2.3.2 implies that the onto-ness in (3.8.2) is reduced to the onto-ness in (3.8.1).

(ii) Next we will prove that $\operatorname{Ker}(t\cdot) = \operatorname{Im} s$ in (3.8.2). Again, $\operatorname{Im} s \subset \operatorname{Ker}(t\cdot)$ is clearly true; so we will prove $\operatorname{Ker}(t\cdot) \subset \operatorname{Im} s$. Let us first consider this question in a neighborhood of $(x_0, 0; \sqrt{-1}(\langle\xi_0, dx\rangle + dt)\infty)$. The hypothesis $tu = 0$ implies that one can find a conoidal neighborhood Γ_j of $U_j \subset S(M \times \mathbf{R})$, $j = 1, 2, \ldots, j_0$, such that $\sqrt{-1}(\langle\xi_0, dx\rangle + dt)\infty \in U_j^\circ$, and one can also find a holomorphic function φ_j in Γ_j so that $tu = \sum_{j=1}^{j_0} b_{\Gamma_j}(\varphi_j)$ holds. Using $\psi_j(x, t)$ and $\xi_j(x)$ such that $\varphi_j(x, t) = t\psi_j(x, t) + \xi_j(x)$, as we noted, one lets

$$\tilde{u}(x, t) = u(x, t) - \sum_{j=1}^{j_0} b_{\Gamma_j}(\psi_j) - \frac{1}{t - \sqrt{-1}0}\sum_{j=1}^{j_0} b_{\Gamma_j}(\xi_j).$$

Clearly, then, $\tilde{u}(x, t) = u(x, t)$ holds in a neighborhood of $(x_0, 0; \sqrt{-1}(\langle\xi_0, dx\rangle + dt)\infty)$, and one also has $t\tilde{u}(x, t) = 0$ since ψ_j and ξ_j are chosen as above. Therefore, by the exactness of (3.8.1), one can find a hyperfunction $v(x)$ such that $\tilde{u}(x, t) = v(x)\delta(t)$ holds. That is, $u = \mathrm{sp}(v(x)\delta(t))$ holds in a neighborhood of that point. Hence, $\mathrm{Ker}\,(t\cdot) \subset \mathrm{Im}\, s$ has been proved. One can prove this in a neighborhood of $(x_0, 0; \sqrt{-1}\langle \xi_0, dx\rangle\infty)$ by using the coordinate transformation $x \mapsto x + at$.

(iii) Lastly, we will prove that s is monomorphic. From the definition of $\hat{\mathscr{C}}_M$, we need to prove two cases.

First, the exactness of the sequence $0 \to \mathscr{B}_{M,x} \xrightarrow{s} \mathscr{C}_{M,(x,0;\sqrt{-1}\,dt\infty)}$ will be proved. We have $\delta(ct) = (1/|c|)\delta(t)$ for $c \neq 0$. If $\mathrm{sp}(u(x)\delta(t)) = 0$ holds in a neighborhood of $(x, 0; \sqrt{-1}(\langle \xi_0, dx\rangle + dt)\infty)$ for some $\xi_0 \neq 0$, then one has $\mathrm{sp}(u(x)\delta(t)) = 0$ for an arbitrary ξ in a neighborhood of $(x, 0; \sqrt{-1}\langle \xi, dx\rangle + dt)\infty)$. On the other hand, $\mathrm{sp}(u(x)\delta(t)) = 0$ holds anywhere by the coordinate transformation $x \mapsto x + at$. Hence, $u(t)\delta(t)$ is a real analytic function whose support is restricted to $\{t = 0\}$. Then one has $u(x)\delta(t) = 0$. The map s, being monomorphic in (3.8.1), implies $u(x) = 0$.

We will prove that

$$0 \to \mathscr{C}_{M,(x;\sqrt{-1}\langle\xi,dx\rangle\infty)} \xrightarrow{s} \mathscr{C}_{M\times\mathbf{R},(x,0;\sqrt{-1}(\langle\xi,dx\rangle+\tau\,dt)\infty)}$$

is an exact sequence. By the coordinate transformation $x \mapsto x + at$, one obtains $u(x)\delta(t) = 0$, including the case where $\tau = 0$. Then

$$u(x) = \int u(x)\delta(t)\,dt$$

implies that s is a monomorphism. The proof of Proposition 3.8.1 is now completed.

As our next topic, we will discuss the concept of a hyperfunction containing holomorphic parameters. By way of introduction, we first recall the notion of a complex conjugate.

Definition 3.8.2. *Let T be a complex manifold. Then a complex manifold $\bar{T}$ is said to be the complex conjugate of T when there exists a homeomorphism $\alpha: T \to \bar{T}$, as topological spaces, such that $f \in \mathcal{O}_{\bar{T}}$ if and only if $\overline{f\circ\alpha} \in \mathcal{O}_T$, where $\overline{f\circ\alpha}(z) = \overline{f(\alpha(z))}$.*

Remarks.

(1) If $\bar{T}$ is the complex conjugate of T, then T is the complex conjugate of $\bar{T}$.

(2) Let $\bar{T}$ be the complex conjugate of T. Then one can embed T into $T \times \bar{T}$ by the map $T \ni \tau \mapsto (\tau, \alpha(\tau)) \in T \times \bar{T}$. Let us denote by T_{real} the case when T is regarded as a real analytic manifold. Then $T \times \bar{T}$ is a complexification of T_{real}. We can identify the tangent space $T(T)$ as a complex manifold with the tangent space $T(T_{\mathrm{real}})$ as a real analytic manifold as follows. Let $X_0 \in T(T)$ and $X \in T(T_{\mathrm{real}})$. Then X_0 and X are identified if and only if $Xf = X_0 f$ holds for an arbitrary $f \in \mathcal{O}_T$.

Next we will describe the above identification via a local coordinate system. Let $(\tau_1, \ldots, \tau_l)$ be local complex coordinates of T, and let $\tau_j = t_j + \sqrt{-1}s_j$, $j = 1, 2, \ldots, l$. Then, for $f \in \mathcal{O}_T$,

$$\frac{\partial}{\partial t_j} f(\tau) = \frac{\partial}{\partial \tau_j} f(\tau) \frac{\partial \tau_j}{\partial t_j} = \frac{\partial}{\partial \tau_j} f(\tau)$$

and

$$\frac{\partial}{\partial s_j} f(\tau) = \frac{\partial}{\partial \tau_j} f(\tau) \frac{\partial \tau_j}{\partial s_j} = \sqrt{-1} \frac{\partial}{\partial \tau_j} f(\tau)$$

holds. Hence, the correspondence should be

$$\frac{\partial}{\partial \tau_j} \leftrightarrow \frac{\partial}{\partial t_j}, \qquad \sqrt{-1} \frac{\partial}{\partial \tau_j} \leftrightarrow \frac{\partial}{\partial s_j} \qquad \text{for } j = 1, \ldots, l.$$

One can also identify the dual spaces $T^*(T)$ with $T^*(T_{\text{real}})$ through the above identification. In terms of coordinates, this is given by

$$d\tau_j \leftrightarrow dt_j, \qquad -\sqrt{-1}\, d\tau_j \leftrightarrow ds_j.$$

In what follows, these identifications are assumed.

Let M be a real analytic manifold, and let X and $\bar{X}$ be complexifications of M such that X and $\bar{X}$ are complex conjugates to each other. Assume that the restriction of the homeomorphism $\alpha: X \to \bar{X}$ to M is an identity map on M. For example, if $M = \mathbf{R}^n$ and $X = \bar{X} = \mathbf{C}^n$, then one may let $\alpha(z_1, \ldots, z_n) = (\bar{z}_1, \ldots, \bar{z}_n)$.

The correspondence $\mathcal{O}_{\bar{X}} \ni f(\bar{z}) \mapsto \bar{f}(z) = \overline{f(\bar{z})} \in \mathcal{O}_X$ induces a sheaf homomorphism $\alpha^{-1}\mathcal{O}_{\bar{X}} \to \mathcal{O}_X$. Hence this induces a sheaf homomorphism $\alpha^{-1}\mathscr{H}^n_M(\mathcal{O}_{\bar{X}}) \to \mathscr{H}^n_M(\mathcal{O}_X)$, which is an isomorphism. Following these preparations, we begin our discussion.

Definition 3.8.3. *For a hyperfunction* $u(x) = \sum_j b(f_j(z)) \in \mathscr{B}_M$, $\overline{u(x)} = \sum_j b(\bar{f}_j(\alpha(z))) \in \mathscr{B}_M$ *is called the complex conjugate of* $u(x)$. *A hyperfunction* $u(x)$ *is said to be a real-valued hyperfunction if* $u(x) = \overline{u(x)}$ *holds.*

Example 3.8.1. δ-function is a real-valued hyperfunction. Since

$$\delta(x_1, \ldots, x_n) = \frac{1}{(-2\pi\sqrt{-1})^n} \sum_{\epsilon} \frac{\epsilon_1 \cdots \epsilon_n}{(x_1 + \sqrt{-1}\epsilon_1 0) \cdots (x_n + \sqrt{-1}\epsilon_n 0)},$$

we have by definition

$$\overline{\delta(x_1, \ldots, x_n)} = \frac{1}{(2\pi\sqrt{-1})^n} \sum_{\epsilon} \frac{\epsilon_1 \cdots \epsilon_n}{(x_1 - \sqrt{-1}\epsilon_1 0) \cdots (x_n - \sqrt{-1}\epsilon_n 0)}.$$

Replace ϵ_i with $-\epsilon_i$ in the above. Then we obtain

$$\overline{\delta(x_1, \ldots, x_n)} = \frac{1}{(2\pi\sqrt{-1})^n} \sum_\epsilon \frac{(-1)^n \epsilon_1 \cdots \epsilon_n}{(x_1 + \sqrt{-1}\epsilon_1 0) \cdots (x_n + \sqrt{-1}\epsilon_n 0)}$$
$$= \delta(x_1, \ldots, x_n).$$

Example 3.8.2. For $\lambda \in \mathbf{C}$, $\overline{(x + i0)^\lambda} = (x - \sqrt{-1}0)^{\bar{\lambda}}$ holds.

If $\bar{X}$ is the complex conjugate of X, then $\alpha: X \to \bar{X}$ naturally induces the antipodal maps a on $\sqrt{-1}SM$ and $\sqrt{-1}S^*M$. It is also plain that we have

$$\text{S.S. } \bar{u} = (\text{S.S. } u)^a = \{(x, -\sqrt{-1}\xi\infty) \in \sqrt{-1}S^*M \,|\, (x, \sqrt{-1}\xi\infty) \in \text{S.S. } u\}.$$

One can also define the complex conjugate of a microfunction as in the above.

Definition 3.8.4. *Let M be an n-dimensional real analytic manifold, and let T be an l-dimensional complex manifold. Let $u(x, \tau, \bar{\tau})$ be a hyperfunction (or a microfunction) on $M \times T$, regarded as an $(n + 2l)$-dimensional real analytic manifold. Then u is said to contain τ as holomorphic parameters, or u depends upon τ holomorphically, if $u(x, \tau, \bar{\tau})$ satisfies the Cauchy-Riemann differential equations $\partial u/\partial \bar{\tau}_j = 0$ for $j = 1, \ldots, l$. In that case, we denote $u = u(x, \tau)$, since $\partial u/\partial \bar{\tau} = 0$.*

Remark. We have $\sqrt{-1}S^*(M \times T) = \{(x, \tau; \sqrt{-1}(\langle \xi, dx \rangle + \langle \zeta, d\tau \rangle + \langle \bar{\zeta}, d\bar{\tau} \rangle)\infty\}$ and $T \times \sqrt{-1}S^*M = \{(x, \tau; \sqrt{-1}\langle \xi, dx \rangle \infty)\}$. Hence, we can regard $T \times \sqrt{-1}S^*M \subset \sqrt{-1}S^*(M \times T)$. Since the principal symbol of the differential operator $\partial/\partial\bar{\tau}_j$ is $\bar{\zeta}_j$, from Theorem 3.4.4, we have S.S. $u \subset \{\bar{\zeta}_1 = \cdots = \bar{\zeta}_1 = 0\} \subset T \times \sqrt{-1}S^*M$.

We will consider a hyperfunction containing τ as holomorphic parameters. Even though we will restrict our discussion to the case $T \subset \mathbf{C}$, the general case $T \subset \mathbf{C}^l$ can be carried out equally well.

Proposition 3.8.2. *Let $T \subset \mathbf{C}$, and let $u(x, \tau)$ be a hyperfunction on $M \times T$ containing a holomorphic parameter. If X and $T \times \bar{T}$ are complexifications of M and T, respectively, then one can choose an open subset D_j of X and an open subset U_j of T such that* $u(x, \tau) = \sum_j b_{D_j \times U_j}(f_j(z, \tau))$ holds.

Proof. First let $u(x, \tau) = \sum_j b(f_j(z, \tau, \bar{\tau}))$. Then one obtains

$$\frac{\partial u}{\partial \bar{\tau}} = \sum_j b\left(\frac{\partial}{\partial \bar{\tau}} f_j(z, \tau, \bar{\tau})\right) = 0.$$

That is, $(\partial/\partial\bar{\tau})f_j$ is a coboundary. Hence, one can express

$$\frac{\partial}{\partial \bar{\tau}} f_j = \sum_k g_{jk}, \qquad \text{where } g_{jk} = -g_{kj}.$$

As in the proof of the Lemma following Proposition 3.2.1, there exists $h_{jk}(z, \tau, \bar{\tau}), j > k$, such that $g_{jk} = (\partial/\partial\bar{\tau})h_{jk}$ holds. For the cases $j < k$ and $j = k$, let h_{jk} be $h_{jk} = -h_{kj}$ and $h_{jk} = 0$ respectively. Then define $F_j(z, \tau, \bar{\tau}) = f_j - \sum_k h_{jk}$. One has $(\partial/\partial\bar{\tau}\ F_j(z, \tau, \bar{\tau}) = 0$. Therefore, one has $F_j(z, \tau, \bar{\tau}) = F_j(z, \tau)$. Furthermore, $\sum_j b(F_j(z, \tau)) = \sum_j b\left(f_j - \sum_k h_{jk}\right) = \sum_j b(f_j) = u(x, \tau)$ holds.

Proposition 3.8.3. *Let a hyperfunction $u(x, \tau)$ depend holomorphically upon τ. If $u(x, \tau_0) = 0$ holds, then there exists a unique hyperfunction $v(x, \tau)$ holomorphically dependent on τ such that $u(x, \tau) = (\tau - \tau_0)v(x, \tau)$ holds.*

Proof. One can let $\tau_0 = 0 \in \mathbf{C}$ without loss of generality. Express $u(x, \tau) = \sum_j b(f_j(z, \tau))$, where $f_j(z, \tau)$ is as in the previous proposition. Then $u(x, 0) = \sum_j b(f_j(z, 0)) = 0$ holds. On the other hand, there is a holomorphic function $h_j(z, \tau)$ so that $f_j(z, \tau) - f_j(z, 0) = \tau h_j(z, \tau)$ holds. Hence, one obtains $u(x, \tau) = \sum_j b(f_j(z, \tau) - f_j(z, 0)) = \tau \sum_j b(h_j(z, \tau))$; i.e. $v(x, \tau) = \sum_j b(h_j(z, \tau))$. Next, we will prove the uniqueness part. Suppose $\tau \cdot u(x, \tau) = 0$. Then $u(x, \tau) = 0$ for $\tau \neq 0$. Hence, $\operatorname{supp} u(x, \tau) \subset \{\operatorname{Re} \tau \geqq 0\}$ holds. One has $d \operatorname{Re} \tau((x, 0)) \neq 0$ and $((x, 0); \sqrt{-1}d \operatorname{Re} \tau(x, 0)\infty) \notin \text{S.S. } u(x, \tau)$. Consequently, by Proposition 3.5.2, $u(x, \tau) = 0$ holds.

Proposition 3.8.4 (Cauchy's Integral Theorem for a Holomorphic Parameter). *Let γ be a closed Jordan curve $\subset T(\subset \mathbf{C})$. Then, for a hyperfunction $u(x, \tau)$ containing a holomorphic parameter τ, $\int_\gamma u(x, \tau)\, d\tau = 0$ holds.*

Note. The above integration $\int_\gamma u(x, \tau)\, d\tau$ means: one first restricts $u(x, \tau)$ on γ and performs the line integral $\oint_\gamma d\tau$.

Proof. Using $f_j(z, \tau)$ in Proposition 3.8.2, express $u(x, \tau) = \sum_j b(f_j(z, \tau))$. Then the proof can be completed by Cauchy's integral theorem for a holomorphic function.

Proposition 3.8.5 (Cauchy's Integral Formula for a Holomorphic Parameter). *Let $\gamma \subset T(\subset \mathbf{C})$ be a closed Jordan curve. Then one has $u(x, \tau) = (1/2\pi\sqrt{-1})\oint_\gamma u(x, \zeta)(1/\zeta - \tau)\, d\zeta$ for a hyperfunction containing a holomorphic parameter τ, where the closed curve γ is taken in the counterclockwise direction once around τ.*

Proof. From Proposition 3.8.3, there exists a hyperfunction v depending holomorphically upon τ such that

$$u(x, \tau + w) - u(x, w) = \tau v(x, w, \bar{w}, \tau)$$

holds. Taking the derivative of this equation with respect to $\bar{w}$ gives us $0 = \tau(\partial/\partial\bar{w})v(x, w, \bar{w}, \tau)$. By the uniqueness assertion in Proposition 3.8.3, one has $\partial v/\partial\bar{w} = 0$. Hence, $v = v(x, w, \tau)$ holds. Therefore one obtains

$$u(x, \zeta) - u(x, \tau) = u(x, (\zeta - \tau) + \tau) - u(x, \tau) = (\zeta - \tau)v(x, \zeta - \tau, \tau).$$

Consequently, Proposition 3.8.4 and the above imply

$$\begin{aligned}\frac{1}{2\pi\sqrt{-1}}\oint u(x, \zeta)\frac{1}{\zeta - \tau}\,d\zeta &= u(x, \tau) + \frac{1}{2\pi\sqrt{-1}}\oint v(x, \zeta - \tau, \tau)\,d\zeta \\ &= u(x, \tau).\end{aligned}$$

Our next aim is to prove the uniqueness of the analytic continuation with respect to a holomorphic parameter of a hyperfunction (and a microfunction). For this purpose, we will prove Proposition 3.8.6, a local version of Bochner's theorem on the envelope of holomorphy of a tube domain. This proposition has many applications.

Proposition 3.8.6 (Komatsu [2]; SKK [1], Chap. I, §3.1). *For* $0 < \epsilon < 1$, *define* G_ϵ *and* F_ϵ *as follows:*

$$\begin{aligned}G_\epsilon \underset{\text{def}}{=} \{(x_1 + \sqrt{-1}y_1, x_2 + \sqrt{-1}y_2) \in \mathbf{C}^2 \,|\, 0 \leqq y_1, 0 \leqq y_2, y_1 + y_2 < 1 \\ \text{and } \epsilon(x_1^2 + x_2^2) + (y_1 + y_2) - \epsilon(y_1^2 + y_2^2) < 1 - \epsilon\}\end{aligned}$$

and

$$F_\epsilon \underset{\text{def}}{=} G_\epsilon \cap \{y_1 = 0 \text{ or } y_2 = 0\}.$$

Furthermore, let U' *be an open set containing* F_ϵ, *and let*

$$U = \{(x_1 + \sqrt{-1}y_1, x_2 + \sqrt{-1}y_2) \in U' \,|\, y_1 < 0 \text{ or } y_2 < 0\}.$$

Then, for any open subset $W \subset \mathbf{C}^n$,

$$\mathcal{O}_{\mathbf{C}^2 \times W}((U \cup G_\epsilon) \times W) \to \mathcal{O}_{\mathbf{C}^2 \times W}(U \times W)$$

is an epimorphism.

Proof. From the long exact sequence of relative cohomology groups, it is sufficient to prove $H^1_{G_\epsilon \times W}(\mathbf{C}^2 \times W, \mathcal{O}) = 0$. Let $\varphi(z_1, z_2) = z_1 + z_2 + \sqrt{-1}\epsilon(z_1^2 + z_1^2)$. Then $G_\epsilon = \{(z_1, z_2) \,|\, \text{Im } z_1 \geqq 0, \text{Im } z_2 \geqq 0, \text{Im}(z_1 + z_2) < 1 \text{ and Im } \varphi < 1 - \epsilon\}$ holds. If one defines

$$K_\epsilon^1 = \{(z_1, z_2) \in \mathbf{C}^2 \,|\, \text{Im } z_1 \geqq 0, \text{Im } z_2 \geqq 0, \text{Im}(z_1 + z_2) \leqq 1, \text{Im } \varphi \leqq 1 - \epsilon\}$$

$$K_\epsilon^2 = \{(z_1, z_2) \in \mathbf{C}^2 \,|\, \text{Im } z_1 \geqq 0, \text{Im } z_2 \geqq 0, \text{Im}(z_1 + z_2) \leqq 1, \text{Im } \varphi = 1 - \epsilon\},$$

then one has $G_\epsilon = K_\epsilon^1 - K_\epsilon^2$. Clearly, $G_\epsilon \subset K_\epsilon^1 - K_\epsilon^2$ holds. One needs to prove $K_\epsilon^1 - K_\epsilon^2 \subset G_\epsilon$. That is, it suffices to prove, "Im $z_1 \geqq 0$, Im $z_2 \geqq 0$, and $\operatorname{Im}\varphi < 1 - \epsilon$ imply $\operatorname{Im}(z_1 + z_2) \neq 1$." One has $\operatorname{Im}(z_1 + z_2) - \epsilon(\operatorname{Im}(z_1 + z_2))^2 \leqq \operatorname{Im}\varphi < 1 - \epsilon$ for $\operatorname{Im} z_1 \geqq 0$, $\operatorname{Im} z_2 \geqq 0$, and $\operatorname{Im}\varphi < 1 - \epsilon$. Hence, if $\operatorname{Im}(z_1 + z_2) = 1$, it would then imply $1 - \epsilon < 1 - \epsilon$, a contradiction.

Since K_ϵ^1 and K_ϵ^2 are compact analytic polyhedrons, one obtains $H^1_{G_\epsilon \times W}(\mathbf{C}^2 \times W, \mathcal{O}) = 0$ by Proposition 2.2.2.

Proposition 3.8.7. *Let* $0 < \epsilon < \frac{1}{2}$. *Define*

$$F_1 = \left\{(x_1 + \sqrt{-1}y_1, x_2 + \sqrt{-1}y_2) \in \mathbf{C}^2 \,|\, y_1 = 0, 0 \leqq y_2 < 1,\right.$$

$$\left. x_1^2 + x_2^2 < 4\left(\frac{1-\epsilon}{\epsilon}\right)\right\};$$

$$F_2 = \left\{x_1 + \sqrt{-1}y_1, x_2 + \sqrt{-1}y_2) \in \mathbf{C}^2 \,|\, y_2 = 0, 0 \leqq y_1 < 1,\right.$$

$$\left. x_1^2 + x_2^2 < 4\left(\frac{1-\epsilon}{\epsilon}\right)\right\};$$

$$G_1 = \left\{(x_1 + \sqrt{-1}y_1, x_2 + \sqrt{-1}y_2) \in \mathbf{C}^2 \,|\, y_1 \geqq 0, y_2 \geqq 0,\right.$$

$$\left. y_2 + (1 - 2\epsilon)(y_1 - 1) < 0,\ x_1^2 + x_2^2 < \frac{1-\epsilon}{\epsilon}\right\};$$

$$G_2 = \left\{(x_1 + \sqrt{-1}y_1, x_2 + \sqrt{-1}y_2) \in \mathbf{C}^2 \,|\, y_1 \geqq 0, y_2 \geqq 0,\right.$$

$$\left. y_1 + (1 - 2\epsilon)(y_2 - 1) < 0,\ x_1^2 + x_2^2 < \frac{1-\epsilon}{\epsilon}\right\}.$$

Then let $F = F_1 \cup F_2$, and let $\mathrm{G} = \mathrm{G}_1 \cup \mathrm{G}_2$. *For a neighborhood* U' *of* F, *define* $U = U' \cap \{y_1 < 0$ or $y_2 < 0\}$. *Then the map*

$$\mathcal{O}_{\mathbf{C}^2 \times W}((U \cup G) \times W) \to \mathcal{O}_{\mathbf{C}^2 \times W}(U \times W)$$

is an epimorphism for an arbitrary open subset $W \subset \mathbf{C}^N$.

Proof. For $\alpha = (\alpha_1, \alpha_2) \in \mathbf{R}^2$, define $G_{\epsilon,\alpha}$, $F_{\epsilon,\alpha}$, and $\tilde{G}_{\epsilon,\alpha}$ as follows: $G_{\epsilon,\alpha} = \{(z_1, z_2) \in \mathbf{C}^2 \,|\, y_1 \geqq 0,\ y_2 \geqq 0,\ y_1 + y_2 < 1$ and $\epsilon((x_1 - \alpha_1)^2 + (x_2 - \alpha_2)^2) + (y_1 + y_2) - \epsilon(y_1^2 + y_2^2) < 1 - \epsilon\}$; $F_{\epsilon,\alpha} = G_{\epsilon,\alpha} \cap \{y_1 = 0$ or $y_2 = 0\}$; and $\tilde{G}_{\epsilon,\alpha} = G_{\epsilon,\alpha} \cap \{x_1 = \alpha_1, x_2 = \alpha_2\}$. Then $F_{\epsilon,\alpha} \subset F$ for $\alpha_1^2 + \alpha_2^2 < (1 - \epsilon)/\epsilon$. Therefore, U' is a neighborhood of $F_{\epsilon,\alpha}$ if U' is a neighborhood of F. Then Proposition 3.8.6 implies that a holomorphic function on U can be extended to one over $U \cup G_{\epsilon,\alpha}$; in particular, over

$U \cup \tilde{G}_{\epsilon,\alpha}$. Let $\tilde{G} \underset{\text{def}}{=} \bigcup_{\alpha_1^2+\alpha_2^2<(1-\epsilon)/\epsilon} \tilde{G}_{\epsilon,\alpha}$. Then $\tilde{G} = \{(z_1, z_2) \in \mathbf{C}^2 | x_1^2 + x_2^2 < (1-\epsilon)/\epsilon,\ y_1 \geqq 0,\ y_2 \geqq 0,\ (y_1 + y_2) - \epsilon(y_1^2 + y_2^2) < 1 - \epsilon\}$. Since a holomorphic function in U can be extended to a holomorphic function in $U \cup \tilde{G}$, and since $G \subset \tilde{G}$, particularly, it can be extended to one in $U \cup G$. This means that

$$\mathcal{O}_{\mathbf{C}^2 \times W}((U \cup G) \times W) \to \mathcal{O}_{\mathbf{C}^2 \times W}(U \times W)$$

is an epimorphism.

The following is an easy corollary from the proposition, which is quite important for applications.

Corollary 1. *A holomorphic function in a neighborhood of* $\{z = (z_1, \ldots, z_n) \in \mathbf{C}^n | |\mathrm{Re}\, z_j| < a$ *for* $j = 1, 2, \ldots, n$, *and* $0 < \mathrm{Im}\, z_1 < b$, $\mathrm{Im}\, z_2 = \cdots = \mathrm{Im}\, z_n = 0\}$ *can be extended to a holomorphic function in*

$$\left\{z \in \mathbf{C}^n \middle| |\mathrm{Re}\, z_j| < \epsilon, \frac{1}{\epsilon}\sqrt{\sum_{j=2}^{n} |\mathrm{Im}\, z_j|^2} < \mathrm{Im}\, z_1 < \epsilon\right\} \qquad \textit{for some } \epsilon > 0.$$

Remark. The above Corollary 1 indicates that $f(z)$ determines an element of $\tilde{\mathscr{A}}$ in a neighborhood of $0 + \sqrt{-1}(\partial/\partial x_1)0$. Hence, from this corollary, when one considers a boundary value of a holomorphic function, one may ignore the higher-order infinitesimals as approaching reals from imaginaries. This seems to be a hidden reason why the microfunction sheaf can be defined on such a geometrically simple structured manifold as $\sqrt{-1}S^*M$, i.e. without using such a notion as a jet of higher order. Of course, this explanation is a kind of afterthought. Sato regarded this fact (Corollary 1) as being obvious when he constructed the theory of microfunctions—the sort of clear intuition that only a genius has!

The following corollary of Corollary 1 is also quite useful.

Corollary 2. *Let* M *be an open subset of* $\mathbf{R}^n$, *and let* ξ_0 *be a non-zero* n*-dimensional real vector. If, for a real analytic function* $f(x)$ *on* M *and* x_0 *on* M, *one has* $f(x + \sqrt{-1}t\xi_0) \neq 0$ *for* $0 < t \ll 1$, $|x - x_0| \ll 1$, *then* $\{z \in \mathbf{C}^n - \mathbf{R}^n | f(z) \neq 0\}$ *is a conoidal neighborhood of* $x_0 + \sqrt{-1}\xi_0 0$.

By these results, we will prove a proposition on the propagation of analyticity.

Proposition 3.8.8. *Let* D *be a connected open subset of* $\mathbf{C}$, *and let* Ω *be an open subset containing* $\{0\} \times D$ *in* $\mathbf{R}^n \times \mathbf{C}$. *Suppose* $u(x, \tau)$ *is a hyperfunction defined on* Ω *containing a holomorphic parameter* τ *such that* S.S. $u \cap \{x = 0\} \subset \Omega \times \sqrt{-1}\Gamma$ *for a properly convex closed set* Γ *in* S^{n-1}.

If $u(x, \tau)$ is a real analytic function in a neighborhood of $x = 0$ and $\tau = \tau_0 \in D$, then $u(x, \tau)$ is real-analytic in a neighborhood of $\{0\} \times D$.

Proof. Since Γ is a properly convex closed set, one can assume, without loss of generality, that $\Gamma \subset V \underset{\text{def}}{=} \{(\xi_1, \ldots, \xi_n) \in S^{n-1} | \xi_1 > |\xi_2| + \cdots + |\xi_n|\}$ after a certain coordinate transformation. One has $\{\tau | |\mathrm{Re}\, \tau| \leqq \alpha$ and $|\mathrm{Im}\, \tau| \leqq \alpha\} \times \{x \in \mathbf{R}^n | |x| \leqq \epsilon\} \subset \Omega$ for some $\alpha(<1)$ and ϵ, since $D \times \{0\} \subset \Omega$. Hence, this allows one to assume $\bar{D} \times \{x \in \mathbf{R}^n | |x| \leqq 1\} \subset \Omega$ to prove the assertion. Since S.S. $u \cap \{x = 0\} \subset \Omega \times \sqrt{-1}V$, Theorem 2.3.4 implies that one can express u as $b_{\tilde{D}}(f(z, \tau))$, where $f(z, \tau)$ is a holomorphic function in $\tilde{D} = \{(z, \tau) \in \mathbf{C}^n \times \mathbf{C} | \tau \in D, |z| \ll 1$ and $\mathrm{Im}\, z_1 > |\mathrm{Im}\, z_2|, \ldots, |\mathrm{Im}\, z_n|\}$. Then we have the following lemma.

Lemma 1. *Let $u(x, \tau) = b(f(z, \tau))$ be defined for $|\tau| < 1$ and be real-analytic in* $\mathrm{Im}\, \tau > 0$. *Then, for some $\epsilon > 0$, $u(x, \tau)$ is a real analytic function in $|\tau| < \epsilon$.*

Proof. First we will prove Proposition 3.8.8 using Lemma 1. In order to prove Proposition 3.8.8, it is sufficient to prove the statement

(A) If $b(f(z, \tau))$ is defined for $|\tau| < 1$ and is real-analytic in a neighborhood of $\tau = 0$, then $b(f(z, \tau))$ is real-analytic in $|\tau| < 1$.

It suffices to prove the following (B) to claim (A).

(B) If $b(f(z, \tau))$ is real-analytic in $|\tau| < a(<1)$, then there exists $\epsilon > 0$ such that $b(f(z, \tau)$ is real-analytic in $|\tau| < a + \epsilon$.

We will derive (B) from Lemma 1. Let $|\tau_0| = a$. Then $b(f(z, \tau))$ is real-analytic in $|\tau| < a$. From Lemma 1, one may assume, after a suitable linear fractional transformation, that $b(f(z, \tau))$ is real-analytic in $U_{\tau_0} = \{\tau | |\tau - \tau_0| < \epsilon_{\tau_0}\}$ for some $\epsilon_{\tau_0} > 0$. Since $\{\tau | |\tau| = a\}$ is compact, one has finitely many τ_i, $i = 1, \ldots, l$, such that $\{\tau | |\tau| = a\} \subset \bigcup_{i=1}^{l} U_{\tau_i}$ holds. Hence, for $\epsilon = \min_i \epsilon_{\tau_i}$, $b(f(z, \tau))$ is real-analytic in $|\tau| < a + \epsilon$. Therefore, it is reduced to prove Lemma 1.

One can let $f(z, \tau)$ be a holomorphic function defined in the domain where $|\mathrm{Re}\, \tau| < 1$, $|\mathrm{Im}\, \tau| < 1$, $|\mathrm{Re}\, z_j| < \epsilon$, $|\mathrm{Im}\, z_j| < \epsilon$ $(1 \leqq j \leqq n)$ and where $\mathrm{Im}\, z_1 > |\mathrm{Im}\, z_2|, \ldots, |\mathrm{Im}\, z_n|$. Since the boundary value $b(f(z, \tau))$ is real-analytic for $\mathrm{Im}\, \tau > 0$, then $f(z, \tau)$ remains to be holomorphic where $\mathrm{Im}\, z_1 \to 0$ (hence $|\mathrm{Im}\, z_j| \to 0$, $j = 2, \ldots, n$). Therefore, for $\mathrm{Im}\, \tau > \delta > 0$, $f(z, \tau)$ is holomorphic so long as $\mathrm{Im}\, z_1 > -\epsilon'$, $|\mathrm{Im}\, z_j| < \epsilon'$ for some $\epsilon' > 0$, $j = 2, \ldots, n$. Then Proposition 3.8.7 implies that, for sufficiently small δ and $\epsilon > 0$, there exists $\epsilon'' > 0$ such that $f(z, \tau)$ is holomorphic for $|\mathrm{Im}\, \tau| < \epsilon''$ and $|\mathrm{Im}\, z_j| < \epsilon''$, $j = 1, \ldots, n$. Therefore, particularly $f(z, \tau)$ is real-analytic where $\mathrm{Im}\, z = 0$ and $|\mathrm{Im}\, \tau| < \epsilon''$. That is, $u(x, \tau) = b(f(z, \tau))$ is real-analytic in $|\tau| < \epsilon''$, completing the proof of Lemma 1. Hence Proposition 3.8.8 has been proved.

We finally obtain the next theorem.

Theorem 3.8.1.

(1) *Let* $u(x, \tau)$ *be a microfunction containing a holomorphic parameter* τ. *For a connected open subset* D *in* $\mathbf{C}$, *let* $u(x, \tau)$ *be defined on* $\tilde{D}_{(x_0, \xi_0)} = \{(x_0, \tau; \sqrt{-1}\langle \xi_0, dx\rangle\infty) | \tau \in D \subset \mathbf{C}\}$. *If* $u(x, \tau) = 0$ *holds in a neighborhood of* $(x_0, \tau_0; \sqrt{-1}\langle\xi_0, dx\rangle\infty)$, *then* $u(x, \tau) = 0$ *in a neighborhood of* $\tilde{D}_{(x_0, \xi_0)}$.

(2) (Uniqueness of Analytic Continuation): *Let* $u(x, \tau)$ *be a hyperfunction containing a holomorphic parameter* τ, *which is defined in a neighborhood of* $\{x_0\} \times D$. *If one has* $u(x, \tau) = 0$ *in a neighborhood of* (x_0, τ_0), $\tau_0 \in D$, *then* $u(x, \tau) = 0$ *in a neighborhood of* $\{x_0\} \times D$.

Proof. The assertion (2) is immediate from Proposition 3.8.8 and from the uniqueness of analytic continuation of a holomorphic function. We will prove (1). Since $u(x, \tau)$ is defined in a neighborhood of $\{(x_0, \sqrt{-1}\xi_0\infty)\} \times D$, there exists an open set V containing $(x_0, \sqrt{-1}\xi_0\infty)$ such that $u(x, \tau)$ is defined in $V \times D$. Furthermore, one may assume that $u = 0$ holds in a neighborhood of $V \times \{\tau_0\}$. For the sake of simplicity, let $V = U \times \sqrt{-1}W$, where U and W are convex. Note that the sheaf $\mathscr{L}_M$ of microlocal operators on $\sqrt{-1}S^*M$ is also flabby, because of the flabbiness of the microfunction sheaf. From this, one concludes that there exists a microlocal operator K such that $K = 1$ and $\operatorname{supp} K \subset V$ in a neighborhood of $(x_0, \sqrt{-1}\xi_0\infty)$. For this K, let $v = Ku$. Then, from the exactness of $\mathscr{B} \xrightarrow{\text{sp}} \pi_*\mathscr{C} \to 0$, one has a hyperfunction g such that $v = \text{sp}(g)$. Then one has $\text{sp}(\partial g/\partial\bar{\tau}) = \partial v/\partial\bar{\tau} = 0$. Hence, $h = \partial g/\partial\bar{\tau}$ is real-analytic in $U \times D$. Let L be a compact convex set contained in U, and let L' be an arbitrary compact convex subset of D. Then it is well known in the theory of partial differential equations with constant coefficients (e.g. Ehrenpreis [1]) that there exists a real analytic function g_0 such that $\partial g_0/\partial\bar{\tau} = h$ holds on $L \times L'$. Define $v = \text{sp}(g - g_0)$. Then, if L is chosen to contain x_0, then by definition $g - g_0$ is real-analytic in a neighborhood of (x_0, τ_0). Since $g - g_0$ has τ as a holomorphic parameter inside $L \times L'$, by Proposition 3.8.8, $g - g_0$ is real-analytic in a neighborhood of $\{x_0\} \times$ (inside L). That is, $u = v = \text{sp}(g - g_0) = 0$ holds in a neighborhood of $\{(x_0, \sqrt{-1}\xi_0)\} \times$ (inside L) for an arbitrary compact convex set L in D. This completes the proof.

CHAPTER IV

Microdifferential Operators

§1. Definition of the Microdifferential Operator and Its Fundamental Properties

The class of microlocal operators, introduced in §4 of Chapter III, is too wide for practical manipulation. For example, in practice it is a very difficult task to compute the composition of two microlocal operators. In this chapter, we will define a microdifferential operator, as a special case of a microlocal operator, and will consider its properties.

The class of microdifferential operators is located between the class of microlocal operators and the class of differential operators. Its appearance has already been shown in Theorem 3.4.3 (Sato's fundamental theorem). According to this theorem, if the principal symbol $p_m(x, \xi)$ of a linear differential operator $P(x, D)$ does not vanish at $(x_0, \sqrt{-1}\xi_0)$, then there exists a microlocal operator as the inverse element. A main topic of this chapter is to define a class among microlocal operators where the inverse exists and where algebraic manipulations can be performed.

Let X be an open subset of $\mathbf{R}^n$, and let $P(x, D) = \sum a_\alpha(x)D^\alpha$ be a differential operator with coefficients in analytic functions, where $\alpha = (\alpha_1, \ldots, \alpha_n)$ is an n-tuple of non-negative integers and $D^\alpha = \partial^{|\alpha|}/\partial x_1^{\alpha_1} \ldots \partial x_n^{\alpha_n}$, as in Definition 3.4.2. We will begin with the computation of the kernel function $K(x, x')\, dx'$ of $P(x, D)$ regarded as a microlocal operator. As we noted in §4 of Chapter III, this is nothing but $(P(x, D_x)\delta(x - x'))\, dx'$. On the other hand, the plane-wave expansion formula (Proposition 3.2.3),

$$\delta(x - x') = \frac{(n-1)!}{(-2\pi\sqrt{-1})^n} \int \frac{\omega(\xi)}{(\langle x - x', \xi\rangle + \sqrt{-1}0)^n}$$

gives, by taking the derivative D_{x_j} of both sides,

$$D_{x_j}\delta(x - x') = \frac{1}{(-2\pi\sqrt{-1})^n} \int \frac{-n!\xi_j}{(\langle x - x', \xi\rangle + \sqrt{-1}0)^{n+1}}\, \omega(\xi).$$

Further differentiation implies

$$D_x^\alpha \delta(x-x') = \frac{1}{(-2\pi\sqrt{-1})^n} \int \frac{(-1)^{|\alpha|}\xi^\alpha(n+|\alpha|-1)!}{(\langle x-x', \xi\rangle + \sqrt{-1}0)^{n+|\alpha|}} \omega(\xi).$$

Hence, we obtain

$$\begin{aligned} K(x,x') &= P(x, D_x)\delta(x-x') \\ &= \frac{1}{(-2\pi\sqrt{-1})^n} \int \sum_\alpha \frac{(-1)^{|\alpha|}a_\alpha(x)\xi^\alpha(n+|\alpha|-1)!}{(\langle x-x', \xi\rangle + \sqrt{-1}0)^{n+|\alpha|}} \omega(\xi) \\ &= \frac{1}{(-2\pi\sqrt{-1})^n} \int \sum_j \frac{(-1)^j(j+n-1)!p_j(x,\xi)}{(\langle x-x', \xi\rangle + \sqrt{-1}0)^{n+j}} \omega(\xi), \end{aligned}$$

where $p_j(x, \xi) = \sum_{|\alpha|=j} a_\alpha(x)\xi^\alpha$. Here $p_j(x, \xi)$ is a homogeneous polynomial in ξ of degree j. A microdifferential operator corresponds to a generalization of the polynomial $p_j(x, \xi)$ to a holomorphic function $p_j(x, \xi)$.

Let us introduce the following function for convenience:

$$\Phi_\lambda(\tau) = \frac{\Gamma(\lambda)}{(-\tau)^\lambda} \qquad \text{for } \tau \in \mathbf{C} - \{\tau \in \mathbf{R} \mid \tau \geqq 0\}. \tag{4.1.1}$$

Since $(-\tau)^\lambda$ is a multivalued function, we choose a branch such that $(-1)^\lambda = 1$ holds for $\tau = -1$. Furthermore, $\Phi_\lambda(\tau)$ cannot be defined at $\lambda = 0, -1, -2, \ldots$ as in the equation (4.1.1), since the gamma-function $\Gamma(\lambda)$ has a pole of order one at $\lambda = 0, -1, -2, \ldots$. But the function

$$(\lambda + m)\Phi_\lambda(\tau) = (\lambda + m)\Gamma(\lambda)/(-\tau)^\lambda = \frac{\Gamma(\lambda + m + 1)}{\lambda(\lambda+1)\ldots(\lambda+m-1)(-\tau)^\lambda}$$

is holomorphic in a neighborhood of $\lambda = -m$ and has the value $\tau^m/m!$ at $\lambda = -m$. Hence, $\Phi_\lambda(\tau) - \tau^m/(\lambda + m)m!$ is holomorphic in a neighborhood of $\lambda = -m$. Since our concern is the singularity with respect to τ, we may take $\tau^m/(\lambda + m)m!$ for $\Phi_\lambda(\tau)$. Then let the value at $\lambda = -m$ be $\Phi_{-m}(\tau)$. When we carry out the computation, we have

$$\Phi_{-m}(\tau) = \frac{-1}{m!}\tau^m\left(\log(-\tau) + \left(\gamma - 1 - \frac{1}{2} - \cdots - \frac{1}{m}\right)\right) \qquad \text{for } m = 0, 1, 2, \ldots,$$

where $\gamma = 0.57721\ldots$ is the Euler constant. We have

$$\frac{\partial}{\partial\tau}\Phi_\lambda(\tau) = \Phi_{\lambda+1}(\tau).$$

Since $\Gamma(\lambda) = \int_0^\infty e^{-t}t^{\lambda-1}\,dt$, we have $\Phi_\lambda(\tau) = \int_0^\infty e^{t\tau}t^{\lambda-1}\,dt$ for $\operatorname{Re}\tau < 0$ and $\operatorname{Re}\lambda > 0$.

We can rewrite $K(x, x')$ in terms of $\Phi_\lambda(\tau)$, as follows:

$$K(x, x') = P(x, D_x)\delta(x - x')$$

$$= \frac{1}{(2\pi)^n} \int \sum_j p_j(x, \sqrt{-1}\xi)\Phi_{n+j}(\sqrt{-1}(\langle x - x', \xi\rangle + \sqrt{-1}0))\omega(\xi).$$

Note that $\Phi_\lambda(\sqrt{-1}\tau)$ is defined for $\operatorname{Im} \tau > 0$. We may omit $+\sqrt{-1}0$ when it is obvious.

Proposition 4.1.1. *Let* $\{a_j(z)\}_{j \in Z}$ *be a sequence of holomorphic functions defined in* $\Omega \subset \mathbf{C}^n$, *which satisfies the conditions* (4.1.2) *and* (4.1.3) *as follows:*

For an arbitrary compact subset K of Ω, *there exists a positive real number* R_K *such that*

$$\sup_{z \in K} |a_j(z)| \leqq (-j)! R_K^{-j} \qquad \text{for } j < 0$$

holds. (4.1.2)

Let K be an arbitrary compact subset of Ω, *and let* ϵ *be an arbitrary positive real number. Then there exists* $C_{\epsilon,K}$ *such that*

$$\sup_{z \in K} |a_j(z)| \leqq \frac{1}{j!} C_{\epsilon,K}\epsilon^j \qquad \text{for } j \geqq 0$$

holds. (4.1.3)

Then, the following (1) *and* (2) *hold:*

(1) *The series* $\sum_{j=-\infty}^{\infty} a_j(z)\Phi_{\lambda+j}(\tau)$, $z \in \Omega$, *is uniformly absolutely convergent in wider sense for* $\{\tau \in \mathbf{C} \mid 0 < |\tau| \ll 1,\ 0 \neq \tau\}$ *and is multivalued for* $\lambda \in \mathbf{C} - \mathbf{Z}$. *Furthermore, it has a pole of order one at* $\lambda = 0, \pm 1, \pm 2, \ldots$.

(2) *The boundary value of* $\sum_j a_j(z)\Phi_{\lambda+j}(\sqrt{-1}\tau)$ *from* $\{\operatorname{Im} \tau > 0\}$ *defines a microfunction containing* $\lambda \in \mathbf{C}$ *as a holomorphic parameter.*

Proof. One has

$$f_1(z, \tau) \underset{\text{def}}{=} \sum_{j=0}^{\infty} a_j(z)\Phi_{\lambda+j}(\tau) = (-\tau)^{-\lambda} \sum_{j=0}^{\infty} a_j(z)\Gamma(\lambda + j)(-\tau)^{-j}$$

$$= \Gamma(\lambda)(-\tau)^{-\lambda} \sum_{j=0}^{\infty} \lambda(\lambda + 1) \cdots (\lambda + j - 1)a_j(z)(-\tau)^{-j}.$$

Since $|\lambda(\lambda + 1) \cdots (\lambda + j - 1)| \leqq |\lambda|(1 + |\lambda|) \cdots (j - 1 + |\lambda|) \leqq (j - 1)!|\lambda|(1 + |\lambda|) \cdots (1 + |\lambda|/(j - 1))$ and $1 + x \leqq e^x$ $(x \geqq 0)$ hold, the last term is less than or equal to $(j-1)!|\lambda|e^{|\lambda|(1+\cdots+1/(j-1))}$. Furthermore, one has $1+\ldots+1/(j-1) \leqq \int_1^{j-1} dx/x = 1+\log(j-1)$. Consequently,

$$|\lambda(\lambda+1)\cdots(\lambda+j-1)| \leqq (j-1)!|\lambda|e^{|\lambda|}(j-1)^{|\lambda|} \tag{4.1.4}$$

holds. Hence, one obtains

$$|\lambda(\lambda+1)\cdots(\lambda+j-1)a_j(z)| < |\lambda|(j-1)^{|\lambda|-1}e^{|\lambda|}C_{\epsilon,K}\epsilon^j$$

for $z \in K$. Therefore, $((-\tau)^\lambda/\Gamma(\lambda))f_1(z,\tau)$ is holomorphic for $z \in \operatorname{Int} K$, the interior of K, and $\epsilon < |\tau|$. Since ϵ and K are arbitrary, consequently $(-\tau)^\lambda f_1(z,\tau)/\Gamma(\lambda)$ is holomorphic for $z \in \Omega$ and $\tau \in \mathbf{C} - \{0\}$. Next we will consider

$$f_2(z,\tau) = \sum_{j<0} a_j(z)\Phi_{\lambda+j}(\tau) = \sum_{j>0} \Gamma(\lambda-j)a_{-j}(-\tau)^{j-\lambda}.$$

This can be rewritten as

$$f_2(z,\tau) = \sum_{0<j<m} \Gamma(\lambda-j)a_{-j}(z)(-\tau)^{j-\lambda} + \sum_{j\geqq m} \Gamma(\lambda-j)a_{-j}(z)(-\tau)^{j-\lambda},$$

where m is a sufficiently large integer. The first term on the right-hand side is a multivalued meromorphic function for $\lambda \in \mathbf{C}$ and $0 < |\tau|$, having a pole at $\lambda \in \mathbf{Z}$. Hence, it is sufficient to prove the assertion for the second term. We have

$$\begin{aligned}\sum_{j\geqq m} \Gamma(\lambda-j)a_{-j}(z)(-\tau)^{j-\lambda} &= \Gamma(\lambda-m)(-\tau)^{-\lambda} \sum_{j\geqq m} \frac{a_{-j}(z)}{(\lambda-m-1)\cdots(\lambda-j)}(-\tau)^j \\ &= \Gamma(\lambda-m)(-\tau)^{-\lambda} \sum_{j\geqq m} \frac{(-1)^m a_{-j}(z)}{(1+m-\lambda)\cdots(j-\lambda)}\tau^j.\end{aligned}$$

For $|\lambda| \leqq m$, $|(1+m-\lambda)\cdots(j-\lambda)| \geqq (1+m-|\lambda|)\cdots(j-|\lambda|) \geqq (j-m)!$ holds. Hence, for $z \in K$ one obtains

$$\left|\frac{a_{-j}(z)}{(1+m-\lambda)\cdots(j-\lambda)}\right| \leqq \frac{j!R_K^j}{(j-m)!}.$$

That is, the second term converges for $z \in \operatorname{Int} K$ and $0 < |\tau| < 1/R_K$.

Remark. Conversely, if $\sum_j a_j(z)\Phi_{\lambda+j}(\tau)$ is uniformly absolutely convergent in wider sense for $0 < |\tau| \ll 1$, then $\{a_j(z)\}$ satisfies the conditions (4.1.2) and (4.1.3).

By virtue of Proposition 4.1.1, we can introduce a special class of microlocal operators. Take $\mathbf{R}^n$ for a real analytic manifold M, and take $\mathbf{C}^n$ for a complexification X of M. Denote the coordinates of the tangent bundle

$T^*X \cong \mathbf{C}^n \times \mathbf{C}^n$ by $(z, \zeta) = (z_1, \ldots, z_n, \zeta_1, \ldots, \zeta_n)$. Let λ be a complex number, and let Ω be an open subset of T^*X.

Definition 4.1.1. *Denote by* $\mathscr{E}^{\infty}_{(\lambda)}(\Omega)$ *the totality of sequences* $\{p_{\lambda+j}(z, \zeta)\}_{j \in \mathbf{Z}}$ *of holomorphic functions defined in* Ω *satisfying the following conditions* (4.1.5) *and* (4.1.6):

$p_{\lambda+j}(z, \zeta)$ *is a holomorphic function defined in* Ω *and homogeneous in* ζ *of degree* $\lambda + j$; *i.e.*

$$\sum_{i=1}^{n} \zeta_i(\partial/\partial\zeta_i)p_{\lambda+j}(z, \zeta) = (\lambda + j)p_{\lambda+j}(z, \zeta) \text{ holds.} \tag{4.1.5}$$

(4.1.6): $p_{\lambda+j}(z, \zeta)$ *satisfies the growth conditions as follows:*

(4.1.6a) *For an arbitrary compact subset* K *in* Ω, *there exists a positive* C_K *such that*

$$|p_{\lambda+j}(z, \zeta)| \leqq C_K^{-j}(-j)!$$

holds for $(z, \zeta) \in K$ *and* $j < 0$.

(4.1.6b) *For an arbitrary compact subset* K *in* Ω *and for an arbitrary* $\epsilon > 0$, *there exists a positive* $C_{K,\epsilon}$ *such that*

$$|p_{\lambda+j}(z, \zeta)| < \frac{1}{j!} C_{K,\epsilon}\epsilon^j$$

holds for $(z, \zeta) \in K$ *and* $j \geqq 0$.

Let

$$P(z, D) = \sum_j p_{\lambda+j}(z, D).$$

Then $P(z, D)$ *is called a microdifferential operator defined in* Ω. *Since the presheaf* $\{\mathscr{E}^{\infty}_{(\lambda)}(\Omega)\}$ *is clearly a sheaf, we denote the sheaf by* $\mathscr{E}^{\infty}_{(\lambda)}$. $\mathscr{E}(\lambda)$ *denotes the sheaf of operators* $P = \sum_j p_{\lambda+j}(z, D)$ *with the property* $p_{\lambda+j} = 0$ *for* $j > 0$. *Furthermore, define* $\mathscr{E}_{(\lambda)} = \bigcup_{j \in \mathbf{Z}} \mathscr{E}(\lambda + j)$. *We also denote* $\mathscr{E}^{\infty}_{(\lambda)}$ *and* $\mathscr{E}_{(\lambda)}$ *for* $\lambda = 0$ *by* $\mathscr{E}^{\infty}$ *and* $\mathscr{E}$ *respectively. Let* $\lambda = \operatorname{ord} P$ *for* $P \in \mathscr{E}(\lambda)$ *and then* λ *is said to be the order of* P. *An element of* $\mathscr{E}_{(\lambda)}$ *is called a microdifferential operator of finite order. An element of* $\mathscr{E}(\lambda)$ *is called a microdifferential operator of order at most* λ. *An element of* $\mathscr{E}^{\infty}_{(\lambda)}$, *which is not in* $\mathscr{E}_{(\lambda)}$, *is said to be a microdifferential operator of infinite order.*

Remark 1. $\cosh(\sqrt{D_1}) \underset{\text{def}}{=} \sum_{n=0}^{\infty} (D_1^n/(2n)!)$ is one of the simplest examples of microdifferential operators of infinite order (in fact, this is a differential operator). Even though the notion of microdifferential operators of infinite order is profoundly related to the structures of holomorphic functions over the field of complex numbers and is a very useful tool in analysis, it is quite difficult to manipulate. Hence we seldom use it in this book. We also mention that a microdifferential operator of infinite order acts on the sheaf of microfunctions as a sheaf homomorphism, as it will be shown below. But it does not act as a sheaf homomorphism on analogues of microfunctions constructed from distributions.

In the above, $P(z, D)$ is called a microdifferential "operator"; in fact, $P(z, D)$ determines a microlocal operator, as will be shown below. Let Ω be invariant under the actions by $\mathbf{R}_+^\times$; i.e. for $(z, \zeta) \in \Omega$ and $c > 0$ one has $(z, c\zeta) \in \Omega$. Then we will prove that $P(z, D)$ determines a microlocal operator on $R(\Omega) = \{(x\sqrt{-1}\xi\infty) \in \sqrt{-1}S^*\mathbf{R}^n \,|\, (x, \sqrt{-1}\xi) \in \Omega\}$. From Proposition 4.1.1, the boundary value of

$$\sum_j p_{\lambda+j}(z, \sqrt{-1}\zeta)\Phi_{\lambda+j+n}(\sqrt{-1}\tau)$$

from $\operatorname{Im}\tau > 0$ defines a hyperfunction. Let $\tau = \langle x - y, \zeta\rangle$, and let $K(x, y, \zeta)$ be the boundary value of $\sum_j p_{\lambda+j}(x, \sqrt{-1}\zeta)\Phi_{\lambda+j+n}(\sqrt{-1}(\langle x - y, \zeta\rangle))$ from $\operatorname{Im}(\langle x - y, \zeta\rangle) > 0$. Then $K(x, y, \zeta)$ is a hyperfunction defined in an open set U containing $\{(x, y, \zeta) \,|\, x = y \text{ and } (x, \sqrt{-1}\zeta) \in \Omega\}$. Furthermore, it is obtained from the holomorphic function defined in $\operatorname{Im}\langle x - y, \zeta\rangle > 0$ and is holomorphic where $\langle x - y, \zeta\rangle \neq 0$. Hence,

$$\begin{aligned} \text{S.S. } K(x, y, \zeta) \subset G = \{&(x, y, \zeta; \sqrt{-1}(\xi, \eta, \rho)\infty \,|\, (x, y, \zeta) \in U, \langle x - y, \zeta\rangle = 0 \\ &\text{and } \xi = -\eta = k\zeta, \rho = k(x - y) \text{ for some } k > 0\} \end{aligned}$$

holds. Then let

$$K(x, y) = \int K(x, y, \zeta)\omega(\zeta).$$

If a point in G satisfies $\rho = 0$, then $x = y$ holds since $k > 0$. Hence,

$$\begin{aligned} Z \underset{\text{def}}{=} &G \cap \{(x, y, \zeta; \sqrt{-1}(\xi, \eta, \rho)\infty) \,|\, \rho = 0\} \\ = &\{(x, y, \zeta; \sqrt{-1}(\xi, \eta, \rho)\infty) \,|\, (x, \sqrt{-1}\zeta) \in \Omega, \\ &x = y, \xi = -\eta = k\zeta, (k > 0), \rho = 0\} \end{aligned}$$

holds. Therefore, the projection from Z to the $(x, y, \sqrt{-1}(\xi, \eta)\infty)$-space is a proper map, and its image is contained in

$$\{(x, y; \sqrt{-1}(\xi, \eta)\infty) \,|\, \xi = -\eta, x = y, (x, \sqrt{-1}\xi) \in \Omega\}.$$

Consequently, $K(x, y)$ is a microfunction defined in $\{(x, y; \sqrt{-1}(\xi, \eta)\infty) \,|\, (x, \sqrt{-1}\xi) \in R(\Omega)\}$ with its support contained in $\{x = y, \xi = -\eta\}$. Hence, $K(x, y)\,dx$ defines a microlocal operator defined over $R(\Omega)$.

Remark 2. The flabbiness of the microfunction sheaf implies that the sheaf $\mathscr{L}$ of microlocal operators is also flabby. Contrary to these flabby sheaves, the sheaf $\mathscr{E}$ of microdifferential operators is a "rigid" sheaf, having uniqueness of continuation. This indicates how great the difference is between these two notions.

In order to establish various formulas in connection with microdifferential operators, we will begin with the next formula.

Proposition 4.1.2.

$$\int_{S^{n-1}} \xi_{1+}^{\lambda_1 - 1} \cdots \xi_{n+}^{\lambda_n - 1} \Phi_{\lambda_1 + \cdots + \lambda_n}(\sqrt{-1}(\langle \xi, x\rangle + \sqrt{-1}0))\omega(\xi)$$
$$= \Phi_{\lambda_1}(\sqrt{-1}(x_1 + \sqrt{-1}0)) \cdots \Phi_{\lambda_n}(\sqrt{-1}(x_n + \sqrt{-1}0)). \quad (4.1.5')$$

Proof. It is sufficient to prove the above formula for the case where x_j is the complex number $x_j + \sqrt{-1}\epsilon_j$ for $\epsilon_j > 0$. If $\operatorname{Im} x_j > 0$ and $\xi_j \geqq 0$, then one has

$$\Phi_{\lambda_1 + \cdots + \lambda_n}(\sqrt{-1}(\langle \xi, x\rangle + \sqrt{-1}0)) = \int_0^\infty e^{\sqrt{-1}t\langle \xi, x\rangle} t^{\lambda_1 + \cdots + \lambda_n - 1}\, dt.$$

Hence, the left-hand side of (4.1.5′) becomes

$$\int_{S^{n-1}} \omega(\xi)\xi_{1+}^{\lambda_1 - 1} \cdots \xi_{n+}^{\lambda_n - 1} \int_0^\infty e^{\sqrt{-1}t\langle \xi, x\rangle} t^{\lambda_1 + \cdots + \lambda_n - 1}\, dt,$$

which can be rewritten as follows:

$$\int_0^\infty t^{n-1}\, dt \int_{S^{n-1}} \omega(\xi)(t\xi_1)_+^{\lambda_1 - 1} \cdots (t\xi_n)_+^{\lambda_n - 1} e^{\sqrt{-1}\langle t\xi, x\rangle}$$
$$= \int_{\mathbf{R}^n} d\xi\, \xi_{n+}^{\lambda_1 - 1} \cdots \xi_{n+}^{\lambda_n - 1} e^{\sqrt{-1}\langle \xi, x\rangle}$$
$$= \prod_{j=1}^n \int_0^\infty \xi_j^{\lambda_j - 1} e^{\sqrt{-1}\xi_j x_j}\, d\xi_j$$
$$= \prod_{j=1}^n \Phi_{\lambda_j}(\sqrt{-1}x_j)$$
$$= \text{the right-hand side of (4.1.5′)}.$$

Using this result, we will give proofs to the formulas (4.1.8) and (4.1.9) of Radon transforms. (Consult Gel'fand, Graev and Vilenkin [1] for the geometric meanings of these equations.) For an arbitrary complex number λ, define a function $\delta_\lambda(x, y)$ on $(\mathbf{R}^n - \{0\}) \times (\mathbf{R}^n - \{0\})$ by

$$\delta_\lambda(x, y) = |x_j|^\lambda \prod_{k \neq j} \delta\left(y_k - \frac{y_j}{x_j} x_k\right)(\pm y_j)_+^{-\lambda - 1} \qquad \text{for } \pm x_j > 0.$$

For $\epsilon_j x_j > 0$ and $\epsilon_k x_k > 0$, $(\epsilon_j, \epsilon_k = \pm 1)$,

$$|x_j|^\lambda \prod_{l \neq j} \delta\left(y_l - \frac{y_j}{x_j} x_l\right)(\epsilon_j y_j)_+^{-\lambda-1} = |x_k|^\lambda \prod_{l \neq k} \delta\left(y_l - \frac{y_k}{x_k} x_l\right)(\epsilon_k y_k)_+^{-\lambda-1}$$

holds. Hence $\delta_\lambda(x, y)$ is well defined as a hyperfunction on $(\mathbf{R}^n - \{0\}) \times (\mathbf{R}^n - \{0\})$. From the definition, the homogeneous degrees of $\delta_\lambda(x, y)$ in x and y are λ and $-\lambda - n$ respectively. One has

$$\operatorname{supp} \delta_\lambda(x, y) \subset \{(x, y) \mid x = ty, t > 0\}$$

and

$$\delta_\lambda(x, y) = \delta_{-n-\lambda}(y, x) = \delta_\lambda(-x, -y).$$

The following lemma is obvious from the definition.

Lemma 1. *Let $u(x)$ be a hyperfunction on $\mathbf{R}^n - \{0\}$ of homogeneous degree λ. Then one has*

$$u(x) = \int \delta_\lambda(x, y) u(y) \omega(y).$$

Lemma 2.

$$\frac{1}{(2\pi)^n} \int_{S^{n-1}} \Phi_{-\lambda}(\sqrt{-1}(\langle x, \xi \rangle + \sqrt{-1}0)) \cdot \Phi_{\lambda+n}(-\sqrt{-1}(\langle y, \xi \rangle - \sqrt{-1}0))\omega(\xi) = \delta_\lambda(x, y). \quad (4.1.6')$$

Proof. One obtains from Proposition 4.1.2.

$$\Phi_{-\lambda}(\sqrt{-1}(\langle x, \xi \rangle + \sqrt{-1}0))\Phi_{\lambda+n}(-\sqrt{-1}(\langle y, \xi \rangle - \sqrt{-1}0)) = \int t_+^{-\lambda-1} s_+^{\lambda+n-1} \Phi_n(\sqrt{-1}(\langle tx - sy, \xi \rangle + \sqrt{-1}0))\omega(t, s).$$

Hence, the left-hand side of (4.1.6′) becomes

$$\frac{1}{(2\pi)^n} \int t_+^{-\lambda-1} s_+^{\lambda+n-1} \omega(t, s) \int \Phi_n(\sqrt{-1}(\langle tx - sy, \xi \rangle + \sqrt{-1}0)\omega(\xi) = \int t_+^{-\lambda-1} s_+^{\lambda+n-1} \delta(tx - sy)\omega(t, s).$$

If, for example, $x_1 > 0$, then

$$\int t_+^{-\lambda-1} s_+^{\lambda+n-1} \delta(tx - sy)\omega(t, s) = \int t_+^{-\lambda-1} \delta(tx - y)\, dt = \left(\frac{y_1}{x_1}\right)_+^{-\lambda-1} \prod_{j \neq 1} \delta\left(\frac{y_1}{x_1} x_j - y_j\right) \frac{1}{x_1},$$

which is equal to $\delta_\lambda(x, y)$.

One easily has the following lemma from the definition.

Lemma 3. *For $g \in GL(n, \mathbf{R})$, one has*

$$\delta_\lambda(gx, gy) = |\det g|^{-1}\delta_\lambda(x, y). \tag{4.1.7}$$

Proposition 4.1.3. *There is a one-to-one correspondence between homogeneous functions $u(x)$ on $\mathbf{R}^n - \{0\}$ of degree λ and homogeneous functions $v(x)$ on $\mathbf{R}^n - \{0\}$ of degree $-n-\lambda$, given by*

$$v(\xi) = \frac{1}{(2\pi)^{n/2}} \int u(x)\Phi_{\lambda+n}(\sqrt{-1}(\langle x, \xi\rangle + \sqrt{-1}0))\omega(x) \tag{4.1.8}$$

and

$$u(x) = \frac{1}{(2\pi)^{n/2}} \int v(\xi)\Phi_{-\lambda}(-\sqrt{-1}(\langle x, \xi\rangle - \sqrt{-1}0)\omega(\xi). \tag{4.1.9}$$

Proof. When $u(x)$ and $v(\xi)$ are given as in (4.1.8), one has

$$\begin{aligned}
&\frac{1}{(2\pi)^{n/2}} \int v(\xi)\Phi_{-\lambda}(-\sqrt{-1}(\langle x, \xi\rangle - \sqrt{-1}0))\omega(\xi) \\
&\qquad = \frac{1}{(2\pi)^n} \int u(y)\omega(y) \int \varphi_{\lambda+n}(\sqrt{-1}\langle y, \xi\rangle + \sqrt{-1}0) \\
&\qquad\quad \cdot \Phi_{-\lambda}(-\sqrt{-1}(\langle x, \xi\rangle - \sqrt{-1}0))\omega(\xi) \\
&\qquad = \int u(y)\delta_\lambda(x, y)\omega(y) = u(x).
\end{aligned}$$

A similar computation can be done when $u(x)$ and $v(\xi)$ are given as in (4.1.9).

Proposition 4.1.4. *Let $P(x, D_x) = \sum_j p_{\lambda+j}(x, D_x)$. Then, for an integer $\mu \geqq n$,*

$$\begin{aligned}
&P(x, D_x)\Phi_\mu(\sqrt{-1}(\langle x, \xi\rangle + p + \sqrt{-1}0)) \\
&\qquad = \sum_j p_{\lambda+j}(x, \sqrt{-1}\xi)\Phi_{\lambda+\mu+j}(\sqrt{-1}(\langle x, \xi\rangle + p + \sqrt{-1}0))
\end{aligned} \tag{4.1.10}$$

holds.

Proof. Let us first consider the case when $\mu = n$. Let

$$K(x, \xi, p) = \frac{1}{(2\pi)^n} \sum p_{\lambda+j}(x, \sqrt{-1}\xi)\Phi_{n+\lambda+j}(\sqrt{-1}(\langle x, \xi\rangle + p + \sqrt{-1}0)).$$

Then, the left-hand side of (4.1.10) equals

$$\begin{aligned}
&\int K(x, \eta, -\langle y, \eta\rangle)\Phi_n(\sqrt{-1}(\langle y, \xi\rangle + p + \sqrt{-1}0))\omega(\eta)\, dy \\
&\qquad = \int K(x, \eta, t)\, dt \int \delta(t + \langle y, \eta\rangle)\Phi_n(\sqrt{-1}(\langle y, \xi\rangle + p + \sqrt{-1}0))\omega(\eta)\, dy.
\end{aligned}$$

On the other hand, one has

$$\delta(t + \langle y, \eta\rangle) = \frac{1}{2\pi}\Phi_1(-\sqrt{-1}(t + \langle y, \eta\rangle - \sqrt{-1}0)).$$

Hence,

$$\int \delta(t + \langle y, \eta\rangle)\Phi_n(\sqrt{-1}(\langle y, \xi\rangle + p + \sqrt{-1}0))\omega(\eta)\,dy$$
$$= \frac{1}{2\pi}\int \Phi_1(-\sqrt{-1}(y_0 t + \langle y, \eta\rangle - \sqrt{-1}0)) \cdot \Phi_n(\sqrt{-1}(\langle y, \xi\rangle + y_0 p + \sqrt{-1}0))\omega(y_0, y)\omega(\eta)$$
$$= (2\pi)^n\delta_n((p, \xi), (t, \eta)).$$

Consequently, the left-hand side of (4.1.10) is

$$(2\pi)^n \int K(x, \eta, t)\delta_n((p, \xi), (t, \eta))\omega(t, \eta) = (2\pi)^n K(x, \xi, p).$$

For the general μ, first note

$$\Phi_\mu(\sqrt{-1}(\langle x, \xi\rangle + p + \sqrt{-1}0))$$
$$= \frac{1}{2\pi}\int \Phi_n(\sqrt{-1}(\langle x, \xi\rangle + t + \sqrt{-1}0))\Phi_{\mu-n+1}(\sqrt{-1}(p - t + \sqrt{-1}0))\,dt.$$

Therefore, one obtains the following:

$$P(x, D_x)\Phi_\mu(\sqrt{-1}(\langle x, \xi\rangle + p + \sqrt{-1}0))$$
$$= \frac{1}{2\pi}\int \Phi_{\mu-n+1}(\sqrt{-1}(p - t + \sqrt{-1}0)) \cdot \left(\sum_j p_{\lambda+j}(x, \sqrt{-1}\xi)\Phi_{n+\lambda+j}(\sqrt{-1}(\langle x, \xi\rangle + t + \sqrt{-1}0)\right)dt$$
$$= \left(\frac{D_p}{\sqrt{-1}}\right)^{\mu-n} \sum_j p_{\lambda+j}(x, \sqrt{-1}\xi)\Phi_{n+\lambda+j}(\sqrt{-1}(\langle x, \xi\rangle + p + \sqrt{-1}0)).$$

Hence the next lemma completes the proof.

Lemma 1. *Let $a_{jk}(t, z)$, $j, k \in \mathbf{Z}$, be a holomorphic function defined in a neighborhood U of $t = 0$ and $z = 0$ such that $\sum_{j,k} a_{jk}(t, z)\Phi_{\lambda+j}(p)\Phi_{\mu+k}(q)$ is uniformly absolutely convergent in a wider sense for $(t, z) \in U$ and $0 < |p|, |q| \ll 1$. Let $u(t, s, x)$ be the boundary value of*

$$\sum_{j,k} a_{jk}(t, x)\Phi_{\lambda+j}(\sqrt{-1}(t - s))\,\Phi_{\mu+k}(\sqrt{-1}s)$$

from $\operatorname{Im}(t-s)$ *and* $\operatorname{Im} s > 0$. *Then, as microfunctions, one has*

$$\int u(t,s,x)\,ds = 2\pi \sum_{j,k} a_{j,k}(t,x)\Phi_{\lambda+\mu+j+k-1}(\sqrt{-1}t)$$

in a neighborhood of $\sqrt{-1}\,dt$.

Proof. Let $a > 0$ be sufficiently small and let

$$I_a(t,x) = \int_{\gamma_a} u(t,s,x)\,ds,$$

where γ_a is a path from $-a$ to a, contained in $\operatorname{Im} s \geqq 0$ and not going through the origin. Then $\int u(t,s,x)\,ds$ is the boundary value of $I_a(t,x)$. On the other hand,

$$I_a(t,x) = \sum_{j,k} a_{jk}(t,x) \int_{\gamma_a} \Phi_{\lambda+j}(\sqrt{-1}(t-s))\Phi_{\mu+k}(\sqrt{-1}s)\,ds$$

holds. We need the following lemma to compute the right-hand side of the above equation.

Lemma 2.

$$\begin{aligned}
&\int_{-a}^{b} \Phi_\lambda(\sqrt{-1}(t-s+\sqrt{-1}0))\Phi_\mu(\sqrt{-1}(s+\sqrt{-1}0))\,ds \\
&\qquad - 2\pi\Phi_{\lambda+\mu-1}(\sqrt{-1}(t+\sqrt{-1}0)) \\
&\quad = -\Gamma(\mu)\sum_{n=0}^{\infty} \frac{\Gamma(\lambda+n)}{n!(\lambda+\mu+n-1)} \cdot \left(e^{(\pi i/2)(\mu-\lambda)} b^{1-\lambda-\mu}\left(\frac{t}{b}\right)^n \right. \\
&\qquad \left. + (-1)^n e^{(\pi i/2)(\lambda-\mu)} a^{1-\lambda-\mu}\left(\frac{t}{a}\right)^n \right).
\end{aligned} \tag{4.1.11}$$

Proof. By virtue of analytic continuation with respect to λ and μ, it is sufficient to consider the case where $\operatorname{Re}\lambda \gg 0$ and $\operatorname{Re}\mu \gg 0$. In that case, one has

$$\begin{aligned}
2\pi\Phi_{\lambda+\mu-1}(\sqrt{-1}(t+\sqrt{-1}0)) = \int_{-\infty}^{\infty} &\Phi_\lambda(\sqrt{-1}(t-s+\sqrt{-1}0)) \\
&\cdot \Phi_\mu(\sqrt{-1}(s+\sqrt{-1}0))\,ds.
\end{aligned}$$

Hence, the left-hand side of (4.1.11) can be written as

$$\left(\int_{-a}^{-\infty} - \int_{b}^{\infty}\right) \Phi_\lambda(\sqrt{-1}(t-s+\sqrt{-1}0))\Phi_\mu(\sqrt{-1}(s+\sqrt{-1}0))\,ds.$$

Here, note

$$\begin{aligned}
&\int_b^\infty \Phi_\lambda(\sqrt{-1}(t-s+\sqrt{-1}0))\Phi_\mu(\sqrt{-1}(s+\sqrt{-1}0))\,ds \\
&= e^{(\pi\sqrt{-1}/2)(\mu-\lambda)}\Gamma(\lambda)\Gamma(\mu)\int_b^\infty (s-t)^{-\lambda}s^{-\mu}\,ds \\
&= e^{(\pi\sqrt{-1}/2)(\mu-\lambda)}\Gamma(\lambda)\Gamma(\mu)\int_b^\infty s^{-\lambda-\mu}\left(1-\frac{t}{s}\right)^{-\lambda} ds \\
&= e^{(\pi\sqrt{-1}/2)(\mu-\lambda)}\Gamma(\lambda)\Gamma(\mu)\int_b^\infty \left(\sum_{n=0}^\infty \frac{\lambda(\lambda+1)\cdots(\lambda+n-1)}{n!}t^n s^{-\lambda-\mu-n}\right) ds \\
&= e^{(\pi\sqrt{-1}/2)(\mu-\lambda)}\Gamma(\lambda)\Gamma(\mu)\sum_{n=0}^\infty \frac{\lambda(\lambda+1)\cdots(\lambda+n-1)}{n!(\lambda+\mu+n-1)} b^{-\lambda-\mu-n+1}t^n.
\end{aligned}$$

The term involving $\int_{-a}^{-\infty}$ can be done similarly. This completes the proof for Lemma 2.

Now we return to the proof of Lemma 1. Lemma 2 implies

$$\begin{aligned}
I_a(t,x) &= 2\pi\sum_{j,k} a_{jk}(t,x)\Phi_{\lambda+\mu+j+k-1}(\sqrt{-1}t) \\
&\quad - \sum_{\substack{j,k,n\\ n\geqq 0}} \frac{\Gamma(\lambda+n+j)\Gamma(\mu+k)}{n!(\lambda+\mu+j+k+n-1)}(e^{(\pi\sqrt{-1}/2)(\mu+k-\lambda-j)} \\
&\quad + (-1)^n e^{(\pi\sqrt{-1}/2)(\lambda+j-\mu-k)})\cdot a_{jk}(t,x)a^{1-\lambda-\mu-j-k}\left(\frac{t}{a}\right)^n. \qquad (4.1.12)
\end{aligned}$$

It remains to be proved that the second-term sum of the above converges absolutely for $0<|a|\ll 1$ and $|t/a|\ll 1$. The second term can be written as

$$\int da \sum_{\substack{j,k,n\\ n\geqq 0}} \frac{\Gamma(\lambda+n+j)e^{(\pi\sqrt{-1}/2)(k-j)}(e^{(\pi\sqrt{-1}/2)(\mu-\lambda)}+(-1)^{n+j+k}e^{(\pi\sqrt{-1}/2)(\lambda-\mu)})}{n!\Gamma(\lambda+j)}$$
$$\cdot a_{jk}(t,x)\Phi_{\lambda+j}(-a)\Phi_{\mu+k}(-a)(t/a)^n. \qquad (4.1.13)$$

For $c \geqq \frac{1}{2}|e^{(\pi\sqrt{-1}/2)(\mu-\lambda)}|, \frac{1}{2}|e^{(\pi\sqrt{-1}/2)(\lambda-\mu)}|$, one has

$$\begin{aligned}
&\sum_{n=0}^\infty\left|\frac{\Gamma(\lambda+n+j)e^{(\pi\sqrt{-1}/2)(k-j)}(e^{(\pi\sqrt{-1}/2)(\mu-\lambda)}+(-1)^{n+j+k}e^{(\pi\sqrt{-1}/2)(\lambda-\mu)})}{n!\Gamma(\lambda+j)}\left(\frac{t}{a}\right)^n\right| \\
&\qquad \leqq c\sum_{n=0}^\infty \frac{(|\lambda|+|j|+n-1)\cdots(|\lambda|+|j|)}{n!}\left|\frac{t}{a}\right|^n = c\left(1-\left|\frac{t}{a}\right|\right)^{-|\lambda|-|j|}.
\end{aligned}$$

On the other hand,

$$\sum_{j,k}|a_{jk}(t,x)|\,|\Phi_{\lambda+j}(-a)|\,|\Phi_{\mu+k}(-a)|\left(1-\left|\frac{t}{a}\right|\right)^{-|j|}$$

converges absolutely. Therefore, the integrand of (4.1.13) is absolutely convergent, which completes the proof of Lemma 1.

By Lemma 1, one can easily complete the proof of Proposition 4.1.4. Let $P(x, D_x) = \sum_j p_{\lambda+j}(x, D_x)$ be a microdifferential operator for $x \in \mathbf{R}^n$, and let $K(x, x')\,dx'$ be the kernel function. For a coordinate system y of $\mathbf{R}^m$, consider

$$K(x, x')\delta(y - y')\,dx'\,dy'.$$

This determines the kernel function of a microlocal operator on $\mathbf{R}^{n+m}$. Furthermore, this is a microdifferential operator, which is equal to $\sum_j r_{\lambda+j}(x, y, D_x, D_y)$, where $r_{r+j}(x, y, \xi, \eta) = p_{\lambda+j}(x, \xi)$. In fact,

$$K(x, x')\delta(y - y') = P(x, D_x)\delta(x - x')\delta(y - y')$$

holds, and Proposition 4.1.4 implies

$$\begin{aligned}
&P(x, D_x)\Phi_{n+m}(\sqrt{-1}(\langle x - x', \xi\rangle + \langle y - y', \eta\rangle + \sqrt{-1}0)) \\
&= \sum_j p_{\lambda+j}(x, \sqrt{-1}\xi)\Phi_{n+m+\lambda+j}(\sqrt{-1}(\langle x - x', \xi\rangle + \langle y - y', \eta\rangle + \sqrt{-1}0)) \\
&= \sum_j r_{\lambda+j}(x, y, \sqrt{-1}\xi, \sqrt{-1}\eta)\Phi_{n+m+\lambda+j}(\sqrt{-1}(\langle x - x', \xi\rangle \\
&\quad + \langle y - y', \eta\rangle + \sqrt{-1}0)).
\end{aligned}$$

Hence, one obtains

$$\begin{aligned}
&R(x, y, D_x, D_y)\delta(x - x')\delta(y - y') \\
&\qquad = \frac{1}{(2\pi)^{n+m}}\int p(x, D_x)\Phi_{n+m}(\sqrt{-1}(\langle x - x', \xi\rangle + \langle y - y', \eta\rangle \\
&\qquad\quad + \sqrt{-1}0))\omega(\xi, \eta) \\
&\qquad = P(x, D_x)\delta(x - x', y - y').
\end{aligned}$$

Henceforth, one regards a microdifferential operator $P(x, D_x)$ on $\mathbf{R}^n$ as one on $\mathbf{R}^{n+m}$.

Proposition 4.1.5. *Let $P(x, D_x) = \sum_j p_{\lambda+j}(x, D_x)$ be a microdifferential operator, let $\{a_j(x)\}_{j\in\mathbf{Z}}$ be a sequence which satisfies the conditions* (4.1.2) *and* (4.1.3), *and let*

$$u(x, \xi, p) = \sum_j a_j(x)\Phi_{\mu+j}(\sqrt{-1}(\langle x, \xi\rangle + p + \sqrt{-1}0)).$$

Then the sequence $\{b_l(x, \xi)\}_{l\in\mathbf{Z}}$, where

$$b_l(x, \xi) = \sum_{l=j+k-|\alpha|}\frac{1}{\alpha!}\left(\frac{1}{\sqrt{-1}}D_\xi\right)^\alpha p_{\lambda+\xi}(x, \sqrt{-1}\xi)(D_x^\alpha a_k(x)),$$

satisfies the conditions (4.1.2) *and* (4.1.3), *and then*

$$P(x, D_x)u(x, \xi, p) = \sum_l b_l(x, \xi)\Phi_{\lambda+\mu+l}(\sqrt{-1}(\langle x, \xi\rangle + p + \sqrt{-1}0)) \quad (4.1.14)$$

holds.

Proof.

$$P(x, D_x)u(x, \xi, p)$$

$$= \frac{1}{(2\pi)^n}\int\sum_{j,k} p_{\lambda+j}(x, \sqrt{-1}\eta)\Phi_{\lambda+n+j}(\sqrt{-1}(\langle x-y, \eta\rangle + \sqrt{-\ }0))$$

$$\cdot\, a_k(y)\Phi_{\mu+k}(\sqrt{-1}(\langle y, \xi\rangle + p + \sqrt{-1}0))\omega(\eta)\, dy$$

holds. The Taylor expansion $a_k(y) = \sum_\alpha (1/\alpha!)(y-x)^\alpha D_x^\alpha a_k(x)$ and

$$(y-x)^\alpha\Phi_{\lambda+n+j}(\sqrt{-1}(\langle x-y, \eta\rangle + \sqrt{-1}0))$$

$$= (\sqrt{-1}D_\eta)^\alpha\Phi_{\lambda+n+j-|\alpha|}(\sqrt{-1}(\langle x-y, \eta\rangle + \sqrt{-1}0))$$

provide one with

$$P(x, D_x)u(x, \xi, p)$$

$$= \frac{1}{(2\pi)^n}\int\sum_{j,k,\alpha}\frac{1}{\alpha!}p_{\lambda+j}(x, \sqrt{-1}\eta)(D_x^\alpha a_k(x))\Phi_{\mu+k}(\sqrt{-1}(\langle y, \xi\rangle + p + \sqrt{-1}0))$$

$$\cdot(\sqrt{-1}D_\eta)^\alpha\Phi_{\lambda+n+j-|\alpha|}(\sqrt{-1}(\langle x-y, \eta\rangle + \sqrt{-1}0))\omega(\eta)\, dy. \quad (4.1.15)$$

When one performs a termwise integration by parts, one obtains

$$P(x, D_x)u(x, \xi, p)$$

$$= \frac{1}{(2\pi)^n}\int\Bigg(\sum_{j,k,\alpha}\frac{1}{\alpha!}(-\sqrt{-1}D_\eta)^\alpha p_{\lambda+j+|\alpha|}(x, \sqrt{-1}\eta)$$

$$\cdot\,(D_x^\alpha a_k(x))\Phi_{\mu+k}(\sqrt{-1}(\langle y, \xi\rangle + p + \sqrt{-1}0))$$

$$\cdot\,\Phi_{\lambda+n+j}(\sqrt{-1}(\langle x-y, \eta\rangle + \sqrt{-1}0))\Bigg)\omega(\eta)\, dy.$$

Therefore, for

$$c_{jk}(x, \eta) \underset{\text{def}}{=} \sum\frac{1}{\alpha!}(-\sqrt{-1}D_\eta)^\alpha p_{\lambda+j+|\alpha|}(x, \sqrt{-1}\eta)D_x^\alpha a_k(x),$$

one has

$$P(x, D_x)u(x, \xi, p) = \frac{1}{(2\pi)^n}\int\sum_{j,k} c_{jk}(x, \eta)\Phi_{\mu+k}(\sqrt{-1}(\langle y, \xi\rangle + p + \sqrt{-1}0)$$

$$\cdot\,\Phi_{\lambda+n+j}(\sqrt{-1}(\langle x-y, \eta\rangle + \sqrt{-1}0))\omega(\eta)\, dy.$$

If one lets

$$f(x, \eta, t, s) \underset{\text{def}}{=} \sum_{j,k} c_{jk}(x, \eta)\Phi_{\mu+k}(\sqrt{-1}(t+\sqrt{-1}0))\Phi_{\lambda+1+j}(\sqrt{-1}(s+\sqrt{-1}0)),$$

then one has the following:

$$\begin{aligned} P(x, D_x)u(x, \xi, p) &= \frac{1}{(2\pi)^n}\int_s^{n-1} f(x, \eta, \langle y, \xi\rangle + p, \langle x - y, \eta\rangle)\omega(\eta)\, dy \\ &= \frac{(\sqrt{-1})^{n-1}}{(2\pi)^n}\int f(x, \eta, t + p, \langle x, \eta\rangle + s)\delta(t - \langle y, \xi\rangle) \\ &\quad \cdot \delta^{(n-1)}(s + \langle y, \eta\rangle)\omega(\eta)\, dy\, dt\, ds. \end{aligned}$$

Let us compute

$$\int \delta(t - \langle y, \xi\rangle)\delta^{(n-1)}(s + \langle y, \eta\rangle)\, dy.$$

The above integral is microlocally (i.e. locally on the cotangent bundle) equal to

$$\begin{aligned} \varphi(t, s, \xi, \eta) \underset{\text{def}}{=} \frac{(-\sqrt{-1})^{n-1}}{(2\pi)^2}&\int \Phi_1(\sqrt{-1}(t - \langle y, \xi\rangle + \sqrt{-1}0)) \\ &\cdot \Phi_n(\sqrt{-1}(s + \langle y, \eta\rangle + \sqrt{-1}0))\, dy. \end{aligned}$$

Then one obtains

$$\begin{aligned} \varphi(t, s, \xi, \eta) = \frac{(-\sqrt{-1})^{n-1}}{(2\pi)^2}&\int \Phi_1(\sqrt{-1}(y_0 t + \langle y, -\xi\rangle + \sqrt{-1}0)) \\ &\cdot \Phi_n(\sqrt{-1}(y_0 s + \langle y, \eta\rangle + \sqrt{-1}0))\omega(\tilde{y}) \\ = (2\pi)^{-2}&(2\pi)^{n+1}(-\sqrt{-1})^{n-1}\delta_{-1}((-t, \xi), (s, \eta)), \end{aligned}$$

where $\tilde{y} = (y_0, y)$. Therefore,

$$\begin{aligned} \frac{(\sqrt{-1})^{n-1}}{(2\pi)^n}&\int f(x, \eta, t+p, \langle x, \eta\rangle + s)\delta(t - \langle y, \xi\rangle)\delta^{(n-1)}(s + \langle y, \eta\rangle)\omega(\eta)\, dy\, ds \\ &= \frac{1}{2\pi}\int f(x, \eta, t + p, \langle x, \eta\rangle + s)\delta_{-1}((-t, \xi), (s, \eta))\omega(\eta)\, ds \\ &= \frac{1}{2\pi}\int f(x, \eta, t + p, \langle x, \eta\rangle + s)\delta_{-1}((-t, \xi), (s, \eta))\omega(\eta, s) \\ &= \frac{1}{2\pi} f(x, \xi, t + p, \langle x, \xi\rangle - t) \end{aligned}$$

holds. Consequently, one has

$$P(x, D_x)u(x, \xi, p) = \frac{1}{2\pi} \int f(x, \xi, t + p, \langle x, \xi \rangle - t)\, dt.$$

Hence, Lemma 1, in the proof of Proposition 4.1.4, implies the equation (4.1.14). We need to prove that (4.1.15) can be integrated by parts. It is sufficient to show that there exist hyperfunctions G_ν, $\nu = 1, \ldots, n$, such that

$$\begin{aligned}
&\Big\{\sum_{j,k,\alpha} \frac{1}{\alpha!} p_{\lambda+j}(x, \sqrt{-1}\eta)(D_x^\alpha a_k(x))\Phi_{\mu+k}(\sqrt{-1}(\langle y, \xi\rangle + p + \sqrt{-1}0)) \\
&\qquad \cdot (\sqrt{-1}D_\eta)^\alpha \Phi_{\lambda+n+j-|\alpha|}(\sqrt{-1}(\langle x - y, \eta\rangle + \sqrt{-1}0))\Big\} \\
&\qquad - \Big\{\sum_{j,k,\alpha} \frac{1}{\alpha!}(-\sqrt{-1}D_\eta)^\alpha p_{\lambda+j}(x, \sqrt{-1}\eta)(D_x^\alpha a_k(x)) \\
&\qquad \cdot \Phi_{\mu+k}(\sqrt{-1}(\langle y, \xi\rangle + p + \sqrt{-1}0)) \\
&\qquad \cdot \Phi_{\lambda+n+j-|\alpha|}(\sqrt{-1}(\langle x - y, \eta\rangle + \sqrt{-1}0)\Big\} \\
&= \sum_{\nu=1}^{n} \frac{1}{\sqrt{-1}} \frac{\partial}{\partial \eta_\nu} G_\nu
\end{aligned}$$

holds. This is because the above equation implies

$$\int \frac{\partial}{\partial \eta_\nu} G_\nu \omega(\eta) = 0.$$

The construction of G_ν is left for the reader. (A formal infinite series expression of G_ν can be found from the following lemma. Then the convergence of the series must be checked).

Lemma. *Let $P(\xi)$ be a homogeneous polynomial of degree m. Then*

$$\begin{aligned}
&a(x)(P(D_x)b(x)) - (P(-D_x)a(x))b(x) \\
&\qquad = \sum_j \frac{\partial}{\partial x_j}\left(\sum_{|\alpha|<m} \frac{(-1)^\alpha (m - |\alpha| - 1)!|\alpha|!}{m!\alpha!}(D_x^\alpha a)P^{(\alpha+\delta_j)}(D_x)b\right)
\end{aligned}$$

holds, where $P^{(\alpha)}(\xi) = D_\xi^\alpha P$ and $\delta_j = (0, \ldots, 0, \overset{j}{\check{1}}, 0, \ldots, 0)$.

Proof. The right-hand side is equal to

$$\begin{aligned}
&\sum_{|\alpha|<m} \frac{(-1)^{|\alpha|}(m - |\alpha| - 1)!|\alpha|!}{m!\alpha!}(D^{\alpha+\delta_j}a)P^{(\alpha+\delta_j)}(D)b \\
&\qquad + \sum_{|\alpha|<m} \frac{(-1)^{|\alpha|}(m - |\alpha| - 1)!|\alpha|!}{m!\alpha!}(D^\alpha a)\left(\sum_j D_j P^{(\alpha+\delta_j)}(D)b\right). \qquad (4.1.15')
\end{aligned}$$

Since the homogeneous degree of $P^{(\alpha)}$ is $m - |\alpha|$,

$$\sum_j D_j P^{(\alpha+\delta_j)}(D) = (m - |\alpha|)P^{(\alpha)}$$

holds. Then the second term of (4.1.15′) becomes

$$\sum_{|\alpha|<m} \frac{(-1)^{|\alpha|}(m-|\alpha|)!|\alpha|!}{m!\alpha!}(D^\alpha a)(P^{(\alpha)}(D)b),$$

and the first term becomes

$$\sum_{0<|\beta|\leq m} \frac{(-1)^{|\beta|-1}(m-|\beta|)!(|\beta|-1)!}{m!}\left(\sum_{\beta=\alpha+\delta_j}\frac{1}{\alpha!}\right)(D^\beta a)P^{(\beta)}(D)b.$$

On the other hand, one has

$$\sum_{\beta=\alpha+\delta_j}\frac{1}{\alpha!} = \sum \frac{\beta_j}{\beta!} = \frac{|\beta|}{\beta!}.$$

Hence, the first term of (4.1.15′) is

$$\sum_{0<|\beta|\leq m} \frac{(-1)^{|\beta|-1}(m-|\beta|)!|\beta|!}{m!\beta!}(D^\beta a)P^{(\beta)}(D)b.$$

Consequently one obtains

$$\sum_{|\alpha|=m} \frac{(-1)^{|\alpha|-1}(m-|\alpha|)!|\alpha|!}{m!\alpha!}(D^\alpha a)P^{(\alpha)}(D)b + aP(D)b$$

$$= aP(D)b - \sum_{|\alpha|=m}\frac{1}{\alpha!}(P^{(\alpha)}(D)(-D)^\alpha a)b = aP(D)b - (P(-D)a)b.$$

We obtain from Proposition 4.1.5 the following fundamental result on the composition of microdifferential operators.

Theorem 4.1.1. *Let $P(x, D_x) = \sum_j p_{\lambda+j}(x, D_x)$ be a microdifferential operator in $\mathscr{E}_{(\lambda)}$, and let $Q(x, D_x) = \sum_k q_{k+\mu}(x, D_x)$ be a microdifferential operator in $\mathscr{E}_{(\mu)}$. Then the composition $R = PQ$ is an element of $\mathscr{E}_{(\lambda+\mu)}$. If $R = \sum_l r_{\lambda+\mu+l}(x, D_x)$, then one has*

$$r_{\lambda+\mu+l}(x, \sqrt{-1}\xi) = \sum_{l=j+k-|\alpha|}\frac{1}{\alpha!}\left(\left(\frac{1}{\sqrt{-1}}D_\xi\right)^\alpha p_{\lambda+j}(x, \sqrt{-1}\xi)\right)$$
$$\cdot (D_x^\alpha q_{\mu+k}(x, \sqrt{-1}\xi)). \tag{4.1.16}$$

Proof. Proposition 4.1.5 implies

$$Q(x, D_x)\Phi_n(\sqrt{-1}(\langle x, \xi\rangle + p)$$
$$= \sum_k q_{\mu+k}(x, \sqrt{-1}\xi)\Phi_{\mu+k+n}(\sqrt{-1}(\langle x, \xi\rangle + p)).$$

Applying Proposition 4.1.5 to $PQ\Phi_n(\sqrt{-1}(\langle x, \xi\rangle + p))$, one has

$$PQ\Phi_n(\sqrt{-1}(\langle x, \xi\rangle + p)) = \sum_l r_{\lambda+\mu+l}(x, \sqrt{-1}\xi)\Phi_{\lambda+\mu+l+n}(\sqrt{-1}(\langle x, \xi\rangle + p))$$

where

$$r_{\lambda+\mu+l}(x, \sqrt{-1}\xi) = \sum_{l=j+k-|\alpha|} \frac{1}{\alpha!}\left(\frac{1}{\sqrt{-1}} D_\xi\right)^\alpha p_{\lambda+j}(x, \sqrt{-1}\xi)(D_x^\alpha q_{\mu+k}(x, \sqrt{-1}\xi))$$

holds. Hence one obtains

$$PQ\delta(x-y) = \frac{1}{(2\pi)^n}\int PQ\Phi_n(\sqrt{-1}(\langle x, \xi\rangle - \langle y, \xi\rangle))\omega(\xi)$$
$$= \frac{1}{(2\pi)^n}\int \sum_l r_{\lambda+\mu+l}(x, \sqrt{-1}\xi)\Phi_{\lambda+\mu+l+n}(\sqrt{-1}\langle x-y, \xi\rangle)\omega(\xi),$$

completing the proof.

Remark. By a similar argument (for a fixed volume element $dx = dx_1 \cdots dx_n$), the conjugate operator $P^*(x, D_x) \equiv \sum_k q_{\lambda+k}(x, D_x)$ of $P(x, D_x) = \sum_j p_{\lambda+j}(x, D_x)$, in the sense of Definition 3.4.3, satisfies

$$q_{\lambda+k}(x, -\sqrt{-1}\xi) = \sum_{k=j-|\alpha|} \frac{(-1)^{|\alpha|}}{\alpha!}\left(\frac{1}{\sqrt{-1}} D_\xi\right)^\alpha D_x^\alpha p_{\lambda+j}(x, \sqrt{-1}\xi).$$

Note that this is a generalization of Lemma 4(2), following Definition 3.4.3, to the case of a microdifferential operator.

Let $\mathcal{O}(\lambda)$ be the sheaf of holomorphic functions on T^*X whose homogeneous degree in the fibre coordinate ζ is λ. Assigning $\sum_j p_{\lambda+j}(x, D_x)$ to $p_\lambda(z, \zeta)$ induces a sheaf homomorphism σ_λ from $\mathscr{E}(\lambda)$ to $\mathcal{O}(\lambda)$. As in the case of a linear differential operator, $\sigma_\lambda(P)$ is said to be the principal symbol of P. Then we obtain the following theorem from Theorem 4.1.1.

Theorem 4.1.2

(i) *If* $P \in \mathscr{E}(\lambda)$ *and* $Q \in \mathscr{E}(\mu)$, *then* $PQ \in \mathscr{E}(\lambda+\mu)$ *and* $\sigma_{\lambda+\mu}(PQ) = \sigma_\lambda(P)\sigma_\mu(Q)$ *hold.*

(ii) *If* $P \in \mathscr{E}(\lambda)$ *and* $Q \in \mathscr{E}(\mu)$, *then* $[P, Q] \in \mathscr{E}(\lambda+\mu-1)$ *holds, and one has*

$$\sigma_{\lambda+\mu-1}([P, Q]) = \frac{1}{\sqrt{-1}}\{p_\lambda(x, \sqrt{-1}\xi), q_\mu(x, \sqrt{-1}\xi)\},$$

which (see §5, Chapter III) *is equal to*

$$\frac{1}{\sqrt{-1}}\sum_{j=1}^{n}\left(\frac{\partial}{\partial\xi_j}p_\lambda(x,\sqrt{-1}\xi)\frac{\partial}{\partial x_j}q_\mu(x,\sqrt{-1}\xi)\right.$$
$$\left.-\frac{\partial}{\partial\xi_j}q_\mu(x,\sqrt{-1}\xi)\frac{\partial}{\partial x_j}p_\lambda(x,\sqrt{-1}\xi)\right).$$

Note that $\mathscr{E}(\lambda)$ is a sheaf of non-commutative rings and that $\mathcal{O}(\lambda)$ is a sheaf of commutative rings. In what follows, we will show how to capture the structures of microdifferential equations via information from the commutative object $\mathcal{O}(\lambda)$. Before this investigation, we will prove the invariance of the class of microdifferential operators under a coordinate transformation.

Let $x = (x_1, \ldots, x_n)$ and $\tilde{x} = (\tilde{x}_1, \ldots, \tilde{x}_n)$ be coordinate systems of X, and let $x = \varphi(\tilde{x})$ and $\tilde{x} = \tilde{\varphi}(x)$. Also let $P(x, D_x) = \sum_j P_{\lambda+j}(x, D_x)$ be a microdifferential operator defined in a neighborhood of $(x_0, \sqrt{-1}\xi_0\infty)$, and denote the point corresponding to $(x_0, \sqrt{-1}\xi_0)$ by $(\tilde{x}_0, \sqrt{-1}\tilde{\xi}_0)$ on $\sqrt{-1}T^*X$. Then we have

$$(\tilde{\xi})_j = \sum_k \frac{\partial x_k}{\partial\tilde{x}_j}(\xi)_k.$$

From the definition,

$$P(x, D_x)\delta(x-y) = \frac{1}{(2\pi)^n}\int K(x, y, \xi)\omega(\xi),$$

where $K(x, y, \xi) = \sum_j P_{\lambda+j}(x, \sqrt{-1}\xi)\Phi_{\lambda+j}(\sqrt{-1}(\langle x-y, \xi\rangle + \sqrt{-1}0))$. Denoting $|\det \partial x/\partial\tilde{x}|$ by $j(\tilde{x})$, we have

$$\delta(x-y) = j(\tilde{y})\delta(\tilde{x}-\tilde{y}),$$

where $\tilde{x} = \varphi(x)$ and $\tilde{y} = \varphi(y)$. Then

$$P(x, D_x)\delta(x-y) = P(x, D_x)\delta(\tilde{x}-\tilde{y})j(\tilde{y})$$

holds. Therefore, we obtain

$$\begin{aligned}P(x, D_x)\delta(\tilde{\varphi}(x)-\tilde{\varphi}(y)) &= j(\tilde{y})^{-1}P(x, D_x)\delta(x-y)\\ &= \frac{1}{(2\pi)^n}\int j(\tilde{\varphi}(y))^{-1}\sum_j p_{\lambda+j}(x, \sqrt{-1}\xi)\\ &\quad\cdot\Phi_{\lambda+j}(\sqrt{-1}(\langle x-y, \xi\rangle) + \sqrt{-1}0)\omega(\xi). \quad (4.1.17)\end{aligned}$$

On the other hand, we can write

$$\langle x-y, \xi\rangle = \langle\varphi(\tilde{x}) - \varphi(\tilde{y}), \tilde{\Theta}(\tilde{x}, \tilde{y}, \xi)\rangle,$$

where $\tilde{\Theta}(\tilde{x}, \tilde{y}, \xi)$ is an n-vector such that

$$\tilde{\Theta}_k(\tilde{x}, \tilde{y}, \xi) = \sum_j \frac{\partial x_k}{\partial \tilde{x}_j} \xi_j.$$

Hence, one can solve $\tilde{\xi} = \tilde{\Theta}(\tilde{x}, \tilde{y}, \xi)$ for ξ, i.e. $\xi = \Theta(\tilde{x}, \tilde{y}, \tilde{\xi})$. Then one has

$$\Theta_j(\tilde{x}, \tilde{y}, \tilde{\xi}) = \sum_k \frac{\partial \tilde{x}_j}{\partial x_k} \tilde{\xi}_k.$$

Since

$$\omega(\tilde{\xi}) = \left|\frac{\partial \tilde{\xi}}{\partial \xi}\right| \omega(\xi)$$

holds for the coordinate transformation $\tilde{\xi} = \tilde{\Theta}(\tilde{x}, \tilde{y}, \xi)$, one obtains from (4.1.17)

$$P(x, D_x)\delta(\tilde{x} - \tilde{y}) = \frac{1}{(2\pi)^n} \int j(\tilde{y})^{-1} \Big(\sum_j p_{\lambda+j}(x, \sqrt{-1}\xi) \cdot \Phi_{\lambda+j}(\sqrt{-1}(\langle \tilde{x} - \tilde{y}, \tilde{\xi}\rangle + \sqrt{-1}0)) \Big) |\partial\tilde{\xi}/\partial\xi|^{-1} \omega(\tilde{\xi}), \tag{4.1.18}$$

where $\xi = \Theta(\tilde{x}, \tilde{y}, \tilde{\xi})$ and $x = \varphi(\tilde{x})$.

Let

$$r_{\lambda+j}(\tilde{x}, \tilde{y}, \sqrt{-1}\tilde{\xi}) = j(\tilde{y})^{-1} |\partial\tilde{\xi}/\partial\xi|^{-1} p_{\lambda+j}(x, \sqrt{-1}\xi).$$

Then one obtains

$$P(x, D_x)\delta(\tilde{x} - \tilde{y}) = \frac{1}{(2\pi)^n} \int \Big(\sum_j r_{\lambda+j}(\tilde{x}, \tilde{y}, \sqrt{-1}\tilde{\xi}) \Phi_{\lambda+j}(\sqrt{-1}\langle \tilde{x} - \tilde{y}, \tilde{\xi}\rangle + \sqrt{-1}0) \Big) \omega(\tilde{\xi}).$$

If one can show that the above integration can be written as

$$\frac{1}{(2\pi)^n} \int \Big(\sum_j \tilde{p}_{\lambda+j}(\tilde{x}, \sqrt{-1}\tilde{\xi}) \Phi_{\lambda+j}(\sqrt{-1}(\langle \tilde{x} - \tilde{y}, \tilde{\xi}\rangle)) \Big) \omega(\tilde{\xi}),$$

then it follows that $P(x, D_x)$ becomes $\sum_j \tilde{p}_{\lambda+j}(\tilde{x}, D_{\tilde{x}})$ by the coordinate transformation. Notice that

$$r_{\lambda+j}(\tilde{x}, \tilde{x}, \sqrt{-1}\tilde{\xi}) = p_{\lambda+j}(x, \sqrt{-1}\xi)$$

holds.

Theorem 4.1.3. *For $j \in \mathbf{Z}$, let $a_{\lambda+j}(x, y, \sqrt{-1}\xi)$ satisfy the growth conditions* (4.1.2) *and* (4.1.3), *and let*

$$p_{\lambda+j}(x, \sqrt{-1}\xi) = \sum_{j=k-|\alpha|} \frac{1}{\alpha!} D_y^\alpha \left(\frac{1}{\sqrt{-1}} D_\xi\right)^\alpha a_{\lambda+k}(x, y, \sqrt{-1}\xi)\big|_{x=y}. \quad (4.1.19)$$

Then one has (i) *and* (ii):

(i) $$\int \sum_j a_{\lambda+j}(x, y, \sqrt{-1}\xi)\Phi_{\lambda+j+n}(\sqrt{-1}(\langle x-y, \xi\rangle + \sqrt{-1}0))\omega(\xi)$$
$$= \int \sum_j p_{\lambda+j}(x, \sqrt{-1}\xi)\Phi_{\lambda+j+n}(\sqrt{-1}(\langle x-y, \xi\rangle + \sqrt{-1}0))\omega(\xi).$$

(ii) *Furthermore, if $a_{\lambda+j} = 0$ for $j > 0$, then $p_\lambda(x, \sqrt{-1}\xi) = a_\lambda(x, x, \sqrt{-1}\xi)$.*

Proof. First let $c_{\lambda+j}^\alpha(x, \sqrt{-1}\xi) = D_y^\alpha a_{\lambda+j}(x, y, \sqrt{-1}\xi)\big|_{x=y}$, and write $a_{\lambda+j}(x, y, \sqrt{-1}\xi)$ as the Taylor expansion

$$a_{\lambda+j}(x, y, \sqrt{-1}\xi) = \sum_\alpha \frac{1}{\alpha!}(y-x)^\alpha c_{\lambda+j}^\alpha(x, \sqrt{-1}\xi).$$

Then one obtains

$$\int\left(\sum_j a_{\lambda+j}(x, y, \sqrt{-1}\xi)\Phi_{\lambda+j+n}(\sqrt{-1}(\langle x-y, \xi\rangle + \sqrt{-1}0))\right)\omega(\xi)$$
$$= \int\left(\sum_{j,\alpha} \frac{1}{\alpha!} c_{\lambda+j}^\alpha(x, \sqrt{-1}\xi)(y-x)^\alpha \Phi_{\lambda+j+n}(\sqrt{-1}(\langle x-y, \xi\rangle + \sqrt{-1}0))\right)\omega(\xi)$$
$$= \int\left(\sum_{j,\alpha} \frac{1}{\alpha!} c_{\lambda+j}^\alpha(x, \sqrt{-1}\xi)(\sqrt{-1}D_\xi)^\alpha\right.$$
$$\left.\cdot\,\Phi_{\lambda+j+n-|\alpha|}(\sqrt{-1}(\langle x-y, \xi\rangle + \sqrt{-1}0))\right)\omega(\xi)$$
$$= \int\left(\sum_{j,\alpha} \frac{1}{\alpha!} c_{\lambda+j+|\alpha|}^\alpha(x, \sqrt{-1}\xi)(\sqrt{-1}D_\xi)^\alpha\right.$$
$$\left.\cdot\,\Phi_{\lambda+j+n}(\sqrt{-1}(\langle x-y, \xi\rangle + \sqrt{-1}0))\right)\omega(\xi).$$

Through the integral by parts with respect to ξ, as in the proof of Theorem 4.1.1, the above integral is equal to

$$\int \sum_\alpha \frac{1}{\alpha!}((-\sqrt{-1}D_\xi)^\alpha c_{\lambda+j+|\alpha|}^\alpha(x, \sqrt{-1}\xi)$$
$$\cdot\,\Phi_{\lambda+j+n}(\sqrt{-1}(\langle x-y, \xi\rangle + \sqrt{-1}0)))\,\omega(\xi).$$

Hence, the class of microdifferential operators is invariant under a coordinate transformation, which completes the proof.

Suppose $P(x, D_x) = \sum_j p_{\lambda+j}(x, D_x)$ belongs to $\mathscr{E}(\lambda)$; i.e. $p_{\lambda+j} = 0$ for $j > 0$. Then the operator $\tilde{p}(\tilde{x}, D_{\tilde{x}})$, obtained from $P(x, D_x)$ by a coordinate transformation $x = \varphi(\tilde{x})$, is an element of $\mathscr{E}(\lambda)$ by Theorem 4.1.3 (ii). Furthermore,

$$\tilde{p}_\lambda(\tilde{x}, \sqrt{-1}\tilde{\xi}) = r_\lambda(\tilde{x}, \tilde{x}, \sqrt{-1}\tilde{\xi}) = p_\lambda(x, \sqrt{-1}\xi)$$

holds. This implies a sheaf homomorphism

$$\sigma_\lambda : \mathscr{E}(\lambda) \to \mathscr{O}(\lambda)$$

defined by $P = \sum_{j \leqq 0} p_{\lambda+j} \mapsto p_\lambda$ is invariant under a coordinate transformation.

We will describe the fundamental properties of a microdifferential operator, which are needed for our main goal in this treatise—the structure theory of microdifferential equations. In the analysis of microdifferential operators, after obtaining a formal solution $\sum_j p_{\lambda+j}(x, D_x)$, it is always a troublesome task to ensure that each homogeneous part $p_{\lambda+j}(x, \xi)$ satisfies the growth conditions (4.1.6a) and (4.1.6b). A relatively convenient method introduced in Boutet de Monvel and Krée [1] is that of a formal norm $N_l^\omega(P; t)$. We will assume $\lambda = 0$ for the sake of simplicity in the following discussion.

Definition 4.1.2. *Let $(x^0, \sqrt{-1}\xi^0\infty)$ be a point on $\sqrt{-1}S^*M$, let ω be a complex neighborhood of $(x^0, \sqrt{-1}\xi^0\infty)$, and let* dim $M = n$. *Suppose that $p_j(z, \zeta)$ is a holomorphic function in ω for a microdifferential operator $P(x, D_x) = \sum_{j=-\infty}^{l} p_j(x, D_x)$, where the homogeneous degree of $p_j(x, \xi)$ in ξ is j. Then the formal norm $N_l^\omega(P; t)$ of $P(x, D_x)$ in ω is a formal sum with respect to t, defined as*

$$\sum_{k,\alpha,\beta} \frac{2(2n)^{-k}k!}{(|\alpha| + k)!(|\beta| + k)!} \sup_\omega |D_z^\alpha D_\zeta^\beta p_{l-k}(z, \zeta)| t^{2k+|\alpha+\beta|}. \tag{4.1.20}$$

Proposition 4.1.6. *If $N_l^\omega(P; \epsilon) < \infty$ holds for any $\epsilon > 0$, then the growth condition* (4.1.6a) *is satisfied. If $\{p_j(z, \zeta)\}_{-\infty < j \leqq l}$ satisfies the condition* (4.1.6a), *then $N_l^{\omega'}(P; \epsilon) < \infty$ for some $\omega' \subset \omega$ and $\epsilon > 0$.*

The proof is almost obvious. In order to prove the latter half of the proposition, use $\sup_\omega |p_j|$ to estimate the derivatives of p_j, via Cauchy's integral formula.

Remark. When there is no fear of confusion, we simply write $N_l(P; t)$ without the superscript ω.

The usefulness of the formal norm is mainly due to the following theorem.

Theorem 4.1.5 (Boutet de Monvel and Krée [1]). *Let*

$$P_1 = \sum_{j=-\infty}^{l_1} p_j^{(1)} \quad \text{and} \quad P_2 = \sum_{j=-\infty}^{l_2} p_j^{(2)},$$

where $p_j^{(1)}$ and $p_j^{(2)}$ are defined in ω. Then

$$N^{\omega}_{l_1+l_2}(P_1P_2; t) \ll N^{\omega}_{l_1}(P_1; t)N^{\omega}_{l_2}(P_2; t) \tag{4.1.21}$$

holds, where for formal power series $A(t)$ and $B(t)$ the notation $A(t) \ll B(t)$ means that $B(t)$ is a majorant series of $A(t)$.

For a proof, see Lemma 1.2 in Boutet de Monvel and Krée [1].

The next theorem and its corollaries will show how powerful Theorem 4.1.5 is.

Theorem 4.1.6. *Let $f(s) = \sum_{s=0}^{\infty} f_k s^k$ be a holomorphic function defined in a neighborhood of $s = 0$. If a microdifferential operator $P(x, D_x)$ of order at most zero is defined in a neighborhood of $(x^0, \sqrt{-1}\xi^0\infty) \in \sqrt{-1}S^*M$, and if $P(x, D_x)$ satisfies*

$$\sigma_0(P)(x^0, \sqrt{-1}\xi^0) = 0, \tag{4.1.22}$$

then $f(P) = \sum_{k=0}^{\infty} f_k P^k$ is a microdifferential operator of order at most zero defined in a neighborhood of $(x^0, \sqrt{-1}\xi^0\infty)$.

Proof. From Theorem 4.1.5, one has

$$N^{\omega}_0(f(P); t) = \sum_{k=0}^{\infty} f_k N^{\omega}_0(P^k; t) \ll \sum_{k=0}^{\infty} f_k N^{\omega}_0(P; t)^k.$$

Then (4.1.22) implies that, for any $\delta > 0$, there can be found ω and $\epsilon > 0$ such that $N^{\omega}_0(P; \epsilon) < \delta$. Hence, from Proposition 4.1.6, the proof follows.

Corollary. *Let $P(x, D_x)$ be a microdifferential operator of finite order. Since the principal symbol $\sigma_m(P)(x, \sqrt{-1}\xi)$ does not vanish in $\Omega \subset \sqrt{-1}S^*M$, there exists a microdifferential operator E defined in Ω such that*

$$PE = EP = I \tag{4.1.23}$$

holds.

Proof. It is sufficient to prove the existence of E at each point in Ω, since such an E satisfying (4.1.23) is also unique locally. Let $q_{-m}(x, \sqrt{-1}\xi) = p_m(x, \sqrt{-1}\xi)^{-1}$, and let

$$R(x, D_x) = I - P(x, D_x)q_{-m}(x, D_x). \tag{4.1.24}$$

Then $R(x, D_x)$ satisfies the condition of Theorem 4.1.6. Hence, $\sum_{k=0}^{\infty} R(x, D_x)^k$ determines locally a microdifferential operator $S(x, D_x)$. Hence, $(I - R)S = I$ holds by the definition of S. This implies that $q_{-m}(x, D_x)S(x, D_x)$ is a right inverse of P. In a similar manner, one can locally obtain a left inverse of P as a microdifferential operator. Since the associative law holds for microdifferential operators, the left and right inverses coincide. Then let $q_{-m}S = E$.

Remark. The following assertion can be proved in the same manner.

Let $P(x, D_x) = (P_{ij}(x, D_x))_{1 \leqq i,j \leqq N}$ be an $N \times N$ matrix of microdifferential operators, and let $m_j = \max_{1 \leqq i \leqq N} \operatorname{ord} P_{ij}$. If $\det(\sigma_{m_j}(P_{ij}))$ does not vanish in $\Omega \subset \sqrt{-1}S^*M$, then there exists an $N \times N$ matrix E of microdifferential operators such that

$$PE = EP = I \tag{4.1.25}$$

holds.

Note. The above assertion can be sharpened. But we will not discuss this further, since in what follows we will not need more than what is stated in the above.

Microlocally speaking, the structure of $Pu = 0$ is trivial outside $V_{\mathbf{R}} = \{(x, \sqrt{-1}\xi\infty) \in \sqrt{-1}S^*M \,|\, \sigma_m(P)(x, \sqrt{-1}\xi) = 0\}$ by the above corollary. Therefore, our focus is on the structure of a microdifferential equation where $\sigma_m(P) = 0$. As we will show in §3 of this chapter, the structures can be described with extreme clarity, even for overdetermined systems. Schematically speaking, we can treat microdifferential equations as if they were algebraic equations. In order to carry this out, some preparative topics are in order. We begin with the analogues of the Späth theorem and the Weierstrass preparation theorem, which are fundamental for the local theory of holomorphic functions of several complex variables. We will not give proofs for these theorems in this book.

Theorem 4.1.7 (Späth Theorem I for Microdifferential Operators). *Let $P(x, D_x)$ be a microdifferential operator of order m such that*

$$(\sigma_m(P)(x, \sqrt{-1}\xi)/\xi_n^p)|_{(x;\xi)=(0;1,0,\ldots,0,\xi_n)} \tag{4.1.26}$$

is a holomorphic function of ξ_n and can never be zero in a neighborhood of $\xi_n = 0$. Then, for an arbitrary microdifferential operator $S(x, D_x)$, there *can be found microdifferential operators* $Q(x, D_x)$ *and* $R(x, D_x)$ *so that the following* (4.1.27) *and division theorem* (4.1.28) *hold in a neighborhood* ω *of* $(x; \sqrt{-1}\xi) = (0; \sqrt{-1}(1, 0, \ldots, 0))$:

$$S(x, D_x) = Q(x, D_x)P(x, D_x) + R(x, D_x). \tag{4.1.27}$$

$\underbrace{[x_n, [\ldots, [x_n}_{p}, R] \ldots] = 0$; *i.e.* $R(x, D_x)$ *can be written as*

$$\sum_{j=0}^{p-1} R^{(j)}(x, D')D_n^j, \text{ where } D' = (D_1, \ldots, D_{n-1}). \tag{4.1.28}$$

Note also that Q *and* R *are determined uniquely, and that* Q *and* R *are of finite order when* S *is of finite order.*

Theorem 4.1.8 (Späth Theorem II for Microdifferential Operator). *Let* $P(x, D_x)$ *satisfy* (4.1.29), *instead of* (4.1.26):

$(\sigma_m(P)(x, \sqrt{-1}\xi)/x_1^p)|_{(x;\xi)=(x_1,0;\xi^\circ)}$ *is a holomorphic function of* x_1 *and can never be zero in a neighborhood of* $x_1 = 0$. (4.1.29)

Then, for an arbitrary microdifferential operator $S(x, D_x)$, *one has a unique presentation of division algorithm as follows*:

$$S(x, D_x) = Q(x, D_x)P(x, D_x) + R(x, D_x). \tag{4.1.30}$$

$\underbrace{[D_1, [\ldots [D_1}_{p}, R] \ldots] = 0$; *i.e.* $R(x, D_x)$ *can be written as*

$$\sum_{j=0}^{p-1} x_1^j R^{(j)}(x', D), \text{ where } x' = (x_2, \ldots, x_n). \tag{4.1.31}$$

See SKK [1], Chap. II, §2.2, for proofs.

As in the theory of several complex variables, a Weierstrass-type division theorem is obtained as a corollary of Theorem 4.1.7.

Theorem 4.1.9. *Let* $P(x, D_x)$ *be a microdifferential operator of order* m *satisfying the condition* (4.1.26). *Then* $P(x, D_x)$ *can be decomposed, in a neighborhood* ω *of* $(x, \sqrt{-1}\xi) = (0; \sqrt{-1}(1, 0, \ldots, 0))$, *as*

$P(x, D_x) = Q(x, D_x)W(x, D_x)$ *uniquely, where* $Q(x, D_x)$ *is invertible in* ω *and* W *has the following form*: (4.1.32)

$W(x, D_x) = D_n^p + \sum_{j=0}^{p-1} W^{(j)}(x, D')D_n^j$, $D' = (D_1, \ldots, D_{n-1})$, *and the order of* $W^{(j)}(x, D')$ *is at most* $p - j$ *such that* $\sigma_{p-j}(W^{(j)})(0, \sqrt{-1}(1, 0, \ldots, 0)) = 0$ *holds.* (4.1.33)

As with the Späth theorem for the local theory of holomorphic functions of several variables, the above theorems provide a fundamental method to normalize microdifferential operators, as one pleases, in the local study of microdifferential equations.

§2. Quantized Contact Transformation for Microdifferential Operators

We have shown in the previous section that the notion of a microdifferential operator is obtained by the microlocalization of a differential operator on $\sqrt{-1}S^*M$. On the other hand, we recognized (in §6 of Chapter III) a bicharacteristic strip as a "carrier" of singularities of solutions of a differential equation, where a glimpse of the interplay contact geometry and differential equations was observed. In this section, we will carry out this program (i.e. "contact transformations of microdifferential operators"). It was probably Maslov [1] in which an attempt was first made to formulate the above idea. But only after Egorov [1] did mathematicians begin to approach this problem—Maslov [1] was not well known outside the USSR and was somewhat lacking in mathematical rigor (perhaps the book was written for physicists). Nowadays this method is crucial for the study of linear differential equations.

Let us begin with some basic notions from contact geometry. Let a complex manifold X be complex-analytic. In the case when X is a real manifold, then T^*X and other notions should be considered over real numbers $\mathbf{R}$; i.e. considering the actions by $\mathbf{R}_+^\times = \{t \in \mathbf{R} | t \gneqq 0$, instead of the actions by $\mathbf{C}^\times = \mathbf{C} - \{0\}$ for the complex category.

Definition 4.2.1. *Let T^*X be the cotangent bundle of a $(2n-1)$-dimensional manifold X. If a 1-dimensional sub-bundle L^* of T^*X satisfies the following condition* (4.2.1), *then (X, L) is said to be a contact structure on X, where L^* denotes the dual bundle of L:*

For a nowhere-vanishing local cross-section ω of L^*, $\omega \wedge (d\omega)^{n-1}$ is non-zero anywhere. (4.2.1)

Note. Since $(f\omega) \wedge (d(f\omega))^{n-1} = f^n \omega \wedge (d\omega)^{n-1}$ holds, the above condition is independent of the choice of ω.

Definition 4.2.2. *When ω satisfies* (4.2.1), *ω is called a canonical 1-form.*

Let us denote $L^* - \{\text{zero sections}\}$ by $\hat{X}$. Then $\mathbf{C}^\times = \mathbf{C} - \{0\}$ acts on $\hat{X}$ as a $\mathbf{C}^\times$-principal bundle over X. One can canonically define a 1-form θ on $\hat{X}$ by $s^*(\theta) = \omega$ for a cross-section s of $\hat{X}$. Then, $(d\theta)^n$ is nowhere zero. $\hat{X}$ is called the symplectic manifold associated with X, and θ is said to be a homogeneous canonical 1-form. Let $f(z)$ be a function defined on $\hat{X}$ (i.e. defined in a neighborhood in $\hat{X}$) satisfying $f(az) = a^m f(z)$ for $a \in \mathbf{C}^\times$ and $z \in \hat{X}$. Then, $f(z)$ is said to be a homogeneous function of degree m, in which case f is said also to be a homogeneous function of degree m on X.

The sheaf of homogeneous functions of degree m on X is isomorphic to $L^{\otimes m}$.

The above definitions are more explicit in the case where X is the cotangent projective bundle P^*Y of an n-dimensional manifold Y; then $\hat{X} = T^*Y - Y$. The classical theorem of Darboux states that a canonical 1-form ω can be written as $\omega = dx_n - p_1\,dx_1 - \cdots - p_{n-1}\,dx_{n-1}$ for a local coordinate system $(x_1, \ldots, x_n, p_1, \ldots, p_{n-1})$, where $(x_1, \ldots, x_n, p_1, \ldots, p_{n-1})$ are said to be canonical coordinates. Note also that in the above case a coordinate system $(x_1, \ldots, x_n, \eta_1, \ldots, \eta_n)$ of $\hat{X}$ can be taken to satisfy $p_j = -\eta_j/\eta_n$ for $j = 1, \ldots, n-1$ and $\theta = \eta_n\omega = \eta_1\,dx_1 + \cdots + \eta_n\,dx_n$, where $(x_1, \ldots, x_n, \eta_1, \ldots, \eta_n)$ are called canonical homogeneous coordinates.

The Poisson bracket, introduced in §5, Chapter III, can be defined intrinsically through a contact structure.

Definition 4.2.3. *The Poisson bracket $\{f, g\}$ of functions f and g on a symplectic manifold $\hat{X}$ of dimension $2n$ is defined as follows:*

$$n\,df \wedge dg \wedge (d\theta)^{n-1} = \{f, g\}(d\theta)^n. \tag{4.2.2}$$

Exercise. Let $(x_1, \ldots, x_n, \eta_1, \ldots, \eta_n)$ be canonical homogeneous coordinates. Then show that the Poisson bracket $\{f, g\}$ can be written as

$$\sum_{j=1}^{n} \left(\frac{\partial f}{\partial \eta_j} \frac{\partial g}{\partial x_j} - \frac{\partial f}{\partial x_j} \frac{\partial g}{\partial \eta_j} \right). \tag{4.2.2$'$}$$

The Poisson bracket $\{f, g\}$ satisfies the following relations, as one can easily see from (4.2.2′).

Theorem 4.2.1.

$$\{f, g\} = -\{g, f\}. \tag{4.2.3}$$

$$\{\{f, g\}, h\} + \{\{g, h\}, f\} + \{\{h, f\}, g\} = 0. \tag{4.2.4}$$

If the homogeneous degree of f is l, and one of g is m, then the homogeneous degree of $\{f, g\}$ is $l + m - 1$. (4.2.5)

Connecting with the Poisson bracket, the Hamiltonian vector field

$$H_f = \sum_{j=1}^{n} \left(\frac{\partial f}{\partial \eta_j} \frac{\partial}{\partial x_j} - \frac{\partial f}{\partial x_j} \frac{\partial}{\partial \eta_j} \right) \tag{4.2.6}$$

is also important. A bicharacteristic strip, introduced in Definition 3.5.1, can be rephrased as the integral curve of H_{p_m}, which is important for the recognition of the notion of a bicharacteristic strip as a concept in contact geometry. Notice also that

$$H_{\{f,g\}} = [H_f, H_g] \left(\underset{\text{def}}{=} H_f H_g - H_g H_f\right) \tag{4.2.7}$$

holds from the definitions.

To summarize the above, a canonical 1-form ω determines a contact structure on X and, furthermore, a homogeneous canonical 1-form θ_X on $\hat{X}$ is naturally defined by ω. Conversely, if a $\mathbf{C}^\times$-principal bundle S is given on X such that a 1-form θ satisfies the following conditions (4.2.8) and (4.2.9), then we will show that a contact structure can be defined on X:

$$(d\theta)^n \text{ does not vanish anywhere on } S. \tag{4.2.8}$$

The 1-form θ is homogeneous; i.e. for $(c, x) \in \mathbf{C}^\times \times S, \theta(cx) = c^r\theta(x)$ holds for some integer r. (4.2.9)

For an arbitrary cross-section s of S, one can take $s^*\theta$ for a canonical 1-form ω. If $r = 1$, then S and $\hat{X}$ coincide. For the general r, $\hat{X} = S^{\otimes r}$ holds. That is, for a canonical homogeneous 1-form θ_X on $\hat{X}$, there exists a map F from the principal bundle S to the principal bundle $\hat{X}$ such that $F(cx) = c^rF(x)$ and $\theta = F^*\theta_X$ hold. We will give an example of this construction of a contact structure.

Example 4.2.1. Let V be a symplectic linear space of dimension $2n$; i.e. there is given a skew-symmetric non-degenerate quadratic form $E(v_1, v_2)$ on a linear space V. Define $X = P(V) = (V - \{0\})/\mathbf{C}^\times$. Then define a 1-form θ on V as $E(v, dv)/2$. Since E is non-degenerate, $(d\theta)^n$ is non-degenerate. For $(c, v) \in \mathbf{C}^\times \times (V - \{0\})$, $\theta(cv) = c^2\theta(v)$ holds. Hence, $P(V)$ has a contact structure induced from θ.

Let us recall some of the basic notions pertaining to a contact structure.

Definition 4.2.4. *Let X and Y be contact manifolds of the same dimension, and let φ be a map from X to Y. If, for an arbitrary canonical 1-form ω_Y on Y, $\varphi^*\omega_Y$ is a canonical 1-form on X, then φ is said to be a contact transformation.*

Example 4.2.2. Let M and N be open subsets of $\mathbf{C}^n$. Consider $\Omega(x, y)$, $x \in M$, $y \in N$, a holomorphic function defined on $M \times N$. Suppose that $\Omega(x, y)$ satisfies the following conditions (4.2.10) and (4.2.11):

The hypersurface $H = \{(x, y) \in M \times N \,|\, \Omega(x, y) = 0\}$ is non-singular; i.e. $\operatorname{grad}_{(x,y)}\Omega(x, y) \neq 0$ holds on H. (4.2.10)

$$\text{On } H, \det \begin{bmatrix} 0 & \dfrac{\partial\Omega}{\partial y_1} & \cdots & \dfrac{\partial\Omega}{\partial y_n} \\ \dfrac{\partial\Omega}{\partial x_1} & \dfrac{\partial^2\Omega}{\partial y_1\,\partial x_1} & \cdots & \dfrac{\partial^2\Omega}{\partial y_n\,\partial x_1} \\ \cdots & \cdots & \cdots & \cdots \\ \dfrac{\partial\Omega}{\partial x_n} & \dfrac{\partial^2\Omega}{\partial y_1\,\partial x_n} & \cdots & \dfrac{\partial^2\Omega}{\partial y_n\,\partial x_n} \end{bmatrix} \neq 0 \text{ holds.} \tag{4.2.11}$$

Then one can define a local isomorphism from P^*M to P^*N via

$$P^*_H(M \times N) = \{(x, y; \xi, \eta) \in P^*(M \times N) \,|\, \Omega(x, y) = 0 \text{ and } (\xi, \eta) = c \operatorname{grad}_{(x,y)} \Omega(x, y) \text{ for } c \neq 0\},$$

as follows. Implicit function theorem implies that the projection $P^*_H(M \times N) \xrightarrow{p} P^*M$ is a local isomorphism. Note that

$$\det\begin{pmatrix} 0 & d_y\Omega \\ d_x\Omega & cd_xd_y\Omega \end{pmatrix} = c^{n-2} \det\begin{pmatrix} 0 & d_y\Omega \\ d_x\Omega & d_xd_y\Omega \end{pmatrix}$$

holds. Similarly, one has a local isomorphism $P^*_H(M \times N) \xrightarrow{q} P^*N$. Hence, p and q induce local isomorphisms $p \circ q^{-1}: P^*N \to P^*M$ and $q \circ p^{-1}: P^*M \to P^*N$. Then, on H, one has $d\Omega = \sum_{j=1}^{n} (\partial\Omega/\partial x_j)\, dx_j + \sum_{j=1}^{n} (\partial\Omega/\partial y_j)\, dy_j \neq 0$. Hence, $p \circ q^{-1}$ and $q \circ p^{-1}$ are clearly contact transformations.

Definition 4.2.5. *The contact transformation obtained as above is called a contact transformation having* Ω *as a generating function.*

Remark. A classical result states that an arbitrary contact transformation can be obtained by the consecutive use of two contact transformations having generating functions. In this sense, a contact transformation having a generating function is a "generic" contact transformation.

Example 4.2.3. The well-known classical Legendre transformation is a contact transformation where the generating function $\Omega(x, y) = x_n - y_n + \sum_{j=1}^{n-1} x_i y_i$. Then the explicit correspondence between (x, ξ) and (y, η) is given by

$$\left.\begin{aligned} x_j &= -\eta_j \eta_n^{-1} && \text{for } j < n \\ x_n &= \langle y, \eta\rangle \eta_n^{-1} && \\ \xi_j &= y_j \eta_n && \text{for } j < n \\ \xi_n &= \eta_n && \end{aligned}\right\} \tag{4.2.12a}$$

and

$$\left.\begin{aligned} y_j &= \xi_j \xi_n^{-1} && \text{for } j < n \\ y_n &= \langle x, \xi\rangle \xi_n^{-1} && \\ \eta_j &= -x_j \xi_n && \text{for } j < n \\ \eta_n &= \xi_n && \end{aligned}\right\} \tag{4.2.12b}$$

Example 4.2.4. A contact structure was defined on $P(V)$ via a skew-symmetric non-degenerate quadratic form E on V in Example 4.2.1. Let

φ be a linear transformation of V with the property $E(v_1, v_2) = E(\varphi v_1, \varphi v_2)$; i.e. it is a symplectic transformation of (V, E). Then a 1-form θ defined on V is invariant under φ. Hence, φ defines a contact transformation of $X = P(V)$.

Definition 4.2.6. *An analytic subset V of a contact manifold X is said to be involutory if the following condition* (4.2.13) *is satisfied*:

$$\text{If } f|_V = g|_V = 0, \text{ then } \{f, g\}|_V = 0. \tag{4.2.13}$$

Remark 1. It is known that the characteristic variety of an arbitrary system of microdifferential equations is involutory (see SKK [1], Chap. II, Theorem 5.3.2). This is extremely fundamental and important for the study of linear partial differential equations. Although the proof given in SKK [1] is transcendental, an algebraic proof of this theorem is available (see Kashiwara [2]).

Remark 2. An involutory subset V of a $(2n\text{-}1)$-dimensional contact manifold X has a codimension which is always less than or equal to n.

Definition 4.2.7. *An involutory (non-singular) submanifold V of X is said to be regular if $\omega|_V \neq 0$ holds everywhere on V.*

Definition 4.2.8. *An involutory analytic subset V of a* $(2n\text{-}1)$*-dimensional contact manifold X is said to be Lagrangian if the codimension of V in X is n.*

Remark 1. A Lagrangian manifold can never be regular in the sense of Definition 4.2.7. It is known that a submanifold Y of codimension n in X is Lagrangian if and only if $\omega|_Y = 0$ holds. From this fact, one can say that a Lagrangian subset is rather exceptional as an involutory subset. Conversely, one should say, this promises that analysis connected with a Lagrangian set is fruitful. Though many interesting topics are in progress, we will not discuss these here (see *Publ. RIMS, Kyoto Univ.*, vol. 12, suppl. [1977], which may be useful for these topics).

Note. A system of microdifferential equations is said to be a holonomic system or a maximally overdetermined system if its characteristic variety is Lagrangian. One of the most fundamental properties of a holonomic system is the finite dimensionality of the (microfunction) solution space; see Kashiwara [1] and Kashiwara and Kawai [1]. Hence one may hope that holonomic systems will be used effectively as a governing principle of functions of several variables, as ordinary differential equations do in one-variable cases. This viewpoint was advocated by M. Sato as early as 1960, in connection with the question of how to characterize fundamental solutions for the Cauchy problem.

Remark 2. By the Jacobi theory, any regular involutory manifold can be expressed as $p_1 = \cdots = p_r = 0$ for some canonical coordinate system (x, p).

Using these geometric preparations, we will quantize a contact transformation, i.e. "a transformation of a microdifferential operator compatible with a contact transformation." By the remark following Definition 4.2.5, one can restrict oneself to consider the case of a contact transformation having a generating function. For the sake of simplicity and convenience, our treatment is within the real analytic category, rather than the complex analytic one.

Theorem 4.2.2. *Let M and N be real analytic manifolds of dimension n. Assume that a real-valued real analytic function $\Omega(x, y)$ defined on $M \times N$ satisfies the conditions* (4.2.10) *and* (4.2.11). *Then, for an arbitrary microdifferential operator $P(x, D_x)$, a microdifferential operator $Q(y, D_y)$ is uniquely determined such that*

$$\int P(x, D_x)\delta(\Omega(x, y))u(y)\,dy = \int \delta(\Omega(x, y))Q(y, D_y)u(y)\,dy \tag{4.2.14}$$

holds for any microfunction $u(y)$. Conversely, if Q is given, then P is uniquely determined so that (4.2.14) *holds. Furthermore, the order of Q is equal to that of P.*

That is, one has

$$\left.\begin{array}{ll} q \circ p^{-1}\mathscr{E}_M(m) \cong \mathscr{E}_N(m), & q \circ p^{-1}\mathscr{E}_M \cong \mathscr{E}_N \\ p \circ q^{-1}\mathscr{E}_N(m) \cong \mathscr{E}_M(m), & p \circ q^{-1}\mathscr{E}_M \cong \mathscr{E}_N, \end{array}\right\} \tag{4.2.15}$$

and the isomorphism

$$q \circ p^{-1}\mathscr{A}_{\sqrt{-1}S^*M} \cong \mathscr{A}_{\sqrt{-1}S^*_N}, \tag{4.2.16}$$

induced by (4.2.15), *is a contact transformation with the generating function* Ω.

Definition 4.2.8. *The isomorphism in* (4.2.15) *is called a quantized contact transformation (with the generating function* Ω).

Remark. Though we will carry out our proof for the sheaf $\mathscr{E}$ of microdifferential operators of finite order, the proof is valid for the sheaf $\mathscr{E}^\infty$ of microdifferential operators of infinite order.

Proof of Theorem 4.2.2. As p and q are defined via the conormal sphere bundle $S^*_H(M \times N)$ of H, we will prove the isomorphism in (4.2.15) via $\mathscr{E}_{M \times N}(m)\delta(\Omega(x, y))$.That is, one is to prove

$$p^{-1}\mathscr{E}_M(m) \cong \mathscr{E}_{M \times N}(m)\delta(\Omega(x, y)). \tag{4.2.17}$$

One of the motivations that led us to this method comes from the fact S.S. $\delta(\Omega(x, y)) = S^*_H(M \times N)$; actually, the characteristic variety of the holonomic system for $\delta(\Omega)$ is (the complexification of) $S^*_H(M \times N)$. (The concept of a holonomic system is used in this proof, though not explicitly. See the remark following the proof of this theorem.) The condition (4.2.10) implies that one can choose a basis $\mathscr{X}_j$, $1 \leqq j \leqq 2n$, for a vector field defined on $M \times N$, so as to satisfy

$$[\mathscr{X}_j, \mathscr{X}_k] = 0 \qquad (j, k = 1, \ldots, 2n) \tag{4.2.18}$$

and

$$\left.\begin{aligned} &\mathscr{X}_j\Omega = 0 \qquad (j = 1, \ldots, 2n-1) \\ &\mathscr{X}_{2n}\Omega = 1. \end{aligned}\right\} \tag{4.2.19}$$

Define $\mathscr{I} = \{P(x, y, D_x, D_y) \in \mathscr{E}_{M \times N} \mid P(x, y, D_x, D_y)\delta(\Omega(x, y)) = 0\}$. Then $\mathscr{I}$ is generated by $\mathscr{X}_1, \ldots, \mathscr{X}_{2n-1}$ and $(\Omega\mathscr{X}_{2n} - 1)$. By the definition of $\mathscr{I}$, one has

$$\mathscr{E}_{M \times N}\delta(\Omega(x, y)) = \mathscr{E}_{M \times N}/\mathscr{I}. \tag{4.2.20}$$

Notice that the common zeros of the principal symbols $\sigma(\mathscr{X}_j)$ and $\sigma(\Omega\mathscr{X}_{2n} - 1)$ of the generators for $\mathscr{I}$ coincide with $S^*_H(M \times N)$. We will denote a point on $S^*(M \times N)$ by $(x, y; \xi, \eta)$, where ξ and η are cotangent vectors corresponding to x and y respectively.

In order to prove the isomorphism in (4.2.17), it is sufficient to prove that, for an arbitrary $A(x, y, D_x, D_y) \in \mathscr{E}_{M \times N}(m)$, $Q_j \in \mathscr{E}_{M \times N}$ and $\tilde{A}(x, D_x) \in \mathscr{E}_M(m)$ can be chosen so that, determining $\tilde{A}$ uniquely,

$$A = \sum_{j=1}^{2n} Q_j R_j + \tilde{A} \tag{4.2.21}$$

may hold, where $R_j(x, y, D_x, D_y)$ are properly chosen generators for $\mathscr{I}$. Since H is non-singular, i.e. condition (4.2.10), then $S^*_H(M \times N)$ is also non-singular and, furthermore, $S^*_H(M \times N)$ is locally isomorphic to S^*M by (4.2.11). Hence, $S^*_H(M \times N)$ can be expressed as

$$\left.\begin{aligned} \eta_j &= p_j(x, \xi), \qquad j = 1, \ldots, n \\ y_j &= q_j(x, \xi), \qquad j = 1, \ldots, n, \end{aligned}\right\} \tag{4.2.22}$$

where $p_j(x, \xi)$ and $q_j(x, \xi)$ are analytic functions for $j = 1, \ldots, n$, and the homogeneous degree of p_j in ξ is 1 and that of q_j in ξ is 0. Therefore, one can let R_j be a microdifferential operator whose principal part is either $\eta_j - p_j$, for $j = 1, \ldots, n$, or $y_j - p_j$ for $j = n + 1, \ldots, 2n$ so that the decomposition in the form of (4.2.21) is possible for the principal part. We will consider next how to choose the terms of R_j of lower order. Since the question is local on $S^*_H(M \times N)$, one can choose analytic functions $a_{jk}(x, y, \xi, \eta)$ and $b_{jk}(x, y, \xi, \eta)$, where $j = 1, \ldots, n$ and $k = 1, \ldots, 2n$, so that one can have

$$\left.\begin{aligned} \eta_j - p_j &= \sum_{k=1}^{2n-1} a_{jk}\sigma(\mathscr{X}_k) + a_{j2n}\sigma(\Omega\mathscr{X}_{2n}), \qquad j = 1, \ldots, n \\ y_j - q_j &= \sum_{k=1}^{2n-1} b_{jk}\sigma(\mathscr{X}_k) + b_{j2n}\sigma(\Omega\mathscr{X}_{2n}), \qquad j = 1, \ldots, n, \end{aligned}\right\} \tag{4.2.23}$$

where the homogeneous degrees of a_{jk} and b_{jk} in (ξ, η) are 0 and -1 respectively. Furthermore, one may assume

$$\det \begin{bmatrix} a_{1,1}, & \ldots, & a_{1,2n} \\ \cdots & \cdots & \cdots \\ a_{n,1}, & \ldots, & a_{n,2n} \\ b_{1,1}, & \ldots, & b_{1,2n} \\ \cdots & \cdots & \cdots \\ b_{n,1}, & \ldots, & b_{n,2n} \end{bmatrix} \neq 0 \tag{4.2.24}$$

holds on $S_H^*(M \times N)$. Next we will find the generators R_j, $j = 1, \ldots, 2n$, for $\mathscr{I}$, satisfying (4.2.21), in the following form (4.2.25):

$$\left.\begin{aligned} R_j &= \sum_{k=1}^{2n-1} A_{jk}\mathscr{X}_k + A_{j2n}(\Omega\mathscr{X}_{2n} - 1) \qquad \text{for } j = 1, \ldots, n \\ S_j &= \sum_{k=1}^{2n-1} B_{jk}\mathscr{X}_k + B_{j2n}(\Omega\mathscr{X}_{2n} - 1) \qquad \text{for } j = 1, \ldots, n, \end{aligned}\right\} \tag{4.2.25}$$

where $S_j = R_{j-n}$, $j = n + 1, \ldots, 2n$, and A_{jk} and B_{jk} are microdifferential operators whose principal symbols are a_{jk} and b_{jk} respectively. By the remark following the corollary of Theorem 4.1.6, (4.2.24) implies that the $2n \times 2n$-matrix

$$\begin{bmatrix} A_{jk} \\ B_{jk} \end{bmatrix}_{\substack{1 \leq j \leq n \\ 1 \leq k \leq 2n}}$$

is invertible. Therefore, one obtains

$$\mathscr{I} = \sum_{j=1}^{n} \mathscr{E}_{M \times N} R_j + \sum_{j=1}^{n} \mathscr{E}_{M \times N} S_j \tag{4.2.26}$$

from $\mathscr{I} = \sum_{k=1}^{2n-1} \mathscr{E}_{M \times N}\mathscr{X}_k + \mathscr{E}_{M \times N}(\Omega\mathscr{X}_{2n} - 1)$. When one lets P_j and Q_j satisfy

$$\left.\begin{aligned} R_j(x, y, D_x, D_y) &= D_{y_j} - P_j(x, y, D_x, D_y) \qquad \text{for } j = 1, 2, \ldots, n \\ S_j(x, y, D_x, D_y) &= y_j - Q_j(x, y, D_x, D_y) \qquad \text{for } j = 1, 2, \ldots, n, \end{aligned}\right\} \tag{4.2.27}$$

then we will prove that R_j and S_j can be chosen to satisfy

$$\left.\begin{aligned}[y_i, P_j] = 0 \qquad \text{for } i, j = 1, \ldots, n\\ [y_i, Q_j] = 0 \qquad \text{for } i, j = 1, \ldots, n\end{aligned}\right\} \tag{4.2.28}$$

and

$$\left.\begin{aligned}[D_{y_i}, P_j] = 0 \qquad \text{for } i, j = 1, \ldots, n\\ [D_{y_i}, Q_j] = 0 \qquad \text{for } i, j = 1, \ldots, n.\end{aligned}\right\} \tag{4.2.29}$$

That is, one wishes to find R_j and S_j so that the following equations should hold:

$$\left.\begin{aligned}R_j = D_{y_j} - P_j(x, D_x) \qquad \text{for } j = 1, \ldots, n\\ S_j = y_j - Q_j(x, D_y) \qquad \text{for } j = 1, \ldots, n.\end{aligned}\right\} \tag{4.2.30}$$

If (4.2.30) holds, then Theorems 4.1.7 and 4.1.8 imply that there exists $\tilde{A}(x, D_x)$ satisfying (4.2.21), in which case the order of $\tilde{A}(x, D_x)$ is at most m, provided that the order of $A(x, y, D_x, D_y)$ is at most m.

We will show that one can choose R_j and S_j so that (4.2.28) and (4.2.29) hold. We will do this by induction on i. That is, we will prove the following $(4.2.28)_k$ and $(4.2.29)_k$ by induction on k:

$$\left.\begin{aligned}[y_i, P_j] = 0 \qquad \text{for } i = 1, \ldots, k, \text{ and } j = 1, \ldots, n\\ [y_i, Q_j] = 0 \qquad \text{for } i = 1, \ldots, k, \text{ and } j = 1, \ldots, n,\end{aligned}\right\} \tag*{$(4.2.28)_k$}$$

and

$$\left.\begin{aligned}[D_{y_i}, P_j] = 0 \qquad \text{for } i = 1, \ldots, k, \text{ and } j = 1, \ldots, n\\ [D_{y_i}, Q_j] = 0 \qquad \text{for } i = 1, \ldots, k, \text{ and } j = 1, \ldots, n.\end{aligned}\right\} \tag*{$(4.2.29)_k$}$$

For the case $k = 0$, the assertion is clearly true. Hence, let $k \geqq 1$ and assume that $(4.2.28)_i$ and $(4.2.29)_i$ hold for $i < k$. By the definition, the principal part of R_k is $\eta_k - p_k(x, \xi)$. Therefore, from Theorem 4.1.17, one can find G_j, H_j, $\tilde{P}_j$, $\tilde{Q}_j$ such that

$$\left.\begin{aligned}P_j = G_j R_k + \tilde{P}_j \qquad \text{for } j = 1, \ldots, n\\ Q_j = H_j R_k + \tilde{Q}_j \qquad \text{for } j = 1, \ldots, n\end{aligned}\right\} \tag{4.2.31}$$

and

$$[y_k, \tilde{P}_j] = [y_k, \tilde{Q}_j] = 0 \qquad \text{for } j = 1, \ldots, n \tag{4.2.32}$$

hold. By the inductive assumption,

$$[y_i, P_j] = [y_i, R_k] = [y_i, Q_j] = 0 \qquad \text{for } i < k \text{ and } j = 1, \ldots, n \tag{4.2.33}$$

holds. Then, the uniqueness part of Theorem 4.1.7 implies

$$[y_i, \tilde{P}_j] = [y_i, \tilde{Q}_j] = 0 \qquad \text{for } i \leqq k \text{ and } j = 1, \ldots, n. \tag{4.2.34}$$

From (4.2.31), one has $\sigma_1(P_k) = p_k(x, \xi)$. Hence, one obtains $\sigma_0(G_k) = 0$, $\sigma_1(\tilde{P}_k) = \sigma_1(P_k) = p_k$. Therefore, $(1 + G_k)$ is invertible by the corollary of Theorem 4.1.6. Then one has

$$\mathscr{I} = \sum_{j=1}^{n} \mathscr{E}_{M \times N}(D_{y_j} - \tilde{P}_j) + \sum_{j=1}^{n} \mathscr{E}_{M \times N}(y_j - \tilde{Q}_j). \tag{4.2.35}$$

A similar argument provides us with

$$[D_{y_k}, \tilde{P}_j] = [D_{y_k}, \tilde{Q}_j] = 0 \qquad \text{for } j = 1, \ldots, n. \tag{4.2.36}$$

Consequently, by replacing R_j amd S_j by $D_{y_j} - P_j$ and $y_j - Q_j$ respectively, $(4.2.28)_k$ and $(4.2.29)_k$ hold. Hence, the induction proceeds, concluding the proof of (4.2.30).

Lastly, we will prove that, for a given A, $\tilde{A}(x, D_x)$ in (4.2.21) is uniquely determined. It is sufficient to prove $A = 0$ for $A(x, D_x) \in \mathscr{I}$. Since R_j and S_j generate $\mathscr{I}$, for $A \in \mathscr{I}$ one has

$$A(x, D_x) = \sum_{j=1}^{n} G_j R_j + \sum_{j=1}^{n} H_j S_j \qquad \text{for } G_j, H_j \in \mathscr{E}_{M \times N}. \tag{4.2.37}$$

Again, by induction, we will prove that G_j and H_j can be chosen to satisfy

$$\begin{aligned} &A = \sum_{j=k}^{n} G_j R_j + \sum_{j=1}^{n} H_j S_j \qquad \text{and} \\ &[y_i, G_j] = [y_i, H_j] = 0 \qquad \text{for } i < k. \end{aligned} \tag{4.2.37$_k$}$$

As before, let $\tilde{G}_j$ and $\tilde{H}_j$ be the remainders when G_j and H_j are divided by R_k respectively. Then one may let

$$[y_k, \tilde{G}_j] = [y_k, \tilde{H}_j] = 0 \qquad \text{for } j = 1, \ldots, n. \tag{4.2.38}$$

Again by the uniqueness part of the division theorem,

$$[y_i, \tilde{G}_j] = [y_i, \tilde{H}_j] = 0 \qquad \text{for } i \leq k \text{ and } j = 1, \ldots, n \tag{4.2.39}$$

holds. Therefore, one obtains the expression

$$A(x, D_x) - \sum_{j=k+1}^{n} \tilde{G}_j R_j - \sum_{j=1}^{n} \tilde{H}_j S_j = T_k R_k, \qquad T_k \in \mathscr{E}_{M \times N}. \tag{4.2.40}$$

Since one may suppose that R_j and S_j are as in (4.2.30), the left-hand side commutes with y_k. Hence the uniqueness of the division implies that $A - \sum_{j=k+1}^{n} \tilde{G}_j R_j - \sum_{j=1}^{n} \tilde{H}_j S_j = 0$. Thus, the induction proceeds. If one repeats the argument for $\sum_{j=1}^{n} \tilde{H}_j S_j$, consequently one obtains $(4.2.37)_n$. From the uniqueness of the division, one has $A = 0$.

Remark. We have obtained the isomorphism in (4.2.15) via the integral transformation (4.2.14) with the kernel function $\delta(\Omega(x, y))$. Note that $Y(\Omega(x, y))$ could have been taken as a kernel function instead of $\delta(\Omega(x, y))$; or, more generally, for a real-valued real analytic function $a(x, y)$, which can never be zero on H, one can use $a(x, y)(\Omega(x, y))_+^\lambda/\Gamma(\lambda + 1)$, $(\lambda \in \mathbf{C})$. Such arbitrariness in the choice of kernel functions indicates that the isomorphism in (4.2.15) is not determined uniquely for a given Ω; that is, an operator transformation cannot be uniquely determined by a geometric transformation. Actually, it is known that the isomorphism is unique up to an inner automorphism by an invertible microdifferential operator of order zero. This is the reason why the terminology "quantized contact transformation" was introduced. We also note that, more generally, we can use any non-degenerate section of a simple holonomic system with its characteristic variety being the conormal bundle of H, instead of $\delta(\Omega(x, y))$ etc., as a kernel function in (4.1.14). It might be better to discuss the problem emphasizing this viewpoint, partly because we can deal with general contact transformations (not necessarily with a kernel function) by that approach, and partly because it is preferable aesthetically. That, however, would require further preparation on the manipulation of holonomic systems; hence we content ourselves here with the above discussion.

Example 4.2.5. We will quantize the Legendre transformation. Let $\Omega = x_n - y_n + \sum_{j=1}^{n-1} x_i y_i$. Then, $(\partial/\partial y_n)^{-1}$ is well defined for $\eta_n \neq 0$ as a microdifferential operator. We have

$$\left(\frac{\partial}{\partial y_n}\right)^{-1}\delta(\Omega) = -Y(\Omega).$$

Hence,

$$x_j\delta(\Omega) = \frac{\delta}{\delta y_j}Y(\Omega) = -\frac{\partial}{\partial y_j}\left(\frac{\partial}{\partial y_n}\right)^{-1}\delta(\Omega) \qquad \text{for } j = 1, \ldots, n-1$$

holds. Since we have

$$x_n\delta(\Omega) = \left(y_n - \sum_{j=1}^{n-1} x_j y_j\right)\delta(\Omega),$$

we obtain the following:

$$x_n\delta(\Omega) = \left(y_n + \sum_{j=1}^{n-1} y_j \frac{\partial}{\partial y_j}\left(\frac{\partial}{\partial y_n}\right)^{-1}\right)\delta(\Omega) = \langle y, D_y\rangle\left(\frac{\partial}{\partial y_n}\right)^{-1}\delta(\Omega).$$

Furthermore,

$$\frac{\partial}{\partial x_j}\delta(\Omega) = y_j\delta'(\Omega) = y_j\left(-\frac{\partial}{\partial y_n}\right)\delta(\Omega) \qquad \text{for } j = 1, \ldots, n-1$$

and

$$\frac{\partial}{\partial x_n}\delta(\Omega) = -\frac{\partial}{\partial y_n}\delta(\Omega)$$

hold. Here, the equation (4.2.14) is equivalent to

$$P(x, D_x)\delta(\Omega(x, y)) = Q^*(y, D_y)\delta(\Omega(x, y)), \tag{4.2.14'}$$

where Q^* denotes the conjugate operator of Q. Consequently, we obtain the correspondence as follows:

$$\left.\begin{aligned}
&x_j = \frac{\partial}{\partial y_j}\left(\frac{\partial}{\partial y_n}\right)^{-1}, && j = 1, \ldots, n-1\\
&x_n = \langle y, D_y\rangle\left(\frac{\partial}{\partial y_n}\right)^{-1} &&\\
&\frac{\partial}{\partial x_j} = y_j\frac{\partial}{\partial y_n}, && j = 1, \ldots, n-1\\
&\frac{\partial}{\partial x_n} = \frac{\partial}{\partial y_n}. &&
\end{aligned}\right\} \tag{4.2.41a}$$

The inverse correspondence can be found as follows:

$$\left.\begin{aligned}
&y_j = \frac{\partial}{\partial x_j}\left(\frac{\partial}{\partial x_n}\right)^{-1}, && j = 1, \ldots, n-1\\
&y_n = \langle x, D_x\rangle\left(\frac{\partial}{\partial x_n}\right)^{-1} &&\\
&\frac{\partial}{\partial y_j} = -x_j\frac{\partial}{\partial x_n}, && j = 1, \ldots, n-1\\
&\frac{\partial}{\partial y_n} = \frac{\partial}{\partial x_n}. &&
\end{aligned}\right\} \tag{4.2.41b}$$

Exercise. Find the above correspondences when the kernel function $\delta(\Omega)$ in (4.2.14) is replaced by $Y(\Omega)$.

We viewed the quantized contact transformation in Theorem 4.2.2 as a correspondence of microdifferential operators. Next we will show that a quantized contact transformation induces an isomorphism on the sheaf of microfunctions. This guarantees that one may use quantized contact transformations freely as one studies the structures of microfunction solutions of microdifferential equations.

Theorem 4.2.3. *Under the same hypotheses as in Theorem 4.2.2, the map* $T:\mathscr{C}_M \to \mathscr{C}_N$, *defined by* $T(u(y)) = \int \delta(\Omega(x, y))u(y)\,dy$, *is an isomorphism.*

Proof. Let

$$I(y', y) \underset{\text{def}}{=} \int \delta(\Omega(x, y'))\delta(\Omega(x, y))\,dx.$$

Then, it is sufficient to prove that I is a kernel function for an invertible microdifferential operator on $\sqrt{-1}S^*M$. By the Weierstrass preparation theorem for a holomorphic function (see Note following Proposition 2.4.3) one may assume that $\Omega(x, y)$ is of the form $x_1 - f(x_2, \ldots, x_n, y)$. We have

$$I(y', y) = \int \delta(f(x_2, \ldots, x_n y) - f_2(x_2, \ldots, x_n, y'))\,dx_2 \cdots dx_n. \tag{4.2.43}$$

From the Taylor expansion, one can find G such that

$$f(x_2, \ldots, x_n, y) - f(x_2, \ldots, x_n, y') = \langle y - y', G(x_2, \ldots, x_n, y, y')\rangle.$$

Hence (4.2.43) becomes

$$\int \delta(\langle y - y', G(x_2, \ldots, x_n, y, y')\rangle)\,dx_2 \cdots dx_n. \tag{4.2.44}$$

Letting $A(y, y', \eta) = |\det(\eta, \partial\eta/\partial x_2, \ldots, \partial\eta/\partial x_n)|^{-1}$, one can rewrite (4.2.44) as

$$\int A(y, y', \eta)\delta(\langle y - y', \eta\rangle)\omega(\eta).$$

Therefore, since $A(y, y', \eta) \neq 0$, Theorem 4.1.3 and the corollary of Theorem 4.1.6 imply that $I(y, y')$ determines an invertible microdifferential operator.

§3. Structures of Systems of Microdifferential Equations

The purpose of this section is to analyze the structure of a general system of microdifferential equations, using quantized contact transformations. In classical mechanics it is an exquisite and fundamental fact that any system can be transformed into an equilibrium system by a canonical transformation (see, for example, Yamanouchi[1]). In terms of analysis, this means that a partial differential equation of the first order can be transformed into a standard form by canonical transformation. The results in this section generalize the above result to the higher order in the linear case. Though we will consider rather restricted systems (see Remarks 4.3.3 and 4.3.4) in this treatise, the essence of our theory will be recognized.

Theorem 4.3.1. *Let $\mathscr{M}$ be an $\mathscr{E}$-Module defined in a neighborhood of* $(x^0, \sqrt{-1}\xi^0\infty) \in \sqrt{-1}S^*M$, *satisfying the conditions* (4.3.1) ~ (4.3.4):

For some left-ideal $\mathscr{I}$ of $\mathscr{E}$, $\mathscr{M} = \mathscr{E}/\mathscr{I}$. (4.3.1)

The zero set $V(J)$ of $J \underset{\text{def}}{=} \bigcup_m \{\sigma_m(P) \mid P \in \mathscr{I} \cap \mathscr{E}(m)\}$ is a non-singular manifold of codimension d in a complex neighborhood of $(x^0, \sqrt{-1}\xi^0\infty)$; *and $\omega|_{V(J)} \neq 0$ also holds, where ω is the canonical 1-form $\sum_j \xi_j dx_j$.* (4.3.2)

The totality of analytic functions, homogeneous in ξ, that vanish on $V(J)$ is J. (4.3.3)

The zero set $V(J)$ is real, i.e. $V(J) = \overline{V(J)}$, where $\overline{V(J)}$ is the complex conjugate of $V(J)$. (4.3.4)

Then, $\mathscr{M}$ can be transformed into the following system $\mathscr{N}$ via a quantized contact transformation

$$\mathscr{N} = \mathscr{E}/(\mathscr{E}D_1 + \cdots + \mathscr{E}D_d). \tag{4.3.5}$$

Remark 4.3.1. Let u be the residue class of $1 \in \mathscr{E}$ at $\mathscr{I}$. Then (4.3.1) may be written as

$$\mathscr{M} = \mathscr{E}u, \qquad \text{where } Pu = 0 \text{ for } P \in \mathscr{I}. \tag{4.3.6}$$

Some readers may consider that (4.3.6) looks more like a system of (micro)-differential equations. However, when one considers a general system (not necessarily with one unknown function) of microdifferential equations, it is more desirable to grasp the system of microdifferential equations as a coherent left $\mathscr{E}$-Module; i.e.

$$0 \leftarrow \mathscr{M} \leftarrow \mathscr{E}^s \leftarrow \mathscr{E}^t$$

is exact, which is more intrinsic. If a system of microdifferential equations is interpreted as an $\mathscr{E}$-Module, then it is clear that an element $\mathscr{H}om_{\mathscr{E}}(\mathscr{M}, \mathscr{C})$ represents a microfunction solution of $\mathscr{M}$ (rigorously speaking, the image of u under an $\mathscr{E}$-homomorphism from $\mathscr{M}$ to $\mathscr{C}$ is a microfunction solution of $\mathscr{M}$). The system $\mathscr{N}$ is sometimes called a de Rham system, or a partial de Rham system.

Remark 4.3.2. Since $\omega|_{V(J)} \neq 0$, then $d \leqq n - 1$ holds in our case.

Before we prove Theorem 4.3.1, a special case of Theorem 4.3.1 will be proved. Note that the proof of Theorem 4.3.1 is essentially reduced to this special case.

Theorem 4.3.2. *Let $P(x, D_x)$ be a microdifferential operator of the first order defined in a neighborhood of* $(x^0, \sqrt{-1}\xi^0\infty) = (0, \sqrt{-1}(0, 0, \ldots, 1)\infty)$

such that the principal symbol $\sigma_1(P)$ is ξ_1. Then, in a neighborhood of $(x^0, \sqrt{-1}\xi^0\infty)$,

$$\mathscr{E}/\mathscr{E}P \cong \mathscr{E}/\mathscr{E}D_1 \tag{4.3.7}$$

holds.

Proof. From Theorem 4.1.9, one can write $P = Q(D_1 - A(x, D'))$, where $D' = (D_2, \ldots, D_n)$ and Q is invertible in a neighborhood of $(x^0, \sqrt{-1}\xi^0\infty)$. Hence, one may assume $P = D_1 - A(x, D')$ to prove this theorem. We will find an invertible microdifferential operator $R(x, D')$ such that $R^{-1}PR = D_1$ and $R = \sum_{k=0}^{\infty} R^k(x, D')$, where R^k, $k = 0, \ldots,$ satisfy the following (4.3.8) and (4.3.9). When such an R exists, it is clear that (4.3.7) holds:

$$R^0 = 1. \tag{4.3.8}$$

$$\left.\begin{aligned} &\frac{\partial}{\partial x_1} R^k(x, D') = A(x, D')R^{k-1}(x, D') \\ &R^k|_{x_1=0} = 0 \qquad \text{for } k \geqq 1. \end{aligned}\right\} \tag{4.3.9}$$

Here, the left-hand sides $(\partial/\partial x_1)R^k(x, D')$ and $R^k|_{x_1=0}$ of (4.3.9) mean the derivative of R^k with respect to x_1 (regarding the microdifferential operator R^k as depending upon the parameter x_1) and the substitution $x_1 = 0$ respectively. The equation (4.3.9) can be written explicitly as

$$R^k(x, D') = \int_0^{x_1} A(s_k, x', D') \int_0^{s_k} A(s_{k-1}, x', D') \int_0^{s_{k-1}} \cdots \int_0^{s_2} A(s_1, x', D')ds_1 \cdots ds_{k-2}\, ds_{k-1}\, ds_k, \tag{4.3.10}$$

where $x' = (x_2, \ldots, x_n)$ and where A is regarded as an operator depending on s_j. Denote the domain of integration of (4.3.10) by V_k. Then the volume of V_k is $|x_1|^k/k!$. The order of R^k is at most zero from the expression (4.3.10). Hence, by letting δ, ϵ and $M > 0$ be as

$$\sup_{|s| \leqq \epsilon} N_0(A(s, x', D'); t) \leqq M < \infty \qquad \text{for } 0 < t < \delta, \tag{4.3.11}$$

the following (4.3.12), obtained from Theorem 4.1.5, implies (4.3.13):

$$\begin{aligned} N_0(R^k(x, D'); t) &\ll \int_{V_k} \cdots \int N_0(A(s_k, x', D') \cdots A(s_1, x', D')\, ds_1 \cdots ds_k \\ &\ll \int_{V_k} \cdots \int \left[\sup_{|s| \leqq |x_1|} N_0(A(s, x', D'); t) \right]^k ds_1 \cdots ds_k. \end{aligned} \tag{4.3.12}$$

$$N_0\left(\sum_{k=0}^{\infty} R^k(x, D'); t \right) \leqq \sum_{k=0}^{\infty} \frac{M^k|x_1|^k}{k!} = \exp(Mx_1) < \infty. \tag{4.3.13}$$

Therefore, $R = \sum_{k=0}^{\infty} R^k(x, D')$ is well defined as microdifferential operator. Furthermore, (4.3.8) and the initial condition in (4.3.9) imply $\sigma_0(R)|_{x_1=0} = 1 \neq 0$. Hence, R is invertible by the corollary of Theorem 4.1.6. On the other hand, by definitions (4.3.8) and (4.3.9) of R^k, one obtains

$$(D_1 - A(x, D')) \circ R = R \circ D_1. \tag{4.3.14}$$

Here, $R \circ D_1$ is the composition of microdifferential operators, not in the sense of $(\partial/\partial x_1)R$ in (4.3.9). In fact, we have $D_1 \circ R = \partial R/\partial x_1 + R \circ D_1$ by definition.

Remark 4.3.3. In the case $\sigma(P) = \xi_1^m$, $m \geqq 2$, rather than $\sigma(P) = \xi_1$, one can prove the similar assertion by replacing $\mathscr{E}$ with $\mathscr{E}^\infty$. That is, the introduction of operators of infinite order allows one to treat the terms of lower orders. Even though it is one of the characteristics of hyperfunction theory to be able to treat such an operator, we restrict ourselves to a consideration of operators of finite order. One reason for so doing is that the treatment of operators of infinite order is at present not so "algebro-analytic" (not without reason—an operator of infinite order is transcendental, depending on the topological properties of the field of complex numbers).

We now begin the proof of Theorem 4.3.1. It is easy to see, because of the condition (4.3.3), that $V(J)$ is involutory. This is a special case of the remark following Definition 4.2.6; when one does not have such an algebraic condition as (4.3.3), the proof is non-trivial. Here is a proof. Let $f(x, \xi)|_{V(J)} = g(x, \xi)|_{V(J)} = 0$. Then one can find P and Q in $\mathscr{J}$ such that $\sigma_1(P) = f$ and $\sigma_1(Q) = g$. Note that $[P, Q] = PQ - QP \in \mathscr{J}$, which implies $\sigma_1([P, Q])|_{V(J)} = 0$. Consequently, since $\sigma_1([P, Q]) = \{\sigma_1(P), \sigma_1(Q)\} = \{f, g\}$, one obtains $\{f, g\}|_{V(J)} = 0$; i.e. $V(J)$ is involutory. Hence, conditions (4.3.2) and (4.3.4) imply that, by a quantized contact transformation, one can write $V(J) = \{\xi_1 = \cdots = \xi_d = 0\}$. By (4.3.3), one can choose $P_1, \ldots, P_d$ in $\mathscr{J}$ such that

$$\sigma_1(P_j) = \xi_j \qquad \text{for } j = 1, \ldots, d. \tag{4.3.15}$$

Furthermore, one may assume $P_1 = D_1$ from the proof of Theorem 4.3.2. We will prove by induction on k ($\leqq d$) that we can take

$$P_j = D_j \qquad \text{for } j = 1, \ldots, k \tag*{(4.3.16)$_k$}$$

for suitable generators of $\mathscr{M}$. Let us assume that $(4.3.16)_k$ holds. Then, by $D_j u = 0$ for $j = 1, \ldots, k$, one may assume

$$P_j = P_j(x, D_{k+1}, \ldots, D_n) \qquad \text{for } j = k + 1, \ldots, d. \tag{4.3.17}$$

Theorem 4.1.9 implies that one has

$$P_{k+1}(x, D_{k+1}, \ldots, D_n) = R(D_{k+1} + Q_{k+1}) \tag{4.3.18}$$

for some microdifferential operator $Q_{k+1}(x, D_{k+2}, \ldots, D_n)$ of, at most, order 0 and an invertible microdifferential operator $R(x, D_{k+1}, \ldots, D_n)$. The assertion on uniqueness in Theorem 4.1.9 indicates that R and Q_{k+1} take the above forms, i.e. depending on some, not all Dj's. Hence, we may begin with $P_{k+1} = D_{k+1} + Q_{k+1}(x, D_{k+2}, \ldots, D_n)$. Thus, one may assume, using $(D_{k+1} + Q_{k+1})u = 0$,

$$P_j = P_j(x, D_{k+2}, \ldots, D_n) \qquad \text{for } j = k+2, \ldots, d. \tag{4.3.19}$$

Then, again by virtue of Theorem 4.1.9, one obtains $P_{k+2} = D_{k+2} + Q_{k+2}(x, D_{k+3}, \ldots, D_n)$. The repeated use of this argument implies

$$\left.\begin{aligned} P_j &= D_j \qquad \text{for } j = 1, \ldots, k \\ P_{k+1} &= D_{k+1} + Q_{k+1}(x, D_{d+1}, \ldots, D_n). \end{aligned}\right\} \tag{4.3.20}$$

Since for $j = 1, \ldots, k$ one has $[D_j, P_{k+1}] = D_jP_{k+1} - P_{k+1}D_j \in \mathscr{I}$, then $[D_j, Q_{k+1}] \in \mathscr{I}$ holds for $j = 1, \ldots, k$. Furthermore, we will show that $[D_j, Q_{k+1}] = 0$. Suppose $[D_j, Q_{k+1}] \neq 0$; then let m be the order of $[D_j, Q_{k+1}]$. Since $[D_j, Q_{k+1}] \in \mathscr{I}$, then $\sigma_m([D_j, Q_{k+1}]) = 0$ on $V(J)$. Therefore, if $q_{k+1,m}$ denotes the homogeneous part of degree m of Q_{k+1}, then on $V(J)$

$$\sigma_m([D_j, Q_{k+1}]) = \{\xi_j, q_{k+1,m}\} = \frac{\partial}{\partial x_j} q_{k+1,m} = 0$$

holds. But $q_{k+1,m}$ does not depend on $\xi_1, \ldots, \xi_d$ by (4.3.20). Consequently, $(\partial/\partial x_j)q_{k+1,m} \equiv 0$, which contradicts the assumption $\sigma_m([D_j, Q_{k+1}]) \not\equiv 0$. Therefore we have $[D_j, Q_{k+1}] = 0$ for $j = 1, \ldots, k$. That is, Q_{k+1} depends only upon $(x_{k+1}, \ldots, x_n, D_{d+1}, \ldots, D_n)$. By the proof for Theorem 4.3.2, one can find an invertible microdifferential operator $R(x_{k+1}, \ldots, x_n, D_{k+1}, \ldots, D_n)$ such that

$$R^{-1}P_{k+1}R = D_{k+1} \tag{4.3.21}$$

holds. One has $R^{-1}D_jR = D_j$ for $j = 1, \ldots, k$, since R commutes with D_j, $j = 1, \ldots, k$. Hence, replacing a generator u of $\mathscr{M}$ by Ru, generators P_j for $\mathscr{I}$ can be chosen so as to satisfy $(4.3.16)_{k+1}$. Hence the induction proceeds.

Next, by making use of $(4.3.16)_d$, we will prove that $\mathscr{I}$ is generated by $D_1, \ldots, D_d$. Suppose $D_1, \ldots, D_d$ do not generate $\mathscr{I}$; then there exists $0 \neq R(x, D_{d+1}, \ldots, D_n) \in \mathscr{I}$. By the definition of $V(J)$, one has $\sigma(R)|_{V(J)} = 0$. Since $\sigma(R)$ does not depend upon $\xi_1, \ldots, \xi_d$, this implies $\sigma(R) = 0$, a contradiction. Hence, $\mathscr{I}$ is generated by $D_1, \ldots, D_d$. That is, $\mathscr{M}$ is isomorphic to the $\mathscr{N}$ given in (4.3.5).

Remark 4.3.4. If operators of infinite order are employed, the condition (4.3.3) is not needed. Since a microdifferential operator of infinite order operates on the sheaf of microfunctions as a sheaf homomorphism, the condition (4.3.3) is not needed for the following theorem (Theorem 4.3.3). We will not go into details here.

Once Theorem 4.3.1 is obtained, the properties of the microfunction solutions of a system of microdifferential equations, which are invariant under quantized contact transformations, can be easily found. One of the most fundamental examples is phrased in Theorem 4.3.3, stating the propagation of solutions along a bicharacteristic manifold. First, we will generalize the notion of a bicharacteristic strip given for a single equation (Chapter III, §5, Definition 3.5.1).

Definition 4.3.1. *Suppose that an involutory submanifold V of $\sqrt{-1}S^*M$ satisfies the conditions* (4.3.2) *and* (4.3.4), *such that* $V = \{(x, \sqrt{-1}\xi\infty) \in \sqrt{-1}S^*M \mid f_1(x, \xi) = \cdots = f_d(x, \xi) = 0\}$. *The bicharacteristic manifold* $b = b_{(x^0, \sqrt{-1}\xi^0\infty)}$, *associated with V passing through $(x^0, \sqrt{-1}\xi^0\infty) \in V$, is defined as the integral manifold of dimension d, going through $(x^0, \sqrt{-1}\xi^0\infty)$, of d Hamilton operators H_{f_j} associated with V, where*

$$H_{f_j} = \sum_{l=1}^{n} \left(\frac{\partial f_j}{\partial \xi_l} \frac{\partial}{\partial x_l} - \frac{\partial f_j}{\partial x_l} \frac{\partial}{\partial \xi_l} \right) \qquad \text{for } j = 1, \ldots, d. \tag{4.3.22}$$

Remark 4.3.5. Since V is involutory, $\{H_{f_j}\}_{j=1}^d$ satisfy the integrability condition. The definition of b does not depend upon the choice of $\{f_j\}_{j=1}^d$.

Theorem 4.3.3. *Let $\mathcal{M}$ be an $\mathcal{E}$-Module satisfying the conditions in Theorem* 4.3.1. *Then one has the following fact* (4.3.23) *in a neighborhood of* $(x^0, \sqrt{-1}\xi^0\infty)$:

The microfunction solution sheaf $\mathcal{S} \underset{\text{def}}{=} \mathcal{H}om_{\mathcal{E}}(\mathcal{M}, \mathcal{C})$ has its support in V and is locally constant along each bicharacteristic manifold. Furthermore, $\mathcal{S}$ is flabby in the direction transversal to bicharacteristic manifolds. (4.3.23)

Proof. Since the assertion of this theorem is invariant under a quantized contact transformation, it is sufficient to prove the case where $\mathcal{M}$ is of the form $\mathcal{N} = \mathcal{E}/(\mathcal{E}D_1 + \cdots + \mathcal{E}D_d)$. For $\mathcal{N}$, (4.3.23) can be proved in a way similar to the proof of the first lemma in §2 of Chapter III.

Remark 4.3.6. It can also be shown that $\mathcal{M}$ is locally solvable on $\sqrt{-1}S^*M$; i.e., for f_j satisfying an algebraic comparable condition, there exists a microfunction solution u such that $P_j u = f_j$, where P_j's are generators of $\mathcal{I}$. More generally, the higher cohomology groups $\mathcal{E}xt^j_{\mathcal{E}}(\mathcal{M}, \mathcal{C})$, $j \geqq 1$, vanish, but we need further algebraic preparation to prove this.

Remark 4.3.7. Having followed this treatise thus far, one may not have any difficulty understanding Theorem 4.3.3. The statement on "the propagation of singularities along bicharacteristic manifolds" (even for $d = 1$), together with the local solvability problem to be discussed later, has been of central interest in the theory of linear differential equations. One had to wait until the advent of microfunction theory to solve this problem in the real analytic category. Even in the C^∞-category, the solution for the case $d = 1$ had not been established before the 1960s. Furthermore, the description using bicharacteristic curves was not thorough enough to use bicharacteristic strips (see §6 of Chapter III). For further historical information on the era of linear partial differential equations before microfunction theory and quantized contact transformations, consult Hörmander [1]. There one may get an idea of what it was like "B.C. (Before $\mathscr{C}$)." As such a difficult problem becomes obvious, one may see progress in mathematics: it is like a view from a mountain peak!

Remark 4.3.8. Theorem 4.3.1 still holds for a complex domain—in which case, (4.3.4) is not needed. In this sense, the theorem is most fundamental. In this book, we defined the notion of a microdifferential operator on $\sqrt{-1}S^*M$. The reader might expect that compositions, transformations, etc. can also be defined for a complex manifold X and the projective cotangent bundle P^*X. In SKK [1], the theory is developed in that way so as to define an operator on a real manifold by the restriction. This method is more universal but less intuitive. Hence, in this book, we gave a less general definition of a microdifferential operator than the one in SKK [1]. Once $\mathscr{E}$ is defined over P^*X, Theorem 4.3.1 can be rephrased as follows: let $\mathscr{M}$ be an $\mathscr{E}$-Module defined in a neighborhood of $(x^0, \xi^0) \in P^*X$ satisfying (4.3.1), (4.3.25), and (4.3.3) where

$$V(J) \subset P^*X \text{ is a non-singular manifold of codimension } d \text{ and } \omega|_{V(J)} \neq 0. \tag{4.3.25}$$

Then (by a quantized contact transformation) $\mathscr{M}$ is isomorphic to $\mathscr{N} = \mathscr{E}/(\mathscr{E}D_1 + \cdots + \mathscr{E}D_d)$.

Using the theory of microdifferential operators developed over P^*X, one can prove this version in a way similar to that shown here. Note that the only geometric fact used for the proof of Theorem 4.3.1 is the following: if $V(J)$ is real and satisfies (4.3.2), then one obtains $V(J) = \{\xi_1 = \cdots = \xi_d = 0\}$ by a real contact transformation. For a complex domain, needless to say, one needs the condition (4.3.25) only, without the condition of $V(J)$ being real, to obtain the above. The rest of the argument is no different for a real domain or for a complex domain.

The above remark (Remark 4.3.8) indicates that the essential part of Theorem 4.3.1 may be viewed as a result for a complex domain. In fact, the condition (4.3.4), which is needed for the case in a real domain, is satisfied only by very restricted equations (although, we may say, it is not such a restrictive condition for equations actually appearing in applications). Thus, we are naturally led to ask the following question. Theoretically speaking, what are "generic" systems of equations for a real domain? First we must ask how far canonical V can be by a real contact transformation. An answer to this question for a system will be given in (4.3.82). To make the essence of the argument apparent, we will treat the case of a single equation.

Theorem 4.3.4. *Let M be an open neighborhood of $x^0 \in \mathbf{R}^n$, and let $f(x, \xi)$ be a real analytic function defined in a neighborhood of $(x^0, \xi^0) \in T^*M - M$, having the following properties* (4.3.26) ~ (4.3.28):

f is positively homogeneous in ξ of degree 1/2; *i.e.* $f(x, c\xi) = c^{1/2}f(x, \xi)$ *for* $c > 0$. (4.3.26)

$$f(x^0, \xi^0) = 0. \tag{4.3.27}$$

$$\frac{1}{2\sqrt{-1}}\{f, \bar{f}\}(x^0, \xi^0) \gneqq 0. \tag{4.3.28}$$

Then there exists a real-valued real analytic function $\Phi(x, \xi, t, \bar{t})$, defined in a neighborhood of $(x, \xi, t, \bar{t}) = (x^0, \xi^0, 0, 0) \in (T^*M - M) \times \mathbf{C} \times \bar{\mathbf{C}}$ ($\bar{t}$ is the complex conjugate of t), satisfying the following three conditions:

$$\Phi(x^0, \xi^0, 0, 0) \gneqq 0. \tag{4.3.29}$$

$\Phi(x, \xi, f(x, \xi), \bar{f}(x, \xi))$ *is positively homogeneous in ξ of degree* 1/2.

(4.3.30)

$$\frac{1}{2\sqrt{-1}}\{f(x, \xi)\Phi(x, \xi, f(x, \xi), \bar{f}(x, \xi)), \bar{f}(x, \xi)\Phi(x, \xi, f(x, \xi), \bar{f}(x, \xi))\} = 1.$$

(4.3.31)

Proof. We will prove a more general statement (C) than the assertion of this theorem, since the proof of (C) is simpler in notations than the one of the theorem.

(C): Let $F(x, \xi, t, \bar{t})$ be a strictly positive-valued real analytic function defined in a neighborhood of $(x^0, \xi^0, 0, 0)$ such that $F(x, \xi, f(x, \xi), \bar{f}(x, \xi))$ is positively homogeneous in ξ of degree 0. Then, one can find a real-valued real analytic function $\Phi(x, \xi, t, \bar{t})$ such that Φ satisfies

$$\begin{aligned}\{f(x, \xi)\Phi(x, \xi, f(x, \xi), \bar{f}(x, \xi)), \bar{f}(x, \xi)\Phi(x, \xi, f(x, \xi), \bar{f}(x, \xi))\}\\ = \{f(x, \xi), \bar{f}(x, \xi)\}F(x, \xi, f(x, \xi), \bar{f}(x, \xi)),\end{aligned} \tag{4.3.32}$$

$\Phi(x^0, \xi^0, 0, 0) > 0$, and $\Phi(x, \xi, f(x, \xi), \bar{f}(x, \xi))$ is positively homogeneous in ξ of degree 0.

The assertion of Theorem 4.3.4 is obtained from (C) for the case $F = 2/\{f, \bar{f}\}$.

Proof of (C). For the sake of simplicity, let $\Psi(x, \xi, t, \bar{t}) = (\Phi(x, \xi, t, \bar{t}))^2$. For a function $g(x, \xi)$ on $T^*M - M$, define Θ and $\bar{\Theta}$ as follows:

$$\left.\begin{aligned} \Theta_g &= \frac{\{g, \bar{f}\}}{\{f, \bar{f}\}} \\ \bar{\Theta}_g &= \frac{\{f, g\}}{\{f, \bar{f}\}}. \end{aligned}\right\} \tag{4.3.33}$$

The differential operators Θ and $\bar{\Theta}$ naturally act on $\Psi(x, \xi, t, \bar{t})$. Then notice that (4.3.32) is reduced to finding a solution $\Psi \gneqq 0$ of the following (4.3.34):

$$\begin{aligned} \Psi(x, \xi, t, \bar{t}) &+ \frac{1}{2} t\left(\frac{\partial \Psi(x, \xi, t, \bar{t})}{\partial t} + \Theta\Psi(x, \xi, t, \bar{t})\right) \\ &+ \frac{1}{2}\bar{t}\left(\frac{\partial \bar{\Psi}(x, \xi, t, \bar{t})}{\partial \bar{t}} + \bar{\Theta}\Psi(x, \xi, t, \bar{t})\right) = F(x, \xi, t, \bar{t}). \end{aligned} \tag{4.3.34}$$

Since this equation is degenerate at $t = \bar{t} = 0$, we will show the existence of a solution of (4.3.34), using a singular coordinate transformation at the origin in the $(t, \bar{t})$-space. That is, one is to find a real-valued function Ω satisfying

$$\Omega(\lambda, t, \bar{t}) \, (\equiv \Omega(x, \xi; \lambda, t, \bar{t})) = \lambda^2 \Psi(x, \xi, \lambda t, \lambda \bar{t}), \qquad \lambda \in \mathbf{R}. \tag{4.3.35}$$

When the dependency of Ω on (x, ξ) is not crucial in arguments, we abbreviate $\Omega(x, \xi; \lambda, t, \bar{t})$ as $\Omega(\lambda, t, \bar{t})$. By (4.3.35), (4.3.34) can be rewritten in terms of Ω as

$$\frac{\partial}{\partial \lambda}\Omega + \Theta(\bar{t}\Omega) + \bar{\Theta}(t\Omega) = 2\lambda F(x, \xi, \lambda t, \lambda \bar{t}). \tag{4.3.36}$$

If the initial value of Ω on $\{\lambda = 0\}$ is given by 0, then Ω, since $\{\lambda = 0\}$ is non-singular for (4.3.36), is determined uniquely from (4.3.36). Furthermore, the complex conjugate $\bar{\Omega}$ of Ω is also a solution, since F is a real-valued function. By the uniqueness theorem of the Cauchy problem, $\Omega = \bar{\Omega}$ holds; i.e. Ω is a real-valued function. Since Θ and $\bar{\Theta}$ are differential operators with respect to (x, ξ), then (4.3.36) implies

$$\left.\frac{\partial \Omega}{\partial \lambda}\right|_{\lambda = 0} = 0. \tag{4.3.37}$$

The initial condition of Ω and (4.3.37) imply that Ω/λ^2 is holomorphic in a neighborhood of $\{\lambda = 0\}$. If one can show that Ω/λ^2 is an analytic function of $(x, \xi, \lambda t, \lambda\bar{t})$, then one can let $\Psi(x, \xi, \lambda t, \lambda\bar{t}) = \Omega/\lambda^2$. To do so, it is sufficient to show that Ω satisfies the following condition on the homogeneity with respect to $(\lambda, t, \bar{t})$:

$$\Omega(c\lambda, c^{-1}t, c^{-1}\bar{t}) = c^2\Omega(\lambda, t, \bar{t}) \qquad \text{for } c \in \mathbf{R} - \{0\}. \tag{4.3.38}$$

Let $\mu = c^{-1}\lambda$, $s = ct$, and $\bar{s} = c\bar{t}$; and let $\tilde{\Omega}(\mu, s, \bar{s}) = \Omega(c\mu, c^{-1}s, c^{-1}\bar{s})$. From (4.3.36) one obtains

$$\frac{1}{c}\frac{\partial}{\partial\mu}\tilde{\Omega} + \Theta\left(\frac{\bar{s}}{c}\tilde{\Omega}\right) + \bar{\Theta}\left(\frac{s}{c}\tilde{\Omega}\right) = 2c\mu F(x, \xi, \mu s, \mu\bar{s}). \tag{4.3.39}$$

Notice that $(\tilde{\Omega}/c^2)(\lambda, t, \bar{t})$ is a solution of (4.3.38), which becomes 0 at $\lambda = 0$. Therefore, by the uniqueness of a solution of (4.3.36), one has

$$\tilde{\Omega}(\lambda, t, \bar{t})/c^2 = \Omega(\lambda, t, \bar{t}). \tag{4.3.40}$$

Hence, (4.3.40) and the definition of $\tilde{\Omega}$ imply (4.3.38). Then, letting $\Psi = \Omega(1, \lambda t, \lambda\bar{t})$, one obtains a solution Ψ of (4.3.34).

From the coefficients of the Taylor expansions of (4.3.36) with respect to λ, one obtains

$$\left.\frac{\Omega}{\lambda^2}\right|_{\lambda=0} = F(x, \xi, 0, 0) > 0. \tag{4.3.41}$$

Hence $\Psi(x^0, \xi^0, 0, 0) \gneq 0$; therefore, one may suppose $\Phi(x^0, \xi^0, 0, 0) > 0$. The homogeneity of $\Phi(x, \xi, f(x, \xi), \bar{f}(x, \xi))$ remains to be proved. Since $f(x, \xi)$ is positively homogeneous in ξ of degree 1/2, it is sufficient to prove

$$\Psi(x, c\xi, c^{1/2}t, c^{1/2}\bar{t}) = \Psi(x, \xi, t, \bar{t}) \qquad \text{for } c > 0. \tag{4.3.42}$$

Then, from (4.3.35), we need to show

$$\Omega(x, \lambda^2\xi, \lambda, t, \bar{t}) = \lambda^2\Omega(x, \xi, 1, t, \bar{t}) \qquad \text{for } \lambda > 0. \tag{4.3.43}$$

Therefore, using (4.3.38), it is sufficient to show

$$\Omega(x, \lambda^2\xi, \lambda, t, \bar{t}) = \Omega(x, \xi, \lambda, \lambda^{-1}t, \lambda^{-1}\bar{t}) \qquad \text{for } \lambda > 0. \tag{4.3.44}$$

By the definition of Θ and $f(x, \xi)$ being positively homogeneous of degree 1/2, (4.3.44) can be proved in the same manner as the proof of (4.3.38). Hence, (C) is proved, which completes the proof of Theorem 4.3.4.

From Theorem 4.3.4, we have the following astonishing result.

Theorem 4.3.5. *Let $P(x, D_x)$ be a microdifferential operator of order m defined in a neighborhood of $(x^0, \sqrt{-1}\xi^0\infty) \in \sqrt{-1}S^*M$. Let the principal symbol $\sigma_m(P)(x, \sqrt{-1}\xi) = f(x, \xi)$ satisfy the conditions*

$$f(x^0, \xi^0) = 0 \tag{4.3.45}$$

and

$$\{f, \bar{f}\}(x^0, \xi^0) \gneqq 0. \tag{4.3.46}$$

Then, the equation $Pu = 0$ *can be transformed into the following equation* $\mathcal{N}$, *defined in a neighborhood of* $(y; \sqrt{-1}\eta) = (0; \sqrt{-1}(0, \ldots, 0, 1))$ *by an invertible (real) quantized contact transformation:*

$$\mathcal{N}: \left(\frac{\partial}{\partial y_1} - \sqrt{-1} y_1 \frac{\partial}{\partial y_n}\right) u = 0. \tag{4.3.47}$$

Proof. By considering $(D_1^2 + \cdots + D_n^2)^{-(2m-1)/4} P(x, D_x)$, one may assume that f is positively homogeneous in ξ of degree 1/2. Furthermore, by considering $\Phi(x, -\sqrt{-1}D_x, f(x, -\sqrt{-1}D_x), \bar{f}(x, -\sqrt{-1}D_x))$, Theorem 4.3.4 implies that one may assume that

$$\frac{1}{2\sqrt{-1}}\{f, \bar{f}\} = 1 \tag{4.3.48}$$

holds, where $f(x, -\sqrt{-1}D_x)$ etc. means $\sum_\alpha a_\alpha(x)(-\sqrt{-1}D_x)^\alpha$ etc. for $f(x, \xi) = \sum_\alpha a_\alpha(x)\xi^\alpha$ etc. Therefore, by a real contact transformation, one may have

$$f = y_1 \eta_n^{1/2} + \sqrt{-1}\eta_1 \eta_n^{-1/2} \tag{4.3.49}$$

in a neighborhood of $(y; \eta) = (0; 0, \ldots, 0, 1)$. Hence, one can assume that the principal symbol of P is $\eta_1 - \sqrt{-1}y_1\eta_n$. Therefore, after a suitable (complex) coordinate transformation, this theorem follows from Theorem 4.3.2.

Notes.

1. $\{f, \bar{f}\} \neq 0$ implies $d_{(x,\xi)}f \not\propto \omega$; i.e. condition (4.3.2) is automatically satisfied.
2. Since Theorem 4.3.2 is a result for a real domain (see Remark 4.3.8), this argument is not rigorous. It is recommended that the unsatisfied reader prove Theorem 4.3.2 for that case. One needs to check the convergence of the infinite series of operators obtained by a successive approximation. See Theorem 2.1.2 and Remark 1 following Theorem 2.1.2, in Chap. II, §2.1, of SKK [1].

Remark 4.3.9. If (4.3.36) in Theorem 4.3.5 is replaced by the condition $\{f, \bar{f}\}(x^0, \xi^0) \lneqq 0$, then the corresponding canonical equation is $\mathcal{N}: (\partial/\partial y_1 + \sqrt{-1}y_1(\partial/\partial y_n))u = 0$ in a neighborhood of $(y; \sqrt{-1}\eta) = (0; \sqrt{-1}(0, \ldots, 0, 1))$.

Remark 4.3.10. Under the assumptions of Theorem 4.3.5, $V = \{f(x, \xi) = 0\}$ and $\bar{V} = \{\bar{f}(x, \xi) = 0\}$ intersect transversally; and $\omega|_{V \cap \bar{V}}$ defines a contact structure on $V \cap \bar{V}$. When $\operatorname{codim}(V \cap \bar{V}) = 2$ and $V \cap \bar{V}$ has the contact structure from $\omega|_{V \cap \bar{V}}$, then one can ask, as an interesting generalization of the above, "What is a canonical equation corresponding to the equation $Pu = 0$?" It is known (Sato-Kawai-Kashiwara [2]), using an argument similar to that given above, that the canonical equation is given by

$$\left(\frac{\partial}{\partial y_1} \pm \sqrt{-1}y_1^k \frac{\partial}{\partial y_n}\right)u = 0. \tag{4.3.50}$$

The nature of the equations of this type was first appreciated in Mizohata [2].

By Theorem 4.3.5, the microlocal study of the solutions of a microdifferential equation is reduced to studying the solutions of a very simple equation, $(\partial/\partial x_1 \pm \sqrt{-1}x_1(\partial/\partial x_n))u = 0$. Next, we will study the structure of the solutions of this special equation by direct computation. The reader should notice that one cannot understand the essence of even such a simple equation without microlocal consideration.

Theorem 4.3.6. *Let* $P(x, D_x) = D_1 - \sqrt{-1}x_1 D_n$, *and let* $Q(x, D_x) = D_1 + \sqrt{-1}x_1 D_n$. *Then there exists a non-zero microlocal operator* $\mathscr{K}$ *defined in a neighborhood of* $(x^0; \sqrt{-1}\xi^0\infty) = (0; \sqrt{-1}(0, 0, \ldots, 0, 1)\infty) \in \sqrt{-1}S^*\mathbf{R}^n$ *such that the sequence*

$$0 \to \mathscr{C} \xrightarrow{Q} \mathscr{C} \xrightarrow{\mathscr{K}} \mathscr{C} \xrightarrow{P} \mathscr{C} \to 0 \tag{4.3.51}$$

is exact.

Hence, in particular, P is solvable in a neighborhood of $(x^0; \sqrt{-1}\xi^0\infty)$, *Q is not solvable, and the image of Q is characterized as the kernel of* $\mathscr{K}$.

Proof. For computational convenience, let us introduce differential operators R and R' defined at $(x, t) \in \mathbf{R}^{n+1}$, where

$$\left.\begin{aligned} R &= P(x, D_x) + \frac{\sqrt{-1}}{2} D_t \\ R' &= P^*(x', D_{x'}) + \frac{\sqrt{-1}}{2} D_t. \end{aligned}\right\} \tag{4.3.52}$$

Here, $P^*(x', D_{x'}) = -\partial/\partial x'_1 + \sqrt{-1}x'_1\, \partial/\partial x'_n$, the conjugate operator of $P(x, D_x)$.

Next, define $\varphi(x, x', t)$ as

$$\varphi = x_n - x'_n + (x_1 + x'_1)t + \frac{\sqrt{-1}}{4}((x_1 - x'_1)^2 + 4t^2)). \tag{4.3.53}$$

Then, by Proposition 2.4.2, $(\varphi + \sqrt{-1}0)^\alpha$ is well defined (see also Example 2.4.3) and, furthermore,

$$\begin{aligned} \mathrm{S.S.}(\varphi + \sqrt{-1}0)^\alpha \subset \{(x, x', t; \sqrt{-1}(\langle \xi, dx\rangle + \langle \xi', dx'\rangle + \langle \tau, dt\rangle)\infty \mid \\ x_1 = x_1', x_n = x_n', t = 0, \\ \xi_1 = \cdots = \xi_{n-1} = \xi_1' = \cdots = \xi_{n-1}' = 0, \\ \xi_n = -\xi_n' = 1, \tau = x_1 + x_1'\} \end{aligned} \tag{4.3.54}$$

holds. Since $R\varphi = 0$ is clearly true, one has

$$R(\varphi + \sqrt{-1}0)^\alpha = 0. \tag{4.3.55}$$

On the other hand, consider the hyperfunction $1/(x + \sqrt{-1}y)$ in Example 2.4.8. Then one has

$$\left(\frac{\partial}{\partial x} + \sqrt{-1}\frac{\partial}{\partial y}\right)\left(\frac{1}{x + \sqrt{-1}y}\right) = 2\pi\delta(x)\delta(y). \tag{4.3.56}$$

Hence, one obtains

$$\begin{aligned} R\left(\frac{(\varphi + \sqrt{-1}0)^\alpha}{t - \frac{\sqrt{-1}}{2}(x_1 - x_1')}\right) &= R\left(\frac{1}{t - \frac{\sqrt{-1}}{2}(x_1 - x_1')}\right)(\varphi + \sqrt{-1}0)^\alpha \\ &= 2\pi\sqrt{-1}\delta(x_1 - x_1')\delta(t)(\varphi + \sqrt{-1}0)^\alpha \\ &= 2\pi\sqrt{-1}\delta(x_1 - x_1')\delta(t)(x_n - x_n' + \sqrt{-1}0)^\alpha. \end{aligned} \tag{4.3.57}$$

Note that, by (4.3.54), $(\varphi + \sqrt{-1}0)^\alpha/(t - (\sqrt{-1}/2)(x_1 - x_1'))$ is well defined as a hyperfunction (see Theorem 3.1.5). Similarly, since $R'\varphi = \sqrt{-1}(x_1 - x_1' + 2\sqrt{-1}t)$, one obtains

$$\begin{aligned} R'&\left(\frac{(\varphi + \sqrt{-1}0)^\alpha}{t - \frac{\sqrt{-1}}{2}(x_1 - x_1')}\right) \\ &= 2\pi\sqrt{-1}\delta(x_1 - x_1')\delta(t)(x_n - x_n' + \sqrt{-1}0)^\alpha - 2\alpha(\varphi + \sqrt{-1}0)^{\alpha-1}. \end{aligned} \tag{4.3.58}$$

Integrating (4.3.57) and (4.3.58) with respect to t gives, respectively,

$$\begin{aligned} P(x, D_x)&\int_{-\infty}^{\infty} \frac{(\varphi + \sqrt{-1}0)^\alpha}{\left(t - \frac{\sqrt{-1}}{2}(x_1 - x_1')\right)}\,dt \\ &= 2\pi\sqrt{-1}\delta(x_1 - x_1')(x_n - x_n' + \sqrt{-1}0)^\alpha \end{aligned} \tag{4.3.59}$$

and

$$P^*(x', D_{x'}) \int_{-\infty}^{\infty} \frac{(\varphi + \sqrt{-1}0)^{\alpha}}{\left(t - \frac{\sqrt{-1}}{2}(x_1 - x_1')\right)} dt$$

$$= 2\pi\sqrt{-1}\delta(x_1 - x_1')(x_n - x_n' + \sqrt{-1}0)^{\alpha} - 2\alpha \int_{-\infty}^{\infty} (\varphi + \sqrt{-1}0)^{\alpha-1}\, dt. \tag{4.3.60}$$

Notice, as an integration of a microfunction,

$$E_{\alpha}(x, x') = \int_{-\infty}^{\infty} \frac{(\varphi + \sqrt{-1}0)^{\alpha}}{\left(t - \frac{\sqrt{-1}}{2}(x_1 - x_1')\right)} dt \tag{4.3.61}$$

is well defined (see Theorem 3.2.1). For the case $\alpha = -1$, one may use the formula

$$\int_{-\infty}^{\infty} \frac{dt}{(a + \sqrt{-1}t)(c + bt + \sqrt{-1}(a^2 + t^2))}$$

$$= 2\pi i \left\{ \frac{Y(a)}{\sqrt{-1}(c - 2\sqrt{-1}ab)} + \frac{Y(\operatorname{Im} \alpha)}{(a + \sqrt{-1}\alpha)\sqrt{-1}(\alpha - \beta)} + \frac{Y(\operatorname{Im} \beta)}{(a + \sqrt{-1}(\beta - \alpha)} \right\},$$

where

$$\begin{cases} \alpha \\ \beta \end{cases} = \frac{\sqrt{-1}b \pm \sqrt{-1}\sqrt{4a^2 + b^2 - 4\sqrt{-1}c}}{2}.$$

Furthermore, a well-known formula

$$\Gamma(-\alpha) \int_{-\infty}^{\infty} (t^2 + A)^{\alpha}\, dt = \sqrt{\pi}\Gamma(-\alpha - \tfrac{1}{2})A^{\alpha + 1/2}, \qquad \operatorname{Im} A > 0 \tag{4.3.62}$$

provides

$$\int_{-\infty}^{\infty} (\varphi + \sqrt{-1}0)^{\alpha-1}\, dt$$

$$= e^{-(\pi/4)\sqrt{-1}} \frac{\sqrt{\pi}\Gamma(-\alpha + \frac{1}{2})}{\Gamma(-\alpha + 1)} \left(x_n - x_n' + \frac{\sqrt{-1}}{2}(x_1^2 + x_1'^2) + \sqrt{-1}0\right)^{\alpha - 1/2} \tag{4.3.63}$$

(cf. Example 3.2.8). Let

$$E(x, x') \underset{\text{def}}{=} \frac{1}{4\pi^2} E_{-1}(x, x') \prod_{j=2}^{n-1} \delta(x_j - x_j'). \tag{4.3.64}$$

Then we can summarize what has been obtained as

$$P(x, D_x)E(x, x') = \delta(x - x') \tag{4.3.65}$$

$$P^*(x', D_{x'})E(x, x') = \delta(x - x') - \frac{e^{-(\pi/4)\sqrt{-1}}}{4\pi}\left(x_n - x_n' + \frac{\sqrt{-1}}{2}(x_1^2 + x_1'^2) + \sqrt{-1}0\right)^{-3/2} \prod_{j=2}^{n-1} \delta(x_j - x_j'), \tag{4.3.66}$$

where equalities hold in a neighborhood of

$$(x, x'; \sqrt{-1}(\langle \xi, dx \rangle + \langle \xi', dx \rangle)\infty) = (0, 0; \sqrt{-1}(dx_n - dx_n')\infty) \in \sqrt{-1}S^*(\mathbf{R}^n \times \mathbf{R}^n).$$

If one lets $\mathscr{K}$ and $\mathscr{H}$ be the microlocal operators having $-(e^{-(\pi/4)\sqrt{-1}}/4\pi) \cdot (x_n - x_n' + (\sqrt{-1}/2)(x_1^2 + x_1'^2) + \sqrt{-1}0)^{-3/2} \prod_{j=2}^{n-1} \delta(x_j - x_j')$ and $E(x, x')$ as kernel functions, respectively, then (4.3.65) and (4.3.66) can be phrased in terms of equations of microlocal operators as, in the neighborhood of $(x; \sqrt{-1}\langle \xi, dx \rangle \infty) = (0; \sqrt{-1}\, dx_n \infty)$,

$$P\mathscr{H} = 1 \tag{4.3.65$'$}$$

and

$$\mathscr{H}P = 1 - \mathscr{K}. \tag{4.3.66$'$}$$

As for $Q(x, D_x)$, let $R = Q^*(x', D_{x'}) - (\sqrt{-1}D_t/2)$, and let

$$R' = Q(x, D_x) - (\sqrt{-1}D_t/2);$$

then one obtains $R\varphi = 0$ and $R'\varphi = \sqrt{-1}(x_1 - x_1' - 2\sqrt{-1})$. In the same manner as above,

$$Q\mathscr{F} = 1 - \mathscr{K}, \tag{4.3.67}$$

and

$$\mathscr{F}Q = 1 \tag{4.3.68}$$

holds, where $\mathscr{F}$ is the microlocal operator defined in a neighborhood of $(0; \sqrt{-1}\, dx_n \infty)$ with the kernel function

$$\frac{1}{4\pi^2}\left(\int_{-\infty}^{\infty} \frac{1}{\left(t + \frac{\sqrt{-1}}{2}(x_1 - x_1')\right)(\varphi + \sqrt{-1}0)}\, dt\right) \prod_{j=2}^{n-1} \delta(x_j - x_j').$$

Hence, for example, one has $Q = Q(\mathscr{F}Q) = (Q\mathscr{F})Q = (1 - \mathscr{K})Q = Q - \mathscr{K}Q$ from (4.3.67), (4.3.68), and the corollary following Definition 3.4.1, concluding $\mathscr{K}Q = 0$. Suppose $\mathscr{K}f = 0$ holds; then (4.3.67) implies $Q(\mathscr{F}f) =$

$f - \mathscr{K}f = f$. Hence $\operatorname{Ker}\mathscr{K} \subset \operatorname{Im} Q$; therefore, $\operatorname{Ker}\mathscr{K} = \operatorname{Im} Q$ holds. Similarly, one obtains $\operatorname{Im}\mathscr{K} = \operatorname{Ker} P$ from (4.3.65′) and (4.3.66′). From (4.3.68) and (4.3.65′), respectively, Q is a monomorphism and P is an epimorphism. This completes the proof of the exactness of (4.3.51).

We will obtain the following decisive theorem by Theorem 4.3.5 (and Remark 4.3.9) and by the above theorem.

Theorem 4.3.7. *Let $P(x, D_x)$ be a microdifferential operator of order m, which is defined in a neighborhood of $(x^0, \sqrt{-1}\xi^0\infty) \in \sqrt{-1}S^*M$, and let $f(x, \xi) = \sigma_m(P)(x, \sqrt{-1}\xi)$. Then one has the following statements:*

(I) *If $f(x^0, \xi^0) = 0$ and $\{f, \bar{f}\}(x^0, \xi^0) \gneqq 0$ hold, then P is epimorphic in a neighborhood of $(x^0, \sqrt{-1}\xi^0\infty)$ and* Ker *P is equal to the image of a microlocal operator $\mathscr{K}$.*

(II) *If $f(x^0, \xi^0) = 0$ and $\{f, \bar{f}\}(x^0, \xi^0) \lneqq 0$ hold, then P is monomorphic in a neighborhood of $(x^0, \sqrt{-1}\xi^0\infty)$, but not epimorphic, and the image of P is equal to the kernel of a microlocal operator $\mathscr{K}$. That is, for the equation $Pu = g$ to be solvable, $\mathscr{K}g = 0$ must hold.*

Remark 4.3.11. Since H. Lewy's sensational discovery in 1957 of the equation

$$\left(\frac{1}{2}\left(\frac{\partial}{\partial x_1} + \sqrt{-1}\frac{\partial}{\partial x_2}\right) - (x_1 + \sqrt{-1}x_2)\frac{\partial}{\partial x_3}\right)u = g$$

without (local) solutions (see Lewy [1]), the solvability of a linear partial differential equation has been a central topic. Now, for a "generic" equation, an answer has been obtained in an ideal form, as in the above theorem. We have treated the geometrically "generic" case. We need to consider a system of equations that does not satisfy the condition (4.3.3) in order to treat the truly "generic" case. With operators of infinite order, an argument similar to that above enables one to handle the case (see Remark 4.3.3). Theorem 4.3.7, together with Theorem 4.3.3, has shown just how useful microlocal analysis is for the study of linear partial differential equations. Even though it is possible to adapt the above equation of Lewy for a canonical form, we choose the canonical form in (4.3.47), which clearly suggests a connection with the contact structure and the following theorem. A microdifferential equation, having (4.3.47) as the canonical form, i.e. the equation satisfying (4.3.46) or with the opposite inequality, is said to be the Lewy-Mizohata type.

As we mentioned in Remark 4.3.10, for the characteristic variety V of P, $V \cap \sqrt{-1}S^*M = V \cap \bar{V}$ has a contact structure. If this fact links Ker P or Coker Q, that would be very interesting. This "expected harmony" beautifully exists, as follows.

Theorem 4.3.8. *Let P, Q, and $\mathscr{K}$ be as in Theorem 4.3.6. Let $M = \mathbf{R}^n$, $N = \{x \in M \mid x_1 = 0\}$, and $Z = \{(x, \sqrt{-1}\xi\infty) \in \sqrt{-1}S^*M \mid x_1 = 0,\ \xi_1 = 0 \text{ and } \xi_n > 0\}$. Z is identified with $\{(x', \sqrt{-1}\xi'\infty) \in \sqrt{-1}S^*N \mid \xi_n > 0\}$ via the projection $N \underset{M}{\times} \sqrt{-1}S^*M - \sqrt{-1}S_N^*M \to \sqrt{-1}S^*N$, where $x' = (x_2, \ldots, x_n)$ and $\xi' = (\xi_2, \ldots, \xi_n)$. Then, one can define sheaf homomorphisms Φ and Ψ over $\Omega = \{(x, \sqrt{-1}\xi\infty) \in \sqrt{-1}S^*M \mid \xi_n > 0\}$ such that*

$$\begin{array}{ccc} \Phi: \mathscr{C}_N & \longrightarrow & \mathscr{C}_M \\ \Cup & & \Cup \\ u(x_2, \ldots, x_n) & \mapsto & -\dfrac{1}{2\pi\sqrt{-1}} \displaystyle\int_{-\infty}^{\infty} \dfrac{u(x_2, x_3, \ldots, x_n')}{\left(x_n - x_n' + \dfrac{\sqrt{-1}}{2} x_1^2 + \sqrt{-1}0\right)} dx_n' \end{array} \tag{4.3.69}$$

and

$$\begin{array}{ccc} \Psi: \mathscr{C}_M & \longrightarrow & \mathscr{C}_N \\ \Cup & & \Cup \\ v(x_1, \ldots, x_n) & \longmapsto & \mathscr{K}v|_{x_1=0}; \end{array} \tag{4.3.70}$$

and then

$$\Psi\Phi = 1: \mathscr{C}_N \to \mathscr{C}_N \tag{4.3.71}$$

and

$$\Phi\Psi = \mathscr{K}: \mathscr{C}_M \to \mathscr{C}_M \tag{4.3.72}$$

hold. Therefore, in particular one has an isomorphism

$$\Psi: \mathrm{Ker}_{\mathscr{C}_M} P \xrightarrow{\sim} \mathscr{C}_N. \tag{4.3.73}$$

Similarly, sheaf homomorphisms $\tilde{\Phi}$ and $\tilde{\Psi}$ can be defined such that

$$\begin{array}{ccc} \tilde{\Phi}: \mathscr{C}_N & \longrightarrow & \mathscr{C}_M \\ \Cup & & \Cup \\ u(x_2, \ldots, x_n) & \longmapsto & \mathscr{K}(u(x_2, \ldots, x_n)\delta(x_1)), \end{array} \tag{4.3.74}$$

$$\begin{array}{ccc} \tilde{\Psi}: \mathscr{C}_M & \longrightarrow & \mathscr{C}_N \\ \Cup & & \Cup \\ v(x_1, \ldots, x_n) & \mapsto & -\dfrac{1}{2\pi\sqrt{-1}} \displaystyle\int \dfrac{v(x_1', x_2, x_3, \ldots, x_n')}{\left(x_n - x_n' + \dfrac{\sqrt{-1}}{2} x_1'^2 + \sqrt{-1}0\right)} dx_1'\, dx_n', \end{array} \tag{4.3.75}$$

$$\tilde{\Psi}\tilde{\Phi} = 1: \mathscr{C}_N \to \mathscr{C}_N, \tag{4.3.76}$$

and

$$\tilde{\Phi}\tilde{\Psi} = \mathscr{K} : \mathscr{C}_M \to \mathscr{C}_M. \tag{4.3.77}$$

Furthermore, one has the isomorphism

$$\tilde{\Phi} : \mathscr{C}_N \xrightarrow{\sim} \operatorname{Coker}_{\mathscr{C}_M} Q. \tag{4.3.78}$$

Proof. Since $\mathscr{K}$ is a microlocal operator, it is almost obvious that Φ, Ψ, $\tilde{\Phi}$, $\tilde{\Psi}$ are sheaf homomorphisms. We will prove (4.3.71). By the definitions,

$$\Phi\Psi(u(x_2, \ldots, x_n))$$

$$= -\frac{e^{-(\pi/4)\sqrt{-1}}}{4\pi} \int \frac{\prod_{j=2}^{n-1} \delta(x_j - x_j')}{\left(x_n - x_n' + \frac{\sqrt{-1}}{2} x_1'^2 + \sqrt{-1}0\right)^{3/2}}$$

$$\cdot \int \frac{u(x_2', x_3', \ldots, x_n'')}{(-2\pi\sqrt{-1})\left(x_n' - x_n'' + \frac{\sqrt{-1}}{2} x_1'^2 + \sqrt{-1}0\right)} dx_n'' \prod_{j=1}^{n} dx_j'. \tag{4.3.79}$$

On the other hand, by the residue computation and (4.3.62), one obtains

$$\int \frac{dx_1'\, dx_n'}{\left(x_n - x_n' + \frac{\sqrt{-1}}{2} x_1'^2 + \sqrt{-1}0\right)^{3/2}\left(x_n' - x_n'' + \frac{\sqrt{-1}}{2} x_1'^2 + \sqrt{-1}0\right)}$$

$$= -2\pi\sqrt{-1} \int \frac{dx_1'}{(x_n - x_n'' + \sqrt{-1}x_1'^2 + \sqrt{-1}0)^{3/2}}$$

$$= -2\pi\sqrt{-1} \frac{\sqrt{\pi}\,\Gamma(1)}{e^{(\pi/4)\sqrt{-1}}\Gamma(3/2)} \cdot \frac{1}{(x_n - x_n'' + \sqrt{-1}0)}$$

$$= -4e^{(\pi/4)\sqrt{-1}} \frac{1}{(x_n - x_n'' + \sqrt{-1}0)}. \tag{4.3.80}$$

When this is substituted into (4.3.79), one obtains $\Psi\Phi = 1$ on Z. We can prove (4.3.72), (4.3.76), and (4.3.77) in the same manner. Next, we will prove (4.3.73). First, notice that $\mathscr{K}^2 = (\Phi\Psi)(\Phi\Psi) = \Phi(\Psi\Phi)\Psi = \Phi\Psi = \mathscr{K}$ by (4.3.71) and (4.3.72). On the other hand, from the exact sequence (4.3.51), for $v \in \operatorname{Ker} P$ there exists $g \in \mathscr{C}_M$ such that $v = \mathscr{K}g$ holds. Therefore, one obtains $\Psi v = \mathscr{K}v|_{x_1=0} = \mathscr{K}^2 g|_{x_1=0} = \mathscr{K}g|_{x_1=0} = v|_{x_1=0}$, which is the isomorphism in (4.3.73). The isomorphism in (4.3.78) can be seen as follows. By (4.3.67), one has

$$\begin{aligned}\tilde{\Phi}u &= \delta(x_1)u(x_2, \ldots, x_n) - Q\mathscr{F}(\delta(x_1)u(x_2, \ldots, x_n))\\ &\equiv \delta(x_1)u(x_2, \ldots, x_n) \bmod Q\mathscr{C}_M.\end{aligned}$$

The flabbiness of the microfunction sheaf (Theorem 3.7.1), Theorem 4.3.7, and Theorem 4.3.8 imply the following theorem.

Theorem 4.3.9. *Notations being the same as in Theorem* 4.3.7, *sheaves* Ker P *in* (I) *and* Coker P *in* (II) *are flabby over* $V \cap \bar{V}\ (= V \cap \sqrt{-1}S^*N)$.

We have restricted our arguments to the case of a single equation. Clever use of Theorem 4.3.1 (regarded as a theorem for a complex domain) enables one to extend the result on the Lewy-Mizohata-type equation to the one on overdetermined systems. Theorem 4.3.8 is very useful for the generalization. We will touch this topic only briefly. Consult §2.3, Chap. III of SKK [1] for details.

Definition 4.3.2. *Let M be a real analytic manifold. Let an involutory submanifold V in a complex neighborhood of* $(x^0, \sqrt{-1}\langle\xi^0, dx\rangle\infty) \in \sqrt{-1}S^*M$ *be written as* $\{(x, \sqrt{-1}\langle\xi, dx\rangle\infty \,|\, p_1(x, \sqrt{-1}\xi) = \cdots = p_d(x, \sqrt{-1}\xi) = 0\}$. *Then the Hermitian matrix*

$$L(x, \xi) = \left(\frac{1}{2\sqrt{-1}}\{p_j(x, \xi), \bar{p}_k(x, \xi)\}\right)_{1 \leq j,k \leq d} \tag{4.3.81}$$

is said to be the "generalized Levi form" of V.

Remark 4.3.12. The number of positive eigenvalues and the number of negative eigenvalues of $L(x, \xi)$ are independent of the choice of defining functions of V. These numbers are also invariant under a real contact transformation. L is called the "generalized Levi form" in an analogy with the terminology used in function theory (e.g. see Hitotumatu [1]). These two notions, however, are closely related (Kashiwara and Kawai [1], Example 3).

Using this notion of a generalized Levi form, the statement on the canonical case of an overdetermined system can be phrased as follows. Let $\mathcal{M}$ be an $\mathscr{E}$-Module defined in a neighborhood of $(x^0, \sqrt{-1}\langle\xi^0, dx\rangle\infty) \in \sqrt{-1}S^*M$ such that $\mathcal{M}$ satisfies (4.3.1), (4.3.2), and (4.3.3). Further, assume that the generalized Levi form of $V(J)$ has p positive eigenvalues and $q\ (= d - p)$ negative eigenvalues at $(x^0, \sqrt{-1}\xi^0)$. Then, $\mathcal{M}$ can be transformed by a quantized contact transformation into the system of equations $\mathcal{N}_p$ (called the $(p, d - p)$-Lewy-Mizohata system):

$$\mathcal{N}_p\colon \begin{cases} \left(\dfrac{\partial}{\partial y_j} - \sqrt{-1}y_j\dfrac{\partial}{\partial y_n}\right)u = 0 & \text{for } j = 1, \ldots, p \\ \left(\dfrac{\partial}{\partial y_j} + \sqrt{-1}y_j\dfrac{\partial}{\partial y_n}\right)u = 0 & \text{for } j = p + 1, \ldots, p + q = d \end{cases} \tag{4.3.82}$$

in a neighborhood of $(y^0, \sqrt{-1}\langle\eta^0, dy\rangle\infty) = (0; \sqrt{-1}\,dy_n\infty)$.

It is an exercise in homological algebra to study the structure of microfunction solutions of the system of equations $\mathcal{N}_p$, if Theorems 4.3.6 and 4.3.8 are employed.

The system (4.3.82) is nothing but the equations that the system $\tilde{\mathcal{N}}$ over $\mathbf{R}^{n+1}$ satisfies on the hypersurface

$$S_p = \left\{ y \in \mathbf{R}^{n+1} \,\middle|\, y_{n+1} - \tfrac{1}{2}\left(\sum_{j=1}^{p} y_j^2 - \sum_{j=p+1}^{d} y_j^2\right) = 0 \right\}$$

in $\mathbf{R}^{n+1}$, where

$$\tilde{\mathcal{N}}: \begin{cases} \dfrac{\partial}{\partial y_j} u = 0 & \text{for } j = 1, \ldots, d \ (d \leqq n-1) \\ \left(\dfrac{\partial}{\partial y_n} + \sqrt{-1}\,\dfrac{\partial}{\partial y_{n+1}}\right) u = 0. \end{cases} \tag{4.3.83}$$

One would expect some links between the solutions of $\mathcal{N}_p$ and the solutions of $\tilde{\mathcal{N}}$, in the sense that one may study the structure of solutions of $\tilde{\mathcal{N}}_p$ from that of $\tilde{\mathcal{N}}$ and vice versa. When the above question is pursued far enough, a question surfaces: "What is the boundary value problem?" A fairly satisfying answer has been obtained in Kashiwara and Kawai [1] and [2], in which the interested reader can find the details.

Let us conclude this book with one final comment. We have obtained the canonical form of each $\mathcal{M}$ in Theorem 4.3.1, providing that the characteristic variety V is real, and in Theorem 4.3.5, providing that $V \cap \bar{V}$ intersects transversally and $V \cap \bar{V}$ has a contact structure, respectively. It is also of theoretical importance to find the canonical form in the rather degenerate case where $V \cap \bar{V}$ is a non-singular involutory submanifold. In this case (assuming $\omega|_{V \cap \bar{V}} \neq 0$) the canonical form is known to be the following partial Cauchy-Riemann system

$$\mathcal{N}: \frac{1}{2}\left(\frac{\partial}{\partial y_{2j-1}} + \sqrt{-1}\,\frac{\partial}{\partial y_{2j}}\right) u = 0 \qquad \text{for } j = 1, \ldots, d, \tag{4.3.84}$$

where $\mathcal{N}$ is considered over a neighborhood of $(y, \sqrt{-1}\langle \eta, dy \rangle \infty) = (0, \sqrt{-1}\, dy_n \infty)$.

Furthermore, a "generic" system $\mathcal{M}$ is isomorphic to the mixture system of the (partial) de Rham system, the Lewy-Mizohata system, and the (partial) Cauchy-Riemann system (SKK [1], Chap. III, §2.4). This result is most fundamental and most exquisite in the local theory of linear partial differential equations.

References

Only those works referred to in the text are listed here. Hence, this bibliography does not include all related works. Since life is finite, one may not need to consult the referenced work unless one has a particular interest.

Atiyah, M.F., R. Bott, and L. Gårding:

[1] "Lacunas for hyperbolic differential operators with constant coefficients I." *Acta Math. 124* (1970), 109–189.

———: [2] "——— II." Ibid. *131* (1973), 145–206.

Bony, J.M.:

[1] "Une extension du théorème de Holmgren sur l'unicité du problème de Cauchy." *C.R. Acad. Sci. Paris, Sér. A 268* (1969), 1103–1106.

———: [2] *Equivalence des diverses notions de spectre singulier analytique.* Sém. Goulaouic-Schwartz, Ecole Polytechnique Expose, 3, Ecole Polytechnique, 1976–1977.

Bony, J.M. and P. Schapira:

[1] *Solutions hyperfonctions du problème de Cauchy.* Lecture Notes in Math. 287, Springer, 1973, pp. 82–98.

Boutet de Monvel, L. and P. Krée:

[1] "Pseudo-differential operators and Gevrey classes." *Ann. Inst. Fourier 17-1* (1967), 295–323.

Bruhat, F. and H. Whitney:

[1] "Quelques propriétés fondamentales des ensembles analytiques-reéls." *Comm. Math. Helv. 33* (1959), 132–160.

Courant, R. and D. Hilbert:

[1] *Methods of Mathematical Physics*, vols. 1 and 2. Interscience, 1953 and 1962 (translation of *Methoden der Mathematischen Physik*, Springer [1924, 1937]). Vol. 2 has been rewritten entirely, but the first edition has its own interest.

Eden, R.J., P.V. Landshoff, D.I. Olive, and J.C. Polkinghorne:

[1] *The Analytic S-Matrix.* Cambridge University Press, 1966.

Egorov, Yu. V.:

[1] "On canonical transformations of pseudo-differential operators." *Uspehi Mat. Nauk 24* (1969), 235–236. In Russian.

Ehrenpreis, L.:

[1] *Fourier Analysis in Several Complex Variables.* Wiley-Interscience, 1970.

Gårding, L.:
[1] "Linear hyperbolic partial differential equations with constant coefficients." *Acta. Math. 85* (1950), 1–62.
Gel'fand, I.M. and G.E. Shilov:
[1] *Generalized Functions*, vol. 1. Academic Press, 1964. (Translation of the Russian original (1959).
Gel'fand, I.M., M.I. Graev, and N. Ya. Vilenkin:
[1] *Generalized Functions*, vol. 5. Academic Press, 1966. The original appeared in 1962.
Grauert, H.:
[1] "On Levi's problem and the imbedding of real-analytic manifolds." *Ann. of Math. 68* (1958), 460–472.
Grothendieck, A.:
[1] *Local Cohomology*. Lecture Notes in Math. 41, Springer, 1967.
Hamada, Y.:
[1] "The singularities of the solutions of the Cauchy problem." *Publ. RIMS, Kyoto Univ. 5* (1969), 21–40.
Hilbert, D.:
[1] *Mathematische Probleme*. Göttinger Nachrichten, 1900, pp. 253–297.
Hörmander, L.:
[1] *Linear Partial Differential Operators*. Springer, 1963.
———: [2] "Uniqueness theorems and wave front sets for solutions of linear differential equations with analytic coefficients." *Comm. Pure Appl. Math. 24* (1971), 671–704.
———: [3] "Fourier integral operators I," *Acta Math. 127* (1971), 79–183.
Hitotumatu, S.:
[1] *Tahensu kaiseki kansuron* [Theory of Analytic Functions of Several Complex Variables]. Baihukan, 1960.
Iagolnitzer, D.:
[1] "Analytic structure of distributions and essential support theory." In *Structural Analysis of Collision Amplitudes*. North Holland, 1976, pp. 295–358.
Iagolnitzer, D. and H.P. Stapp:
[1] "Macroscopic causality and physical region analyticity in *S*-matrix theory." *Commun. Math. Phys. 14* (1969), 15–55.
John, F.:
[1] "The fundamental solution of linear elliptic differential equations with analytic coefficients." *Comm. Pure Appl. Math. 3* (1950), 273–304.
———: [2] *Plane Waves and Spherical Means Applied to Partial Differential Equations*. Interscience, 1955.
———: [3] "Continuous dependence on data for solutions of partial differential equations with a prescribed bound." *Comm. Pure Appl. Math. 13* (1960), 551–585.
Kashiwara, M.:
[1] "On the maximally overdetermined system of linear differential equations I." *Publ. RIMS, Kyoto Univ. 10* (1975), 563–579.
———: [2] Unpublished. Consult, also, Guillemin, V.W., D. Quillen, and S. Sternberg: "Integrability of characteristics." *Comm. Pure Appl. Math. 23* (1970), 39–77.

Kashiwara, M. and T. Kawai:
[1] "On the boundary value problem for elliptic systems of linear differential equations I." *Proc. Japan Acad. 48* (1972), 712–715.
———: [2] "——— II." Ibid. *49* (1973), 164–168.
———: [3] "Micro-hyperbolic pseudo-differential operators I." *J. Math. Soc. Japan 27* (1975), 359–404.
———: [4] "Finiteness theorem for holonomic systems of micro-differential equations." *Proc. Japan Acad. 52* (1976), 341–343.
Kawai, T.:
[1] "On the theory of Fourier hyperfunctions and its application to partial differential equations with constant coefficients." *J. Fac. Sci. Univ. Tokyo 17* (1970), 467–517.
———: [2] "Construction of local elementary solutions for linear partial differential operators with real analytic coefficients: (I) The case with real principal symbols." *Publ. RIMS, Kyoto Univ. 7* (1971), 363–397.
Kawai, T. and H.P. Stapp:
[1] *Micro-local Study of the S-Matrix Singularity Structure.* Lecture Notes in Phys. 39, Springer, 1975, pp. 38–48.
Komatsu, H.:
[1] "Resolution by hyperfunctions of sheaves of solutions of differential equations with constant coefficients." *Math. Ann. 176* (1968), 77–86.
———: [2] "A local version of Bochner's tube theorem." *J. Fac. Sci. Univ. Tokyo, Sect. IA, 19* (1972), 201–214.
Lax, P.D.:
[1] "Asymptotic solutions of oscillatory initial value problems." *Duke Math. J. 24* (1957), 627–646.
Leray, J.:
[1] "Problème de Cauchy IV." *Bull. Soc. Math. France 90* (1962), 39–156.
Lewy, H.:
[1] "An example of a smooth linear partial differential equation without solutions." *Ann. of Math. 66* (1957), 155–158.
Malgrange, B.:
[1] "Faisceux sur des variétés analytiques réelles." *Bull. Soc. Math. France 83* (1955), 231–237.
Maslov, V.P.:
[1] *Theorie des perturbations et méthodes asymptotiques.* Gauthier-Villars, 1972. The original appeared in Russian in 1965.
Matsushima, Y.:
[1] *Differentiable Manifolds.* Marcel Dekker, 1972. The original was published in Japanese by Shōkabō (1965).
Mizohata, S.:
[1] "Some remarks on the Cauchy problem." *J. Math. Kyoto Univ. 1* (1961), 109–127.
———: [2] "Solutions nulles et solutions non-analytiques." Ibid. *1* (1962), 271–302.

Morimoto, M.:

[1] *Edge of the Wedge Theorem and Hyperfunctions.* Lecture Notes in Math. 287, Springer, 1973, pp. 41–81.

———: [2] Unpublished. See, also, Morimoto, M.: "Support of hyperfunction and singular support (Sato's conjecture and sheaf $\mathscr{C}_{N|X}$)." *Kyoto daigaku sūrikaiseki-kenkyūsho-kokyūroku 168* (1972), 28–59. In Japanese.

Nakanishi, N.:

[1] *Ba no ryoshiron* [Quantum Field Theory]. Baihukan, 1975. In Japanese.

Oshima, T. and H. Komatsu

[1]: *Ikkai henbibun hōteishiki* [Partial Differential Equations of First Order]. Iwanami koza kisosūgaku, kaisekigaku (II) (iii). Iwanami shoten, 1977. In Japanese.

Petrowsky, I.G.:

[1] "On the diffusion of waves and the lacunas for hyperbolic equations." *Mat. Sb. 17* (1945), 289–370.

Sato, M.:

[1] "Theory of hyperfunctions, I. II." *J. Fac. Sci. Univ. Tokyo 8* (1959–1960), 139–193, 387–437.

———: [2] *Recent Development in Hyperfunction Theory and Its Applications to Physics.* Lecture Notes in Physics 39, Springer, 1975, pp. 13–29.

Sato, M., T. Kawai, and M. Kashiwara:

[1] (referred to as SKK [1]) *Microfunctions and Pseudo-differential Equations.* Lecture Notes in Math. 287, Springer, 1973, pp. 265–529.

———: [2] "On the structure of single linear pseudo-differential equations." *Proc. Japan Acad. 48* (1972), 643–646.

Schwartz, L.:

[1] *Théorie des distributions.* Hermann, 1950–1951.

Suzuki, F.:

[1] "On the global existence of holomorphic solutions of the equation $\partial u/\partial x_1 = f$." *Sci. Rep. Tokyo Kyoiku Daigaku, Sect. A., 11* (1972), 253–258.

Vladimirov, V.S.:

[1] *Methods of the Theory of Functions of Many Complex Variables.* The MIT Press, 1966. The original appeared in Russian in 1964.

Yamanouchi, T.:

[1] *Ippan rikigaku* [General Dynamics], 3rd ed. Iwanami shoten, 1965. In Japanese.

Yosida, K.:

[1] *Bibunhōteishiki no kaihō* [Methods of Solving Differential Equations], 2nd ed. Iwanami shoten, 1978. In Japanese.

Index

JAPANESE INDUSTRIAL HISTORY

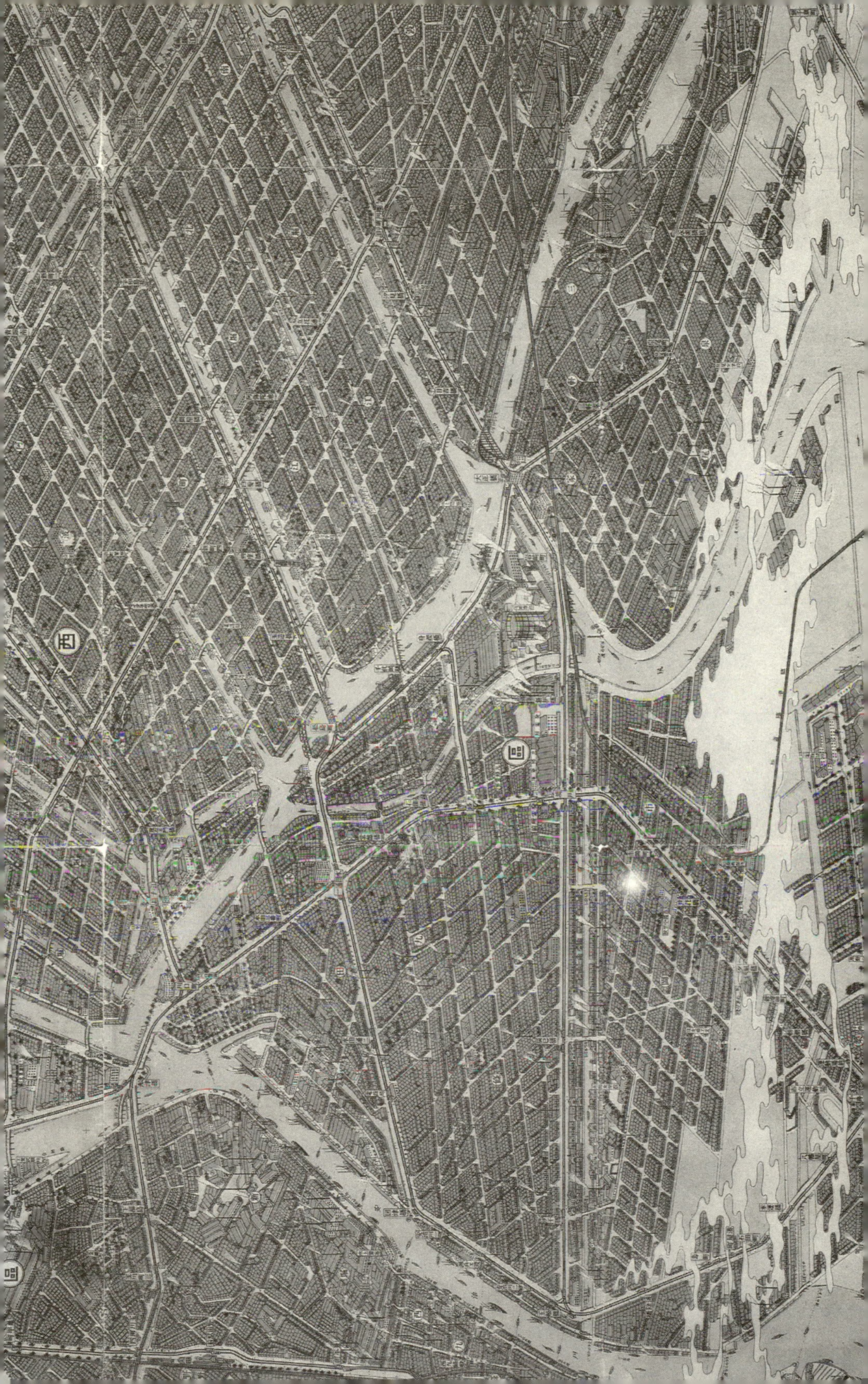

JAPANESE INDUSTRIAL HISTORY

Technology, Urbanization, and Economic Growth

Carl Mosk

M.E. Sharpe
Armonk, New York
London, England

Frontispiece:
Osaka in 1924: The Mock Painted Picture of the Great Osaka [Portion]
Courtesy of Osaka City Museum.

Library of Congress Cataloging-in-Publication Data

Mosk, Carl.
Japanese industrial history : technology, urbanization, and economic growth / Carl Mosk
p. cm.
Includes bibliographical references and index.
ISBN 0-7656-0700-X (alk. paper)
1. Industries—Japan—History. 2. Corporations—Japan—Growth—History. 3. Industrialization—Japan—History. 4. Urbanization—Japan—History. 5. Technology—Japan—History. 6. Japan—Commerce—History. 7. Japan—Economic conditions. 8. Japan—Social conditions. I. Title.

HC462 .M637 2001
338.0952—dc21 00-059511

Printed in the United States of America

The paper used in this publication meets the minimum requirements of
American National Standard for Information Sciences
Permanence of Paper for Printed Library Materials,
ANSI Z 39.48-1984.

∞

BM (c) 10 9 8 7 6 5 4 3 2 1

For Donna

Contents

List of Maps, Figures, and Tables

Maps

Figures

Tables

Preface

This book combines economic history and economic geography. A word is in order about my approach to both subjects, and my reason for combining them in this study.

Economic history is both history and economics. As history, economic history attempts to put us in the past, it attempts to explain how and why events occurred in the past, and it attempts to put the past in perspective. Some history also tries to draw inferences about the present or future. As economics, economic history attempts to shed light on the validity of theories by providing empirical evidence with which the theory can be critically examined. Thus, insofar as theory is useful for prediction, economic history can help us make predictions. It is my belief that this study concerning Japanese industrial history does provide a general framework for interpreting Japan's economic development that can be also usefully applied to other countries. It is possible that my thesis about infrastructure-driven growth in Japan may be applicable to other countries. Perhaps it may be even useful in deriving policies for economic development in these nations.

Unfortunately, most economics and economic history seems to be cast in a geographic vacuum. One has no sense of where industrialization took place or why it took place in certain areas. This is true of much of the literature on Japan's economic history, for instance. This inattention to geography is unfortunate, in my opinion. Industrialization tends to occur in some regions of a country and not in others. This matters, not only because concentration generates scale economies, but also because ecology shapes the location of economic activity, thereby affecting the course of economic history. With my focus in this book upon Osaka and the Tōkaidō industrial belt, I attempt to systematically address the question of how and why geographic concentration mattered in Japan's economic development.

Thus, this book asks much of the reader. It asks the reader to be economist, historian, urban specialist, and geographer. I can only hope that the effort elicited of the reader will provide some measure of intellectual sustenance and will stimulate discussion about the importance of cities, technology, and industrialization in both developing and currently developed economies.

Acknowledgments

Some books have a long gestation, and some have a short gestation. This book has a very long gestation.

Growing up in the hills of Berkeley, I would often spend lazy afternoons lying in the grass and gazing out over a magnificent panorama that included San Francisco, the Golden Gate Bridge, and the rolling hills of Marin County. I even started a novel that was to revolve around the history of a great industrial seaport. But I put it aside. Little did I know that many decades later, I would become a student of San Francisco's great sister city in Japan, the industrial seaport Osaka.

Then, as a graduate student, I had the privilege of working with Henry Rosovsky at Harvard University and with a group of scholars at the Institute of Economic Research at Hitotsubashi University, especially Umemura Mataji. From these scholars I learned much about Japan's economic development and about the importance of long swings. Indeed, Ohkawa Kazushi, Umemura Mataji, and Minami Ryōshin have published widely on long-swing development in Japan. I imbibed this knowledge, and for many years used it as a key concept around which to organize my lectures about Japanese economic development. Now, finally, in my own way, I have at last come to terms with the long swing in Japanese economic history.

During the late 1990s, I became increasingly interested in the relationship between urbanization and industrialization. During the summer of 1997, with the generous support of Tom Havens, I started doing research on the topic at the East Asian Library at Berkeley. I also was allowed to use the City and Regional Planning Library at the university. I am grateful to the staffs at these two libraries for assisting me in my research efforts.

Taking up sabbatical leave in the fall of 1997, I was fortunate enough to have the opportunity to become a Visiting Scholar at the International Research Center for Japanese Studies in Kyoto. I am grateful to Hayami Akira for sponsoring my stay at the center. Arriving in the early fall, I decided to under-

take a study of the economic and demographic development of a major Japanese city. After I weighed the pros and cons of choosing either nearby Osaka or Kyoto, I elected to choose the former. I believe that I made the right choice.

The staff at the International Research Center for Japanese Studies provided wonderful service, securing books about Osaka from libraries all over the country. I am especially grateful to Ayabe Fusako, Danmoto Sachiko, Imai Yoshiko, Kawakita Emiko, Nakamura Yoshiyuki, Suezawa Reikko, Tachiuri Toshmasa, Tsuruta Sachi, and Yoshioka Yōko, who helped me locate and order books. In the Osaka area, I also secured materials at the Osaka City Hall, the Osaka Prefecture Library, and the Osaka City Museum. I am grateful to their staffs for assisting me in my studies. Finally, I made a very fruitful trip to Tokyo, where I used the Bunken Centaa at the Institute of Economic Research at Hitotsubashi University. I am very grateful to my friend Saito Osamu of the Institute, who arranged for me to use the materials and who provided me with much useful advice.

In Victoria, I have benefited much from discussions with my colleagues in the Economics Department. Conversations with Ian King, Merwan Engineer, and David Sconnes clarified my thinking about endogenous growth theory, and discussions with Kenneth Avio and Gerald Walters helped me crystallize my thinking about economic evolution. Finally, wide-ranging discussions with Yehuda Kotowitz stimulated my thinking on an extremely wide range of issues in economics too numerous to recount here. I also want to express my gratitude to Ole Heggen of the Geography Department, who prepared the figures and maps that grace this volume. He is a true professional.

I have a debt to two editors: Roger Haydon at Cornell University Press and Sean Culhane of M.E. Sharpe. Roger gave me much valuable advice and arranged for readers to comment on earlier versions of the manuscript that has become this book. I am very grateful to Roger and to the anonymous readers who read the manuscript thoroughly and made many detailed comments. Finally, Sean made an extremely helpful suggestion when he gave the book the title it now has, thereby suggesting I broaden the focus of the study so it would not be mainly focused upon Osaka.

It goes without saying that, while I have listened and responded to advice from these various quarters, I have stubbornly gone off to make something of all of this material on my own. For this reason, I take full responsibility for all defects that may remain in the present volume.

Part I

Water and Wood

1

Infrastructure, Technology, and Geography

A Layered Infrastructure

At first it seems chaotic, incongruous, at war with itself.

Is it because Osaka is so diverse? Is it because within Osaka's environs are factories large and small; sprawling markets and minuscule retail outlets; powerful companies whose business extends into the farthest reaches of the planet and suppliers of elegant fabrics to a select few; port facilities servicing global shipping and underground shopping malls; wide rivers and narrow canals; small, wood-framed residential houses on narrow, winding streets and towering concrete apartment complexes abutting congested boulevards? Or is it because this diversity is characteristic not just of the metropolitan area as a whole, but of most of the local districts within it? In other words, is it the density of the diversity—the packing in of disparate activities in small spaces—that overwhelms us?

Or does Osaka seem chaotic because of its sheer scale? Is it the volume of the far-flung plumes of lurid gas spilling out of the great *kombinato* complexes composed of petrochemical and iron and steel plants sandwiched together, which brings home to us the gigantic scale of the industrial capacity ringing the shores of Osaka Bay? Or is it the multistoried amalgamation of subway stations, department stores, shopping arcades, restaurants, and electric railroad terminals that drives home the sense of economic activity carried out on a vast scale but within a narrow compass?

Or is it the layering of infrastructure, the piling up of one type of infrastructure on another, which seems to plays havoc with our sense of normal order? The surest way to experience this layering of infrastructure is to walk

the streets of Osaka's city center. For instance, start your journey at the huge Nanba railroad station in the southern reaches of the city's center and proceed northward to the equally imposing railroad complex at Umeda that lies at the northern apex of Osaka's innermost core.[1]

Commence walking at Nanba station, terminus of the Nankai Railroad and of the Nanba Cat that whisks the foreign visitor in from the artificial island in Osaka Bay that is the stunning Kansai International Airport. Nanba is a massive piece of infrastructure, for it serves as the multistoried transport hub connecting the Nankai Railroad lines above ground to the subway lines several stories below. Sandwiched in between are shopping centers and a major department store. And running out from the subway entrances in all directions is a Byzantine-like shopping mall, where one can browse for hours, sampling everything from electronic gadgetry to books and rice crackers.

Walk north, departing from the Nanba Station by crossing the street and plunging into the Sennichimae and Dōtonbori shopping arcades. If the Nanba Station symbolizes the piling on of land transport and retail infrastructure, Sennichimae and Dōtonbori exemplify the piling on of land-based and water-based infrastructure. The district draws its character from the Dōton canal and the street of *kabuki* theaters and movie houses fronting onto the ancient waterway. Now, walk eastward along the canal, which for centuries proudly served as a major thoroughfare for shallow-hulled wooden boats bearing goods and passengers, and which today sports festive jets of water. If you proceed far enough along the canal, you will reach a dark, noisy point where it makes a sharp turn northward. Overhead is one of the branches of the Hanshin expressway system, erected on mighty concrete pillars thrust into the waterway. One form of infrastructure has been piled on top of another.

Now not only are disparate forms of infrastructure piled vertically on top of one another within Osaka's core, they also run parallel to each other. Consider walking northward from the Nanba/Sennichimae/Dōtonbori area toward the city's heart at Yodoyabashi. One can either proceed along Midosuji Boulevard, which is wide and spacious enough to accommodate buses, or one can make use of an old narrow street, Shinsaibashi. Let us opt for the latter, a narrow street that has been transformed into a covered shopping arcade by restricting daytime traffic to walkers and human pulled carts, *niguruma*, which are used to transport parcels and bundles, and to haul away garbage and refuse. This is a perfect example of Osaka's ancient infrastructure. Thus, it is fitting that it leads the walker into Senba, where Osaka's early modern merchant elite held sway and where wholesalers, *tonya*, continue to ply their trade. But just a short block away, and running parallel to this quaint little street cum arcade, is the wide Midosuji boulevard, along which moves a seemingly endless phalanx of trucks, buses, and automobiles.

Along the rest of one's walk—through Yodoyabashi and Kitahama, the financial core of Osaka bristling with banks and security houses, and across the Yodoyabashi Bridge and Nakanoshima, a small island dominated by Osaka's imposing city hall—one comes upon the layering of infrastructure time and time again. Subways and train stations abut rivers once used as thoroughfares; expressways pass over wide boulevards running parallel to narrow alleyways and meandering streets. And equaling the Nanba Station complex of stations, shopping arcades, and department stores is the Umeda/Osaka Station complex that lies just north of Nakanoshima. A vast array of underground malls fed by a serpentine labyrinth of old winding shopping streets lies at the end of our walk. In short, in Osaka's core, modern infrastructure is layered on top of ancient infrastructure.

Paralleling the layering of new infrastructure upon ancient infrastructure evident in the great cities of Japan's industrial belt like Osaka is the historical layering of infrastructure creation underlying Japan's modern industrial history. Employing canals, opening banks, shoring up river embankments, and dredging harbors and the rivers emptying into them was typical of the beginnings of infrastructure investment underlying Japan's first great industrial spurt in the late 1880s and 1890s. Several decades later, construction of electric railroad lines and their stations in places like Umeda and Nanba proved to be crucial for the second industrial spurt during the second decade of the twentieth century. Building subway lines and wide, paved boulevards accommodating buses and trucks was fundamental to the infrastructure construction buttressing Japan's third industrial spurt, that of the 1930s. Finally, the construction of the expressways, underground shopping arcades, and *kombinato* was intimately tied up with the industrial spurt between 1953 and 1969. In short, the physical layering of infrastructure evident within Osaka parallels the historical layering of infrastructure that supported four great waves of industrial expansion in Japan between the late 1880s and the late 1960s.

Underneath the apparent chaos of Osaka as exemplified by its layered infrastructure, does there reside a coherent logic—the logic of Japan's industrial expansion and economic development? Is the logic of chronological layering in Japanese industrial history the hidden order underlying the physical layering that is so dramatically evident in Osaka? And if so, what is the historical logic of Japan's industrial expansion? And why is infrastructure creation so important for that logic?

Infrastructure-Driven Growth and the Long Swing

The importance of investment in infrastructure for Japanese industrialization and economic growth should not be underestimated. Indeed, it is one of the

main tenets of this book that Japanese industrialization has been infrastructure driven.

To appreciate the crucial role played by infrastructure capital formation in Japanese industrial history, it is essential to understand that there has been a distinct rhythm to economic growth and industrialization in Japan over the period 1887–1969, during which Japan caught up with the major Western economies in terms of income per capita and labor productivity. Industrial output and economic growth were especially rapid during the periods 1887–1897, 1911–1919, 1930–1938, and 1953–1969. Within the literature on Japanese economic development, these phases of especially intense industrial expansion and income growth are known as upswings of long swings.[2] A long swing is a pattern of growth in gross domestic product (and related variables like the growth in disposable income per capita and in domestic fixed capital formation) in which the rate of output growth rises, reaches a peak, declines, reaches a trough, and then begins to rise once again. Each long swing consists of two phases: an upswing phase that initiates the long swing, and a downswing phase that brings it to a close. Relative to average growth over the entire long swing, growth rates during the upswing exceed the long-swing average, and growth rates during the downswing fall short of the long-swing average. Using this long-swing framework, the following long swings and phases can be demarcated historically. The first long swing occurred between 1887 and 1904, the upswing taking place between 1887 and 1897, and the downswing between 1897 and 1904. The second long swing had two upswing phases, the first between 1904 and 1911 when the focus of capital formation was on infrastructure, and the second between 1911 and 1919 when capital buildup was concentrated in the industrial sector. This surge sustained over fifteen years was followed by a deep downturn phase lasting from 1919 until 1930. In 1930, the third long swing was initiated, its logic being broken by preparations for total war during and after 1938. The upswing of a fourth long swing commenced in the middle 1950s and lasted until the end of the 1960s.

While the existence of long swings has been exhaustively documented and commented upon in the literature, their interpretation has been much debated. Some scholars have interpreted these long swings in terms of private-sector domestic fixed capital formation. They emphasize capital accumulation and savings, adopting an investment-led growth theory of Japanese development. Others favor an interpretation based upon trade. They argue that Japanese development is best understood as export led (at least one interpretation combines the private domestic investment and export demand–oriented interpretations). And interpretations in terms of credit creation and the expansion and performance of the banking sector have also been ad-

vanced.[3] In this book, I argue that the logic of long-swing development in Japan is far-reaching because long swings are simultaneously innovation waves, structural change waves, periods when domestic infrastructure investment and industrial expansion interact systematically, and periods in which Japan's involvement with global economic and geopolitical circumstances has certain salient and important characteristics. Each long swing has a distinct focus in terms of technological and organizational change. Each long swing involves a definite transformation in the structure of factor and product markets. And during each long swing, infrastructure capital formation is crucial for industrial expansion, in some cases preceding the industrial expansion and in at least one case occurring simultaneously with it.

In order to provide the reader with a road map for the analysis we are embarking upon, it is useful to state the four main hypotheses involving long swings that I advance in this study.

The first hypothesis involves the character and importance of long swings. Japan industrialized through a sequence of long swings. Underlying each long swing was a wave of innovation, imitation, and creative destruction associated with new technologies, fresh methods for harnessing and distributing energy, changes in the composition of raw materials demanded, and novel products brought to market. Innovation promoted investment in both infrastructure and manufacturing. Success in innovation generated excess profits in the enterprises adopting the innovation, which in turn encouraged imitation and capital investment in the sector of the economy making extensive use of innovations. By encouraging imitation and high rates of capital formation in innovation-using enterprise, higher profits accruing to the activities employing the innovations promoted increase in the share of the economy employing the innovation. This is diffusion. By the same token, differential rates of profits tended to undermine the position of those enterprises not employing the innovation. This is creative destruction. Now each long swing was also associated with structural changes involving the flow—or lack of flow—of capital and labor resources and knowledge between the sectors of the economy adopting the innovations and those that largely ignored the innovations basic to the long swing. Because the distribution of income and opportunities between these two sectors was tied up with the flow of resources between them, the potential for domestic social unrest was intimately linked to structural change inherent in the long swings.

The second hypothesis concerns the relationship between infrastructure investment and industrial expansion. Investment in infrastructure—physical, human capital enhancing, and financial—laid the groundwork for industrial expansion phases of long swings. It established the groundwork on the supply side by reducing labor costs, by facilitating the mobilization of capital

for capital accumulation in manufacturing and mining, and by chopping the costs of raw materials and energy faced by the factory managers. Moreover, learning how to build, utilize, and maintain infrastructure created technological spillovers benefiting industrial concerns. In short, infrastructure expansion reduced the supply costs for industrial activity. In turn, protracted industrial expansion put pressure on the underlying infrastructure, thereby creating demand for further expansion in the scale and quality of infrastructure. Pressure took a variety of forms, including price surges, tightening of labor markets, destabilization of credit markets due to excessively buoyant expectations and speculation, and limits imposed on growth of demand related to the growth of Japan's exports and imports and to its geopolitical relations with other countries.

The third hypothesis concerns Japanese government as coordinating and facilitating agent. Japan's national and local governments played a crucial role in creating infrastructure. In the case of some types of infrastructure, central and local governments fully shouldered the burden of constructing infrastructure. And in the case of other types of infrastructure, governments acted as facilitating or coordinating agents, working together with private-sector companies in carrying out infrastructure construction projects. Moreover, governments empowered local groups of citizens to initiate and implement the reworking and modernizing of certain kinds of infrastructure.

The fourth hypothesis states that long-swing development in Japan followed a domestic logic tempered by global circumstances. The process of domestic innovation, imitation, and creative destruction realized through capital formation in infrastructure and industrial plant and equipment provided the basic impetus for economic growth in Japan. However, Japan did not develop in a vacuum. In many cases, domestic innovation in technology was based on the study of Western models and on the adaptation of those models to Japanese circumstances and resource constraints. In some cases, innovation took the form of literal imitation of Western models through reverse engineering. In other cases, innovations were hybrids, combining foreign and Japanese methods. Moreover, changing global economic and geopolitical circumstances had an important impact on Japan's trade growth. And, in turn, Japan's increasingly voracious appetite for raw materials and its burgeoning presence in foreign markets for goods and services had an important impact on the other great economic and political powers.

With these propositions as framework for further discussion, consider Table 1.1. This table provides a summary description of the three long swings of the prewar period. Note that underlying each long swing are important technological and organizational innovations. And note that for each long swing, I provide a name describing important structural features of the economy during

the long-swing period. I call the first the "balanced growth" long swing, the second the "transitional growth" long swing, and the third the "unbalanced growth" long swing. What do I mean by "balanced," "transitional," and "unbalanced" growth? With these terms I describe the relationship between those sectors of the economy adopting the innovations crucial to each long swing, and those sectors of the economy eschewing the vast majority of the innovations. Balanced growth refers to a situation where both sectors grow (at roughly comparable rates), not only in terms of output and productivity, but also in terms of demand for capital and labor services and their capacity to attract these services from the owners of the factors of production. In balanced growth, the two sectors are roughly on the same footing. Unbalanced growth refers to a situation where one sector grows much more rapidly than the other, and this high-growth sector enjoys access to certain types of labor and capital services from which the other sector is effectively barred. That is, unbalanced growth is dualistic growth in which labor and capital markets are segmented. Transitional growth refers to growth leading from a balanced-growth economy phase toward an unbalanced-growth economy.

Table 1.1 and the four hypotheses accompanying it give us a historical framework for analyzing Japanese industrial history. Thus we have a setting in historical time within which we can carry out our analysis. But, as yet, we do not have a geographic setting.[4] And the discussion of layering of infrastructure, scale, and density of diversity within Osaka points to the importance of geographic concentration. So, at this juncture, we turn our attention to formulating a hypothesis concerning geographic concentration that will guide us in our subsequent analysis.

Geographic Concentration

The geography Japan enjoyed in the 1880s when it commenced the first long swing reflected both natural endowments and economic developments in the three centuries prior to 1880. Let us briefly review these historical changes and their impact on the economic geography of Japan before serious industrialization got under way.

Our concern is with the period known in the historical literature as "early modern." Typically, early modern Japan is defined as the period from 1600 to 1868. It is a period when rural demilitarization and almost total isolation from the West conditioned the political economy and economic growth potential of the country. It was also a period when a nationwide system of infrastructure was laid down and markets expanded, giving rise to far-reaching specialization and division of labor, and an organic—or natural—economy prospered and reached a comparatively high level of technological and orga-

Table 1.1

The Long-Swing Framework

Phases	Infrastructure: Physical	Infrastructure: Human capital	Infrastructure: Financial	Agriculture	Manufacturing	Energy/ raw materials	Trade and international relations	Economic geography
Balanced-Growth Long Swing								
Upswing of 1887–1897 Downswing of 1897–1904	Dredging of ports and building breakwaters; shoring up river banks. Steam railroad track laid on modest scale.	Growing enforcement of 1872 Fundamental Code of Education; by 1904, 94 percent of children attend compulsory elementary school. Western medicine embraced. Civil Code of 1898 created.	End of experiment with national banks. Private banks proliferate. The *zaibatsu* emerge in banking. Bank of Japan introduces over-loan policy.	Diffusion of best-practice traditional techniques from southwest to northeast. Expansion in use of fertilizers and new seed varieties. Land reclamation and rural road improvement occurs.	Factory system introduced and Commercial Code passed in 1890, revised in 1899. Steam, coal, and gas lighting used in textile mills. Cartel created in modern textile industry.	Coal using steam power tapped on a growing scale. Demand for iron ore and other minerals increases.	Japan successfully negotiates end of unequal treaties with Western powers. Japan secures Taiwan and indemnity after Sino-Japanese War; shifts from silver to gold standard.	The Tōkaidō region emerges as Japan's nascent industrial belt or core region. Osaka becomes the economic core of the core. City planning for Tokyo initiated.

Transitional-Growth Long Swing								
Upswing of 1904–1911 (Upswing 1)	Hydroelectric power generation expands. Steam railroad nationalization and electric railroads begin.		Reorganizing of *zaibatsu* commences as holding companies with multi-subsidiaries.	Productivity growth in agriculture slows as diffusion process reaches completion.	Mining law and factory act pass diet. Import substitution policies for shipbuilding and railroads introduced.	Electricity emerges as important new power source.	Russo-Japanese War occurs. Japan secures special rights in Korea. Korea becomes part of empire in 1910.	Dominance of the Tōkaidō belt in manufacturing, and Tokyo and Nagoya emerge as important industrial centers.
Upswing of 1911–1919 (Upswing 2)	Huge buildup of private sector plant and equipment.	Higher education revamped. University Ordinance of 1918.	Ongoing restructuring of *zaibatsu*. New *ziabatsu* emerge.	Rice Riots of 1918 occur. Import of foodstuffs from empire stepped up.	Labor markets in heavy industry internalized; labor boss system weakened.	Use of internal combustion engine expands. Unit drive system encouraged by electrification.	Collapse of European trade due to World War I embargoes. Japan gains markets in Asia.	Growing gap in income per capita between industrial belt and periphery (remainder of country).

(continued)

Table 1.1 *(continued)*

Phases	Infrastructure: Physical	Infrastructure: Human capital	Financial	Agriculture	Manufacturing	Energy/ raw materials	Trade and international relations	Economic geography
Transitional-Growth Long Swing (continued)								
Downswing of 1919–1930	Roads paved, widened, and extended. Power grids created. Great Kantō earthquake reconstruction occurs.	Imperial University system expands, and industrial research institutes proliferate.	After financial crisis of 1927, banking system revamped. Growing concentration in big-five banks takes place.	Domestic agriculture stagnates as empire expands output; landlord-tenant unrest occurs.	Light industry contracts during 1920s; heavy industry expands slowly. Mechanization of small plants occurs.	Synthetic fibers introduced and improved upon. Cartels operate in various industries.	Anglo-Japanese alliance ends. British pound–based gold standard fails, tariffs raised, and autarky grows.	City planning extended to big six cities of the industrial belt. Land conversion policies systematically applied.
Unbalanced-Growth Long Swing								
Upswing of 1930–1938	Airports built. Massive expansion in heavy industrial plant and equipment occurs.	Industrial and commercial vocational schools proliferate.	*Zaibatsu* begin to issue stock to public for holding companies, and *shinzaibatsu* proliferate.	Domestic stagnation continues; Diet fails to pass Land Reform bill.	Heavy industry expansion tied to military buildup. Machinery sector grows rapidly.	After Manchurian Incident (1931–1933), Japanese military seizes foreign policy initiative.	Applications of internal combustion engine to aviation occurs. Optics and radar advanced.	Morbidity and mortality fall in big industrial cities; Tokyo emerges as core of industrial belt.

nizational sophistication.[5] By an organic or natural economy, I mean an economy whose major sources of energy are the natural resources of wind, water flow, rain, fire, and physical exertion by humans and domesticated animals (an inorganic economy is one that relies upon stocks of latent power pent up in fossil fuels or in petroleum reserves). We shall divide the early modern period into two subperiods: the period from about 1600 to 1720, which I characterize as a period of extensive growth; and the period from 1720 to 1868, which I dub a period of intensive growth.[6] By extensive growth, I have in mind growth in output that reflects growth in the scale of population, land under cultivation, and use of resources. By intensive growth, I mean growth in productivity of labor that reflects improvements in skills, better education and organization of production, and a rise in the ratio of capital to labor.

To clarify the distinction between intensive and extensive growth, let Y be output, P be population, and y be output per person (thus $y = Y/P$). Adopting the convention that $G(x)$ stands for the growth rate of a variable x, we can write:

$$G(Y) = G(P) + G(y). \qquad (1)$$

For instance, a 1 percent rise in output could be due to a 1 percent rise in population and a 0 percent rise in population; or it could reflect a 0.5 percent rise in population and a 0.5 percent rise in income per head. With this equation, I define extensive growth as growth in output due solely to growth in the scale of the economy. Thus, during extensive growth, $G(Y) = G(P)$. By contrast, under intensive growth, $G(Y) = G(y)$. To be sure, characterizing growth between 1600 and 1720 as completely due to expansion in the scale of the economy and its population, and growth between 1720 and 1868 as completely due to improvements in labor productivity, is extreme. But it is my assertion that this description is broadly true, and that it is helpful in understanding how Japan's economic geography developed prior to the 1880s.

The foundation stone of extensive economic growth was political and involved demilitarization of the countryside. The "pax Tokugawa" brought down the curtain on a period of bitter internecine conflict in which rival warlords (*daimyō*) vied for regional power and fought bloody protracted battles with one another. Rural villages were caught up in these regional conflicts. Indeed, *samurai* (military retainers to *daimyō*) typically resided in villages and carried on farming in addition to readying themselves for warfare. Under the system of federal rule systematized by the Tokugawa overlords during the course of the seventeenth century, *daimyō* and their attending *samurai* were relocated to castle towns in the fiefs over which the *daimyō* were given authority (each fief had one castle town that became its adminis-

trative center). In order to prevent *daimyō* from falling under the sway of, and forming alliances with, foreign powers, the central government (the *bakufu* or *shogunate*) imposed an isolationist policy visa-à-vis the West, restricting contact to a small colony of Dutch traders who resided in the island of Dejima in Nagasaki harbor. Finally, the *bakufu* developed a system of monitoring the warlords by requiring their attendance in Edo, the capital of the *bakufu*, on an alternating basis.

The economic consequences of pax Tokugawa were immense. During the first century of *bakufu* rule, both land- and water-based infrastructures were expanded, and villages were increasingly knit together by roads and irrigation canals. Why? Castle town construction and the building of roads, along which *daimyō* were required to travel on their way to Edo for compulsory attendance at the court of the *bakufu*, generated a massive building boom. And ending local conflicts between villages by removing the restive *samurai* from their rural environs helped bring to an end protracted squabbling over water rights, and the arbitrary diversion of water brought from rivers along irrigation canals by aggressive villages riding roughshod over their neighboring villages. Water was crucial to an agrarian economy that, by dint of geography and climate, was heavily dependent on rice cultivation. For rice was not only the principal source of calorie intake in early modern Japan. It was also the basis for taxation in most fiefs. In a typical fief, the *daimyō* remunerated his *samurai* with allotments of rice. Thus, elimination of local conflicts stimulated investments by the central *bakufu* authorities, fief administrations, and local villages in improving the distribution and management of water resources. Dikes were constructed along riverbanks, irrigation canals were dredged, and the direction of river flow was straightened out in places. The fief and *bakufu* authorities had a vested interest in such improvements. By increasing the amount of land that could be cultivated through the extension of irrigation networks, they increased the amount of rice output from which they could draw rice taxes. An important by-product of the taming of water flow in the larger rivers was expansion of water transportation by shallow flat-bottomed boat (*kobune*).

In short, infrastructure grew under early Tokugawa rule in three key dimensions: the urban building boom; the carving out of road networks; and the improvement and extension of rivers, canals, and irrigation ditches.

The economic implications of this buildup in infrastructure were as far-reaching as the investment in the infrastructure itself. The amount of land that farmers could efficiently exploit increased. And because the number of mouths sustainable on the land was roughly proportional to the ratio of land in rice cultivation, population increased as well. A combination of larger populace and improved transportation increased the size of markets for goods

and services. The geographic circumference of the area that a typical producer or distributor could service ballooned. Thus, extensive growth resulted. The population grew, rice production and agricultural output soared, and the scope of markets expanded.

Associated with the buildup of infrastructure, the widening scale of agricultural output, and the growing efficiency of transportation and distribution was the emergence of a core economic zone, the Tōkaidō region, indicated by the shaded region in Map 1.1. (The geographic extent of the Tōkaidō region in early modern Japan was somewhat more modest than what is indicated here, because the shaded area in this map represents the industrial belt that the Tōkaidō region of early modern Japan eventually became.) As can be seen, the Tōkaidō region lies along the Pacific Ocean seaboard of Japan's main island, Hōnshū. And—although it is not apparent in the map, which regrettably, fails to contain information about Japan's topography—the core region consists of three huge plains linked together by coastal and inland transport networks. The three plains are the Kantō plain spreading out from Tokyo, the Nobi plain fanning out from Nagoya, and the Kinki or Kinai plains and basins centered on Osaka Bay and the Yodo River basin, which stretches from the environs of Lake Biwa and Kyoto to Osaka Bay.

The heart of the core during the early Tokugawa period was the Kinai region, which was blessed with relatively fertile soils, abundant rainfall, and a comparatively benevolent climate. By dint of these natural advantages, rice cultivation buttressed by production of fruit, vegetable, and fiber crops flourished in the Kinai. With the expansion of coastal commerce that was carried along roads and especially on ships (which circled the main island of Honshū and which transported goods into the *bakufu* capital of Edo from the Kinai and elsewhere), innovations involving the introduction of new crops pioneered in the Kinai were diffused to Edo and its environs. Under *bakufu* rule, the geographic reach of the core expanded northward. Originally concentrated in the Kinai and in the Yodo River basin, the economic core of the country spread northward and eastward, so that it eventually came to encompass Edo, the Kantō plain. The two great metropolitan centers directly controlled by the *bakufu*—Osaka on the Inland Sea and Edo at the mouth of the Kantō plain—marked out the southern and northern poles of the core, the most economically sophisticated zone, of the extensive growth economy.

But extensive growth ran into limitations. At the most basic level, the limits to growth were physical. Given the technologies developed in Tokugawa Japan, diminishing returns to the expansion of irrigation networks and land improvement set in. It became more and more expensive—in terms of resources—to convert land from other uses into paddy fields. And as rice cultivation reached its upper limits, so did the volume of taxes that most *daimyō*

Map 1.1 **The Prefectures and Regions of Japan, the Tōkaidō Industrial Belt, and the Six Big Cities**

The Six Big Cities

a. Tokyo d. Kyoto
b. Yokohama e. Osaka
c. Nagoya f. Kobe

Tōkaidō Industrial Belt

Regional boundaries

Prefecture boundaries

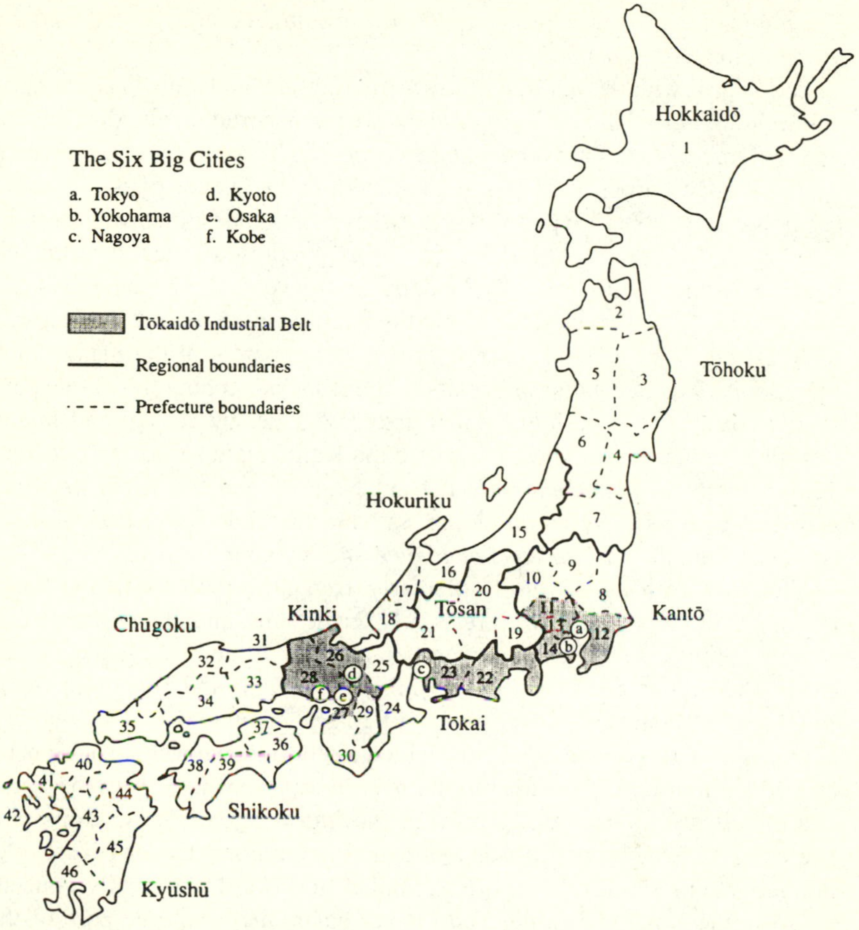

Regions and Prefectures

HOKKAIDŌ	KANTŌ	HOKURIKU	TŌKAI	CHŪGOKU	KYŪSHŪ
1 Hokkaido	8 Ibaraki	15 Niigata	22 Shizuoka	31 Tottori	40 Fukuoka
	9 Tochigi	16 Toyama	23 Aichi	32 Shimane	41 Saga
TŌHOKU	10 Gumma	17 Ishikawa	24 Mie	33 Okayama	42 Nagasaki
2 Aomori	11 Saitama	18 Fukui		34 Hiroshima	43 Kumamoto
3 Iwate	12 Chiba		KINKI	35 Yamaguchi	44 Oita
4 Miyagi	13 Tokyo	TŌSAN	25 Shiga		45 Miyazaki
5 Akita	14 Kanagawa	19 Yamanashi	26 Kyoto	SHIKOKU	46 Kagoshima
6 Yamagata		20 Nagano	27 Osaka	36 Tokushima	
7 Fukushima		21 Gifu	28 Hyogo	37 Kagawa	
			29 Nara	38 Ehime	
			30 Wakayama	39 Kochi

and the *bakufu* could extract from the villages they controlled. As a result, the number of *samurai* retainers who could be sustained by fief authorities reached limits as well. In short, extensive growth ran out of steam.

These limits did not spell the end of economic growth itself. Indeed, it continued because the expansion of markets under extensive growth spawned specialization and division of labor and the proliferation of new crops. That is, larger markets promoted innovation. Innovations took many different forms, but were especially associated with the introduction of new kinds of crops—*mikan* or mandarin-style oranges and cotton are good examples—and the systematic development of craft or proto-industry. Proto-industry took a variety of forms. It included the manufacture of implements and tools for agriculture, the weaving of baskets, the processing of foods (soy sauce and rice wine, sake, for instance), the spinning of cotton and silk thread, and the weaving of the threads into garments. Diffusion of these innovations raised the productivity of labor and also increased the demand for skills, including the managerial skills of keeping accurate accounts. Thus, the spread of proto-industry in the countryside increased the demand for physical capital accumulation and for education in rural Japan.

As extensive growth ran into limits and gave way to intensive growth, the competitive advantage (in terms of spawning innovations and attracting investments in infrastructure and manufacturing establishments) of the great metropolitan centers of the Tōkaidō core were weakened somewhat. For instance, in the Osaka area, expansion of proto-industry tended to take place in the rural districts contiguous to the city rather than in the city itself. The reasons are manifold. But one salient reason is that proto-industry made heavy use of agricultural products as inputs into production, and it relied on natural sources of energy such as wind, fire, and water power, which were potentially abundant in many villages, especially in the Kinai region with its productive farming sector. A second factor was born out of the nature of the political economic system put into place by the *bakufu* during the period of intensive growth. Under normal circumstances, taxes were paid upon rice production. Diversifying into crops other than rice and earning larger and larger proportions of total farm income from crops and activities other than rice production reduced the proportion of total income paid out in tax. Finally, the nature of the skills required in craft industry favored rural villages. Much of the skill involved intimate knowledge of the raw materials. Farmers who produced silk cocoons, raw cotton, tea, fruit, and the like naturally knew a great deal about raw materials that were used in proto-industry.

In short, the Tōkaidō region constituted the economic core—the region in which the most sophisticated economic activity was concentrated—of early modern Japan. During intensive growth, distribution and marketing was es-

pecially well developed in the great metropolitan centers of this core region, Osaka and Edo. Osaka was favorably located because it was in the rich Kinai area and because it lies at the mouth of the Yodo River basin and on the relatively calm Inland Sea. Edo was favored because it lies on the great Kantō plain at the mouth of great rivers that flow into Tokyo Bay and because it was the capital city of the *bakufu*. Under intensive growth, the competitive advantage of these two great metropolitan areas was eroded somewhat. Proto-industry flourished in their hinterlands, especially in the rural villages near Osaka. But Osaka and Edo (which was to become Tokyo) continued to be great centers for distribution and marketing under intensive growth.

Industrialization and the opening up of Japan to thoroughgoing commercial interaction with the West reversed the erosion of competitive advantage enjoyed by the great cities of the Tōkaidō. Once Japan was forcibly opened up to the West in the 1850s and jettisoned its feudal form of government in the late 1860s, the competitive advantage of the great metropolitan cities of the Tōkaidō was immensely strengthened. A wave of innovations swept over the great centers of the Tōkaidō as Japanese began to systematically adopt and emulate Western forms of infrastructure and Western practices and equipment in manufacturing. The great metropolitan centers on the three great plains of Honshū became magnets for investment in both infrastructure and in industry.

As a result, three great seaport/industrial/financial complexes emerged in prewar Japan: Osaka/Kobe on the Inland Sea, Nagoya at the mouth of the Nobi plain, and Tokyo/Yokohama/Kawasaki in the environs of Tokyo Bay. The fifth main hypothesis of this book concerns the role of these great metropolitan centers of the Tōkaidō in the interaction of infrastructure and industry basic to the long swing.

Thus the fifth hypothesis concerns geographic concentration of industrial activity and infrastructure buildup. The efficiency of infrastructure investment in terms of generating spurts in industrial activity, and in terms of promoting innovation within industrial enterprises, was enhanced by the concentration of infrastructure investment and manufacturing in the Tōkaidō industrial belt, especially in the three big seaport/industrial/financial complexes, Tokyo-Yokohama, Nagoya, and Osaka-Kobe. Geographic concentration promoted scale economies in marketing new products, in the securing of raw materials, in the exporting of manufactured goods, and in the supplying of energy. Geographic concentration also led to a sharp divergence between the economic fortunes of the great metropolitan districts of the Tōkaidō core and the (mainly rural) periphery of the country.

The core/periphery distinction is important for an understanding of the economic geography of Japan's early modern economic development and its

industrial history after 1880. But the core/periphery distinction is not the only important geographic distinction we need to make in appreciating the relationship between the economic geography of Japan's economic growth and its industrial development through the long swings. Where and why the economic heart of the core is located is also important; that is, analyzing the locus of the core of the core is important. The reason is that the core of the core did not remain fixed over the course of the three prewar long swings. As can be gleaned from Table 1.1, the core of the core in the first long swing was Osaka. But after the first long swing, there was a strong pull toward Tokyo and away from Osaka. By the end of the upswing of the third long swing, Tokyo had emerged as the core of the core. Why did this happen?

The Giants Compared

The shift from an Osaka-centered economy to a Tokyo-centered economy is one of the most important themes in the economic geography of modern Japan. My sixth hypothesis bears on the question of why this shift occurred.

Thus the sixth hypothesis concerns the core of the core. The core of Japan's industrial belt core shifted over the course of Japan's industrialization. During the first long swing, Osaka was the core of the core. But, during the course of the second and third long swings, Tokyo's capacity for creating physical, human-capital enhancing, and financial infrastructure grew relative to Osaka's capacity. Because Tokyo's relative competitive advantage in infrastructure creation soared, Osaka's relative competitive advantage deteriorated. The regional economies of both Tokyo and Osaka expanded greatly in absolute terms, but in relative terms, the Tokyo-centered regional economy expanded more vigorously. By the third long swing, Tokyo's supremacy in political, cultural, and economic affairs—known as Tokyo-centrism—was fully established. Tokyo's competitive advantage in creating infrastructure was due in part to ecological factors, and in part to the fact that it was the political capital of the nation.

What do we mean by ecology, and why did it matter? Ecology entails a variety of characteristics: climate, soil type, topography, the frequency and magnitude of natural disasters, the relative calm of ocean and river waterways, and so forth. Ecology is heavily shaped by nature, but it is not exclusively affected by it. Human agency interacts with nature to shape the ecological constraints facing individual regions. Indeed, in contemporary Japan it is difficult to find any locality that has not been shaped by the purposeful activity of the human race. Canals, huge landfill projects in harbors and bays, river embankments thrown up, contoured mountain slopes, paths

and roads are all examples of the impact of human agency on natural environment. Consider the ecological contrast between Tokyo-Kantō and Osaka-Kinai, which prevailed in the 1870s. I set out the main features of that contrast in Table 1.2. As is apparent from the items contrasted, some of the differences are clearly attributable to nature—the prevalence of flooding in the Osaka area and the prevalence of earthquakes in the Tokyo area—but some, like the settlement pattern and the type of livestock used, involve human agency. To be sure, humans may have responded to natural constraints in establishing one type of settlement pattern and not another, but in any event, human agency was involved in determining the actual ecological constraints facing the two regions.

That the geological and morphological structure of Japan is important for the contrast between the Osaka and Tokyo districts is evident from their soil conditions and natural disaster patterns. The main island of the Japanese archipelago, Honshū, consists of northern and southern halves divided by a great depressed zone, the Fossa Magna of Naumann, which stretches from the Pacific Island to the Sea of Japan. The mighty mountain chain that runs on a northeast/southwest axis, as if it is a backbone for the island, is cut across by the Fossa Magna rift that runs along an axis almost vertical to the mountain range. And nestled within the basins carved out of this huge fissure is a cluster of volcanoes known as the great Fuji volcano chain. Because the Kantō plain lies just north of the Fuji volcanic chain, it is situated in the shadow of a host of volcanoes, and this is the source of the geological instability visited on the Kantō plain. Earthquakes are frequent in the Kantō, and occasionally they are devastating. To be sure, the Osaka region is also occasionally rocked by great quakes. But, in comparison with Tokyo-Kantō, the environs of Osaka are tremor-free. However, what nature gives with one hand, it takes away with the other. For Osaka has been the victim of continual assaults from water and air. Typhoons and flooding have played havoc with the greater Osaka area. In short, Tokyo's main natural enemies are earthquakes (and hence conflagrations). Osaka's assaults have mainly been from wind, rain, and inundations.

And humans have responded to these constraints. Consider the topography of the two regions during the heyday of the organic or natural economy. Because volcanic ash pockmarked the brown soils in the Edo environs, the arable fields surrounding the *bakufu*'s capital were poor in quality. It was difficult for farmers to carry out paddy rice production in many parts of the Kantō plain. Hence, dry field cultivation dominated agricultural production in the hinterland of Edo. And much of the region was left in wasteland, or in forest where the elite hunted for game. By contrast, the Kinai region around Osaka was extremely fertile; and because rainfall was prevalent there, it was

Table 1.2

Ecological Contrast: Kinai-Osaka and Kantō-Tokyo

Item/district	Kinai-Osaka	Kantō-Tokyo (Edo)
Climate	Mild winter (few days below freezing) except in Kyoto and Japan Sea district, hot summer.	Mild winter (around three to four days below freezing), strong northwesterly winds, hot summer.
Geology and soil	South and west of Fossa Magna, alluvial lowlands surrounded by diluvial borderlands; Yodo River formed from rivers flowing out of Kyoto, Lake Biwa, and Nara basins flows into Osaka Bay; red soils with little volcanic ash.	North and east of Fossa Magna, alluvial lowlands surrounded by diluvial terraces; Tone, Ara, and Tama rivers cross plain, empty in Tokyo Bay; Tokyo on low flood lands and delta of Sumida River; in vicinity of many volcanoes, brown soil laden with volcanic ash, infertile and deficient in basic minerals.
Frequent natural disasters	High frequency of flooding; occasional earthquakes.	High frequency of earthquakes; occasional flooding.
Tokugawa period agricultural productivity/ population density	Relatively high agricultural productivity, extensive double cropping of dry fields; high population density.	Relatively low agricultural productivity, and relatively low population density.
Agricultural land use	Rice fields dominant; land also supports cotton production and other industrial crops.	Dry fields dominant (vegetables, fruits, mulberry, barley, potatoes, tea, and tobacco).
Settlement pattern	Small compact rural agglomerations, rectangular in shape, surrounded by moats and hedges.	Rural agglomerations irregular in shape, typically surrounded by forest.
Livestock	Cows dominant.	Horses dominant.

extremely well suited for rice cultivation. Hence, densely settled villages cultivating rice flourished in the hinterland of Osaka. Because land was relatively productive in the Osaka area, very little acreage was left undeveloped in its hinterland. In short, at the time when industrialization began, the productivity—and therefore price—of arable land in the environs of Tokyo was considerable less than the productivity—and therefore price—of land in the hinterland of Osaka.

Now the cost of converting dry field and wasteland to residential, industrial, or infrastructure usage is far less than the cost of converting paddy field to such uses. There are several reasons for this. Paddy field land is generally more valuable and expensive than dry field land. It embodies more agricultural infrastructure in the form of drainage and irrigation ditches. And, because in traditional Japanese rice cultivation, the flow of irrigation water was usually managed by the village acting as a collective entity, it was more difficult for real estate developers to buy out individual farms since conversion of farmland affected the water distribution patterns of those households that did not wish to abandon farming. In short, with industrialization, both Tokyo and Osaka expanded outward, gobbling up hinterland for housing, factories, roads, and other forms of infrastructure. And in this outward expansion, Tokyo had a clear competitive advantage over Osaka. Conversion of land for wide roads accommodating trucks and buses was an important element in this competitive advantage on the land.

Moreover, because fires were common in Edo, the authorities in the capital decreed that wide boulevards cum firebreaks should be cut through the city. By contrast, Osaka was a welter of narrow streets and canals upon which goods were shipped in flat-bottomed boats. For this reason, within the city cores, Tokyo also had an advantage in infrastructure based on the land. For in the Tokyo of the 1880s there were already in place spacious boulevards that would accommodate horse-drawn trains, and, later on, trucks and buses.

The competitive advantage that Tokyo possessed in developing infrastructure based on land was a major factor in Tokyo's emergence as the economic core of the core. But the ease of land conversion on the periphery of Tokyo, and the existence of a network of wide roads in the center of the city from the outset of industrialization, were not the sole factors favoring Tokyo over the long run. Because Tokyo was the capital, the Bank of Japan and the Ministry of Finance were headquartered there. Thus, Tokyo had a competitive advantage in attracting banks and other financial intermediaries. Moreover, the Tokyo-based Ministry of Education was naturally inclined to favor education in the capital over education elsewhere. For instance, it was easier for the ministry to monitor and regulate higher education in the capital than in distant reaches of the country. Thus, the University of Tokyo became the

apex of the imperial university system during the prewar period. In short, Tokyo had a competitive advantage in the development of land-based physical infrastructure, human capital–enhancing infrastructure, and financial infrastructure. During the first long swing, Osaka's advantages in water-based infrastructure and proto-industrial economy skills prevailed, and Osaka was the core of the core. But as the economy passed through the second and third long swings, the relative importance of land-based infrastructure, education and industrial research, and financial intermediation increased. Thus, the relative attractiveness of the Tokyo region improved. The pull toward Tokyo was inexorable. By the late 1930s, Tokyo-centrism was firmly established.

The Approach

This book began with concrete observations. Contemporary Osaka is home to a remarkable density of diversity, a massive scale of economic activity, and a richly layered infrastructure. It is the burden of this volume that, underlying the phenomena I have called attention to, are fundamental aspects of Japanese industrial development.

I have elaborated six hypotheses that I believe capture the most important features of that development. These hypotheses revolve around the character and importance of long swings as innovation waves and as carriers of structural change; the interaction of capital formation in infrastructure and industry; the importance of government as coordinating and facilitating agent in infrastructure investment; the interaction of a domestic logic of development with global economic and geopolitical forces in shaping the long-swing pattern; the importance of geographic concentration of manufacturing and infrastructure in the Tōkaidō industrial belt; and the geographic shift in the locus of the core of the core, from Osaka to Tokyo. The evidence supporting my six hypotheses makes up the bulk of chapters 2 through 7 of this book. In the appendices to this chapter, I provide the reader with a bridge to the literature in economics underlying my theory of long swings as innovation waves, and my theory of infrastructure-driven growth.

In chapter 2, I discuss extensive and intensive growth in early modern Japan, the development of the Tōkaidō core, and the emergence of Osaka as the core of the core. I highlight the interaction of investment in infrastructure and expansion in agriculture and proto-industry in this discussion. This chapter provides the background for chapter 3, which deals with the first long-swing period and the development of Osaka as the "Manchester of the Far East."

Chapters 4 through 7 focus upon innovation and the interaction of infrastructure and industrial development during the second and third long swings, when Japan passed through the transition from balanced growth to unbal-

anced growth. In chapter 4, I review the main contours of the transitional-growth and unbalanced-growth long swings, and I make some observations about the interaction of global economic and geopolitical circumstances with the internal logic of long-swing development in Japan. Chapter 5 takes up infrastructure investment in greater detail, focusing on infrastructure development in the nascent Tōkaidō industrial belt and, within that belt, in the Osaka and Tokyo regions. Chapter 6 turns to industrial expansion. And as with chapter 5, it deals in detail with Osaka and the struggle between the two cities over which region was to be the core of the core. Chapter 7 looks at the interaction of government and the private sector in developing infrastructure. The interaction of government and private sector in Osaka, and the pull toward Tokyo, are the central themes. Taken together, chapters 4 through 7 constitute the heart of the book in the sense that evidence in support of all six hypotheses is systematically laid out in these four chapters.

Finally, chapter 8 serves as a conclusion. I also use the chapter as a vehicle for making a few observations about the fourth long swing between the early 1950s and the late 1960s. I close by bringing the reader back to contemporary Osaka, a city that at first glance seems to have reaped the whirlwind of chaos.

Appendix A.1: Innovation Waves, Kondratieff Waves, and Long Swings

It is important to distinguish between innovation waves, Kondratieff waves, and long swings. The innovation wave is a theoretical concept attributable to Joseph Schumpeter (1964). Innovations—the setting up of new production functions by entrepreneurs—stimulate imitation because entrepreneurs secure supernormal profits by exploiting inventions, creating new products, and so forth. As a result of imitation, excess profits are eroded as the innovation is diffused. Moreover, producers who cling to older methods or older products are marginalized. Thus, creative destruction takes place as some producers are driven out of business.

Kondratieff waves and long swings are empirical phenomena. A Kondratieff wave is a long wave of about fifty years in duration. It is a wave in prices. In the upswing, prices tend to rise initially, reach a peak, and then fall; in the downswing, prices initially fall, reach troughs, and then begin to recover. Average price levels are greater over the upswings than they are over the downswing. Typically, the Kondratieff waves are measured using moving averages of prices. Long swings are shorter waves than Kondratieff waves. They are usually two decades in length and are measured in terms of growth rates of output and growth rates of, and levels in, fixed domestic

capital formation. During upswings of long swings, growth rates for income and capital formation rise and then reach peaks. Downswings are initiated by declines in growth rates. Like Kondratieff waves, upswings have above-average growth rates and capital formation rates, and downswings below-average growth rates and capital formation rates. And like Kondratieff waves, long swings are usually measured with moving averages for the variables involved. Because the Juglar cycle is about four to seven years in length, both Kondratieff waves and long swings are usually measured using moving averages that average out the Juglar movements. Thus, in the case of Japan where Juglar cycles are about seven years in length, long swings are usually based on seven-year moving averages (in the case of aggregate measures like gross domestic product, but swings for specific sectors may be based on averages for shorter periods of time as they are in this study). The concept of the long swing is due to Kuznets (1971), who illustrates how he used the idea in much of his research.

Schumpeter (1964) argued that Kondratieff waves are innovation waves. In his dating, the first Kondratieff took place between 1787 and 1842. The key innovations he focused on were the new technologies of the industrial revolution, especially the use of coal, improvements in iron and steel production, the harnessing of steam power, and the development of the factory system. England was the first country to go through this transformation, which involved shifting from an organic to an inorganic economy (cf. Wrigley [1987, 1988]). Schumpeter associated the second Kondratieff with the railroadization of the globe, and dated it from 1842 to 1897. Other innovations associated with this wave include the growing use of steam power in water transportation, the use of steel rail in building railroad track networks, and the emergence of large vertically and horizontally integrated companies exploiting economies of scale in production and of scope in marketing. The dating for the third Kondratieff is 1897 to 1938. Schumpeter argued that it was driven by an innovation wave centering around the use of electricity and the internal combustion engine and involved the paving of roads and the development of synthetic chemicals.

It is reasonably clear that Schumpeter's interpretation of Kondratieff waves assumed that the new innovations would come in a small cluster of core or advanced countries that were at the forefront of technological innovation. But the number of countries in the core gradually expanded over time. Thus, France, Belgium, the Netherlands, and the United States began to industrialize during the first Kondratieff. Germany, Switzerland, Denmark, Norway, Sweden, Italy, Australia, Canada, Japan, and Russia began industrialization during the second Kondratieff and Argentina, Turkey, Brazil, and Korea during the third. Since, by borrowing and adapting the

innovations and technologies developed elsewhere, later-developing countries could potentially grow more rapidly than the earlier developers had, the ranks of the core countries grew over time. This makes it difficult to determine the scope of the data required to measure a Kondratieff. Which countries are to be included? And are hybrid technologies developed in later-developing countries like Japan considered innovations driving Kondratieffs?

Walt Rostow (1980, 1998) rejects the identification of Kondratieff waves with innovation waves. I agree with his position on this issue. Rostow also works with a stage theoretic concept for individual countries. According to his framework, countries prepare for the takeoff into sustained growth, pass through the takeoff, and then experience the drive to maturity. Rostow (1963) argues that leading sectors are key to each phase. There is a superficial similarity between the theory that I offer and Rostow's stage theory, insofar as my interpretation of long swings rests on the idea that certain sectors are especially important in specific long swings. For instance, in the balanced-growth long swing, textiles was unusually important; in the second long swing, chemicals, electricity, and railroads were especially important; and in the third long swing, the airplane industry played an crucial role. But my theory and Rostow's are basically different. I emphasize infrastructure; I reject the idea of a universal theory of stages applicable to all countries; and I focus on long-swing dynamics that Rostow does not emphasize.

This brings me to the long swing. From the work of Simon Kuznets (1971) and Brinley Thomas (1973), it is apparent that long swings are intimately tied up with capital formation. Indeed, because capital and population flowed from Great Britain to the United States during the nineteenth century, the long swings in the United States and Great Britain were out of phase with one another. During upswings in Great Britain, capital stayed at home, wages rose, and labor stayed at home as well. During downswings in Great Britain, capital flowed to the United States, where it earned a higher rate of return, pushing up wages in the United States and encouraging a flow of labor from Great Britain to the United States. Capital formation in the United States was very sensitive to rates of population increase stemming from immigration to the country. Thus, during upswings, immigration and population soared, which encouraged even more capital accumulation to accommodate the increasing numbers of people. Capital formation in infrastructure (e.g., railroads) was an important element in this "population sensitive" capital formation.

In the case of my theory for Japan, innovation waves, Kondratieff-like waves, and long swings all overlap. Because Japan was initially a follower country, there was a large backlog of foreign technology that Japanese inno-

vators could draw upon in coming up with new products, new techniques, and new sources of energy and mechanical power. Hence, Japan proceeded rapidly through innovation waves. Since each innovation wave was associated with high rates of growth in fixed-capital formation, the waves overlapped with long swings driven by capital formation. And because prices were driven up rapidly during upswings, the long swings also have a Kondratieff-like character. But, while the overlap of the three types of processes holds for Japan, there is no reason to think this holds for other countries. In particular, as I emphasize throughout this book, long swings were accompanied by domestic structural change and by changes in the relationship of Japan to other countries. It is doubtful that these same circumstances are or were duplicated by other countries.

Appendix A.2: Sources of Growth Accounting and the Residual Factor

What explains growth rates of output? How much is due to increases in labor supply? To improvements in labor quality, increases in capital stock, improvements in the quality of capital? How important are scale economies? What role does expansion in trade play? And what is the pure effect of "technological catch-up" and of innovations involving technological catch-up? By the pure effect of catch-up or innovation, I mean the effect on growth of these variables independent of capital accumulation. For instance, in my argument, excess profits in the sector or sectors employing the innovations encourages imitation and, hence, high rates of accumulation in those sectors. This "accumulation effect" of innovation is different from the "pure" impact of innovation.

Growth accounting offers an answer this question. At the core of growth accounting is the following equation:

$$G(Y) = [\alpha G(K) + \beta\, G(L)] + G(R) \qquad (2)$$

where Y is output, L is the labor force, K is capital, R is the "residual factor," a is the share of output accruing to labor, b is the share of output accruing to capital, and $G(x)$ stands for the growth rate of a variable x. It should be noted that theoretically labor and capital are measured in the same terms as output is measured, that is, as flows over a period (usually an annual flow). Thus, if properly measured, the variable "L" is the flow of labor services over the period for which income Y is measured, and K is the flow of capital services over the identical period. Also, it should be noted that, if we let w stand for wages, and r for the return on capital, then

$$\alpha = [wL]/Y; \text{ and } \beta = [rK]/Y. \tag{3}$$

Since the sum of *wL* and *rK* must equal total output *Y*, it follows that:

$$\beta = 1 - \alpha \tag{4}$$

Thus, growth accounting involves "parceling out" the growth rate of total output into the growth rate of combined inputs—$\alpha G(K) + \beta G(K)$—and the remainder, the "residual" factor.

It should be noted that it is possible to derive the growth equation (2) from a Cobb-Douglas production function with constant returns to scale:

$$Q = A\, K^{\alpha} L^{\beta}, \text{ with } \alpha + \beta = 1 \tag{5}$$

in which, as before, α is labor's share in output, β is capital's share, Q is output, and A is an index of organizational and technological progress. But is this all that A is an index for? What about scale economies? In any event, it is not necessary to start with the Cobb-Douglas production function in order to derive the growth-accounting equation (2). One can simply assume that the contributions to growth in output are to be estimated as they are in equation (2) and proceed on that arbitrary assumption. For a thoroughgoing discussion of the relationship between production functions and growth-accounting estimates, the interested reader is asked to consult the article by Hulten (2000).

In theory, empirical growth accounting is straightforward. In practice, it is difficult. How does one measure the growth in labor? How does one take into account improvements in the quality of labor services? How does one adjust capital stock figures to get capital service flow figures that take into account the rate at which capital is utilized? How much weight do we give to (1) structural shifts from low productivity to high productivity sectors; (2) scale economies; (3) the fruits of investment in research and development; (4) the fruits of investment in infrastructure that improves the efficiency with which capital and labor are combined to produce output; (5) the benefits of trade, including widening of domestic markets through export and exposure to new products and foreign competition; and (6) "pure" organizational and technological changes that do not involve direct investment? The greatest difficulty involves accurately taking into account all of those factors accounting for growth that economic agents actually invested in. If one invests directly in improvements in, or if one directly uses the flows from, the factors of production, one is making use of variables that are "endogenous" to the economy. If these are not fully accounted for, the estimating procedure given by equation (2) will tend to overestimate the impact of the "residual" factor,

thereby exaggerating the "pure" effects of "exogenous" improvements in knowledge and organization.

Much empirical work has gone into estimated contributions of growth. For Japan, the estimates of Kazushi Ohkawa and Henry Rosovsky (1973) and Edward Dennison and William Chung (1976) are an essential starting point for research into this area. Angus Maddison (1987) has reworked these estimates for Japan and has come out with the following contributions to output growth for Japan between 1913 and 1950. Of the total growth rate of output of 2.24 percent over the period, he attributes 94.2 percent to growth in factors that can be directly accounted for. Hence, the residual accounts for 5.8 percent of total output growth over this period. Of the nonresidual factors accounting for growth, he attributes 16 percent (of total output increase) to increase in labor quantity, 27 percent to improvements in labor quality, 34 percent to increase in nonresidential capital quantity, and 20 percent to improvements in capital quality. Labor hoarding contributes negative 25 percent. Scale economies contribute 3 percent. And structural change contributes 28 percent. In short, his estimates suggest that investments in labor and capital quality, and the sheer impact of industrialization that moved workers from less productive to more productive activities, played a major role in prewar growth. From the point of view of this study, I believe Maddison's estimates support my notion that infrastructure improvements that improved labor quality and helped to mobilize capital for industrial activity played an extremely important role in Japan's growth through the second and third long swings. However, since Maddison does not take into account the economic geography of growth, it is hard for me to accept his estimates for scale economies. By failing to account for the scale economies stemming from geographic concentration, he denigrates the role of scale economies, in my opinion.[7] In any event, his estimates suggest that Japanese growth during the period of the second and third long swings was almost exclusively due to endogenous factors.

Appendix A.3: Economic Evolution and Endogenous Growth

In the previous appendix, I briefly discussed the distinction between endogenous and exogenous factors in economic growth. The literature on economic growth has gone through three phases, and in each phase different weights (or interpretations) have been placed upon the relative importance of endogenous and exogenous factors. In the early literature from the 1950s and 1960s, estimates of the growth equation (2) given in the previous appendix suggested that most growth was due to "manna from heaven" or exogenous technological and organizational change. Then, in the late 1960s, the argument was advanced that improvements in capital and labor quality—

improvements that were actually invested in—might account for most of the growth in output not accounted for by increases in the quantity of capital and labor. Thus, the argument was advanced that growth might be more endogenous than was suggested by the original growth estimates.

A more recent literature argues for endogenous growth along different lines (for details and extensive discussion, see Grossman and Helpman [1994], Hulten [2000], Romer [1994], and Solow [1994]. This literature emphasizes spillovers from investments in capital made with specific enterprises. Thus, let each firm operate with a constant returns-to-scale Cobb-Douglas production function as given in equation (5), and assume that a 1 percent increase in the plant and equipment of the firm raises firm output by α percent and creates spillovers that raise the output of all other firms by β percent. Then, assuming the index of technology and organization is just equal to the accumulated spillovers, one gets for the aggregate production function:

$$Q = K^{\beta}(K^{\alpha}L^{\beta}), \text{ where } \alpha + \beta = 1. \tag{6}$$

In this case, there are no diminishing returns to investment in capital because $Q = KL^{\beta}$. Thus growth is endogenous and accumulation of capital will cause it to continue.

In the infrastructure-driven growth theory I propose in this work, the main source of spillovers is investment in infrastructure capital. These create technological advantages and drive down costs for raw materials, energy, and the quality-adjusted unit cost of the factors of production, thereby stimulating growth. And according to this logic, the growth that results is largely endogenous. It should be noted that David Aschauer (1989) provides evidence suggesting that public-sector infrastructure investment was important for stimulating economic growth in the United States.

And so the perspective adopted here emphasizes endogenous factors. It also emphasizes evolutionary factors. In order to understand the significance of this assertion, it is necessary to make a few remarks about the evolutionary approach to economics.

The evolutionary approach adopted in this study rests upon a large literature. Discussing it in detail is superfluous. However, a few key points made in the literature are worthy of mention here. Of the various evolutionary approaches developed, my approach draws most heavily from that developed by Joseph Schumpeter (1964) and by Thorstein Veblen.[8] Geoffrey Hodgson (1993) provides a useful overview of the views of these two pioneers of evolutionary economics. He notes that Schumpeter's theory emphasizes the clustering of innovations. This clustering is akin the idea promoted by some theorists of Darwinian evolution that random variation takes place in waves,

a clustering of new species bursting upon the scene suddenly. Joel Mokyr (1990) also stresses that technological evolution may take place historically in clusters, the rates of innovation rising and then falling. This analogy to biological evolution underlies my interpretation of upswings of long swings as periods of unusually intense innovation. The biological analogy comes into my account in my stress upon the spawning of hybrids as a vehicle for generating innovations.

Hodgson (1993) contrasts the Schumpeterian vision of evolution proceeding through periods of intense species creation followed by periods of weeding out of less fit species with Veblen's theory of cumulative causation. According to Veblen's theory, organizational rules, social routines, and technologies evolve through mutual interaction. I draw upon this notion with my theory of the interaction of infrastructure investment with technological change in industry. Infrastructure investment involves institutional changes. For instance, the socialization of labor occurs partly within the educational system. Hence, increasing the educational certification required of workers shapes the social routines and norms on the shop floor. It should be noted that economic evolution through cumulative causation is a form of endogenous growth. The impetus for organizational and technological changes arises from the ongoing interaction of social norms, business routines, and technological breakthroughs.

The notion of diffusion of innovations through a process akin to a weeding out of less fit species also plays a role in my theory. It is useful to employ some elegant and simple algebra developed by Richard Nelson and Sidney Winter (1982, 238–40) to illustrate how diffusion works. Consider diffusion of an innovative technique that is pitted against an older competitor technique. Assume that profit-oriented firms are the agents that accumulate the capital used to employ the new technique. And assume that these profit-making enterprises plow back all of their profits into further capital accumulation. Let the subscript N denote new technique and O the old technique; let K_N and K_O be the capital applied to using the old and the new techniques respectively; let P be the price of output; let r be the interest rate, the unit cost of borrowing capital; let l be labor input per unit of output, which is the inverse of labor productivity; and let w stand for wages. Finally, let us assume that labor and capital markets are not segmented. That is, let us assume firms using old techniques and firms using new techniques are able to compete on an equal footing—at identical prices—for the factors of production, and sell their products in the same output markets at the identical unit price P. Finally, assume that capital-labor ratios are the same for old and new techniques alike.

Under these assumptions, the only difference between employing an old or a new technique arises from the superiority of the new technique with

respect to unit costs. That is:

$$l_N < l_O. \tag{7}$$

In this case, profits generated by adopting the new technique exceed those accruing to producers sticking to the old technique. Assume those enterprises clinging to the old technique secure returns on capital that just "break even." That is, their rate of return on capital just equals the cost of borrowing funds to purchase capital. Then, the enterprises using the new technique make returns per unit of capital exceeding the "break even" level. Now, suppose investment is directly proportional to profits. Thus:

$$\Delta K_N / K_N = \lambda\ (P - r - wl_N); \qquad \text{(7a),}$$

and

$$\Delta K_O = \lambda\ (P - r - wl_O). \qquad \text{(7b)}$$

From these equations, it can be seen that investment in plant and equipment exploiting the new technique yields higher rates of return, thereby stimulating higher rates of capital formation than does investment in capital tied up with the old technique.

Over time, diffusion takes place. The new technique is more profitable and, for this reason, drives out the old technique. Various scenarios generate this result: differential rates of capital favoring the new technique; wages rising due to competition for labor, which renders unprofitable old-technique production that was just "breaking even" before wages drifted upward, and so forth.

As is demonstrated by concrete examples presented in chapters 3, 5, and 6 of this book, a simple diffusion model of this sort does capture important features of the diffusion of technological and organizational innovations. But it misses some fundamental aspects of the process. One deficiency with the approach is the fact that it ignores barriers to diffusion associated with shortages in the supply of labor willing or able to work with the new technique or capital. Another deficiency has to do with the fact that under unbalanced growth, labor and capital markets are segmented. That is, wages paid to workers employed in production that uses the new technique are higher than wages paid to those working with the older technique; and the borrowing costs for capital acquisition are lower for the innovative firms than for those clinging to older methods of production. In short:

$$r_N < r_O; \text{ and } w_N > w_O \tag{8}$$

In such an environment, the logic of capital accumulation and the logic of diffusion of new techniques differs from the story told by equations (7a) and (7b). And so does the story involving the diffusion process. Emphasis shifts away from the role of wages as a mechanism for dooming old techniques, and it shifts toward differential gains in labor productivity. The sector using new techniques pays higher wages. But it is generously compensated by securing labor willing and able to acquire the skills required of the innovative technique. In sum, diffusion of new techniques under conditions of unbalanced growth differs from the type of diffusion process occurring under balanced-growth conditions. In the latter case, a model of the form given by equations (7a) and (7b) is broadly applicable. In the former case, the story revolves around the high rates of productivity gain accruing to innovators.

Notes

1. The walk from Namba to Umeda is not long. I did it on a fine November day in about three hours. For intensive photograph collections illustrating places in Osaka at various periods of its history, see Okamoto (1978) and Ōsaka-shi (1989). For walking guides to Tokyo that emphasize the role of the historic past in shaping contemporary vistas in the city, see Ashihara (1986) and Hidenobu (1995).

2. For details on the long swings, see chapters 3 and 4. For graphs illustrating the wave-like character of the long swings in Japan over the period between the late 1880s and the 1970s, see Minami (1986, 1994), Ohkawa and Rosovsky (1973), and Ohkawa and Shinohara with Meissner (1979).

3. For various interpretations of long swings in Japanese economic development, see the chapters in Ohkawa and Shinohara with Meissner (1979).

4. On Japanese geography, see Collcutt, Jansen, and Kumakura (1988) and Trewartha (1945).

5. For the distinction between organic and inorganic economies, see Wrigley (1988). I am also indebted to Brinley Thomas for pointing out this distinction in his Royer lectures delivered to the Department of Economics at the University of California, Berkeley, during the late 1970s.

6. Jones (1988) also makes a distinction between intensive and extensive growth. However, his use concepts and mine differ somewhat.

7. On the importance of regional concentration for economic growth, see inter alia Ausbel and Herman (1988), Boserup (1981), Evans (1985), Henderson (2000), and Hirsch (1993).

8. For details concerning the theories of Veblen, see Rutherford (1994).

2

Under *Bakufu* Rule

Extensive Economic Growth

Before the Japanese economy became industrial, it became proto-industrial. And, before its proto-industrial (craft) production sector flourished in a phase I call intensive economic growth, Japan passed through a protracted phase of extensive economic growth.

Extensive economic growth involved a massive increase in the size of the agricultural economy (especially in the magnitude of rice output), the population, and the scope of markets. Due to the surge in the size of markets, specialization and division of labor were spawned, out of which emerged technological and institutional innovation. New seed varieties and tools were introduced into farming. Merchant houses and shippers pioneered the regularly scheduled transport of fertilizers, processed foods like soy sauce, cooking oils, and rice wine. Furthermore, a brisk commerce developed in cotton and silk fabrics that hitherto had been mainly produced by individual households for their own internal consumption. Markets expanded because the population grew and output grew. But markets also expanded because the physical infrastructure supporting trade and distribution was vastly improved during the period of extensive growth. Moreover, markets expanded because the financing of shipping and investment in land reclamation, the introduction of new agricultural crops, and craft production grew in sophistication and scale. Indeed, infrastructure investment was an important component of aggregate demand during the period of extensive growth, and it paved the way for much of the expansion in output that occurred during that period.

That the Japanese economy became increasingly sophisticated and diverse under conditions of extensive and intensive growth should not blind us to the fact that it remained a natural, organic economy.[1] From the early sev-

enteenth century until the 1870s when steam power and coal began to fuel manufacturing production, the Japanese economy operated under the constraint of negative feedback in the sense that expansion in the scale of output was sharply limited by the available land area, in particular by the amount of vegetation that could be produced on the surface of the land, and by the volumes of wind and water circulating above, on, and under the surface of that land. In a negative-feedback economy, increases in the production of calorie-yielding vegetation won through land reclamation, irrigation, and drainage stimulate increases in population. As increasing numbers press up against available surface-level resources, the costs to further extension of growth soar. And in a negative-feedback economy, resources must be poured into maintaining an infrastructure constantly ravaged by the natural forces of wind, rain, flooding from ocean and river, and earthquakes, for the infrastructure of organic economies is mainly constructed from timber, rock, soil, and vegetation. In short, there were important limits to growth operating within Japan's organic economy.

However, the existence of limits to growth in the organic economy did not spell the end of growth. It continued but in a new guise. With the innovations attending vigorous expansion in the output of the proto-industrial sector, diversification into new types of production permitted economic progress—albeit slow progress—to continue in the face of limits to extensive growth. Nevertheless, as long as an economy fails to exploit the vast reserves of energy stocked up under the soil or in reservoirs of water, as long as it is an economy based on natural forces rather than fossil fuels, the pace of productivity gain is strongly limited by the specter of negative feedback, by the amounts of energy that an individual worker can harness, and by the costs of maintaining infrastructure.

Infrastructure construction was crucial to extensive growth. As an example, consider two of the most striking achievements of extensive growth: the creation of a huge market for the selling and distribution of rice; and the emergence of a core region knitting together far-flung regional markets through its commercial activities, namely the Tōkaidō. The buildup of Osaka perfectly illustrates both achievements. The Osaka of extensive growth was designed to be a water city. In many ways, it was akin to seventeenth-century Venice or Amsterdam. The goal of the *bakufu* in encouraging merchants to gather there was to create a national rice market, where rice—especially from central and southwestern Japan—was to be bought and sold in large quantities.[2] The authorities planned, and secured labor and resources for creating, a vast network of canals in the great delta of the Yodo River where it drains the rivers of the Kinai region into the Inland Sea. Plying the checkerboard of canals were wooden flat-bottomed boats carrying passengers and

goods. Thus was the Osaka of extensive growth born. Osaka's surge to prominence was rooted in its handling rice, which not only served as the main medium for collecting taxes but was also the medium for reimbursing the *samurai* military class.

Paradoxically, it was the drive to demilitarize that informed the policy of taxing agricultural villages in terms of rice, and then redistributing most of the fruits of that taxation to the *samurai*. To prevent local conflicts from erupting in the countryside, *samurai* were forcibly relocated into castle towns controlled by the *daimyō* to whom they owed fealty. Since the *samurai* were no longer able to farm, the *daimyō* taxed the peasantry and distributed substantial amounts of that revenue to his military retainers. Thus, Osaka, bristling with warehouses for rice and canals for moving it in boats, was a potent symbol for the politics of pax Tokugawa.

The political system that nurtured Osaka's rise to preeminence is known as the *bakuhan* system. And the construction of the word itself—combining the Chinese character for *baku* (short for *bakufu* meaning "tent government" and referring to the central authority of the *shogun*) with that for *han* (meaning a fief or domain controlled by a *daimyō*)—provides us with an important clue as to how the political system of early modern Japan operated. For the word "*bakuhan*" indicates that power was shared between regional authorities, the *daimyō*, and the central ruler, the *shogun*. Thus it was a federal system of a sort.

Testifying to its vital importance as the place where fiscal resources were extensively handled and marketed, Osaka fell under direct *bakufu* rule. Indeed, like the *shogun*'s own castle town of Edo in the Kantō plain, Osaka was administratively controlled by the central government. Indeed, Toyotomi Hideyoshi—who, in the closing decades of the sixteenth century, managed to become establish hegemonic control over all of Japan and fashioned the policy of demilitarizing the countryside—ordered the construction of a menacing castle at Osaka from which he intended to rule. But, in the struggle to take over Hideyoshi's status as *shogun* after his death, an alliance of warlords loyal to Tokugawa Ieyasu, whose fief was in the Kantō plain, managed to defeat its enemies and strip away the authority of *shogun* from the Toyotomi dynasty. The decisive battle occurred at Sekigahara in 1600, and after that battle the Tokugawas established their capital in Edo. When *samurai* loyal to Hideyoshi's son gathered in Osaka castle during the second decade of the seventeenth century, the *bakufu* attacked their forces, and in the process partially destroyed the great edifice. After this victory, the *bakufu* established direct rule over Osaka.

Thus, Osaka was a divided city. On one hand, it was a castle town of sorts, where *samurai* and administrative representatives of the fiefs that maintained warehouses in the environs of the city resided. At the same time, it was home

Map 2.1 **Osaka During the Tokugawa Period**

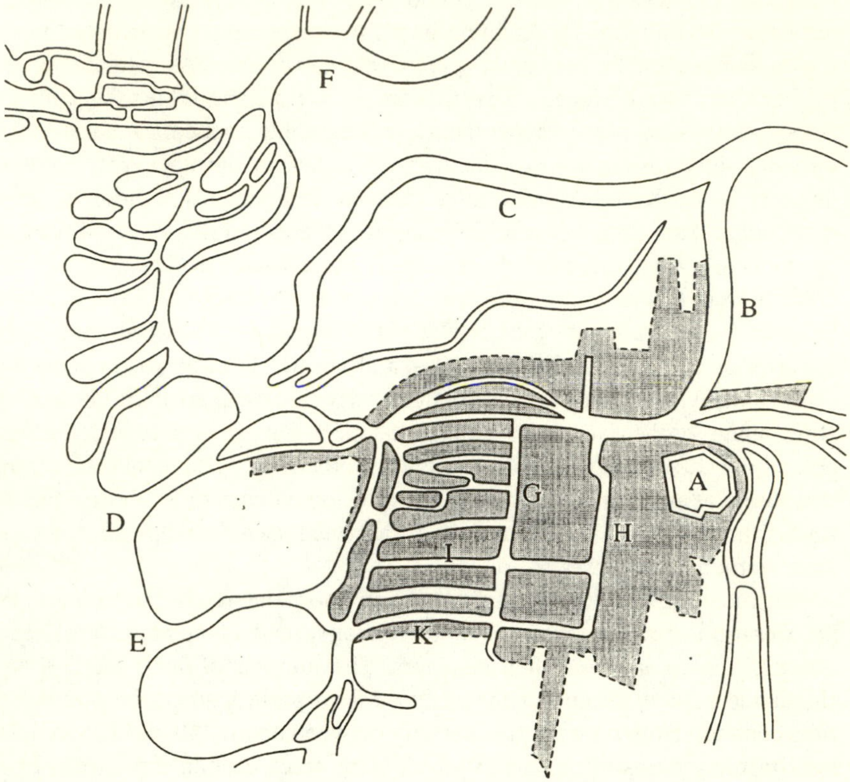

A Osaka Castle
B Yodo River
C Nakatsu River
D Mouth of Aji River
E Mouth of Shirinashi River
F Kanzaki River
G Nishiyoko Canal
H Higashiyoko Canal
I Naga Canal
K Dōton Canal

Sangō

Note: The shaded area labeled Sangō includes the districts occupied by the samurai. These districts occupied by the samurai are not formally part of Sangō.

to an equally impressive assemblage of powerful and wealthy merchants. It was formally divided into several distinct districts: a collection of neighborhoods occupied by *samurai*; a district where outcasts resided; and three groupings of residential quarters mainly occupied by merchants and artisans, each district controlled by a city magistrate (the term "*Sangō*," which is occasionally used to describe Osaka during the early modern period, refers to the three groupings under the direct control of the city magistrates). A rough sense of the layout of the great metropolis is given in Map 2.1. Thus Osaka was simultaneously an important military district, a political and administrative center, and a mighty concentration of commercial and marketing activity.

Building cities and connecting these nodal points together with roads, or along rivers and the ocean coastline, was one major aspect of early modern Japan's infrastructure buildup. The building boom was especially concentrated in the region that became the economic core of Tokugawa Japan. As the national rice market, Osaka was designed to be in the core of the emerging Tokugawa economy. And—while being precise about the exact geographic boundaries of the core is difficult—it is reasonable to also include within the core the other great metropolis of the Kinai, Kyoto. Moreover, the immediate hinterlands of these two great metropolises, the Yodo River basin connecting them, and its hinterland were all economically advanced, partly because agricultural output per hectare was especially high in this region. Thus, Osaka, Kyoto, the Yodo River basin, and their hinterlands, were all part of the core. But the core did not run just through the Kinai. It also ran northward from Osaka along the Tōkaidō road connecting Osaka and Kyoto to Edo.[3] For the Tokugawa's political capital at Edo, staked out at the mouth of the rivers debouching into Tokyo Bay, emerged as the northern pole of the core, just as Osaka became the southern pole. But connective infrastructure was not limited to the core. It also joined core to periphery. For instance, the individual fiefs created roads that were used for both administrative and commercial purposes.

The most important facet of the building boom outside of the core was the proliferation of castle towns, *jōkamachi*, that served as the residential headquarters for *daimyō* and their retainers. The number of such castle towns fluctuated somewhat during the early modern period, since the number of fiefs changed from time to time, varying between about 240 and 295 in number. In many cases, these castle towns were erected upon a pre-Tokugawa foundation. Originally, many had been staked out as market towns, their sites reflecting natural advantages as transportation nodes.[4] Under the federal structure that developed with *bakufu* rule, castle towns took on a rich combination of administrative and commercial functions. None were formally called cities—the Chinese character for city, *shi*, was not applied to any conurbations until the 1880s—nor were they under some form of municipal government rule as were some cities in the West at this time. Indeed, under the standard administrative model of the *bakuhan* system, a castle town was administered by one (or several) *machi bugyō*, municipal administrators appointed by the ruler of the domain governing the castle town. In practice, neighborhood (*chō*) associations played an active role in managing urban affairs, organizing religious events, and assembling and recruiting fire brigades to fight conflagrations.[5] In sum, the construction of castle towns created a network of regional administrative cum commercial nodes connected by roads, rivers, canals, and seacoast, so that core was linked to periphery, and points in the periphery were linked to one another.

The networking of nodal points in core and periphery was partly achieved through the building of dikes and embankments along rivers. Moreover, systematic effort was made by national and regional authorities to straighten out river flows. But the greatest effort of the central authority was put into constructing the Gokaidō road network, consisting of 5 main roads, 8 auxiliary roads, and almost 250 post stations. The investment in this infrastructure creation perfectly illustrates the connection between the political goal of demilitarization and infrastructure construction in early modern Japan. Key to the pax Tokugawa was the idea of a balance of power checking the ambitions of individual *daimyō*. By monopolizing all intercourse with foreign powers, and by introducing an isolationist policy vis-à-vis the Western powers (the Dutch being the sole exception), the *bakufu* preventing individual fiefs from coming under the sway of, and forming alliances with, foreign powers. Moreover, by decreeing that each *daimyō* maintain a residence in Edo and attend the *shogun*'s court on an alternating basis—a policy known as *sankin kōtai*—the *bakufu* could carefully monitor the activities of its potential rivals for power. As part of the policy, the *daimyō* were required to leave their children and wives in Edo on a full-time basis, thereby making them virtual hostages of the *bakufu*. And in order to adequately monitor the comings and goings of the *daimyō* and their retainers along the Gokaidō network of roads leading toward Edo, the *bakufu* built checkpoints along the roads where travelers were subject to search.

The political logic of road construction is clear, but economic benefits also flowed from the infrastructure so created. Commercial and recreational travel took place along these roads. Thus, the scope of markets was enlarged and enriched.

For linking nodal points, road construction was probably more important than the revamping of rivers. Nevertheless, the economic impact of reconfiguring rivers was surely tremendous. To understand why working on the flow paths of rivers was so important, the nature of Japan's annual rainfall must be mentioned. It tends to be concentrated in two comparatively short periods: the early summer when monsoons sweep across the country, and the autumn when typhoon storms buffet the archipelago with rampaging winds and drenching rain. Because Japan is mainly mountainous and because the rivers tend to drop down through steep mountain gorges at precipitous angles, the volume of water flowing through rivers varies tremendously from season to season. During periods of intense rainfall, rivers swell. This generates tremendous churning, which throws up silt, erodes land, and causes flooding. At the same time, the water was potentially valuable, provided it could be harnessed to paddy rice production in the form of irrigation flow. Thus, the problem was how to control water flows and how to distribute

them so that the land area under irrigation could be maximized. Coordination between local communities was essential for harnessing water for irrigation purposes and for flood control. And demilitarization of the countryside under *bakufu* rule created the conditions necessary for coordination. By squelching local conflicts over water rights, by promoting the reworking and shoring up of river banks, with mud and rock-strewn levees, and by dredging river beds so that flat-bottomed wooden boats loaded with timber and foodstuffs could ply rivers, the nascent *bakuhan* system gave a huge push to expansion of agricultural production.

As a result of these riparian infrastructure projects, extensive swaths of land were converted from dry field usage to paddy field production. New paddy fields, *shinden*, were created through the building of irrigation ditches. Also, they were created through piling up masses of land in areas hitherto covered by water, namely at lakes, rivers, and the shores of the ocean. Thus, paddy acreage expanded at a heady rate, increasing by 70 percent between 1450 and 1600, and by an additional 140 percent between 1600 and 1720.[6] In short, if the building of massive administrative, market, and transportation hubs like Osaka and Edo constituted the most dramatic achievement of early modern Japan's infrastructure buildup, the creation of *shinden* through improvements in the management of water was the main engine behind extensive economic growth.

With the increased capacity of the land to sustain human numbers, population growth took off. Lack of credible data for Japan prior to 1720 makes it impossible to put an exact figure on the increases in rice production and the associated expansion of human numbers between 1600 and 1720. Point estimates for population growth vary between 0.3 percent and 1.0 percent.[7] But it is clear that the scale of agricultural output vastly increased, and as the capacity to feed people improved, the number of mouths swelled. From the point of view of supply, land reclamation, extension of the irrigation system, control of flooding, all contributed to growth in the amount of food. And from the point of view of demand, the creation of a nationally integrated market with the linking of nodes in core and periphery, by making it possible for producers to market surpluses, increased demand for foodstuffs.

The restructuring of the Yodo River basin in the Kinai usefully illustrates the interrelationship between infrastructure improvements in core and periphery. The Yodo River is formed from the confluence of the Katsura and Kamo Rivers (which cut through the city of Kyoto); the Uji River, which flows out of Lake Biwa; and the Kizu River, which empties out of the Nara basin. Fed by these rivers flowing out of the mountains, the mighty Yodo makes its way into the Inland Sea at Osaka Bay. In shoring up the banks of the Yodo, in redirecting it by removing meanders, and in carving from it

drainage and irrigation ditches, the Tokugawa-era *bugyō* administrators were naturally drawn into structuring anew the configuration of the river, especially where it emptied into the Inland Sea at Osaka.

Thus, the huge engineering project resulting in the constructing of a checkerboard pattern of canals within Osaka was part and parcel of the reworking of the Yodo River. As can be seen from Map 2.1, much of Osaka during the early modern period consisted of islands created by channeling out canals throughout the area. In the heart of Osaka—the area bordered by the Nishiyoko Canal on the West, the Higashiyoko Canal on the east, the Dōton Canal on the south, and the tributaries emanating from the Yodo River on the north—was the principal concentration of merchant and artisan residents.[8] Crammed into the slim sliver of land perched just north of this warren of canals and known as Nakanoshima (literally "Central Island") were many of the warehouses of the *daimyō*. Concentrated into this small space were vast fiscal reserves of the *bakuhan* system, handled by the bailiffs employed by fiefs. In short, this key piece of geography in the heart of the Tōkaidō core was a tiny product of the great infrastructure buildup on the Yodo River, which helped knit together core and periphery.

Moreover, creating *shinden* that was so essential to stimulating economic advance in the periphery played a crucial role in the reconfiguring of the Osaka region. Consider Map 2.1. The area lying west of the shaded district on the map consists of a series of island like masses of land divided by rivers. The rivers flowing through this region are the Kanzaki, Nakatsu, Aji, Shirinashi, and Kiso. And the masses of land are landfill, *shinden*. Some of these were created by the force of nature as tree branches trapped in the floor of the river served as gathering points for mud and silt rushing down the Yodo River and the smaller rivers debouching into Osaka Bay. And some of these *shinden* were the deliberate creation of contractors employed by merchants and government authorities interested in creating new agricultural fields that they could rent to farmers. In short, methods of infrastructure creation reconfiguring the periphery were also being utilized to fashion the key nodal points in the core.

Now, the massive building boom of the sixteenth and seventeenth centuries that ushered in extensive growth put unprecedented pressure on Japan's natural-resource base. An excellent example is the pressure upon forest reserves in the Kinai and Inland Sea districts, stemming from the building of grand wooden temples, shops, and residences in Osaka and Kyoto.[9] In appreciating the impact that the building of *bakufu*-controlled metropolitan centers and *daimyō*-controlled castle towns had upon the resource base of early modern Japan, it is important to keep in mind that capital depreciates. And the infrastructure of Tokugawa Japan depreciated rapidly because of

earthquakes, fires, and flooding. Moreover, cutting down forests created ecological problems. While four-fifths of Japan's land area consists of hills and mountains that were originally covered in forests, the soil on most of the steep hillsides is thin. Hence, the soil often loses nutrients through leaching and mudslides accompanying the torrential monsoon and typhoon rainstorms sweeping across the island archipelago. Thus, denuding a mountainous region eroded soil. And soil erosion led to the silting of rivers. As those forests in the immediate vicinity of the great metropolitan centers of the Kinai and the Tōkaidō were deforested, loggers increasingly had to penetrate distant and precipitous mountain reaches. Eventually, expensive methods involving sending down logs along mountainsides and transporting logs in the form of rafts were widely developed. But this raised the costs of securing additional timber. Thus, as with land reclamation, infrastructure construction and maintenance ran into diminishing returns and escalating unit costs. In sum, extensive growth ran up against limits to growth.

How did the economy respond to these limits? The most decisive factor involved population. It appears that limits to fresh expansion of arable land checked population growth. Indeed, at the national level, population growth virtually ceased after the 1720s. But policy also played a role in coping with the crisis. For instance, fiefs began to curtail the wholesale exploitation of forests, and they began to undertake reforestation programs. Moreover, the crisis gave a fillip to technological and organizational innovations. For instance, a system of linking chutes fashioned out of peeled logs was developed to slide logs down mountainsides. Splash dams feeding into wood-processing sites were built to facilitate the transport of logs along rivers.

As the technology for getting timber out of the periphery improved, the ranks of the merchant community based in the periphery swelled. Indeed, imitation of merchant practices pioneered in the core facilitated the growth of an increasingly sophisticated cadre of rural-based merchants and traders.

Limits to growth also spelled an end to the proliferating ranks of the number of *samurai* that had accompanied extensive growth.[10] As the number of *samurai* supported by rice stipends peaked, so did the economic advance of many of the castle towns. Indeed, during the eighteenth and nineteenth centuries, many castle towns entered into a period of stagnation or even decline. In short, the curtain on extensive growth was brought down during the early eighteenth century in most regions of Japan. Key symptoms of this climacteric were a checking of population increase, the encouraging of systematic management of resources used in infrastructure construction, and the clamping down on the creation of new "slots" for the *samurai* warrior-bureaucrats.

But, as limits to extensive growth were reached, a period of intensive

economic growth was ushered in. At the root of the intensive growth was improvement in the quality of the factors of production, innovations spawned by the specialization and division of labor attending expansion in the size and scope of markets, and creative response to the limits to extensive growth. The average worker became more productive because he or she now worked with a superior tool, or was healthier and stronger than his or her ancestors, or had mastered a new technique, or was more efficiently able to take advantage of the winter season by concentrating on making craft products for market. In short, intensive growth was forged in the crucible of extensive growth.

Intensive Economic Growth

By intensive economic growth, I mean improvements in the quality of the factors of production; growing commercialization in the organization of production and marketing; technological and organizational innovation; and increased diversity in the use of raw materials and in the commodities and services supplied to market. Let us consider some concrete examples. A growing stock of threshing frames and plows, an expanded water-transport base consisting of flat-bottomed wooden ships and sailing vessels, the diffusion of silk-cocoon raising in the mountains of central Honshū and the growing of raw cotton in the Kinai, the nurturing of plum orchards and orange groves in the Kii peninsula near Osaka, the processing of dried fish remains for fertilizer along the northern reaches of the Tōkaidō region—all are illustrations of the innovations and the increasing diversity of production in intensive-growth Japan. A noteworthy feature of these examples is the strong interrelationship between agriculture and the craft or proto-industrial goods proliferating during the latter part of the Tokugawa period. Consider the following concrete examples of inputs into proto-industry: raw silk generated from cocoons that feed on mulberry plants; raw cotton serving as an input into the spinning sector; rice as an ingredient in the manufacture of sake rice wine. Because most of the inputs for proto-industry were produced in agricultural villages, it was natural that manufacturers exploiting these inputs also sprang up in the countryside.

In short, the principal locus for intensive growth was the countryside. The organic sources of energy (wind, water, and fire) were relatively abundant in rural districts, and so were the raw materials. Thus, farm households began to diversify, reducing their dependence on farming and farming alone. As craft production opportunities proliferated in rural villages, parents increasingly invested in training their children in a diverse range of skills. And, because the sites for producing material inputs exploited in proto-industry tended to be farming villages, the skills essential for working with the raw

materials were readily transferred from agriculture to manufacturing. Thus, those regions like the Kinai that enjoyed bumper agricultural yields were also natural spots for the gestation and diffusion of proto-industrial activities. The existence of good infrastructure supporting widespread marketing of output also promoted craft production in the Kinai. The infrastructure in the Tōkaidō/ Kinai core was superior to the infrastructure in the periphery; thus, innovation and diffusion of proto-industrial activity tended to be concentrated there as well.

Pioneering quantitative research concerning Japanese proto-industrialization undertaken by Osamu Saito (1985, 209–11) bears out the superiority of the Kinai in both agriculture and in the new areas of production.[11] In Saito's geographical decomposition, the term "Kinai, narrowly defined" refers to the five fiefs in the vicinity of Osaka and Kyoto; and the term "Kinai, surrounding area" refers to eight more domains in the area. According to Saito's estimates for the 1870s, population density, urbanization, and agricultural output per *chō* were far higher in the Kinai than they were in the rest of Japan.[12] For instance, population density in the "Kinai, narrowly defined" was 15.2 persons per *chō*, urbanization levels were 35 percent, output of cereals in yen per *chō* was sixty, and output of industrial crops was twenty-nine yen per *chō*. For the "Kinai, surrounding area," these figures were, respectively, 9.7, 13 percent, 54 yen, and 8 yen. For Japan as a whole, the respective averages were 7.8 persons per *chō*, 16 percent urbanization, 36 yen per *chō* for cereals, and 7 yen per *chō* for industrial crops. Figures for the Kantō region surrounding Edo are roughly comparable to those for Japan as a whole. The superiority of the Kinai in terms of agriculture—especially in terms of the density of rice fields—and industrial crops was definitely associated with superiority in terms of the index of village industrial activity.[13] For instance, Saito's estimate for the index in the "Kinai, narrowly defined" is .474; for the "Kinai, surrounding area," it is .148; and for Japan as a whole, it is .204. Of the regions for which Saito gives figures, the only regions outside of "Kinai, narrowly defined" enjoying levels for the index of village industrial activity exceeding that for the country as a whole are Eastern Japan (with a level of .236) and Kantō West (with a level of .376). And, like "Kinai, narrowly defined," the region known as "Kantō West" was part of the Tōkaidō.

In short, proto-industry flourished in the rich agricultural environs of the Kinai, and to a lesser extent in the remainder of the Tōkaidō. That is, the rural sector of, or contiguous to, the core became the seat of proto-industrialization during intensive growth.

Now, it should not be forgotten that, unlike extensive growth that involved population increase and an expansion in the scale of activity, intensive growth

was primarily associated with enhancement in the level of income per head. That population stagnated at the national level reflected the fact that extensive growth had run into limits, and the fact that, as a rule, proto-industrialization in Japan did not promote demographic increase.[14] In fact, the national picture showing stagnation in population growth masks some important regional differences. It is evident from recent research that in some regions of the country, population growth continued unabated, and in other districts, decline and atrophy set in. However, in the locales where population growth lost its impetus, it seems that the principal cause was plummeting rates of reproduction within marriage, coupled with delays in getting married.[15] In most areas of the country, and especially in the Kinai, proto-industrialization tended to cause marriage ages to rise for females. For with jobs springing up in nearby craft-producing centers, farm families found themselves increasingly tempted to dispatch their daughters to manufacturing concerns for a year or two. Or, farm families found themselves investing in their own spinning wheels and weaving frames, holding onto their daughters for an extra couple of years before arranging for their marriages. To be sure, there were exceptions to this proposition. There were communities north of the Kantō where, prior to the diffusion of proto-industrialization, the progeny of the landless were consigned to the ranks of the celibate. But as demand for craft workers boosted their earnings prospects, marriage became a possible outcome for these formerly hapless villagers.[16] In general, however, intensive economic growth put a damper upon demographic increase.

An important impetus for growth in rural craft activity was the increase in the ranks of the rural village merchants. Many of these merchants imitated the practices developed by urban merchants, especially in Osaka, who pioneered accounting and bookkeeping, the development of futures markets, and the provision of insurance for long-distance voyages.[17] The wholesaler merchant, or *tonya*—pronounced *toiya* within Osaka—increasingly diversified out of rice into a wider variety of activities in the great metropolitan centers of the Tōkaidō. That the community of Osaka *tonya* had decisively shifted away from mainly handling rice is indicated by the fact that by the 1670s, there were over fifty types of *tonya* plying their trade in the great metropolis at the mouth of the Yodo River. By then, Osaka-based merchants were active in handling cotton, oils, paper, firewood, charcoal, fish, fowl, swords, copper smelting, and so forth. Indeed, two of the families that eventually created the famed financial cliques, the *zaibatsu*, of the post-1870 period, the Mitsui and the Sumitomo families, were active in non-rice-related activities in the Osaka region during the sixteenth century. *Tonya* developed their own *shinden*, rented land, ferreted out raw materials in the hinterland, and assiduously expanded the geographic scope of their trade networks. For

instance, as logging penetrated deeper and deeper into the periphery, specialist houses in Osaka and Kyoto financed the cutting and transporting of timber on a large scale.

Of particular interest to the Osaka merchant community was the management of the putting-out production of cotton textiles. In putting out, the merchant provides the raw material to the households or individuals doing spinning, weaving, or dying, and pays for the work in stages. The Kinai was especially well suited for cotton production, the environment—soil quality, climate—being ideal for spinning and weaving as well as for the growing of the cotton itself. Thus, the merchants, first in the core metropolitan centers and later on in the periphery, were in the forefront in encouraging the diffusion of proto-industrial activity. Typically, the merchant houses in the conurbations of the Tōkaidō were the innovators in developing markets for the new products, and the merchants of the periphery imitated them.

And imitation gave rise to competition and conflict. For as wealthier farmers cum rural merchants emulated practices of the urban merchant houses, competition between the two groups naturally arose. The powerful merchant houses of Osaka did not view the growing competition with equanimity. They turned against the imitators. In doing so, they exploited the financial distress visited upon the political authorities due to the shift from extensive to intensive growth. For as the growth trajectory in rice production leveled off, so did the growth trajectory in the fiscal resources flowing into fief and *bakufu* coffers. For this reason, the political authorities were willing, on occasion, to heed the clamor of the Osaka *tonya*, selling them monopoly and oligopoly rights within certain lines of business. The authorities periodically issued licenses guaranteeing exclusive marketing rights within specialized lines of business to designated merchant associations—*kabu nakama*—in exchange for funds. Wielding these licenses, the *kabu nakama* struck back at their competitors. In effect, by issuing these licenses, the government was encouraging the creation of cartels.

But it was not a unified bureaucracy that agreed to the licensing of cartels. On the one hand, sale of the licenses brought in needed funds to the *bakufu*. This revenue supplemented rice taxes whose volume had leveled off as extensive growth ran out of steam. On the other hand, permitting cartels tended to stifle competition, discouraging enterprise in villages and generating opposition by consumer groups. For this reason, the *bakufu* was increasingly drawn into factional disputes within its own bureaucracy as well as in the community of merchants. Osaka merchant groups were vocal in arguing for their prerogatives. For instance, the Osaka political elite was especially vocal in defending the interests of the cartel created in the cotton industry, the *wataya nakama*.

In becoming embroiled in the affairs of cartels forged among the merchant elite in the great metropolitan centers like Osaka, the *bakufu* established an important precedent that was followed by post-Tokugawa governments: The *bakufu*—and especially its agents appointed the great cities like Osaka—helped to monitor the activities of the cartel members, thereby keeping them "honest." By monitoring the members of the cartels, the authorities kept cartel members from abandoning the cartel in favor of potential rival coalitions, thereby keeping the cartels strong. In short, the relative deterioration of the fiscal fortunes of the fief and central governments whose fiscal base in rice production failed to keep pace with growth in the economy as a whole encouraged the growth of cartels receiving official blessing.

Regional Competitive Advantage

Intensive growth reworked the geographic structure of Japan's economic core, shifting the heart of the core away from the great metropolitan centers of the Tōkaidō and toward the rural hinterlands of these centers. In particular, intensive growth favored Osaka's hinterland in the Kinai region. If we think of competitive advantage in a "zero-sum" sense, so that gain in competitive advantage by one region necessarily comes at the expense of regions that compete against it, then we can say that innovation in intensive-growth Japan favored the competitive advantage of Osaka's hinterland. The entire Tōkaidō economy grew as a result of proto-industrialization, but some regions grew more rapidly than others.

In appreciating the economic logic of the geographic shift under intensive growth, it is important to keep in mind the wellsprings of intensive growth. The key factors that supported it were (1) technological progress and innovation induced by the specialization and division of labor arising from the expansion in size of markets during extensive growth; (2) per-worker productivity increase due to investing in improving the education and skills of workers; and (3) reliance on hitherto underutilized resources, like labor services during the winter season when the demands placed upon farm families were minimal.[18] As noted before, this type of growth was realized in enhanced output per worker rather than in a larger and burgeoning populace. For this reason, it is misleading to measure a region's changing ("zero-sum") competitive advantage in terms of its demographic experience during intensive growth.

In short, extensive growth and intensive growth were profoundly different. Under the former, but not the latter, the demographic experience of a region corresponded to its economic experience. The point is worth stressing because an influential thesis advanced by Thomas Smith (1988) assumes

that regions failing to grow demographically in the eighteenth and nineteenth centuries were undergoing some form of economic decline, at least in relative terms.[19] Noting that many castle towns declined in size during the latter half of the Tokugawa period, and that in general the country tended to reverse the trend toward increased urbanization established in the seventeenth century, Smith (1988) argues that economic status of most conurbations plummeted during the later Tokugawa period because (1) the income distribution shifted in favor of the rural villager and against the *samurai* who populated the castle towns; (2) a lack of international trade led to the atrophy of seaports like Osaka; and (3) proto-industry tended to expand in farming areas, farmers having an especially strong incentive to shift into craft production because they lessened their tax burden by diversifying away from rice.

It is true that the position of the *samurai* was eroding as extensive growth gave way to intensive growth. And it is true that the demographic growth of seaports like Osaka was limited by the prohibition on most international commerce. But it does not follow that the economic core of the country staked out during the period of infrastructure buildup and extensive growth was sapped of its vigor, losing competitive advantage to the periphery. Indeed, the core of the early Tokugawa period economy was the Kinai, especially the conurbations Osaka and Kyoto, and the Yodo River basin. This was the core of the core during the seventeenth century, and it continued to be the core of the core in the eighteenth and nineteenth centuries. To be sure, cities like Osaka did experience population decline, but this reflected a shift in the locus of growth from Osaka proper to its hinterland. Osaka—especially the city and its hinterland—continued to flourish, especially in terms of per capita output.

However, there is no doubt that Osaka proper did enter a period of demographic decline under intensive growth. Consider the figures for the period 1721–1855 marshaled in Table 2.1. It can be seen that Osaka's failure to grow in human numbers, so evident in the table, was characteristic of the Kinai region as a whole. In terms of population growth—as opposed to growth in per capita income—the Kinai district was not dynamic during the era of intensive economic growth. By contrast, both Edo and the Kantō region show signs of greater demographic vigor than do Osaka and its Kinai environs. Indeed, while Osaka went into a precipitous decline in population during the eighteenth century, Edo—which was already a city of a half million residents in 1740—continued to surge forward in terms of human numbers, despite the fact that nearly ninety major fires broke out in Edo and its environs between 1601 and 1868.[20] But it would be misleading to conclude from these figures that the economy of Osaka and its Kinai environs went into a tailspin while that of Edo flourished. As noted before, what counted during intensive growth was not growth in population but growth in income per capita.

Table 2.1

Long-Run Population Growth: Kinai-Osaka and Kantō-Edo, 1721–1872

A. Indices of population (1798 = 100) for the Kinai and Kantō regions and for all Japan, 1721–1872.

	Kinai[a]			Kantō[b]			
Year	Narrowly defined	Surrounding domains	Total	North	South	Total	Japan
1721	109.8	108.2	108.9	129.7	112.0	117.7	102.3
1756	105.9	105.5	105.7	123.6	109.9	114.7	102.4
1786	99.6	102.9	101.5	103.7	99.1	100.6	98.5
1804	98.5	99.8	99.2	97.7	99.3	98.7	100.6
1828	102.5	104.5	103.6	94.1	102.7	99.9	106.7
1846	97.6	102.6	100.4	93.6	106.1	102.0	105.7
1872	98.8	115.5	108.1	116.5	120.1	119.0	129.3

B. Estimates of population (1,000s) for Edo (Tokyo) and Sang) (Osaka), 1740–1855.

	Approximate date			
City	1740	1830	1840	1855
Edo	501.0	546.0	529.8	574.0
Sangō	403.7	374.3	341.5	321.2

Sources: Hayami (1996), Table 3; Shinshū Ōsaka-shi Shi Hensan Iinkai (1989), *Ōsaka-shi shi*, vol. 4, pp. 199, 200, and 205; and Saito (1987): p. 130.

Notes: [a]Kinai narrowly defined consists of the five domains in the immediate environs of the Kyoto, Nara, and Osaka conurbations. The surrounding domains are eight in number. The area encompassed by the total of thirteen domains is roughly equivalent to the region known today as the Kinai region.

[b]Within Kantō there are three domains in the north and five domains in the south.

What explains the diverging fortunes of Osaka and Edo under intensive growth? Two explanations are compelling. One involves the relative degree of economic competition from the hinterlands of the conurbations; the other, the nature of their labor markets. In emphasizing the flourishing of proto-industry in the environs of Osaka, I have touched upon the first point already. As for the second point, it has been established by a number of scholars that the labor markets of Osaka and Edo operated differently.[21] One of the most important innovations pioneered by the merchant houses concentrated in Osaka was the internal labor market in which a new worker was recruited into the house at a very young age, and internally promoted and trained exclusively by the house. A typical recruit entered the house as a penurious apprentice (*detchi*), aspiring to be promoted to clerk (*tedai*), and eventually chief clerk (*bantō*). Saito (1987) argues that this system of slow promotion

tended to lead to late marriage and low rates of population increase. By contrast, according to both Gary Leupp (1992) and Osamu Saito (1987), there was a large "casual" labor market in Edo, which was populated by common laborers and servants. These workers moved around from one employer to the other in a fluid fashion and were not constrained by formalized promotion ladders. Thus, they did not put off marriage until they were promoted. As a result of having a large pool of fluid workers, Edo had a stronger demographic impulse than did Osaka. But Osaka, and especially the greater Osaka region including its hinterland, had a stronger economic growth pulse than did Edo.

Tokugawa Infrastructure in Decline and Crisis

While Japan did not completely seal itself off from the Western powers—the Dutch settlement at Dejima Island served as a window onto the West, and the Dutch community as a conduit through which the *bakufu* informed itself about some of the technological advances made in the West—the diffusion of Western technology into Japan was severely limited under *bakufu* rule. Thus, as an increasing number of Western European economies made the transition to an inorganic economy and embraced and railroads—adopting the steam engine for use in textile factories and transportation, the Darby and puddling processes in iron and steel production, and the flying shuttle and spinning jenny—the gap between Japan and the West widened. As luck would have it, the Dutch were laggards in harnessing the new technologies of the inorganic economy. Thus, Japan's window on the West failed in two dimensions: it was too small, and its conduit to Western advances was inappropriate for appreciating the gap that had emerged between itself and the West.

Because the new industrial and transportation technologies had important military applications, the widening gap between Japan and the West was perhaps most telling to the *bakufu* advisors and the *samurai* in the fiefs in the fields of warships, cannon, and guns. Indeed, as the gap widened, the capacity of Japan to ward off Western threats to its sovereignty deteriorated. Hence, during the 1850s, Japan found it impossible to stave off the Western powers that wanted to open it up.[22] The Western powers were interested in opening up Japan for a variety of reasons: to secure coaling stations for their fleets and merchant marines; to secure new markets for goods and fresh sources of supply of resources; and to bring more and more of Asia into the international system of diplomacy based on the Western concept of the nation-state.

Between the 1850s, when the Western powers pried Japan open, and 1868, when the *bakufu* completely collapsed, Japan passed through a period of transition. During that transition, the financial, human-capital enhancing, and

physical infrastructures of the *bakuhan* system were continually and harshly tested and found wanting. As the credibility of the physical and institutional bases of Tokugawa infrastructure was made a mockery of, the credibility of the entire political system was called into question. The *bakuhan* system collapsed because its entire infrastructure was increasingly viewed as inadequate. Correcting one or two deficiencies was not enough. If Japan was to mount effective resistance and response to the Western challenge, root and branch reform was needed.

But while it was evident to most of the Japanese, who became aware of the dimensions of the Western threat, that root and branch reconstruction of infrastructure was desirable, particular groups and individuals had a vested interest in resisting reforms. Osaka's financial fate under transition well illustrates the themes of infrastructure inadequacies and the fact that reform seriously undermined vested interests.

Under *bakufu* rule, Osaka's money supply operated largely on the basis of silver. Indeed, silver was heavily mined in Japan during the Tokugawa period, and sold to the Dutch in exchange for pelts and other commodities. The Sumitomo fortune was heavily tied up with its interest in silver smelting. The rest of Japan operated on a money system in which three metals—gold, silver, and copper—were jointly used for coinage. The *bakufu* monopolized coinage but allowed fief governments to issue paper currency, a practice that even it eventually adopted.

During the negotiations between the Western powers and Japan that took place throughout the 1850s—talks that revolved around Japan's conceding extraterritorial rights in selected treaty ports including Osaka, coaling stations, and permission to carry on commerce—diplomatic negotiations over the establishment of a medium of exchange that would both serve Japan's domestic needs and the needs of the foreign community became especially contentious.[23] Because the Mexican silver dollar was being widely utilized by Westerners active in Asia, the Western negotiators pressed for conversion of the Japanese coinage system to a form consistent with the metallic content standard in the Mexican silver dollar.

The *bakufu* eventually acceded to the Western demands concerning coinage. In doing so, the government threw into jeopardy the financial positions of various communities throughout the country, and especially the merchants in Osaka. For the West operated with an international exchange rate between silver and gold, according to which one ounce of gold purchased fifteen ounces of silver. But within the sealed off economy of Tokugawa Japan, one ounce of gold was exchanged for five ounces of silver. As a result of this discrepancy in exchange rates, arbitrage carried out by Western interests—Westerners purchasing gold with silver in Japan, which they sold on Western

markets for silver with which they bought more gold in Japan—was devastating to those Japanese holding substantial amounts of silver. Since Osaka's merchant community held substantial portions of its assets in the form of silver, the new *bakufu* policy wreaked havoc on the merchants of Osaka.

In short, in attempting to meet the challenge of dealing with the Western powers, the central government rapidly lost legitimacy. The transition when this loss of legitimacy created sufficient momentum for reform and effective military revolt against the *bakufu* was protracted. But there is something wonderfully symbolic about the year 1868, when the old regime finally collapsed in ruins. This is especially true in the annals of Osaka, for in that year, flooding created havoc in the city. Flooding had been a periodic occurrence under *bakufu* rule whenever incessant rainfall swelled the waters of the mighty Yodo River. In 1868, waters once again spilled over the embankments so carefully built up along the great river during centuries of *bakufu* rule.[24] And, in that year, foreigners entered the city to take up residence in the Kawaguchi district according to the terms laid out in the treaty port agreement of 1858.[25] In short, the inadequacy of the Tokugawa regime and its infrastructure was made painfully apparent in the chaos brought on by nature and foreign power alike.

The civil war dooming the *bakufu* also created chaos in the Osaka of 1868. In the aftermath of the feeble attempt by the *shogun*'s armed forces from Osaka and the Kinai, fighting broke out in the streets of the city and in the environs of Hideyoshi's menacing castle. In the confusion in and around the massive structure, an uncontrollable conflagration broke out and rapidly spread, turning much of the original structure into smoldering ruins. The very infrastructure that had fostered extensive and intensive growth seemed to be crumbling.

In light of the superior technology marshaled by the West, Tokugawa infrastructure seemed woefully inadequate. Rather than being a point of pride, the infrastructure that had been painstakingly constructed during extensive growth at the key nodal points in the core of the country—in Osaka and in Edo especially—was becoming a source of shame for Japan as it was being forcibly opened up under a treaty port system that humiliated it. But what was to replace that ancient infrastructure? In the chaotic conditions surrounding the burning of Hideyoshi's castle, in the loss of life and property taking place in the wake of the failure of the Yodo River embankments to contain the restless churning of the swollen waterway, who could view Japan's future prospects with equanimity? Could the people of Osaka do anything but gaze in terror as the infrastructure of the old order perished in flames and floods? Could they do anything but stare in numbing fear as they were seemingly plunged into a new world whose logic and shape they could only vaguely and dimly discern?

Notes

1. I draw this discussion of negative and positive feedback from Wrigley (1987, 1988). For a discussion of the pre-industrial Dutch economy as an example of a highly developed and efficient organic economy functioning under conditions of negative feedback, see de Vries and van der Woude (1997).

2. For a recent English-language account of Osaka during the early modern period, 1600–1868, see McClain and Wakita (1999).

3. On the Tōkaidō road and the other great road networks knitting Tokugawa Japan together, see Vaporis (1994).

4. See Rozman (1973).

5. See, for instance, Ito (1996) and pages 184–88 in Shinshū Ōsaka-shi Hensan Iinkai (1990), *Ōsaka-shi shi*, vol. 4. Sally Hastings (1995) provides useful material concerning neighborhood associations in Tokyo, during *bakufu* rule and afterwards.

6. The figures appearing in the text are taken from Totman (1993, 149). For an extensive discussion of the manifold methods used to create *shinden*, see Kikuchi (1996).

7. Hanley and Yamamura (1977, 43–44).

8. In principle, a river in Osaka is and was differentiated from a canal within the city by a designation following the name of the river or canal: "*kawa*" or "*gawa*" stands for river, and "*hori*" or "*bori*" stands for canal. Throughout this book, in giving the names of canals or rivers, I translate the designation for river or canal. Thus I give "Aji River," rather than "Ajigawa." I make one exception to this practice. In the opening section of the book, in referring to the district of Osaka known as "Dōtonbori," I use the phrase that appears in guidebooks rather than the phrase "Dōton Canal."

9. On Japan's forests, see Totman (1985, 1993, and 1995). On the impact of fires in the great conurbations on the demand for wood in Tokugawa Japan, see McClain (1994).

10. See Murakoshi (1995). For a discussion of dynamics of Tokugawa population among various social classes, see Hanley and Yamamura (1977), Hayami (1993, 1996, 1997), and Saito (1985, 1987). Also see the references cited in note 15 for this chapter given below.

11.The term "proto-industrialization" was originally developed to describe pre-industrial craft production in Western Europe and England.

12. A *chō* is a unit of land area equivalent to 2.45 acres.

13. The ratio of land area in paddy to dry fields for the "Kinai, narrowly defined" was 3:2 and for the "Kinai, surrounding area" was 3.7. For Japan as a whole, the ratio was 1.4; for Eastern Japan, 1:1; for Kantō East, 1:1; for Kantō West, 0:6; for Central Japan, 2:2; and for Western Japan, 1:6.

14. In parts of Western Europe, proto-industrialization did stimulate population increase. Perhaps it also did so in some districts of Japan, especially in the areas west of the Kantō and in eastern Japan.

15. For a discussion of recent research findings in Japanese historical demography, see Hayami (1996, 1997), Hayami and Miyamoto (1997), Mosk (1983, 1996), Saito (1987), and Smith (1988).

16. See Hayami (1997) and Mosk (1996, chap. 3).

17. On the sophisticated practices of Osaka merchant houses, see Hirschmeier and Yui (1995), McClain and Wakita (1999), Morikawa (1992), and Saito (1987).

18. On investment in education and skills, see Mosk (1995a, 1996).

19. See Smith (1988, 15–49). Smith's thesis is based upon comparison between pre-industrial Europe and pre-industrial Japan. He notes that Japan's proto-industrial economic development was associated with a falling off of urbanization and atrophy in the economic vitality of many cities. By contrast, in Europe proto-industrialization and urbanization moved in tandem with one another. In addressing the question of why proto-industry developed in rural Japan rather than in the cities themselves, Smith fails to note that there was little change in the geographic locus of the core region in Japan. He also neglects to note that Japanese population densities in arable districts of rural Japan tend to be higher than in comparable districts of proto-industrial Europe. For this reason, the relative attractiveness of Japan's great cities as locales for exploiting scale economies in marketing was less than it was in Europe.

20. Osaka proper not only lost markets to competition from communities within its own hinterland. Osaka also lost markets to Edo, which had been dependent on Osaka for many of its goods during the seventeenth century, but which increasingly began to supply itself as the eighteenth century wore on. On the shift in production of goods to Edo and the Kantō, see Miyamoto and Uemura (1997). That Edo became less dependent upon Osaka and its environs, however, does not mean that the Osaka-centered economy declined. Indeed, it remained the core of the Tōkaidō core by expanding production and marketing throughout the periphery of Japan, as well as within the Kinai region (which was blessed with high per capita income) itself.

21. On the contrast between the labor markets of the two conurbations, see Leupp (1992) and Saito (1987). On the demographic expansion of the cities of Osaka and Tokyo after 1883, see chapter 3.

22. On the opening of the country, see Totman (1993, 465–551).

23. For a detailed account of the currency crisis, see Frost (1970).

24. In terms of the ecological contrast between Osaka/Kinai and Tokyo/Kantō highlighted in Table 1.2, it is interesting to note that a series of earthquakes struck the Kantō area during the 1854–55 period when Commodore Matthew Perry of the United States was making his historic visit to Japan. A quake in 1854 in Izu caused a tsunami that wreaked havoc on the Russian squadron anchored in Shimoda Harbor. And, in the following year, a much more violent eruption caused substantial damage to Edo, killing perhaps as many as 100,000 persons. It was this later earthquake that was associated in many woodblock prints with the tremor set off by the Americans.

25. On the treaty port system and its impact on economic development, see Elvin (1974a, 1974b) and Hoare (1994).

3

Manchester of the Far East

The Balanced-Growth Long Swing

Building upon a proto-industrial and agricultural base reinvigorated by sweeping infrastructure investment, Japan embraced innovations in the use of steam power, coal, and the factory system. As a result, Japan's economy passed through the first long swing between the late 1880s and the middle of the first decade of the twentieth century.

Compared to the slow, steady growth of output under extensive and intensive economic growth, the pace of expansion in production during the first long swing was unparalleled in Japan's history up to this point. But if expansion was rapid, its contour was not linear. Growth spurted, then faltered. As can be seen from Table 3.1, during the decade after 1887, output grew at an annual rate of 3 percent. It climbed throughout the years 1887–1894, peaking in 1894 with a growth of 5.0 percent.[1] From this peak it commenced its descent, bottoming out in the trough of the downswing with a growth of 0.1 percent in 1902. Per capita income also surged forward rapidly and then lost its forward momentum. Disposable income per head enjoyed an annual average growth of 2.7 percent between 1887 and 1897, the growth rate rising to a peak of 4.1 percent in 1891, after which it fell, reaching a trough in 1902 at –0.9 percent. In the long-swing rhythm of growth established at the close of the nineteenth century, the contour of economic expansion mimics a wave: growth speeds up, peaks, declines and reaches a trough, and then picks up once again.

Investment in new infrastructure that exploited the technologies developed in Europe and North America during Japan's lengthy isolation from the West fueled the expansion. Indeed, the crucial role played by capital formation in new infrastructure can be seen in various figures provided in Table 3.1. Consider first the growth rate of fixed domestic capital formation (panel B.1).

Table 3.1

The Balanced-Growth Long Swing

A: Per annum growth in income (gross national product [GDP]; gross national expenditure [GNE]), population (P), disposable income per capita (DYPC). Share of GDP arising from agriculture (A%) and manufacturing (M%). Percentage of gainfully employed population not engaged in farming or fishing (PLFNFF).

Phase	GDP	GNE	P	DYPC	A%	M%	PLFNFF
1887–1897 (U)	3.0%	3.1%	1.0%	2.7%	37.6%	8.5%	31.8%
1897–1904 (D)	2.2	1.8	1.2	0.1	34.0	11.0	34.1

B.1: Capital stock and capital formation. Percentage of GNE in: gross domestic fixed capital formation (GDFCF%); gross national savings (GNS%); net lending to rest of world (NL%); net national savings (NNS%); consumption of fixed capital (CFC%); private saving (PS%); government saving (GS%); and military capital formation (MIL%). Per annum growth rate for gross domestic fixed capital formation (GGDFCF).

	% of GNE								
Phases	GDFCF%	GNS%	NL%	NNS%	CFC%	PS%	GS%	MIL%	GGDFCF
1887–1897 (U)	9.8%	14.8%	0.4%	5.4%	9.4%	4.6%	0.8%	1.1%	5.6%
1897–1904 (D)	11.2	12.6	–1.1	3.9	8.7	5.0	–1.2	2.1	2.1%

B.2: Capital stock and capital formation. Percentage of gross domestic fixed capital formation (including military capital formation) in private primary industry (PRPRI%) and private nonprimary industry (PRNPRI%); private residential construction (PRRES%); government

(G%); and military capital formation (MIL%).

	Private			Government	
Phases	PRPRI%	PRNPRI%	PRRES%	G%	MIL%
1887–1897 (U)	12.1%	8.9%	35.2%	43.8%	2.6%
1897–1904 (D)	10.1	9.6	31.8	48.5	6.0

B.3: Capital stock and capital formation. Per annum growth rates of capital stock. Growth rates for the major components of gross capital stock: livestock and plants (LSPL); producer's durable equipment (PDUREQ); total gross capital stock excluding residential (NRESC); and residential buildings (RESB). Growth rates for components of nonprimary-sector capital stock: public works (PW); railroads (RR); utilities (UTIL), and nonresidential buildings (NRB).

	Gross capital stock				Nonprimary-sector capital stock			
Phases	LSPL	PDUREQ	NRESC	RESB	PW	RR	UTIL	NRB
1887–1897 (U)	1.1%	7.2%	2.5%	1.1%	3.1%	11.9%	30.7%	3.6%
1897–1904 (D)	1.1	8.4	3.0	0.8	3.7	2.9	23.1	3.4

(continued)

Table 3.1 *(continued)*

C.1: Corporate finance and banking. Companies: number (COM); percentage of companies that are partnerships (P%), limited partnerships (LP%), or joint-stock companies (JS%). Paid-up capital per company (in 1,000 yen) in partnerships (PUCP), in limited partnerships (PUCLP), and in joint-stock companies (PUCJS) in 1,000 yen. Total corporate-paid up capital (TPUC) in million yen.

	Companies				Paid-up capital per company (1,000 yen)			
Year	COM	P%	LP%	JS%	PUCP	PUCLP	PUCJS	TPUC (million yen)
1896	4,596	7.5%	36.3%	56.2%	43	19	139	397
1905	9,006	14.2	39.0	46.8	47	16	203	975

C.2: Corporate finance and banking. Per annum growth rates for bank offices (BOG); net worth of banks (NWBG); bank deposits (BDEPG); and bank loans (BLOANG). Bank offices per 10,000 population (BDEN). Total borrowing of banks from the Bank of Japan as a percentage of total funds employed by banks for national banks (NBNBOR), and for all ordinary banks (OBNBOR).

	Per annum growth rates					Borrowing from Bank of Japan	
Phases	BOG	NWBG	BDEPG	BLOANG	BDEN	NBNBOR	OBNBOR
1887–1897 (U)	18.2%	13.2%	21.9%	21.0%		31.0%	12.5%
1897–1904 (D)	5.7	7.2	14.1	10.3	0.69	—	9.0%

D.1: Labor. Labor force per annum growth rates: overall labor force (LFG), agricultural labor force (ALFG), and nonagricultural labor force (NALFG). Per annum growth rates for the capital/labor ratio (KLRG); and for labor productivity in agriculture, nominal labor productivity (NOMLPG) and real labor productivity (RLPG) [based on gross value added per worker]. Sex ratio (females/100 males) for agricultural labor force (SRALF) and per annum growth rate for percentage of labor force in nonagriculture (NALF%G).

Phases	Labor force, per annum growth rates			Agricultural labor productivity				
	LFG	ALFG	NALFG	KLRG	NOMLPG	RLPG	SRALF	NALF%G
1887–1897 (U)	0.6%	0.1%	1.7%	1.9%	7.0%	1.4%	87.0	1.1%
1897–1904 (D)	0.5	0.1	1.2	2.4	5.5	1.8	86.8	0.7

D.2: Labor. Growth in wages. Per annum growth in nominal w n agriculture (NWAMG); females in agriculture (NWAFG); males in manufacturing (NWMMG); and females in manufacturing (NWMFG). Per annum growth rates in real wages for males in agriculture (RWAMG); females in agriculture (RWAFG); males in manufacturing (RWMMG); and females in manufacturing (RWMFG). Per annum growth rates in wage differentials (agriculture wage/manufacturing wage) males (DMG) and females (DFG).

Phases	Nominal wages, per annum growth				Real wages, per annum growth				Wage differentials	
	NWAMG	NWAFG	NWMMG	NWMFG	RWAMG	RWAFG	RWMMG	RWMFG	DMG	DFG
1887–1897 (U)	7.7%	9.0%	5.6%	5.8%	2.6%	3.9%	0.5%	0.8%	2.2%	–3.0%
1897–1904 (D)	3.2	4.1	5.2	4.4	–1.4	–0.4	0.7	–0.1	–2.0	0.3

E: Prices. Growth rates for major price indices: GNE deflator (GNEDG); consumer price index (CPIG); investment goods price index (IGPIG); agricultural goods price index (AGPIG); manufactured goods price index (MGPIG); and commerce-service price index (CSPIG).

Phases	GNEDG	CPIG	.GPIG	AGPIG	MGPIG	CSPIG
1887–1897 (U)	5.3%	5.1%	4.2%	5.6%	4.3%	4.2%
1897–1904 (D)	4.7	4.6	2.2	3.7	4.1	5.7

(continued)

Table 3.1 *(continued)*

F.1: Trade. Percentage of exports that are merchandise (EXM%), services (EXS%) and factor income from foreign sources (EXFS%); and percentage of imports that are merchandise (IMM%), other (IMO%), and factor income paid abroad (IMFA%). Terms of trade (export price index/import price index), levels and per annum growth rates (TTL, TTG). Per annum growth rates for commodity trade, exports (ECOMG) and imports (ICOMG).

	Percentage of exports			Percentage of imports			Terms of trade		Commodity trade growth	
Phases	EXM%	EXS%	EXFS%	IMM%	IMO%	IMFA%	TTL	TTG	ECOMG	ICOMG
1887–1897 (U)	90.9%	8.4%	0.7%	91.6%	5.7%	2.7%	131.0	2.1%	7.8%	12.7%
1897–1904 (D)	85.6	13.1	1.3	85.1	12.5	2.5	145.7	0.5	6.4	5.5

F.2: Trade. Trade balance: net exports/trade ratio (NX/T); net decrease in specie held abroad/trade ratio (NDS/T); and export of gold/trade ratio (GX/T). Percent of exports and imports that are primary commodities (EPRI%, IPRI%). Ratio of exports to gross domestic fixed capital formation (XCFR).

	Net trade position ratios			Primary commodities in trade		Export/capital formation ratio (XCFR)	
Phases	NX/T	NDS/T	GX/T	EPRI%	IPRI%	Level	Per annum growth
1887–1897 (U)	–5.2%	0.0%	–2.0%	40.4%	29.8%	37.8%	2.2%
1897–1904 (D)	–9.3	–2.1	0.5	27.8	43.0	52.2	5.3

Sources: Miyamoto (1984), pp. 53–54; Nakamura (1983), p. 206; Ohkawa and Shinohara with Meissner (1979), pp. 180–81, 184–89, 251–53, 256–63, 278–91, 293–95, 302–4, 323–29, 331–35, 354–59, 373–75, and 387–93; and Suzuki (1980), pp. 10–1.

Notes: Most figures given represent averages of seven-year moving averages for the periods given in the table (this is particularly true for growth rates based on seven-year moving averages of the underlying series). In some cases, the averages are for periods shorter than the upswing and downswing phases. "U" stands for upswing and "D" for downswing. A "–" indicates that the figure cannot be computed.

During the upswing, it grew at an average annual rate of 5.6 percent, its growth peaking at 11.5 percent in the last year of the upswing. From this peak, growth in capital formation slowed a bit, reaching a trough in 1902 of 10.8 percent. Now, as can be seen from panel B.2 of the table, a substantial share of the increment to capital stock was due to expenditures by government. Moreover, as can be seen from panel B.3, of the capital stock created outside of the primary sector (mainly agriculture, forestry, and fishing), generous portions took the form of investment in utilities and railroads. But the importance of infrastructure creation to the first long swing was not restricted to formation of social overhead capital. Indeed, as panels C.1 and C.2 make evident, the development of incorporation and the growth of a nascent banking sector were essential to growth during the first long swing. Thus, infrastructure investment was integral to expansion in demand throughout both the upswing and the downswing of the first long swing.

In the table, I describe the first long swing as a balanced-growth long swing. By this term, I have two criteria in mind. Growth was balanced because both the Western technology sectors and the traditional Japanese technology-using sectors grew. Growth was balanced because both sectors competed successfully for the factors of production, labor, and capital, and competed on a level playing field. That the sector employing traditional technology grew—measured in terms of both primary production and agriculture—is evident from panel A of Table 3.1 (where it can be seen that the percentage of output flowing out of agriculture drops by only a small amount during the entire swing period); that it was successful in attracting capital resources is evident from panel B.2 (where it is shown that private capital formation in primary activities exceeded that in nonprimary activities during both upswing and downswing phases). And that agriculture was able to generate productivity growth and to compete for labor services is evident from panels D.1 and D.2. To be sure, as the labor market for females, who constituted the bulk of the labor force in light manufacturing during all four long swings between 1887 and 1969, tightened with the expansion in demand for workers in textiles, the agricultural wage fell relative to the manufacturing wage. But the labor forces in both manufacturing and agriculture grew during both upswings and downswings, as farming households freed up female workers for industrial employment by hanging onto male workers and by increasing hours of work for the workers who remained.[2] Indeed, during the upswing of the first long swing, growth in wages for agriculture outstripped growth in wages for manufacturing by a considerable margin.

The multifaceted nature of infrastructure improvements stimulated growth in both primary production and in manufacturing and services for many reasons. The development of banking facilitated capital formation in both sec-

tors. Promoting compulsory education for all socioeconomic groups and vocational education for the rural elite stimulated the diffusion and adaptation of best-practice traditional technologies from high-productivity to low-productivity districts. Investing in schools of higher learning aimed at teaching Western concepts in engineering and science to a (largely) nonrural elite improved the capacity of managers in factories to work with foreign machinery. Promoting Western medicine in schools of higher learning bolstered knowledge about infection and applications of the germ theory of disease.[3] Building roads and improving the flow of traffic on waterways and coastline expanded the size of markets for both farmers and for producers of manufactures, whether they relied upon proto-industrial technologies or were embracing the new steam-driven machinery imported from the West. Finally, building infrastructure—especially dredging ports and constructing breakwaters to accommodate large steam-driven ships—promoted trade, thereby stimulating demand for the products of both primary and nonprimary sectors and opening up foreign supplies of raw materials and machines.

The impact of infrastructure construction on trade must not be overlooked. Japan's trade grew rapidly at the close of the nineteenth century, and much of this growth stemmed from the improvement in port facilities. To be sure, because Japan was able to secure gold reserves in the form of a war indemnity won in the Sino-Japanese war, it was able to go onto the gold standard in 1897, and this also gave a push to both exports and imports. But accommodating large ships that plied the high seas was essential to developing the trading capacity of an island nation. In any event, the surge in Japan's trade was remarkable: While world trade expanded at a growth rate of 3.0 percent between 1886 and 1890, Japan's trade grew at 11.4 percent; while world trade expanded at a growth rate of 1.8 percent between 1891 and 1905, Japan's trade expanded at a growth rate of 9.9 percent.[4] And, reflecting the nature of balanced growth in which both primary and nonprimary production grew strongly, Japan's export of primary products played a key role in its ability to import. As can be seen from panel F.2 of Table 3.1, primary products were essential to exporting, especially during the upswing of the long swing. However, a clear shift away from exporting raw materials is evident in the sharp drop in dependence upon primary product exports recorded during the transition from the upswing to the downswing phases of the first long swing.

Now, tightening of supply and demand in labor markets, and limits on the expansion of trade (due to running negative balances in the balance of payments, and to constraints upon the capacity of ports to handle shipments), both helped bring to an end the first upswing phase and ushered in the downswing between 1897 and 1904. But pressure of expansion upon infrastructure played a role in areas other than in the port capacity. Pressure on banking

and the money supply played a role. This was especially true because during the upswing, price levels rose rapidly. As can be seen from panel E of Table 3.1, the upward drive in prices was especially pronounced for agricultural products and the consumer price index. The latter began to rise until its growth peaked at 8.1 percent in 1894, after which it tapered off, reaching a trough of 3.2 percent annual growth in 1901. In short, the upswing ushered in a downswing because pressure mounted on the infrastructure underlying the expansion. The capacities of physical infrastructure, the financial sector, and the labor market to supply the services for continuing rapid growth were exhausted. During the downswing, the growth in the demands placed upon infrastructure slackened, and continuing infrastructure investment helped pave the way for the second long swing.

This overview of expansion in output during the close of the nineteenth century provides us with a general understanding of the mechanics of the balanced-growth long swing. To probe its dynamics in more detail, let us turn to an examination of industrialization and economic growth within the Tōkaidō, and within the key conurbations of Osaka and Tokyo.

From Wards to City

In the wake of the *bakufu*'s collapse, Osaka, which lay at the geographic heart of Japan's proto-industrial economy, became the flourishing center of the emerging industrial economy of the nation. Indeed, Osaka became the core of the entire East Asian regional economy, earning for itself the sobriquet "Manchester of the Far East."[5]

Osaka's early dominance in Japan's industrial affairs was no accident. The regional focus of the economy parallels its structure. Under balanced growth, the agricultural/proto-industrial sector was tightly integrated with the new inorganic technology-using manufacturing sector. The new inorganic economy was focused around light industry and infrastructure, and the traditional economy was dependent on the latter sector and involved in generating light manufacturing output. Thus, the new steam- and coal-based sector carved out a market niche, and that niche coexisted with the one already serviced by the proto-industrial/agricultural sector.

Coexistence of the two niches was favored by the fact that product markets were highly segmented. Steam-power-driven, coal-using factories concentrated on making Western-style goods that could be sold internationally as well as domestically. The proto-industrial sector relying on wood and water continued to sell traditional Japanese-style clothing, broadly known as *kimono*. And coexistence was favored by the fact that labor—mainly young farm girls—flowed back and forth between the two worlds, being paid com-

parable wages in both endeavors. Thus, under this type of coexistence, it was natural that Osaka, which was the center of the proto-industrial economy, became the center of the new inorganic industrial economy.

In becoming the Manchester of the Far East, Osaka also became a legally defined city. Under a series of administrative experiments carried out by the new Meiji government that was committed to grafting the unitary nation-state model of Western Europe onto the system of government inherited from the Tokugawa period, Osaka's governance was transformed. First, Osaka became a coalition of four nominally independent wards (*ku*). Then, the wards were amalgamated to form a city (*shi*).

Now, grafting Western precedents onto existing Japanese practice was hardly limited to the sphere of government. Equally radical was the importing and adapting of techniques and organizational forms pioneered in Europe, England, and the Atlantic economy after the mid-nineteenth century. These techniques were harnessed to infrastructure construction, to agriculture, to energy production, and to manufacturing. As the experiments in grafting Western practice upon Japanese tradition proliferated, the great conurbations that constituted the core of the Tōkaidō core during extensive growth reemerged as the most dynamic locales in industrializing Japan. Why?

Several factors contributed to Osaka and Tokyo reestablishing primacy in economic innovation. Under the treaty port system, foreigners were allowed to congregate in only a few locales, and both Tokyo and Osaka were places where Westerners resided and carried on business. Moreover, population was densely settled in the Kinai and elsewhere along the Tōkaidō, and there was a rich supply of merchants in this region who possessed the connections and resources for undertaking nationwide trade and distribution. Finally, artisans and workers possessing skills in proto-industrial craft industries were concentrated in the hinterlands of the great conurbations.

These arguments do not explain why industry tended to concentrate in the immediate environs of the great cities of the Tōkaidō, rather than in their more distant hinterlands. After all, under extensive growth, the hinterlands of Osaka and Edo had flourished. Now, to a certain extent, the hinterlands of ancient Osaka and Tokyo did continue to boom under industrialization. However, because these hinterlands were eventually brought under the administrative control of the two giant metropolises, these former rural districts ended up as parts of the cities that they had once bordered. So city growth through boundary changes partly explains why Osaka and Tokyo emerged as industrial dynamos.

While this theory of incorporating hinterland is valid, it is not a full explanation. For within their old boundaries, Osaka and Tokyo became key industrial centers during the first long swing. A key factor in the reestablishing of

supremacy by Osaka and Tokyo is domestic and foreign trade, which, as noted earlier, expanded rapidly. The natural deepwater ports of Kobe and Yokohama emerged as key centers of trade as soon as the country was opened up since foreigners wanted to reside in ports that they could access with steamships possessing deep hulls. Since Osaka was situated in the immediate vicinity of Kobe, and Tokyo was not far from Yokohama, Osaka and Tokyo were favored by the location of two of the great natural deepwater ports. By contrast, Nagoya had no natural deepwater port in its immediate environs. Hence, it did not emerge as a major manufacturing center before the railroad network developed to a point allowing it to effectively carry on both international and domestic trade.

In appreciating the importance of trade for the growth of industry in Osaka and Tokyo, the overriding importance of coal as a source of energy for the factories dependent upon Western technology must be appreciated. Unlike England's coal deposits that were heavily concentrated in Wales and the Midlands, Japan's coal deposits were not located in areas well suited for industrialization. Thus, exploiting coastal shipping to bring coal to the industrial centers proved more cost-effective than locating manufacturing near coal deposits. In short, switching from an organic to an inorganic economy strengthened the competitive advantage of seaports relative to their hinterlands.

The presence of foreigners in the treaty ports was certainly a factor in the emergence of Osaka and Tokyo as innovating centers.[6] For instance, the consumer demonstration effect was operating. The presence of foreigners garbed in Western clothing and employing a plethora of materials, such as brick and machines like those used for steam power, spurred imitation among the Japanese who came into frequent contact with them. Imitation usually was a highly creative process because it involved creating new market niches within Japan, and it involved the spawning of hybrid products that combined Western and Japanese elements. Brick building construction for government offices, arc lighting for places of amusement, gas lighting for factories, steel reinforcements for river embankments, concrete blocks for breakwaters, lighthouses to direct coastal shipping—all are examples of Western concepts that were imitated and adapted by the Japanese, especially in the great conurbations like Osaka and Tokyo.

Thus, infrastructure embodying Western concepts and machines tended to gravitate toward the great centers of the Tōkaidō. As a result, those Japanese eager to exploit the new infrastructure and to learn how it worked so they might adapt it tended to concentrate it these centers. Therefore, Osaka and Tokyo once again emerged as the geographic centers of Japan's surging wave of innovation. However, Osaka was in a particularly advantageous position. Within its hinterland was a vigorous proto-industrial economy. Thus,

farm households possessing an intimate knowledge of raw cotton, silk, and the manufacturing of wooden implements abounded in the region. In short, the human capital base supporting light manufacturing was unusually rich in the Osaka environs. Moreover, because Japan was an island nation, the earliest and initially most cost-effective applications of steam power to transport were in water transportation. And Osaka—blessed with canals, access to the Inland Sea and the Yodo River, and nearness to the deepwater port of Kobe—was especially well suited to exploit the new Western technologies. Thus, Osaka enjoyed economies of scale in innovation: The potential applications of steam power in potentially profitable endeavors were unusually numerous within the conurbation, and the labor force with which to exploit these applications was densely concentrated within its hinterland. In short, capacity to innovate is the most important factor behind Osaka's emergence as the core of the newly industrializing Tōkaidō core.

The role of scale economies in promoting innovation in early industrializing Japan, and the importance of government as a mobilizing agent for this innovation, cannot be overestimated. A disparate set of examples of innovations involving government—the postal system, banking, imperial rituals, and the architectural designs employed in building official buildings—may help the reader grasp these points.

One of the most striking characteristics of government-inspired innovation during the early Meiji period is the remarkable degree of hybridization brought on by restless experimentation with various Western models.[7] For instance, British models were used for the navy, and for the telegraph and postal systems. French institutions were originally used as models for the army and for the primary school system. But in both cases—the first involving emulation of the British and the second imitation of the French—initial experiments were jettisoned. The Meiji government eventually decided to follow the American model for primary schooling and the German model for the army. In the case of banking, the Japanese government first tried out an American system of national banks, only to abandon this effort and to adopt a Belgian-style central bank system.[8] One consequence of this practice of ongoing experimentation was the spawning of models drawing from the practices of various different countries. Models created in this manner were often hybrids, differing fundamentally from their progenitors.

One explanation for restless experimentation lies in scarcities. The number of foreigners living in Japan when the country was initially opened up was minuscule. And the number of foreign technical experts whom the new government could afford to recruit abroad and bring to Japan was also small.[9] Consider the case of innovations involving telegraph and postal systems. The fact that British technicians had been employed by the Lighthouse De-

partment was a crucial factor in adapting British practice in communications. Nevertheless, in the case of the postal service, the standard British practice of the day was to rely upon steam-driven railroads. Now, in the early Meiji period, the railroad network was in its infancy—a national line of twenty-nine kilometers connecting Shinbashi in Tokyo to Yokohama was opened in 1872, and a second line of thirty-three kilometers linking Osaka to Kobe was completed two years later—and thus it was impossible to use railroads for deliverying mail during this period. An interesting hybrid emerged, one that spliced earlier British practice onto contemporary British practice, and the amalgam onto Chinese practice. Earlier British methods using horse-drawn carriages were used to join some nodal points to one another. The human-pulled cart or rickshaw that had been developed in China—in 1870, under the Japanese name *jinrikisha*, it began to carve out a widening niche in the Japanese market—was used to supplement this form of land transport. And these two forms of inorganic economy transport were linked, in relay fashion, to steam-driven transport using steam railroads where available, and steamships.

The use of steamships for mail delivery is noteworthy for a variety of reasons. It illustrates the point that during the first long swing, seacoast transport was far more cost effective than railroad transport. And, it underlines the importance of government's contracting out services to the private sector for the emergence of large private companies operating initially in the infrastructure sector where so much of balanced-growth long swing investment was taking place. The company securing the contract for transporting the mail along Japan's seacoast was Mitsubishi, which was eventually to become one of prewar Japan's most powerful financial cliques, known as *zaibatsu*.[10] It was the Yubin Kisen Mitsubishi Kaisha whose fortunes soared with the postal delivery business. It was out of fear of this company's becoming a monopoly that the Japanese government developed a subsequent strategy of encouraging the growth of competing oligopolies, large enough to exploit economies of scale but locked in competition, albeit competition among the few. This illustrates the way Japanese government has acted as a coordinating or facilitating agent during the modern period, working in concert with mammoth companies whose activities it stimulates through the awarding of lucrative contracts.

Geographic scale economies, especially those offered by the Tōkaidō core and the existence of huge conurbations within that core, also played a role in building up Japan's postal system. Consider the rapid increase in postal service activity in Osaka, as evidenced by the volume of deliveries within the city.[11] Setting the index of postal deliveries in 1890, the index (which stood at 59.7 in 1885) jumped to 180.4 in 1895. The rapidity of this increase is

remarkable. To be sure, the postal system was being developed during this system. But the main factor underlying the rapid expansion within Osaka is concentration of population in the great conurbations of the Tōkaidō belt. Establishing a nationwide postal system involved considerable fixed overhead cost, in total and on a per capita basis. But the per-person fixed cost for building up the infrastructure of mail delivery within the Tōkaidō was far less.

The proliferation of banks in Osaka offers another useful example of how hybridization and scale economies interacted in the first long swing and the buildup to it.[12] The early Meiji government found itself enmeshed in the twin problems of creating a currency that was convertible to foreign currencies at fixed exchange rates—a problem theoretically solved when Japan went on the gold standard in 1897—and of creating financial intermediaries that could satisfy the demand for credit creation in the newly industrializing economy. The government first tried out an American-style system of national banks whose collateral served as the base for note issue. Disappointed with the fruits of this experiment, the government veered off into new directions, adopting a combination of British and continental models under which the Bank of Japan became the lender of last resort (it was created in 1882 and most of its policy-making initiative was given over to the powerful Ministry of Finance). As part of this reform program, the government also established a series of specialized banks—savings banks, commercial banks, and banks concentrating on foreign trade and the handling of foreign currencies—that eventually took over many of the functions originally given to the national banks. The national banks themselves eventually became ordinary private commercial institutions, joining the ranks of already chartered private banks like the Mitsui Ginkō of the emerging Mitsui *zaibatsu*. With its merchant community already well versed in sophisticated financial operations, and feeling the demand for credit in its burgeoning industrial economy, Osaka emerged as a principal center of the new banking industry. Between 1877 and 1893, forty banks were established within Osaka, fourteen of which had been national banks founded during the first period of financial experimentation.

Hybridization also informed the concept of the Meiji monarchy. In fashioning a symbol of Japanese nationality and of the nation-state—one that, like the money supply, would meet both domestic and international requirements—the architects of Japan's modern imperial institutions grafted atavistic Shinto rituals and shrines onto a European concept of monarchy.[13] In effect, the government created two capitals for the country, both in the Tōkaidō: a western capital at Kyoto in the Kinai, and an eastern capital in Tokyo. In the ancient capital Kyoto, the emperor appeared in court robes, beardless, above politics and godlike. In the modern capital Tokyo, the

emperor's regalia was symbolic of the general march of civilization and indicative of authority as an internationally recognized head of state.[14]

The diffusion and modification of methods in architecture for official government buildings affords yet another illustration of the subtle interplay of Western models with Japanese tradition.[15] Active elsewhere in Asia, Thomas Waters, a British architect, received a number of important commissions during the Meiji period. Waters' designs left a strong neoclassical imprint upon the monumental architecture of Meiji Japan. For instance, Waters designed the great mint in Osaka (the largest in the Far East). The complex that was the mint consisted of two structures: a low foundry block with a Tuscan portico elevated upon a podium as primary motif for the façade; and a two-story reception pavilion fashioned in brick and stucco, covered in Japanese tiles, and decked out with a Tuscan-style veranda. In order to secure the brick for the mint construction, Waters had to introduce brick-making technology in its entirety, importing a brick kiln from Hong Kong.

Bankrolled by another massive Tōkaidō infrastructure project, the building of a fireproof brick thoroughfare for the Ginza district of Tokyo (whose central area, a full ninety-five hectares, had been gutted in one of Tokyo's perennial conflagrations), Waters went on to design a succession of brick-producing facilities in various locales in Japan. Waters' Bricktown in Ginza, which consisted of 916 fireproof buildings boasting continuous colonnaded facades and screening terraces, and his mint in Osaka inspired a generation of Japanese master craftsmen. These craftsmen imitated Waters' neoclassical principles in design, but because they were trained to work in wood, they tended to graft Tokugawa-style wood construction methods onto Waters-style neoclassical lines. In short, another hybrid was being created.

Thus Meiji innovation involved a complicated overlapping of various Western models upon one another, and upon inherited Japanese practice. A good example of this principle in the arena of government is afforded by the administrative system introduced during the late nineteenth century. Modeled on the Prussian system of local government, the administrative system that evolved through a succession of experiments does seem, at first glance, to reject core principles of Tokugawa administration. For instance, the new administrative system attached paramount importance to the nation-state and—because many of the functions of government were concentrated in the hands of national ministries—was rooted in a unitary state model. But because the Meiji experiments always placed lower-level jurisdictions directly under the umbrella of higher-level jurisdictions, the vertical principle of hierarchy integral to *bakuhan* government continued under a new guise. In this sense, Meiji practice was compatible with Tokugawa practice. The fiefs were abolished and out of these fiefs were created seventy-five prefec-

tures. Three of these prefectures—the great Tōkaidō conurbations of Tokyo, Osaka, and Kyoto that had been under direct *bakufu* rule during the Tokugawa period—retained their special status by being designated as *fu*. The remaining seventy-two prefectures were designated as *ken*. In any event, whether *fu* or *ken*, prefectures were created by amalgamating fiefs. Thus, the system of prefectures was overlaid on the system of fiefs.

Now, under the fief form of administration, urban castle towns and rural villages functioned as separate administrative entities under the jurisdiction of a *daimyō*. The new Meiji system was compatible with this arrangement in the sense that under each prefecture were separate urban and rural districts known as *ku*. Rural *ku* were subdivided into villages (*mura*) and towns (*machi*). Thus, in its initial phase, local administrative structure was realized in the form of seventy-five prefectures that were subdivided into rural and urban districts subject to further subdivision.

While the new system had the considerable virtue of being fairly compatible with the Tokugawa legacy, it was remarkably inefficient in terms of wasting staff. Reducing the costs of staffing the various layers of bureaucracy was a major reason for many of the reforms introduced during the Meiji period. As these reforms occurred, the ancient fief identities became more and more diluted in successive waves of amalgamation. For instance, through a series of consolidations, the number of prefectures was reduced to forty-seven. Moreover, the number of villages and towns dropped drastically during the late 1880s. For example, more than 7,000 villages and towns disappeared between 1874 and 1886; and between 1888 and 1889, another 56,000 were extinguished. In this way, the number of villages and towns was reduced to around 16,000 by 1889, the year in which a city code was finally enacted and the European concept of municipal government made its initial appearance. In that year, thirty-six cities (*shi*) were created, many of them the amalgamation of the smaller *ku* districts created earlier.

Thus, municipal government was introduced as part of a rationalization program designed to streamline local government and to cut the number of bureaucrats. Osaka, born as a municipal entity in April 1889, is a perfect example of this phenomenon. Four wards—Higashi-*ku* (Eastern Ward), Nishi-*ku* (Western Ward), Kita-*ku* (Northern Ward), and Minami-*ku* (Southern Ward)—were merged to form Osaka-*shi*. The remainder of Osaka-*fu* (Osaka prefecture, which was enlarged from its initial 1871 boundaries through the program of reducing the number of prefectures by joining contiguous prefectures) was subdivided into an urban (*shi*) portion, which at the time consisted of only two cities, Osaka and Sakai, and a rural (*gun*) portion, which was further subdivided into villages (*mura*) and towns (*machi*).[16] Thus, the main impetus leading to the creation of Osaka as a formally defined city was

wholesale nationwide rationalization of local government. The contrast with Europe is striking. For unlike European cities, many of which had won independence before the nation-state was created, Japanese conurbations of the Tokugawa era were administrative creatures of the *bakuhan* system. Thus, their fate as administrative units did not come from within their own development, but rather was imposed upon them by administrative reforms aimed at restructuring government as a whole.

To be sure, rationalizing government took place according to an impeccable logic that was embedded into the way the system functioned. Key to that logic is the principle that smaller units are neatly arranged as branches extending off larger units. Overlapping jurisdictions are ruled out. Thus, to reduce the number of subdivisions within an area, the authorities simply amalgamate some or all of the separate divisions into larger units. For this reason, municipal expansion through an increase in the size of the land area and population formally incorporated into a city's boundaries, is readily accommodated within the logic of subdivision. For example, in expanding its boundaries a city can either incorporate one or several contiguous counties (*gun*), thereby obliterating them in the process. Or, alternatively, it can carve out chunks of neighboring *gun* in the form of groupings of villages and towns, bringing these under its administrative umbrella and thereby shrinking the number of subdivisions the neighboring *gun* administers. From the viewpoint of the prefecture's organizational chart, municipal incorporation simply reconfigures the maze of branches, thereby affecting the number or sizes of the branches and the identity of the larger unit from which they branch out.

Osaka's modern history provides ample illustrations of the way municipal expansion takes place under the logic of Japan's governmental organization. As can be seen from Map 3.1, Osaka city expanded through three administrative—as opposed to landfill—expansions. In 1897, as a by-product of commencing the massive port construction project that made Osaka one of Japan's greatest deepwater harbors, the city absorbed portions of Nishinari-*gun* and Higashinari-*gun*. That is, Osaka absorbed an area that had been part of its hinterland during the Tokugawa period. Then, in 1925, it took over the remainder of the two counties it had cannibalized in 1897, thereby growing tremendously in size: The city now stretched from boundaries demarcated by the Kanzaki River on the west to the Yamato River on the southwest.[17] Generous chunks of Osaka's former hinterland were now directly incorporated into it. Finally, a smaller incorporation occurred after World War II. In short, during the course of the first and second long swing, Osaka absorbed substantial chunks of the hinterland where proto-industry had once flourished under *bakufu* rule.

Thus, Osaka's modern history is inseparable from the histories of the two

Map 3.1 **Territorial Expansion of Osaka, 1889–1955**

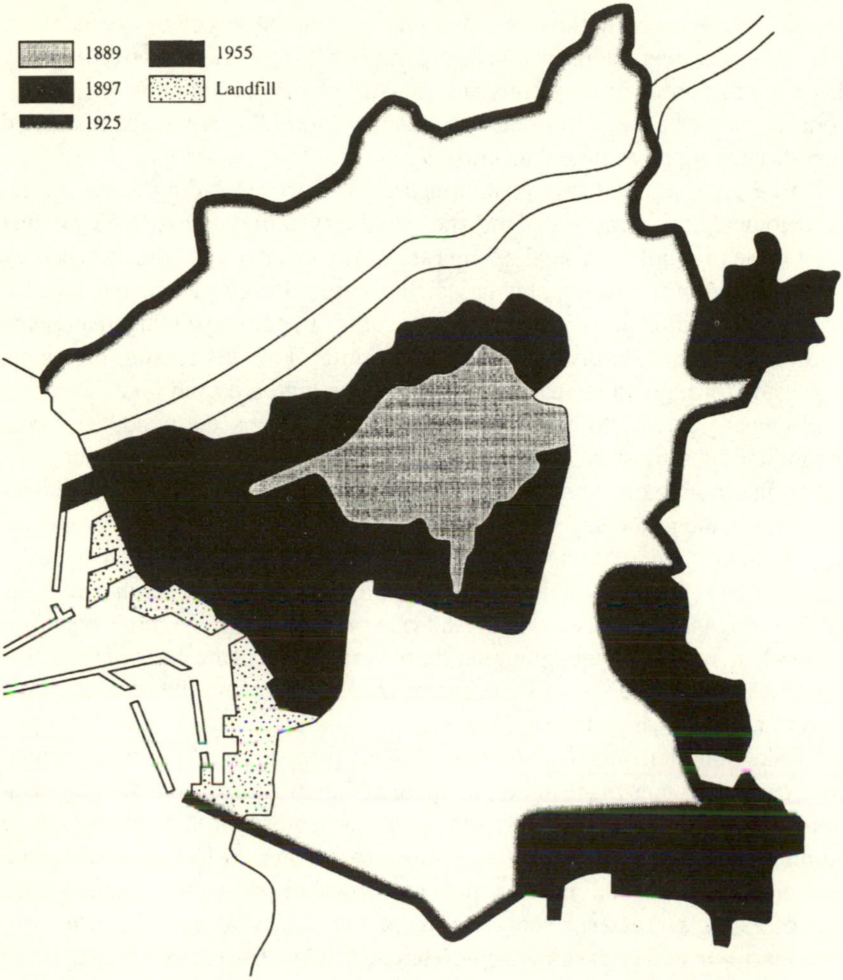

counties, Higashinari-*gun* and Nishinari-*gun*, which once bordered it and which it eventually absorbed. Indeed, many of the factories that made Osaka so famous as an early industrializing conurbation for East Asia were located along the waterways running through the two counties. And as the population concentrations and industrial zones within greater Osaka (including these two counties) proliferated, the city grew in terms of density of settlement inward from its waterways, and outward from the center that was the original Osaka of Tokugawa times.

Map 3.2 **Territorial Expansion of Tokyo, 1889–1932**

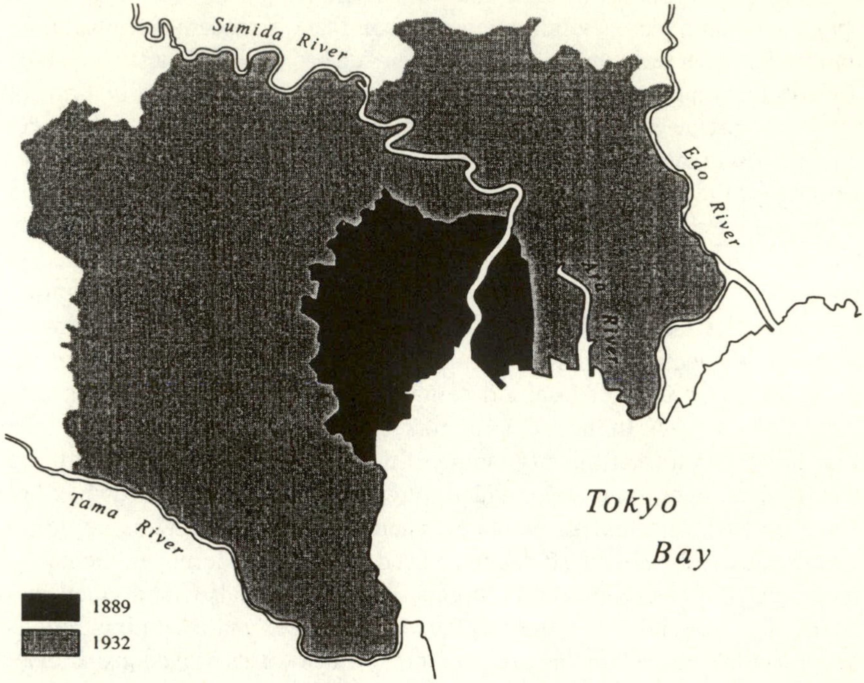

Dividing the city into zones defined in terms of the date when the zone became part of the city is useful for analysis of commuting and settlement patterns discussed in chapter 7. Thus is helpful at this juncture to define these zones here, where we can readily consult Map 3.1. Perusal of the map shows that the prewar expansion of the city amounted to encircling with ever larger concentric rings the original core, that is, the Osaka of the Tokugawa era. I shall refer to these three rings as core or center, middle ring, and periphery. The core is the area that legally became the city of Osaka in 1889. The middle ring is the zone added on in 1897 as the harbor project was getting off the ground. And the periphery is the area incorporated in 1925, as the city pushed out to the Kanzaki River boundary with neighboring Hyogo prefecture on the northwest and to the Yamoto River boundary shared with Sakai city on the east.

Tokyo's expansion is less clearly defined in terms of waterways and port development. True, like Osaka, it commenced its growth during the early Tokugawa period as a nodal point in a massive delta where rivers and ocean converge. However, as can be seen from Map 3.2, during the half century after Tokyo became a legally recognized metropolis in 1889, it expanded in one and

only one large incorporation, the 1932 administrative reorganization. In that expansion, Tokyo gobbled up land that was mainly lying out to the west, in the sprawling Kantō plain. Befitting the fact that Tokyo's modern manufacturing buildup has been less tightly linked to water than has Osaka's industrial history, Tokyo's growth in terms of land area has been less clearly defined in terms of water than has the growth of its great rival at the southern reaches of the Tōkaidō.

It goes without saying that internal administrative reorganization has accompanied the population increase occurring within these two metropolitan giants. At the most superficial level, the administrative system for each city has evolved in the sense that either each individual ward has expanded in terms of population, or it has been subject to subdivision and reorganization of boundaries. A common pattern has been for internal reorganization to occur simultaneously with incorporation of hinterland. But, as Table 3.2 demonstrates in the case of Osaka, internal reorganization has taken place at other times as well. In the 1925 incorporation, the area encompassed by the four original wards of the city (some of these had expanded in size during the 1897 incorporation) was divided into eight wards.[18] The wards had become bloated with human numbers. Their ward administrations could no longer effectively meet the demands placed upon them. Moreover, the newly incorporated areas brought into the city in 1925 were also subdivided into wards, five in total. Thus the city now had thirteen ward administrations under its umbrella. Even this division into thirteen wards proved insufficient to handle Osaka's burgeoning population. In 1932, the city was further subdivided, this time into fifteen wards. In short, proliferation of wards has accompanied Osaka's demographic surge to gargantuan proportions.

This also holds for Tokyo. Tokyo, originally spread out over a far larger land area than Osaka—in 1889, Tokyo's land area was 81.2 square kilometers, a figure far exceeding Osaka's original 15.3 square kilometers—commenced its existence as a municipality with thirteen wards. In its gigantic 1932 incorporation, four hundred square kilometers were added to the city's geographic extent. And, in order to manage this new region, the number of wards in Tokyo increased from thirteen to thirty-five.

Thus, Japan's two largest metropolises added on residential population, housing, factories, and infrastructure, through incorporation as well as through accretion within fixed boundaries. As can be seen from panel B of Table 3.2, population increase was greatest in the periphery zone of the city, and least pronounced in its core. In the case of Osaka, the impetus for growth tended to be directed to the margins of the periphery. Population concentrations nibbled away at the edges of the periphery, as the periphery filled in from the expanding center and middle ring of the city outward, and from the outermost boundaries of the periphery inward.

Innovation during the last three decades of the nineteenth century moved by fits and starts. Experiments were tried, then modified or abandoned, and other experiments were undertaken. Most of these experiments involving creating hybrids, as the practices of various Western nations were plumbed for models and concepts, and these were overlaid on existing Japanese institutions. Path dependence was at work in this evolution. Old habits, norms, and the rationale buttressing them die slowly. As the hybrid system of administration that developed before and during the first long swing evolved, the past continued to haunt the present and thus reached out to mold the future. Thus, in the Japan of the 1930s, traces of Tokugawa principles of administration peppered the country. A good example is afforded by the way administrators were selected throughout the prewar period. Mayors of most cities in Japan were elected, but the Ministry of Home Affairs appointed governors for the prefectures. Now, recall from chapter 2 that during the Tokugawa period the conurbation of Osaka was under direct *bakufu* rule. Thus, following the Tokugawa logic that great cities like Osaka are especially important to the central authorities, the Ministry of Home Affairs also selected the mayor for Osaka. To be sure, the mayor had to work with a city assembly elected by the citizens of the city to whom the franchise was extended. But the old principle of central control was staunchly conserved with this practice.

Agriculture in Balanced Growth

One of the most important consequences of the centralization of government accomplished through the early reforms of the Meiji era was the impetus given to growth in the traditional, organic sector of the economy. The agricultural/proto-industrial sector benefited from many of the changes wrought during the period of restless experimentation that generated the first long swing. It benefited from the investments in roads, port construction, and the gradual building of a nationwide steam railroad network. And it benefited from spread of elementary schooling and the founding of vocational and technical schools that specialized in agricultural technology. Nevertheless, most of the improvements driving expansion in agricultural output during balanced growth were variations on existing past practice developed in pockets of the country under *bakufu* rule. Better irrigation, improved seed varieties, and more generous utilization of fertilizers were the hallmarks of agricultural productivity growth during the Tokugawa period. These were also the hallmarks of agricultural productivity growth during the first long swing. What was new was the system of government and the way it taxed the fruits of farming.

By contrast, during the century of and after the industrial revolution in England—that is, between 1750 and 1850—the productivity advances won

Table 3.2

Osaka and Tokyo: Resident Population, 1889–1925

A. Total estimated resident population (1,000s) and land area (square kilometers).

Item	Tokyo[a] 15-*ku* area (1889 boundaries)	Tokyo[a] 35-*ku* area (1932 boundaries)	Osaka[b] Original 4-*ku* area (1889 boundaries)	Osaka[b] Enlarged 4-*ku* area (1897 boundaries)	Osaka[b] 13-*ku* area (1925 boundaries)
Area	81.2	550.9	15.3	55.7	178.9
Population, 1883	885.5	n.e.	351.2	479.2	548.1
Population, 1890	1,207.3	n.e.	476.4	641.2	729.9
Population, 1909	1,623.1	n.e.	n.e.	1,204.6	1,405.5
Population, 1920	2,173.2	3,350.6	635.2	1,253.0	1,768.3

B. Osaka: center, middle ring, and periphery, 1883, 1895, and 1920.

Date	Total population (1,000s) Center	Total population (1,000s) Middle ring	Total population (1,000s) Periphery	Percentage of expanded Osaka population (%) Center	Percentage of expanded Osaka population (%) Middle ring	Percentage of expanded Osaka population (%) Periphery
1883	351.2	128.0	68.9	64.1	23.3	12.6
1895	488.7	206.6	111.3	60.6	25.6	13.8
1920	635.2	617.8	515.3	35.9	35.0	46.2

C. Sex ratio, Osaka: center, middle ring, periphery (males/100 females).

Date	Expanded Osaka	Center	Middle ring	Periphery
1883	n.e.	119.1	n.e.	n.e.
1925	107.4	119.9	116.3	107.4

Sources: Nihon Naikaku Tōkei-kyoku (1906), various tables; Nihon Naikaku Tōkei-kyoku (various dates) Taishō kunen *Taishō jūyonen kokusei chōsa fu ken hen chōsa hōkoku*; (for 1920 and 1925); Nihon Naikaku Tōkei-kyoku (1992) as edited by A. Hayami; and Tokyo Hyakunen Henshū Iinkai (1979), various pages.

Notes:

[a]In the original city—including the area not legally defined as Tokyo *shi* (Tokyo City) before 1889—there were 15 *ku* (wards). In 1932 an additional 20 *ku* lying to the north, west, and south of the original 15 *ku* area, were added. See Map 3.2.

[b]The term "Expanded Osaka" refers to the city with its boundaries subsequent to the 1925 incorporation of those portions of Nishinari *gun* and Higashinari *gun* not incorporated into Osaka city earlier, that is, the year 1897. A brief summary of the administrative changes underlying the formation of prewar Osaka city are as follows: (a) Before creation of the legally recognized entity Osaka *shi* (Osaka City) in 1889, the area that became Osaka City consisted of four *ku* (wards)—Higashi (East), Nishi (West), Kita (North), and Minami (South). These were joined together to form Osaka *shi* and were surrounded by three *gun* (counties)—Nishinari, Higashinari, and Sumiyoshi. In 1896 Sumiyoshi was absorbed into the other two counties, and in 1897 the original four *ku*, and hence Osaka *shi* as a whole, absorbed portions of Nishinari and Higashinari *gun*. During the 1925 incorporation of those portions of Nishinari and Higashinari *gun* not absorbed earlier, the area added into the city in 1897 was administratively separated from the four *ku* into which it had been absorbed in 1897, and four new *ku* were created: Konohana, Minato, Naniwa, and Tenōji. I refer to the area encompassed by these four wards as the "middle ring" and to the area in the original four *ku* area of 1889 as the "center." Out of the area incorporated into Osaka *shi* in 1925, five new wards were created, namely Nishiyodogawa, Higashiyodogawa, Higashinari, Sumiyoshi, and Nishinari *ku*. I call the area encompassed by these five new wards the "periphery" of Osaka *shi*. In 1925 there were a total of thirteen wards in Osaka, and in 1932 population increase in some districts occasioned a further reorganization. Two new *ku* were carved out of the thirteen *ku*: Taishō (mainly taken out of Minato *ku*) and Asahi *ku* (mainly taken out of Higashinari *ku*). See Map 3.1.

In order to estimate the populations of Tokyo and Osaka prior to the first comprehensive national de facto census count of 1920, I used estimates of resident populations either for each ward of what became Tokyo and Osaka in 1889 or of each city taken as a whole. In order to estimate resident population (*genjū jinkō*), one adds the population whose household registers (*koseki*) are in the given administrative unit one is interested in to the net flow of registered population in and out of the unit (the so called *kiryū* population). In order to estimate the populations of "center," "middle ring," and "periphery" for Expanded Osaka, I combined the following data: resident population estimates for Osaka city or for the wards that became Osaka city, figures running back to 1883; and figures on Nishinari, Higashinari, and Sumiyoshi *gun* back to 1883. I calculated the percentage of the three combined *gun* area lost to Osaka city in 1897 (65 percent) and assumed this ratio applied all the way back to 1883. In this way I could allocate the combined *gun* into the middle ring and periphery and thereby secure an estimate for Expanded Osaka resident population all the way back to early Meiji.

n.e. = not estimated.

were mainly of a different type. These advances were mainly fueled by rationalizing farm boundaries (known as enclosure), and by applying mechanical power to agriculture. To be sure, better roads and drainage ditches played a role in Britain, just as they did in Japan. But it was the tractor, the harvester, the reaper, the chemical industry, and the railroad that remade Western farming. Indeed, the chemical industries of Europe and North America began churning out relatively inexpensive synthetic fertilizers that vastly improved the productivity of land, and of the farmers tilling and weeding it. Thus, Western agriculture was revamped through better infrastructure, improved access to markets, and especially through the systematic substitution of capital in the form of machinery for labor known as mechanization.

This type of technological progress was of limited applicability to early industrializing Japan. This assertion does not imply that innovators in Japan ignored Western developments in the field of farming. To the contrary, a nascent chemical industry sprang up in Japan, partly because it was able to establish a flourishing niche for synthetic fertilizers through an aggressive marketing campaign. Moreover, on the wide plains of Hokkaidō—which, reflecting its harsh northern climate, remained a virtually unsettled frontier prior to the 1870s—it proved feasible to exploit American-style mechanization and to plant wheat, rye, and barley in profusion. In addition, the lure of rationalizing rice production through the wholesale readjustment of plots proved infectious in a number of locales throughout the country. For instance, a governor of Ishikawa prefecture became so enamored with the possibilities of enhancing productivity with the creation of larger-sized plots that he founded a model farm of three hectares in his district to tout the benefits of readjusting plots. This experiment proved so successful that a rich landlord emulated it, convincing the villagers in the areas where he owned land to collectively reorganize their scattered arable fields.[19] In short, on a highly selective basis, Japanese farmers did manage to exploit some Western innovations.

However, the most important influence on Japanese agriculture exercised by Western models was the model of the centralized nation-state, for, during the Tokugawa period, agricultural productivity varied widely between regions. It varied partly because fief boundaries acted as barriers to the diffusion of best-practice technique. In general, agricultural productivity in western Japan, which reaped the benefits of a benevolent climate and the (relative) absence of volcanic ash in the soil, was high. For instance, in many parts of southwestern Japan, it proved possible for farmers to double-crop rice and/or to raise side crops of vegetables. Moreover, in the Kinai, with its highly developed proto-industrial sector, farm households generated a rich profusion of industrial crops.

Thus, the main thrust of Meiji agricultural policy was to obliterate these

regional differentials in agricultural productivity through the systematic encouragement of diffusion of best-practice technique. The government sought out those farmers whose methods were deemed exemplary. It subsidized speaking tours for these veteran farmers, so that they could relay their ideas and methods to communities throughout the country. It established experimentation stations where new seed varieties and innovative methods of planting were developed. And it encouraged villagers to drain their rice fields after the fall harvest, so that side crops could be put in. In short, as a result of centralization, government emerged as a coordinating and facilitating agent for the agricultural sector. The main beneficiaries of this activity were the villages in the Kantō and in northeastern Japan.

As a result of the systematic diffusion of traditional agricultural technology from western Japan to eastern Japan, the yawning gap in agricultural productivity that existed between the region surrounding Tokyo and that surrounding Osaka closed. It did not altogether disappear. Ecological differences between the two regions continued to have an impact on the nature of crops produced in the two districts. These points can be gleaned from the figures assembled in Table 3.3. Apparent from panel A of the table is the sustained growth in farming in the northern reaches of the Tōkaidō, especially in the environs of Tokyo. A striking contrast between Tokyo and Osaka prefectures is revealed by the figures on number of farm households. In Tokyo prefecture the number sharply increases. In Osaka prefecture the number plummets. This bears testimony to the fact that there was more undeveloped, or underdeveloped, land in the Tokyo-Kantō area than there was in the Osaka-Kinai region.

The fact that agriculture and proto-industry grew more vigorously in the environs of Tokyo than it did in the environs of Osaka does not erase the fact that Osaka was situated in a region with unusually valuable rice paddy land. Tokyo's hinterland continued to be dominated by relatively low productivity dry field farming, even after the balanced-growth swing when diffusion of best-practice techniques raised productivity throughout the Kantō and in the more northeastern reaches of the country. Ecology continued to be important, even in the face of important technological advances.

Now, because of these longstanding ecological differences, as the cities of Osaka and Tokyo pushed out into their hinterlands, they confronted radically different challenges involving the conversion of land. Osaka pressed out into high-productivity/high-rent paddy fields. It pushed out into land that embodied centuries of investment in the form of irrigation canals and drainage ditches. By contrast, as the densely settled fringes of Tokyo encroached upon agricultural communities, they impinged upon land that was primarily given over to dry field farming. This land was not especially productive, nor

Table 3.3

Farming and Agricultural Land Use in Several Regions of Japan, and in Osaka and Tokyo Prefectures, 1880–1939

A. Farm households

	Indices of farm households (1880–1882)						Percentage of all Japanese farm households in Tōkaidō		
		Tōkaidō							
Period	Japan	Total	North	Tokyo	South	Osaka	Total	North	South
1880–1889	99.7	98.9	99.7	100.5	98.3	98.4	23.6%	8.8%	7.6%
1900–1909	100.1	93.6	97.9	184.8	90.0	89.1	22.3	8.6	7.0
1920–1929	100.0	91.9	97.0	162.6	83.6	78.8	21.3	8.5	6.4
1930–1939	101.8	88.3	96.0	152.5	80.0	73.0	20.7	8.3	6.1

B: Agricultural land use, shares of national land, relative land productivity and price, tenancy and cow/horse ratio[a]

Item/region	Japan	Kantō	Tokyo	Kinki	Osaka
Arable land rate (PLAR)	15.3%	29.1	23.8	14.9	33.6
Paddy land rate (PADLR)	54.1%	44.3	21.6	76.5	81.5
Percent of national cotton land (PCF)	100%	20.5	0.3	10.8	5.9
Relative paddy land price (RPADP)	100	100.4	117.6	124.9	145.4
Relative cotton field productivity (RPCF)	100	67.2	56.9	146.2	175.4
Arable land density (ALDEN)	766.4	874.3	4,086.4	1,582.2	2,607.7
Tenancy (TENR)	32.8%	42.2	39.8	38.0	59.3
Cow-horse ratio (CHR)	83.6	16.2	122.9	1,729.3	8,703.9

Sources: Nihon Naikaku Tōkei-kyoku (various dates), *Nihon teikoku tōkei nenkan*, various tables; and Nihon Naikaku Tōkei-kyoku (1930), *Nōgyō chōsa kekka hōkoku. Shōwa yonen:* various tables.

Notes: [a]The variables in the panel are defined as follows: PLAR = Percentage of land in district that is arable in 1929; PADLR = Percentage of arable land that is in paddy in 1929; PCF = Percentage of total national cotton land in district in 1901; RPADP = Price of a hectare of paddy land relative to national paddy land price = 100 in 1901; RPCF = Output of an hectare of cotton field land relative to national average = 100 in 1901; ALDEN = Population per .01 hectare of arable land in 1929; TENR = Percentage of cities, towns, and villages having arable land with percentage of ownership of cultivable land under 50 percent in 1929; CHR = Ratio of cows to 100 horses in 1901.

was it very expensive. In short, expansion in Osaka faced higher costs for land conversion than did expansion in Tokyo. The impressive improvements in Kantō agriculture so integral to balanced growth muted these differentials, but did not eliminate them.

Physical Infrastructure

To the Japanese who came into regular contact with the coal- and steam-based technology employed by Westerners, applying the technology to an ancient infrastructure that had fallen into disrepute during the death throes of the Tokugawa period must have surely seemed attractive. After all, infrastructure creation had laid the groundwork for extensive growth during the Tokugawa period. Would it not help propel Japan into industrialization?

In the treaty port of Osaka were all of the ingredients essential to harnessing a coal- and steam-based technology to the revamping of infrastructure. Demand for it was unusually vigorous in Osaka and its neighbor Kobe. Most important, the new steam technology was easily applied to water transport, and rivers, canals, and ocean permeated the great conurbation of Osaka. With steam-driven equipment, it was possible to carry out the dredging of rivers, building breakwaters and erecting dikes and embankments for waterways on a scale hitherto undreamed of in Japan. In addition, the supply of technical expertise was also available. Under the treaties creating extraterritorial rights for certain Western nations, Osaka's Kawaguchi district was given over to a small but knowledgeable group of Europeans and Americans, the Dutch and British playing an especially important role in the community's affairs.

A substantial number of these foreign residents were in Osaka because they were employed as specialists in, or technical advisors for, enterprises managed by the government. For instance, the mint designed by Waters employed foreign engineers, gold and silver experts, printers, testers, and machinists during the Meiji period. Of these residents in the Kawaguchi district, perhaps the three who contributed the most to the harnessing of Western technology to rework the infrastructure of Osaka were three Dutch engineers deeply versed in the design of deepwater ports and the waterways emptying into them. Over the three decades from 1873 to 1901, these three worked assiduously to plan and direct the reorganization of water flow in the lower Yodo River basin and the building of the port of Osaka.[20]

Both the native and foreign communities of Osaka had, from the very first year of Meiji in 1868, been intrigued by the idea of harnessing the new technology in the revamping of the waterways of the city and its environs. The main technical hurdle that had to be cleared was the natural depth of the harbor. It was too shallow to accommodate typical ocean-going vessels, whether

they be wooden sailing ships or steamships with metal hulls. Lighters—shallow wooden hulled boats propelled by oar or sail—were used to transship passengers and freight arriving or departing from the deepwater harbor in Kobe to or from Osaka. But Osaka Bay itself was not a natural deepwater port. The flow down the Yodo River brought with it immense volumes of silt, especially during and immediately after the torrential rainstorms that swept across the country during the months of early summer and autumn. With the construction of a railroad line linking Osaka to Kobe, transshipment also took place on the land. Nevertheless, the business community of Osaka, perhaps recalling the days when Osaka was a great seaport serviced by a veritable armada of ships, believed that Osaka needed a deepwater port.

Buttressing the lobbying campaign for an improved port in Osaka was the threat of regular flooding at the mouth of the Yodo River, for flooding and the dredging of the Yodo River were interconnected. Both problems could be solved by building breakwaters and piers, dredging, constructing spillways and floodways, driving down metal piles to shore up the banks of rivers, and reconfiguring the channels through which the Yodo River empties out into Osaka Bay. For this reason, the Osaka community had an argument that favored their city in the competition between Osaka and Tokyo for construction of a deepwater harbor.[21] After all, fires and earthquakes were the most pressing ecological dangers facing Tokyo. Thus, Bricktown was built in the Ginza district of Tokyo.

Exploiting the threat of flooding, the business community of Osaka embarked on a long and eventually successful effort—punctuated by delays due to the Sino-Japanese war, and to demands placed on the small number of professionals with practical experience in carrying out port improvements by projects elsewhere in the country—to win approval of the national ministries allocating resources for port construction. The scarcity of trained professionals who understood the nature of water flow in the rivers and natural harbors of Japan was a major constraint. The national government's effort at building Western-style ports commenced with the debacle in Nobiru in northern Japan during the 1870s. In carrying out construction, the government relied upon foreign exports schooled in building breakwaters but lacking an appreciation for the tidal conditions at Nobiru. As a result, underwater work dragged on, and many of the structures and foundation stones, designed to buttress the breakwater, were dashed to pieces in the violent currents.[22] Learning from this bungled attempt, the government now embarked on a more cautious program, carefully studying the sites and assembling a team of experts who canvassed a wide range of possible risk factors before approving projects. The deliberate pace for approving new port improvement schemes is indicated by the fact that it was only in 1889 that construction began on

two huge breakwaters—each three to six meters in width and confining between them an anchorage of five hundred hectares—in Yokohama, which was Japan's principal port after 1859.[23]

Hence, Osaka waited. But tragedy tipped the scales in its favor. In 1885, the Yodo River once again broke through its levees, submerging more than 57,000 houses and leaving more than 200,000 victims in its wake. In the aftermath of this disaster, the lobbying campaign picked up steam. Bolstering it was the position of the military that wanted better port facilities in Osaka so it could dispatch capital ships from anchorage there. And so, in 1897, at the close of the upswing of the first long swing, the national government commenced the massive project of reconfiguring the Yodo River flow and building two giant breakwaters and a gigantic iron pier in Osaka Harbor. The project was huge and was not fully completed until 1929.

As with most of the other technical and organizational innovations of Japan during the first long-swing period, the Osaka project was a hybrid, mixing Tokugawa practices with Western methods and machinery. Consider two of the key elements in the project: excavation of waterways and landfill. These two elements were in evidence during the early seventeenth century when the canals of Osaka were carved out and new fields, *shinden*, were thrown up in the silted mouth of the Yodo River. To the timeworn practices of Tokugawa infrastructure construction were added a host of innovations developed in the West: steam power for dredging and pile driving; concrete blocks for building breakwaters; and iron for making the huge pier that jutted out into Osaka Bay at Tenpozan.

In marshaling the resources to realize this complex project, the national government worked in concert with the private sector and with the municipality of Osaka. For instance, warehouse construction at the Sakurajima docks sheltered by the northern breakwater was undertaken by the great Osaka-based Sumitomo *zaibatsu*. And the city built infrastructure complementing the project. It opened up a city-managed electric tramway line in 1903 with two legs: a leg connecting the old city core to the Tenpozan pier, and a leg running through the area built up through landfill that now effectively sealed off the flow of the Aji and Shirinashi rivers from one another. In short, government was acting as both direct investor in infrastructure construction and as coordinator and facilitator.

A major goal of the project was to stop flooding on the Yodo River. Work to this end was completed by 1910. A wholly new channel for the Yodo River's stately entrance to Osaka Bay was excavated. Today, this heavily fortified channel is known as the Shinyodo River, "*shin*" meaning "new." And the Kema Dam and Kema Floodway were also completed as components of this project.

Once the problem of flooding was solved, and the direction that silting on the river would naturally take was known, a channel running north of the northern breakwater was constructed so that the silt would not spill into the deepened harbor. The harbor itself was defined in shape by breakwaters consisting of more than 54,000 blocks of concrete. A north jetty more than 3,000 yards long encompassed the mouth of the Aji River, which was itself the extension of rivers passing through the center of the city and flowing around Nakanoshima (where most of the fief warehouses had been located during the heyday of the *bakufu*). A south jetty almost 5,000 yards long sheltered the mouths of the Shirinashi River and a canal cutting northward from the Kizu River. Thus the myriad of canals in Osaka were directly linked to the sheltered harbor.[24]

It goes without saying that expanding Osaka's capacity to import raw materials and export finished goods increased the volume of trade going through both Osaka and Japan. Indeed, it is one of the arguments of this book that the expansion of Japanese trade heavily depended upon the port construction projects that were integral to the first and second long swings. While the project in its entirety was not completed until the 1920s, it proved possible to use some of the facilities earlier. Indeed, as early as 1912, the mayor commissioned a special committee to study the feasibility of using the partially completed harbor facilities for trade. The Sakurajima docks, partially sheltered from the open waters of the Inland Sea, had already begun doing a brisk business. Now, the central pier at Tenpozan also entered the fray. To grasp the impact of the port on trade going through Osaka Harbor, consider the following figures: In 1912, when Sakurajima accounted for 80.7 percent of the trade flow in Osaka Harbor, the index of trade (with the level in the year 1914 set at 100) was 90; by 1916, when Sakurajima's share had dropped to 65.3 percent because more and more trade was going through the main pier at Tenpozan, the index of trade had reached a figure of 155.5.[25] After 1912, Osaka entered into direct rivalry with the two other great ports of the Tōkaidō, Kobe and Yokohama.

How rapid was the buildup of Osaka Harbor trade? Consider the volume of Osaka's trade as a percentage of Yokohama's trade. In 1873–1882, the percentage was 2.7 percent (Kobe's trade as a percentage of Yokohama's was 26.2 percent in that year); by 1923–1932, Osaka's percentage had jumped to 55.7 percent (Kobe's percentage was 124.9 percent in that year).[26] Combining the percentages of Osaka and Kobe in 1873–1882 gives 29.0 percent; and by 1923–1932, the combined percentage has jumped to 180.6 percent. Osaka's port project had vastly increased the capacity of the southern apex of the Tōkaidō to carry on trade.

With port, rivers, and canals fully linked, a vast labyrinth of water was

created in Osaka. Without appealing to visual aids, it is impossible to convey to the reader the sheer architectural grandeur of the port and its feeder waterways. On the cover of this volume is a portion of a reproduction of a famous map of Osaka, commissioned in 1924, after the harbor construction was fully completed and before the hinterland in Higashinari-*gun* and Nishinari-*gun* was administratively absorbed into the city proper.[27] The degree to which prewar industrial Osaka depended on water for the movement of passengers and goods can be gleaned from the following remarkable fact. A map drawn up during the 1920s shows that of the forty-odd major distribution points handling specified types of raw materials and manufactured goods—that is of the places designated for loading and unloading coal, fertilizer, raw cotton, iron and steel, cotton thread, cement, fresh fish, glass, timber, rice, and so forth—all but one point lies either on a waterway or at the ocean's edge.[28]

Under *bakufu* rule, Osaka had been a "water city," for it was conceived in terms of canals and rivers. Under prewar industrialization, Osaka continued to be a water city. Indeed, before Osaka Harbor construction was initiated, Osaka's burgeoning empire of factories had already sprawled out along the waterways running through the city and its hinterland. The major steam-driven textile mills of the early Meiji period shot up upon *shinden* landfill on or near the Yodo, Kanzaki, or Aji rivers, and in the area south of the Dōton Canal, where plants line the Kizu River. Chemical plants were nestled in and along the Shinyodo and Yodo rivers. Metalworking plants proliferated in the vicinity of Sakurajima and near the mighty warehouse complex in the environs of the Shirinashi and Kizu rivers.[29]

With the completion of the port project in 1929, Osaka's destiny on the water was fully realized. During the heyday of the proto-industrial economy, its infrastructure made it a mighty pre-industrial Venice or Amsterdam of the Far East. And, by the 1930s, bristling with breakwaters and a huge iron dock that handled the massive ships plying the Pacific, it had become an industrial Venice of the Far East.

Transportation

Infrastructure and the means of transportation are two sides of the same coin. They complement each other. Under *bakufu* rule, goods (and passengers) circulated through the maze of canals and rivers on small boats (*kobune*), passengers walked through the unpaved streets, and goods were transported in freight wagons (*niguruma*) that were usually pulled by humans. The roads were narrow, and the canals and rivers were relatively wide.

With the harnessing of Western technology, both infrastructure and the means of transportation were transformed. But the pace of transformation

was measured. Transportation on the water was most strongly affected during the balanced-growth long swing. Transportation on the land was eventually revolutionized by the railroad, but building a railroad network took time. Indeed, it was not until the upswing of the second long swing that the impact of railroad infrastructure became significant. Thus, for the most part, the means of land transportation under early industrialization was not terribly different than it was during the Tokugawa period. To be sure, the human-pulled passenger vehicle, the *jinrikisha*, was a novel innovation. Originally developed in China, where it was known as the rickshaw and where it was built with wooden wheels, it was introduced into Japan in the 1870s and was soon outfitted with rubber wheels, thus becoming a hybrid born out of the rickshaw and the Western bicycle. But human-pulled wooden freight carts, *niguruma*, continued to be the main vehicle used for transporting goods.

For most places and peoples in Japan during the first long swing—especially for the inhabitants of isolated villages situated deep in the valleys or on the slopes of massive mountains—the predominant means of transportation on the land was foot or *niguruma*, and on the water the small wooden boat, the *kobune*. Penetration of the railroad, or of steam technology in general, in the region outside of the Tōkaidō was extremely limited. Consider the figures given in Table 3.4 on the percentages of vessels registered in the Tōkaidō in the closing decades of the nineteenth century. As can be seen from panel A of the table, outside of the northern district of the Tōkaidō, the inroads made by steam power were limited. Indeed, as far as registered vessels are concerned, it appears that wind- and human-powered water transport continued to play an extremely important role in most of Japan, even in the industrializing southern district of the Tōkaidō. However, a word of caution is in order here. The figures refer to registration of vessels. It is quite possible that use of vessels powered by steam in the Inland Sea was far more common than is suggested by these figures.

As the first long swing spread across the Tōkaidō, the demand for transportation vehicles of all sorts soared. More freight was being hauled. More passengers were traveling for business and leisure. The burst of demand for traditional means of ground transportation is clearly seen in Figure 3.1, which gives five-year moving averages for the number of *niguruma* being used in the country. Growth is brisk during the 1870s when trade with the West began to expand. Upswings during the late 1880s and mid-1890s are also evident in the figure. This timing fits the long-swing pattern.

How did the northern and southern districts of the Tōkaidō respond to the surge in demand for transportation? In order to gauge this impact, let us focus on vehicles used for short-haul transport. In particular, the figures in panel B of Table 3.4 for *jinrikisha*, *niguruma*, and *kobune* give us important

Figure 3.1 **The Growth of *Niguruma* in Japan, 1877–1900** (Five-Year Moving Averages)

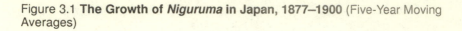

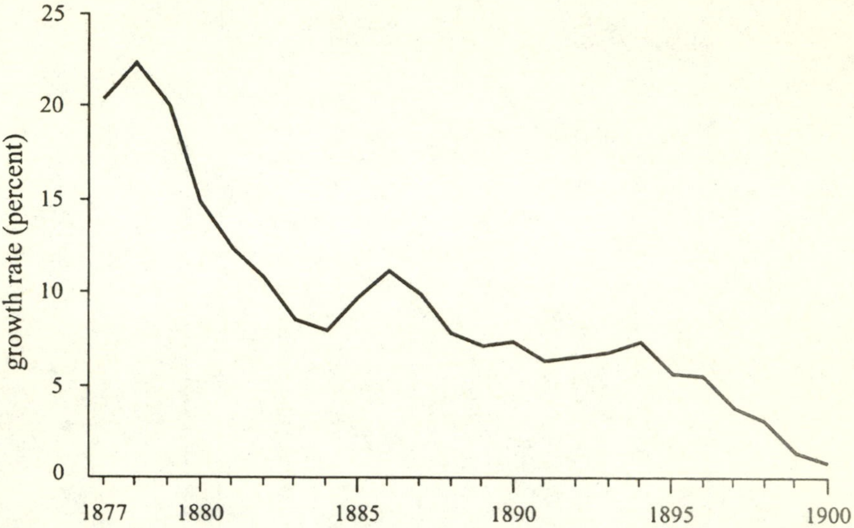

clues. As can be gleaned from the table, the surge in manufacturing activity affected the buildup in transport vehicle stocks in two radically different ways. Land transport vehicle usage expanded by leaps and bounds in Tokyo. And in Osaka, both water and land transport vehicle usage soared. Moreover, because Osaka was a water city, the ratio of land transport vehicles to water transport vehicles was far lower in Osaka than it was in Tokyo. The ecological contrast between the two cities, and their differing comparative advantages in terms of infrastructure, was clearly reflected in the means of transportation.

Financial Infrastructure: Merchants, Entrepreneurs, and the *Zaibatsu*

The infrastructure-driven growth hypothesis emphasizes three aspects of investment in infrastructure: investment in physical infrastructure like roads, electrical power grids, and port facilities; investment in schools, public health measures, and industrial research institutes that enhance human capital; and investment in financial infrastructure.

The growth of banking is one of the most salient characteristics of Japan's first long swing. Evidence on this score appeared in the first two sections of

Table 3.4

Transportation in Japan, in the Tōkaidō, and in the Prefectures of Osaka and Tokyo: 1881–1900

A. Infrastructure and infrastructure support: percentage in Tōkaidō and its subdistricts, and Osaka/Tokyo ratio (Tokyo = 100)[a]

	Urban population/ workers			Space: residential, storage, product			Vehicles/ ships			
District,[b] or ratio	UP	LAB	CARP	RLS	MANS	WHS	CC	SAILSS	SAILSL	STEAML
Tōkaidō	68.7%	26.9	24.3	24.0	53.1	22.7	49.3	73.1	51.6	81.3
Tōkaidō, north	37.6	11.0	10.4	9.9	22.9	8.5	23.0	30.3	28.1	72.0
Tōkaidō, middle	5.8	7.5	6.0	6.7	8.5	5.0	11.1	2.6	7.6	1.3
Tōkaidō, south	25.3	8.5	7.9	7.3	21.8	9.9	15.2	40.2	16.0	8.0
Osaka/Tokyo	37.5	70.8	43.3	67.3	83.7	114.5	55.3	132.9	43.0	11.8

B. Transportation vehicles: Osaka/Tokyo ratio (Tokyo = 100) and ratio of land to water transportation, Tokyo and Osaka[c]

	Osaka/Tokyo ratio				(jin + nig)/kb	
Years	jin	nig	jin + nig	kb	Tokyo	Osaka
1881–1885	74.1	87.4	83.5	111.0	503.4	377.4
1896–1900	48.8	65.1	61.7	116.2	911.8	484.4

C. Per annum growth rates based on five-year moving averages for various periods, land and water transport; Japan, Tokyo, and Osaka[c]

Period	jin + nig			nig		kb	
	Japan	Tokyo	Osaka	Tokyo	Osaka	Tokyo	Osaka
1882–1885	6.8%	7.7%	5.9%	7.8%	8.2%	–0.1%	1.9%
1886–1889	7.9	6.9	2.2	7.5	3.6	3.7	2.5
1890–1883	6.8	4.3	4.5	5.9	3.8	–1.0	0.7
1894–1887	6.1	3.3	3.4	3.3	3.2	2.2	2.8

Sources: Nihon Naikaku Tōkei-kyoku (1882–1904) *Nihon tōkei nenkan* (various dates), various tables; and Umemura, Takamatsu and Itoh (1983), various tables.

Notes: [a]The percentages in this panel refer to percentages of national totals located in the designated districts of the country. The variables are as follows: UP = urban population; LAB = unskilled laborers; CARP = carpenters; RLS = residential living space; MANS = area of manufacturing establishments; WHS = warehouse capacity (space); CC = cargo carts (*niguruma*); SAILSS = small sailing ships; SAILSL = large sailing ships; STEAML = large steam ships.[b]The districts of the Tōkaidō consist of the following prefectures: North—Saitama, Chiba, Tokyo, and Kanagawa; Middle—Shizuoka and Aichi; South—Kyoto, Osaka, and Hyogo.[c]The variable names are as follows: jin = *jinrikusha*; nig = *niguruma*; jin + nig = sum of *jinrikusha* and *niguruma*; kb = *kobune*; and (jin = nig)/kb = ratio of "jin + nig" to "kb" with "kb" set at 100.

this chapter. The proliferation of credit-creating institutions testifies to several facets of balanced growth in Japan. It testifies to the role of the government in experimenting with a variety of innovations designed to encourage the creation of financial intermediaries linking savers to investors. It also testifies to the strength of the increase in demand for funds, flowing from a rapid rise in fixed-capital formation within both the agricultural and manufacturing sectors during the upswing of the long swing. And it testifies to the weakness of equity (stock and bond) markets. That is, it testifies to a strong reliance on banks for the financing of investment. Finally, it leads to the question of whether there was a dearth of entrepreneurs who were able to build their businesses through internal accumulation of profits that they plowed back into more capital accumulation during the first long swing.

In order to understand how and why investment banking emerged as the engine for mobilizing capital in early industrializing Japan, we must examine, albeit briefly, the Janus-faced nature of entrepreneurial activity during the first long swing. For, on the one hand, giant combines emerged—the *zaibatsu*—as engines of capital accumulation and entrepreneurial drive. On the other hand, minuscule businesses headed up by entrepreneurs who had been active in the proto-industrial economy also proliferated in droves as the first long swing got under way.

What is a *zaibatsu*? The term combines two Chinese characters, "*zai*," meaning "financial," and "*batsu*," meaning "clique" or "faction." In the Japanese literature, the term "*zaibatsu*" refers to a collection of diversified enterprises under the ownership of a single family or extended family consisting of a main line and branch lines.[30] In the early wave of *zaibatsu* formation, which roughly coincided with the first long swing and the buildup to it, most of the *zaibatsu* heads were known as "political merchants" because they had strong connections with the Meiji oligarchs. Indeed, the Meiji oligarchy favored the interests of powerful merchants—as had the *bakufu* during the periods of extensive and intensive economic growth—and helped many *zaibatsu* acquire assets during the 1880s by selling off government-owned companies to them at cheap prices. But not all *zaibatsu* took this form. Some, like Sumitomo, which had emerged in the environs of Osaka during the late sixteenth century (Sumitomo's "well frame" logo was created in 1590) already owned mines and built their empires from a base in raw materials supplying industries like mining.

During the first long-swing period, the *zaibatsu* focused their attention on infrastructure development, trade and transportation, raw materials production, and banking. Most stayed away from manufacturing. Mitsui expanded from its base in banking, and then diversified into mining and trade. The Yasuda *zaibatsu* also emerged from a base in banking and finance. Ōkura

was involved in trade, and Mitsubishi in shipping and postal delivery. Sumitomo and Furukawa built their initial fortunes around mining.

The *zaibatsu* began as closed enterprises (e.g., partnerships), and gradually, as family capitalism gave way to managerial capitalism, shifted toward becoming open enterprises (e.g., joint stock companies). Institutional innovations played a role in the evolution of these diversified combines. For instance, implementation of the Commercial Code in the early 1890s encouraged the *zaibatsu* to set up either limited partnerships (*gōshi gaisha*), or unlimited partnerships (*gōmei gaisha*) for the enterprises they controlled.

In thinking about the early *zaibatsu* as family-managed enterprises, it is important to differentiate between those that emerged out of ancient Tokugawa merchant houses mainly based in Osaka, Kyoto, or Edo, and those that sprang up after 1868. During the first long swing, the first type of *zaibatsu* were managed by chief clerks (*bantō*), acting in the name of the family and its assets. The second type were usually directly managed by the entrepreneur and his—there were no female heads of *zaibatsu*—sons. Mitsui is a good example of the first type. During the Tokugawa period, the Mitsui house was active in the dry goods and money exchange businesses. It adopted a system whereby its assets were held in a family partnership, and the actual direction of affairs was placed in the hands of a trusted chief clerk. Mitsubishi is an example of the second type. Its founder, Iwasaki Yaturo, was born into a *samurai* house and became an unusually active entrepreneur during the early Meiji period. He directed the family business until his demise. In short, many of the early *zaibatsu* were not really family-managed businesses in the sense that many of the early British factories of the industrial revolution were family managed. In fact, to some degree, they had already taken on a managerial form. But this was not true of all the *zaibatsu*, for those founded by entrepreneurs during the Meiji period did not really make the transition to a managerial form until the founding head died.

The *zaibatsu* form was born out of scarcities—scarcities in entrepreneurial talent, scarcities in knowledge of foreign languages, and scarcities in physical capital. Within a fully developed "model" *zaibatsu* was a financial cluster (banks, trust and insurance companies), a general trading company (*sōgō shōsha*), raw material suppliers, and industrial enterprises. By pooling entrepreneurial drive among the affiliated enterprises, *zaibatsu* secured scale economies in making business decisions. By pooling technical expertise, they were able to make informed technical decisions essential to innovation, especially in the technology- and capital-intensive heavy industries that became important during the second long swing. And, by sharing access to financial resources, they were able to balance their portfolios. Older line, already profitable, enterprises provided flows of profits that could be tapped

by the financial cores of the *zaibatsu* in securing "venture" capital for new lines of business that did not yet have a well-established market niche. Thus, the *zaibatsu* form was important for stimulating capital accumulation in innovative areas that had not yet developed proven track records in profitability. Risk and return are negatively correlated. New risky ventures were subsidized by less risky, older enterprises. By diversifying its portfolio across various lines of business, the *zaibatsu* were usually able to simultaneously move into innovative lines of activity and avoid going under in bankruptcy. The key here is exploiting scale economies.

Most of the *zaibatsu* eventually reached a form in which powerful banks were at the core. These banks became the great investment banks of prewar Japanese industrialization. Why did this happen? To some extent, it happened because equity markets are difficult to set up. Brokerage houses, rating institutions, auditing firms, scrutiny by government watchdogs, and so forth must exist if the public is to be able to trust the information upon which it makes investment decisions. The capital market during the first long swing was thin. This constrained the growth of equity financing.

Government policy played a role as well. The government adopted a strategy of operating as a coordinating agent. It wanted to facilitate the emergence of powerful companies that could invest in both infrastructure development and industrial activities. It realized that scarcities existed, and it encouraged the growth of *zaibatsu* that exploited scale economies, thereby overcoming scarcities. For this reason, the Bank of Japan continued to use the policy of overloan that it employed to shore up national banks before they were phased out.[31] Under the overloan policy, the Bank of Japan allowed certain banks—like the powerful *zaibatsu* banks that eventually were heavily involved in providing the capital for railroads, shipbuilding, chemicals, and iron and steel—to run "negative reserves" at the Bank of Japan. That is, instead of requiring that these banks maintain positive reserves (thereby being net lenders to the central bank), so that they could meet the threat of potential runs on their liquid assets by drawing upon their reserves, the Bank of Japan, acting as a lender of last resort, permitted them to become net borrowers. Basically, the Bank of Japan was insuring the credibility of the big investment banks. Since the public knew this, they tended to flock to these banks. And so did entrepreneurs with ideas for innovations.

Now, since the Bank of Japan was located in Tokyo, and because maintaining good relations with the central bank was crucial to the *zaibatsu* banks, there was a natural tendency for the *zaibatsu* to establish their headquarters in Tokyo. For this reason, Japan's financial infrastructure tended to gravitate toward the capital. Why did industrial activity fail to follow this gravitational pull? Why did it concentrate in Osaka? Part of the reason lies in the fact that

some *zaibatsu* did keep their headquarters in Osaka. Sumitomo is a good example. But the main reason is that during the first long swing, and during early phases of the second long swing, investment in factories and plant and equipment for manufacturing came from small proprietors, many of whom emerged from the proto-industrial sector.

The importance of proto-industrial production for incipient Japanese industrialization should not be overlooked. Consider the figures in Table 3.5. As can be seen, it is only in machinery and printing—the former sector not having much competition from the proto-industrial sector, and the latter servicing a niche of better-educated Japanese, many of whom were concentrated in the Tōkaidō—that the region outside of the Tōkaidō fails to enjoy a substantial share of total national output. Proto-industrial producers, who predominated in the periphery outside the Tōkaidō core, were very important for generating manufacturing output during the first long-swing period.

Proto-industrial producers were not always opposed to using Western technology. In many cases, they played a key role in adopting Western techniques. They carried out the "proto-industrialization" of Western technology, thereby rendering it compatible with the conditions in the Japanese labor and product markets. And Osaka, which had been the center of the greatest geographic concentration of Tokugawa proto-industry, naturally became a magnet for merchants and entrepreneurs active in this "proto-industrialization" of Western technology movement.

Let us consider some concrete examples for the Osaka area. Johzen Takeuchi (1991) gives a remarkable account of proto-industrialization of Western technology for several industries active in Osaka during the first long swing: shell buttons, brushes, and knit fabrics.[32] He argues that in the case of all three industries, initial attempts to establish literal replications of Western mass production methods failed dismally. For instance, in 1888, Matsumoto Jutaro, president of the one hundred and thirtieth National Bank in Osaka, imported from the United States a complete factory system for manufacturing brushes. Petty producers quickly imitated his production methods. The proto-industrial producers eventually secured a dominant position in the brush industry, because they were more effective in training and motivating workers. In the case of shell button manufacturing, small producers learned how to make buttons from Western-managed concerns active in Osaka. The proto-industrial plants imitated the technology and then improved on it, both in terms of technical engineering and of organizing the flow of activity on the shop floor.

In sum, to some degree capital accumulation in Japan took on a bifurcated shape during the balanced-growth long swing. On the one hand, a small number of larger industrial enterprises and investors in large infrastructure

Table 3.5

Japanese Manufacturing, 1870–1900: Dominance of the Tōkaidō; Predominance of the Prefectures of Osaka and Tokyo

A. Output of manufacturing industries, circa 1890. Shares of national output in Tōkaidō and its subdistricts; and Osaka/Tokyo ratios (with Tokyo = 100) for industries grouped according to whether Tōkaidō share is above or below fifty percent[a]

	Tōkaidō share above 50%				Tōkaidō share under 50%						
District[b]	PRINT	CER	WOOD	MACH	TEXAP	OILW	CHEM	METAL	PAPER	FOOD	MIS
Tōkaidō	71.3%	52.8%	53.1%	91.3%	36.2%	30.4%	36.1%	12.9%	31.5%	32.9%	29.8%
Tōkaidō, north	30.4	9.6	7.7	5.2	9.3	4.9	7.6	2.0	14.5	10.3	3.8
Tōkaidō, middle	4.5	12.6	0.2	0.5	5.1	6.4	2.7	0.2	4.5	8.5	15.9
Tōkaidō, south	36.4	30.6	45.1	85.6	21.8	19.1	25.8	10.7	12.5	14.1	29.8
Osaka/Tokyo	94.7	169.0	1,419.7	1,322.4	465.8	21,345	365.1	961.3	42.9	159.2	216.4

B. Osaka, 1870–1897: Banks and spinning companies [capital in ( )] and factories [employment in ( )][c]

	Banks				Factories			
District	National	Private	Saving	Spinning Companies	Textiles	Machinery	Match	Glass
Higashi *ku*	9 (1,090)	11 (1,320)	3 (500)	—	1 (15)	—	2 (345)	1 (38)
Minami *ku*	—	5 (930)	—	—	2 (94)	4 (230)	5 (838)	3 (97)
Nishi *ku*	3 (370)	9 (550)	—	—	2 (2,325)	2 (56)	6 (249)	1 (n.a.)
Kita *ku*	1 (100)	1 (500)	—	1 (185)	13 (520)	9 (449)	7 (1,717)	20 (446)

Nishinari *gun*	—	—	1 (300)	1 (500)	9 (268)	1 (15)	3 (103)	1 (20)
Higashinari *gun*	—	—	3 (130)	5 (4,600)	7 (612)	1 (68)	6 (310)	2 (217)

Sources: Shinshū Ōsaka-shi Shi Hensan Iinkai (1991), *Ōsaka-shi Shi*, vol. 5, pp. 326–78; and Umemura, Takamatsu, and Itoh (1983), various tables.

Note: [a]Manufacturing industries are as follows: PRINT = printing and publishing; CER = ceramics; WOOD = wood products; TEXAP = textiles and apparel; OILW = oil and waxes; CHEM = chemicals; METAL = metal products; PAPER = paper products; FOOD = food products = beverages plus seasonings, seafood, tobacco; MIS = miscellaneous. Figures are based on averages in *yen* for the years 1889, 1890, and 1891.

[b]For the three subdistricts of the Tōkaidō, see notes to table 3.4.

[c]For banks and spinning companies figures in parenthesis are the sum of paid-in capital at the date of founding of the institution in 1,000 *yen* for all institutions of a given type in the district. For factories, the figure in parenthesis is the total number of workers for all factories in the district. "n.a." indicates figures are not available.

projects gravitated toward investment bank financing. Indeed, since development of banking was one of the key elements of infrastructure buildup during the first long swing, the very growth of huge banks was an integral part of the infrastructure expansion. On the other hand, most of the new industrial enterprises active during this period came out of the preexisting proto-industrial sector that was heavily concentrated in the Osaka region. These companies had modest capital needs and did not have to rely on the big investment banks that sprang up with the emergence of the *zaibatsu*. Since agricultural projects were usually small, resembling most of the new industrial projects in magnitude, growth was balanced. Typical investors in agriculture and manufacturing competed on a common playing field.

Factories and Mechanization

While the proto-industrialization of Western technology was important in the adaptation of Western technology in prewar Japan, there were sectors in which it did not play an important role. In these sectors, enterprises using Western technology carved out a niche that was distinct from that occupied by the proto-industrial producers within the sector. Rather than go head to head, the two types of firms coexisted by operating in different worlds and in different ways.

A classic example is the cotton textile industry employing British mules, power looms, ring frames, jennies, and steam engines.[33] Centered in Osaka during the first long swing, it coexisted with a proto-industrial sector based on the putting-out system. The proto-industrial producers tended to stick to making traditional *kimono* that they had been producing during the Tokugawa era. The Western technology–using sector focused on manufacturing cloth and clothing that competed with imports from India that flooded the markets of the great metropolises of the Tōkaidō with the opening of the country to trade. In particular, British companies were especially active in bringing in Indian cloth, much of it produced in British-managed factories in India. Indian-style cloth was suitable for making Western clothing that was turned out on wide looms, but it was not suitable for making Japanese *kimono* that was woven on narrow looms. Thus, in the cotton textile industry, the conditions needed for the coexistence of two very different types of producers were well established.

The ecology of Osaka and its environs made it an especially attractive locale for producing both types of textiles. Osaka—or rather the fields of Higashinari-*gun* and Nishinari-*gun* close to rivers along which raw cotton, thread, and finished cloth could be readily shipped—was favored because raw cotton was produced in the hinterlands of the metropolis, and damp

breezes flowing off the Pacific carried moisture into the environs.[34]

Both foreign and native entrepreneurs actively entered the Western-style cotton textile industry in greater Osaka. Particularly innovative was Osaka Cotton Mills, which commenced operations in 1884 with more than 10,000 steam-powered spindles. The company introduced electrical lighting that it generated itself so that it could run shifts day and night, thereby economizing on scarce capital. Because the skills required for satisfactory—if not exemplary—performance of tasks were acquired relatively quickly by freshly recruited workers, the industry was able to draw upon young farm girls. Thus, wages for females in agriculture basically set the floor for wages in light manufacturing.[35]

Enjoying an elastic supply of relatively cheap labor, the Osaka-based cotton textiles industry became competitive with importers of Indian cloth in a relatively short time. Thus, by the early 1890s, producers in the Osaka area became sufficiently versed in the techniques of cotton manufacturing that they no longer needed to rely on domestic sources of raw cotton. Indian staple cotton was superior to domestic staple, but it was somewhat costly because it was brought in from overseas. However, by the middle of the first long swing, the industry was so efficient that it could afford to use the imported raw material.

While steam-powered mechanized spinning and weaving became one of the most dynamic sectors of manufacturing within Osaka during the first long swing, proto-industrial production continued to flourish there as well. For instance, it was especially entrenched in the Sennan district of Osaka prefecture, where hand-operated narrow looms were utilized in making *kimono* for the domestic market.[36] Why did the producers in Sennan refuse to switch over to mechanized production?

The nature of steam technology played a major role in keeping the two sectors distinct. To be tapped effectively, steam power was generated in large central steam engines harnessed to weaving frames and looms through pulleys and straps. Small sheds could not make use of steam engines since the fixed cost of running the engine could not be spread over a large number of machines. Only in large factories was the unit cost of steam power low enough to render it competitive with hand production. Technology kept the industry sharply bifurcated.

Electrification, which was the crucial new energy source for the second long swing, changed this. Electrical motors operate on the unit drive principle. Each motor can be switched on or off independently of the other motors. Thus, electrification did lead to the mechanization of proto-industrial style production in the Sennan district. Compare the following figures on power looms and workers employed in Sennan in 1906 (before electrical

power was introduced into the area) with the figures for 1912 after villages in the area were supplied with power. The figures given compare the number of looms and workers concentrated in factories and in rental (*chinbata*) operations. The latter were commonly used in the putting-out organization of production (since no power looms were used in *chinbata* operations in Sennan in either 1906 or 1912, I do not give figures for that category here):[37]

	Factories			Rental loom operations	
Years	Power looms	Hand looms	Workers	Hand looms	Workers
1906	—	790	820	11,471	11,780
1912	9,227	5,415	5,415	833	843

Electrification collapsed some of the barriers keeping small proto-industrial operations in a wholly separate niche from large Western technology using factories. Indeed, under the force of mechanization associated with electrification, labor-intensive operations largely vanished in the Osaka region.

Thus, Osaka's competitive advantage during the first long swing lay in the fact that its economy stood on two legs: proto-industry and steam-powered production for light manufacturing. But its competitive advantage in exploiting steam power was not limited to cotton textiles and food processing. Osaka also became a nascent shipbuilding center during the first long swing. Under *bakufu* rule, Osaka had been a major transport hub, and hence craftspeople building and repairing the flotilla of tiny wooden boats used in the Inland Sea and on the Yodo River congregated in the environs of the city. With the opening up of the Inland Sea to Western-style sailing ships and steam-propelled vessels, many of these craftspeople turned their attention to learning how to build and repair the hulls, masts, and cabin interiors of these ships.

The brisk growth in shipping in the Inland Sea, and more generally along Japan's coastlines, created a burgeoning demand for shipbuilding. Reacting to the national government's encouragement of a merger between Japan's two major shipping lines, creating the mammoth Nippon Yusen Kaisha (N.Y.K.), and, spurred on by the interest of the Sumitomo *zaibatsu* in securing a stake in the rapidly expanding coastal trade, fifty of the small shipbuilding and ship-using firms concentrated in the Osaka area banded together. They merged in 1884 to form the Osaka Shosen Kaisha (O.S.K.), which proceeded to go into head-to-head competition with the N.Y.K.

Under the subsidization policy in place until the latter half of the 1890s, individual enterprises could apply for government subsidies and the two companies vied for them.[38] The policy reflected the division of Japan's coastal trade into two distinct niches: one involved transporting people and goods from

one locale in Japan to another; and the other involved transoceanic traffic. The former was largely in the hands of Japanese companies, and the latter in the hands of Western enterprises. Why did Western companies control the latter niche? The main answer is personnel. Japanese crews could not handle the long-distance voyages. They were not trained to do so. Only after decades of training cadres of Japanese seamen, captains, and engineers in government-managed schools—that is, only until the 1890s—were Japanese companies prepared to compete with Western companies in the long-distance transoceanic business. When they reached the point when they could compete, the government changed its subsidization policy, and began to support activity along specific designated route lines rather than specific companies.[39]

Reflecting the fact that until the 1890s, Japanese companies concentrated on coastal shipping, the shipbuilding industry that emerged in Osaka during the first long swing focused its efforts on producing vessels of medium size and vessels having wooden hulls. These were suitable for the domestic coastline business. The fact that this early phase of Western-style shipbuilding involved building boats with wood favored builders in the Osaka area. Workers versed in constructing small wooden boats were plentiful in the environs of Osaka, where *kobune* were still heavily relied upon for transportation. In short, just as with textiles, Osaka's strength in the proto-industrial sector created a competitive advantage in the innovative sector employing Western technology extensively.

Osaka's dominance in shipbuilding during early Japanese industrialization is the subject of Table 3.6. This table provides data on the location of dockyards where the 626 ships registered in the Japanese merchant marine had been constructed. As can be seen from the figures, an extremely large percentage of the steamships produced in domestic yards came from the Settsu district—Settsu refers to Osaka and Kobe—and almost every one of those ships was built out of wood. The larger ships with metal hulls were almost exclusively the product of foreign dockyards, mostly located in the United Kingdom.

In understanding the evolutionary dynamics of the shipbuilding industry, it is important to keep in mind that during the initial phase of Western shipbuilding (when Japanese workers were not yet comfortable working to Western specifications and were not proficient in welding and manipulating Western tools), productivity and profitability in the Western-oriented dockyards was low. Indeed, in comparison to the enterprises turning out *kobune*, the Western ship-producing dockyards were far less profitable during the period 1877–1880.[40] But after 1885, as Japanese workers learned how to do their tasks more efficiently, the Western-style yards became increasingly profitable. In the 1885–1890 period, their profitableness falls slightly short of that registered for the *kobune*-oriented enterprises; and by 1902–1908, the large capital-intensive firms producing Western-style ships are the most profitable

Table 3.6

Steamers in the Japanese Merchant Marine, 1898

A. Classified by locale of registration (626 ships)[a]

	Number of ships [percentages in ( )]			Average registered tonnage		Average horsepower	
Hull type	Total	Outside Settsu	Inside Settsu	Outside Settsu	Inside Settsu	Outside Settsu	Inside Settsu
Wood	397 (100)	275 (69.3)	122 (30.7)	81.3	27.2	110.3	26.9
Nonwood[b]	229 (100)	128 (44.1)	101 (44.1)	1,193.5	753.0	179.1	118.7

B. Classified by locale of construction/shipyard of origin (626 ships)

	Percentage made in:				Average tonnage			Average horsepower		
		Japan[a]			Made in Japan			Made in Japan		
Hull type	Total	Outside Settsu	Inside Settsu	Abroad[c]	Outside Settsu	Inside Settsu	Made abroad	Outside Settsu	Inside Settsu	Made abroad
Wood	100	38.3	60.0	1.7	79.0	94.0	204.0	27.0	27.0	36.4
Nonwood[b]	100	5.7	12.2	82.1	341.8	292.9	1,151.5	54.7	81.4	169.9

Sources: Japan Department of Communications, Mercantile Merchant Marine Bureau (1898), various pages.
Notes: [a]The "Settsu" category includes the following designations: Settsu, Osaka, Kobe, and Hyogo.
[b]Nonwood includes the following categories: iron, steel, combined iron and steel, and composite.
[c]This category includes a small number of shipyards of unknown designation.

in the shipbuilding field. An excellent example of how Japanese learned to work with Western technology in the shipbuilding and related sectors is afforded by the story of Nishiyama Unotsuke.[41] Nishiyama went to work for the Osaka Iron Works, Shipbuilding, and Dock Company established by a British entrepreneur, E.H. Hunter. Under Hunter's guidance, Nishiyama learned the intricacies of iron and steel making, and he eventually left Hunter's yard to start his own small firm near the banks of the Aji River in Osaka.

This last set of points brings me to an observation about the constraints imposed on upswings that are propelled forward by innovations in Japan. Innovating sectors often find their capacity for expansion checked by inadequate supplies of skilled labor. During downswings and phases of intense infrastructure buildup (which includes founding and staffing schools and academies devoted to training prospective workers), supplies of labor have increased in response to the surges in demand that occurred earlier.

In sum, during the first long swing, the Tōkaidō began to take shape as Japan's industrial belt, and, within that belt, Osaka emerged as the core of the core. Osaka's dominance in industrial affairs is remarkable because, as can be seen from panel A of Table 3.5, it is apparent in almost every type of manufacturing activity, paper and printing excepted, Osaka far outstripped Tokyo. Compare the following figures, for 1890, on steam horsepower and on workers in factories with 300–499 workers, and in factories with 500 workers or more:[42]

	Steam horsepower		Workers	
Factory size	Osaka	Tokyo	Osaka	Tokyo
300–499 workers	3,419	855	9,152	5,069
500 workers or more	30	304	1,630	1,449
Total	3,449	1,159	10,782	6,518

If one is referring to Osaka's preeminence in industrial matters, it was certainly aptly named "Manchester of the Far East."

Osaka Triumphant

But does the nickname "Manchester of the Far East" adequately describe Osaka's role in Japanese industrialization and economic growth? How similar were the key industrial cities of early industrializing Japan to the key industrial cities of early industrializing Great Britain? How similar was the urban system of nascent industrializing Japan to that which developed during Great Britain's first century of industrialization? How similar were the industrial belts that emerged in these two great island nations?

Likening Osaka of the first long swing to the great cotton town of southeast Lancashire, Manchester, seems natural. Like the Manchester of the British Industrial Revolution, Osaka was home to grand steam-powered cotton and spinning factories. Like Manchester, Osaka had become a grimy city with myriads of smokestacks belching out black fumes. Like Manchester, Osaka had a voracious appetite for coal. Like Manchester, which depended on waterways for securing its raw cotton—the Duke of Bridgewater Canal and the Irwell River—Osaka relied upon its canals and rivers for shipping raw materials and finished products. Manchester depended upon the nearby seaport of Liverpool; Osaka depended upon Kobe close at hand. And, like Manchester, Osaka was home to a commercially oriented elite that had to cope with a political elite residing in a distant national capital.[43]

To put this comparison in context, it is useful to compare the geographic contours of the industrial belts in the two nations. Map 1.1 provides us with the geography of Japan's industrial belt, the Tōkaidō. And Map 3.3 provides us with a similar illustration for Great Britain. Befitting its shape (the spread in the southern region around London and the narrowing in the northwest near Liverpool), the British industrial belt of the industrial revolution is known as the "Industrial Coffin." Manchester is situated in the northwestern sector of the "Industrial Coffin." Like the Manchester/Liverpool complex that serves as the northwestern pole of the British industrial belt whose southeastern pole is London, the capital, the Osaka-Kobe complex serves as the southwestern pole of the Tōkaidō industrial belt. The parallel between the two cities is certainly intriguing.

Moreover, both cities played a key role in promoting the mechanization of manufacturing. For instance, of the three hundred–odd Boulton-Watt steam engines put into use in Great Britain between 1775 and 1800, over a third were employed in Lancashire. Of steam engines used in British textile concerns, over 80 percent were situated in Lancashire.[44] In short, there are compelling parallels between the two metropolitan centers and their industrial belts.

But, more thorough consideration concerning cites and industrial belts reveals glaring differences. At the root of these differences is diversity. Osaka's economy was far more diverse than was Manchester's at similar phases of industrial development. Being more diverse, Osaka's economy made greater contributions to economic development in Japan than did Manchester in Great Britain.

Consider the geographical locus of the great industrial centers in the two countries. As can be seen by comparing Map 1.1 with Map 3.3, most of the great conurbations of the Industrial Coffin lie inland. In comparison, the Tōkaidō is ocean-oriented. Five of the six major metropolitan centers of the Tōkaidō are situated along the seacoast. Two are in and around the Kantō plain,

Map 3.3 **Principle Towns of England, Wales, and Scotland Inside and Outside of the "Industrial Coffin" Counties**

and two lie near the mouth of the Yodo River draining the basins and plains of the Kinai. During the heyday of Japan's organic economy, infrastructure that made heavy use of access to the water developed along the coastline joining the Kantō to the Kinai, and the area became the administrative and commercial core of the country. Since the economy did not make use of coal at the time, securing it played no role in the formation of this core region or of the infrastructure tying its nodal points together. The main coal deposits in Japan are situated in the southern island of Kyūshū and in the northern island of Hokkaidō. They are far apart in terms of geography, and they are not close to the geographic center of the country.[45]

By contrast, the network of canals and waterways that linked together major nodes in Britain's Industrial Coffin was largely a by-product of the industrial revolution itself. Most of the British canals were constructed during the eighteenth and nineteenth centuries so that boats could move coal, iron ore, brick, and limestone around the Industrial Coffin.[46] Moreover, the location of coal deposits played a key role in the creation of an industrial zone slicing across England along the Liverpool-London axis. Collieries were in abundance within this zone. Indeed, a key reason underlying the rise of the southeast Lancashire cotton textile complex around Manchester was the fact that coal was being actively mined in this region. Of the 2,500 collieries operating in Britain in 1855, about 360 were concentrated in Lancashire, and these Lancashire mines generated about 14 percent of Britain's total tonnage of coal.[47] In short, coal was intimately tied up with the formation of the Industrial Coffin. But it was not a factor in the formation of the Tōkaidō industrial belt.

Comparing the buildup of Osaka and Manchester prior to, and during, industrialization also reveals important differences between the two conurbations and their roles in the economic development of their respective nations. As a key nodal point in the commerce of Tokugawa Japan, Osaka had become a huge population center prior to industrialization. Osaka's population during the seventeenth century hovered around 4,000 individuals. By contrast, Manchester did not become a heavily populated city until the Industrial Revolution. In 1801, Manchester had a population of only 75,000 people. By 1841, it had soared to 235,000 thousand people.

And Osaka was diverse. Its merchant houses specialized in trade and financial matters. It handled rice, vegetables, fish, fertilizer, cotton, silk, and swords on a grand scale. Craft production flourished in Osaka, and its merchants were active in putting-out production throughout the Kinai region.

Once Japan was opened up to trade with the West, Osaka became diverse in new ways. Because Japan began to industrialize a full century after Britain did, Japanese innovators could draw upon a far richer menu of techniques, machines, and sources of energy than could the innovators in Britain

during the period 1750–1840. To give a few illustrations: Osaka textile mills generated their own thermal electricity in the 1880s; and steamships with screw propellers played an important role in transporting goods during Osaka's early industrialization. Osaka was a more important innovator than Manchester because it had a greater opportunity to innovate.

For these reasons, I believe Osaka and the other great conurbations of the Tōkaidō played a more important role in Japan's early industrialization than did Manchester and the cities of the Industrial Coffin in Britain's early industrialization. Population size and role in promoting economic development are certainly not the same, but the distribution of city sizes in the two countries does provide us with an interesting way of ferreting out the relative importance of major cities in the two countries.

Consider the rank-size distribution of cities in the core industrial and periphery regions of the two countries. The rank-size distribution is characterized as follows. Rank the cities of a country in terms of their population sizes. The rank-size distribution relates the population sizes of the cities to their order in the ranking. Typically, the relationship is expressed in terms of natural logarithms. The following equation captures the association:

$$\log(P_i) = \sigma_0 + \sigma_1 [\log(i)] + \varepsilon. \qquad (9)$$

In this equation, $\log(x)$ is the logarithm of variable x, P_i is the population size of the city with ith rank, i is the rank in the order, and ε is the error term. The steeper the slope—that is, the larger in absolute value is the coefficient σ_1, which is negative in sign—the more populous are the largest cities relative to the smaller cities. In short, the larger in absolute value is the coefficient on the logarithm of a city's rank, the more concentrated is a region's population, the more dominant are the largest cities in the region.

Comparing the rank-size distributions of cities in the industrial belts of Britain and Japan during early industrialization, it is apparent that compared to the British Industrial Coffin, Japan's T)kaid) was heavily concentrated. This can be seen graphically for Japan in Figure 3.2, showing the rank-size distribution within the T)kaid) to be very steep. Indeed the absolute value of the coefficient estimate σ_1 is 1.81. But in Japan's periphery, the region outside of the industrial belt, the relationship is far flatter. The absolute value for σ_1 is 0.46. Consider England, Wales, and Scotland. In 1801, the absolute value for the estimate of σ_1 in the Industrial Coffin core is 1.2; and the estimate for the periphery outside of the industrial core is 0.9. In short, during early industrialization, the overwhelming dominance of the six big conurbations of the Japanese industrial belt stands in stark contrast to the urban structure of Britain's Industrial Coffin.

Figure 3.2 **Rank-Size Distributions for the Cities in Two Regions of Japan, 1903: Tōkaidō Cities and Cities Outside of the Tōkaidō**

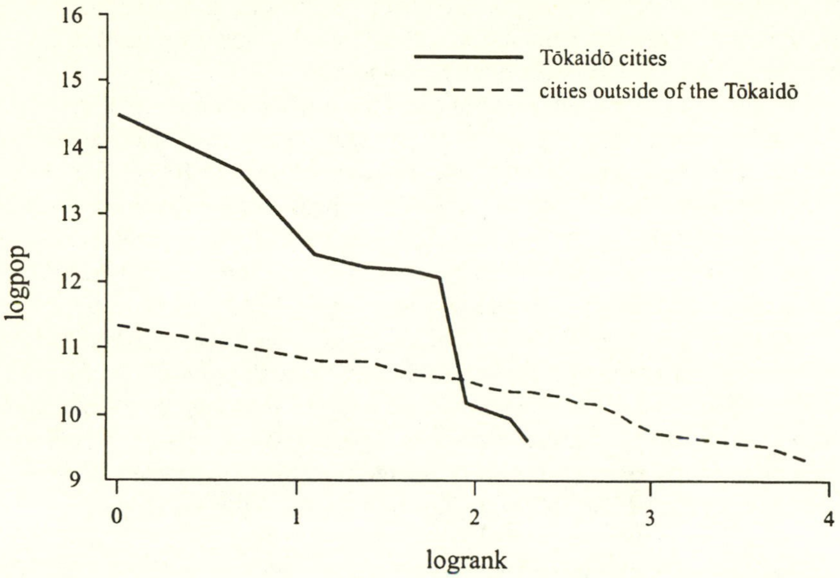

Thus, while the analogy between Osaka and Manchester was an analogy that naturally occurred to Westerners who saw Osaka's emergence as a major textiles manufacturing metropolis during the first long swing, it is flawed. In terms of overall impact on industrialization and innovation, Osaka's role was far greater than Manchester's. Early industrializing Osaka was far more diverse in its economic and political functions than was Manchester. The rank-size distribution calculations testify to very important differences between the early industrialization of Great Britain and Japan.

Indeed, as earthshaking as was innovation in Great Britain's Industrial Revolution, Meiji Japan was rocked with an even greater wave of innovations. As we have seen in this chapter, these innovations were hardly restricted to the industrial sector and to the technology of energy production. Innovations abounded in political organization, financial infrastructure, education and learning, architecture, agriculture, and physical infrastructure. During the first long swing, the infrastructure of Japan was thoroughly transformed.

In the great wave of innovation sweeping over Meiji Japan, Osaka emerged as the core of the new industrial core. Building upon its strengths in water-based infrastructure and proto-industrial knowledge, Osaka became the center of industrial innovation in Meiji Japan.

Thus, factories began to spring up in Osaka and its hinterland. Snapping up land, entrepreneurs set up production in the new Manchester of the Far East. And as the demand for land soared, so did its price. Consider the ratio of the price of land used for commercial, residential, and industrial purposes—known in Japanese as *takuchi*—to paddy land.[48] Between the early 1880s and the end of the century, this ratio falls in Tokyo prefecture. For instance, the ratio for Tokyo prefecture in 1880–1884 is 3.4; and, by 1895–1899, it has dipped to 2.58. The buying up of land by industrial concerns and builders of residential housing was desultory. By contrast, the Osaka land market was hot. The ratio of *takuchi* to paddy in Osaka prefecture in 1880–1884 was 1.43; by 1895–1899 it had reached 2.13. And this was in a prefecture where paddy land was extremely valuable.

Moreover, in absolute terms, Osaka's per hectare *takuchi* price surpassed that of Tokyo at the close of the century. In the general inflation of the upswing of the long swing, *takuchi* prices surged forward with unusual vigor in Osaka. But a mere twenty years earlier, things were vastly different. During the 1880–1884 period, the average value of *takuchi* in Osaka was 55 percent of that in Tokyo. During 1895–1899, the figure was 102 percent. Osaka's economy was booming, and this was reflected in the prices for land.

So, in the balanced-growth long swing of 1887–1904 that paved the way for Japan's industrialization, Osaka emerged as the center of the new innovating sector of the economy. But the balanced-growth long swing provided the foundations for the transitional-growth long swing. And infrastructure that buttressed the transitional-growth long swing differed in significant ways from the infrastructure supporting the balanced-growth long swing. Infrastructure erected on the land—electricity and electric railroads—was key to the second long swing. At the turn of the new century, Japan entered into an age of electricity and steel. In that world, the competitive advantage in developing new infrastructure enjoyed by Tokyo outstripped that of Osaka. A shift toward a Tokyo-centric world began.

At the turn of the new century, Osaka's preeminence in industrial matters was overwhelming. It was so overwhelming that guidebooks to Japan safely assumed that Osaka's status as Manchester of the Far East would remain unshaken as Japan's industrial progress continued apace. But glory achieved is seldom glory retained. The glory garnered by Osaka was garnered for the moment only. Osaka's glory was fated to be transitory.

Notes

1. It must be kept in mind that I use seven-year moving averages for most of the variables describing the long swings. For discussion on this point, see Appendix A.1

and the discussions concerning long swings in Minami (1986, 1994), Ohkawa and Rosovsky (1973), and Ohkawa and Shinohara with Meissner (1979).

2. On this point, see the discussion in Mosk (1995a, chap. 2; 1996, chaps. 4 and 5).

3. Mosk (1995a, Chap. 2; 1996, Chaps. 4 and 5; 2000a) discusses the importance of investment in education for Japanese economic development and for the prevention of infectious mortality, and provides details concerning the evolution of the various layers of the educational system. Mosk (1996) emphasizes the positive feedback of improvements in health and physical strength upon work capacity.

4. These figures are taken from table 4 of Mosk (2000a).

5. The term "Manchester of the Far East" appears in many accounts of Osaka's industrial history. I have not been able to discover who first coined the term. Perhaps it was one of the British nationals who lived in Osaka's Kawaguchi district that was provided to the foreign community under the terms of the treaties guaranteeing extraterritorial rights to nationals of some Western powers.

6. On the economic advantages enjoyed by treaty ports, see Elvin (1974a, 1974b), Hoare (1994), Howe (1981b), and Murphey (1974).

7. For a general discussion of early Meiji institutional experimentation, see Westney (1987). For a discussion of foreign models used by the Meiji era Navy in Japan, see Evans and Peattie (1997).

8. The choice of competing national models may have reflected bureaucratic infighting, one faction favoring country A and the other country B. On bureaucratic sectionalism and factionalism, see Samuels (1983).

9. On foreign technical experts employed in Meiji Japan, see Jones (1980).

10. For the emergence of the Mitsubishi *zaibatsu*, see Morikawa (1992) and Wray (1984). The nature of balanced-growth long-swing *zaibatsu* as either political merchants or as developers of infrastructure, especially in mining, is discussed later in this chapter. The evolution of the *zaibatsu* through the second and third long swings is discussed in chapter 6.

11. See page 498 in Shinshū Ōsaka-shi Shi Hensan Iinkai (1991), vol. 5.

12. See Goldsmith (1983), Patrick (1967), and Westney (1987) for more extensive discussions of early Meiji banking.

13. See Fujitani (1996) for details.

14. Some members of the Meiji oligarchy favored making Osaka the capital of the country. But, since Osaka was close to Kyoto, and since striking a geopolitical balance between eastern and western Japan was important, the advocates for Tokyo prevailed over the advocates for Osaka.

15. I draw upon Stewart (1987) here.

16. The *gun* units were stripped of their administrative functions in 1926. They continue to exist only insofar as they demarcate boundaries.

17. A series of incorporations proposed in the early 1940s would have led to a fourth major incorporation. According to the plan floated at the time, a number of surrounding cities—Sakai-*shi*, Toyonaka-*shi*, and Suita-*shi*—would have been absorbed into Osaka. The plan was rejected. After the war, Osaka city incorporated a few contiguous villages and towns in a relatively modest expansion of boundaries.

18. In the table, estimates of population for Tokyo and Osaka prior to 1920 appear. These estimates are not based on de facto census counts, because the national government did not carry out a nationwide census until 1920. For the years between the early 1870s and the first census taken in 1920, population counts are based upon tabulations worked up by local government offices from

household registration books filed in the offices. In principle—since individuals are legally required to register changes of domicile within a short period after they have moved residence from one locale to another—it is possible to obtain population totals for local areas like Osaka and Tokyo by adding the net additions of residents during a time period to population estimates for the locale at the commencement of the time period. Net additions to population are registered births plus inmigration, minus deaths and outmigration. For details on Japan's household registration system, see Mosk (1983).

19. See Sawamoto (1981, 123–24) for details. The early Meiji experiments in arable land reorganization helped foster the Cultivable Land Reorganization Law of 1899. This law had a tremendous impact on city planning. See the discussion in chapter 7. The Land Tax Reform of 1873, which abolished paying taxes in kind by requiring payment in money (taxes now dependent upon the capitalized land value owned by an individual), was also derived from Western models.

20. See Shinshū Ōsaka-shi Shi Hensan Iinkai (1991), *Ōsaka-shi shi*, vol. 5, pp. 396 ff.

21. The business community in Tokyo, under the leadership of Shibusawa Eiichi, made out a case for building a deepwater port in Tokyo Bay during the Meiji period. However, they did not advance their case with the fervor of the business community of Osaka.

22. See Sawamoto (1981, 125 ff.).

23. The Yokohama port improvements were completed in 1917.

24. Shinshū Ōsaka-shi Shi Hensan Iinkai (1991), *Ōsaka-shi shi*, vol. 5, various pages. Subsequent improvements involved extending the breakwaters northward and southward. To some extent, these improvements were undertaken because a typhoon damaged many sections of the original structure in 1934. Osaka was also active in securing national funding for a waterworks that used metal pipes. This project was completed in 1897. Part of the rationale for this public works project was an outbreak of cholera in 1886. Cholera often broke out in port cities.

25. See pages 474–75 in Shinshū Ōsaka-shi Shi Hensan Iinkai (1991), *Ōsaka-shi shi*, vol. 5.

26. See Tōyō Keizai Shinpōsha (1935).

27. I am grateful to Osaka City Museum for the right to reproduce their copy of this famous privately commissioned map.

28. See map 3 in Ōsaka-shi (1980), *Meiji-Taishō Ōsaka-shi shi*, vol. 3.

29. For a remarkable map displaying the water oriented locus of manufacturing in prewar Osaka, see Shinshū Ōsaka-shi Shi Hensan Iinkai (1996), *Ōsaka-shi shi*, vol. 10, map 10. This volume is devoted to maps.

30. On the *zaibatsu*, see Hirschmeier and Yui (1995), Morikawa (1992), Nakamura (1983), Okochi and Yasuoka (1984), and Wray (1984). On main line and branch line families in the Japanese stem family system, see Mosk (1983, 1996).

31. For a description of how the overloan policy works, see Suzuki (1980). For some measures of the degrees of overloan during the first long swing, see panel C.2 of table 3.1.

32. See Takeuchi 1991 for the shell button business (36–41); for the brush industry (58–61); and for the knit fabric industry (86–95).

33. For very detailed treatments of the importation and adaptation of Western technology in the textiles industry, see Kiyokawa (1995) and Minami and Makino (1995a, 1995b). Kiyokawa emphasizes the importance of technically trained engineers for the

decisions that led the cotton textile industry to switch from mule to ring frame spinning. He also emphasizes the sharing of technical information among the major spinning concerns for the rapid diffusion of ring frames in the industry.

34. From panel C of Table 3.2, it is apparent that the male-to-female sex ratio is far lower in the periphery of Osaka than it is in the core of the city. This ratio reflects the fact that light industry relied heavily on female labor input, and light industry was heavily concentrated in the periphery of the city. By contrast, the merchant houses, which mainly recruited male workers as apprentices, tended to be located in the older part of the city, that is, in the core.

35. On this point, see Mosk (1995a, chapter 2).

36. See Abe and Saito (1988).

37. These figures are taken from Abe and Saito (1988, 154).

38. The discussion about shipping in this section, and elsewhere in this volume, draws heavily from the accounts provided in Chida and Davies (1990), Wray (1984), and Yui (1985).

39. For discussion of various systems of subsidies used to support shipping and shipbuilding in prewar Japan, see Chida and Davies (1990), Wray (1984), and Yui (1985).

40. For estimates of profits, bearing out these assertions, see Ohkawa and Otsuka (1994, 48).

41. See Nishiyama (1997).

42. These figures are taken from Tōkyō Hyakunen Henshū Iinkai (1979, 378–9). The underlying data do not appear to be complete. Therefore, it is likely the figures cited here are underestimates of the true magnitudes.

43. It should be noted that there are a number of interesting parallels between Japan and Great Britain as countries. For instance, both countries are island nations lying off the Eurasian continent. And within their geographic zones—Great Britain on the west and Japan on the east—each nation was the first to industrialize.

44. See Smith (1968, 155 ff.). Of the more than 97,000 power looms utilized in England and Wales in 1835, 64 percent were located in Lancashire.

45. See Trewartha (1945, 87). In 1936, mines in Kyūshū accounted for 71 percent of the coal output in Japan proper (excluding its empire in Korea, Taiwan, and Manchuria). Mines in Hokkaidō generated about 23 percent of the coal output of Japan proper in that year.

46. See Hadfield (1968, 21, 34, 144 ff.). For instance, the Duke of Bridgewater constructed a canal with the principal aim of shipping coal from mines on his own estate to Manchester.

47. See Smith (1968, 148). The climate in Manchester, like that of Osaka, was ideal for cotton spinning. The spinning districts were built up along the slopes facing west, where they benefited from the damp breezes flowing off the Atlantic. Moreover, Manchester was close to Liverpool, which was a major port handling raw cotton brought in from the southern states of the United States across the Atlantic.

48. Figures on land prices are drawn from Nihon Naikaku Tōkei-kyoku (1882–1902), *Nihon teikoku tōkei nenkan*. The term *takuchi* refers to land used for commercial, residential, and industrial purposes. Typically, *takuchi* excludes wasteland and land used in agricultural cultivation. I go into the issue of land speculation and rising land prices in greater detail in chapter 5, where I deal with speculation accompanying escalating land values.

Part II

Electricity and Steel

4

The Transitional-Growth and Unbalanced-Growth Long Swings

The Logic of Internal Development and the Impact of Global Economic and Geopolitical Change

The first long swing laid the foundation for Japan's industrialization. During the second and third long swings, the nation industrialized. Industrialization—defined as a growth in the proportion of output flowing from, and labor force engaged in, manufacturing and mining—is a trend phenomenon. Thus, the trend toward industrialization was an ongoing concomitant of Japan's moving through transitional growth and unbalanced-growth long swings.

Industrialization was not the only trend accompanying the second and third long swings. Others were also important. The number of companies incorporated as either partnerships, limited partnerships, or joint-stock enterprises surged forward. Paid-in capital in the corporate sector jumped, as did paid-in capital in the banking sector. And, most important of all, the structure of the economy changed. During the second long swing, the economy made a decisive shift toward unbalanced growth.

As a result of this shift, the economy became increasingly dualistic. Sectors enjoying high levels of innovation grew rapidly in terms of output and employment. These sectors included key branches of infrastructure (electricity, electrical railroads, and the extension and improvement of roads) and of manufacturing (especially heavy manufacturing like iron and steel, metal making, and machinery manufacturing). At the same time, other sectors grew far more slowly. Agriculture and light manufacturing experienced desultory growth. Moreover, segmented markets for labor and capital services emerged. Innovating sectors secured capital on relatively inexpensive terms; lagging

sectors paid higher interest rates than did the dynamic branches of infrastructure and industry. The reverse was true for labor: The unit cost of labor was far lower in light manufacturing and agriculture than it was in heavy industry. For this reason, the relative cost of the factors of production—as measured by the ratio of wages to the unit cost of capital—was very different in the two worlds. In innovating branches, the ratio was high, and in lagging branches it was low.

Thus, trends and swings overlap. But they overlap in another way as well. A number of authors have pointed to the possibility that trend acceleration occurred in Japan.[1] In quantitative terms, trend acceleration refers to a situation in which the trend rate of growth accelerates over time. For instance, consider the annual average rates of growth of gross domestic product over successive upswings of the long swings.[2] In the first upswing between 1887 and 1897, gross domestic product grew at a rate of 3.0 percent; during the second upswing of the second long swing—the one associated with manufacturing expansion—gross domestic product grew at 4.4 percent; and during the upswing of the third long swing, it grew at 4.6 percent. Thus, the rate of growth during upswings increased over each successive long swing. This is trend acceleration in output growth.

Trend acceleration also appears in rates of capital formation. Consider growth rates for fixed domestic capital formation over successive upswings. In the upswing of 1887–1897, the rate of growth of fixed domestic capital formation was 5.6 percent; in the first upswing of the second long swing between 1904 and 1911, the growth rate was 6.1 percent; during the second upswing of the second long swing between 1911 and 1919, it was 6.9 percent; and during the upswing of the 1930s, it was 10.6 percent. This acceleration in accumulation also shows up in the percentage of gross national expenditure going into gross domestic capital formation. In 1887–1897, this figure is 9.8 percent; in 1904–1911, it is 13.2 percent; in 1911–1919, it is 16.2 percent; and in 1930–1938, it is 18.5 percent. Thus, the rate of accumulation rises with each successive long swing upswing. This is trend acceleration in capital formation.

Trend acceleration can be viewed as the outcome of a strictly endogenous infrastructure-driven economic-growth process.[3] By "endogenous," I mean two senses of the word: the sense that the logic of growth is internal to the Japanese economy itself, and the sense employed in appendix 3 of chapter 1. In the latter sense of the word, endogenous growth means growth stemming from capital accumulation and spillovers resulting from that accumulation. In my theory of infrastructure-driven growth, investment in infrastructure spills over into the industrial sector of the economy, thereby promoting technological and organizational progress induced by accumulation of infrastructure.

One of the key linkages inducing trend acceleration from the internal logic of domestic growth is the savings rate. If the rate rises because income and income per head rise—either responding to higher levels of income or to growth in income—then growth in income (and income per head) induces a rise in the savings rate and an increase in domestic fixed capital formation.[4] The greater the rate of saving, the greater the capacity of the economy to innovate. This is because a larger pool of savings makes it easier for governments and private firms to borrow. Thus, higher savings encourages a brisker increase in the volume and quality of the infrastructure capital stock, thereby promoting a more rapid pace to the incidence of spillovers. And, by reducing the interest rate, higher savings helps entrepreneurs more readily borrow in order to acquire new plant and equipment.[5] In short, the pace of, and diffusion of, innovations is linked to accumulation. A positive response of savings rates to income growth buttresses trend acceleration.

Of course, assuming that income growth induces a rise in savings and, hence, increased domestic capital accumulation presupposes that institutions exist to channel funds from savers to investors. It is in recognition of the crucial role played by financial intermediaries in facilitating this matching of interests, simultaneously satisfying savers and investors, that I stress the importance of the expansion of the banking sector in infrastructure development.

In short, there is considerable evidence supporting the position that the Japanese economy evolved endogenously. However, even the most ardent supporter of endogenous growth must recognize that Japan did not develop in a vacuum. As we have seen, during the first long swing, Japan depended on other countries for technology, and it sought out models for banks, government agencies, and commercial law in foreign lands. Thus, it is necessary to briefly review some of the ways the international community affected Japan during the course of the second and third long swings. I will focus on three international tendencies affecting the performance of the Japanese economy during the period 1904–1938: the breakdown in the gold standard, the rise in protectionism, and a drift toward autarky; growing international competition between three competing ideologies for organizing economies and societies in the West, namely, competition between fascism, communism, and liberal democracy; and geopolitical changes.[6] These three sets of influences are historically intertwined. And, historically, they are intimately tied up with Japan's growing interdependence with its empire, an empire that it began to build up in stages after 1895. It secured Taiwan as a result of its victory in the Sino-Japanese War, and it formally annexed Korea into its empire in 1910, having secured international recognition for its "special interests" in Korea upon defeating Russia in the Russo-Japanese War.

The breakdown in international trade and the worsening of the interna-

tional geopolitical situation began in 1914 with the onset of World War I.[7] During the war itself, the belligerent countries embargoed trade and attacked each other's merchant marines. After the war, the major powers were unable to reestablish a viable mechanism for managing exchange rates. Path dependence was strong. The British pound–based gold standard had worked prior to 1914. Countries attempted to get back onto the system after 1919. But trying to go back on at the prewar par levels caused problems. In Japan, the government responded by promoting deflation throughout most of the 1920s, thereby hoping to get back on the gold standard at the prewar par with a domestic price level at which Japanese exporters could effectively export. Reparations, war debt, and unpaid commercial loans compounded the travails the powers found themselves in. Countries had different priorities. Viewed as a nonzero sum game, the great powers failed to make the compromises needed to get the old system going. Underlying the failure to agree upon rules for adjusting exchange rates was the fact that the United States was emerging as the dominant economic power. And the United Kingdom—which had dominated production and trade between 1800 and 1850—was in decline. To avert crisis, the world probably should have moved to a U.S. dollar–based gold standard. But this did not happen until the end of World War II.

Tariffs also complicated matters. There was some excess capacity at the close of World War I, especially in shipbuilding. After the war, unemployment soared in a number of countries, including the United States and Japan, which emerged as the economic winners of the great conflict as trade gradually shifted from the Atlantic toward the Pacific. Tariffs and other forms of protection seemed to be an appealing way of bolstering home demand in order to cut the ranks of the unemployed. The American Smoot-Hawley Tariff of 1930 was the culmination of a decade of failed international negotiations over tariff reduction.

The result of a breakdown in the gold standard based on the British pound, and the failure to secure international tariff restraint, was a drift toward autarky and the emergence of semiclosed trading blocs. After the Ottawa Conference of 1932, the British Commonwealth went onto a system of Commonwealth Preference, with sterling as a clearing currency. The United States and much of Latin America operated with the dollar. Japan went back on the gold standard in 1930 at par. Its gold reserves exhausted in late 1931, it went off gold at the end of the year to pursue a yen bloc strategy for its empire. France, Belgium, and the Netherlands stayed on gold until the mid-1930s, then followed the other major countries in abandoning it.

The League of Nations had failed to ward off the rise of protectionism and the collapse of the international exchange rate regime. After 1931, it

failed to stop wars of aggression, wars of colonial expansion. The Japanese military, increasingly impatient with the failures of internationalism of the 1920s, moved first, steadily taking over Manchuria between 1931 and 1933 in the Manchurian Incident. Emboldened by Japan's audacity, Italy moved against Ethiopia. Germany supported Franco in the Spanish Civil War. Then, Hitler embarked upon a calculated program of aggression against Germany's neighbors.

Underlying the drift to global warfare after 1931 was the confrontation of three competing ideologies after World War I: fascism, communism, and liberal democracy. Nationalism and Wilsonian self-determination fanned the flames erupting on the international arena as the ideologies competed in country after country through the setting up of national parties affiliated with each international movement.

European fascist ideology had a strong egalitarian bias. It was critical of capitalism and of class divisions that supposedly emanated from private market–oriented industrialization. In this sense, it shared views with communism. Indeed, both ideologies were rooted in nineteenth-century social movements that resisted the growth of private markets and that focused public attention upon the possibility of a growing concentration of wealth in the hands of large capitalist enterprises and their owners. Nevertheless, fascism and communism—or particularly the version of socialism in one country espoused by Stalin and his faction within Russia—advocated radically different analyses of the failures of capitalism and radically different policies as to how to promote economic development.

As a matter of policy, fascism leaned toward the creation of corporations directly managed by government that would eliminate class conflict by bringing together labor and capital in order to coordinate and maximize production in the interests of the state. The rights of unions were severely restricted under corporatist legislation in Italy and Germany. But the state took a less active role in clipping the wings of big business. Thus, to a communist ideologue, fascism was an ideological expression of monopoly capitalism. Still, for all of their differences, communism and fascism shared a common political logic. Both were totalitarian, both suppressed opposition through a policy of arbitrary state action that operated above the law with secret police–implemented domestic terror. In short, in principle, fascism promoted equality. In practice, it favored business interests and suppressed the interests of some groups, including unions.

Whether Japan was fascist in the sense of corporatism and totalitarianism has been much debated. Most scholars seem to reject the idea that it became fascist in the European sense. Rather, the prevailing view seems to be that Japan became ultranationalist during the 1930s, contemptuous of Western-oriented indi-

vidualism and liberal democracy, but not corporatist and not totalitarian. To be sure, carrying out its imperialist program during the 1930s imposed costs on its own economy in the form of military expenses and the diversion of investment funds to building up capital in the empire. And expansion into China during the 1930s led to increased friction with the United States and Great Britain.

It is important not to overly emphasize the role of ideology in international affairs. When ideology and geopolitical interests clashed, geopolitical interests often prevailed. Briefly, in 1939 and 1940, it appeared that an entente of Germany, Italy, the USSR and Japan would stare down the United Kingdom, the United States, and Canada. But the German-USSR alliance failed to hold. Japan joined the Axis with Germany and Italy in hopes of seizing British, Dutch, and American colonial interests in Asia; with the aim of securing a relatively free hand in China; and in anticipation of signing a nonaggression treaty with Russia. But in 1941, Germany betrayed Japanese strategic interests by attacking Russia, and Japan betrayed German interests by attacking the United States. Short-run geopolitical convenience undermined ideological purity. When it suited their interests, fascist and communist dictators were more than willing to forge alliances of convenience.

In sum, the imperial strategy adopted by Japan underwent a wrenching shift during the early 1930s. Before 1930, it was devoted to securing a buffer zone on the continent, and it was premised upon confronting the two great continental powers of East Asia, Russia, and China. After 1931, the imperial strategy became extremely expansionist and aggressive, encouraging confrontation with the two great global powers of East Asia, the United States, and Great Britain.

How did this international tumult affect Japan's trade? There is no doubt that due to the drift toward autarky and global geopolitical confrontation, world trade growth faltered during the period between the world wars. And, since Japan's trade growth is strongly influenced by world trade growth, it slowed down as well.[8] Over the period 1886–1915, world trade grew by 3.3 percent, and Japan's trade by 9.1 percent; during 1921–1938, world trade grew by 1.4 percent, and Japan's trade grew by 6.7 percent. In both periods, Japan's trade growth exceeded world trade growth in absolute magnitude by approximately 5 percent. Thus, the slowdown in world trade did have some impact on Japan. But, because Japan was able to secure markets in Asia during World War I that hitherto had been dominated by European powers, it was able to expand its trade more rapidly than other countries could. Moreover, it expanded trade within its own empire, especially after the rice riots of 1918. In the wake of the riots that exploded over the high price of rice, the government stepped up its efforts to encourage growth in agricultural productivity in Taiwan and Korea. The goal in increasing agricultural production in the colonies was to beef up exports of rice and other foodstuffs to Japan proper. Moreover, for Manchuria

and northern Korea, Japan secured hydroelectric power and raw materials for manufacturing.[9] In addition, Japanese enterprises aggressively expanded exports to Taiwan, Korea, Manchuria, and northern China.

The deterioration in the international situation impinged upon Japan's ability to secure foreign technology. This was especially true for technologies that had obvious applications to manufacturing military hardware. After 1931, Japanese companies found it increasingly difficult to secure licensing arrangements for certain sensitive technologies. Thus, totally indigenous research and development was stepped up.[10]

In sum, on first consideration the case for a wholly endogenous theory of Japanese industrialization and economic growth seems plausible. But, because Japan interacted with other countries through its trade, importing of foreign technology, and military activities abroad, it is impossible to accept a completely endogenous theory. Thus, we must accept the fact that international events exogenous to Japan did play a role in shaping the long swings that we observe in the historical record. To state this is not to reject the idea of an endogenous logic to Japanese economic development. Indeed, the infrastructure-driven-growth interpretation of long swings is a theory based on a strictly endogenous logic. Incorporating international factors into an understanding of the dynamics of long swings is tantamount to arguing that while the logic of infrastructure-driven growth is the primary determinant of Japan's growth path, the specific historical growth path that the economy traversed has been influenced by events external to the country.

Physical Infrastructure and Industrialization in the Transitional-Growth Long Swing

The long swing initiated in 1904 and completed in 1930 is the most important long swing in Japan's economic development. To be sure, during this swing the Japanese economy did not achieve the dramatic growth that it did between 1953 and 1969 when gross national expenditure increased at annual rates of almost 10 percent.[11] However, something more important happened. Systematic electrification of the country occurred. In the wake of electrification, the wholesale mechanization of manufacturing took place. And unbalanced growth commenced.

The contours of the second long swing, the transitional growth long swing, are captured in Table 4.1. These contours are classic. The upswing begins with infrastructure buildup associated with electrification and the construction of intra-city electrical railroad lines. In the table, this period—between 1904 and 1911—is designated as the first upswing phase (U1). Electrification slashed energy prices, and it encouraged mechanization because electrical motors were run on the unit drive system. Thus, the infrastructure

Table 4.1

The Transitional-Growth Long Swing

A. Per annum growth in income (gross national product [GDP]; gross national expenditure [GNE]; population [P]; disposable income per capita [DYPC]. Share of GDP arising from agriculture [A%] and manufacturing [M%]. Percentage of gainfully employed population not engaged in farming or fishing [PLFNFF].

Phases	GDP	GNE	P	DYPC	A%	M%	PLFNFF
1904–1911 (U1)	2.6%	2.3%	1.2%	0.6%	31.9%	13.6%	35.9%
1911–1919 (U2)	4.4	4.1	1.3	4.1	28.7	17.9	41.4
1919–1930 (D)	2.5	2.5	1.4	0.8	22.3	19.8	50.1

B.1. Capital stock and capital formation. Percentage of GNE in: gross domestic fixed capital formation (GDFCF%); gross national savings (GNS%); net lending to rest of world (NL%); national savings (NNS%); consumption of fixed capital (CFC%); private saving (PS%) with private corporate savings as a percent of GNE in ( ); government saving (GS%); and military capital formation (MIL%). Per annum growth rate for gross domestic fixed capital formation (GGDFCF).

	Percent of GNE								
Phases	GDFCF	GNS	NL	NNS	CFC	PS	GS	MIL	GGDFCF
1904–1911 (U1)	13.2	12.0	–2.7	3.4	8.5	4.0 (0.3)	–0.6	1.9	6.1
1911–1919 (U2)	16.2	20.1	3.2	11.0	9.1	6.2 (0.9)	4.8	1.7	6.9
1919–1930 (D)	18.7	15.4	–0.8	5.3	10.1	1.4 (–0.1)	3.9	1.7	1.4

B.2. Capital stock and capital formation. Percentage of gross domestic fixed capital formation (including military capital formation) in private primary industry (PRPRI%) and in private nonprimary industry (PRNPRI%); in private residential construction (PRRES%); in government (G%); and in military capital formation (MIL%).

	Private			Government	
Phases	PRPRI%	PRNPRI%	PRRES%	G%	MIL%
1904–1911 (U1)	8.6%	13.9%	30.6%	46.9%	5.3%
1911–1919 (U2)	6.1	18.4	31.4	44.1	3.7
1919–1930 (D)	5.3	18.1	28.5	48.2	6.8

B.3. Capital stock and capital formation. Per annum growth rates of capital stock. Growth rates for the major components of capital stock: livestock and plants (LSPL), producer's durable equipment (PDUREQ), total gross capital stock excluding residential (NRESC), and residential buildings (RESB). Growth rates for components of nonprimary sector capital stock: public works (PW), railroads (RR), utilities (UTIL), nonresidential buildings (NRB), general machinery (MAC), tools and fixtures (TOOL), and vehicles (VEH).

	Gross capital stock				Nonprimary capital stock						
Phases	LSPL	PDUREQ	NRESC	RESB	PW	RR	UTIL	NRB	MAC	TOOL	VEH
1904–1911 (U1)	2.1%	8.7%	4.3%	0.8%	3.9%	6.3%	26.3%	4.2%	n.a.	n.a.	10.3%
1911–1919 (U2)	2.1	10.1	5.2	1.2	3.5	4.5	13.8	3.5	5.1%	5.9%	19.2
1919–1930 (D)	2.6	4.8	4.4	1.1	4.6	3.5	9.6	4.3	5.3	5.3	13.8

C.1. Corporate finance and banking. Companies: number (COM); percentage of companies that are partnerships (P%); limited partnerships (LP%); or joint-stock companies (JS%). Paid-up capital per company (in 1,000 yen) in partnership (PUCP), in limited partnerships (PUCLP), and in joint-stock companies (PUCJS) in 1,000 yen. And total corporate paid-up capital (TPUC) in million yen.

	Companies				Paid-up capital per company (1,000 yen)			
Year	COM	P%	LP%	JS%	PUCP	PUCLP	PUCJS	TPUC
1905	9,006	14.2%	39.0%	46.8	47	16	203	975
1910	12,308	20.3	38.9	40.8	56	20	248	1,481
1915	17,149	17.8	40.2	41.8	60	20	258	2,167
1920	29,917	15.7	30.0	54.2	123	42	456	8,238
1925	34,345	15.1	33.6	51.1	171	63	543	11,160

(continued)

Table 4.1 *(continued)*

C.2. Corporate finance and banking. Per annum growth rates for bank loans (BLOANG), bank deposits (BDEPG), and paid-in capital (BPICG). And share of big five banks in total paid-in capital (B5SH) and ratio of bank borrowing from Bank of Japan relative to total funds employed by banks (NBOR).

	Per annum growth rates				
Phases	BLOANG	BDEPG	BPICG	B5SH	NBOR
1904–1911 (U1)	17.3%	19.0%	4.0%	8.4%	3.6%
1911–1919 (U2)	2.8	4.8	12.1	14.9	2.8
1919–1930 (D)	1.2	7.4	4.4	19.7	4.5

D.1. Labor. Labor force per annum growth rates: overall labor force (LFG); agricultural labor force (ALG); nonagricultural labor force growth (NALFG); and percentage of labor force in nonagricultural employment (NAL%G). Per annum growth rates for labor productivity in agriculture and manufacturing: nominal labor productivity in agriculture (ANOLPG); real labor productivity in agriculture (ARLPG); nominal labor productivity growth in manufacturing (MNOLPG); and real labor productivity in manufacturing (MRLPG). Sex ratio (females/100 males) for agricultural labor force (SRALF).

	Labor force, per annum growth rates				Labor productivity, per annum growth rates				
Phases	LFG	ALG	NALFG	NAL%G	ANOLPG	ARLPG	MNOLPG	MRLPG	SRALF
1904–1911 (U1)	0.5%	–0.1%	1.5%	1.0%	4.5%	2.3%	3.9%	3.4%	86.6
1911–1919 (U2)	0.7	–1.4	3.5	2.8	12.2	3.0	11.4	2.6	83.4
1919–1930 (D)	0.8	–0.2	1.8	1.8	–1.8	0.8	0.9	3.9	83.3

D.2. Labor. Growth in wages. Per annum growth in nominal wages for males in agriculture (NWAMG); for females in agriculture (NWAFG); for males in manufacturing (NWMMG); and for females in manufacturing (NWMFG). And per annum growth rates in real wages for males in agriculture (RWAMG); for females in agriculture (RWAFG); for males in manufacturing (RWMMG); and for females in manufacturing (RWMFG). Per annum growth rate in wage differentials (agriculture wage/manufacturing wage) for males (DMG) and for females (DFG).

	Nominal wages, per annum growth				Real wages				Wage differentials	
Phases	NWAMG	NWAFG	NWMMG	NWMFG	RWAMG	RWAFG	RWMMG	RWMFG	DMG	DFG
1904–1911 (U1)	3.0%	2.6%	3.4%	4.2	0.03%	–0.3%	0.5%	1.3%	–0.5%	1.7%
1911–1919 (U2)	12.1	11.6	14.2	13.6	4.0	3.4	6.0	5.4	–1.4	1.5
1919–1930 (D)	–1.3	–1.0	3.0	0.5	–0.8	–0.5	3.6	1.0	–4.5	2.0

D.3. Labor. Labor's share in nonprimary income, total (LST) and in corporate income (LSCOR). Share of corporate income in total income (CYS).

Phases	LST	LSCOR	CYS
1904–1911 (U1)	62.4%	72.2%	40.8
1911–1919 (U2)	67.5	75.8	47.8
1919–1930 (D)	60.9	66.9	45.8

E. Prices. Growth rates for major price indices: GNE deflator (GNEDG); consumer price index (CPIG); investment goods price index (IGPIG); agricultural goods price index (AGPIG); manufactured goods price index (MGPIG); and commerce-service price index (CSPIG).

Phases	GNEDG	CPIG	IGPIG	AGPIG	MGPIG	CSPIG
1904–1911 (U1)	3.0%	2.9%	1.0%	2.1%	1.8%	3.4%
1911–1919 (U2)	9.3	8.2	10.1	8.8	8.6	7.7
1919–1930 (D)	–0.7	–0.6	–2.2	–2.6	–3.3	2.3

F.1. Trade. Percentage of exports that are merchandise (EXM%), services (EXS%), and factor income from foreign sources (EXFS%); and percentage of imports that are merchandise (IMM%), other (IMO%); and factor income paid abroad (IMFA%). Terms of trade (export price index/import price index) levels and per annum growth rates (TTL, TTG). Per annum growth rates for commodity trade, exports (ECOMG), and imports (ICOMG).

	Percentage of exports			Percentage of imports			Terms of trade		Commodity trade growth	
Phases	EXM%	EXS%	EXFS%	IMM%	IMO%	IMFA%	TTL	TTG	ECOMG	ICOMG
1904–1911 (U1)	81.8%	15.5%	2.7%	76.7%	13.1%	10.3%	143.5	–1.5%	7.3%	3.8%
1911–1919 (U2)	79.0	17.2	3.8	83.3	8.4	8.4	136.9	1.8	5.2	7.1
1919–1930 (D)	77.6	16.7	5.7	86.4	10.2	3.5	156.0	–1.0	5.7	5.3

(continued)

Table 4.1 *(continued)*

F.2. Trade. Trade balance: net exports/trade ratio (NX/T); net decrease in specie held abroad/trade ratio (NDS/T); and export of gold/trade ratio (GX/T). Percent of exports and imports that are primary commodities (EPRI%, IPRI%). Ratio of exports to gross domestic fixed capital formation (XCFR).

	Net trade position ratios			Primary commodities in trade		Export/capital formation ratio	
Phases	NX/T	NDS/T	GX/T	ERPI%	IPRI%	Level	growth
1904–1911 (U1)	–7.0%	–1.6%	–0.2%	21.8%	49.8%	59.5%	0.9%
1911–1919 (U2)	5.2	–1.7	–1.4	16.3	59.0	73.8	0.2
1919–1930 (D)	–2.6	1.1	–0.8	9.9	56.7	70.2	4.0

Sources: See sources to Table 3.1.
Notes: Most figures given represent averages of seven-year moving averages for the periods given in the table (this is particularly true for growth rates based on seven-year moving averages of the underlying series). In some cases, the averages are for periods shorter than the upswing and downswing phases. "U" stands for upswing, and "D" stands for downswing. "N.a." stands for "not available."

construction boom worked on the supply side of the economy. By reducing production costs in manufacturing, it set in motion the second phase of the upswing, occurring between 1911 and 1919 and designated as the second upswing phase in the table (U2). It is important to keep in mind that the industrial expansion of the 1910s began before World War I commenced. Thus, the fillip to exports, and the encouragement of import substitution, transmitted to the economy by the breakdown in trade among the European powers, did not initiate the second upswing phase. Rather, its roots lie on the supply side. To be sure, demand growth stemming from exports and import substitution helped. Nonetheless, supply changes were more fundamental.

Growth became excessive. It could not be sustained. To a certain extent, it could not be sustained because World War I ended. The conclusion of hostilities allowed exporters in the European powers to reenter the markets that they abandoned during the conflict. And they were able to reenter Japan's market, undercutting Japanese domestic producers of machinery and capital goods. But growth also ran into limits imposed by infrastructure.

Infrastructure imposed limits in a variety of areas. The demand for skilled labor and technically trained managers—engineers—soared during the industrial boom of the 1910s. So, labor supply was one limiting factor. Another was credit creation. Prices soared during the second upswing. In a period of rapid inflation or deflation—that is, in a period of price instability—wholly rational individuals begin to wonder whether the price changes will continue or not. They are rational but not fully informed. So, they follow theories propounded by optimists or pessimists who purport to know what is going on. Expectations of optimistic and pessimistic camps diverge. The greater the instability, the greater is the divergence in expectations between the camps. Appreciating the fact that individuals have differing opinions, speculators enter the fray. There is a wedge to profit by buying cheap and selling high, and they act accordingly. Speculation fuels credit creation that is only loosely associated with economic fundamentals. Banks make speculative loans, and some of these fail. The possibility for panics and runs on banks lurks around the corner.

The capacity of physical infrastructure to sustain the manufacturing boom was also strained. Electrical delivery systems were pushed to their limits. Brownouts occurred. Commuting railroads became overcrowded. Land prices were pushed up, encouraging speculation in land. Since land was needed for expansion of infrastructure, speculation in land created problems. Potential sellers of land held off selling it in the anticipation of further inflation in the value of their asset.

The economy entered the downswing. During this phase, from 1919 to 1930, infrastructure was revamped and expanded. The educational system was completely overhauled.[12] Banks failed, and the banking system was completely

reorganized in the aftermath of the financial crisis. And physical infrastructure construction went on at a heady pace. Hydroelectric power grids were expanded and consolidated. Train lines were extended and rolling stock was added on. And, most important, road paving and building took off in response to the growing demand for surfaces that could handle buses and trucks. The economy began to absorb innovations involving the internal combustion engine.

In short, the transitional-growth long swing was classic for two reasons. It was classic because it incorporates a phase of industrial expansion sandwiched between two phases of infrastructure expansion. The logic of infrastructure-driven growth is especially evident in this second long swing. It is classic because the booms and busts are unusually clear, especially in the series on prices. That is to say, downswings and upswings have dramatically different characteristics. Let us briefly review these.[13]

During the first upswing (U1), the peak growth for output was in 1908, when gross domestic product grew at 3.7 percent. Peak growth for gross domestic product originating in manufacturing, however, peaked in 1911, at 7.7 percent. Upswings due to infrastructure construction usually generate slower growth in the overall economy than do upswings due to manufacturing activity. The reason is that infrastructure generates potential service flow, but until demand for that service flow picks up—as it does during manufacturing booms—the service flow recorded in the national accounts is modest. Thus, industrial upswings record greater growth rates for overall output because they involve expansion in the service flow from the infrastructure that they rely upon and expansion in industrial output itself. Thus, as manufacturing output growth picked up toward the tail end of the first upswing, the impetus to overall growth picked up. Since infrastructure was being put into place during this phase, price increases were modest. The peak for growth in the consumer price index was in 1908, when it reached 4.1 percent. The investment goods price index was especially quiescent. Its growth peaked at 1.7 percent, during 1908. In short, the first upswing phase of the second long swing was one of relative price stability.

Stability vanished during the second upswing phase. The rate of growth in gross domestic product picked up, peaking at 5.8 percent in 1917. Growth in gross domestic product from manufacturing, however, peaked earlier, in 1912 at 9.3 percent. Indeed, the fact that manufacturing output growth peaked before World War I is significant. It points to the importance of infrastructure improvements put into place before 1911 for the surge in manufacturing production. The European conflict was crucial to sustaining the upswing in manufacturing, but its deepest roots lie in domestic infrastructure investment. Prices skyrocketed. Growth in the consumer price index peaked at 15.0 percent in 1915. A year later, the investment goods price index reached its peak growth of 17.5 percent. Speculation began to play a substantial role as the economy overheated.

The labor market tightened during both upswing phases. Nominal wage increases virtually moved in tandem with nominal labor-productivity increases. Labor's share in income rose.

Never before had the Japanese economy reached the dizzying heights of growth and inflation that it did during the war years. And for the first time, it sustained export surpluses on a regular basis. But the infrastructure was now pressed to the limit, and the war ended, thereby wiping out the possibility for further current account surpluses in trade. The upswing gave way to a deep downswing.

During the downswing, overall growth slowed to a crawl. The trough for growth in gross domestic product was reached in 1927 when the economy grew at 1.5 percent. Growth in gross domestic product from manufacturing reached a trough earlier, in 1921, when it grew at 1.8 percent. The upward surge in prices was reversed. Prices fell. Growth in the consumer price index reached a trough of negative 4.7 percent in 1928; growth in the investment goods price index hit a trough of negative 7.6 percent in 1927. The depth of the downswing was profound, especially as measured in terms of prices.

Labor's share in income followed the opposite course it had traversed during the manufacturing upswing. It dropped. Nominal and real wages for females dropped because the downturn in light manufacturing was especially deep.

In sum, the second long swing was classic in its contours. The upswings were strong, the downswing deep. Phases of intense investment in infrastructure alternated with a phase of intense investment in manufacturing. Moreover, during the second long swing, the economy made a clear and wrenching shift toward unbalanced growth. The growth of agriculture virtually ceased, but manufacturing growth soared. Increasingly, capital was flowing to industry. For the male labor force—in manufacturing mostly concentrated in the new rapidly expanding heavy industries—gaping wage differentials developed and accelerated. Wages in agriculture for males plummeted relative to those in industry.

The Unbalanced-Growth Long Swing

The investment in roads, electrical power grids, and electric railroad expansion of the 1920s paved the way for a massive expansion of manufacturing during the 1930s. During the upswing of the third long swing, output growth recorded prewar peaks: gross domestic product grew at 7.0 percent in 1935 and gross domestic product in manufacturing grew at 12.5 percent. The quantitative characteristics of this long swing are captured in Table 4.2. Comparing the characteristics of this long swing with the transitional-growth long swing that preceded it reveals important structural differences between the two swings. These must be acknowledged in any theory that claims to offer a comprehensive interpretation of the long-swing process.[14]

Table 4.2

The Unbalanced-Growth Long Swing

A. Per annum growth in income (gross national product [GDP]; gross national expenditure [GNE]; population [P]; disposable income per capita [DYPC]. Share of GDP arising from agriculture (A%) and manufacturing (M%). Percentage of gainfully employed population not engaged in farming or fishing [PLFNFF].

Phase	GDP	GNE	P	DYPC	A%	M%	PLFNFF
1930–1938 (U)	4.6%	5.0%	1.3%	2.7%	19.0%	27.0%	54.0%

B.1. Capital stock and capital formation. Percentage of GNE in: gross domestic fixed capital formation (GDFCF%); gross national savings (GNS%); net lending to rest of world (NL%); net national savings (NNS%); consumption of fixed capital (CFC%); private saving (PS%) with corporate savings as a % of GNE in ( ); government saving (GS%); and military capital formation (MIL%). Per annum growth rate for gross domestic fixed capital formation (GGDFCF).

	% of GNE								
Phase	GDFCF%	GNS%	NL%	NNS%	CFC%	PS%	GS%	MIC%	GGDFCF
1930–1938 (U)	18.5%	15.0%	–0.2	5.2%	9.8%	7.1% (–1.0%)	–1.9%	3.8%	10.6%

B.2. Capital stock and capital formation. Percentage of gross domestic fixed capital formation (including military capital formation) in private primary industry (PRPRI%) and in private nonprimary industry (PRNPRI%); in private residential construction (PRRES%); in government (G%); and in military capital formation (MIL%).

	Private			Government	
Phases	PRPRI%	PRNPRI%	PRRES%	G%	MIL%
1930–1938 (U)	5.0%	17.7%	27.3%	50.0%	8.3%

B.3: Capital stock and capital formation. Per annum growth rates of capital stock. Growth rates for the major components of capital stock: livestock and plants (LSPL), producer's durable equipment (PDUREQ), total gross capital stock excluding residential (NRESC), and residential buildings (RESB). Growth rates for components of nonprimary sector capital stock; public works (PW), railroads (RR), utilities (UTIL), nonresidential buildings (NRB), general machinery (MAC), tools and fixtures (TOOL), and vehicles (VEH).

	Gross capital stock				Nonprimary capital stock						
Phase	LSPL	PDUREQ	NRESC	RESB	PW	RR	UTIL	NRB	MAC	TOOL	VEH
1930–1938 (U)	2.1%	4.6%	4.0%	1.2%	5.5%	0.3%	6.8%	3.9%	8.0%	8.4%	13.8%

C.1: Corporate finance and banking. Companies: number (COM); percentage of companies that are partnerships (P%); limited partnerships (LP%); or joint-stock companies (JS%). Paid-up capital per company (in 1,000 yen) in partnerships (PUCP), in limited partnerships (PUCLP), and in joint-stock companies (PUCJS) in 1,000 yen. Total corporate paid-up capital (TPUC) in million yen.

	Companies				Paid-up capital per company (1,000 yen)			
Year	COM	P%	LP%	JS%	PUCP	PUCLP	PUCJS	TPUC
1930	51,910	16.4%	46.2%	37.4%	323	200	440	19,633
1935	84,146	19.5	52.8	27.7	184	137	439	22,352
1939	85,122	17.9	43.0	39.0	214	168	499	34,025

C2: Corporate finance and banking. Per annum growth rates for bank loans (BLOANG), bank deposits (BDEPG), and paid-in capital (BPICG). Share of big five banks in total paid in capital (B5SH) and ratio of bank borrowing from Bank of Japan relative to total funds employed by banks (NBOR).

	Per annum growth rates				
Phase	BLOANG	BDEPG	BPICG	B5SH	NBOR
1930–1938 (U)	1.2%	7.4%	–3.0%	27.5%	1.0%

(continued)

Table 4.2 *(continued)*

D.1: Labor. Labor force per annum growth rates: overall labor force (LFG), agricultural labor force (ALG), nonagricultural labor force growth (NALFG), and percentage of labor force in nonagricultural employment (NAL%G). Per annum growth rates for labor productivity in agriculture and manufacturing: nominal labor productivity in agriculture (ANOLPG); real labor productivity in agriculture (ARLPG); nominal labor productivity growth in manufacturing (MNOLPG); and real labor productivity in manufacturing (MRLPG). Sex ratio (females/100 males) for agricultural labor force (SRALF).

	Labor force, per annum growth rates			Labor productivity, per annum growth rates					
Phase	LFG	ALG	NALFG	NAL%G	ANOLPG	ARLPG	MNOLPG	MRLPG	SRALF
1930–1938 (U)	1.2%	–0.3%	2.4%	1.2%	6.4%	1.1%	8.7%	5.5%	84.2

D.2: Labor. Growth in wages. Per annum growth in nominal wages for males in agriculture (NWAMG); for females in agriculture (NWAFG); for males in manufacturing (NWMMG); and for females in manufacturing (NWMFG). Per annum growth rates in real wages for males in agriculture (RWAMG); for females in agriculture (RWAFG); for males in manufacturing (RWMMG); and for females in manufacturing (RWMFG). Per annum growth rate in wage differentials (agriculture wage/manufacturing wage) for males (DMG) and for females (DFG).

	Nominal wages, per annum growth				Real wages				Wage differentials	
Phase	NWAMG	NWAFG	NWMMG	NWMFG	RWAMG	RWAFG	RWMMG	RWMFG	DMG	DFG
1930–1938 (U)	–0.9%	–0.5%	0.1%	–1.5%	–2.2%	–1.8%	–1.2%	–2.9%	–1.2%	–0.5%

D.3. Labor's share in nonprimary income, total (LST), and in corporate income (LSCOR). Share of corporate income in total income (CYS).

Phase	LST	LSCOR	CYS
1930–1938 (U)	66.9%	60.9%	45.3%

E. Prices. Growth rates for major price indices: GNE deflator (GNEDG); consumer price index (CPIG); investment goods price index (IGPIG); agricultural goods price index (AGPI); manufactured goods price index (MGPIG); and consumer-service price index (CSPIG).

Phases	GNEDG	CPIG	IGPIG	AGPIG	MGPIG	CSPIG
1930–1938 (U)	1.9%	2.2%	3.4%	5.0%	3.0%	1.3%

F.1. Trade. Percentage of exports that are merchandise (EXM%), services (EXS%) and factor income from foreign sources (EXFS%); and percentage of imports that are merchandise (IMM%) and factor income paid abroad (IMFA%). Terms of trade levels (export price index/import price index), and per annum growth rates (TTL, TTG). Per annum growth rates for commodity trade, exports (ECOMG), and imports (ICOMG).

	Percentage of exports			Percentage of imports			Terms of trade		Commodity trade growth	
Phase	EXM%	EXS%	EXFS%	IMM%	IMO%	IMFA%	TTL	TTG	ECOMG	ICOMG
1930–1938 (U)	76.0%	17.6%	6.4%	78.9%	16.6%	4.5%	120.4	–4.6%	9.9%	3.7%

F.2. Trade. Trade balance: net exports/trade ratio (NX/T); net decrease in specie held abroad/trade ratio (NDS/T); and export of gold/trade ratio (GX/T). Percent of exports and imports that are primary commodities (ERPI%, IPRI%). Ratio of exports to gross domestic fixed capital formation (XCFR).

	Net trade position ratios			Primary commodities in trade		Export/capital formation ratio	
Phase	NX/T	NDS/T	GX/T	ERPI%	IRPI%	Level	Per annum growth
1930–1938 (U)	–0.5%	0.3	3.3%	6.9%	61.1%	104.9%	0.5%

Sources: See sources for Table 3.1.
Notes: Most figures given represent averages of seven-year moving averages for the period given (this is particularly true for growth rates based on seven-year moving averages of the underlying series). In some cases, the average is for a period shorter than the period indicated (i.e., some averages are for 1930–1936, some for 1930–1937, etc.).

Change in the international context in which the upswings of 1911–1919 and 1930–1938 occurred accounts for some of the differences that are apparent from comparing Tables 4.1 and 4.2. Military spending is far more important in the third long swing than it was in the second long swing. The third long swing was never completed due to the onset of mobilization for total war in 1938. The trade balance is positive during the second upswing of the second long swing. This is due to World War I. During the upswing of the third long swing, exports rise rapidly, but the trade balance is not positive. Indeed, export demand is very important to total overall demand growth during the upswing of the third long swing. Exports actually exceed domestic fixed capital formation during the upswing. In both upswings, there were barriers to continuing expansion of exports. In the boom of the 1910s, the constraint was the end of the war; in the 1930s, the constraint was the division of the globe into distinct trading blocs that pursued autarky policies, excluding imports.

The unbalanced-growth upswing was relatively stable. Prices were far less explosive during the upswing of the third long swing than they were between 1911 and 1919. The consumer price index reached a peak growth rate of 7.6 percent in 1936; in the same year, the investment goods index hit a peak growth rate of 9.1 percent. Pressure of rising investment goods prices upon capital formation was less apparent during the upswing of the 1930s than it was in the upswing of the 1910s. Perhaps this reflects the fact that inefficient producers of capital goods were driven into bankruptcy during the downswing of the 1920s, wiped out by renewal of competition from European suppliers. During the 1930s, efficient producers constituted a larger share of the industrial enterprises in operation.

The third long swing is definitely an unbalanced-growth long swing. Throughout the period, the percentage of output from manufacturing and mining exceeds that generated within agriculture. The second long swing was simply transitional.

The most significant difference between the second and third long swing, however, involves the operation of labor markets. During the upswing of the unbalanced long swing, labor productivity growth far outstrips wage growth. Indeed, real wages drop for both males and females throughout the upswing, while nominal and real labor productivity grows, especially in manufacturing. As a result, labor's share drops to a relatively low average level. This stands in marked contrast to the upswing of the 1910s, when labor's share rises, with nominal labor productivity growth exceeded by wage growth. In my opinion, this difference in the behavior of the labor market reflects the impact of infrastructure restructuring during the 1920s upon the labor market of the 1930s. Education expanded drastically during the 1920s. Thus, the

supply of skilled workers expanded vigorously throughout the period of slack demand for their services.[15] Supply of fresh school graduates continued to shift out during the 1930s, although the military draft cut into these. The extent of the outward shift in supply was not fully exhausted by the expansion of demand. Moreover, workers were better fed and had more physical stamina. They could, and did, work longer hours.[16] Enhanced work capacity was especially prominent in the case of females laboring in agriculture, who increasingly replaced males called up for military service or recruited in the ranks of industrial workers. Figures on the sex ratio (females per one hundred males) in agriculture testify to this. The ratio falls until the beginning of the third long-swing upswing. Then, it commences to rise.

In short, there is no universal pattern to the long swings. During the first long swing, infrastructure and industry expand together, and as they do so, agriculture expands as well. During the second long swing, there is a clear demarcation of infrastructure investment and industrial investment phases. Infrastructure expands, paving the wave for the industrial expansion that takes place during the second-upswing phase of the long swing. The industrial expansion puts pressure on infrastructure, which is substantially revamped and expanded during the downswing phase. This lays the groundwork for the industrial upswing of the uncompleted third long swing. There is a clear structural shift in the economy during the second long swing. The economy moves from being balanced to being unbalanced. During the third long swing, growth is wholly unbalanced.

To some degree, differences between the long swings reflect the trend toward industrialization and the fact that the pace of innovation was far more rapid in manufacturing than it was in agriculture. That is, the differences rise as endogenous features of the growth process. But this is not the entire story. The changing international situation plays a role as well. The impact of the world's geopolitical and trade climate is reflected in the performance of trade, for instance. It is also reflected, to a degree, in the emergence of unbalanced growth. Because the government was intent on developing agricultural output in the colonies in order to solve potential food supply difficulties, it invested resources in developing Korean and Taiwanese agriculture at the expense of domestic farming. So, it is difficult to separate international influences from those inherent in growth itself.

In sum, Japan industrialized during the second and third long swings, and infrastructure paved the way for the two great massive expansions of industrial activity, during the 1910s and 1930s. To better understand how infrastructure development and expansion of factory production interacted, we need to embed the details of the story within a specific geographic context. As we already know, prewar Japanese infrastructure buildup and manufac-

turing expansion were highly concentrated in the Tōkaidō industrial belt, and especially in the Osaka-Kobe, Nagoya, and Tokyo-Yokohama districts. We now turn our attention to a recounting of this detailed story.

Notes

1. See Minami (1994) and Ohkawa and Rosovsky (1973).

2. It must be stressed that these figures refer to seven-year moving averages, and not to annual growth rates based upon gross domestic product for successive years.

3. For other interpretations of trend acceleration, see Minami (1994) and Ohkawa and Rosovsky (1973). I strongly reject the notion that the slow and—for some subperiods, deeply negative—growth of the Japanese economy between 1938 and 1953 is a downswing phase of the third long swing. And, by the same token, I reject the notion that the postwar upswing of the period 1953–1969 is a simple continuation of a process of growth initiated with the first long swing of 1887–1904. In some ways, it can be viewed as a logical extension of a growth process begun at the end of the nineteenth century. Indeed, in my view it can be viewed as a fourth upswing set in motion by investment in, and reworking of, infrastructure during the period leading up to 1945 and 1953. But there are strong elements of discontinuity between the prewar and postwar periods. Thus, I view the so called miracle growth, or convergence, long swing of the 1953–1969 period as initiating a growth process that differs in many dimensions from that initiated in the 1880s and completed in the late 1930s. On this point, see my discussion in chapter 8.

4. For the argument linking the savings rate to income growth, see Ohkawa and Rosovsky (1973), especially chapter 6. Also, see Minami (1994, 155–64). A number of theoretical arguments can be advanced to explain why savings rates rise as income (and especially per capita income) rise. For instance, people develop consumption patterns based on their normal expected levels of income. As income rises, they may not anticipate that the rise is permanent. They may expect that a downturn will shortly occur, wiping out their gains. Hence, they may not immediately adjust their spending patterns upward. Again, households may have target levels of assets relative to income that they wish to meet. As household income rises, the households will attempt to increase their assets. If they are not able to secure financing for the purchase of assets from banks or other financial intermediaries, they are forced to save in order to purchase the desired assets.

5. For tests of the sensitivity of investment to interest rates, see Minami (1994, 134–37).

6. The text in the remainder of this section draws heavily from Mosk (2000a).

7. Japan was nominally a belligerent country, being drawn into the war because of the Anglo-Japanese alliance. In practice, Japan was only marginally involved in the conflict. However, its military took advantage of the war to wrest away some German-held islands in the Pacific from Germany. And its industrialists took advantage of the breakdown in the capacity of European exporters to ship their goods to Asia, by expanding their markets for goods and services throughout Asia and the Middle East. Japanese liner traffic increased tremendously during this period. Moreover, Japanese producers effectively operated behind high tariffs during the war, since European suppliers of machinery could not ship their products to Japan. Japanese importers relied mainly on American sources for capital equipment, thereby driving

up the price of imported machines and equipment. Thus, World War I promoted import substitution in Japan.

8. On this point, see Minami (1994, 186–91).

9. See Schumpeter (1940).

10. For research and development encouraged by the Japanese Imperial Navy, see Evans and Peattie (1997). Some of the papers in Goto and Odagiri (1997) deal with indigenous research and development and technological innovation in Japan.

11. See estimates given in Ohkawa and Shinohara with Meissner (1979).

12. See Mosk (1995, chapter 2).

13. It should be emphasized that the statistics on peaks and troughs discussed below are based on seven-year moving averages.

14. For example, Ohkawa and Rosovsky (1973) emphasize the importance of labor market tightening in their interpretation of long swings. But the behavior of labor markets that they emphasize—one that works through a shrinking of wage differentials between agriculture and manufacturing during upswings and a widening in downswings, and that depends upon labor's share in nonagriculture falling during upswings—is not universally recorded for the long swings. For instance, labor's market share in nonagriculture rose during the second upswing of the transitional-growth long swing, but fell during the upswing of the unbalanced-growth long swing.

15. On outward shifts in labor supply during the period between the world wars, see Mosk (1995, 2000a).

16. On gross nutrition, net nutrition—net of the demands placed upon it by disease and physical work—and the impact of improvements in net nutrition upon work capacity in Japan, see Mosk (1996, 2000a).

5

Infrastructure

Electricity and Railroads

An inorganic niche employing steam and coal cradled within an economy that fundamentally remained organic and proto-industrial gave birth to Japanese industrialization. However, it was electricity, not steam, that made Japan into a mighty industrial power.

In order to supply electricity from local thermal power plants, to transport it from remote hydroelectric generating stations at dam sites erected within deep mountain ravines, electricity must be sent across space. And to do this, massive investment in high-voltage power lines and huge metal pylons is required. The costs are huge. But looming equally large are the prospective benefits, especially when the number of consumers of power at any given distribution node is large. The more numerous the potential users, the larger the consumer base over which the fixed costs—of building dams, generating facilities, power lines, and transforming stations—are spread. Thus, the larger the potential market, the lower are unit costs.

The most obvious economic benefits are the increases in output fueled by capital accumulation in electrical generation and supply. By cutting down costs of energy and promoting mechanization, electrification renders factories more productive. But there are also important technological spillovers. Engineers trained in the intricacies of power generation possess skills that are readily transferred to industrial concerns. Setting up schools specializing in (and programs within institutions of higher learning focused upon) electrical engineering addresses potential bottleneck problems in both the infrastructure and industrial sectors.

In short, there is a close connection between innovation in infrastructure

and innovation in industry during the second long swing. Of course, demand for electrical power comes from sources other than factories. Indeed, industrial demand may be insufficient to create the economies of scale justifying widespread investment in electrical power utilities. It is for this reason that cities are important. The greater the concentration of consumers of electrical power for lighting and running machinery, the larger are the scale economies that can be tapped. Thus, the great metropolitan centers of the Tōkaidō like Osaka played a crucial role in the buildup of electrical power capacity in Japan. These cities were the wellsprings for powerful scale economies. But they were more than that. Their governments played an active role in investing in utilities and/or working together with large private suppliers of power. In short, city governments were investors in, and facilitating agents for, private-sector involvement in the creation of electrical power infrastructure.

What type of private suppliers did city governments interact with in developing electricity? They were companies providing lighting. But, more important, they were railroad companies. Key to the first upswing of the second long swing is the overlap of railroadization and electrification.

Since railroad construction began almost as soon as the *bakufu* was overthrown, why is it reasonable to claim that railroadization and electrification overlapped? It is quite true that the early Meiji government was keen on developing a nationwide nationalized railroad network as a way of stimulating commerce and as an integral part of its program of military buildup. For instance, the early Meiji government employed more than ninety British technicians in designing a network for the country, and it began to actively purchase British locomotives in order to achieve its goal.

Despite these grandiose plans, railroad construction—carried out either directly by the national authorities or by investors granted licenses from the government—proceeded at a desultory pace. For instance, in 1883–1884 (1883 marks the first year that private lines were completed), only about 260 miles of track had been laid down. Combining national and private lines, and indexing the total mileage of track for 1883–1884 at one hundred, the index for 1883–1890 is 263.3; for 1891–1900 is 1038.2; and for 1901–1905 is 1755.2.[1] By 1907, there were a mere 5,013 miles of track in Japan. While not inconsiderable, it amounted to little more than a first step.

Why was growth in railroad-track mileage so lethargic prior to the onset of the second long swing? The answer is simple: competition from sailing ships and steamships. Japan is an island nation. Moreover, during the first long swing, with the exception of Kyoto, all of its big cities lay on the seacoast. Four of these five metropolises—Tokyo, Yokohama, Osaka, and Kobe—were either natural deepwater seaports or were situated but a short distance away from natural deepwater ports. Moreover, as explained in

chapter 3, locating near coal was not important for the concentration of industry in the Tōkaidō industrial belt. For instance, in the case of the British Industrial Coffin, situating production near coal was crucial. Thus, a network of industrial towns spreading across the midlands of England sprang up during the late eighteenth and early nineteenth centuries. Linking these towns that were principally situated inland by a railroad network had a compelling economic logic behind it. By contrast, in early industrializing Japan, it was the steamship that was crucial to the knitting together of the industrial belt.

Indeed, steamships were of great significance in the first long swing for a variety of reasons. They carried passengers and freight. They created demand for deepwater ports and breakwater construction. Their use stimulated demand for industrial enterprises, which during Meiji could more easily cope with the difficulties of building ships with wooden hulls than with the problems of building steam railroad rolling stock. Granted, craftspeople trained in working with wood were able to work on wooden freight cars and the wooden interiors of passenger cars. But locomotives were a different matter altogether. In 1887, of the ninety-seven locomotives employed in Japan, ninety-five had British origins.[2]

Two events during the first decade of the twentieth century transformed this state of affairs: nationalization of most of the steam railroad lines and widespread electrification of the Tōkaidō industrial belt. Throughout the period between the early 1870s and the 1890s, nationalizing private steam lines had been a preoccupation of many politicians and bureaucrats in the national railroad administration. Thus, a year after the Imperial Diet opened in 1890, the government introduced a bill buying out many of the private lines that had been constructed before that date.[3] Passage of the legislation did not occur instantaneously. But, in 1906, a nationalization bill did become law.

By the close of 1906, the national railroad system had gathered in its hands approximately 65 percent of all track that had been laid in Japan up to that point. Further nationalization actions were undertaken in 1907–1908, and at the same time the railroad bureaucracy was reorganized. The cabinet set up the Tetsudōin (Railroad Department), raising it to full ministry status as the Tetsudōshō (Ministry of Railroads) in 1920. Under the terms of the legislation, seventeen private railroad companies were immediately bought out (at prices that many critics deemed inflated). In nationalizing steam railroads, the government left open the possibility of continued private-sector involvement in railroad construction. Some private steam railroad lines escaped being swallowed up by the Railroad Department. Private enterprises were allowed to offer intercity transport services, and municipalities were allowed to enter the intercity tramway business.

Nationalizing the railroads created powerful scale economies that stimulated research and development and innovation in the machinery industry. The new national railroad system was sufficiently flush with resources that it could afford to implement a massive reverse engineering project for locomotives.[4] In 1909, the government promulgated a national self-sufficiency policy. It followed this up in 1912 with specific directives creating a designated factory system policy. Under these guidelines, the Japan National Railroads imported four steam locomotives from four different foreign suppliers—American, British, and German companies provided the rolling stock—and directed technicians to painstakingly study the manufacture of every one of the components. Working from this knowledge, a design for a domestic locomotive emerged. Japan National Railroads asked a small number of favored companies to produce the individual items, while it served as a clearinghouse for the project as a whole. It tested the parts, returned defects, requested changes and modifications, and took responsibility for final assembly. In short, it operated as a coordinating or facilitating agent.

By creating scale economies, nationalization fostered import substitution and the diffusion of precision machining. Later on, during the 1920s, as the industry moved into electrical locomotive manufacture, Japan National Railroads once again served as a clearinghouse for designated companies. It brought together designers from Hitachi, Shibaura, Mitsubishi Denki, and Kawasaki Shipyards in a coordinated program that involved designing motors, fashioning air brakes, and the like. The effort paid off. By the 1920s, the rolling stock industry joined shipbuilding and electrical machinery in becoming one of the top three subsectors (in terms of value of output) of the burgeoning machinery industry.

As important as nationalization was to creating a derived demand for machinery production, electrification played an even more decisive role in the expansion of the railroad system in Japan. To be sure, prior to World War II, electrification of railroads was almost exclusively associated with short distance lines, especially with regional intercity lines and intracity lines like that initiated by Osaka City in 1903 as part of its contribution to the port development project. Still, electricity revolutionized all trains, regardless of whether their locomotives ran under steam or electrical power. For instance, electrical lighting of cars and stations and communication lines facilitated travel at night. Night travel was especially important in a densely inhabited country like Japan, where rail lines typically cross roads. Moving freight at night minimizes disrupting traffic crossing tracks, traffic crossing tracks mainly moving during daylight hours.

The building of railroads wholly dependent on electrical power revolutionized both railroad transport and the electrical power sector. Compared to

steam power, electrical power—especially hydroelectric power (in a pre-nuclear power era)—is relatively cheap to produce in Japan. The country is mountainous. Annual rainfall is plentiful. Moreover, trains powered by electricity do not pollute the air as steam does. Furthermore, electrically driven trains can be used as subway trains. Running electrified trains created synergies. Private companies active in the electrical railroad sector usually also sold electricity to local communities bordering their tracks, and used the proceeds from sale of electricity to help defray costs. This was especially important for their revenue flow during the period when they were just offering passenger service and had not yet established a reliable customer base for rail service.

In short, during the first upswing of the transitional growth long swing, the railroad and electrical power sectors expanded in tandem. Statistics conveying the massive expansion of these two sectors after 1900 appear in Table 5.1. Regarding the buildup in electrical power-generating capacity, it must be kept in mind that from the early 1880s, thermal power was being used in Japan, albeit on a modest scale. Initially, private companies concentrated in the great conurbations of the Tōkadō provided the thermal electricity. So, as in the West, electrical power generation got its start by exploiting coal. However, the lag between the introduction of thermal-generated electricity and the harnessing of water flow for electrical power creation was far shorter in Japan than it was in North America or Europe.[5] For instance, in 1887, only five years after New York became the world's innovator in offering electrical lighting, Tokyo Dentō (Tokyo Electric Light) began providing it. Following Tokyo's lead, private utility companies proliferated through the years of the first long swing. Most of them utilized Edison equipment imported from the United States.[6]

The importance of the private sector for early electrical power generation in Japan should not be overlooked. The sprouting up of relatively small enterprises providing power during the first long swing testifies to this. But Japan paid a price for this early reliance on private initiative. Uncoordinated private activity can, under certain circumstances, undermine potential scale economies. For instance, in nineteenth-century Britain, private developers built the steam railroad network. They were largely unconstrained by overarching national standards or guidelines. Hence, companies often ignored what others were doing in terms of setting width—gauge—for track. As a result, at some junctions, trains running on one company's set of tracks could not run on the tracks entering the same junction that were laid down by other companies. This made merger of lines more costly. Companies created out of mergers were forced to undertake ripping up some track and laying replacement for it, or they had to maintain two sets of rolling stock.

In the Japanese case, the problem involved the type of current, and the number of cycles for that current, adopted by the various companies. Kobe Dentō introduced direct current in 1888; Osaka Dentō brought in alternating current in 1889; Kyoto Dentō and Nagoya Dentō settled upon direct current in 1889, as did Yokohama Dentō in 1890. Osaka and western Japan ended up using current with sixty cycles. But Tokyo and eastern Japan settled on fifty cycles. At a subsequent point, the government commissioned studies to look into the cost of establishing one single standard for current. But it concluded that the costs of conversion were too excessive to justify the effort. Thus, to this day, the country is divided into incompatible electrical zones.

Investment in electrical utilities was an interesting facet of infrastructure investment during the first long swing, but the scale of power generation was actually minuscule. For instance, Osaka Dentō started up by running eleven arc lights in Nishi Dōtonbori (near the Dōton Canal). In 1891, it opened up a separate facility at Nakanoshima. The total number of lights that it serviced from those two stations was less than eighty.[7] Of course, as demand for electrical power soared, and as Japanese engineers became familiar with electrical machinery, the scale of thermal power generation took off. For instance, in 1911, on the banks of the Aji River, Osaka Dentō brought into service the Ajigawa generating plant. For a time, this was the largest generating plant in all of East Asia.[8]

In any event, it was the harnessing of hydroelectric power that gave a decisive push to the Japanese power industry as the new century dawned. For instance, the number of thermal power stations and hydroelectric-generating stations in Japan in 1903 were twenty-one and nine, respectively. By 1912, at the close of the first upswing of the transitional growth long swing, the number of thermal power stations had increased to 147, but the number of hydroelectric stations had surged forward much more vigorously, reaching a total of 199.[9]

Thus, a combination of thermal and hydroelectric power fueled the industrial upswings of the 1910s and 1930s. Both sectors were competitive. As can be seen from Figure 5.1, until the Pacific War began to cut into Japan's fuel supplies, generation of both thermal and hydroelectric power experienced impressive growth. Growth was especially salient during the two manufacturing upswings. Because infrastructure had been laid down in the decade prior to each industrial upswing—in the 1904–1911 period leading up the 1911–1919 boom, and in the 1919–1930 period leading up to the 1930–1938 boom—supplies of electricity responded to surges in demand with relatively high elasticity.

Private enterprises providing lighting were not the only firms entering the

Table 5.1

Electricity and Railroads: Japan, 1900–1941[a]

A. Railroads: Operation kilometers, rolling stock for national railroads, and rolling stock kilometers, indices (1920–1921 = 100).

	Operation—kilometers				Rolling stock (national railroad only)			Rolling stock—kilometers		
Period	National	Regional	Street	Total	Locomotives	Freight cars	Total	Passenger	Freight	Total
1872–1879	0.7	—	—	0.5	0.8	1.6	0.6	n.a.	n.a.	n.a.
1880–1889	3.3	8.4	—	3.9	1.7	3.7	1.7	n.a.	n.a.	n.a.
1890–1899	9.4	84.8	—	23.7	5.7	9.0	5.0	3.1	6.1	3.1
1900–1909	35.6	104.4	—	45.1	30.0	32.6	29.7	16.7	20.5	16.7
1910–1919	85.1	66.7	92.5	78.6	77.1	79.5	82.1	59.9	64.1	59.9
1920–1929	115.9	141.8	113.2	120.9	113.5	119.3	114.2	94.8	110.3	94.8
1930–1941	159.4	207.4	109.1	162.6	128.0	138.1	143.9	114.7	105.9	114.7

B. Railroads: Electricity generated, electric lights, electrified cars and electric lights in cars, indices (1920–1921 = 100).

Period	Electricity generated	Electric lights	Wattage-lights	Electrified cars	Electric lights in cars
1910–1914	30.4	30.0	47.8	46.1	40.6
1915–1919	93.8	63.9	77.5	75.0	68.3
1920–1924	103.3	128.7	122.6	126.1	124.8
1925–1929	143.4	232.0	247.2	179.9	178.3
1930–1934	647.0	320.1	391.6	179.6	208.3
1935–1939	724.2	450.1	610.9	190.4	235.8
1940–1941	1,494.2	555.0	732.2	210.1	269.8

C. Electrical lights and total electricity generated and supplied for Japan and the Tōkaidō and its subdistricts, indices (1920–1921 = 100). Percentages of national lights and electricity generated within Tōkaidō, and percentages of Tōkaidō total in subdistricts.

	Indices [1920–1921 = 100]					Percentage in Tōkaidō; Percentage of Tōkaidō in subdistricts			
		Tōkaidō[b]					Percentage of Tōkaidō in:[b]		
Period	Japan	Total	North	Middle	South	Percentage in Tōkaidō	North	Middle	South
C.1 Number of electrical lights									
1900–1909	2.4	3.8	3.8	1.9	4.8	65.2	38.6	11.6	49.8
1910–1919	41.7	51.4	54.4	39.1	54.8	54.5	42.4	14.5	43.2
1920–1929	127.2	119.7	123.7	116.3	117.4	38.6	41.6	20.4	38.1
1930–1937	163.1	160.4	174.7	149.8	151.3	40.2	44.0	19.4	36.6
C.2 Electricity generated and supplied									
1915–1919	70.4	74.7	79.5	70.0	70.0	45.3	53.9	10.5	35.6
1920–1929	164.6	182.7	134.4	205.2	237.2	47.6	39.0	12.8	48.2
1930–1937	339.6	386.7	270.8	503.4	500.2	48.9	34.5	14.7	50.8

Sources: Nihon Naikaku Tōkei-kyoku (various dates), *Nihon teikoku tōkei nenkan*, various tables; Nihon Tetsudō-shō (various dates), *Tetsudō tōkei shiryō tetsudōin tōkei zuhyō*, various tables; and Minami (1965), various tables.

Notes: [a]Original figures on electricity flow generated and supplied are in kilowatt-hours.

[b]Subdistricts are defined in terms of the following groups of prefectures: North—Saitama, Chiba, Tokyo, and Kanagawa; Middle—Shizuoka and Aichi; and South—Kyoto, Osaka, and Hyogo.

Figure 5.1 **Hydroelectric and Thermal Power Generation: Growth Rates in Japan, 1916–1960** (Five-Year Moving Averages)

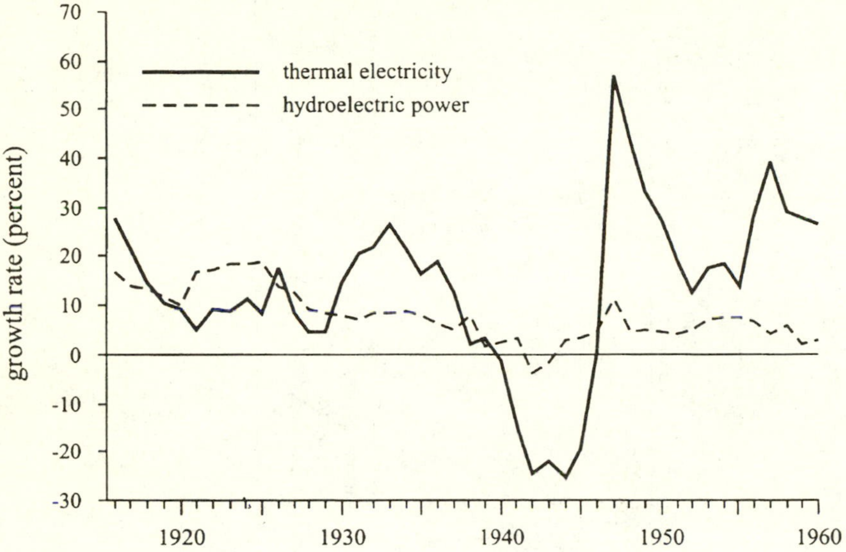

burgeoning field of power generation. At the beginning of the second long swing, a new type of supplier to electricity entered the fray. This was the intercity or intracity railroad or tramway company.[10]

Even prior to electrification, a number of companies in Japan had been offering intercity and intracity service with horse-drawn tramways. For instance, following the lead of companies in the United States (mainly concentrated in larger metropolitan centers) that had begun offering horse drawn transport service prior to the Civil War, a company opened up a similar service in Tokyo in 1885. Tokyo was a natural place to introduce the business. It had a copious population; and wide boulevards, built during the Tokugawa era and extended thereafter, cut through its environs. Having wide streets to run horse-pulled tramways on was a virtual necessity. The horses dunged the roads, creating a noxious stench and health hazards. Needless to say, Osaka's checkerboard of narrow streets was incompatible with the use of horse-drawn tramways.[11]

Since horse-drawn railroads created hazards to health, entrepreneurs and city officials were keen to find replacements. Unfortunately, steam power–driven tramways were not a viable option. In close quarters, steam pollutes. Engines give off sparks and cinders that can set off conflagrations. And the noise can be deafening up close.

With the introduction of cheap electricity, electrical-powered tramways

offered a viable alternative to horse-driven trains. Pollution and fires were not a real issue. And electrical service could be effectively delivered along wires even on relatively narrow streets. Thus, as electrification spread across the United States, the conversion of horse-drawn tramways to electrical tramways occurred in its wake. By 1890, around 1,000 miles of American street railroad track—about a sixth of total urban tramway track—was handling electrical service. After this, electrification really took off. By 1903, 98 percent of American intracity lines were run with electricity. Following the American lead, the Tokyo-based horse tramway company, Tokyo Tetsudō (Tokyo Railway) switched over to electrical power in 1903. Indeed, it was not even the first railroad in Kantō to offer electric rail service. A year earlier, the Keihin line, running between Tokyo and Yokohama, opened up electric intercity service.

Tokyo's innovations in electrical rail service stimulated a wave of imitation throughout the Tōkaidō. Osaka soon became a regional center for electric railways. Not only was Osaka the preeminent industrial center of Japan, it was also—by dint of geography—the natural hub for a network of lines linking together a host of major conurbations in the Kinai. Lying east of Kobe, south of Kyoto and Nara, west of Nagoya, and north of Sakai and Wakayama, Osaka was perfectly positioned to be the centerpiece of the electric railroad network for the Kinai. Steam railway lines already joined a number of these cities to Osaka prior to the early 1900s. Now, emulating the Kantō innovations, those companies offering steam service that escaped nationalization scrambled to switch over to electricity. In addition, new start-up companies jumped in.

With Osaka at its geographic core, a powerful regional electrical railway oligopoly emerged between 1905 and 1914. Coming into Osaka from Kobe on the west was the Hanshin line. A second line, linking Kyoto to Osaka, was developed by the Hankyū Railroad. The company building this line had originally been involved in creating rail connections running into the mountainous resort towns of Takarazuka and Minō from Kobe. Foreign residents of Kobe frequented the spas and entertainment centers clustered in Takarazuka and Minō. Reflecting its original base of operations in Minō, the company that eventually became Hankyū was originally known as the Minōarima. A third line, the Keihin, also joined Osaka to Kyoto—unlike the Hankyū line that ran down the western bank of the Yodo River—forging a link between the two great conurbations along the eastern bank of the Yodo River. Joining Nara to Osaka via track laid down in the eastern reaches of Osaka prefecture was the Osaka Kidō Railway. This company eventually merged with enterprises running lines from Nagoya to Osaka, becoming the Kintetsu Railway. From the south and west came an important complex of lines: Nankai run-

ning out of Wakayama City; Hankai, joining Sakai to Osaka; and Kōya, linking Osaka to the lush verdant mountains of the Kii peninsula. These three companies merged to form the powerful Nankai company.[12]

As the intercity electric railroad oligopoly took shape, a hub-and-spoke pattern emerged with Osaka at the hub. The suburban lines radiated out from the hub like spokes on a wheel.

The terminuses of these lines radiating out from Osaka became key nodes within Osaka. Indeed, the core of Osaka after the first upswing of the transitional-growth long swing—by which I mean the functional core of the city and not the core, or center, as defined in Table 3.2—is basically the area contained within the region defined by an imaginary line joining the terminuses of the major suburban lines to one another. At the northern perimeter of this emerging core were the terminuses for the Hanshin, Hankyū, and Keihan Railways. In Umeda and close to Osaka Station through which passed the Tōkaidō mainline of the Japan National Railroad went up stations for the Hanshin and Hankyū Railroads. And, to the east, within easy walking distance of the Umeda/Osaka Station complex, went up the terminal for the Keihin company. This station was located at the very edge of the Senba district, in proximity to Nakanoshima and Kitahama where Osaka's financial and administrative might was concentrated. Along the eastern periphery of the core, at Uehonmachi, was constructed the terminal for the Osaka Kidō line.[13] At the southern perimeter of the core, slightly south of the Dōton Canal, a cluster of stations that eventually were incorporated in the Nanba complex appeared.

Sweeping around the contour of much of this core was the track of the Japan National Railroad. This track took the shape of a broad arc, running from the complex of docks and warehouses at Sakurajima through Osaka Station and then along the eastern periphery of the core past Osaka Castle. Today this line is the famous loop line of Osaka. Nor was the national government the only government drawn into providing service within the emerging core of the metropolis. Osaka's municipal government also became involved.

Why was the city's government pulled into the vortex of rail lines defining the city's core? While specific historical circumstances undoubtedly help account for the city government's involvement, there were a number of general considerations that surely played a role. By building a tramway system linking together the key nodes on the periphery of the core, the city brought order and coherence to the hub-and-spoke pattern that was developing. And the city improved its leverage as a coordinating agent for the emergent railway oligopoly. By directly controlling one important piece of the transit puzzle, the municipal authorities gained bargaining power that they wielded in their dealings with the private companies and the bureaucracy of Japan National Railroad.

For these reasons, Osaka's municipal authorities plunged into the busi-

ness of building and managing a growing web of tramway lines. They began by constructing the line connecting the pier at Tenpozan to the city's core. This was a piece of the massive harbor construction project. Building upon this initial foray into the tramway business, the city expanded its system in three distinct phases.[14] The first phase, between 1903 and 1910, was carried out in conjunction with the laying out of the hub/spoke pattern created by the building of the suburban electric train lines. During this phase, the city built two lines that crisscrossed the core, transecting it in a crosslike pattern. Along the north/south axis, the city constructed a line joining Umeda to Tennōji through Nanba. The city also built a line running roughly vertical to this line. As a result of this construction, the city-managed system ended up with a total of 12.8 kilometers of track and three lines. During the second building phase—between 1911 and 1915—the city created an additional eighteen lines, pushing out service into the middle ring, into the portion of the city incorporated in 1897. With the additional 43.2 kilometers of track laid down in this second phase, the system managed by the city brought within its nexus nodal points on the key rivers and canals. Finally, in the third phase, the city added on seven more lines.

So, by 1926, one year after it incorporated the periphery, the city managed a tramway system with eighty-eight kilometers of track. This network of rail knitted together most of the land- and water-based transport systems penetrating the city's core and middle ring. Only the newly incorporated periphery, brought under the city's aegis in 1925, was not serviced by this system.

The transportation of Osaka between the world wars was a remarkable hybrid. Steam railroads, electric railroads and tramways, *kobune*, *jinrikisha*, and *niguruma*, were all widely used within its boundaries. Ancient and new transport jostled up against one another, complementing each another. Moreover, the infrastructure that these diverse modes of transport utilized was layered. Waterways and narrow roads constituted the original backbone of the city's infrastructure; train and tramway lines were laid on top of it.

Once it embraced electrical trams, the city of Osaka was inexorably drawn into the business of supplying electric power. Indeed, most of the private electric railway companies did both. Spurring the city on were repeated difficulties encountered in negotiating compensation contracts with Osaka Dentō, the private monopoly that provided the city proper with most of its electricity. In 1923, the city bought out the company, and set up its own bureau to supply electrical power to its tramway system, and to the factories and residential housing units under its jurisdiction.[15] It should be stressed that this statement applies to the city proper as defined by its boundaries in 1923. In the periphery that it formally incorporated in 1925, private companies—the Ujigawa company, and the Keihan, Hanshin, and Nankai railways—com-

peted in providing electricity. As in so many other areas of infrastructure development, private enterprises and public bureaus shared the market.

The form of competition developing in railroads and electrical power generation and delivery was competition among the few. Competition was between a few large companies and government agencies. That is, market was structured as an oligopoly. But just because competition is among a few does not mean it is not fierce. For example, consider the growth of passenger traffic flow in and out of Osaka between 1912 and 1926 in terms of an index of passenger flow.[16] Setting the flow figures for 1912 at a base of 100 for each railroad company, the figures for 1926 on the various railroad lines are as follows: Japan National Railroad had 256; Hanshin had 429; Hankyū had 729; Keihan had 259; Nankai had 542; and Osaka Kidō (Kintetsu) had 860. As can be seen, in those fourteen years, some lines managed to increase their flow of riders eightfold; others were not able to even triple their passenger flow. From the figures given here, it would appear that competition was especially keen along the Kobe-Osaka-Kyoto axis. Competition in the south and east seems to have been less intense.

The fact that the market for suburban train service in and around Osaka was extremely competitive probably meant that profits were squeezed. To be sure, entrepreneurs and investors anticipated making a substantial profit from their endeavors. However, when a number of parties expected this to be the case—as happened here—they all jumped into the business at the same time.

In such a feverish boom of construction, excessive investment may occur. While it is difficult to say whether investment was excessive in the sense that profits were pushed to subnormal levels over the short term, it is clear that there was a real boom between 1900 and 1914 in the Kinai region. This can be seen from Figure 5.2. What is especially striking about the Kinai area (which is labeled "Tōkaidō South" in the figure) is the exceptional rapidity of growth of electric rail service. By comparison, growth in the Tokyo region is anemic.

Of course, growth was less rapid in the Kantō because, prior to the development of electrical railroads, horse-drawn trains had operated in this region. Thus, converting these to electricity did not require a great deal of new investment. Track had already been laid. Rolling stock had already been acquired. The great advantage early development of horse-drawn railways afforded to the Tokyo-Yokohama region is indicated by the following figures. In 1908, of the eighteen companies providing intercity or intracity rail service in Japan, the two great Kantō companies—Tokyo Tetsudō and Keihin—possessed 68 percent of all the cars, and used 32 percent of the total track held by these companies. By contrast, that year the two enterprises of the Kinai controlled a total of 11 percent of the cars, and 14 percent of the track.[17]

Figure 5.2 **Passengers on Electric Railroads: Growth Rates for Japan and for the Northern and Southern Districts of the Tōkaidō, 1902–1932**
(Five-Year Moving Averages)

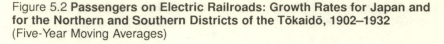

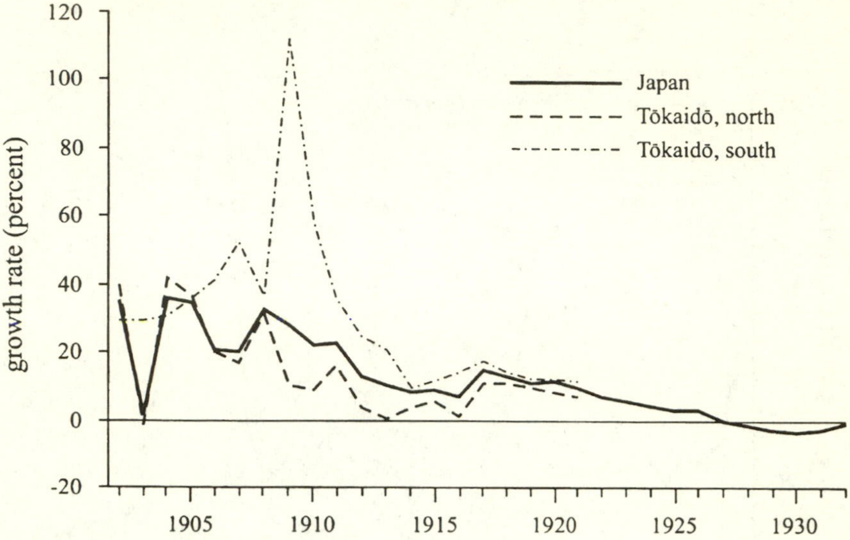

In short, the Kantō had a substantial advantage in developing railroad transport. This advantage became very apparent during the first upswing phase of the transitional-growth long swing. It was to become even more apparent during the downswing phase of the transitional-growth long swing.

Lags and the Pressure of Industrial Expansion on Infrastructure

In the historical logic of infrastructure-driven growth, phases in which infrastructure investment is especially intense lay the groundwork for surges in industrialization. In turn, these surges in manufacturing put pressure on the infrastructure. As a result, the profitability of investing in new infrastructure rises, and resources flow into the sector. This ushers in a phase of renewal, expansion, and improvement in infrastructure. It expands. Then, after a lag, demand for the service flow from the infrastructure expands because the level of industrial activity surges forward. At this point, after a lag, infrastructure expands once again. In short, lags are built into the logic of infrastructure-driven growth.

Now, the pressure that industrial booms exercise on infrastructure takes several forms. In the case of private enterprise–operated infrastructure, ex-

Figure 5.3 **Japanese National Railroads: Growth Rates for Rolling Stock and Kilometers Operated, 1875–1915** (Five-Year Moving Averages)

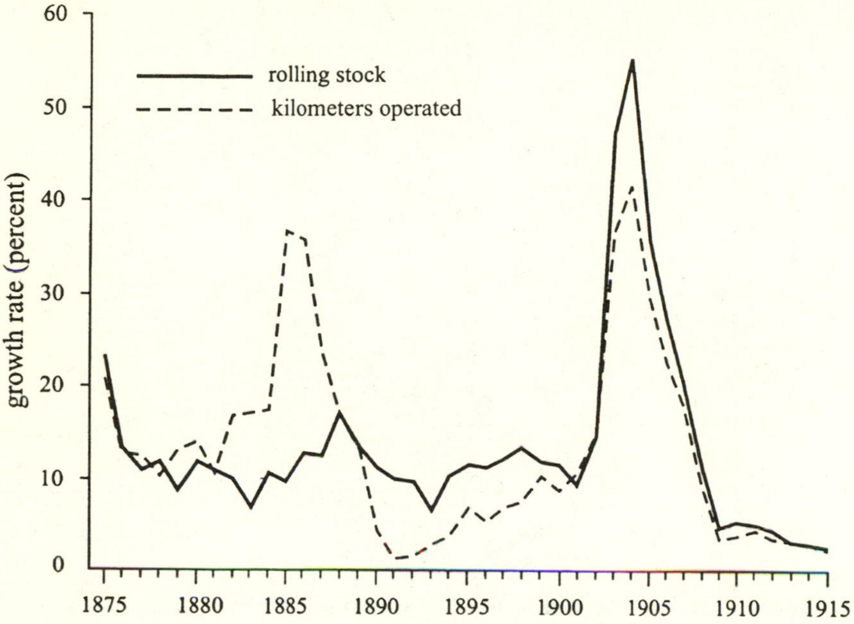

pected profits are the key. As the demand for their services picks up, the suppliers of the services earn greater revenue. Their capacity to issue stock or borrow from banks improves. And they are better equipped to plow back profits into more capital accumulation. In the case of publicly managed companies in the infrastructure sector—like the Osaka City–managed tramway that enjoyed a monopoly position within the core of the city—pressure took a slightly different form. As city-managed trains became excessively packed and overcrowded, the public complained. Newspapers ran editorials calling for improvements. Politicians took notice. The city responded by expanding the scale and reach of its system.

Perhaps the simplest way to conceive of the lags inherent in the timing of infrastructure building and its usage is in terms of stocks and flows. First, in anticipation of future flows, stock—capital—grows. Then, flow—services—pick up. Finally, once again, in response to the surge in flow, stock increases.

This type of pattern was ubiquitous in Japan's economic growth up until the early 1970s. Consider, as a case in point, the pattern for the national railroad system of Japan displayed in Figure 5.3. Buildup of rolling stock during the 1880s is followed by a surge in growth (measured in terms of

Figure 5.4 **The Hankyū and Hanshin Railroads: Growth Rates for Passengers (Hankyū) and Rolling Stock (Hanshin), 1908–1937**
(Five-Year Moving Averages)

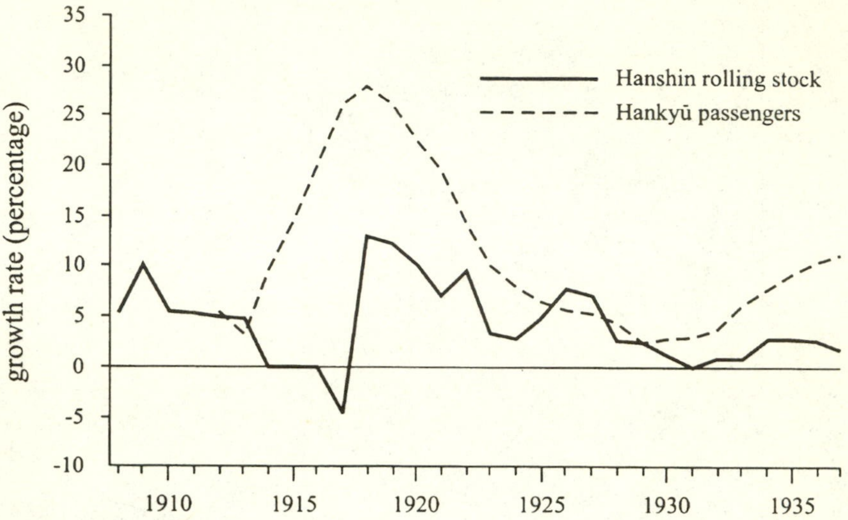

kilometers operated) during the first manufacturing surge of the 1880s. The period when nationalization occurred seems to be an exception to this rule. In fact, it is the exception that proves the rule. For, in the case of nationalization, a base of existing customers already using the lines came with acquisition of the track and rolling stock for the lines taken over.

Another example, this time for the private sector, illustrates my point in a slightly different way. Consider Figure 5.4 giving figures on growth of passenger flow for the Hankyū company, and growth of rolling stock for Hanshin. Hankyū and Hanshin were bitter rivals. As demand on one line picked up, the other expanded its rolling stock, thereby offering less crowded service and greater frequency of service. In a competitive industry, the lag mechanism is very powerful.

The logic of stock/flow adjustment with lags applies to the Osaka City–managed tramway system as well. Consider the pattern depicted in Figure 5.5, which gives growth rates for city-managed ground transportation between 1908 and 1936 (because the city's legal boundaries changed in 1925, the growth rates are given for two subperiods—1908–1921 and 1925–1936—separately). Growth in passenger flow picks up during the manufacturing boom periods of the 1910s and 1930s, and it is precisely during these periods that growth in kilometers operated is languid.

Figure 5.5 **Osaka City-Managed Ground Transportation: Growth Rates for Passengers and Kilometers Operated, 1908–1936** (Five-Year Moving Averages)

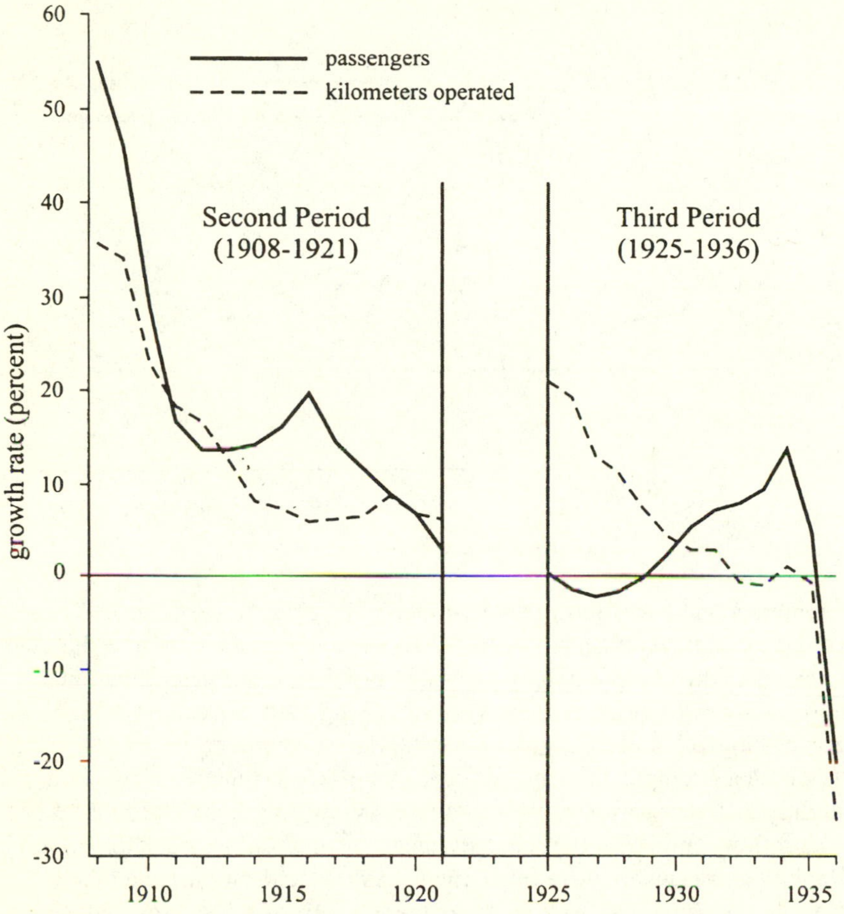

Thus, through patterns of boom and bust, Osaka's transportation infrastructure was built up. How does this buildup compare with that taking place in Tokyo and the Kantō region? We can address this question by systematically comparing the two districts of the Tōkaidō. In Table 5.2, comparison of panels B and C reveals that there are salient differences between the infrastructure expansion taking place in the northern and southern extremes of the industrial belt. Growth in Tokyo and Kantō is far less jerky than is growth in the Osaka region, which is subject to violent fluctuation.

Tokyo's lead in the new revolutionary land-based infrastructure reflected

both the relative advantages and disadvantages of its environs. Compared to Osaka, Tokyo had a decided advantage in land-based infrastructure. For instance, it already had wide boulevards and thus could accommodate horse-drawn tramway traffic. However, it was at a disadvantage on the water. Befitting its reputation as a Venice of the Far East, Osaka had a decisive advantage in water-based infrastructure. And, as panel A of Table 5.2 makes evident, Osaka continued to rely on its rivers and canals throughout the prewar period. Boat traffic did not increase as dramatically as did traffic on railroads. Nevertheless, as the city's population and the scale of economic activity boomed, so did the volume of activity flowing on Osaka's water network. Indeed, in 1902, Osaka granted a company the exclusive rights to manage a water taxi/water bus system within the city.

The relationship between use of water transport and use of land transport is interesting. As can be seen from Figure 5.6, passenger flow on both land and water tended to move in wavelike patterns. To some extent, the waves move together. This reflects the fact that during periods of rapid industrialization, demand for both picked up. And they complemented one another to a degree: a commuter might board a water bus first and then transfer to a tramway, for instance. To a degree—because water transport and land transport were also substitutes for one another—use of the two modes of transport moved in opposite directions. Thus, during the downswing of the 1920s, growth of ground transportation faltered, but growth in water transport actually picked up.

That Osaka and the Kinai were falling behind Tokyo and Kantō in electric train infrastructure had widespread ramifications. It affected speeds of innovation in a variety of endeavors, for electrical railway enterprises were innovators in both transportation and in electrical power distribution. Thus, Tokyo did manage to move ahead of Osaka in the revolutionary new technologies underlying the second long swing.[18]

However, as the electrical power industry evolved and began to take advantage of scale economies, the linkage between railways and electricity weakened. Key to the scale economies in the prewar Japanese electrical power sector was hydroelectricity. In order to reap the profits that delivering this new form of energy offered, private enterprises rushed into the market, especially in the wake of the huge industrial expansion of the 1910s. The most important thrust of electrical power–generating infrastructure construction in the period between the world wars was the creation of three massive power grids delivering energy to the Tōkaidō industrial belt. A Keihin grid delivered power to Tokyo and Yokohama from dams in the Japanese alps north of the Tenryū River. And from rivers and lakes in the alps south of the Tenryū River was tapped the power for two other massive grids, the Chūbu grid,

Table 5.2

Intracity Passenger Transportation by Train, Electric Street Railroad, and Boat in Osaka and Tokyo, 1900–1940

A. Osaka: Indices with 1920–1921 = 100 for passenger flows on boats, trains, and electric street railroads (ESC) and percentage on boats (PerBoat).

Second period in city's history				
	Indices			
Period	Boat	Train	ESC[a]	PerBoat
1900–1904	1.7	52.8	0.3	5.2%
1905–1909	16.8	69.0	4.7	14.3
1910–1914	50.0	86.2	33.2	13.4
1915–1919	68.8	89.0	68.9	9.4
1920–1924	100.7	114.7	108.1	8.9

Third period in city's history				
	Indices			
Period	Boat	Train	ESC	PerBoat
1925–1929	123.3	201.1	116.8	9.4%
1930–1934	123.0	197.9	115.2	9.4
1935–1938	202.1	451.5	107.2	11.8

B. Osaka: Indices with 1920–1921 = 100 for stocks and passenger transportation flows for electric street cars[a].

Second period in city's history				
	Stocks[b]		Flows[c]	
Period	Track	Cars	Pass-Kilo	Daily
1903–1910	14.3	10.6	8.7	6.6
1911–1915	57.6	67.6	57.4	39.4
1916–1920	84.3	86.9	82.9	81.1
1921–1924	113.1	120.8	107.0	110.6

Third period in city's history				
	Stocks[b]		Flows[c]	
Period	Track	Cars	Pass-Kilo	Daily
1925–1930	132.3	133.0	110.2	115.2
1931–1935	142.6	115.4	117.4	94.8
1936–1940	146.2	115.0	136.3	125.6

C. Tokyo: Number of stations on, and indices (1920–1921 = 100) for mileage of track (Track) and passenger flow (Pass) for electric street railroad.

First period in city's history				Second period in city's history			
		Indices[d]				Indices[d]	
Period	Number of stations	Track	Pass	Period	Number of stations	Track	Pass
1905–1909	246	52.8	40.8	1932–1937	406	118.0	71.2
1910–1914	324	77.5	53.0				
1915–1919	330	91.7	72.5				
1920–1924	406	103.4	98.4				
1925–1929	481	112.4	104.3				
1930–1931	406	116.0	80.6				

Sources: Ōsaka-shi Yakusho (various dates), *Ōsaka-shi tōkeisho*, various tables; and Tōkyō-shi Yakusho (various dates), *Tōkyō-shi tōkei nenpyō*, various tables.

Notes: [a]The Osaka electric street car system was inaugurated in 1903.

[b]The term "Track" stands for kilometers of track in operation, and the term "Cars" stands for number of cars in operation.

[c]The term "Pass-Kilo" stands for passenger-kilometers, and the term "Daily" stands for daily passenger flow.

[d]The figure given for Pass for 1905–1909 is actually only for the year 1909.

Figure 5.6 **Ground- and Water-Based Transportation in Osaka City: Growth Rates for Passengers on Water and on Steam and Electric Trains, Buses, and Subways, 1908–1934** (Five-Year Moving Averages)

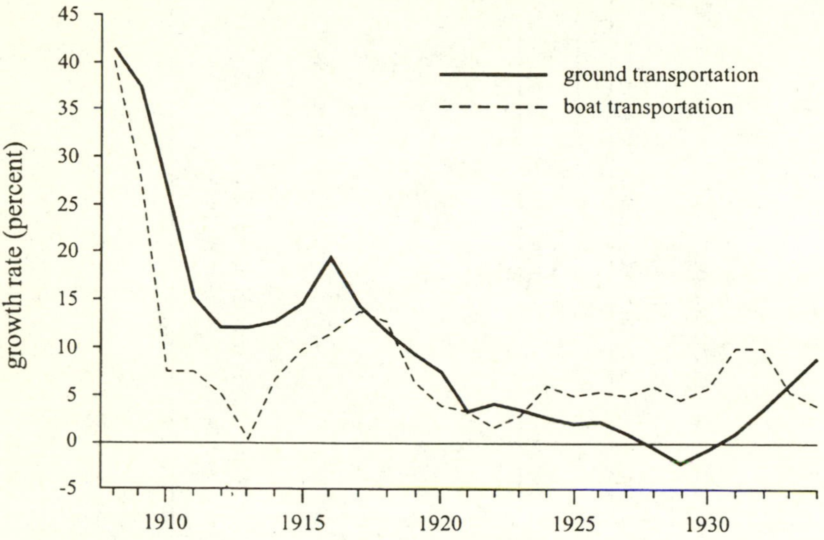

providing power to Nagoya and the Nobi plain, and the Keihan grid, which supplied electricity to Osaka and Kobe.[19]

In order to secure funding for these huge projects, electrical power companies entered into a wave of mergers. They also cut costs through rationalization. For the Keihan and Chūbu grids of the southwestern Tōkaidō, the most important companies coming out of the welter of buyouts, mergers, and amalgamations were Daidō and Nihon Denryoku.[20] In 1923, Daidō unveiled its first major power generating station in the mountains at a site along the Kiso River, which flows into Ise Bay near Nagoya. The new Daidō grid extended a high-voltage delivery line into Osaka. In 1936, Daidō opened up a second line to Osaka. Daidō was not the only company to jointly provide electricity to Nagoya and Osaka. In 1924, Nihon Denryoku constructed a trunk line feeding power to Nagoya. It eventually extended the line southward along the Tōkaidō, to Osaka. Rationalization and mergers substantially reduced the number of enterprises active in the field. Exploiting scale economies was key to this rationalization and merger wave. Competition among the few had become competition among the very few.

As Japan girded itself for total war in the late 1930s, military cabinets carried out an even more sweeping reorganization of the power industry. In 1938, the cabinet put forward legislation nationalizing the enterprises in the

electrical power field. With the Denryoku Kanrihō (Electrical Power Management Law), the private firms were absorbed into nine government-controlled regional power companies. And the Kinai region came under the jurisdiction of a huge Kansai company.[21]

In sum, pressure on infrastructure was key to the alternating rhythm whereby industry expanded, then infrastructure expanded, then industry once again expanded. And, beginning with the upswing initiating the transitional-growth long swing, the focus of the new physical infrastructure was upon land-based infrastructure. Now this shift toward the land put pressure on land prices, especially in the great conurbations of the Tōkaidō.

Land Prices and Speculation

Land development and infrastructure development are intimately related in modern Japan. Indeed, this linkage already existed under *bakufu* rule. For instance, during the Tokugawa period, securing a regular flow of water for irrigation enhanced land value. Building roads meant that hitherto isolated regions now secured access to wider markets. This, too, enhanced land values.

During the second long swing, propinquity to an electric railroad line became strongly linked to land value. Indeed, Japan went through a virtual revolution in land prices during the first upswing phase of the transitional-growth long swing. This revolution is not just apparent in land prices. It also shows up in the proliferation of land development companies that accompanied escalation in land prices in the vicinity of the new intercity railroads.

At the close of chapter 3, I noted that a concerted rise in the price of residential and commercial real estate within Osaka accompanied the upswing of the balanced-growth long swing. As is apparent from Table 5.3, the proliferation of factories within Osaka pushed up *takuchi* prices so vigorously that, by the turn of the century, a hectare of *takuchi* commanded a price about five times that of a hectare of paddy land. In the country as a whole, the ratio of *takuchi* land to paddy land was about unity, so Osaka's ratio was already far above that.

With hindsight, it is clear that the surge in residential and commercial land prices in late-nineteenth-century Osaka was modest. For, as can be seen from panel C of Table 5.3, in a relatively short period of time—between 1910 and the outbreak of World War I—Osaka's *takuchi* prices really took off. Indeed, the ratio of *takuchi* to paddy suddenly jumped to forty. It hovered around that level during the industrial boom of the World War I years, and then resumed its upward course during the 1920s.

Table 5.3

Land and Relative Land Prices: Osaka and Tokyo Prefectures and Osaka City, 1900–1939

A. Private land use in three categories: Paddy field, dry field, and *takuchi*. Indices of land area in each category and ratios of paddy land area to dry field land area (Pad/Dry) and *takuchi* to dry field area (Tak/Dry)[a].

	Tokyo prefecture					Osaka prefecture				
	Indices, 1920–1921 = 100			Ratios [base = 100]		Indices 1920–1921 = 100			Ratios [base = 100]	
Period	Paddy	Dry Field	*Takuchi*	Pad/Dry	Tak/Dry	Paddy	Dry Field	*Takuchi*	Pad/Dry	Tak/Dry
1900–1904	114.2	98.3	79.9	42.2	24.4	97.6	122.5	77.7	324.2	37.2
1905–1909	111.9	98.9	85.1	41.1	25.8	99.1	114.9	81.0	350.9	41.4
1910–1914	106.2	97.7	91.1	39.4	28.0	101.0	105.8	88.0	386.4	48.8
1915–1919	101.8	97.3	95.8	38.0	29.5	100.1	80.2	94.4	647.8	87.6
1920–1924	106.5	98.0	91.2	39.4	27.9	99.4	104.4	89.1	424.2	55.4
1925–1929	93.1	95.8	110.5	35.3	34.6	98.4	96.7	120.8	414.2	73.3
1930–1934	84.6	90.9	143.1	33.8	47.4	95.3	89.4	147.4	433.2	96.7
1935–1938	77.2	86.8	166.0	32.3	57.3	91.8	83.6	164.7	446.7	115.6

B. Relative land prices: Ratio of land prices, Osaka prefecture/Tokyo prefecture, base = 100, 1911–1938.

	Type of land				Type of land		
Year	Paddy	Dry field	*Takuchi*	Year	Paddy	Dry field	*Takuchi*
1911	122.8	206.4	121.1	1924	122.8	188.1	99.1
1912	122.9	203.6	100.1	1929	123.1	170.7	100.0
1918	122.2	192.6	97.5	1934	115.3	110.2	102.5
1920	122.7	192.3	97.5	1938	120.5	94.9	87.5

C. Osaka City, indices for land use [1920–1921 = 100], and ratios of per hectare land price relative to paddy land price = dry field price/paddy land price (Dry/Pad) and *takuchi* price/paddy land price (Tak/Pad)[b].

	Land area, indices 1920–1921 = 100 expanded Osaka City			Land price ratios [paddy = 1]					
				City before 1925 expansion		Combined *gun*		Expanded Osaka City	
Period	Paddy	Dry Field	*Takuchi*	Dry/Pad	Tak/Pad	Dry/Pad	Tak/Pad	Dry/Pad	Tak/Pad
1901–1904	108.9	159.1	62.0	0.60	5.3	0.41	0.93	0.45	4.2
1905–1909	109.9	146.0	68.4	0.59	5.1	0.35	0.70	0.40	3.9
1910–1914	108.1	121.1	80.1	0.58	40.2	0.36	1.95	0.40	27.5
1915–1919	103.9	110.0	89.5	0.45	38.9	0.36	2.54	0.36	29.1
1920–1924	98.2	96.4	104.7	0.55	47.5	0.36	2.58	0.38	27.4
1925–1929	90.4	83.2	124.9	—	—	—	—	0.58	29.8
1930–1934	78.3	69.4	150.3	—	—	—	—	0.68	40.0
1935–1939	69.9	54.7	166.7	—	—	—	—	0.80	42.3

D. Osaka City, per hectare land price ratios for 1926: Center, middle ring, and periphery[c].

Ratio [base = 1]	Osaka City	Center	Middle ring	Periphery
Dry field price/paddy price	0.58	0.50	0.48	0.58
Takuchi price/paddy price	25.3	66.5	19.9	6.7

Sources: Nihon Naikaku Tōkei-kyoku (various dates) *Nihon teikoku tōkei nenkan*, various tables; Ōsaka-fu (various dates), *Ōsaka-fu tōkeisho*, various tables; Ōsaka-shi Yakusho (various dates), *Ōsaka-shi tōkeisho:* various tables; and Tōkyō-shi Yakusho (various dates), *Tōkyō-shi tōkei nenpyō*, various tables.

Notes: [a]*Takuchi* private land consists of land used for housing, retail and wholesale, warehouses, office buildings, factories, and private schools and religious organizations, etc.

[b]The term "expanded Osaka City" refers to the city with its boundaries after the 1925 incorporation of those parts of Nishinari and Higashinari *gun* not incorporated in 1897. The term "combined *gun*" refers to the area in the two *gun* incorporated in 1925.

[c]Paddy price is for the city as a whole. The dry field and *takuchi* prices are for the center, middle ring, and periphery separately. For definitions of "center," "middle ring," and "periphery," see the notes to Table 3.2.

As is apparent from panel D of Table 5.3, the strong upward push to residential and commercial land prices was not felt uniformly throughout the environs of the city. In the newly incorporated periphery of the city in 1926, the ratio was around seven. By comparison, in the core it had reached an astronomical level of 66.5. Why did land prices particularly escalate in the core of the city? If the cause were population growth and the proliferation of factories, the periphery—and not the core—should have experienced the greatest upward pressure on land prices. For, as was shown in chapter 3, population and manufacturing establishments tended to increase the most rapidly in the periphery. Thus, a direct linking of population pressure to land prices is not possible.

A more plausible account links the sharp run-up in land prices to the buildup in infrastructure that was taking place during the first decade of the twentieth century within Osaka. This new infrastructure was having an especially powerful impact on Osaka's core, because the proliferation of suburban railways running into the core, vastly increased the daytime (but not residential or nighttime) population working within it.

Railroad and electrical power infrastructure was being built in all of the major Tōkaidō conurbations during the first decade of the twentieth century. Hence, it is not surprising that *takuchi* prices rose in all of the major cities of Japan's industrial belt. This result is apparent from Figure 5.7. However, the magnitude of the movement and the violence of the fluctuations were more pronounced in Osaka than elsewhere in the Tōkaidō.

In Osaka, the sharp fluctuation was really a speculative bubble. During the 1910s, soaring land prices fed an inflation that was already substantial due to nationwide increases in the consumer price index. In the wake of the industrial boom, nominal prices for land dropped. Nationwide deflation was rampant in all quarters. But the relative price of *takuchi* land in Osaka fell only slightly, working its way up and down in a sawtooth pattern. Then, in the later 1920s, it resumed its upward march.

Price instability created diverging expectations about future price movements, and this opened the door for speculation. A clear sign that land speculation was becoming important in Osaka's real estate market once the upward surge in *takuchi* land prices commenced is the proliferation of land development companies in the region. In 1911, three companies entered the market. Eight more joined in 1912, six more in 1914, eight more in 1917, and twelve in 1918. In the year the industrial downturn began, 1919, a whopping thirty-four new entrants crowded into the real estate business.[22] The two great upswings of the transitional-growth long swing had transformed Osaka's land market forever.

Figure 5.7 **The Relative Per-Hectare Price of *Takuchi* to Paddy Land (*Takuchi* Price/Paddy Land Price Ratio): Expanded Osaka City (Inclusive of Nishinari-*gun* and Higashinari-*gun* Before 1925) and Japan, 1902–1932** (Five-Year Moving Averages)

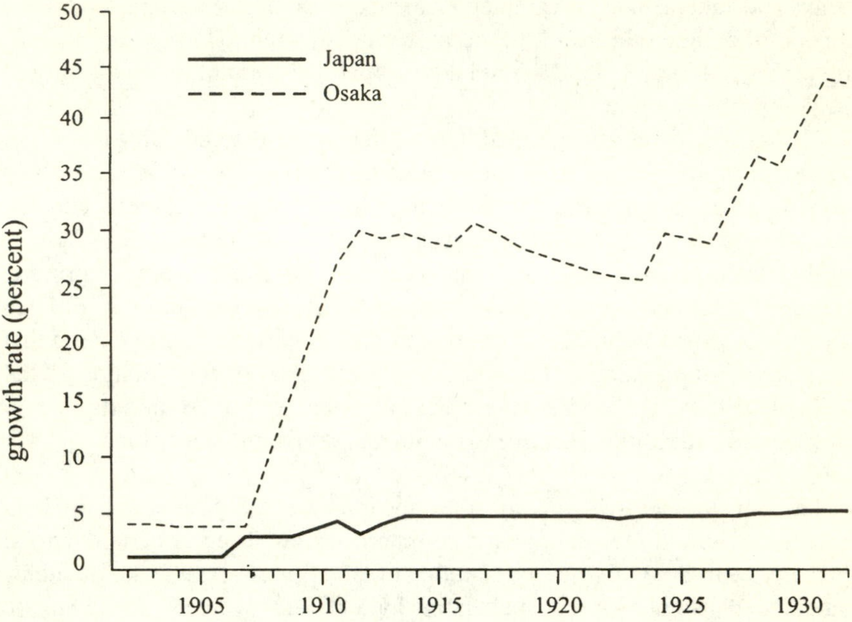

In the forefront of Osaka's land development were the electric railroads. Is this surprising? The suburban railway companies appreciated the fact that after they opened stations in an area, land prices in the vicinity of the station soared. Thus, the companies "internalized the positive external benefits" that they created when they opened up new stations. They bought up land in the area surrounding the locale of the prospective station before they actually constructed the station facility. By buying the land at an early date, they were able to scoop it up at a cheap price. However, once they opened up the station for service, they either sold the land at a high price or developed rental housing on their holdings. This pattern is very clear in the real estate development maps that various suburban railway companies in the Osaka area have published as part of their company histories.[23]

The strategy of diversifying into real estate and electrical power generation is the strategy associated with the *zaibatsu*. Indeed, the railway companies of the Tōkaidō belt are classic examples of regional *zaibatsu*. And, as

zaibatsu, they diversified into more than real estate. They branched out into leisure activities. Consider the Hankyū company. In 1906, it operated a park in Minō. In 1912, it opened up a hot springs resort and entertainment complex at Takarazuka. At the same time, it opened up railroad service between Osaka and Takarazuka. A year later, Hankyū opened up a station in Toyonaka, north of the Shinyodo River. In the year following that, Hankyū began selling off the holdings of land that it had long before acquired in the vicinity of the prospective station.

The railway firms also jumped into the department store business. For, as more and more persons used their trains to commute to work in the core districts of the great metropolitan centers like Osaka, the greater was the flow of persons through the terminal stations. And, of course, after work, or on their lunch breaks, these commuters would want to do some shopping. The commuters constituted a kind of captive audience. Thus, the railway companies joined the ranks of the five mighty department store chains that were starting up business throughout the great cities of the Tōkaidō in the wake of the first upswing of the second long swing. In short, regional *zaibatsu* proliferated as a result of the infrastructure buildup of the transitional-growth long swing.

The development of railway lines spearheaded the conversion of land from agricultural to residential and commercial uses. The public understood this very well. So, with the skyrocketing of land prices in the wake of suburban rail expansion, a kind of speculative logic became well entrenched in the rural communities that were being steadily transformed into bedroom communities for the great conurbations of the industrial belt. Do not sell right away. Let prices rise, and you will secure greater capital gains. Thus, in the short run, escalating land prices actually discouraged conversion of land from farming to residential or commercial purposes.

Local and national governments were well aware that rampant speculation could slow the conversion of land. They appreciated the fact that there were many other barriers slowing conversion of farming land, especially in districts—like those surrounding Osaka—where paddy field production was entrenched. For paddy field villages were cohesively held together by a strong mutual interest in the securing of, and the internal distribution of, irrigation water for the community as a whole. Running roads, sewer lines, and water pipes into these villages—that is, laying down the infrastructure necessary for building up factories, dormitories, and residential housing—could potentially wreak havoc on the older infrastructure embodied in irrigation and drainage ditches.

Speculation only compounds the problem. Put in modern infrastructure. Land prices rise. This threatens the collective integrity of the village.

Realizing the difficulties, the national ministries came up with an ingenious solution. Under the Arable Land Reorganization Law of 1899—a law that had originally been put forward as a tool for rationalizing the layout of paddy fields and upland—the national government came up with a scheme for providing subsidies to local communities that agreed to put forward land reorganization projects linked to putting in modern infrastructure. With the proliferation of bedroom communities in the hinterlands of the great metropolises of the Tōkaidō, the Arable Land Reorganization Law became a major vehicle for carrying out land conversion.[24]

Wielding the Arable Land Reorganization Law served the government's purposes in two ways: It cut the direct costs it had to shoulder in building infrastructure; and, by taking into account the initiative and interests of the local communities themselves, resistance to improvements was minimized. If at least two-thirds of the landowners holding at least two-thirds of the land within a locality agreed upon a plan for reorganizing their holdings and for putting in the infrastructure, they were invited to establish a land reorganization committee (*kōchi seiri kumiai*). What was the incentive on the part of the community to form a committee? The carrot was partial funding for the project. The government expressed its willingness to facilitate conversion by offering subsidies to the committees.

Thus, spurred on by speculation in land, the process of land conversion proceeded with the assistance of the national government. During and after the initiation of the first upswing of the transitional-growth long swing, land conversion picked up momentum throughout the Tōkaidō. The process went far more smoothly in the environs of Tokyo than it did in Osaka's environs. To see how the pace of land conversion proceeded in these two great conurbations and their hinterlands, see panel A of Table 5.3 and Figures 5.8 and 5.9. These displays and the table make apparent that the two regions were experiencing completely different fates as land conversion took place.

In Osaka, the tendency was to convert dry field arable first, and only later the more valuable paddy fields. The sole—albeit prominent—exception to this proposition was the spurt in dry field growth in Osaka prefecture during the latter half of the 1910s and the early 1920s. During the 1920s, retrenchment in light industry was widespread in Osaka. Hence, dry fields may have been put in because land that had been used as *takuchi* land was actually put back into agriculture. For Tokyo, a pattern that was virtually the reverse of Osaka's held sway. Paddy fields—generally far inferior to those surrounding Osaka—were converted first. Again, there is one striking exception. During the early 1920s, the Kantō district was rocked by a huge earthquake, the Tokyo-Yokohama earthquake of 1923. While acknowledging these exceptions, it is the contrast in land conversion that should be emphasized.

Figure 5.8 **The Conversion of Agricultural Land in Osaka Prefecture: Growth Rates for Paddy and Dry Field Land, 1902–1935**
(Five-Year Moving Averages)

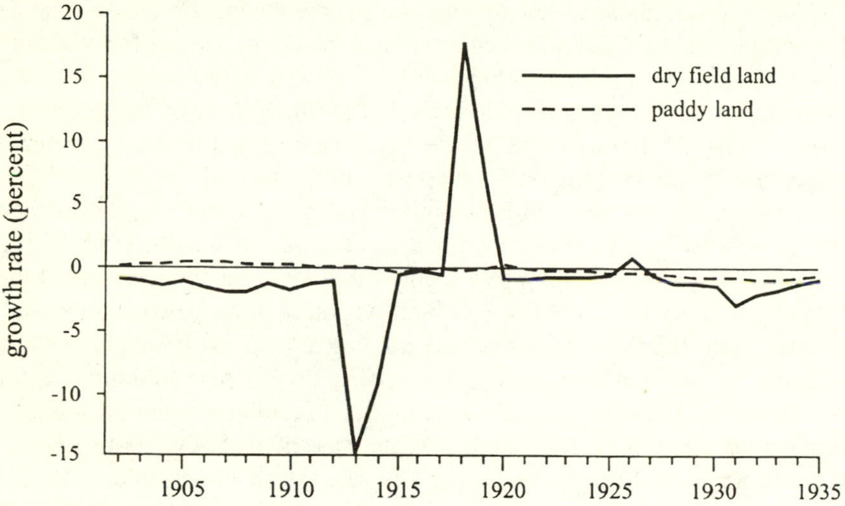

Figure 5.9 **The Conversion of Agricultural Land in Tokyo Prefecture: Growth Rates for Paddy and Dry Field Land, 1902–1935**
(Five-Year Moving Averages)

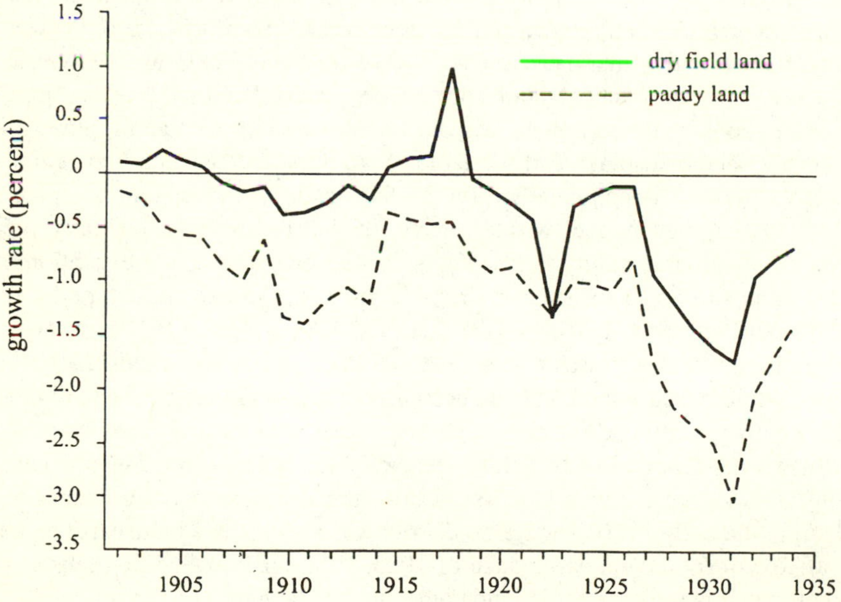

Ecological factors help explain why land conversion proceeded differently in the two districts. In the vicinity of Tokyo were fields of dubious value for farming. A substantial share of the arable around Tokyo was in dry fields. By contrast, Osaka was surrounded by highly productive and valuable paddy fields. Thus, the whole process of land conversion was far more laborious, painful, and costly in Osaka than it was in Tokyo. Given the relative scarcity and poor quality of paddy land in the hinterland of Tokyo, rice-producing farmers in that district were less likely to be able to extract large speculative gains out of the sale of their land. Holding onto it in anticipation of future gains in price was not an especially attractive option. In the periphery of Osaka, the reverse was the case. Paddy was all over the place. Farmers had a reasonable expectation of extracting large capital gains by holding onto the land, especially during periods when the demand for land conversion was strong. In short, as the two great cities pushed out into their hinterlands, Tokyo had tremendous advantages over Osaka.[25]

Strange to say, natural disasters also contributed to Tokyo's competitive advantage in land conversion. Experiencing the great Kantō earthquake actually gave a strong push to land conversion within the Tokyo region. In the aftermath of the great earthquake and the ensuing fires that engulfed the region, huge sections of both Tokyo and Yokohama had to be rebuilt. In order to cope with the disaster, both the national and municipal authorities took on special powers. The national government took responsibility for fifteen designated districts within Tokyo. Tokyo's city government took responsibility for another fifty, and, Yokohama City took charge of thirteen districts under its jurisdiction.[26] The authorities carrying out reconstruction were empowered to expropriate some land for public use. As a result, about 10 percent of the land in Tokyo and its hinterland fell into the hands of government. The city used much of its newly acquired land to expand the existing road network and to lay down new sewer lines and water pipes. In short, the same type of natural disaster that had shaped Edo's physical infrastructure was shaping that of twentieth-century Tokyo as well.

Trucks, Buses, and Roads

During the second long swing, the internal combustion engine became an important source of power in Japan. Harnessed to modes of ground transport, its use revolutionized the movement of goods and passengers. Buses and trucks came into their own. And during the third long swing, the internal combustion engine, modified for use in airplanes, continued to revolutionize transportation in Japan.

Following the lead established in Europe and North America, motorized

vehicles made their appearance in Japan at the turn of the twentieth century. Imports of buses commenced in 1902 when the city of Hiroshima placed an order for an American-made vehicle.[27] Gradually, automobiles began to make their appearance throughout the large cities of Japan, and especially in the conurbations of the Tōkaidō. Their numbers swelled especially in Tokyo. Is this surprising? After all, Tokyo had wide roads. To be sure, the roads were not yet paved, but they were wide and convenient for automobile traffic. On his trip to Japan to design the Imperial Hotel in Tokyo, Frank Lloyd Wright waxed effusively about Tokyo's wide boulevards: "Teeming, enormous area is fascinating Yedo. A vast city channeled with wide bare-earth streets swarming with humanity their interminable length."[28] It was clear Tokyo had a huge advantage in the harnessing of the internal combustion engine to ground transport.

Tokyo's ecology contributed to its being an innovator in bus transport in other ways. The great Kantō earthquake of 1923 had torn up miles of track, thereby rendering them virtually useless. The Japanese were aware of the use Europeans had put automobiles to—for taxicabs and ambulances during World War I—and they quickly seized on the bus as a mode of transport superior to the train for moving populations through densely inhabited areas subject to violent earthquakes. Bus routes could be changed relatively easily; train routes could not. Earthquakes were less likely to knock out bus service. And, in event of an emergency, one could not rely upon slow-moving *jinrikisha* and *niguruma*. A similar logic applied to using trucks for transport of freight.[29]

The surge in inner city land prices in the great conurbations of the Tōkaidō during the upswings of the second long swing contributed to the demand for buses and trucks. For, as factories and service industries crowded into these great cities, land prices in city cores soared. This made living in the central cores of cities extremely expensive. Increasingly, workers sought housing on the periphery of the growing conurbations, or in the communities springing up along the electric intercity suburban rail lines. Thus, a strong fillip was given to the demand for buses and trucks that could traverse wide swaths of the burgeoning residential districts located within the great metropolises and their hinterlands.

The use of motorized vehicles increased dramatically after 1915, that is, about the time that the forceful surge in land prices was being felt. As can be seen from Table 5.4, there was a substantial growth in use of cars, buses, and trucks in Osaka and Tokyo, especially Tokyo, around this time. Perusal of a listing of bus companies that started up service during the period between the world wars makes this clear. Many of the private bus lines were founded by suburban train companies servicing their stations and land development projects. There was an especially dramatic buildup of suburban bus lines

Table 5.4

Cars, Buses, and Trucks in Osaka and Tokyo, 1907–1940

A. Cars, buses, and trucks in Osaka and Tokyo: Indices with 1920–1921 = 100 for total vehicles in each city, and ratio of Osaka total to Tokyo total with Tokyo = 100.

	Years					
Item	1907–1909	1910–1914	1915–1919	1920–1924	1925–1929	1930–1932
Index for Tokyo City	0.6	7.6	34.4	124.8	338.5	535.6
Index for Osaka City	0.0	1.7	33.6	146.7	508.2	837.2
Osaka/Tokyo ratio	0.0(%)	6.9	28.7	37.1	48.0	50.6

B. Buses and electric street railroads in Osaka City: Indices with 1920–1921 = 100 for kilometers of operation (KOP), number of cars, daily operating kilometers (DOPK), and passengers per day (PASPD) for electric street railroads, and for buses and electric street railroads combined, 1921–1940.

	Electric street railroads				Buses and electric street railroads combined			
	KOP	Cars	DOPK	PASPD	KOP	Cars	DOPK	PASPD
1921–1924	113.1	120.8	107.0	110.6	113.1	120.8	107.0	110.6
	Third period in city's history							
1925–1930	132.3	133.0	110.2	115.2	199.5	157.5	127.7	120.0
1931–1935	142.6	115.4	117.4	94.8	363.1	200.4	188.1	116.9
1936–1940	146.2	115.0	136.3	125.6	383.9	258.6	229.3	177.5

C. Buses and electric street railroads in Tokyo City: Indices with 1920–1921 = 100 for electric street railroad passengers (ESRPAS), and for passengers on buses and street railroads (ESRBUSPAS); and percentage of total passengers on buses and street railroads who are on buses (PBUS).

	First period in city's history						Second period
Item	1905–1909	1910–1914	1915–1919	1920–1924	1925–1929	1930–1931	1932–1937
ESRPAS	40.8	53.0	72.5	98.4	104.3	80.6	71.2
ESRBUSPAS	40.8	53.0	72.5	98.4	123.3	99.7	105.8
PBUS	—	—	—	—	15.4	19.2	32.4

Sources: Same as sources for Table 5.2.

feeding into Tokyo during the 1920s. And, after it incorporated Higashinari-*gun* and Nishinari-*gun*, Osaka introduced, in 1927, a city-managed bus transport system that was specifically targeted at the newly incorporated region of the city.[30] In short, growth in demand stemming from the huge industrial

expansion of the 1910s was a powerful factor promoting rapid diffusion of motorized vehicles in Japan.

However, supply as well as demand was operating in shaping the diffusion of the new form of transport. Compare the pace of diffusion of cars, buses, and trucks in Osaka with that in Tokyo. For instance, as can be seen from panel A of Table 5.4, diffusion was quicker in Tokyo than it was in Osaka. For instance, during the 1925–1929 period, the total number of registered vehicles in Osaka was less than 50 percent of that for Tokyo. And Tokyo's lead is especially surprising in light of the fact that—in those years—Osaka was a far bigger metropolis than was Tokyo. Osaka had just gone through its major incorporation, and Tokyo and not yet swallowed up its hinterland to the west. Moreover, as is apparent from panels B and C of Table 5.4, bus service cut into tramway service far more rapidly in Tokyo than it did in Osaka. By the early 1930s, as bus service in Tokyo and its suburbs surged, electric tramway service started to fall. The new was driving out the old. This decline in Tokyo's tramway service is especially striking because it continued even after 1932. In that year, Tokyo's population increased greatly in its great incorporation. By contrast, in post-1925 Osaka, electric tramway service swung up and down. It reached new heights during the 1930s.

In short, Tokyo was outstripping Osaka in the use of motorized vehicles. And it was doing so despite the fact that more factories were concentrated in Osaka than in Tokyo, and despite the fact that—between 1925 and 1932—Osaka was actually more populous than Tokyo. This is even more remarkable in light of the fact that prices for land in the core of Osaka had skyrocketed to dizzying heights during the first upswing of the transitional-growth long swing. Thus, the pressure on residential housing with Osaka was especially acute. The demand for transportation that made convenient commuting from the city's periphery to its core was especially pronounced in Osaka. Since all of these arguments point to stronger demand in Osaka than in Tokyo, it would seem that supply constraints must have been very different in the two locales.

What were these supply constraints? The most obvious factor is the road network. Tokyo had a compelling advantage in roads, especially in wide roads. From Table 5.5, it can be seen that, in 1930, Osaka had almost double the length of roads that Tokyo had but less than half the road area that Tokyo enjoyed. Osaka's roads tended to be extremely narrow. And incorporation of the suburban periphery in 1925 only worsened Osaka's plight. For the paddy fields in the Osaka area were very valuable, and hence farmers had converted as much land as they could to paddy, leaving the scantiest amounts for foot traffic and for the pulling and pushing of carts. Now, in its incorporation of suburbs in 1932, Tokyo also experienced a reduction in average road width as well. Thus, it is evident that Osaka's problem with roads antedated its big

incorporation in 1925. Indeed, as the figures for roads in turn-of-the-century Tokyo and Osaka reveal, Osaka had always had narrower roads than Tokyo.[31]

Indeed, the fact that Osaka's roads were narrow had been a barrier to the diffusion of earlier forms of land transportation. Horse-drawn tramway service had never been a factor in Osaka's pre-1900 transportation, precisely because its roads were so narrow. The narrowness of its streets had even been a barrier to hanging out banners from shop fronts and residences. Between 1869 and 1887, Osaka prefecture issued nineteen regulations restricting the rights of shopkeepers and homeowners to hang out signs, banners, and other objects over the streets.[32]

So roads had acted as a constraint upon Osaka for a long time. Now, the character of its roads was slowing the diffusion of the internal combustion engine within the city. This is clear from Table 5.5, but graphs convey the point even more strongly than do the numbers in the table. As can be seen from comparing Figures 5.10 and 5.11, the development of wide road supply followed completely different trajectories in the two metropolises. In Tokyo, wide road growth was minimal. In Osaka, it is evident that a surge in vehicles put pressure on supply. In the wake of this surge in demand for wide roads, governments—city, prefecture, and nation—responded to this pressure by expanding the wide road network. These violent supply-and-demand fluctuations are not evident in the case of Tokyo. In Tokyo, a scarcity of wide roads does not appear to have been a factor in shaping the diffusion of motorized vehicles.

Because there were more wide roads in Tokyo, it had more cars, trucks, and buses during the 1920s. Therefore, pressure to pave the existing roads was greater in Tokyo than in Osaka. Indeed, by 1938, 52 percent of Tokyo's roads were paved. By comparison, the percentages for the other great conurbations of the Tōkaidō were lower: in Yokohama, the figure was 19 percent; in Nagoya, 20 percent; in Osaka, 45 percent; in Kyoto, 15 percent; and in Kobe, 42 percent. Moreover, 8.5 percent of Tokyo's area was given over to roads. By comparison, 7.6 percent of Osaka's areas were in roads and only 3.7 percent of Kyoto's area was used for roads.[33]

By 1938, Osaka was catching up to Tokyo in terms of paved roads. But the constraint on wide road supply was proving telling, not only for the short run but also for the long run. In comparison to Tokyo's endowment of physical infrastructure, Osaka's was inadequate. Moreover, developing new infrastructure was more costly in Osaka than it was in Tokyo. Timing was not a trivial issue. For instance, after land prices in the core of the great conurbations of the Tōkaidō had shot up during the upswings of the second long swing, the cost of acquiring property and rights of way so that roads could be widened in the center of the cities skyrocketed. So, Tokyo's advantage in already having wide boulevards in the center of the city was of great significance.

Table 5.5

Roads in Osaka and Tokyo, 1898–1938

A. Population, land area, total road length, total road area, and average road width—Osaka and Tokyo cities, 1900 and 1930.

City/date	Population[a]	Area (sq. kilometer)	Roads[b] Area (sq. meter): A	Roads[b] Length (meter): B	Roads[b] Average width: A/B
Osaka, 1900	695,297	55.7	2,220,414	404,966	5.5
Osaka, 1930	2,453,573	178.9	10,496,519	2,533,873	4.1
Tokyo, 1900	1,339,726	81.2	6,515,033	894,022	7.3
Tokyo, 1930	2,070.913	81.2	14,598,007	1,362,816	10.7

B. Growth rates for "wide" road length; growth rates for cars, buses, and trucks combined; and ratio of "wide" roads length, expanded Osaka relative to official Tokyo[b].

Period	Average annual growth rate of "wide" length: Tokyo	Average annual growth rate of "wide" length: Osaka	Average annual growth rate for cars, buses, and trucks combined: Tokyo	Average annual growth rate for cars, buses, and trucks combined: Osaka	"Wide" road-length ratio Osaka/Tokyo (base = 100)
1900–1904	0.23%	2.36%	—	—	67.3
1905–1909	1.07	3.63	n.a.	n.a.	75.0
1910–1914	0.38	–0.24	45.5	79.9	81.8
1915–1919	0.35	4.02	38.5	52.9	82.7
1920–1924	0.63	4.35	20.6	27.8	103.4
1925–1929	4.47	2.38	17.8	20.7	113.0

C. Roads in Osaka's periphery: Indices of road length and percentage distribution by road type in Nishinari-*gun* and Higashinari-*gun* combined, 1901–1924[b].

	Indices of road length (1920–1921 = 100)				Percentages (of total road length)		
Period	National	Prefectural	Local	Total	National	Prefectural	Local
1901–1904	53.4	97.1	61.7	63.7	3.1	9.8	87.1
1905–1909	53.4	97.1	66.2	67.7	2.9	9.2	87.9
1910–1914	49.5	83.0	62.1	62.9	2.5	8.0	82.5
1915–1919	57.3	89.7	66.8	67.9	3.2	8.6	88.2
1920–1924	106.7	157.7	100.8	104.7	3.8	9.6	86.6

D. Roads in Osaka: Indices (1920–1921 = 100) for area and length; and percentages of total area of roads within the city that are city roads, are "narrow," and are paved, 1900–1937[b].

	Indices					Percentages of total area		
	Length							
		City (local) managed roads						
Period	Total in city	"Wide"	Narrow	Total	Area of all roads in city	Percentage in city-managed roads	Percentage that is in narrow roads	Percentage that is paved
	Second period in city's history							
1900–1904	59.2	59.7	15.8	56.8	43.9	87.0	1.9	—
1905–1909	65.5	65.8	26.4	63.2	48.9	87.6	2.8	—
1910–1914	75.9	75.4	46.7	73.5	60.9	87.7	4.3	—
1915–1919	83.8	79.2	107.5	81.1	76.5	87.7	9.0	—
1920–1924	103.2	103.1	99.4	102.9	106.1	81.9	6.5	—
	Third period in city's history							
1925–1929	346.6	157.4	2,904.4	342.4	182.1	82.4	57.2	5.1
1930–1934	337.6	155.8	2,782.5	332.8	205.2	81.7	56.3	8.7
1935–1937	340.4	n.a.	n.a.	n.a.	246.6	85.3	n.a.	13.9

(continued)

Table 5.5 *(continued)*

E. Roads in Tokyo: Indices of area (1920–1921 = 100); average width (meters); and percentages of total road length by type, 1899–1938[b].

	Indices of road area (1920–1921 = 100)			Average width (meters)		Percentage of total road length in types of city roads	
Period	National	Prefectural	Total	City roads	All roads	"Wide"	"Narrow"
First period in city's history							
1899–1904	61.3	16.2	73.3	7.1	7.4	95.4	0.6
1905–1909	79.4	17.8	82.7	7.7	8.1	95.4	0.6
1910–1914	100.6	19.2	92.8	8.1	8.6	95.5	0.6
1915–1919	98.2	19.2	94.1	8.2	8.7	95.6	0.6
1920–1924	100.0	101.6	101.2	8.4	9.1	91.2	0.0
1925–1931	113.5	125.8	128.4	9.0	9.8	n.a.	n.a.
Second period in city's history							
1932–1938	318.2	876.9	500.4	5.1	5.7	n.a.	n.a.

Sources: Same as sources for Table 5.3.

Notes: [a]1900 population figures are actually estimates for 1895 and in the case of Osaka are based on boundaries established in 1897.

[b]Roads are classified according to whether they are constructed and maintained by the national, prefectural, or local (city, town, or village) authorities. In general, prefectural roads were supposed to be equal to or greater than three *ken* in width (one *ken* = 1.82 meters), prefectural roads were equal to or greater than two *ken* in width, and local roads were either classified as "wide"—one *ken* or over in width—or "narrow," that is, under one *ken* in width. For the purposes of this table, I define "wide" roads as those that are one *ken* or over in width. The term "expanded Osaka City" refers to the city with its boundaries as expanded in the incorporation of 1925 (cf. Table 5.3 and notes to the table). "Paved" refers to both pavement by brick and by asphalt. Panel D refers to Osaka within its officially defined boundaries.

n.a. = not available.

Figure 5.10 **"Wide" Road Length and Cars, Buses, and Trucks: Growth Rates in Expanded Osaka City (Inclusive of Nishinari-*gun* and Higashinari-*gun* Before 1925), 1902–1933** (Five-Year Moving Averages)

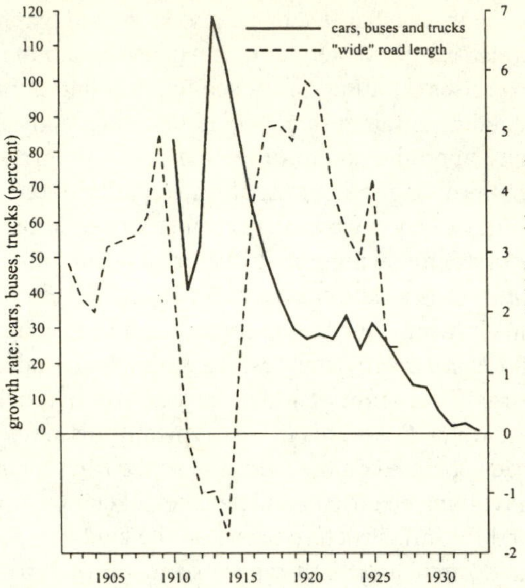

Figure 5.11 **"Wide" Road Length and Cars, Buses, and Trucks: Growth Rates in Tokyo City, 1905–1929** (Five-Year Moving Averages)

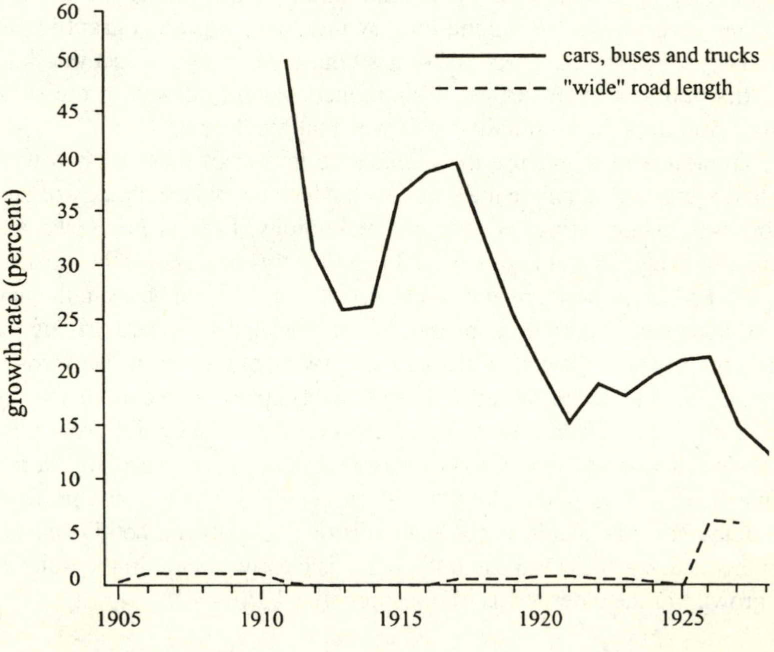

The Geographic Pull Toward Tokyo

Economic geography matters for economic development. And one of the reasons geography matters is that it shapes, and is shaped by, infrastructure. Ecology shapes infrastructure, which, in turn, shapes economic growth. The resulting growth exercises a pull upon subsequent buildup in infrastructure.

How environmental constraints shape economic geography over the long run depends critically upon the nature of technological progress. As the first and second and then third long swings transformed the Japanese economy, the nature of the technology of energy production and transportation was radically altered. During the first long swing, the diffusion of steam power and steamships was key. Water transportation and water-based infrastructure was favored during this long swing. But, in the second and third long swing, there was a decisive shift toward energy sources, machines, and modes of transportation that mainly used land and not water. To be sure, hydroelectric power was generated from water flow. But it was typically distributed across the land. Electric railroads operated on the land, and so did trucks, buses, and cars. Thus, innovations involving electricity and the internal combustion engine created a strong demand for infrastructure based on the land.

Within the Tōkaidō industrial belt where so much of Japan's industrial might was concentrated, Osaka had a competitive advantage in developing infrastructure that used water. By contrast, Tokyo had a competitive advantage in developing infrastructure that used land. For this reason, as the first long swing gave way to the second long swing, there was a geographic pull toward Tokyo. And as the second long swing gave way to the third long swing, this pull was even further strengthened. Each pull was in the same direction. And thus the cumulative pull was that much greater.

The simplest way to capture the cumulative impact of these pulls toward Tokyo and Kantō is to examine the speed with which the old technology of land transport was disappearing. By the old technology, I mean *jinrikisha* and *niguruma*. In Table 5.6 and Figure 5.12, I examine this process for the Tōkaidō as a whole, and for its northern and southern districts.[34] As can be seen, the new modes of transport—railroads, subways, buses, trucks, cars—were driving out the old forms. Up until the end of the second upswing of the transitional-growth long swing, the older forms of transport grew with expansion in overall activity. However, after the industrial upswing of the 1910s when they did grow positively, their ebb was decisive. Growth was always negative, especially during the downswing of the 1920s. Growth was especially negative in the northern district of the Tōkaidō, that is, in the Kantō district. Tokyo's competitive advantage on the land vis-à-vis Osaka and the Kinai is clearly evident in the trajectories of growth for the older forms of transport after 1920.

Figure 5.12 ***Jinrikisha* and *Niguruma* Combined: Growth Rates in the Northern and Southern Districts of the Tōkaidō, 1902–1935** (Five-Year Moving Averages)

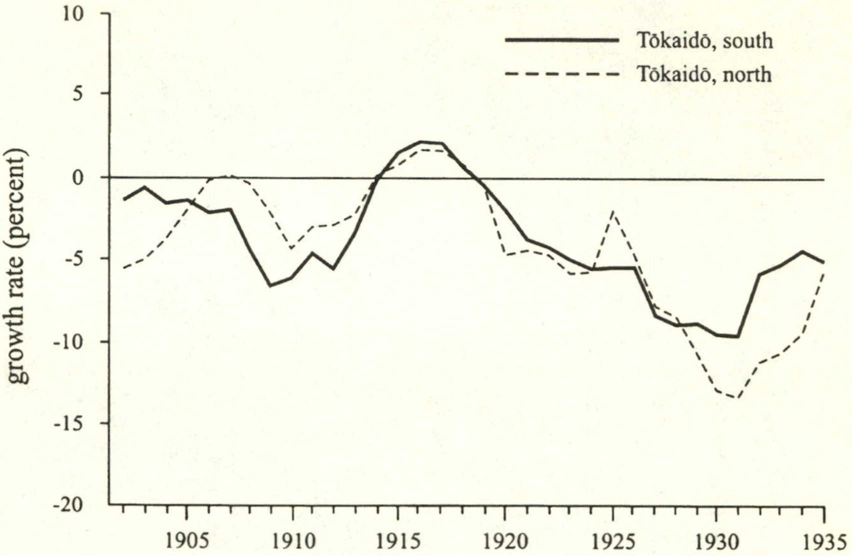

Physical infrastructure construction, and the expansion of energy delivered and transported using this infrastructure, paved the way for the two great industrial expansions of the 1910s and 1930s. Electricity and the electric railroad were key to the first industrial expansion; and roads, motorized vehicles, and electric power grids were key to the second industrial expansion. The main focus of this infrastructure construction was on investing in structures and equipment that used, or were situated upon, the land. And because Tokyo had a competitive advantage in developing infrastructure rooted on the land, there was a strong geographic pull toward Tokyo that occurred during the second and third long swings.

The Tokyo-centrism that characterizes contemporary Japan is due to a variety of factors. Infrastructure is basic to it. And—while Tokyo also developed a competitive advantage over Osaka in financial and in human capital–enhancing infrastructure after the beginning of the second long swing—Tokyo's advantages in twentieth-century physical infrastructure investment are perhaps the single most important factor of all. In terms of physical infrastructure creation, the contest between Osaka and Tokyo over which metropolis was to be the core of the core boils down to the following statement: Osaka won the battle on the water, only to lose the war on the land.

Table 5.6

Old and New Transportation: The Pace of Conversion in Japan, in the Tōkaidō, and in Osaka, 1900–1938

A. *Jinrikisha*, *niguruma*, and *kobune* in the Tōkaidō: Indices (1920–1921 = 100), 1900–1938[a].

	Jinrikisha				*Niguruma*				*Kobune*			
		Tōkaidō				Tōkaidō				Tōkaidō		
Period	Japan	Total	North	South	Japan	Total	North	South	Japan	Total	North	South
1900–1909	162.6	188.3	161.8	207.4	64.1	71.0	69.3	67.1	266.6	184.8	210.0	131.7
1910–1919	111.9	113.0	99.3	116.0	85.6	88.5	89.1	84.9	168.4	132.3	146.5	100.0
1920–1929	70.5	69.8	61.0	68.9	99.3	97.5	92.9	96.9	86.8	83.8	76.0	86.9
1930–1938	23.3	17.9	14.8	18.0	74.2	57.3	50.4	63.1	77.5	70.3	65.4	69.7

B. Cars, Buses, and Trucks in the Tōkaidō: Indices (1920–1921 = 100) and percentages in Tōkaidō, and within Tōkaidō in north and south, 1913–1937[a].

	Indices				Percentages		
		Tōkaidō				Of Tōkaidō in district	
Period	Japan	Total	North	South	Tōkaidō as percentage of Japan	North	South
1913–1919	24.9	27.6	30.0	23.4	79.8	73.7	22.0
1920–1929	348.2	303.1	287.4	291.4	61.9	61.8	28.1
1930–1937	1,418.3	1,271.0	1,108.4	1,304.9	62.1	57.0	28.5

C. Indices (1920–1921 = 100) for transportation vehicles in expanded Osaka City, 1900–1935[b].

Vehicle type	Period						
	1900–1904	1905–1909	1910–1914	1915–1919	1920–1924	1925–1929	1930–1935
Kobune	98.5	83.2	81.3	91.8	100.1	97.3	77.1
Jinrikisha	320.7	263.4	144.0	106.5	86.5	44.2	21.1
Niguruma	63.5	74.7	87.1	96.5	98.2	86.5	n.a.
Cars, buses, and trucks	—	—	1.7	33.6	146.7	508.2	861.9

Sources: Nihon Naikaku Tōkei-kyoku (various dates), *Nihon teikoku tōkei nenkan*, various tables; Ōsaka-fu (various dates), *Ōsaka-fu tōkeisho*, various tables; and Ōsaka-shi Yakusho (various dates), *Ōsaka-shi tōkeisho*, various tables.

Notes: [a]The "north" district of the Tōkaidō consists of the prefectures of Saitama, Chiba, Tokyo, and Kanagawa and the "south" district of the prefectures of Kyoto, Osaka, and Hyogo. See Table 3.4 and notes to that table.

[b]Expanded Osaka City is defined as the area inclusive of the incorporation of 1925. See Table 5.3 and notes to that table. The series for cars, buses, and trucks commences in 1902.

Notes

1. The figures are drawn from Ramseyer and Rosenbluth (1995, 120). Ramseyer and Rosenbluth and Ericson (1996) are fairly dismissive of the contribution of Japan's railroads to early Japanese economic growth. For a positive view of the railroad's contribution, see Minami (1965, 1986, 1994).

2. See Ericson (1996, 375–77) and Ramseyer and Rosenbluth (1995, 120). Ericson argues that it was not until 1912 that trains pulled ahead of boats for the conveyance of domestic freight.

3. See Ramseyer and Rosenbluth (1995, 127 ff.). Ramseyer and Rosenbluth argue that Japanese politicians were prisoners of a pork barrel logic in dispensing funds for new railroad construction. They were chiefly concerned with serving the interests of the constituents of their own districts. That this was true does not undermine the argument concerning the importance of scale economies in the Tōkaidō industrial belt for stimulating railroad construction: Construction of the intercity and intracity electric railroad lines that was a key part of the infrastructure buildup of the 1904–1911 period was undertaken by private parties and not by the national government, which lavished its attention on the steam railroad network.

4. The account here relies heavily upon Sawai (1985, 1998).

5. It should be kept in mind that railroadization occurred far earlier than did electrification in North America and the more advanced countries of Europe. In Japan, the major expansion of railroads took place simultaneously with electrification, reflecting the fact that Japan was catching up technologically with the West at a rapid pace. For further comments on this point, see appendix A.1 in chapter 1.

6. For this paragraph, I draw heavily upon Kansai Denryoku Kabushiki Gaisha (1978, 3 ff.).

7. See Kansai Denryoku Kabushiki Gaisha (1978, 9).

8. According to Trewartha (1945, 273–74), Osaka and Kobe depended upon thermal electricity to a great extent, while Tokyo and Yokohama relied more upon hydroelectric power. To some extent, this dependence reflects geography. The distance from the Japanese alps, where most of the generating stations were located, to the Kantō region was far shorter than the distance from the alps to the Kinai district.

9. See Kansai Denryoku Kabushiki Gaisha (1978, 17).

10. My account of the intercity electrical railroad and tramway is heavily influenced by the account of McKay (1976) for Europe.

11. I have been able to find only one reference to a horse-driven tramway in Osaka. This reference is to a line operating from the front of Tenōji Nishimon. The company is called Osaka Basha Tetsudō (Osaka Horse Railway). See Hanshin Denki Tetsudō Kabushiki Gaisha (1985, 583). It appears that the line was insignificant, for its name does not surface in any of the standard works concerned with railroad development in the Osaka area.

12. For details on these lines, see the accounts in Hanshin Denki Tetsudō Kabushiki Gaisha (1959), Hanshin Kan Modanizumu Ten Jikkō Iinkai (1985), Keihan Shinkyū Kō Denkitetsu Kabushiki Gaisha (1959), Kinki Nihon Tetsudō Kabushiki Gaisha (1960), Mizutani (1994), Nankai Denki Tetsudō Kabushiki Gaisha (1978), Nawa (1929), Takechi (1980), and Taku (1936a). Osaka's financial community was heavily involved in raising the funds bankrolling these enterprises.

13. Keihan's original terminus in Osaka was to the east of its present site at

Yodoyabashi. In contemporary Japan, Kintetsu (originally Osaka Kidō) has its terminus in Osaka at Nanba.

14. See the discussion in Ōsaka-shi Denki-kyoku (1990, 7 ff.). Also see Ōsaka-shi Kōtsū-kyoku (1980a, 48 ff.). Some lines managed by the city—nineteen in total—were constructed as part of special projects outside of those conceived in the distinct phases of expansion discussed in the text.

15. In 1902, Osaka City began experimenting with compensation contracts. It awarded monopolistic rights to run the city's water taxi system to a firm. In exchange, the company paid the city a portion of its revenues, paying compensation for its right to provide the service. In the course of buying out Osaka Dentō, the city allowed a private company, Daidō, to acquire some of the physical assets that had been owned by Osaka Dentō. Daidō secured the generating plant on the Aji River, for instance.

16. These figures are taken from Hanshin Denki Tetsudō Kabushiki Gaisha (1985, 189).

17. See Hanshin Denki Tetsudō Kabushiki Gaisha (1985, 9).

18. Tokyo took the lead in putting in subway lines. It completed its first line in 1927. Osaka's first subway line, connecting Umeda to Shinsaibashi, was not completed until 1932.

19. See Trewartha (1945, 105).

20. For this point, and for the discussion in the text that follows the making of this point, I rely on Kansai Denryoku Kabushiki Gaisha (1978, 37). Daidō started out as Nagoya Dentō in 1887. It emerged as a separate corporate entity when the assets of Osaka Dentō were divided up in 1918. Afterward, during the 1920s, Daidō merged with Osaka Sōden, Hokuriku Denka, and Osaka Denki (the former Osaka Dentō).

21. After the war, this company was denationalized and became a private enterprise. During the period 1938–1945, other mergers guided by the hand of the national government occurred. For instance, in the field of intercity railroads, some lines were merged in order to create scale economies. The Kintetsu railway was basically formed through the merger of a variety of smaller firms during this epoch.

22. See Shinshū Ōsaka-shi Shi Hensan Iinkai (1993), *Ōsaka-shi shi*, vol. 6.

23. See, for example, the maps in Keihan Shinkyū Kō Denkitetsu Kabushiki Gaisha (1959), and in Kinki Nihon Tetsudō Kabushiki Gaisha (1960).

24. See the discussion in chapter 3, and in chapter 7 below. Bestor (1989) discusses applications of the law to land conversion in the hinterland of Tokyo. The legislation went through a number of revisions during the period between the world wars.

25. The most dramatic and compelling example of Tokyo's advantage in expanding out into cheap dry field and wasteland is afforded by the construction of the Chuo line running from central Tokyo out to the western reaches of the prefecture. In sharp contrast to the railway lines that crisscross the Kinai, the Chuo line track runs straight. Because land was so cheap on the Kantō plain, it was possible for the government to buy up huge swaths along a straight line. But in the Kinai, the tracks weave, reflecting the fact that when they were being laid out, the gradient in land prices was steep. Thus the companies tended to buy up the cheapest land and did not heed the fact that they would be creating a meandering path for the resulting rail line. I am grateful to Saito Osamu for this point.

Disputes between landlords and tenants over tenant rights may have slowed conversion of land in the Osaka area as well. Tenancy was especially prevalent in the Osaka environs (cf. panel B of Table 3.3). On disputes over land rights, see Mosk (1996, 2000a).

26. See Nakagawa (1929, 112–3). In the reconstruction following the earthquake of 1923, Tokyo substantially expanded the area of the city given over to roads, largely by widening existing thoroughfares.

27. See Nihon Noriai Jidōsha Kyōkai (1957, 16).

28. This quotation appears in Stewart (1987, 86).

29. Ōshima (1992, 95) provides interesting data on the use of trucks and trains for moving military freight in the Japan of 1936. For short hauls, freight was almost exclusively moved by truck. Long-distance hauls were mainly accomplished by railroad.

30. On the development of private bus lines, see Nihon Noriai Jidōsha Kyōkai (1957). Testifying to growing interest in bus lines during the 1920s, Japan National Railroad carried out a nationwide survey of persons using its train stations in 1925. The purpose of this survey was to determine which stations could be, and should be, served by bus lines. Within Osaka, private bus service commenced in 1924. The "blue line" company began operating a fleet of eighty Model T Fords in that year. In 1926, the city entered the business with an eye to serving the forty-five villages and towns that it had incorporated a year earlier. In 1940, under the exigencies of wartime rationalization, the city bought out the blue line company. See Ōsaka-shi Kōtsū-kyoku (1980b, 8 ff.).

31. Prewar roads were of three types: relatively wide national roads; moderately wide roads managed by the prefectures; and narrow local roads that were managed by cities, towns, or villages. See the notes to Table 5.5 for details.

32. See Ishida (1987, 150).

33. See page 88 in Ōsaka-shi (1954), *Shōwa Ōsaka-shi shi. Keizai hen*: vol. 2.

34. In constructing the index, I computed separate indices for *jinrikisha* and *niguruma*. I weighted each of these indices by one-half. Then, I added the resulting weighted figures together to generate a composite index.

6

Factories

The Product Cycle

Innovation and rapid diffusion of innovation underlie Japan's industrial success. This point is key to the ongoing debate about whether or not government intervention in the Japanese economy has helped or hindered industrial expansion. When government policies have encouraged innovation and its diffusion, they have assisted the growth of industry. When they have discouraged innovation or its diffusion, they have stifled industrial expansion.

Now, in speaking of government policies, one must be specific. What government are we discussing? What type of policies are we considering? Are we taking into account the possibility that policies and practices adopted by the government (or various levels of government) actually emanate from the enterprises in the industries for which the policies are drawn up? In the case of "bottom up" policy or administrative procedure—that is, policies that emerge out of the industries themselves—is it accurate to refer to the resulting policies as creations of the government sector?

The viewpoint developed in this volume leads to a number of conclusions about these questions. Let us briefly discuss them here. In the remainder of the chapter, I will provide concrete examples supporting the assertions. The first conclusion is that policies affecting industry are made at all levels, both national and local. In particular, the role of municipal governments should not be ignored. The second conclusion is that the most important policies developed by governments in Japan have not been those directly linked to industry. Rather, they have been policies involving investment in, and operation of, the infrastructure—financial, human capital–enhancing, and physical—that has supported innovation and its diffusion in manufacturing. In short, the most important industrial policies are: stimulating credit creation

by shoring up banking and encouraging household saving; developing the educational system so that demand for skilled labor does not substantially outstrip supply; and building the physical infrastructure that supplies energy, supports domestic and international trade, and promotes innovation through spillovers.

What about policies directly targeted at manufacturing? At the national level, one must consider the following: direct subsidies, either for a specific industry or for a designated subgroup of enterprises within the industry; creation of standards or sharing of information; imposition of tariffs on competing imports; exchange rate policy that has an impact on the terms of trade; and various types of coordinating and facilitating policies aimed at rationalizing production, removing gluts from the domestic market, and so forth. At the local level, it is clearly impossible to impose tariffs and use monetary policy to affect the real exchange rate. However, it is possible to implement the other policies. Of these, the coordinating and facilitating function has been most productive, in my opinion. This is because the most effective policies for directly promoting industrial activity have been those that share both "bottom up" and "top down" elements.

It is important not to lose sight of the point that diffusion of innovation is key to industrial development in Japan. Much of the innovation in modern Japan has taken place because firms and individuals have embraced innovation on their own, without the direct encouragement, knowledge, or cajoling of government. In certain cases, one can see that government has had a positive influence. In other cases, it would appear that the opposite is the case. How this all adds up is difficult to assess. Perhaps it is impossible.

One simple way of trying to get an overview of the role that government may have played in Japan's industrial history is to examine the product cycle: The relationship between the net trade position of a sector in the economy—that is, exports minus imports for the sector—and the development of the sector. The idea is that an innovating sector that borrows many of the ideas for innovation from enterprises in other countries (perhaps importing the machinery to implement these innovations from these same companies) starts out at a disadvantage relative to the enterprises it is borrowing from. So, at first, the innovating sector is not competitive with the sector in the foreign country from which it borrows. Imports exceed exports. Foreign producers undercut the domestic market. But as the innovations are diffused throughout the innovating sector—and as workers and managers become more efficient at working with the new techniques through repetition and experimentation (known as learning by doing)—the domestic sector is able to accomplish thoroughgoing import substitution. Imports fall; exports rise.

Eventually, the sector becomes a net exporting sector. Finally, the sector becomes backward in the sense that the pulse of innovation shifts to newer sectors. It goes into decline, perhaps because other, less developed countries take it up. And so, once exports and production fall, imports rise. Once again, the sector is in a net importing position.

To summarize: In the product cycle, there are four phases. A sector starts out in a net import position. Then, import substitution occurs. This is followed by a phase when exports exceed imports, and the country is a net exporter. Finally, the sector goes into decline, loses its grip on the domestic resources of capital and labor, and eventually even loses its grip on the domestic market. Imports surpass exports.

Let us relate the product cycle to key subsectors of manufacturing over successive phases of the prewar long swings. In Table 6.1, I provide relevant figures for textiles—the classic light-manufacturing sector that was especially innovative during the first long swing—and two heavy subsectors, chemicals and machinery. In making the transition from balanced to unbalanced growth, the engine of Japanese manufacturing shifted from the light industries to the heavy industries. Machinery is the classic subsector of this period. It is also a major source of growth during the postwar upswing between 1953 and 1969.

Has the product cycle been closely linked to the upward and downward movements in output growth and capital formation associated with the successive phases of the long swings? Or has it been basically shaped by trends? From panel A of Table 6.1, it is evident that the absolute growth of textile production and exports moves with upward and downward swings in the growth rate during the long swings. The picture for imports is less clear. As for the phases of the product cycle, it appears that trend is far more important than is fluctuation over the long swings. Indeed, from panel B it is clear that the ratio of net exports to domestic production steadily rises over time. The sole exception is the downswing phase of the transitional growth long swing, when the net export ratio drops somewhat from the level established for 1911–1919. Thus, during the balanced-growth long swing, textiles successfully managed import substitution. And during the transitional-growth long swing, the industry increasingly staked out a position as a net exporter.

The importance of textiles production for Japanese manufacturing output throughout the prewar period should not be underestimated. As panel B of Table 6.1 shows, the share of manufacturing output attributable to textiles varied between 25 percent and 30 percent. There is a steady trend toward increasing shares over time as well. Thus, domestic and international demand for textiles was of great importance to Japan's industrial growth, and to its success in exporting manufactures.

Table 6.1

The Pace of the Product Cycle in Prewar Japanese Manufacturing

A: Growth rates for domestic production (DPG), exports (EXG), and imports (IMG) in subsectors of manufacturing (textiles, chemicals, machinery), and in manufacturing as a whole (calculations based on seven-year moving averages for figures given in 1934–1936 prices).

	Textiles			Chemicals			Machinery			Manufacturing		
Long-swing phases	DPG	EXG	IMG	DPG	EXG	IMG	DPG	EXG	IMG	DPG	EXG	IMG
1887–1897 (U)	9.4%	11.5%	3.6%	4.4%	10.1%	15.5%	11.3%	76.3%	14.5%	5.5%	10.9%	9.8%
1897–1904 (D)	0.7	6.1	–1.2	4.3	7.8	11.9	12.9	40.9	12.3	2.1	7.9	3.1
1904–1911 (U1)	6.9	8.1	–5.3	5.0	7.3	6.7	11.7	14.7	1.6	5.0	8.2	1.2
1911–1919 (U2)	7.7	6.8	2.8	6.6	5.9	14.2	15.5	12.8	3.3	7.4	6.3	6.7
1919–1930 (D)	6.0	6.3	7.0	8.2	4.9	5.6	2.6	12.3	0.3	4.9	6.2	4.7
1930–1938 (U)	5.9	7.2	–4.2	13.0	6.3	2.7	19.4	24.3	0.5	9.0	10.1	3.9

B: Composition of domestic manufacturing output (% of manufacturing in textiles [Tex%]; in chemicals [Chem%]; and in machinery [Mac%]). And levels and per annum growth rates for the ratio of net exports/domestic product (Tex = textiles; Chem = chemicals; Mac = machinery; and Man = manufacturing).

	Percent of manufacturing output			Net exports/domestic production, level				Net exports/domestic production, growth rate			
	Tex	Chem	Mac	Tex	Chem	Mac	Man	Tex	Chem	Mac	Man
1887–1897 (U)	24.3%	8.7%	1.6%	0.04	0.5	–21.2	–0.4	–46.5%	–0.8%	2.8	3.2
1897–1904 (D)	25.4	9.1	3.3	1.3	0.3	–18.6	–0.4	16.3	–12.3	–9.7	–9.3
1904–1911 (U1)	25.5	9.6	6.0	2.1	0.2	–8.8	–0.1	5.6	14.6	–9.6	11.3
1911–1919 (U2)	27.5	9.4	10.9	3.0	0.1	–3.2	0.6	0.1	464.6	–9.9	3.3
1919–1930 (D)	29.3	10.1	11.1	2.7	–1.4	–2.6	0.2	0.3	1.0	–5.1	192.9
1930–1938 (U)	29.5	14.3	13.7	3.1	–0.4	–0.1	0.9	2.0	–257.1	22.6	9.5

Sources: Ohkawa and Shinohara with Meissner (1979), pp. 302–4 and 323–29.

In terms of *rates* of growth heavy manufacturing subsectors had the distinct edge. This is especially true for machinery. With the sole exception of the downswing of the 1920s, machinery production growth rates are consistently above 10 percent, while growth rates for textiles generally fall below 8 percent (the upswing of the balanced-growth long swing being the sole exception). Along with shipbuilding and railroad rolling stock manufacturing, machinery is extremely dynamic, especially during the transitional growth and unbalanced-growth long swings.[1]

It is interesting to compare the product cycle for machinery to that for textiles. As Table 6.1 shows, during the first long swing, machinery is a significant importer of foreign production. But there is a clear trend toward import substitution. By the upswing of the third long swing, net imports relative to domestic production is insignificant. So, machinery is following the path laid out by textiles. It is also apparent that the pace at which manufacturing proceeds through the product cycle is linked to the upswings and downswings in the overall growth of gross domestic product. During industrial upswing phases—1887–1897, 1911–1919, and 1930–1938—import of machinery soars. Thus the growth of import substitution is slowed down. During downswings, the reverse is the case. This pattern is very different from that evident for textiles. And the reason is important: Most of the machinery produced during the prewar period is used as capital goods in manufacturing. Machines are used to generate output. Thus, as industrial production soars during upswings, the demand for machinery picks up. If domestic production of machinery fails to satisfy this voracious demand, manufacturers seeking equipment must seek out suppliers in other countries. Import substitution in machinery production reduces this dependence on the foreign sector. For this reason, there is an inverse relationship between trends in the growth of imports and the growth of domestic production for the machinery sector. Over time, import growth declines; domestic product growth increases.

How important is government policy for the product cycle? And what policies are important? To shed some light on this question, let us briefly review salient features of innovation in three sectors of Japanese manufacturing: Textiles, shipbuilding, and railroad rolling stock manufacture. The questions we will address are as follows: Did government programs specifically targeted at the sector and designed to promote import substitution play a significant role in the development of these sectors? How important were these programs? To what extent did the programs emanate from the "bottom up"? Or, compared to these approaches, was the government's role in creating infrastructure more important?

Let us first consider textiles. During the 1880s, the national government attempted to create industry wide guilds (*dōgyō kumiai*) in textiles. The

government's stated reason was to encourage the import of Western technology in the industry. Officials reasoned that scale economies in securing and adapting foreign technology were crucial. Unlike a single company, an industry-wide guild could raise the funds required for bankrolling missions seeking out technical information in foreign countries.

The government's effort failed. The fact that the industry was profoundly divided stymied it. On one side were a few large steam-powered factories; on the other side were myriads of small producers, organized under the putting-out system. Merchants were dominant in the latter sector. They resisted the government's efforts since they felt they might lose their market niche if they joined an industry-wide guild. Organized into an overarching structure that could potentially market their output, the cottagers on whom they depended might find alternative ways of securing raw materials and selling product. In any event, the cottage industry was scattered around in farming villages. It was very fragmented, and thus hard to bring under a single umbrella.

However, the large steam-using enterprises were interested in forging cooperative arrangements among themselves. They had an incentive to do so: exploiting scale economies in ferreting out Western technology and organizational methods in cotton spinning and weaving. Thus, they banded together to form their own guild, the *boseki rengōkai*. Basically, it operated as a cartel for the large integrated spinning and weaving enterprises. Because the great majority of these companies had established their factories within the environs of Osaka, it was easy for their executives to hold meetings. They set up study groups and sent fact-finding missions abroad. And they were innovative in other areas. They established rules for jointly curtailing production when gluts developed in the market.[2] In short, during the first long swing, the government's effort to organize the key industrial sector, cotton textiles, fell flat. A cartel-like structure did develop within one subsector of the industry, but this happened because the firms believed creating a cartel was in their interests.

Government-inspired import-substitution polices during the first long swing were successfully implemented in the shipbuilding sector, however. Under the aegis of the Shipbuilding and Navigation Promotion Law of 1895, the government attempted to reduce the dependence of shipping companies on foreign suppliers of ships with large steam engines and steel hulls. The government's aim was to create a merchant marine that could meet the demands of transoceanic voyages and that was domestically supplied with vessels. Under this policy, the increased volume of shipping handled by Japanese trading companies translated into increased orders for domestic shipyards. As a result of World War I embargoes on shipping suffered by the European

powers, Japan emerged as the third-largest maritime nation in the world. A case in point is the Osaka Shosen Kaisha (O.S.K.) operating out of Osaka.[3] During the course of hostilities, it managed to open up twelve ocean routes and five near seas services. As the O.S.K. and other shipping lines expanded the scope and scale of activity, Japan's dockyards saw their business boom. This was especially true during the upswing of 1911–1919.

Just as World War I spurred demand for domestic shipbuilding, so conclusions of hostilities brought on retrenchment. Two factors were decisive in putting the brakes on expansion of the industry: the resuming of international competition in shipbuilding and transoceanic trade, and the Washington Agreement on Naval Disarmament that took effect in 1921. The latter agreement set limits on the number of, and capacity of, capital ships permitted to Great Britain, the United States, and Japan.[4] After 1930, shipping and shipbuilding revived. During the upswing of the unbalanced-growth long swing, new orders multiplied rapidly.

How successful was import substitution in the industry? In 1919, fifteen steamships with a gross tonnage of 947 tons were imported. In that year, 323 steamships were produced by domestic dockyards. The gross tonnage of the vessels manufactured domestically was almost 640,000 tons. In short, in the last year before international competition in shipbuilding resumed in earnest, imports were insignificant relative to domestic production. But with the resumption of international trade, the number of imported steamships jumped. In 1925–1929, twenty-two steamships with a gross tonnage of around 65,000 tons were imported. Domestic production consisted of 104 vessels with a gross tonnage of about 69,000 tons. In short, although Japanese yards had learned how to work with iron and steel, they had not yet achieved full import substitution. Japan's naval buildup during the 1930s renewed the demand growth they had experienced during World War I. Import substitution promoted by the government was a mixed success in this industry.

In the case of railroad rolling stock, the case for government-sponsored import substitution is stronger. In chapter 5, we discussed the reverse engineering project initiated by Japan National Railroad under the rubric of national sufficiency. As a result of this policy, imports did drop. For instance, in 1920, Japan imported rolling stock amounting to about 4 million yen; by 1930, this figure had dropped to less that 9,000 yen.[5]

In short, in heavy industry there is a stronger case to be made for government intervention aimed at stimulating import substitution. But, even in the heavy industries, growth in demand was crucial to growth in production, and hence to the pace of learning by doing in the industries. Demand for railroad rolling stock was strong in the 1920s because domestic railway companies were building up their stocks during that period; and it was strong because

there was demand in Japan's colonies and in the Manchurian railroad network that it had secured special rights over.

In the case of machinery, the surge in demand is certainly paramount. Electrification played a major role in stimulating demand because it increased the level of mechanization in factories large and small. And demand growth was also strong because heavy industries were more capital-intensive than light industries, and they required more machinery. To move into the nascent heavy industries was to innovate. Successful innovation garnered a healthy flow of profits. Meeting surging growth in demand for machinery from the heavy industries was key to machinery sector expansion. Policy had little to do with this.

Thus, structural change driven by private profit-oriented innovation was far more important than industrial policy for the growth of the industrial sector. The rush into the heavy industries depended very heavily on electrification because, in the heavy industries, small firms engaged in doing subcontracting for larger firms proliferated during the industrial upswing of the 1910s. This would not have happened had small factories not mechanized, and mechanization would not have happened without electrification. It was use of electrical motors that made economical the mechanization of small factories. Consider the figures given in Table 6.2. The rapidity of the diffusion of electrical motors is readily apparent.[6] By 1930, over 80 percent of all motors used in Japan were electrical. Use of electric motors and mechanization were strongly correlated. For instance, among small factories—those with labor forces ranging from five to twenty-nine workers—less than a quarter were mechanized in 1909. But as a result of the buildup of electrical infrastructure and power-generating capacity, mechanization of small factories soared. In 1930, the percentage of small factories that were mechanized jumped to over 75 percent.

The rapid expansion of heavy industry in Japan during the second long swing transformed the social nature of production. It did so by shifting the weight of industrial employment away from textiles, food products, and the other subsectors of light manufacturing toward heavy industry. In Osaka, the shift toward heavy industry was especially pronounced because, during the 1920s, employment in light manufacturing actually declined in the city and its environs. By contrast, employment in heavy industry in and around Osaka grew during the 1920s, and expanded especially rapidly during the 1930s. Consider the trajectories for employment growth in light and heavy manufacturing for Osaka and Tokyo prefectures depicted in Figures 6.1 and 6.2. As can be seen from the changes in employment, the structural shift toward heavy industry was overlaid on the long-swing pattern. Within the two greatest conurbations of the Tōkaidō, light manufacturing experienced a much sharper downturn than did heavy manufacturing.

Skill requirements for blue-collar workers in light and heavy industry generally differed. In light manufacturing, work was extremely repetitive and a new worker could acquire a reasonable level of skill fairly quickly. Monitoring effort and output was relatively easy. For instance, girls in the textile mills were routinely tested as to the speeds that they could run weaving and spinning machines. Thus, departing workers could be easily replaced. By contrast, in heavy industry, technology was sophisticated and innovation was more frequent. New products were constantly being developed. The nature of tasks was in constant flux. A worker in heavy industry was expected to master a broad range of skills, and to switch assignments frequently.[7] Turnover was a problem in heavy industry since departing workers usually took valuable skills with them, and replacements had to be trained at considerable cost.

Thus, light industry contented itself with a labor force largely consisting of young girls recruited out of the countryside. A typical factory recruit in textiles worked one or two years, and then returned to the farming village from which she had been recruited. Turnover was high in this industry. By contrast, in heavy industry, turnover was relatively low: firms sought to hire males eager to be trained and employed for long periods of time. Workers tended to stay with enterprises in heavy industry.

Thus, as manufacturing swung toward heavy industry, the industrial labor force increasingly became masculine. The trend away from a labor force principally composed of women is evident for Osaka in the figures assembled in Table 6.3. In judging the trends indicated by these figures, it should be kept in mind that the underlying data are for the city of Osaka defined in terms of its official boundaries. Thus, workers in factories located in the region that eventually became the zone of the city I call the periphery zone (namely, Higashinari-*gun* and Nishinari-*gun* combined) are not included in the estimates given here for the years prior to 1925. This is important because textile mills and proto-industrial enterprises in textiles, which mainly employed females, were mainly concentrated in this periphery zone. For this reason, the proportion of the labor force that is female actually jumps for the city with its official boundaries in the year incorporation of the periphery takes place. Prior to incorporation, there is a clear trend toward a masculine labor force.

The impact of the proliferation of subcontracting accompanying the shift from light to heavy industry upon average factory size is made evident in Table 6.2. There is a clear downward trend prior to 1925. In 1925, average factory size drops to an even lower level, presumably because proto-industrial producers of yarn, cloth, and clothing are very active in the periphery.

Table 6.2

Light and Heavy Industry: Factory Size and Regional Concentration, Japan, the Tōkaidō, and the Prefectures of Osaka and Tokyo, 1909–1940[a]

A. Average factory size (workers per factory): Japan, the Tōkaidō, and subdistricts of Tōkaidō.

	Light industry[b]				Heavy industry[c]			
		Tōkaidō				Tōkaidō		
Period	Japan	Total	North	South	Japan	Total	North	South
1923–1925	38.5	38.0	40.1	42.7	122.7	42.1	39.6	45.6
1936–1938	24.1	23.4	22.6	25.1	128.7	37.1	38.1	34.7

B. Percentage of manufacturing workers and output in regions: of Japan in Tōkaidō; of Tōkaidō in subdistricts; and ratios of Osaka to Tokyo (prefecture) with Tokyo base = 100.

	Workers				Output (in value terms)			
Period or average of individual years	Percentage in Tōkaidō	% of Tōkaidō in: North	% of Tōkaidō in: South	Ratio of Osaka to Tokyo	Percentage in Tōkaidō	% of Tōkaidō in: North	% of Tōkaidō in: South	Ratio of Osaka to Tokyo
1909, 1914	53.2%	30.5%	49.7%	110.6	67.9%	27.6%	59.6	136.0
1919–1922	54.6	32.0	47.7	117.1	61.5	32.7	50.6	128.0
1939–1940	60.0	41.6	38.6	77.0	62.9	44.8	39.5	81.2

C. Powered factories as % of all factories, and composition of total manufacturing power capacity by type of engine (percentage of all power capacity), Japan.

Powered factories: % in factory size group				Composition of manufacturing power capacit			
	Number of workers						
Year	5–29	30–99	100 and over	Year	Steam engines and turbines	Internal combustion engine	Electric motors
1909	20.5%	69.7%	88.4%	1910	80.6%	6.6%	21.8%
1919	54.3	88.6	99.3	1920	31.3	3.4	58.9
1930	80.1	95.3	99.6	1930	15.6	1.2	81.8

Sources: Minami (1986), Tables 5.11 and 5.12 (pp. 127 ff.); and Nihon Tsūsan-shō Daijin Kanbo Chōsa Tōkei-kyoku (1961), *Kōgyō tōkei 50 nen shi*, various tables.

Notes: [a]All figures on factories are for factories with five workers or more. For prefectures in Tōkaidō and subdistricts, see Tables 3.4 and 5.6. [b]Light industry consists of food products, textiles, wood products, printing, and ceramics. [c]Heavy industry consists of chemicals, metal working, and machinery. [d]The Osaka/Tokyo ratio sets the Tokyo base equal to 100.

Figure 6.1 **Factories (with Five Workers or More) in the Light-Industrial Sector: Growth Rates for Employment in Osaka and Tokyo Prefectures, 1925–1937** (Five-Year Moving Averages)

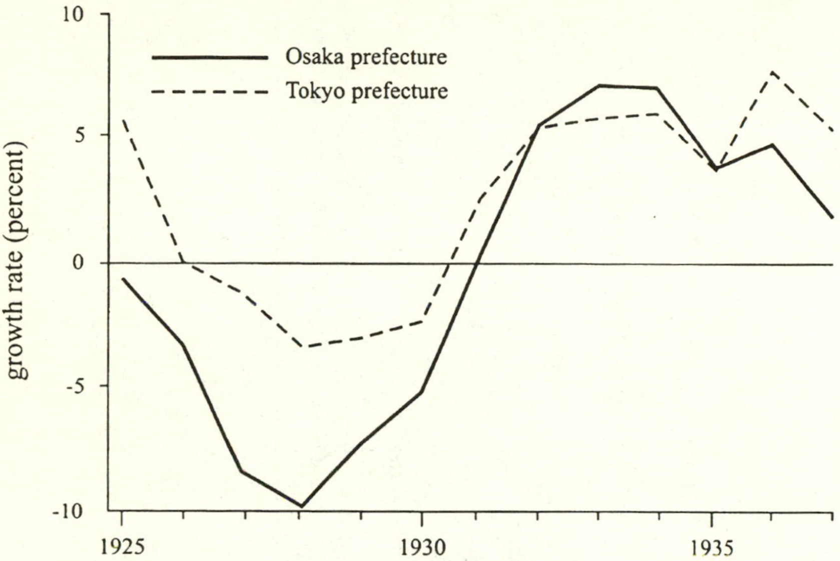

Figure 6.2 **Factories (with Five Workers or More) in the Heavy-Industrial Sector: Growth Rates for Employment in Osaka and Tokyo Prefectures, 1925–1937** (Five-Year Moving Averages)

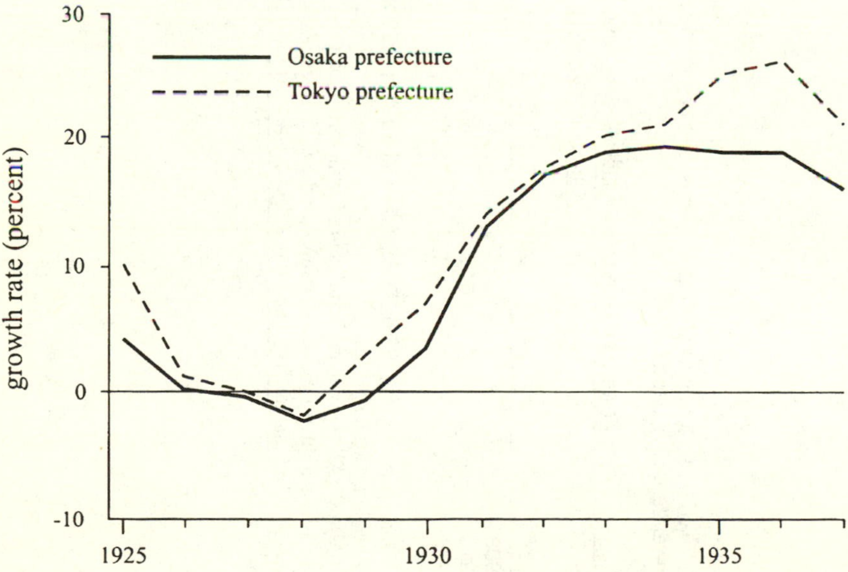

Table 6.3

Factories and Factory Workers in Osaka, 1900–1925[a]

	Indices			Average factory size (workers per factory)		Sex ratio for factory workers	Distribution of employment by subsector of manufacturing (percentage of all workers)		
	Workers								
Period	A series	B series	Number of factories	A series	B series	males/100 females)	Textiles	Machinery	Chemicals
1900–1904	89.9	139.0	108.7	10.6	6.0	128.2	54.5%	14.6%	25.9%
1905–1909	112.1	207.7	145.8	9.9	6.7	133.6	44.5	17.3	22.7
1910–1914	146.2	277.6	232.2	8.1	5.7	199.4	44.0	19.3	15.9
1915–1919	223.7	613.0	399.7	7.2	7.4	264.2	28.2	32.1	12.0
1920–1924	204.1	549.3	431.2	6.0	6.0	298.5	23.9	33.7	11.5
1925	290.0	780.2	1,029.3	3.6	3.6	172.6	38.1	27.7	15.4

Sources: Ōsaka-shi Shakai-bu (1996), *Ōsaka-shi shakai-bu chōsa hōkosho*, vol. 1: pp. 309–29.

Note: [a]All statistics are for Osaka defined in terms of its official boundaries at the time the data was collected. Thus the large jump in the indices for number of workers and factories in the year 1925 reflects the incorporation of Nishinari-*gun* and Higashinari-*gun* in that year. The term used in the source for "worker" is *shokkō*. The source provides two series on total workers in Osaka, series that differ rather markedly before 1915 but not thereafter. The series I dub the "A" series is for workers classified by gender, and the series I dub the "B" series is for workers classified by subsector of manufacturing within which they are employed. Thus the figures on sex ratios of workers were secured from the A series and the figures on distribution of employment by subsector are from the B series.

In sum, government played very little role in directly assisting the import substitution experienced by the textile industry during the first long swing. Its industrial policies directly targeted at subsectors of manufacturing played some role in the development of heavy industry, however. In the subsector specializing in railroad rolling stock, the reverse engineering project undertaken by the Japan National Railroad working in conjunction with a few large manufacturers was relatively successful. However, much more important than its undertaking policies specifically targeting certain industrial endeavors was its general support for infrastructure buildup. Electrification, for instance, paved the way for a shift toward heavy industry. And electrification had other implications that ultimately shaped the capacity of government to carry out industrial policy. A striking example of this is offered by the textile industry during the interwar era.

The Proto-Industrialization of Manufacturing and the Industrialization of Proto-Industry

One of the salient characteristics of innovation in Osaka during the first long swing is the harnessing of Western technology to proto-industry. Abandoning Western forms of shop floor organization in factories using Western machinery and producing Western-style products, and replacing these with proto-industrial forms, was important in a number of fields, including the manufacture of shell buttons, brushes, and knit fabrics.[8] I have referred to this as the proto-industrialization of industry. It was very common in Osaka and its environs because the community of craftspeople with proto-industrial skills was so very large at the end of the Tokugawa period.

Proto-industrial shops using Western techniques and machines occupied a middle ground between the ancient organic economy and the new inorganic economy. In general, prior to electrification, they mainly operated with human labor input. The machines used in the shops did not operate with inanimate sources of energy.

Widespread electrification during the transitional-growth long swing changed this. Electrification weakened the competitive position of those who clung to labor-intensive techniques in manufacturing. Increasingly, small factories switched from producing for merchants, *tonya*, to producing output for larger companies. In short, the petty producer was less and less likely to be a proto-industrial producer, and more and more likely to be a subcontractor.

Even in the branch of textiles where *tonya* maintained a strong position in the niche supplying Japanese-style clothing to the domestic consumer—the *tonya* had successfully fought off the efforts of the government to force them

into industry-wide guilds during the first long-swing period—electrification was having an impact.[9] With electrification, most small factories mechanized, driving out handloom production that was now no longer cost effective. Many of the merchants and wholesalers responded by setting up their own factories. Fragmentation in the industry dissipated.

The government—intent on rationalizing manufacturing wherever it could in the wake of the onset of the downswing of the 1920s—took the opportunity afforded by consolidation in textiles to reintroduce an industrial policy it had earlier abandoned. It reopened its campaign for forging guilds in which government and private enterprises would jointly coordinate the marketing and acquisition of raw materials. The resulting guilds, whose numbers proliferated during the era between the world wars, took the form of *kōgyō kumiai* (industrial associations). The Ministry of Commerce and Industry authorized industrial associations to undertake a variety of functions. These included ensuring minimum quality standards, securing equipment and machines for the common use of affiliated companies, and purchasing raw materials in bulk for distribution to member firms. In short, the government encouraged the industry as a whole to form a huge cartel. This cartel was able to exploit scale economies in marketing and in acquisition of capital equipment and raw materials that individual enterprises could not secure.

Rationalization was a major feature of textiles throughout the downswing of the transitional-growth long swing. For instance, the large integrated spinning and weaving enterprises adopted high drafting, and they streamlined mass production methods. They also introduced scientific cotton blending that allowed them to produce a strong raw thread (unlikely to break while being spun) with a mix of expensive and inexpensive staple cotton. By mixing fibers from expensive and inexpensive staples, the companies created a thread that was cheaper than that made exclusively with expensive staple, but still had the strength of expensive staple in the sense that it was unlikely to break under spinning. In addition, the larger firms practiced coordinated curtailment of operations to keep gluts from developing in the market.

Rationalization also helped smaller producers ride out the downturn of the 1920s. Many diversified out of producing only Japanese-style clothing and entered the Western clothing market. In short, they began to produce for both the domestic and the international market.

Thus, rationalization in textiles was an outgrowth of innovations undertaken by the large integrated spinning and weaving concerns, and of industrial policy sponsored by the government. But industrial policy worked only because of electrification. Thus, it is difficult to give too much credit to industrial policy in the rationalization movement of the interwar years. But regardless of its sources, rationalization was important for the industry. Indeed, it was a

major factor behind the fact that the export capacity of the industry continued to improve during the period between the wars. The ratio of exports to domestic production continued to rise largely because of rationalization.

But rationalization also undercut the institutional arrangements that held together the proto-industrial sector. In this sense, the trend toward rationalization occurring because of the convergence of technological change and industrial policy was eroding the very foundations of proto-industrial activity in Japan.

Education and the New Technological Imperative

During the second long swing, the new capital-intensive heavy manufacturing sector emerged as the engine of Japan's industrial growth. As a result, the demand for labor educated in science, engineering, and technical fields like accounting increased. How could supply be increased to meet this surge in demand?

Government was active in addressing the problem of demand outstripping supply. In the wake of the massive industrial upswing of the 1910s, national funding for the imperial universities and research laboratories increased. Engineering was a field given special priority in this massive infusion of resources. As in so many other areas of infrastructure construction, the government worked in coordination with the private sector. For instance, the powerful *zaibatsu* joined forces with the Ministry of Education and the Ministry of Commerce and Industry in putting together the funding for engineering research that was most likely to yield benefits to industry. And at the local—prefecture and municipal—levels, governments were also active sponsoring the creation of research centers that could meet the needs of small and medium-sized companies.[10]

At the national level, the ministry most active in responding to the crisis in higher education occasioned by the industrial boom of the 1910s was the Ministry of Education. This ministry was responsible for the national school system that encompassed elite imperial universities, national higher technical schools (*senmongakkō*), and national vocational schools (*jitsugyō gakko*). Moreover, it set standards for, accredited, and allocated student slots for both universities and technical schools and vocational schools managed by local governments and by private parties. In 1918, responding to the concerns about a future shortage of engineering and technical school graduates, the Ministry of Education appointed an extraordinary commission concerned with revamping the higher educational system. This commission recommended that the Imperial University system—that got off the ground in 1886 with the founding of Tokyo University (later Tokyo Imperial University)—

be expanded. It also recommended that a combination of private and public multiple-faculty universities be created to supplement the Imperial University system. It argued that the more prestigious and successful *senmongakkō* be elevated to universities (*daigaku*), taking on the "university" appellation that had hitherto been largely reserved for the imperial universities. Working from the recommendations of the commission, the Ministry of Education also began to increase the number of departments within each of the imperial universities, especially in the technical fields.

Local communities became actively involved in the competing for the planned expansion in the infrastructure of higher education. Osaka, for instance, was becoming increasingly concerned about the fact that it lagged behind Tokyo in institutions of higher learning. After all, Tokyo enjoyed an advantage not only because it was the national capital of modern Japan, but because it had also been the capital of early modern Japan. Under *bakufu* rule, the shogun had managed academies of higher learning in Edo, and these were amalgamated to form Tokyo University. Under the recommendations put forward by the commission on higher education, Osaka was to be the site for a new imperial university. Osaka Imperial University was created in a manner similar to the way Tokyo University had been created. A variety of technical schools set up in Osaka at earlier dates were fused together to form the original faculties of Osaka Imperial University.[11] In 1933, the Ministry of Education founded within Osaka Imperial University a faculty of engineering. It accomplished this by incorporating under the umbrella of Osaka Imperial University the Osaka Technical University that had been founded in 1896. With the stroke of a pen, a faculty with eight industrially oriented departments—mechanical engineering, chemical technology, zymotechnology (science of fermentation), metallurgy, shipbuilding, electrical engineering, and applied science—joined Osaka Imperial University.

Osaka's securing of an imperial university hardly occurred in a political vacuum. Giants of the industrial community of Osaka carried on a protracted lobbying campaign on behalf of the project, cajoling and pressuring the Ministry of Education. Particularly active was a coalition of powerful business interests, including C. Itoh and Co., Kubota Ironworks, and the Sumitomo *zaibatsu*. This coalition established a League for the Establishment of Scientific and Industrial Research that acted as a lobbying group. The league, offering funds it had accumulated from its membership as an enticement, was eventually successful in getting the Ministry of Education to create an Institute for Scientific and Industrial Research at Osaka Imperial University. This institute was modeled upon the Institute of Physical and Chemical Research already established in Tokyo.[12]

The Ministry of Education was not the only national institution active in

the field of industrial education. The Ministry of Commerce and Industry was also active in setting up research institutes. For instance, it established the Osaka Industrial Research Institute that carried out industrial experiments in its laboratories. The Electrochemical Laboratory also set up facilities in Osaka, Indeed, there were seven national research institutes operating in Osaka in 1943, according to a nationwide survey.

Zaibatsu and other large companies based in Osaka were involved in securing these national educational and research facilities. But Osaka was also home to myriads of small and medium-sized businesses whose markets were regional and whose clout was local. These companies tended to work with local governments rather than the national ministries. A fruit of the interaction of the local business community and local government in the field of industrial research was the Osaka Prefecture Research Institute. This institute not only carried on its own program of research. It also devoted considerable resources to addressing the needs of small and medium-sized companies in the prefecture. It answered practical questions concerning factory guidance and diagnosis by correspondence, telephone, and on-site visit. For instance, in 1936, it responded to more than 2,000 requests concerning concrete industrial problems, an example being how to properly do industrial welding. Moreover, it carried out extensive surveys of company practices, more than 950 in 1936.

Municipal governments were also active. For instance, the Osaka Municipal Technical Research Institute, founded in 1916 and massively expanded after the industrial boom of the 1910s had run its course, carried out basic and applied research. In the chemical field, its technicians and scientists conducted basic research in organic, inorganic, and electrical chemistry. Furthermore, in the applied fields, it carried out studies concerning machine efficiency, materials, and architectural problems. Like the Osaka Prefecture Research Institute, it responded to requests for diagnosis of factory inefficiency, charging user fees for its services rendered.

In sum, in the wake of the transitional-growth long swing, the great conurbations of the Tōkaidō were fast becoming centers for heavy industry in Japan. Thus, the demand for industrial education and research was felt with special force in these giant metropolitan areas.[13] It is in this field that Tokyo had a tremendous advantage. It was the seat of the national government. Tokyo Imperial University was the crown jewel of the educational system in the scientific and technical fields. Thus research institutes tended to spring up in the capital. For instance, in 1923 a survey was carried out of the 162 research institutes active in Japan. Of the institutions surveyed, forty-two were located in Tokyo. Only seventeen were located in Osaka, which was still the greatest industrial center in the country.[14] But the growing concen-

tration of industrial research capacity in Tokyo was enhancing the status of the capital as a place to locate factories in the heavy industries where research was crucial. Tokyo's human capital–enhancing infrastructure was superior to that of Osaka. And as unbalanced growth took over from balanced growth, this advantage became increasingly important.

The Evolution of Banking and the *Zaibatsu*

The intense pressure placed upon infrastructure by the transitional-growth long swing was also felt in the field of banking and finance. Just as with physical and human capital–enhancing infrastructure, government stepped in to rework that financial infrastructure. Just as with physical and human capital infrastructure, government coordinated its efforts with the powerful oligopolies active in the field. Thus the evolution of banking and finance was strongly influenced by the interaction of government ministries and the great financial cliques, the *zaibatsu*.

During the first long-swing period and the buildup to that swing, banking was fragmented. Most banks were unit banks. Generally banks did not have branches. A bank was synonymous with its single branch, which was its headquarters. This suited the needs of Japanese industry since most enterprises had modest funding needs, fixed capital requirements being relatively limited in light manufacturing.[15] But with the growth of heavy industry, the demand for massive loans necessary for building large integrated chemical plants, huge blast furnaces, and lengthy assembly lines in electrical machinery production increased. Fragmented banking was becoming an obstacle to the emergence of heavy industry as the engine of industrial growth.[16]

Thus the financial sector became increasingly concentrated. The trend toward concentration began even before the end of the first long swing, in 1901, when the number of commercial banks reached its historical maximum, at a total of 1,867. With the growing demand for funding of both large infrastructure projects—intercity electrical railways, electric power grids—and of plant and equipment for heavy industry, there were strong incentives to merge and amalgamate banks. As can be seen from panel C.2 of Table 4.1, the share of the big five (*zaibatsu* affiliated) banks in the paid-in capital of the banking sector began to rise after the onset of the second long swing. By 1911–1919 it was around 15 percent.

But the strongest pressure to reorganize banking came after the downswing of the transitional-growth long swing commenced. Credit had become overextended during the speculative atmosphere of the World War I boom. Financial crises occurred in 1920, 1922, 1923 (associated with the great Kantō earthquake), and 1927 when the Bank of Taiwan collapsed. Bank runs took place,

and a large number of banks failed. The government fostered amalgamations in order to shore up the financial sector. Then, in 1927, it abolished the old Banking Law. A new law came into effect in January 1928. It required that banks have a minimum of capital. Of the approximately 1,400 banks in Japan, about half were disqualified. Thus concentration took on growing force. By 1931, there were less than 700 banks; by 1936, the number was just a bit above 400. Governmental policy was accelerating tendencies that were brought on by the structural transformation of the economy itself.[17]

Concentration in the big five banks also picked up steam. By the upswing of the unbalanced long swing, over 27 percent of paid-in capital in the banking sector was in the hands of the big five banks. Concentration in banking was associated with concentration of financial resources in the *zaibatsu*.

Thus, the second and third long swings were associated with concentration of economic power in the *zaibatsu*. With the transition for oligarchy rule to democracy and the party system, the country's most powerful business interests, the *zaikai*—epitomized by the Industry Club of Japan formed by the major *zaibatsu* in 1917—consolidated its political power by becoming the preeminent financing base for the two main political parties, the Seiyukai and the Minseito (the Mitsui *zaibatsu* favored the Seiyukai, and the Mitsubishi *zaibatsu* favored the Minseito.) Because the *zaibatsu* had done very well during the upswings of the transitional-growth long swing, diversifying into heavy industry and aggressively expanding their role as trading and shipping empires, they were committed to internationalism. They supported cooperation with the Western powers, and were not enthusiastic about imperialism, insofar as it meant autarky. For this reason, the military cabinets of the 1930s were hostile to the old-line *zaibatsu*. The military governments favored the newer industrial combines known as *shinzaibatsu* (literally "new" zaibatsu) like Nissan and Nichitsu that were active in developing infrastructure and manufacturing in Korea and Manchuria.[18] Nevertheless, after 1938, the old-line *zaibatsu* enterprises in the heavy industries grew by leaps and bounds because of the increase in demand for war-related material, airplanes, transportation vehicles, and the like. Even though the *zaibatsu* lost political muscle during the 1930s, their enterprises occupied key places within the burgeoning heavy industry sector.

As the *zaibatsu* diversified their operations and became increasingly involved in heavy industry with its voracious appetite for physical plant and equipment, they changed their organizational forms.[19] They increasingly adopted a holding company/multisubsidiary form of organization in which the headquarters/holding company was treated as a partnership into which the family controlling the *zaibatsu* placed most of its assets. And the individual companies were turned into joint-stock subsidiaries. After 1931, un-

der political pressure from the ultra-nationalist right wing and under economic pressure because of their growing need for funds due to the massive investment requirements of the upswing of the 1930s, the *zaibatsu* even opened up their holding companies to public ownership, turning these into joint-stock companies. In short, the *zaibatsu* form of organization evolved as the structure of the economy evolved. Pressure due to industrial expansion reworked the institutional infrastructure of the *zaibatsu*.

Just as with the banking sector, some *zaibatsu* failed during the downswing of the 1920s. An example is the Suzuki *zaibatsu*. Suzuki had formed during the second upswing, the industrial upswing, of the transitional-growth long swing. It was one of a number of new *zaibatsu*—Kuhara, Iwai, Murai, Matsukata, and Nomura also joined the ranks of the *zaibatsu* during this period—that took off during the boom conditions of the 1910s. Suzuki was active in shipbuilding in the Osaka-Kobe area. After the shipbuilding industry went into the doldrums in the post–World War I period, Suzuki *zaibatsu* became increasingly dependent upon loans from the Bank of Taiwan to shore up its asset position (it was also adversely affected by the great Kantō earthquake). In the financial panic of 1927, Mitsui Bank cut off its supply of call money to the Bank of Taiwan, and the Bank of Japan refused to come to the rescue. As an Osaka-based *zaibatsu*, Suzuki was not popular with the Tokyo-based financial elite, including the Tokyo-based Mitsui and Mitsubishi *zaibatsu*. When the Bank of Taiwan failed, so did the Suzuki *zaibatsu*.

In short, during the downswing of the 1920s, there was systematic weeding out of shaky financial institutions that had overexpanded during the heady days of the 1910 industrial boom. Government, working together with the powerful *zaibatsu* that controlled the big banks, restructured the financial infrastructure during the 1920s largely by eliminating fragmentation in the sector. By the early 1930s, it was ready to handle the massive expansion in demand for financial resources underlying the capital accumulation of the 1930s.

A Rich Profusion: Osaka Factories, Large, Medium, and Minuscule

Osaka's strength lay in the overlay of Western technology upon a richly developed proto-industrial economy. It was this strength that had made it the Manchester of the Far East. It was precisely because of this that Osaka's economy was highly diverse. Tiny putting-out and subcontracting operations jostled up against giant *zaibatsu*-managed factories producing iron and steel, ships, and railroad rolling stock. *Tonya* plied their trade in the shadow of the stock market and the mighty *zaibatsu* banks and trust and insurance companies.

Thus, by World War I, Osaka had become home to a highly diverse economy. According to statistics reported in the 1920 census and reported in panel A of Table 6.4, industrial employment and service-sector employment were roughly comparable. Service sector employment was geographically concentrated in the city's core where the old merchant houses continued to operate their businesses. Thus, in the 1920 census, which antedates the incorporation of 1925, commercial activity is perhaps exaggerated, at least for Osaka as a whole, including its periphery. Even within its pre-1925 boundaries, Osaka was a remarkably diverse place. It was a major transit hub; it was a commercial giant; and, increasingly, it was home to factories large and small, factories in light industry and factories in heavy industry.

Analyzing employment within the city's three zones—core, middle ring, and periphery—reveals that Osaka had become diverse within each of its three main districts. Data from the 1930 census reported in panel B of Table 6.4 demonstrate this point (the figures are for daytime population; thus they do not take into account the fact that many of the residents of the periphery commuted into the core to work in the service sector). Crammed into the core were factories, department stores, ancient houses for merchants and their apprentices, banks large and small, huge railway stations, and government agencies. Osaka's diversity ran deep.

However, Osaka's greatest diversity lay in the way the city functioned as a home for manufacturing production, as a home for factories of all sizes and types. Photographs of Osaka taken during the first three decades of the twentieth century make this evident.[20] Lining the banks of rivers, spewing out grimy fumes and clouds of blackened smoke from chimneys piercing the skyline, Osaka's factories were of every size and type. Within the city was the integrated spinning and weaving plant, and myriads of tiny hovels in which workers carried on putting-out production at the bequest of merchants. And within Osaka were factories producing some of the largest steel ships in the world, and tiny establishments producing small wooden-bottom boats used for plying its ancient network of canals.

Under the force of electrification and with the cajoling of government, proto-industrial production was gradually disappearing, giving way to mechanized factory production year after year. So Osaka was gradually losing some of its diversity and, of course, its original competitive advantage as a locus for industrial and commercial activity, as proto-industrial forms of production and distribution waned. How diverse was the city's industrial economy between the wars? A detailed survey conducted in Osaka City in 1933 provides us with a wealth of information bearing on this question. Consider the figures given in Table 6.5. The incredible diversity of Osaka's world of factories can be gleaned from panels A through D of the table. In the population

Table 6.4

Employment in Osaka Prefecture and in Osaka City, 1920 and 1930

A. Composition of total employed population by primary occupational attachment (both sexes), 1920[a].

	Sector (percentage of resident employed population in occupation/industry)				
			Services		
Locale	Primary	Manufacturing	Commerce	Total	Facilitating
Osaka prefecture	15.1%	43.4%	25.5%	32.4	7.2%
Osaka City, official	0.9	44.6	34.8	43.0	9.3
Expanded Osaka City	2.9	48.1	30.7	38.2	8.7

B. Composition of labor force by sector (both sexes), 1930[b]

			Manufacturing			Services			
Locale/sector	Primary	Mining	Light	Heavy	Total	Commerce	Professions	Total	Facilitating
Osaka prefecture	10.3%	0.2%	20.0%	13.5%	34.3%	33.4%	7.9%	41.6%	12.9%
Osaka City	1.6	0.2	21.0	15.8	37.8	38.1	7.8%	46.3	13.4%
Osaka City, center	0.3	0.1	20.3	8.4	29.7	53.8	7.4	61.6	7.9
Osaka City, middle ring	0.4	0.2	19.1	19.4	39.3	34.2	7.1	41.6	17.9
Osaka City, periphery	3.6	0.2	23.2	18.2	42.6	29.9	8.7	38.9	13.5

Sources: Nihon Naikaku Tōkei-kyoku (various dates), *Kokusei chōsa hōkoku, fu ken hen* [Census] (for 1920 and 1930), various tables.

Notes: [a]By "Expanded Osaka City" in 1920, I mean the official city combined with the Nishinari and Higashinari-*gun* area. Service sector employment includes domestic services. By "facilitating industry" for 1920, I mean the "transportation" occupation. [b]For light and heavy industry categories, see Table 6.1 and notes to that table (I did not classify "miscellaneous"). Here "facilitating" industry includes construction, utilities, transportation, and communications. Domestic service is within services, but is not included in either commerce or professions. For center, middle ring, and periphery, see Table 3.2 and notes to that table.

of factories enumerated in the city, over 65 percent employed fewer than five workers. Nevertheless, large factories—less than 1 percent of the factories had more than one hundred workers and thus the percentage of factories that were large was very small—generated a significant share of the output, around 40 percent of the total value added in manufacturing.

Diversity is not only apparent in the distribution of production units according to scale of operation. It is also apparent in the range of industries. About 50 percent of manufacturing value added was generated by three industrial subsectors: machinery, chemicals, and metalworking. And about 50 percent came from various forms of light industry.

How diverse was manufacturing activity within each of the three zones of the city? Did each zone specialize in certain types of industries? As can be seen from Table 6.5, there were differences between the three zones in terms of the type of factories clustered within them. Textiles and ceramics tended to concentrate in the periphery. There was a tendency for metalworking and machinery to cluster in the middle ring. Printing establishments were usually found in the core.

Transportation costs were certainly a factor shaping the clustering of factories within the various zones. For instance, metalworking plants had a voracious appetite for heavy bulky goods like coal and iron ore that, at the time, were most efficiently transported by boat. Hence, these factories crammed in along, or within the vicinity of, the banks of the Aji River. Since metalworking establishments were clustered in this district, users of metal products also located in the same environs. Thus, many machinery-making establishments also located near the Aji River banks. Farther out from the core of the city, located on the rivers running through the city's periphery, were the spinning and weaving establishments that had played such an important role during the first long swing. And in their shadow were located countless tiny putting-out establishments and cottage operations that also produced textiles.

How important was putting out and other atavistic practices from the Tokugawa past to Osaka's industrial economy between the world wars? With electrification and governmental industrial policy eating away at the foundations of proto-industry, was it not being driven out of the market, a victim of the creative destruction integral to waves of innovation? There is no way to get a completely accurate picture of how entrenched the proto-industrial sector was in Osaka during the 1920s and 1930s. Finding a full set of proxy variables for measuring proto-industry is extremely difficult. However, we can draw some inferences from the factory survey of 1933 for in that survey are questions about apprentices, and about sources of income net of raw materials costs (income net of raw materials costs is value added). Included in the reported sources of income are figures on *kōchin* payments.[21] Proto-industrial enterprises usually depended upon apprentices, and they were usually remunerated with *kōchin* payments (which consti-

tuted the payments made by the merchant who contracted out putting-out work). Hence, since *kōchin* payments are made net of raw materials—the raw materials being supplied by the merchant—the ratio of *kōchin* payments to total net value added provides us with a reasonable proxy for continued dependence upon putting-out contracts. Using figures on dependence on apprentices and on *kōchin*, what can we say about the incidence of proto-industrialization in interwar Osaka?

Putting out was very important for the smallest factories. As can be seen from panels D and E of Table 6.5, small factories—those with one worker or two or three—depended very heavily upon *kōchin* payments for their revenue. In general, it was in textiles where *kōchin* remained well entrenched. But even in wood products and machinery manufacturing, it existed.

Interestingly enough, relying upon apprenticeship was not correlated with dependence upon *kōchin*. Recruiting workers through apprenticeship systems was not common practice in the industries with high ratios of *kōchin* payments to value added. Rather, apprenticeship was still well entrenched in the food products subsector of manufacturing, which had a low *kōchin*-to-income ratio. Most of the food-products enterprises relying upon apprenticeship catered to a domestically oriented market and continued to employ methods pioneered during the Tokugawa period. Apprentices learned their trade by working under masters who themselves had learned the business as apprentices. Thus, dependence upon apprenticeship is really a measure of the vestiges of the guild system that prevailed in many craft trades and merchant houses during the Tokugawa period. Putting out provided an alternative to the guild system during the Tokugawa period. Thus, the two types of contracting arrangements had been substitutes, not complements, during extensive and intensive early modern growth. And this state of affairs persisted during the period between the world wars.[22]

During the period between the wars, proto-industrial practices persisted in the great industrial metropolis of Osaka. But they were surely declining. In a survey of factories carried out later, there are no data about *kōchin*. Perhaps the incidence of putting-out was disappearing so quickly that the officials responsible for the questionnaire did not even bother to inquire about it.

Proto-industrialization was disappearing because rationalization of industry was gaining force, especially after the end of the second upswing of the transitional growth long swing. Rationalization was part and parcel of the reworking of infrastructure integral to the downswing of the 1920s. Banking mergers were promoted; the educational system was expanded to meet the demand for engineers and technically trained blue-collar workers in heavy industry; road networks were being built; and hydroelectric power grids were being extended. The industrial upswing of the 1910s had put tremendous pressure on infrastructure. Now government, working in tandem with the private sector, was responding to this pressure.

Table 6.5

Osaka Factories: The 1933 Survey

A. Distribution of factories and of factory output (value added) according to factory size[a].

District	Factory size (number of workers per factory): 1	2–4	5–10	11–30	31–100	101–1,000	1,000–
Percentage of all factories within district							
Osaka City	19.4%	46.6%	23.1%	7.7%	2.5%	0.6%	0.05%
Center	16.7	48.8	25.6	7.0	1.6	0.3	0.03
Middle ring	19.2	48.1	22.3	7.6	2.5	0.6	0.07
Periphery	21.8	43.4	22.1	8.5	3.6	0.9	0.04
Percentage of all output (value added) of factories in district							
Osaka City	0.5%	6.5%	8.9%	14.0%	20.0%	33.2%	17.1%
Center	0.6	18.6	14.2	18.3	12.8	24.9	10.7
Middle ring	0.4	3.6	7.2	12.9	19.9	28.3	27.7
Periphery	0.5	3.3	8.0	13.0	24.0	43.1	8.2

B. Composition of factory output (value added) by subsector of manufacturing within districts[a,b].

Zone/subsector	Textiles	Ceramics	Wood products	Printing	Food products	Metal-working	Machinery	Chemicals
Osaka city	13.0%	7.8%	3.3%	4.6%	7.8%	23.9%	16.8%	18.0%
Center	8.3	2.3	3.6	18.8	10.3	17.8	8.9	12.8
Middle ring	8.3	3.2	4.1	0.9	6.3	32.3	21.4	16.2
Periphery	20.7	15.7	2.2	1.9	8.3	16.8	15.0	22.7

C. Composition of factory output (value added) according to district for subsectors[a,c].

Zone/subsector	Textiles	Ceramics	Wood products	Printing	Food products	Metal-working	Machinery	Chemicals
Osaka ity	100%	100%	100%	100%	100%	100%	100%	100%
Center	11.4	5.3	19.6	75.1	23.6	13.3	9.4	12.7
Middle ring	28.4	18.2	55.1	8.5	36.1	60.2	56.8	39.9
Periphery	60.2	76.6	25.3	16.5	40.3	26.6	33.8	47.5

D. Intensity of putting-out activity, POI (*kōchin*/value added ratio); sex ratio of employees, SRE (males/100 females); percentage of employed workers who are family workers or apprentices, RFWA; and average factory size (average number of workers per factory), Osaka City [a,d].

Item/subsector	Textiles	Ceramics	Wood products	Printing	Food products	Metal-working	Machinery	Chemicals
POI	25.6%	1.3%	10.8%	5.6%	1.0%	5.6%	9.8%	1.2%
SRE	96	841	2,119	1,162	382	1,311	1,720	5,565
RFWA	22.9%	12.4%	50.0%	22.0%	49.1%	20.4%	28.2%	10.6%
AFS	11.1	21.9	4.0	11.5	4.8	10.5	8.8	16.5

E. Intensity of putting-out activity, POI (*kōchin*/value added ratio), in Osaka and its zones according to factory size[a,d].

	Factory size (number of workers in factory)						
District	1	2–4	5–9	10–29	30–99	100–999	1000–
Osaka City	66.6%	21.4%	18.1%	8.9%	6.8%	4.5%	10.7%
Center	68.9	12.2	18.5	9.2	7.0	4.6	0
Middle ring	63.4	32.1	17.6	7.8	7.1	3.8	7.2
Periphery	68.4	35.3	18.4	9.8	6.5	4.9	16.2

Sources: Osaka-shi Yakusho (1935), *Osaka-shi kōgyō chōsa. Shōwa hachinen*, various tables.

Notes: [a]Figures are for all factories including factories run as secondary sources of income, for instance, by farmers. Examination of the data suggests that the total number of factories excluding secondary-income-source operations is approximately equal to the total number of factories, and that most of the secondary-source-income operations are those with only one, or at most four, workers, these often involving putting-out production. For statistics on putting-out activity, see panels D and E of this table. [b]The totals of the percentages do not add to 100 percent within each district because I do not include the "miscellaneous" subsector of manufacturing. [c]In some cases, due to rounding off of decimals, percentages may not add to 100 percent. [d]See text for a discussion of putting out and of the *kōchin* variable.

The industrial policy of the interwar years was focused upon rationalizing industry. Policies included promoting industry-wide associations that exploited scale economies in marketing and in acquiring raw materials. Another policy was management of coordinated cutbacks in output across all enterprises. Thus, eliminating fragmentation was key to rationalizing industry. For this reason, proto-industrial practices tended to wither in the face of rationalization, although they did not disappear altogether.

Osaka's competitive advantage had been in overlaying industrialization on a proto-industrial platform, integrating one with the other. So, as proto-industrialization waned under the force of technological change and industrial policy, Osaka's competitive advantage was eroded. More important, the reworking of physical, human capital–enhancing, and financial infrastructure was also eroding Osaka's competitive advantage, especially compared to Tokyo's competitive advantage. Tokyo's strengths were in land-based infrastructure (roads, railroads, and electricity), formal education and research, and banking and finance. Most of the powerful *zaibatsu* maintained their main headquarters in Tokyo. In Tokyo, their banks could easily stay in touch with the Bank of Japan and the Ministry of Finance; in Tokyo, they could readily find out what plans for rationalization were afoot in the Ministry of Commerce and Industry. In Tokyo, they could recruit the top students from Tokyo Imperial University, and they could keep up on new developments issuing from the research institutes clustered in the capital. Moreover, it was far cheaper to build wide roads that could handle trucks and buses in Tokyo than it was in Osaka.

In short, because it had a competitive advantage in developing the new infrastructure crucial to the second and third long swings, Tokyo edged out Osaka as the core of the Tōkaidō core during the late 1930s. This is apparent from Table 6.2. Some writers have argued that the growing importance of the military in creating demand for industrial production was key to the decisive shift toward Tokyo-centrism apparent in the last years of the 1930s. This is doubtful. As perusal of Figures 6.1 and 6.2 demonstrate, both light and heavy manufacturing tended to grow more rapidly in Tokyo prefecture than in Osaka prefecture throughout the 1920s and the 1930s. Thus, Osaka's loss of preeminence was long coming.

Differential rates of land conversion testify to Tokyo's surge to preeminence in economic affairs. Consider the growth of *takuchi* land depicted in Figure 6.3. As can be seen, the nonagricultural economy of Tokyo was plunging ahead at a far brisker pace than was Osaka's. In absolute terms, all locales with the Tōkaidō were growing economically. But the northern end of the Tōkaidō was growing considerably more rapidly than were the other regions within it.

Figure 6.3 ***Takuchi* Land in Osaka and Tokyo Prefectures: Growth Rates, 1902–1935** (Five-Year Moving Averages)

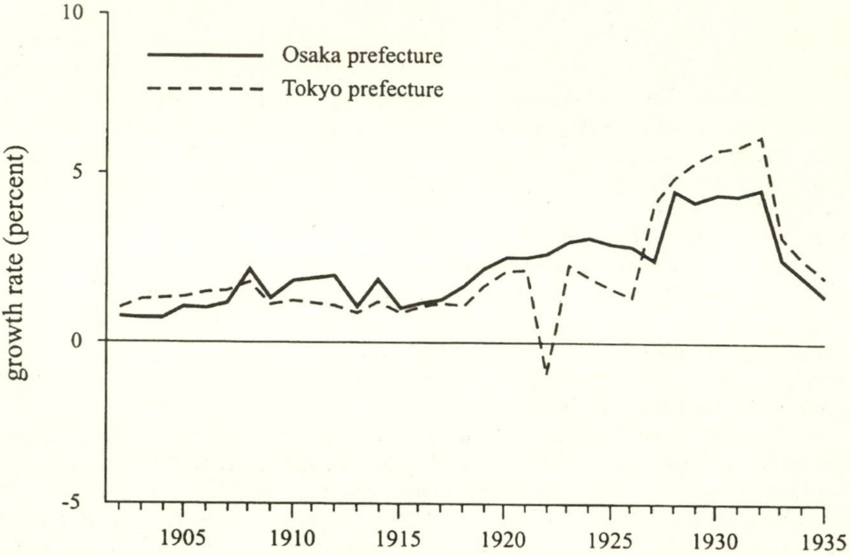

In the heart of Tokyo was the Imperial Palace, the great national ministries clustered around Hibiya Park, Tokyo Imperial University, and the Bank of Japan. This world of business and bureaucracy was intent on rationalizing infrastructure and industry for the purpose of national aggrandizement. Tokyo was girding for total war. It was a hard, dry place, cut through by wide paved boulevards congested with trucks, buses, and cars. It was imperial. By comparison, in Osaka's core was Senba, a world of merchant houses and *tonya*. In physical terms, Osaka was defined in terms of its network of ancient canals that cut through it in a great crisscross pattern. It was a world of colorful bridges and narrow unpaved streets plied by *niguruma* and *jinrikisha*. It appeared to look back to its proto-industrial past, which was being steadfastly denigrated by the force of technological change and the new wave of rationalizing industrial policies.

Economic evolution was undermining the usefulness of Osaka's ancient network of canals, just as it was playing havoc with the utility of proto-industrialization. History is replete with examples of how an initial advantage becomes, at a later juncture, a disadvantage. If there is a general lesson to be learned from Osaka's fall from preeminence, perhaps it lies in the cold logic of that proposition.

Notes

1. For the discussion of the machinery sector, I draw heavily from Sawai (1989). I do not discuss the product cycle for chemicals in the text. It was strongly affected by the development of Japanese investment in its empire, especially in Korea. Molony (1990) discusses the activity of Japanese combines in Korea. They exploited hydroelectric power, which was heavily generated in northern Korea, in the making of fertilizers and other types of chemicals utilizing the electrochemical process. Since imports of chemicals from Korea contribute to the total import figures underlying Table 6.1, a substantial portion of the imports are actually fruits of production of Japanese companies brought into Japan proper.

2. In my opinion, the origins of postwar industrial policy lie in the creation of this cartel. See the discussion in Ohkawa and Rosovsky (1973) and in Uriu (1996). Ramseyer (1996) interprets the behavior of the original cotton textiles cartel in terms of the labor market supply conditions that it faced.

3. On the origins of the O.S.K., see chapter 3. For this section of the text, I draw upon Chida and Davies (1990).

4. Despite the restrictions placed on the size and characteristics of Japan's Imperial Navy, improvement and refitting were allowed under the terms of the agreement. For instance, in 1928–1931, naval orders accounted for 48 percent of hours worked by private shipyards. See Chida and Davies (1990, 46). On the Washington Agreement, the renewals of that agreement, and Japan's ultimate decision to cease to participate in the arrangements between the three great naval powers of the Pacific, see Evans and Peattie (1997).

5. These figures are based on nominal figures given in Sawai (1985, 94 and 146). I deflated the nominal figures by a price index for manufactured goods given in Ohkawa and Shinohara with Meissner (1979).

6. In emphasizing the importance of electrification for mechanization, I follow Minami (1965, 1986, 1994).

7. For a detailed discussion of dualism in prewar Japanese labor markets, see Mosk (1995, 24–88).

8. See chapter 3.

9. For this section of the text I draw heavily upon Allen (1940) and Seki (1956). During World War I, a massive run-up in yarn prices accompanied the general inflation of the period. In 1920, the bottom fell out of the market, and prices dropped precipitously. Many cotton yarn traders and exporters were driven into bankruptcy. Larger spinning and weaving enterprises, many of which had accumulated massive levels of profits during the World War I boom, managed to stay afloat during the retrenchment that spread through the industry during the 1920s. It should be noted that the spinning and weaving companies that were members of the *boseki rengōkai*, which is discussed in the first section of this chapter, adjusted to the depressed market conditions of the 1920s through coordinated cutbacks in production, and through coordinated curtailments of operation. Doane (1998) discusses some of the cooperative arrangements that knit together firms with industries or between industries, giving a wide range of examples.

10. Mosk (1995a, chapters 2 and 4) develops the argument that the Ministry of Education has elastically responded to changes in the demand for technically trained graduates in manufacturing and industry.

11. For this discussion of the founding and expansion of Osaka Imperial University and of research institutes in the Osaka area, I rely upon Sawai (1997). The fact that both applied and basic research in science and engineering was being extensively carried out in prewar Japan at these universities and research institutes testifies to the importance of indigenous research and development in innovation within Japan. Regarding the role of domestic research and development for facilitating the import and adaptation of foreign technology, see Wakasugi (1997).

12. The League ended up assisting the government in raising funds for the institute. Another example of private enterprises subsidizing research at Osaka Imperial University is the grant given by Osaka Electric Co. Ltd. to the Engineering Department of the university for the creation of an academic chair in welding engineering.

13. For the emergence of Nagoya as a major industrial center, see Mosk (2000b).

14. The figures here are drawn from Sawai (1997). Sawai indicates that he believes that the survey undercounted the number of industrial research institutes active in the country. Of the seventeen enumerated within Osaka Prefecture, ten specialized in chemicals.

15. Patrick (1984) argues that during the pre-1905 period, the informal financial sector—consisting of money lenders, rotating credit cooperatives, and bankrolling of investments by friends and relatives—was as important (in terms of the flow of funds channeled through these institutions) as was the new financial sector consisting of banks, the postal savings system, and insurance companies. That is to say, the old financial institutions that had been the backbone of the proto-industrial economy coexisted with the new financial institutions, just as proto-industrial manufacturing firms coexisted with large integrated spinning and weaving enterprises employing mules, jennies, and power looms. After 1905, however, the informal sector went into decline in terms of the proportion of intermediation it carried out, as opposed to that carried out by banks, trust companies, and the like.

16. For the discussion of banking, I rely heavily upon Adams (1964), Goldsmith (1983), Nakamura (1983), Patrick (1967, 1984), and Suzuki (1980). For my discussion of the evolution of the *zaibatsu*, I rely upon Bisson (1954), Hadley (1970), Hirschmeier and Yui (1995), and Morikawa (1992).

17. Takafusa Nakamura (1983) emphasizes the importance of the balance of payments in Japan's trade as a primary determinant of the expansion or contraction of the money supply. That is, he emphasizes factors that, to some extent, are exogenous to the Japanese economy because they are shaped by international developments external to Japan. He argues that the Bank of Japan and the Ministry of Finance were heavily constrained in their capacity to regulate the money supply. The capacity of these organizations to freely set monetary growth targets were also limited by the fact that the government sector either ran budget surpluses (in which case, it was a net saver) or budget deficits (in which case it was a net lender) on its fiscal account. For instance, during the periods 1911–1919 and 1919–1930, government savings were positive, but during 1930–1938, they were negative. See panel B.1 in Tables 4.1 and 4.2.

18. For details concerning Nichitsu, which was heavily involved in the chemicals industry and in fertilizer production, see Molony (1990).

19. Changes in the tax code also played a role in these institutional changes.

20. For photographs, pictures, and illustrations of Osaka and its environs during the period between the world wars, see Nishiyama (1997), Okamoto (1978), and Osaka-shi (1989). For pictures of Tokyo, see Oi (1992). Tokyo's advantage in having wide

roads, and Osaka's dependence on water transport, can be clearly seen by comparing the photographs for the two cities.

21. I am grateful to Saito Osamu for his assistance in interpreting this survey, and for advising me as to the correct interpretation of the *kōchin* concept.

22. The male/female sex ratios given in Table 6.5 are interesting. The levels recorded for this ratio show that males vastly outnumbered females in manufacturing employment in the Osaka of the 1930s. The only exception to this rule is the textiles subsector. In this industry, the number of male workers was about equal to the number of female workers.

7

Cities

A Social Fabric Under Pressure: Inequality and Well-Being

During the transition from balanced to unbalanced growth, the social landscape of Japan was transformed. It was transformed because, under dualistic growth, innovating sectors grew rapidly and noninnovating sectors hardly advanced at all. It was transformed because segmentation in factor markets developed. Innovating industries had access to labor and capital services that were unavailable to laggard sectors. And it was transformed because of the growing concentration of assets: *zaibatsu* and landlords possessed land, equipment, and structures. Tenant farmers and blue-collar workers did not. In short, the distribution of opportunities and the distribution of rewards—in the form of income flows and wealth—was becoming increasingly unequal as balanced growth gave way to unbalanced growth. This disrupted society.[1]

Most of the social tension unleashed in Japan after the upswing of the second long swing was due to the domestically driven economic evolution of the country, and the accompanying switch from balanced to unbalanced growth. International factors played a role as well. Importing foodstuffs from Korea and Taiwan undercut domestic agriculture. The breakdown in trade and the drift toward autarky hurt exporters of raw silk and textiles alike. The United States, increasingly protectionist after the collapse of the stock market in 1929, forced voluntary export restraints upon Japan's cotton textile industry. Exports of raw silk to the United States plummeted after 1930. Commonwealth Preference was implemented in the British Commonwealth after the Ottawa Conference of 1932. As a result, Japan lost markets in Canada, the United Kingdom, Africa, and Asia.

The face-off of opposing ideologies in Europe had an impact on the domestic intellectual agenda in Japan. The incipient left wing in Japan was

emboldened by the apparent success of communism in Russia. Fascism appealed to ultranationalists in Japan who wanted to promote rationalization of industry under military or direct imperial rule. Domestic social tensions were reflected through an intellectual prism that was partly Western and partly Japanese in origin. This complicated the domestic dialogue concerning emerging social inequities.

It is important to keep in mind that the transition from balanced to unbalanced growth did not trigger an increase in the absolute level of poverty in Japan. To the contrary, the standard of living improved for the population as a whole. Nor does it appear to have deteriorated for any particular socioeconomic group. Figures on the height and weight of schoolchildren, factory workers, and military recruits bear out the proposition that the biological standard of living—defined in terms of anthropometric measures like height, weight, and chest girth—improved for all groups during the 1920s and 1930s. Now, the biological standard of living depends heavily upon net nutrition, that is, upon gross nutrition net of the calories, nutrients, and food reserves drained off by fighting off disease and by physical exertion.[2] During the interwar period, gross nutritional intake improved; the incidence and severity of infectious disease eased somewhat; and, due to a growth in school attendance among youth, the incidence of calorie-burning physical work among children and youth declined. Thus, there is really no evidence suggesting that there was a pronounced decline in per capita welfare for the population as a whole or for large subpopulations of the population. Indeed, infant mortality rates plummeted during the 1930s, and life expectancy rose.

Improvement in levels of welfare and the biological standard of living was greater for some groups than others.[3] That is, the relative standard of living changed. Gaps between some groups and other groups widened. The most important gap was between the Tōkaidō and the rest of the country, the periphery. In the industrial belt, and especially in the great conurbations of the industrial belt like Osaka, Nagoya, and Tokyo, per capita incomes soared with the expansion of heavy industry in these locales. Incomes in agriculture and proto-industry did not appreciate very much. Thus, the gap between the great metropolitan centers of the Tōkaidō and the rest of the country widened. Moreover, there were long-standing differentials in the biological standard of living between northeastern Japan and western and central Japan. In general, populations in central Japan enjoyed the best biological standard of living, and those in northeastern Japan the worst. These geographic differentials between northeast and central/southwestern Japan were long-standing.[4] In short, there were important geographic differences in the standard of living, and these appear to have widened during the period between the world wars. And they widened largely because conditions improved more in the

great conurbations of the industrial belt than in the remainder of the country.[5]

In short, under the transition from balanced to unbalanced growth, the social fabric was being ripped apart along geographic lines. Rural villages and towns were being left behind; the great cities of the Tōkaidō were moving forward by leaps and bounds. Moreover, within cities and rural areas, the distribution of wealth was increasingly becoming an issue because of the rise in urban land prices relative to other prices. Land was a source of collateral for bank loans. Thus, those persons and institutions possessing land in abundance were favored in terms of getting access to capital.

Moreover, the animosity of renters toward owners was compounded by the fact that as the price of urban land soared, so did the rental value of that land. Thus there was a natural tendency for urban rents either to rise or to not fall relative to the prices for other goods and services. And urban land prices naturally affected rural land prices, especially in the immediate hinterlands of large cities, for farming land in the periphery of large cities could be converted to *takuchi* purposes. Thus, there was upward pressure on the relative price of farming land in the Tōkaidō, especially in the vicinity of the great conurbations.

The situation was especially galling for the tenant farmer in the immediate periphery of a great industrial metropolis. As agricultural productivity growth dried up after 1910, the role of landlords as facilitating agents for improvements in agriculture was disappearing. Landlords were no longer making much of a contribution, but they were collecting rents. Due to the expansion of the gigantic metropolitan centers, there was constant upward pressure on land prices and, hence, upon land rent.

Adding to the woes stemming from a rending of the social fabric due to changes in the relative status of different groups and regions within the country were two other social problems: unemployment and underemployment, and environmental degradation. With the downturn of the 1920s, the unemployment or underemployment of labor became a festering issue, and, at the same time, the ill effects of pollution and harm to the natural environment were becoming increasingly evident.

Japan's ability to cope with the pressures on the social fabric brought on by the upswings of the transitional-growth long swing is in no small measure due to government policies formulated at both the national and local levels. To some degree, a bias in government policy facilitated society's response to the pressure. The bias was toward solving problems by building and improving infrastructure. Thus, especially at the local level, government managed to kill two birds with one stone. In order to cope with social disruption, government mobilized investment in infrastructure. And, by creating new infrastructure, it encouraged further industrial expansion. In turn, by increasing the demand for labor, industrial expansion helped take pressure off the labor market.

Mobilizing infrastructure to tackle social issues is not uncommon. However, the degree to which Japanese cities relied upon this method is exceptional. The infrastructure so created is a common-pool resource. It is open to use by all parties, but it is subject to crowding. Because it is open to all parties, it differs from policy responses targeted at particular groups, such as those designated as impoverished. Thus, emphasizing common-pool resources means de-emphasizing targeted programs. There is a cost to heavily relying upon the common-pool approach.[6]

The fact that Japanese municipal administrations opted for a common-pool resource approach is not simply a matter of theoretical interest. It has consequences: Taking a common-pool resource approach opens up competition for the streams of benefits flowing from the resource. It encourages competition. It appeals to self-interest, for some parties may reasonably expect to benefit more than others by the creation of the resource. Hence, it encourages government to involve private parties in creating the resource, as they expect to extract unusually generous benefits from doing so. Thus, government tends to adopt a facilitating or coordinating role when it develops common-pool resources.

While cities like Osaka had already experienced lulls in growth before the downswing of the 1920s (growth was slow between 1897 and 1904), the problems posed by the earlier downturn were not nearly as acute as those arising during the 1920s. The first major incident sparking problems that required solution during the 1920s was the rioting set off in 1918 by spiraling rice prices. In 1918, the Rice Riots spread across cities and countryside alike. And rising costs for foodstuffs only exacerbated pressure upon urban household budgets that were being already squeezed by escalating rents. Added to these problems was a surge in unemployment that began in 1919. For instance, in Osaka, the unemployment rate for casual workers in 1919 was 8.5 percent; by 1925, it had increased to 12.9 percent.[7] Moreover, fresh school graduates in the city were also being adversely affected. In 1923, the unemployment rate for fresh school graduates was 24.3 percent; by 1925 it was 26.5 percent.[8] The problem was serious.

How did the city respond to the surge in unemployment? It set up a department to survey conditions. It established youth exchanges and youth employment centers. It built public bath houses and simple functional cafeterias to serve the indigent unemployed. In addition, it started up an unemployment relief program, putting some of the unemployed to work in public work projects.[9] Most of these measures were premised on the assumption that intervention was required for only a short period of time, that is, on a temporary basis. But two programs involved building infrastructure that would be long-lasting: setting up job exchanges and the unemployment relief pro-

gram. Under the latter program, the city upgraded and expanded its sewer lines and water distribution system. For instance, in 1927, the length of the city's sewer lines was about 549,000 meters. By 1935, the length had jumped to almost 962,000 meters. And in 1939, it exceeded 1.2 million meters.

Dealing with the city's unemployed labor force did not occur in a political and ideological vacuum. Union membership and labor unrest both climbed as the ramifications of the 1920s downturn worked its way through the economy. After 1921, May Day attendance figures increased, peaking at almost 9,500 participants in 1932. Union membership also climbed throughout the 1920s and early 1930s. In 1926, about 38,000 persons were in Osaka-based unions. By 1934, the figure had increased to over 68,000 people.

To be sure, in terms of sheer magnitude of participants, interwar rural unrest involving disputes between landlords and tenant farmers far overshadowed the labor unrest of the great cities of the Tōkaidō. But rural unrest was fragmented. In large cities, there are scale economies in the marshaling of protest. It was easier to mount threats that might potentially threaten domestic tranquility, thereby gaining national attention in the great metropolitan centers like Osaka and Tokyo.

Now, because unemployment and underemployment was linked to the long-swing pattern, they largely ceased to be problems during the upswing of the 1930s. While unemployment vanished off the list of pressing social issues during the industrial upswing of the third long swing, ecological devastation tended to worsen as industrial production began to soar once again. Two examples of ecological devastation for which interwar Osaka became famous (perhaps notorious is a more appropriate term) were air pollution and subsidence. The latter involved the sinking of the earth's surface due to extensive pumping of groundwater from the soil beneath the earth's surface.

Air contamination became a pressing issue as early as the first long swing. Indeed, the Japanese term for pollution—*kōgai*—was coined by the Osaka press to describe the grime and soot-filled smoke emitted from the coal-burning factories in the metropolis. By the 1930s, Osaka's air had become so noxious that the city imposed regulations on emissions.[10] It did so because its own City Health Experimentation Stations, which had been systematically sampling air quality in the environs, called out for controls. By then Osaka had acquired a new nickname. It was already known as a "city of water" and the "Manchester of the Far East." Now, it became known as a "city of smoke."

Unlike air pollution that assaults our senses in manifold ways—ugly skies, thick smog, and burning irritation on the eyes—the subsidence problem is not so immediately apparent. As the porous aquifers beneath the earth's surface are depleted, the soil is depleted. As a result, it becomes

more compact. Eventually, buildings begin to sink or tilt, and the pavement and foundations begin to crack.

In Osaka, surface flooding accompanied subsidence. Ground levels sank below the water levels in the canals and rivers near them, resulting in flooding. As factories sank under the water, all that protruded were smokestacks. By the industrial upswing of the 1910s, the fact that the city had a major problem on its hands was widely appreciated. Around 1930, the city began to collect data on the decline in the upper reaches of the water table. To secure the data, it established hundreds of measurement sites throughout the city. Addressing the problem took more time. Indeed, it was not until after World War II that an adequate understanding of the causes of subsidence, and remedies for counteracting it, were developed.

In sum, prior to the development of a regulatory regime that gave producers incentives to prevent or reduce the emission of pollutants, and to practice conservation with respect to water drawn from beneath the earth's surface, Osaka's factories treated the environment as they would any public good. As a result, over the six decades between the late 1870s and the late 1930s, Osaka became an increasingly ugly and fetid city. Its canals and rivers were littered with refuse and chemicals. Its air became gray and noxious. And its surface area creased and cracked because of subsidence. Ecological deterioration accompanied the disruption visited upon the social fabric.

Center, Middle Ring, and Periphery

Complicating the city's response to problems of social disruption and environmental deterioration was its changing demography. As the city's daytime, working population grew, the proportion of workers who commuted into the city from suburbs also grew. Osaka's burgeoning population of nonresident workers did not have the same incentive to confront the city's growing list of difficulties, as residents did. Thus the city was becoming a captive of the demographic wedge being driven between its daytime and nighttime populations.

With the proliferation of intercity railway lines between 1905 and 1914, the ratio of daytime population to nighttime population began systematically to rise. By World War I, Osaka was a massive transport hub for an increasingly interconnected Kinai region. This hub was served by radial train lines linking Osaka to all the other major cities in the southern Tōkaidō. And along each of these lines, suburban communities sprang up. These became bedroom communities for Osaka.

Residents of Osaka were fleeing the city because of skyrocketing land prices, pollution, and other symptoms of disruption plaguing the metropolis. With the massive incorporation of 1925, many of these nonresidents of Osaka

proper once again became residents. As land prices continued to rise throughout the 1930s, growing percentages of Osaka's daytime population moved their households into the suburban developments springing up along the railway lines. Of course, households were not the only parties driven out of the city by escalating land prices. Factories were also being driven out of the city as land became increasingly expensive. If it was factory employment that brought workers into the city, factories fleeing the city tended to reduce the city's daytime and nighttime populations.

How close a relationship is there between factory employment and population in Osaka? The overriding importance of factories for population growth in Osaka is evident in Figure 7.1. Especially after the manufacturing boom of 1911–1919 got under way, the close association of the series for population and factory employment is clear. Both surge during the 1910s. Both experience a pronounced downturn during the early 1920s. And both expand in the mid-1920s. In short, by the second long swing, Osaka was a fully developed industrial city in the sense that the city's demography was heavily shaped by its manufacturing activity.

The demography of region or city is determined by its natural growth as well as by its pull on migrants to the area. The components of Osaka's population increase are displayed in Table 7.1. Comparing the demography of pre-1925 Osaka (with its official boundaries) to the demography of "expanded Osaka" inclusive of Higashinari-*gun* and Nishinari-*gun* reveals some interesting results concerning the reasons the population grew more vigorously in the periphery than it did in the center and in the middle ring. The major reason is inmigration. In the periphery, inmigration rates are high. But they are very low in the two inner zones (reflecting flight to the suburbs and factory closings. The rates are actually negative between 1920 and 1924).

But migration is not the only factor accounting for the periphery's strong demographic expansion. The excess of fertility over mortality—which is the natural rate of increase—is far greater in the periphery than it is in the center and middle ring. Why is this difference so large in the periphery, and so small (indeed, even negative in some periods) in the center and middle ring? It is apparent that one factor is fertility. In the periphery it is fairly high; in the center of the city it is low. Why is this? From panel B of Table 7.1, it is clear that differences in the composition of the population by gender underlie the differences in fertility rates. Males exceed females everywhere in the city. The excess of males is especially pronounced in the center of the city, where the merchant houses were clinging to their apprenticeship system. Apprentices usually married only after promotion to more exalted status within the enterprise. Hence, single apprentices living in housing provided by the merchants clustered in the city's inner core.

Table 7.1

Population Increase in Osaka, 1897–1938

A. Population increase rate (PIR); Crude birth rate (CBR); Crude death rate (CDR); natural rate of increase (NRI); implied net in-migration rate (NIMR); and infant death rate (IDR): official Osaka City and—prior to 1925—Nishinari-*gun* and Higashinari-*gun* combined (all rates per 1,000), 1897–1938[a].

Rate	1897–1899	1900–1904	1905–1909	1910–1914	1915–1919	1920–1924	1925–1929	1930–1934	1935–1938
				Osaka City (official boundaries)					
PIR	58.3	38.7	32.7	34.2	21.8	–14.7	33.0	24.8	51.3
CBR	20.4	19.5	19.7	18.4	20.1	28.2	30.4	29.8	26.0
CDR	23.7	20.9	20.4	17.1	19.2	21.9	18.8	12.8	10.9
NRI	–3.2	–1.4	–0.6	1.3	0.9	6.4	11.6	12.8	10.9
NIMR	61.6	40.1	33.3	32.8	20.9	–21.0	21.5	12.0	40.4
IDR	303.4	251.0	233.4	256.4	253.4	218.4	158.1	132.5	104.8
				Nishinari-*gun* and Higashinari-*gun* combined					
PIR	n.a.	34.6	78.4	91.3	98.5	80.5	—	—	—
CBR	n.a.	34.7	32.4	31.4	35.5	37.9	—	—	—
CDR	n.a.	28.1	25.8	25.4	29.8	27.2	—	—	—
NRI	n.a.	6.6	6.6	6.0	5.6	10.7	—	—	—
NIMR	n.a.	28.0	71.8	85.3	92.9	70.2	—	—	—

B. Zones of expanded Osaka City: Indices for population (1895 = 100); percentage of population in zone (%); and sex ratio of population (males/100 females), 1895–1935[b].

	Indices (1895 = 100)			Percent of Osaka population			Males/100 females		
	Center	Middle	Periph.	Center	Middle	Periph.	Center	Middle	Periph.
1895	100	100	100	60.6	25.6	12.6	n.e.	n.e.	n.e.
1925	127.8	342.4	703.6	n.e.	n.e.	n.e.	119.9	116.3	107.4
1935	137.6	453.9	1,240.1	22.5	31.4	46.2	121.8	119.1	107.6

Sources: Nihon. Naikaku Tōkei-kyoku (various dates), *Kokusei chōsa hōkoku, fu ken hen* (1925, 1930, and 1935); and sources used in Table 5.3.

Notes: [a]Rates are per 1,000 population except for infant death rate, which is the rate per 1,000 live births. The figures for Nishinari-*gun* and Higashinari-*gun* combined for 1900–1904 are actually for 1901–1904. N.a. = not available or not estimated.

[b]For definitions of "expanded Osaka City" and of the zones of Osaka, see Table 3.2. "Middle" stands for "middle ring," and "periph." for periphery. n.e. = not estimated. Due to rounding, percentages may not add to 100 percent.

Figure 7.1 **Population and Factory Employees in Osaka City, 1900–1925, Indices with 1900 = 100**

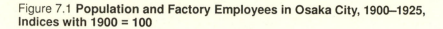

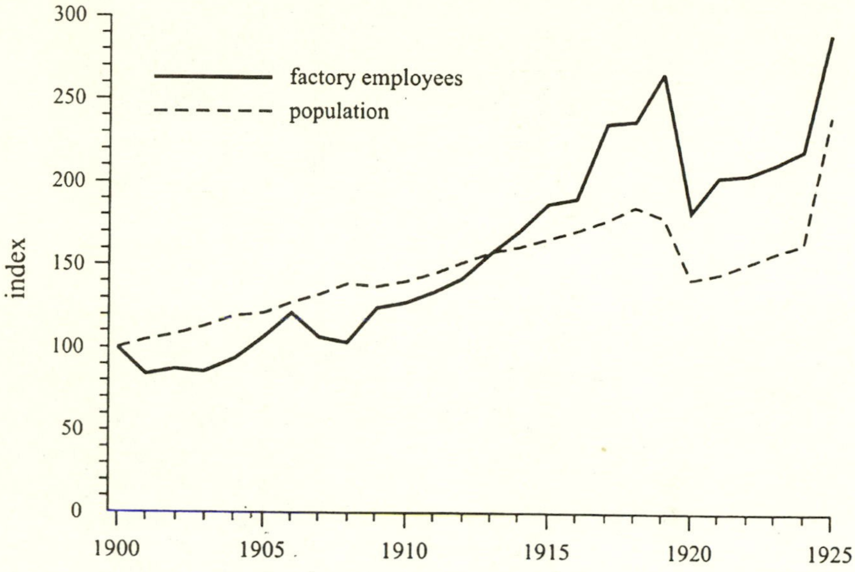

Viewing the demography of Osaka through the prism of the life cycle helps us to understand how the city's demography worked. According to the life cycle perspective, age is a major determinant of demographic behavior. That is, age structure is a key factor shaping vital rates. The age structure of Osaka's population varies tremendously between the zones. At the city's center, growth rates for the young adult population aged fifteen to thirty-four are substantial. However, growth rates for the population over age thirty-four are low, even negative. In particular, for the age group thirty-five to thirty-nine, growth rates are negative.

In the natural life cycle of an apprentice recruited into a merchant house, a young man typically enters the house in his teenage years. He resides with the house until he is promoted and is rewarded with a regular salary that permits him to marry. Accordingly, he is able to marry, and upon marriage moves out of the merchant's residence into lodgings elsewhere in the area. He now becomes a commuter.

Of course, not all young men fleeing the center of the city for its periphery were former apprentices moving through the life cycle. Some were workers fleeing escalating rents. Some were following their employers, who were pulling out of the core of the city where land prices were being driven upward. Indeed, many factory workers moved with the factories

that employed them. It was not uncommon for single workers to live either in dormitories supplied by their employer or in boarding houses near their place of employment.

In short, fertility was influenced by the employment structure of the metropolis and its land use patterns. What about mortality? As can be gleaned from panel A of Table 7.1, mortality risks in Osaka (defined in terms of its official boundaries) between 1897 and 1924 were high. In particular, the infant death rate (IDR) was substantial.[11] The factors shaping this mortality were manifold. But one of the main culprits was the city's water supply. Although the city had constructed a series of filtration ponds near Sakuranomiya (northeast of the city's heart at Nakanoshima), where it cleaned and processed water drawn from the Yodo River before dispensing it to the city's populace through a labyrinth of metal pipes, it had done so at a time when the population base it was serving was far smaller than it was after the first upswing of the transitional-growth long swing commenced. This system had been designed to handle only the needs of the city within its original pre-1897 boundaries. As the population of the metropolis and especially of its as yet unincorporated periphery soared, the authorities cobbled together a makeshift system of sewer lines from agricultural drainage ditches that ran through the farmland in the middle ring and periphery. These fetid waterways became the breeding ground for infection. With few flush toilets in the rapidly expanding middle ring of the city, and with reliance on waterways and ground wells for drinking water, is it surprising that loss of infant life was substantial?

Once the city incorporated its periphery in 1925, it set about addressing its water sanitation problem. It built a new water processing facility and extended the length of sewer and water distribution lines using unemployment relief workers. In mounting a campaign to cope with its burgeoning population, the city marshaled the assistance of local community organizations that sprang up within it. Particularly helpful were the local health cooperatives (*eisei kumiai*) that assisted the municipal authorities in disposing of fecal waste, much of which was sold off as night soil to farmers in the area.[12] Thus, in coping with its increasing numbers, the city focused upon building infrastructure.

As a result of population pressure and escalating land prices, Osaka's nighttime population began "hollowing out." The ratio of daytime population to nighttime population drifted up above a ratio of unity, especially in the center of the city. Commuting into the city became a way of life for increasing numbers of workers. In order to better understand this phenomenon, which was going on all over the Tōkaidō and spurred on by the Ministry of Home Affairs, the Cabinet Bureau of Statistics used the 1930 census as

a vehicle for collecting information on interprefecture migration and daytime commuting.[13]

From the figures in the 1930 census, it is possible to gauge how extensive commuting was in Osaka between the world wars. Using the census tables, I have constructed a set of figures on daytime to nighttime populations and on commuting flows for Osaka as a whole and for the three zones within it. These figures appear in Table 7.2, and show that the city's center was most decisively affected by commuting. In the center, the ratio of population flowing in to population flowing out was over four-to-one for males, and over three-to-one for females. In the early morning, workers came pouring in to the city through the Nanba terminus in the south of the core, and through the Umeda/Osaka Station complex in the north of the core. These workers staffed the banks, merchant houses, retail outlets, government offices, and department stores that were densely crowded in between Nanba and Umeda. Commuting flows for the periphery were the mirror opposite of those for the center. The periphery was a net supplier of daytime population to other locales, mainly the city's core. Only in the middle ring was there a balance. Inflows and outflows were approximately equal, and the ratio of daytime to nighttime population was about unity.

So, Osaka's core was increasingly coming to resemble the commercial center standard for large modern conurbations in North America and Europe. Still, Osaka's core did retain some special features. For instance, as is apparent from panel C of Table 7.2, in the core of the city labored a large number of people whose place of work was their place of residence. Presumably many of these individuals were apprentices working in the merchant houses and proto-industrial food processing businesses crammed into the city center. According to the 1930 census, the average number of members within a household was 4.9 in the core, and only 4.3 in the periphery.[14] The percentage of households with at least nine members was 26.5 percent in the center, 13.2 percent in the middle ring, and 10.4 percent in the periphery. The proportion of population residing in houses with at least six rooms is consistent with these figures. The proportion is 20.3 percent in the core, 8.6 percent in the middle ring, and 9.3 percent in the periphery. This pattern differs from the one that characterized most North American cities in the early twentieth century. In most North American cities, smaller households tended to cluster in the center, and larger households pushed out into the periphery. In interwar Osaka, the reverse was the case.

Osaka's interwar core was Janus-faced. On the one hand, it provided sites for massive department stores and imposing banks. On the other hand, within its environs clustered *tonya* and merchant houses whose lineages extended back to the early Tokugawa period.

Table 7.2

Commuting Students and Workers: Osaka and Tokyo, 1930

A. Ratio of weekday daytime to night time population in Osaka and its zones: males, females, and total[a].

Group/zone	Osaka City	Center	Middle ring	Periphery
Male	104.1	135.6	100.9	86.3
Female	100.3	107.6	98.7	97.3
Total	102.3	122.9	99.9	91.6

B. Ratio of weekday daytime to nighttime population in Tokyo City and Tokyo prefecture[b].

Tokyo City	Tokyo City, center	Tokyo City, middle ring	Tokyo prefecture	Tokyo prefecture, not in Tokyo City
114.6	156.7	103.6	100.6	91.9

C. Ratio of weekday flow of commuting students and workers coming into Osaka or its zones relative to those going out (IF/OF), and ratio of those working at home relative to those working in the same zone as their residence but not working at home (WAH/WOH)[c].

	IF/OF (inflow/outflow)			WAH/WOH (work at home/outside home)		
Zone	Male	Female	Total	Male	Female	Total
Osaka City	124.8	107.3	121.5	116.9	76.1	103.0
Center	442.0	315.3	413.9	200.7	132.5	177.2
Middle ring	106.0	69.9	98.9	92.7	66.3	84.8
Periphery	36.6	43.9	37.9	90.3	53.5	76.8

Sources: Nihon Naikaku Tōkei-kyoku (1935–1936) *Shōwa gonen kokusei chōsa hōkoku, jūgyō no basho*, various tables.

Notes: [a]In all ratios, the value of the base is set at 100.

[b]I define the center of Tokyo as the four "inner" wards of Kōjimachi, Kanda, Nihonbashi, and Kyobashi. The other eleven wards I call the "middle ring." In ratios, the value of the base is 100.

[c]In all ratios, the value of the base is set at 100. WAH/WOH measures the degree to which individuals working within their ward of residence actually work at their residence.

Housing

Osaka's population grew most vigorously in the periphery, and therefore the demand for new housing was concentrated there. Since population increase was especially rapid during the manufacturing expansions of the 1910s and 1930s, demand for new housing stock surged during those periods as well. In a market dominated by profit-oriented housing developers and factory owners catering to

the needs of their workers by providing dormitory housing—that is, in a market operating according to private market principles—demand and supply interacted through the price mechanism. However, it took time to acquire land, convert it, and build the housing. Therefore, demand and supply interacted with a lag. Particularly in densely inhabited Japan, and especially in Osaka because it was surrounded by valuable paddy fields, lags were protracted. Demand surged, housing prices and rents rose, and then supply increased.

The key variable linking demand to supply is the vacancy rate. Other things being equal, the lower the vacancy rate, the greater is upward pressure on rents. Declining vacancy rates induce increases in housing stock with a lag. The efficiency of Osaka's housing market—in terms of supply responding to demand—is demonstrated by Figure 7.2. As can be seen, movements in housing growth lag behind movements in demand growth, for which the vacancy rate is a proxy. Demand growth is inducing supply growth.

However, despite efficient coordination of supply and demand through vacancy rates and price, Osaka's housing market was increasingly becoming a political issue, a source of discontent that spilled over into the arena of social protest. Indeed, it was during the downswing years of the 1920s when vacancy rates were on the rise that the number of arbitration and mediation cases handled by Osaka's municipal authorities soared. These cases were brought under the aegis of the Leased Land and Tenant Arbitration Law, and year after year throughout the 1920s they increased in volume. For instance, there were 544 cases brought for tenant arbitration in 1924; by 1929, almost 3,000 cases were brought in.[15] Lease land arbitration also increased. In 1924, forty-three cases were handled by the authorities; by 1929, this number had increased almost sixfold.

Why did housing become such a festering issue during this period? Why did landlord and tenant unions proliferate in cities like Osaka at the same time that confrontations between landlord and tenant unions were fanning the flames of rural discontent? The most telling reason is land. With the upswings of the transitional-growth long swing, a boom in land was set off, with land prices soaring in the centers of the great Tōkaidō conurbations that were increasingly becoming magnets for heavy industry and blue-collar workers.

Other factors were undoubtedly at work as well. The labor market fell into the doldrums during the downswing of the 1920s. Lags were a problem. Vacancy rates had plummeted during the manufacturing boom of the 1910s. And housing stock only caught up with the earlier demand surge during the 1920s.

A fourth factor was housing quality.[16] As can be seen from Table 7.3, the rear tenement building constituted a significant share of Osaka's housing stock. These were typically dark, grim places opening up onto narrow walkways. Moreover, because rear tenements were being thrown up especially on the periphery where the city's growing numbers were locating housing, their

Figure 7.2 **Housing Stock Growth and Housing Vacancy: Rates for Osaka City, 1907–1935** (Five-Year Moving Averages)

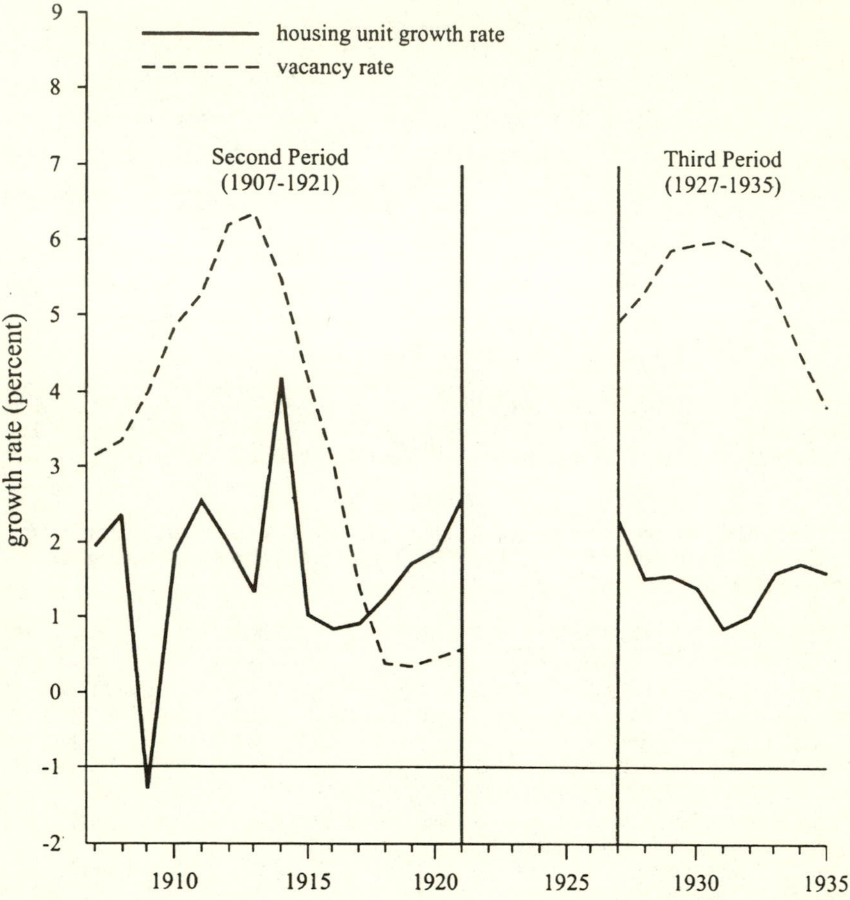

share in the city's housing stock was on the rise. As panel B of the table shows, housing developers were attempting to elastically expand the supply of housing units by increasing the number of floors in their structures. They were trying to expand upward rather than outward, treating the lack of land availability as their biggest bottleneck. Developers were busy demolishing low-level structures, and replacing them with multistoried apartment buildings and dormitories. This type of construction was especially prevalent in the city's middle ring and center, where land prices were the highest. In the periphery, where land was far cheaper, developers favored the single-floor tenement building facing out onto a back alley.

Table 7.3

Osaka Housing: Stocks and Flows, 1905–1938

A. Indices of housing units and of per capita daily water supply (1920–1921 = 100) and percentage of housing units that are (rear) tenements and percentage that are vacant.

	Indices (1920–1921 = 100)				Percent of housing units	
	Housing units			Per capita daily water		
Period	Front	Rear	Total	supply	Rear (tenements)	Vacant
	Second period in city's history					
1905–1909	81.3	81.5	81.4	12.8	35.8%	3.2%
1910–1914	91.4	80.8	87.6	35.6	32.7	6.2
1915–1919	97.5	93.7	96.1	50.1	34.8	1.4
1920–1924	105.9	102.1	104.5	110.8	34.9	1.0
	Third period in city's history					
1925–1929	185.4	244.8	206.6	82.3	42.3	4.6
1930–1934	213.6	265.7	232.2	87.5	40.9	5.2
1935–1938	249.0	237.5	257.8	90.3	37.9	3.2

B. Vacancy rate (VACR), demolition rate (DEMR), gross new housing construction rate (GNHCR), net new housing construction rate (NNHCR), and percentage of new housing that is two stories or more (PNHTS): Osaka and its zones, average values of annual figures for 1929–1932[a].

Variable/zone	Osaka City	Center	Middle ring	Periphery
VACR	5.4%	3.1%	4.6%	7.1%
DEMR	6.0	8.3	7.0	4.5
GNHCR	24.8	16.0	21.8	31.5
NNHCR	18.8	7.7	14.8	27.0
PNHTS	72.1%	79.4%	91.1%	61.5%

Sources: Ōsaka-shi Yakusho (various dates), *Ōsaka-shi tōkeisho*, various tables; and Ōsaka-shi Shakaibu (1996), various tables.

Notes: [a]The DEMR, GNHCR and NNHCR rates are per 1,000 housing units. NNHCR = GNHCR–DEMR.

Regardless of whether it was multi-level or single-level, virtually all of Osaka's housing stock was made of wood. These wooden structures easily caught on fire. Thus, fear of conflagration naturally brought municipal regulators into the housing market. How dependent was prewar Osaka on wood as a building material? Two surveys carried out in connection with Osaka's city planning—the first conducted in 1922 and the second in 1925—divided the city's buildings into three categories, A, B, and C. A-type structures were wood frame, wooden, wood frame with brick, or earthen. B-type structures were brick, stone, or concrete. And C-type structures were built with steel frames (sheeting

being made of brick, stone, or metal). In both surveys, 98.8 percent of housing fell into category A. And in both years, 97.6 percent of shops fell into category A. Only in the case of factories did the proportion in category A dip below 95 percent.[17] In short, Osaka was highly combustible. And for this reason, the city was becoming increasingly concerned about the proper maintenance and upkeep of the building stock.

Nonetheless, slums do not appear to have been a major problem in Osaka, nor in any of the other great conurbations of the Tōkaidō. A survey carried out by the Ministry of Home Affairs in 1925 of cities with over 500,000 residents reveals that only 2.5 percent of the housing units were considered slum units. Fires remained an issue. Thus, the threat of fire hanging over the metropolis was a cudgel that tenant unions could and did wield in their battles with landlord unions and with government bureaucrats.

Of course, residential housing is not the only type of structure concentrated within neighborhoods. This statement holds almost everywhere. But, it was especially true in giant cities of the Tōkaidō like Osaka. This is because shops, restaurants, and markets were everywhere in interwar Osaka, Tokyo, and Nagoya. It was the rare neighborhood that was without a shop.

A remarkable survey conducted by Osaka City in 1935 of its shops and shopping arcades provides us with quantitative documentation on the penetration of shops into ordinary residential neighborhoods. In panels B and C of Table 7.4, I give figures for the *chō* and *chōme* neighborhoods into which the wards (*ku*) were divided for the purpose of identifying subdistricts within their ward boundaries.[18] As can be seen from panel B of the table, the number of *chō* or *chōme* having resident populations within them—some consisted of areas that had little other than warehouses or factories within them—varied tremendously from ward to ward.

The characteristics of households with the neighborhoods also vary widely from ward to ward. Yet, despite this variation in the size and complexity of household units within these neighborhoods, there is a remarkable uniformity across the wards as far as the number of average per capita ratio of retail shops to population is concerned. This point can be seen from a study of the figures on the ratio of retail shops to population (the variable RSPP in panel C of Table 7.4) for the *chō* and *chōme*. In some wards, there are neighborhoods with extremely high ratios—these are neighborhoods with shopping arcades in them—but in general shops are situated everywhere. Some of the wards are primarily industrial, and some of them are primarily commercial, and some of them are primarily residential. Despite this, the average density of shops is pretty much the same in all of them.

Table 7.4

Shops and Households in the *Chō* and *Chōme* of Osaka: The 1935 Survey

Shops per 100 resident population and percentage of Osaka population and shops within zones[a].

	Shops per 100 population			% of population and shops by type within zone				
					Shops			
Zone	Wholesale	Retail	Both w/sale and retail	Population	Total	W/sale	Retail	W/sale/ retail
Osaka City	0.39	2.5	0.24	100%	100%	100%	100%	100%
Center	1.18	2.6	0.58	22.4	31.4	67.2	23.5	53.5
Middle ring	0.40	2.6	0.22	31.4	31.9	27.9	32.8	29.1
Periphery	0.04	2.3	0.09	46.2	36.7	4.9	43.7	17.5

B. *Chō* and *Chōme* in the wards: number of *Chō* and *Chōme* in ward (CHON); average and maximum populations (PAV, PMAX); and average and maximum household size (AHSAV, AHSMAX)[b].

Zone (ward)	CHON	PAV	PMAX	AHSAV	AHSMAX
Center (C): Kita	179	1,395	5,290	5.3	34.8
C: Higashi	205	819	3,596	6.6	21.6
C: Nishi	176 (177)	747	5,827	5.9	14.2
C: Minami	92	241	811	5.1	10.1
Middle Ring (MR): Konohana	113 (115)	1,891	13,598	13.2	861
MR: Minato	222 (227)	1,438	4,904	4.6	10.0
MR: Taishō	85 (87)	331	4,170	4.8	11.2
MR: Tenōji	68	1,830	6,996	5.0	16.7
MR: Naniwa	83	1,821	6,017	4.9	7.1
Periphery (P): Nishiyodogawa	65 (68)	2,918	22,124	4.8	11.5
P: Higashiyodogawa	184 (186)	1,233	14,192	4.9	15.5

P: Higashinari	100 (122)	2,985	26,175	4.6	5.5
P: Asahi	59 (65)	2,884	25,562	4.6	9.8
P: Sumiyoshi	245 (246)	1,115	17,542	4.9	63.4
P: Nishinari	211	965	16,734	4.6	5.7

C. *Chō* and *Chōme* in the wards: total shops per 100 households (SPH); Total shops per 100 population (SPP); wholesale shops per 100 population (WSPP); and retail shops per 100 population (RSPP): Average (AV) and maximum (MAX) within *Chō* or *Chōme*[b].

	SPH		SPP		WSPP		RSPP	
Zone (ward)	AV	MAX	AV	MAX	AV	MAX	AV	MAX
C: Kita	27.4	1,150	4.3	45.5	1.0	27.7	2.9	45.5
C: Minami	39.1	200.0	6.5	200.0	2.4	11.3	3.5	200.0
C: Nishi	33.4	174.1	5.9	32.9	2.2	21.0	2.6	9.4
C: Minami	26.0	76.0	4.5	10.1	1.1	5.5	2.6	7.7
MR: Konohana	16.8	571.4	3.5	105.6	1.1	98.6	2.3	27.3
MR: Minato	10.9	64.3	2.4	11.9	0.1	2.5	2.2	11.9
MR: Taishō	12.7	50.0	2.7	10.0	0.2	2.0	2.3	10.0
MR: Tenōji	12.5	38.3	2.7	8.0	0.2	1.0	2.2	7.6
MR: Naniwa	22.1	96.9	4.5	18.0	0.9	16.1	3.2	12.8
P: Nishiyodogawa	9.7	47.4	2.1	9.4	0.04	0.3	2.0	9.0
P: Higashiyodogawa	8.8	47.3	1.9	9.4	0.04	0.6	2.1	57.1
P: Higashinari	12.3	42.0	2.6	8.5	0.7	0.5	2.4	8.4
P: Asahi	8.9	44.2	2.0	10.5	0.03	0.5	2.0	10.1
P: Sumiyoshi	10.9	172.4	2.3	31.3	0.03	0.9	2.2	31.3
P: Nishinari	12.7	70.0	2.8	13.0	0.04	1.4	2.6	12.6

Sources: Ōsaka-shi Yakusho (1937), *Ōsaka-shi tempo bunpai chōsa hōkoku*, various pages.

Notes: [a]Population is resident population, not daytime population. Wsale = Wholesale shop, Wsale/Retail = shop combining wholesale and retail functions.

[b]Average and maximum are the average and maximum for the *chō* and *chōme* within each ward. Figures in ( ) in the CHON refer to total number of *chō* and *chōme*, including those with no resident population or no available data.

This uniformity for retail shopping does not hold for wholesale shopping. As can be seen from panel A of Table 7.4, wholesale operations are heavily concentrated in the city center. The department stores and the *tonya* are there. The dense canal network and the great electric train terminal complexes are there. Transport is good, and foot traffic immense. The same principle applies to why gas stations tend to cluster at the same intersections, why antique stores generally are situated next to each other, why wholesale shops locate in the city's core. Buyers know where to come. They come where there are a lot of possibilities, where they are likely to find what they are looking for. Hence, sellers tend to cram into the same area where other sellers are situated.

Thus, within interwar Osaka was a remarkable richness of markets and shops. Included in this diverse mixture of shopping venues were gleaming department stores built by national chains and regional transportation cum land developer *zaibatsu*, and tiny wooden shops crammed into huge shopping arcades like Shinsaibashi near Nanba Station.[19]

Of all of the markets crammed into Osaka between the wars, the most interesting was the one set up and managed by the city itself. This market was located in Fukushima ward and was situated along the banks of the Aji River near Nōda.[20] In my opinion, the construction of this market is a paradigm for the way national and local governments in modern Japan have responded to social disruption: build infrastructure. For the market's origins lie in the Rice Riots of 1918. In that year, the Ministry of Home Affairs sent out directives to municipalities, encouraging them to obtain land for the building of central wholesale markets. The notion was that by improving the efficiency of distribution, local government–managed central markets would drive down food prices within their environs. The ministry offered subsidies for building markets. Osaka was the first city to secure approval for a subsidy. Approval came in 1925, and the market was opened for business in 1931.

In the logic used by the bureaucrats of Osaka City and the Ministry of Home Affairs hangs a tale. It was a logic harking back to the days of Hideyoshi who, intent upon staving off internecine conflicts, invited merchants from Fushimi and Sakai to assemble in Osaka so that a national rice market could be created. Government best solves social problems by encouraging market activity, and building infrastructure is a way of accomplishing these goals.

Local Government as Fiscal and Coordinating Agent

The setting up of the central wholesale market in Fukushima illustrates the priority given by Japanese authorities to the building of infrastructure. It also illustrates how Japanese government at all levels acts as a coordinating agent,

bringing together private enterprises and individuals to truck, barter, and carry on market exchange.

Governments have a variety of options open to them. They can mimic military organizations, using direct command and control techniques. After 1938, the Japanese national government attempted to do this. Governments can use fiscal and monetary approaches. Of course, at the local level, only fiscal approaches are possible once a central banking system has been established. They can use transfers, taxing one set of parties in order to secure resources to give to other parties. They can adopt the regulatory mode, setting standards and imposing zoning. And they can use the coordinating or facilitating approach, in which they work with private enterprises. Examples of coordinating or facilitating activity include mediation and arbitration, industrial policy and the monitoring of cartels, and the like.

Of this menu of options, Japanese cities like Osaka have tended to emphasize two: direct fiscal intervention through spending on infrastructure buildup, and coordination. Why have they favored these two approaches?

Part of the explanation for the bias favoring spending on infrastructure and coordination lies in fiscal constraints. Some of these constraints were imposed on local governments by the architects of the Meiji government, who wanted to centralize fiscal decision making in national ministries. Before the American occupation following World War II, budgetary authority was largely centralized in Tokyo. Local governments were seriously hampered in their ability to raise taxes. Consider the following figures concerning taxes collected in Osaka between 1931 and 1940.[21] The figures represent percentages of all taxes, local and national, collected within the city during the years indicated):

	National	Prefecture Taxes			City Taxes			
Years	Total	Total	National surtax	Special	Total	National surtax	Prefecture surtax	Other
1931–1935	67.1%	12.0%	7.5%	4.4%	20.9%	5.2%	13.6%	2.1%
1936–1940	81.8	13.7	5.7	2.1	10.5	3.7	5.2	1.5

The national government collected the lion's share of taxes within Osaka. By controlling the purse strings, national ministries secured leverage over local government decisions and priorities. Thus, because infrastructure construction was a definite priority of the national government, it automatically became a priority of Osaka's municipal government.

The national government's emphasis on infrastructure makes sense in terms of its desire to promote economic growth and industrialization. The *bakufu* had set in motion extensive and intensive growth with its program of building up conurbations, roads, and waterways. After Japan opened up to the West in the 1850s, it began to systematically refashion its infrastructure to accommodate the new sources of energy and transportation that flooded into the country.

Infrastructure construction is both creative and destructive. In order to build new infrastructure, old infrastructure must be dismantled, refurbished, or perhaps completely destroyed. Doing this takes time, and eats up resources and capital, both economic and political. The value of privately held land is closely linked to the infrastructure in its immediate environs. Villagers need irrigation ditches for irrigating their paddy fields. They fight to make sure that widening of roads does not eliminate these ditches. Merchants resist giving up valuable floor space so that the city's road network can be expanded. In short, in a densely inhabited country like Japan—and especially in the densely inhabited conurbations of the Tōkaidō—the process of converting from one type of infrastructure to another is arduous and expensive. Since local economic interests were intimately tied up with infrastructure improvements, the national authorities favored schemes that moved direct responsibility for implementing improvements down to the local level.

With these points in mind, consider the overview of government expenditure given in Table 7.5. It is apparent from the figures in the table that the relative importance of government capital formation varied over the phases of the long swings (this can also be seen from Tables 3.1, 4.1, and 4.2). During periods when infrastructure investment was dominating economic growth, government outlays exceeded those coming from the private sector. During phases when manufacturing was the engine of expansion, the reverse was the case.[22]

Also apparent in these figures is a decided shift from central government outlays to local government outlays. This shift does not really have anything to do with a change in the capacity of local government to generate tax revenue. Rather, it reflects the fact that the national ministries were providing more and more subsidies to local governments for infrastructure improvements. During the first long swing, improving and constructing ports were key to infrastructure buildup. This was most efficiently done by national organizations that could mobilize scarce technical expertise and physical machinery for the massive projects. During the second and third long swings, the focus of infrastructure investment was on intercity electric railroads and intracity tramways, road networks, and subways. Carrying out these projects was generally best done at the local level by private companies and municipal and prefecture governments.[23]

National subsidies to local governments were important in funding local government investment in infrastructure. Local governments also undertook direct fiscal responsibility for some projects, or at least for portions of some projects. From panel C of Table 7.5, it is apparent that local governments entered capital markets during the second long swing to raise resources. And, as panel D demonstrates, these governments mainly raised resources in order to improve human capital and physical infrastructure. Not surprisingly, the emphasis on types of infrastructure invested in varied with time. For instance, in 1880–1899, about 24 percent of local government spending was on public works, and about 20 percent was devoted to education.[24] By 1920–1934, the proportion going to public works had fallen below 17 percent, and the proportions going to education, social welfare and sanitation, and support for industry and electricity and gas had all risen. Education now absorbed 24 percent of funds; social welfare and sanitation, 7 percent; promotion of agriculture, commerce, and industry, 5 percent; and electricity and gas, 6 percent. Successive waves of industrial expansion had put pressure on infrastructure, and therefore local government priorities had changed with the changing demands placed on it.

Municipal Planning: Theory and Reality

City planning was born out of pressure applied to infrastructure. Like the hybrid transportation systems and hybrid factories that jumbled up Western concepts and techniques with Japanese conditions and practices, the Japanese city planning movement was a mixture of Western innovation and Japanese tradition.

At a general level, city planning has been around for a long time. Ancient Rome had regulations concerning the building of structures.[25] In Tokugawa Japan, *bugyō* administrators took responsibility for coordinating neighborhoods in preventing fires and in securing well water for drinking and for fighting conflagrations. However, the new city planning movement sweeping the West after the 1820s was different from the earlier types of city planning that existed. Its aim was to cope with the disruptions to health and the violations of community space due to industrialization.

Within Europe, the character of the city planning movement varied from country to country.[26] In England, city planning evolved out of the mid-nineteenth-century public health movement, which was attempting to prevent the spread of infectious diseases like cholera within the industrial metropolises of the Industrial Coffin. Gradually, as the flight to the suburbs gathered force, the focus of the movement changed. At the turn of the century, the Garden City and New Town movements—whose goal was creating networks of model cities that mixed manufactures, services, and farming in a balanced way—took over center stage in the English setting.

Table 7.5

Central Government, Local Government, and Osaka City Government Expenditures, 1885–1940

A. Marginal contributions to growth in effective demand of government consumption expenditure (GCE), gross domestic fixed capital formation (GDFCF), private (domestic) capital formation (PCF), government (domestic) fixed capital formation (GFCF), and ratio of net capital imports to gross capital formation (RNCIGDF): Japan, 1885–1940.

	Ratio of increment of component of gross national expenditure to total increase in gross national expenditure				Ratio of capital imports to GDFCF[a]	
Period	GCE	GDFCF	PCF	GFCF	Period	Ratio
1888–1900	12.6	17.9	7.8	10.1	1885–1900	0.4 (9.9)
1900–1910	15.3	30.5	18.4	12.1	1901–1910	19.2
1910–1920	6.4	27.9	20.1	7.8	1911–1920	–5.9
1920–1930	26.0	7.5	–6.9	14.4	1921–1930	4.2
1930–1938	14.8	45.3	25.6	19.7	1931–1940	–5.1

B. Percentage of government expenditures on central government; on local government (prefecture, city, *gun*, town, and village); and on cities only.

	Period								
Level of government	1895–1899	1900–1904	1905–1909	1910–1914	1915–1919	1920–1924	1925–1929	1930–1934	1935–1939
Central	69.0%	65.3%	72.2%	63.9%	65.0	56.5	49.2	46.6	53.9
Local	31.0	34.7	27.8	36.1	35.0	43.5	50.8	53.4	46.2
City	3.7	5.7	6.4	10.9	9.7	13.3	21.4	22.1	18.7

C. Surplus or shortfall of revenue and proceeds of bond flotations, six big cities and Osaka, 1914–1929 (million yen)[b].

Group of cities, city/year	1914	1921	1922	1923	1924	1925	1926	1927	1928	1929
	Surplus or shortfall in revenue (million yen)									
Six big cities	–10	–52	–83	–88	–84	–136	–220	–442	–224	–207
Osaka	–3	–19	–23	–27	–26	–37	–50	–196	–52	–49
	Proceeds of bond flotations									
Six big cities	10	68	95	102	120	143	251	431	184	177
Osaka	1	32	16	36	29	50	51	171	28	28

D. Osaka city government: General administrative expenditures (GAE) and expenditures for city-managed companies (CMCE), 1930s[c].

Periods	Percentage of GAE expended on: City planning projects	Percentage of GAE expended on: Education	Ratio of CME to GAE, base = 100 (CMCE/GAE)	Transportation company expenses as % of CMCE	Transportation company expenses as % of GAE
Average for 1930–1932	15.9%	21.3%	86.0	66.8%	57.4%
Average for 1935–1937	13.9	22.8	83.2	65.8	54.7

Sources: Minami (1994), pp. 140 and 153; Nakamura (1983), pp. 167 and 170–71; and Ōsaka-shi (1952), *Shōwa Ōsaka-shi shi. Gyōsei hen,* table 4.

Notes: [a]The % in ( ) is for imports of capital inclusive of reparations from the Sino-Japanese War.

[b]The six big cities are Tokyo, Yokohama, Nagoya, Kyoto, Osaka, and Kobe.

[c]Transportation company expenses are a subcategory of CMCE. The ratio of these expenses relative to GAE, therefore, is calculated for purposes of comparing the burden of city-managed transportation relative to general administrative expenses and its subcategories like city planning projects and education.

In France, city planning was heavily shaped by the refashioning of the sewer and road networks of Paris under the direction of Baron Haussmann and his colleagues. Paris, hitherto a warren of winding narrow streets, was transformed by Haussmann's bold plan to create wide radial boulevards that cut through the city, creating star-like figures in some areas. Haussmann's innovations laid the groundwork for the development of modern transport in France. This promoted industrialization. At the same time, it stymied street fighting in the city. It also gave an aura of imperial grandeur to the capital. In Germany, systematic use of zoning emerged as a powerful new innovation for carrying out city planning. Zoning vested powerful regulatory power in the hands of municipal technocrats who divided up cities systematically into districts, some reserved for residences, some for factories, some for leisure and recreation.

As innovations in city planning proliferated in Europe, the role of city planning was widened. Increasingly, the city planner was expected to make decisions about building new infrastructure, develop zoning, and take measures to ensure that public health was adequately maintained.

In one sense, the development of city planning in Japan reflects the sequence with which various Western innovations made their way into the country.[27] In the early Meiji period, the opening up of ports to commerce with the Western powers brought on outbreaks of cholera. Cities turned to Western concepts to fight these outbreaks. Laying down metal pipes and providing adequate drinking water and sewage removal became priorities during the first long swing. Later, in preparation for the promulgation of laws on local government in 1889, the national authorities issued a special ordinance for the capital. This was known as the Tokyo Municipal Ordinance. Under this set of directives, Tokyo's infrastructure development was put under the direct control of the Ministry of Home Affairs. Strongly influenced by Haussmannization, the Ministry used Paris as a model for building up a capital that would be a symbol of modern civilization but would also have a touch of imperial grandeur. Thus, they set about improving upon, and extending, the grand boulevards and thoroughfares that had crisscrossed Edo during the heyday of the *bakufu*.

Wholesale suburban development in the Tōkaidō following the first upswing of the transitional-growth long swing redirected the focus of the city planning movement once again. Now, the Ministry of Home Affairs was abuzz with discussions of the Garden City movement and of the influential volume penned by Ebenezer Howard, *Garden City*.[28] According to the view that sees Japan's city planning movement evolving step by step as hitherto untapped elements of the international city planning movement were discovered and imported into the country, the decision of the Ministry of Home Affairs to

extend the coverage of the ordinances worked up for Tokyo to the other five big cities of the Tōkaidō appears to be a natural extension of imitation of the West that is seemingly prevalent throughout Meiji Japan. In other words, an elite bureaucracy became aware of developments abroad, sifted through them, established priorities, and thereby shaped the city planning movement in Japan. This view argues that the Japanese city planning movement is rooted in Western innovations and approaches to administration.[29]

I would like to suggest an alternative view here. While it is true that Japan's city planning is a hybrid, and as such does incorporate Western innovations, it is a hybrid with diverse origins, not all of them coming from city planning as practiced in early modern Japan, or in the industrializing West. To be sure, the Ministry of Home Affairs was keeping tabs on fresh developments in city and regional planning as it was evolving in Europe and in North America. Nevertheless, the real roots of Japanese city planning lie in the practical difficulties of converting infrastructure. They are mainly tied up with developing infrastructure suitable for accommodating Western modes of transportation. In what seems to be a bizarre twist, the vehicle that has served as the main engine of city planning in Japan—land reorganization (*kukaku seiri*)—emerged out of a law passed in 1899 for improving farming, namely the Arable Land Reorganization Law. This law was originally designed to enhance the productivity of farming by promoting the consolidation of minuscule plots into larger plots, effectively accommodating the use of draft animals.

Japanese city planning did have Western origins. However, the main germ was not Haussmann in Paris. It was not Ebenezer Howard. It was not German zoning. Rather, it was the model of enclosed farms central to agriculture in the West. And its main use was to building infrastructure for modernizing transportation. Its flavor was rationalization of land use for expansion of industry. In this sense, creating greenbelts and promoting radical zoning initiatives for carving out of the middle of metropolises beautiful natural oases were not really seriously put on the agenda.

For if one reviews the city planning documents worked up by the Osaka authorities in preparing for an evaluation of its plans by the committees commissioned by the Ministry of Home Affairs, one is struck by two things. The focus of Osaka's planning is almost exclusively construction projects. Moreover, the main mechanism suggested for securing the land required for the projects is *kukaku seiri*. What is conspicuous by its absence is a grand design, an overriding vision and rationale, articulation of a set of ethical or aesthetic concerns, and the like. Rather the focus is completely practical. It smacks of the rationalization that deeply informs infrastructure reorganization and expansion during the downswing of the transitional-growth long swing. Projects are proposed on a piecemeal basis.

The project-by-project character of Osaka's city plans is best summarized by examining how it was evaluated in terms of reaching targets for individual projects. Prior to 1939, Osaka adopted for review by the Ministry of Home Affairs three plans. The first plan was worked out in 1919 and approved in 1921. It established targets for improving waterways and sewers, avenues and cemeteries, and parks. This plan was completed in its entirety over a decade period commencing in 1921. The second plan, upon which actual work commenced in 1932, was completed at about a 60 percent rate within two years. The third plan, adopted in 1937, was formulated as Japan began its massive military campaign in China, and it was barely implemented before the Pacific War got under way, shattering the possibility for completing any targets. If we look at specific projects, completion rates for the prewar period are as follows: Osaka station area renovation had a completion rate of 70 percent; building avenues and cemeteries had a completion rate of 100 percent; creating parks and green areas had a completion rate of over 30 percent; sewer and water improvement had a completion rate of slightly more than 30 percent; and the subway construction project was 15 percent completed.[30]

Priority was given to developing roads in the earliest city planning developed for Japan as a whole. Indeed, a law on road development was promulgated in 1919, the year city planning was codified. This focus on road construction was natural because of the pressure put upon infrastructure due to the industrial expansion of the 1910s. It was natural because the first city planning carried out within Japan was exclusively for Tokyo, and developing roads was always a priority in Tokyo. It had been a priority in Edo because of earthquakes and fires. For the same reason it was a priority in early Meiji Japan. The harnessing of the internal combustion engine to urban transport only increased the interest of the authorities in widening existing roads and extending the road network in the capital.

In Osaka, which embarked upon systematic city planning after the escalation in central city real estate prices, bureaucrats had to contend with difficulties that those making plans for Tokyo did not have to wrestle with. Existing streets in Osaka were narrow. And the cost of securing land to widen the roads was horrendous. For this reason, much of Osaka's urban planning expenditures were devoted to converting the narrow Midosuji Street joining Umeda to Nanba into a spacious boulevard. This project dragged on until 1937. In that year national roads number sixteen, the Naniwa-Sumiyoshi Boulevard, and number twenty-six, the Nanba-Sakai Boulevard, were also opened up for traffic.[31] Widening roads in the Osaka area was extremely costly. Thus, the national government played a key role in ensuring that serious progress was made.

Osaka's government not only faced severe fiscal constraints in building

the infrastructure key to its interwar city plans. It also found itself often hemmed in politically. The city was caught between the power of the Ministry of Home Affairs above and the power of local community groups like the *chōkai* below. Because local community groups could hold up infrastructure construction by actively opposing it and refusing to sell land that its members owned, local municipal administrations had to proceed gingerly. This is one reason they actively encouraged the creation of, and courted, Land Reorganization Associations.

The idea behind Land Reorganization Associations was simple. Local Land Reorganization Associations—*kukaku seiri kai* in Japanese—received resources for, and were empowered to make, improvements. The main government funding for these projects took the form of grants-in-aid made by the Ministry of Home Affairs. The association members themselves were usually expected to come up with part of the funding themselves. In this way, little groups improved streets, built pocket parks, and secured adequate sewage and fresh water services.[32] The idea was that the market would compensate the members of the association for their effort and their funding. Once the infrastructure was put into place and improved, the prices of the real estate that association members could command on the market would rise. Thus, working through Land Reorganization Associations accomplished two goals, one political and the other economic. Actively soliciting the cooperation and ideas of association members staved off political opposition to infrastructure improvement. In addition, it reduced the direct financial burden shouldered by government.[33]

Improving the front of Osaka station illustrates well the realities of *kukaku seiri*. The actual renovations took place over a ten-year period, from 1932 to 1942. But prior to this, five years were devoted to a planning phase, when *kukaku seiri* associations wrestled with debates over how to proceed. Who was to surrender land? How much was each party supposed to sacrifice for the common good? The process ate up precious time.

In effect, Osaka's city government became hostage to local neighborhood politics and big business interests. Under the logic of *kukaku seiri*, it was forced into a coordinating role, working with the local electric railway *zaibatsu*, the Japan National Railroads, and the owners of small shops in the vicinity. Hemmed in on all fronts—by regional *zaibatsu*, national ministries, and neighborhood associations—the ingenuity of city bureaucrats was taxed to the limit.

For this reason, the administrations of the major cities of Japan attempted to increase their political leverage by reaching out to one another. For instance, in 1927, Osaka hosted the National City Problems Conference for representatives of the six big cities.[34] But such efforts did not vastly enhance

the positions of the administrations of the great conurbations of the Tōkaidō. Their best option was to rely upon coordinating private interests. Osaka was no exception.

One of the casualties of relying upon coordinating private interests was zoning. To effectively implement zoning, a city's administration must assert itself through comprehensive and binding regulation. It must impose binding constraints on private agents. And it must enforce its dictates through fines and penalties. Osaka's city administration was in no position to do this.

Under the city's putative zoning plan—which was laid out in its planning documents—the city's center was to be given over to commercial establishments. The remainder of the city was divided into two districts: one basically industrial, the other basically residential. The industrial district was to encompass the entire port complex. It was supposed to stretch from the northwestern sector of the city along a broad band fronting onto the Shinyodo River and the harbor all the way to the boundary with Sakai City to the southeast. In addition, a swath of land running along the core of the city's northern rim (near Hideyoshi's ancient castle) was also given over to industrial purposes. The remainder of the city not assigned to commerce or industry was to be for residential purposes. Thus, a strip of land in the northwest of the city, and most of the southwestern reaches of the city, were to be set aside for housing. At least this was the plan.

But reality was different. Let us compare the planned concentration of factories in the city's plans with the actual concentration of factories as it developed over the period 1933–1939. As can be seen from Table 7.6, the city was not successful in keeping factories from sprouting up in those parts of the periphery set aside for housing. Indeed, from panel A of the table, it is apparent that the number of factories actually declined in the region earmarked for industrial development, and the number of factories rose in the area supposedly set aside for residential use. The main reason for this is the proliferation of very small factories in those neighborhoods set aside for housing under the plan. The figures in the table make this clear. If we restrict our concern to large factories (with thirty workers or more), the actual reality is closer to what the planners intended. In any case, reality and theory were out of line. Osaka's power to zone was severely constrained because it relied on coordinating private interests.

In sum, as Japanese city planning evolved, it absorbed concepts and techniques pioneered in the West. It combined these with indigenous practices. As a result, it innovated. A hybrid was created. The use of *kukaku seiri* is testimony to the power of an evolutionary process in which innovation, drawing upon disparate and unexpected sources, brought private and public interests together in an economically and politically viable fashion.

Table 7.6

The Movement of Factories in Osaka, 1933/1939: Center, Middle Ring, and Periphery

A. Net additions or losses in number of factories between 1933 and 1939 (1939 value minus 1933 value) according to factory size (number of employed workers per factory) and zone of city[a].

Factory size group/zone	Osaka City	Center	Middle ring	Periphery (and planning sub-district) Total	Residential	Industrial
All factories	–7,507	–3,824	–3,981	+288	+834	–546
1–4 workers	–4,060	–1,260	–2,223	–577	+541	–1,118
5–9 workers	–4,494	–2,183	–1,743	–568	–48	–520
10–29 workers	+689	–297	+18	+958	+246	+712
30–99 workers	+230	–62	-63	+355	+82	+273
100–999 workers	+110	–19	+13	+116	+12	+104
1,000 workers & up	–2	–3	–3	+4	+1	+3

B. Net additions or losses, 1933/1939, in number of factories in textiles, metal working, and machine/tool making: factories with 30–99 workers and factories with 100–999 workers[a].

Sub-sector/ zone	Osaka City	Center	Middle ring	Periphery Total	Residential	Industrial
	Factories with 30–99 workers					
Textiles	+11	+1	–4	+14	+9	+5
Metalworking	+64	–9	–17	+90	+12	+78
Machine/toolmaking	+265	–3	+20	+248	+68	+180
	Factories with 100–999 workers					
Textiles	+3	0	0	+3	+2	+1
Metalworking	+35	–1	+3	+33	+3	+30
Machine/toolmaking	+84	+1	+9	+74	+10	+64

Sources: Ōsaka-shi Yakusho (1935), *Ōsaka-shi kōgyō chōsa, Shōwa hachinen*, various tables; and Ōsaka-shi Yakusho (1940), *Ōsaka-shi kōgyō chōsa, Shōwa jūyonen*, various tables.

Notes: I divide the periphery into two subdistricts according to the zoning designations laid out in Osaka City planning: the district set aside for residential development (which I approximate with Sumiyoshi and Higashinari wards combined) and that set aside for industrial/factory development and/or residential development (the remaining wards within the periphery).

Tokyo Triumphant

Tokyo and Osaka have been rivals for a long time. In the early seventeenth century, an army coming out of Edo and the Kantō plain attacked the *samurai* retainers defending the interests of Hideyoshi's line, thereby destroying much of Hideyoshi's great castle in the successful effort to bring Osaka un-

der direct *bakufu* rule. In the 1920s, the combined machinations of the *zaibatsu* based in Tokyo managed to bring down the Bank of Taiwan and the Suzuki *zaibatsu* based in Osaka. The rivalry has been intense. It has been bitter. It has been fought out in the military and political fields. And it has been fought out in economic matters.

In the decades following the Meiji Restoration, Osaka successfully built upon its proto-industrial base. During the first long swing, it became the Manchester of the Far East. Meanwhile, Tokyo's fortunes seemed to be deteriorating. The capital presented such a poor spectacle that many high-ranking officials actually contemplated building a new capital elsewhere. Indeed, a plan was even floated for creating three capitals in Japan. Kyoto would be the imperial capital; Osaka would be the western capital; and Tokyo would be the eastern capital.[35]

Gradually Tokyo's situation improved. True, a plan put forward by its business community for a deepwater port failed to secure funding. Nevertheless, industry began to spread along the banks of Sumida and Ara rivers, in the so-called downtown *shitamachi* area of the metropolis, and along Tokyo Bay.[36] Still, the industrial buildup in Osaka far outdistanced that in the capital. Osaka's advantages in waterways, proto-industrial skills, and financial acumen supported its preeminence in economic affairs.

With the second and third long swings, the demand for new kinds of infrastructure soared. The physical infrastructure demanded extensive use of land. It included electrical power grids, electric railroads and tramways, and roads. The new human capital–enhancing infrastructure centered on institutions of higher learning and industrial research centers. And the new financial infrastructure centered on big city banks, the Bank of Japan, and the *zaibatsu*. In all of these areas, Tokyo had a significant competitive advantage over Osaka. Hence, because of the close association between infrastructure creation and industrial expansion, there was a strong pull toward Tokyo-centrism in economic affairs after the new century dawned.

Tokyo's early lead in city planning also bolstered its competitive advantage. From three decades of experimentation with city planning between 1889 and 1919 in Tokyo sprang many of the key ideas underlying the city planning law of 1919.[37] Tokyo was the crucible from which emerged Japan's modern practices of city planning.

Finally, Tokyo drew an arrow out of Osaka's quiver. In the late 1930s, as landfill went into the Kawasaki area lying between Tokyo and Yokohama, Tokyo's harbor was deepened, and it was connected directly to the harbor in Yokohama by a great canal. As a result, the mighty Keihin port complex in and around Tokyo Bay emerged. Onto this *shinden* went steel mills, shipbuilding yards, and factories in other key heavy industries.

And so, as innovator and imitator in infrastructure, Tokyo surged ahead of Osaka in the economic realm. As a result, the geographic center of Japan's burgeoning economy—the heart of the Tōkaidō core—was pulled northward from the Kinai to Kantō, from Osaka to Tokyo. Six decades earlier a city in decline, Tokyo had emerged triumphant, staking out its claim to be the political, intellectual, and economic center of the nation. Tokyo-centrism was firmly and definitively established during the third long swing that carried Japan to economic heights no one would have dreamed of a mere fifty years earlier. Japan was now a great industrial and military power, and Tokyo was the political capital of a rapidly growing empire on the Asian mainland and over the waters of the Pacific.

Notes

1. For details on many of these points, see Mosk (1996, 2000a). See Szreter (1997) for a general theory as to why disruption accompanies industrialization and for illustrations of the notion in Great Britain in the eighteenth and nineteenth centuries.

2. For the net nutritional hypothesis, see Mosk (1996).

3. Minami (1998) and Mosk (2000a) reach rather different conclusions about whether relative inequality increased or decreased in Japan between the world wars. Using estimates of Gini coefficients for household income and figures on wage differentials, Minami concludes that inequality increased. Employing figures on the biological standard of living, Mosk argues that socioeconomic differentials either did not widen or actually shrank during the period between the wars.

4. The fact that fertility rose in northeastern Japan between the early 1900s and the mid-1920s may have caused the differentials in the biological standard of living to increase. For details, see Mosk (2000a).

5. In early-twentieth-century Japan, age-standardized mortality was higher in urbanized and industrial prefectures than it was in rural agricultural prefectures. Thus, as Mosk and Johansson (1986) demonstrate, there was an inverse relationship between per capita income and mortality in the prefectures of the country. Later, due to improvements in the application of the germ theory of disease to public health, and due to exploiting scale economies in providing public health services in major urban areas, mortality fell more dramatically in urban Japan than it did in rural Japan, during the second long-swing period. By 1930, per capita income and age-standardized mortality were largely independent of one another at the level of the prefectures. And, by 1950, age-standardized mortality was far lower in cities and industrial prefectures than it was in rural prefectures. For this reason, the gap in income between rural and urban areas was also, increasingly, becoming a gap in health. Therefore, the income gap was increasingly being associated with a gap in overall welfare.

6. For a discussion of common-pool resources, see Ostrom (1990). Common-pool resource stocks yield flows of benefits that are appropriated by groups and individuals. Once appropriated, these flows are not available to other persons or institutions. There is rivalry in consumption. Public goods are different. There is no rivalry in the consumption of public goods (a standard example is national defense). On the ideology of Japanese approaches to public sector intervention in social matters, see Garon (1997), which emphasizes the role of moral suasion campaigns in Japan.

7. See pages 10–11 in Ōsaka-shi Shakai-bu (1996), *Ōsaka-shi shakai-bu chōsa hōkokusho*, vol. 27.

8. The figures are from Harada (1968, 135). Most of the unemployed fresh school graduates came from the vocational schools, but a significant number were graduates from *senmongakkō*. The figures on unemployment probably understate the actual incidence of unemployment. Many of those laid off were young girls who had come to Osaka to work in the mills. Once they were laid off, most of them probably returned to the farms from which they came.

9. See pages 34–37 in Ōsaka-shi Shakai-bu (1996), *Ōsaka-shi shakai-bu chōsa hōkokusho*, appendix volume. Between 1919 and 1942, Osaka carried out 126 surveys. Slightly more than half of these dealt with labor issues. By comparison, Tokyo, whose bureau for social affairs carried out eighty-nine surveys during the same period, showered less attention on labor problems. It devoted less than a third of its surveys to these matters. Sugihara and Tamai (1996, 56–57) discuss social welfare programs set up in Osaka during the interwar period. In 1919, the private Ohara Institute for Social Research was established in Osaka. Its agenda was the study of social problems involving wages, population pressure, suicide, and so forth.

10. See pages 84–85 in Shinshū Ōsaka-shi Shi Hensan Iinkai (1993), *Ōsaka-shi shi*: vol. 7.

11. Figures for Nishinari-*gun* and Higashinari-*gun* are from prefecture yearbooks. These volumes do not provide figures on infant deaths, despite the fact that the yearbook for Osaka City does provide figures for infant deaths within its boundaries. In general, infant death rates in pre-industrial Japan appear to be fairly low. Cf. Mosk (1983, 1996). For this reason, I consider the infant mortality rates for pre-1925 Osaka to be quite high.

12. See page 543 of Ōsaka-shi (1955), *Shōwa Ōsaka-shi shi. Shakaihen*. Until the 1920s, when farmers began to rely more heavily upon chemical and synthetic fertilizers, most of Osaka's fecal waste was sold to nearby farming communities for night soil.

13. Japanese officials told me that the 1930 census was the most accurate and comprehensive in Japanese history.

14. These figures are taken from Table 29 (pages 386–415) in Nihon Naikaku Tōkei-kyoku (1935) *Shōwa gonen kokusei chōsa hōkoku. Fu, ken hen*, vol. 29.

15. These data are drawn from *Ōsaka jutaku nenpō* reprinted in the Ōsaka-shi Shakai-bu (1996) series.

16. On slum areas in Osaka, see Sugihara and Tamai (1996). National programs aimed at subsidizing housing programs developed after landlord-tenant disputes and confrontations proliferated in the great cities of the Tōkaidō. The national government began to make direct grants and low interest-rate loans to tenant unions.

17. See page 458 in Ōsaka-shi (1954) *Shōwa Ōsaka-shi shi. Keizaihen*, vol. 2 (vol. 2 of *Keizaihen*). Between 1926 and 1936, there was a substantial decline in the proportion of warehouses fashioned from wood.

18. Whether the term "*chō*" or "*chōme*" is used appears to be arbitrary. I am grateful to Saito Osamu for this point.

19. The first subway line completed in Osaka, in 1933, had a terminus at Shinsaibashi.

20. The author recently visited the market and secured some useful literature concerning its history and the scope of its present-day operations. Cf. Ōsaka-shi

Chūo Oroshiuri (1997). Needless to say, one does not see *niguruma* there now. But, in photographs taken in the 1930s, it is apparent that wooden freight carts were parked in massive arrays along the edges of the market. Today, trucks take up the designated parking areas.

21. See page 25, and the pages following, in Ōsaka-shi (1954) *Shōwa Ōsaka-shi shi. Keizaihen*, vol. 1 (vol. 1 of *Keizaihen*). National taxes include income taxes, taxes on corporate profits and interest income, and so forth. Special—other—prefecture taxes include taxes on houses, boats, telephone pole use, and so forth.

22. It should be kept in mind that import of capital from foreign countries played very little role in Japan's economic development. Indeed, Japan was a capital exporter to its empire during the 1920s and 1930s. Securing war indemnities did play a role in Japan's acquiring foreign financial resources, however.

23. On the responsibilities of national, prefecture, municipal, town, and village governments for road maintenance and construction, see chapter 5.

24. These figures are taken from Ohkawa, Shinohara, and Umemura (1966, 192–93).

25. Concerning city planning in Rome, see Robinson (1992).

26. For the development of city planning in Europe, see Benevelo (1971), Hohenberg and Lees (1985), and Ladd (1990).

27. For my discussion of the history of city planning in Japan, I draw upon Fujimori (1982), Honjo (1971), Ishida (1987), Nihon Naimu Daijin Kanbō Keikaku-ka (1988), Ōsaka Toshi Kyokai (1933), Tōkyō Kukakusei Gyō Shi (1987–1988), and Watanabe (1993).

28. The Ministry of Home Affairs had Howard's book translated into Japanese. See Watanabe (1993, 45).

29. The decision to extend city planning to the six big cities was made in 1918. In 1923, the City Planning Law was extended to another twenty-five cities. See Nihon Naimu Daijin Kanbō Toshi Keikaku-ka (1988, 15 ff).

30. See Ōsaka-shi Kai Jimu-kyoku Chōsa Kai (1970), 41 ff.). Also see the references cited in footnote 27 immediately above, and Ōsaka-shi (1955) *Shōwa Ōsaka-shi shi. Gyōseihen*.

31. Osaka's government also invested a lot of resources in paving its roads. As a result, by 1938, around 45 percent were actually paved.

32. In some cases, government agencies provided full subsidy for projects that they approved.

33. For contemporary applications of the method, see Hebbert (1994), Honjo (1971), Osaka City Government, City Planning Department (1997), and Ōsaka-shi Kensetsu-kyoku (1995).

34. Samuels (1983) emphasizes the importance of horizontal linkages in the formulation of local government policy in Japan. This emphasis differs from the one given by Steiner (1965), who stresses the importance of "top down" relationships.

35. See Fujitani (1996) and Stewart (1987).

36. See Map 3.2 in chapter 3.

37. Osaka advanced its own proposals in the debates leading up to the promulgation of the city planning law. To a degree, Osaka's views were accommodated in the legislation. See Watanabe (1993, 158 ff.).

8

Conclusions

The Argument Restated

The law of unintended consequences is always with us. When the Western powers broke Japan open during the 1850s, it was with the aim of securing coaling stations on the Asian rim of the Pacific and of finding new markets for goods and services. They did not expect to awaken a sleeping giant. Without the slightest intention of doing so, Western penetration of Japan unleashed one of the most remarkable chains of innovation the world has witnessed. As a result, within eight decades, Japan had emerged as one of the world's great industrial and military powers.

Why was Japan able to successfully mount a response to the Western challenge? Why was the country able to accomplish something that no other country outside of the European cultural area had yet managed to achieve? Why did wave after wave of innovation sweep across Japan and not across other regions of the globe confronted by the Western powers during the nineteenth century?

The argument laid out in this book addresses these questions. It is simple. Japan was able to innovate because of its infrastructure. Modern economic growth in Japan has been infrastructure driven. Economic development occurred in Japan because infrastructure investment laid the groundwork for industrial expansion that in turn created a demand for more infrastructure, for new infrastructure, and for ancient infrastructure transformed. Innovation occurred when infrastructure was built. This innovation spilled over into the industrial sector. Innovation transformed industry. The resulting expansion in industrial activity put pressure on the infrastructure sector. It responded through innovation. Growth fed on growth.

The Japan of the 1850s was agrarian and proto-industrial. Proto-industry and agriculture were supported by a well-developed infrastructure. Great cities, castle towns, and smaller market towns were knit together by a network of roads and waterways. The merchant communities in conurbations like Osaka were well versed in marketing and in financial dealings. They had developed accounting, futures markets, insurance for ocean-going voyages, and sophisticated methods of handling a variety of currencies and media of exchange. Industrial skills were proliferating, especially in proto-industry. Education was being advanced among both the *samurai* bureaucrats and the peasantry, especially the wealthier peasantry who were diversifying into proto-industrial craft activities.

Infrastructure and proto-industry were especially developed in the Tōkaidō region. Most of the Tōkaidō lay along the narrow belt of land on the Pacific seaboard of Japan's main island of Honshū stretching from Tokyo Bay and Edo in the north to Osaka and the Inland Sea in the south. Also in the Tōkaidō was the Kinai, encompassing the great Yodo River basin that emptied out into Osaka Bay and drained the waters of Lake Biwa and the rivers of Kyoto into the Inland Sea. Several great seaports in the Tōkaidō handled the shallow wooden craft that circulated around the main island of Japan. These included Edo, political capital of the *bakufu*; Osaka, capital of Japan's merchant community; and Nagoya. Edo and Osaka were huge conurbations, alive with merchants, craftspeople and artisans, and *samurai*.

When the Western powers secured extraterritorial rights in Japan after 1858, they naturally wanted to maximize their market penetration. Thus, they negotiated for treaty ports in the Tōkaidō where population was concentrated and infrastructure was good. When their nationals took up residence in Edo and Osaka, they brought with them their steam-driven ships and their ideas for harnessing steam power and fossil fuels to transport and for manufacturing. Japanese nationals, seeing the utility of the Western technology for building both national wealth and military power, began to take an active interest in harnessing it on their own terms. Thus, spawning hybrids—partly Western and partly Japanese—commenced in earnest. This spawning of hybrids was naturally concentrated in the Tōkaidō.

Osaka, center of a vibrant proto-industrial community and home to a host of market-savvy merchants, became one of the chief centers for this wave of innovation. Osaka specialized in harnessing the innovations for industrial production and for transportation. With its maze of canals and its easy access to Kobe, a natural deepwater port, Osaka was able to rapidly harness steam power and the latent energy in coal to water transport. It then aggressively lobbied for a deepwater port and secure harbor. Artisans and proto-industrial entrepreneurs were active in Osaka, and thus it soon emerged as a center for

industrial innovation as well. Innovating held out the potential reward of high profits. Thus, entrepreneurs were naturally attracted to using and adapting the techniques for energy production and the machinery developed in the West. The steam-driven factory took root in Osaka.

By contrast, Tokyo became the center for political innovation. Grafting Western concepts of government, commerce, and banking onto traditional Japanese institutions, Tokyo spearheaded a revamping of Japan's educational and administrative infrastructure. The Bank of Japan was created, and specialized banks like the Yokohama Specie Bank were set up. The growth of private banking was encouraged, partly because the new government issued bonds to the former *samurai* and encouraged them to invest these bonds in banks. In doing away with the old, the government in Tokyo was determined to develop policies that made the transition to a new economic world as painless as possible for those with a vested interest in the old regime.

Hence, innovation in Meiji Japan took a variety of forms. There was innovation in industry, innovation in politics and government. With this innovation, the old was pushed aside in order to make room for the new. Innovation abounded in all areas of economic and social life. To refute the old argument that the Japanese were simply imitators of the West, one simply needs to point to the following phenomena. A new form of enterprise emerged, the *zaibatsu*. The proto-industrialization of manufacturing took place. Hybrids were spawned in profusion: the *jinrikisha*; the postal delivery system relying on a diverse mixture of Western and Japanese forms of transportation; classical Greek and Roman architecture wedded to Japanese craft in woodworking; and an imperial court that oscillated between ancient tradition in Kyoto and Western imperial tradition in Tokyo.

As a result of these innovations, Japan passed through the first long swing. It was a balanced-growth long swing in the sense that industry, agriculture, and infrastructure all developed simultaneously. For the first time in its history, Japan's economy grew rapidly. Between 1887 and 1897, output and output per head grew at an average annual rate of around 3 percent. The rate of capital accumulation was also high, nearly 6 percent. Eventually industrial expansion outpaced the infrastructure supporting it. The economy slowed its growth, and a downswing occurred between 1897 and 1904.

Now the infrastructure sector responded. Economic growth picked up with the boom, spearheaded by the intercity electrical railways and their program of offering electricity to surrounding communities as well as railway service. Population had increasingly crammed into the great conurbations of the Tōkaidō during the first long swing. Therefore, the demographic base supporting intercity railways and electrical power grids was well established. Unit costs for providing railway services and electricity were low in the densely populated giant conurbations of the nascent industrial belt.

Electrification revolutionized manufacturing. By promoting the unit drive system, it spurred on the mechanization of manufacturing in large and small factories alike. As a result, subcontracting—which was rooted in proto-industrial putting out—proliferated.

Electrification also created demand for electrical machinery. Japanese companies, attracted by the profits they could earn in the new field of machinery production, jumped into the market. The breakdown in international trade due to the onset of World War I helped them expand their domestic markets. Thus, a massive industrial upswing took place between 1911 and 1919. During this period, overall output grew at over 4 percent and manufacturing at over 7 percent. As the upswing gathered force, the demand for transport grew. Japan began importing automobiles and trucks to meet this demand. The internal combustion engine and electricity were the revolutionary new sources of mechanical power underlying the innovations of the second long swing.

The industrial upswing of the 1910s put tremendous pressure on the physical, human capital–enhancing, and financial infrastructure. Prices soared during the upswing, especially for land in the cores of the great Tōkaidō conurbations. This stimulated speculation and overexpansion of credit. The renewal of imports of machinery and capital goods after World War I cut into the profits of domestic suppliers whose efficiency did not match that of foreign enterprises. When the downturn began in the aftermath of World War I, banks began to fail in droves. Demand for engineers and technically trained blue-collar workers in heavy industry had also soared during the upswing. Thus there was pressure on the institutions of higher learning and on the vocational schools. At the end of the industrial upswing, there was also excess demand for more electricity, and for roads along which buses and trucks could run.

Government, working in a coordinating role with powerful national and regional *zaibatsu* and with local community groups, responded to the pressure imposed on infrastructure. Roads were built. City planning was extended to all of the great conurbations of the Tōkaidō, and then shortly thereafter to larger groups of cities. Electrical power grids were expanded. Higher education was revamped and the Imperial University system was expanded. Banks were merged and consolidated with the government's encouragement. All of this took place during the downswing of the second long swing, between 1919 and 1930.

The second long swing was a transitional-growth long swing. During the course of the industrial upswing of the 1910s, heavy industry emerged as the engine of growth. Labor and capital markets became increasingly dualistic and segmented. Agriculture began to stagnate, in part because the diffusion of best-practice traditional technique had been exhausted—in part because

the government experimentation stations had not yet completely the development of new seed varieties, and in part because, after 1918, the government encouraged a massive stepping up in the import of rice and other raw foods from its empire in Korea and Taiwan.

Growth became unbalanced. Heavy manufacturing, thirsty for workers who could master skills needed for working with the sophisticated technologies and machinery it was putting into place, offered relatively high wages and secured capital at relatively low interest rates. Light manufacturing, proto-industry and agriculture paid low wages. Because innovation was heavily biased toward the heavy industries now, it was in these sectors that profits were likely to be generous. Hence, suppliers of capital favored the emerging heavy industries, and the sectors that were key to the first long swing found themselves increasingly marginalized in the capital market.

As a result, differentials—in income and in opportunities—between rural and urban areas, between Tōkaidō core and periphery, widened. Moreover, because land prices in urban areas and along the tracks laid down by the intercity railways had shot up during and after the first upswing of the second long swing, there was also a growing gap in terms of wealth between those who owned land and those who did not. The social fabric was threatened. Government was drawn into mediating disputes between landlords and tenants, in the countryside and in the great metropolitan centers. Unemployment and underemployment became a social issue during the 1920s. Governments put the unemployed collecting relief to work on projects extending the infrastructure.

Renewed industrial growth came to the rescue of the authorities. The reworking of infrastructure during the downswing of the 1920s paved the way for a huge industrial expansion during the 1930s. Applications of the internal combustion engine to aircraft were important. Once again, a new mode for harnessing energy was central to the upswing.

Tokyo-centrism also took definitive shape during the third long swing. With the beginning of the second long swing, Tokyo's competitive advantage in developing the infrastructure crucial for the technologies of second and third long swings began to show up in disparate fields. Tokyo and the Kantō had decided advantages over Osaka and the Kinai in electrical power generation, intercity and intracity electric railway development, and creating a road network usable by motorized vehicles. Tokyo had a competitive advantage in banking and finance, largely because it was the capital and the Ministry of Finance and the Bank of Japan were located there. In addition, Tokyo had an advantage in technical education and industrial research, mainly because the Ministry of Education and Tokyo Imperial University were located in its environs. These advantages in creating infrastructure became

increasingly important as the transitional-growth swing gave way to the unbalanced-growth swing. By the late 1930s, Tokyo had become the economic, intellectual, and political capital of the country.

In looking back over the story we have just recounted, it is evident that there is a strong endogenous logic to Japan's growth. Indeed, trend acceleration occurred over the course of the long swings in the sense that during successive upswings, growth rates increased. For instance, during the upswing of the 1911–1919 period, gross domestic product grew at 4.4 percent; and during the upswing between 1930 and 1938, it grew at 4.6 percent. Moreover, the rate of domestic fixed capital formation growth (and the percentage of aggregate demand due to capital formation) grew over successive upswings. Savings rates fed on income growth. Restructuring of financial infrastructure ensured that increased savings would flow into the hands of eager investors. Investment in infrastructure had spillovers in industry. As a result of the subsequent surge in industrial production, gross domestic product and per capita income rose. This raised savings, thereby facilitating further capital accumulation in the infrastructure sector.

Thus Japanese growth is basically endogenous in two senses of the word: Capital accumulation is crucial for growth, and the process of growth is largely shaped by domestic factors. The most important long-run trends at work—industrialization, infrastructure expansion, proliferation of corporations, improvement in human capital, proliferation of types of energy sources—are at work because of the internal logic of development according to which growth begets growth.

Nevertheless, Japan did not industrialize in a vacuum. Geopolitical, security, and military concerns played some role in her economic growth in all phases of the three prewar long swings. In order to secure Western renunciation of the unequal treaties imposed on it, and to create a buffer zone on the mainland, Japan fought two wars with the great land powers of East Asia, China and Russia, during the first long swing and the beginning of the second. From the victories that Japan eked out in these conflicts, it secured war indemnities, foreign exchange, and territory. By 1910, Japan's formal empire included Taiwan and Korea, and her informal empire extended into Manchuria and parts of China.

Between 1910 and 1930, the system of international trade broke down. The gold standard collapsed and tariff walls were erected. As more and more countries opted for autarky, Japan felt increasingly isolated. Lacking raw materials, the drift toward autarky and the creation of trade blocs was especially threatening. If Japan could not export, it could not import raw materials. Indeed, it was always possible that countries might cut off their export of raw materials to Japan. For instance, Japan imported petroleum and scrap

steel from the United States. Would the United States threaten to cut off these supplies? Thus, securing control over a large Asian empire that could supply Japan with raw materials and absorb its manufactures increasingly seemed to have an appeal. This was especially the case for Japan's army and navy, which were, quite naturally, concerned about securing adequate supplies of petroleum, rubber, iron ore, and coal.

After 1931, frustrated with the breakdown in international trade and the drift toward autarky, and perhaps spurred on by the desire to stave off internal social conflict by waving the bloody flag of patriotism, Japan's leaders stepped up their effort to expand Japan's Asian empire. The army gobbled up Manchuria. Then, in 1937, it began a full-fledged invasion of northern China. Continuing Japanese conquest in China was anathema to the two great oceanic powers of the Pacific, the United States and the United Kingdom. Thus, as Japan became increasingly intent on solving its problems in the international trade and social spheres through aggressive actions in China, it inched closer and closer to conflict with the United States and Great Britain.

In short, throughout the prewar period, military and geopolitical concerns played a role in Japanese development. Thus, during the late 1930s, fueled by a drift toward total war, government spending on the military soared. In 1930–1938, capital formation in the military sector accounted for over 8 percent of total capital formation. Japan now had the economic capacity to flex its military muscle and make the world take notice. The United States and the United Kingdom looked on Japan's actions with growing concern. It is doubtful that, in awakening the sleeping economic giant that was Japan in the 1850s, either power anticipated that the island archipelago off the Chinese mainland would, within a mere eight decades, become a great industrial and military power that threatened their vital interests in Asia.

The Rain of Fire

Now it was Japan's turn to awaken a sleeping giant. German gains against England, France, and Holland in 1940 bolstered the contention of those in the military who believed war with the United States was inevitable. For decades, the Imperial Japanese Navy had been preparing for conflict against the U.S. Navy. The faction urging on war against the United States argued that if Japan could cripple the American fleet, it could sweep Western colonialism out of Asia and substitute Japanese imperialism. A bill passed in the U.S. Congress outlaying funds for rapid buildup of the Pacific Ocean and Atlantic Ocean fleets seemed to shorten the time frame within which the Japanese military could effectively move. Thus, in late 1941, Japan attacked the United States at Pearl Harbor.

Japan's strategy was to defeat the American navy in a few set battles involving large battleships. Japan's military hoped that the Americans, defeated in several key confrontations, might sue for peace, thereby giving Japan a free hand to continue its imperial expansion in Asia. The last thing Japan wanted was a war of attrition for it was understood that in a war of attrition, the United States would eventually prevail. The American economy was far larger than was Japan's; its capacity to produce bombers and battleships was far greater than Japan's; and its command of raw materials far outran that in Japanese hands. Nonetheless, the Pacific War settled into a long war of attrition.

By early 1945, Japan was in full retreat and the United States was preparing for its assault on the island archipelago itself. The Americans planned a land invasion of Okinawa and Kyūshū, and they expected to carry out a widespread aerial campaign, for lurking in the wings was the Boeing B-29, the Superfortress. Originally designed to be a long-range bomber that could be used against Germany, fleets of B-29s were being flown to Guam and the Marianas. The United States was poised to launch the systematic firebombing of the Tōkaidō.

Japan's industrial capacity was highly vulnerable to aerial attack. Because it was so heavily concentrated in the strip of land running along the Pacific coastline of Honshū from the Kantō to the Kinai, it could be readily targeted. Moreover, the buildings in the conurbations of the Tōkaidō were mainly fashioned out of wood. That Japan's industrial capacity was vulnerable to aerial assault was apparent to the military commands of both nations as early as April 1942. In the Doolittle Raid of 1942, B-25B bombers, dispatched from the carrier *Hornet*, attacked military installations and factories in Tokyo and Yokohama. They sent a very loud message to the Japanese high command.[1]

Flying out of Guam and Saipan, bombing bays bristling with over 2,000 tons of napalm and oil incendiary bombs, thundering in low over the targeted area at night when Japanese air defenses were least effective, an armada of over three hundred B-29s reached Tokyo on March 9, turning its *shitamachi* industrial district between the Sumida and Ara Rivers into a raging inferno. With temperatures rising rapidly, a wild conflagration spread out across a four-by-three mile section of the capital. Forty percent of Tokyo's buildings were wiped out. The devastation was horrendous: 80,000 persons killed; more than 160,000 persons wounded. Bodies littered bridges, roads, and canals, charred and torched beyond recognition.

Even before the flames in Tokyo could be squelched, the American air command turned its attention southward along the Tōkaidō, pounding Nagoya less than twenty-four hours later. Osaka came next. On the night of March

14, the Superfortresses reached the city, burning a nine-square-mile patch of devastation through the industrial section of the metropolis. Sweeping across the harbor, blowing up docks and warehouses, setting ablaze factories along the Aji River, blasting into ruins most of Senba, Nanba, and Umeda, tearing apart structures along the Kizu and Shirinashi Rivers, the American onslaught systematically smashed Osaka's industrial might and the infrastructure bedrock upon which it rested. Three days later, the bombers revisited the Kinai to lay waste to Kobe. In less than two weeks, the heart of Japan's industrial capacity—twenty-nine square miles of the Tōkaidō—had been torn to shreds.

In the spring of 1945, in the year of Japan's darkest hour, Osaka passed through the rain of fire. And, in those searing flames, Osaka's infrastructure was forever transformed.

Infrastructure Reworked Once Again

After Japan was forced open in the 1850s, it radically revamped its infrastructure. Subsequent reworking of infrastructure—during the first upswing of the transitional-growth long swing and during the 1920s—was far-reaching but was not nearly as thoroughgoing as was that occurring during the first long swing and the buildup to that first long swing. In the aftermath of the Pacific War, Japan's infrastructure was radically reworked once again. Indeed, the postwar restructuring of Japan's infrastructure during the buildup to, and parallel with, the industrial upswing between 1953 and 1969 is, in many ways, comparable to that happening during the early Meiji period. First and foremost, it was a radical restructuring because so many institutions and policies adopted in the period leading up to the war were jettisoned. With its defeat, and prodded by the American occupation authorities, Japan changed its direction in fundamental ways.

It is because Japan's infrastructure was radically reworked after 1945 that I believe the discontinuity brought on by its defeat in the Pacific War was profound. For this reason, I reject the idea that 1938–1953 marks a downswing phase of the third long swing.[2] Rather, I believe that 1938–1945 was a period so distorted by war that it is improper to call it a downswing of an economic growth process; and I believe that 1945–1953 marks a period of infrastructure reconstruction that set the stage for the upswing of the 1950s and 1960s.

Discontinuity comes from two sources: domestic and international. Let us briefly consider each.

On the domestic front, infrastructure was thoroughly reworked. Consider factor markets. Labor markets were reformed. Collective bargaining and unionization were permitted. A labor standards act was passed. The labor

boss system was outlawed. The higher educational system was dramatically expanded, and compulsory education was extended through middle school. Institutions of education, higher as well as lower, were opened up to women.[3]

Capital markets were transformed as well. The *zaibatsu* holding companies were dissolved.[4] Land reform was carried out. Landlordism virtually ceased to exist as a result of the terms set out in the land reform. In short, the problem of a highly unequal wealth distribution that plagued Japan during the period after 1910 was addressed in the capital market reforms. Moreover, the banking system was reorganized. And the ratio of financial assets to total income was dramatically changed, in part due to inflation that also helped wipe out the compensation paid to former landlords in the land reform.[5]

Much physical infrastructure had been destroyed as a result of the bombing of Japan. It was rebuilt. Ironically, the destruction of the buildings in the centers of the great metropolises was a boon to those officials intent on improving the infrastructure in the great cities of Japan. For in the wake of the bombing, property values in city centers plummeted. Thus, city governments could snap up land for widening roads, and extending the road network, at low prices.

On the international front, changes were also fundamental. Most important, Japan lost its empire. Its capital market was no longer drained by the investing done in Taiwan, Korea, and Manchuria. Food was no longer imported from the empire. Indeed, the government addressed the problem of income distribution by protecting domestic agriculture and by buying up rice at a subsidized price that guaranteed to the marginal farmer an income comparable to that earned by a typical blue-collar worker. Japan also adopted a constitution limiting its capacity to engage in military actions outside of its boundaries. As a result, military spending plummeted as a proportion of gross national expenditure.

At the general international level, the 1950s ushered in a new era of expanding trade among the capitalist countries. Successive rounds of the General Agreement on Tariffs and Trade cut tariffs. The Bretton Woods system created a gold standard backed by the United States dollar that persisted until the early 1970s, after which a period of more flexible exchange rates was ushered in. Moreover, much technology had been "pent up" during the period 1930–1950 because of the Great Depression and the world war. This technology was more easily diffused than the technology that had focused around raw-materials extraction in the nineteenth century (the United States had a strong comparative advantage in developing technology involving raw materials). Hence, a large number of countries caught up with the United States during the period 1950–1970. Convergence of Western European countries and Japan toward levels of productivity comparable to that in the United

States occurred because of the expansion in trade and the diffusion of the new technologies.[6] Hence the Japanese upswing between 1953 and 1969 is really a convergence long swing. It is part and parcel of an international movement sweeping across the world of countries that had become highly industrial prior to 1940. It mainly involved catching up with the United States. Catching up with each other was less important.

Still, despite the discontinuities that I have just listed, there are important continuities between the prewar and postwar periods that should not be ignored. Without attempting to be exhaustive, I wish to briefly touch upon six of these: continuity in the shift from water- to land-based infrastructure; continuity in the dominance of the Tōkaidō core in Japanese technological and institutional innovation (included here is continuity in Tokyo-centrism); continuity in the process through which metropolises "hollow out" their nighttime populations; continuity in the proliferation of satellite cites clustering around giant metropolitan centers; continuity in the use of *kukaku seiri* as a tool for carrying out land conversion in the environs of great cities; and, finally, continuity in the way the hierarchy of cities has evolved.

Consider infrastructure. In Table 8.1, I give figures on Japan's roads and on estimated indices for domestic transportation flows. As can be seen from panel A of the table, between 1947 and 1970 there was a marked increase in the use of automobiles, trucks, and buses within Japan. Within Osaka, the increase in trucks and buses (registered in the city) was especially striking. Osaka's experience is captured in Figure 8.1. From Table 8.2, it is apparent that until the early 1970s, most of the expansion in Osaka's motor vehicle stock involved trucks. Increasingly, the city's businesses were depending on trucks to transport raw materials and finished goods.

In order to handle this massive expansion in motor vehicle traffic, Osaka's city government extended its road network and expanded the width of many of its roads. The fact that the American aerial bombing had destroyed most of the buildings and structures in the center of the city—thereby reducing the value of the land they stood on to extremely low levels—assisted the city government in its effort to carry out in the aftermath of war, a task that was so difficult to carry out prior to the Pacific conflict.

Nevertheless, most of the roads in contemporary Osaka are still narrow. They were narrow before the war and they are narrow to this day. To be sure, Osaka has its wide boulevards and highways. However, Osaka is not Tokyo; the Kinai is not Kantō. Perusal of photographs of Osaka's streets taken during the 1950s and 1960s gives the impression of incredible, debilitating congestion. As a result of this congestion, the city eventually closed down its tramway system, which was contributing to the clogging of its arteries.

In lieu of the tramway, the city contrived to create an extensive subway

Table 8.1

In the Aftermath of War: The Expansion of Japanese Transportation and Transportation Infrastructure, 1947–1970

A. National, prefectural, and local roads: indices for length (1965 = 100), and percentage of total length that is paved (PPAV), and percentage of total length that is impassable by cars (PIMP), 1947–1970[a].

	National roads		Prefectural roads			Local roads		
Year	Index	PPAV	Index	Paved (PPAV)	PIMP	Index	Paved (PPAV)	PIMP
1947	33.8	18.1%	39.0	3.7%	46.4%	88.1	n.a.	76.8%
1960	89.0	32.7	78.1	7.6	8.3	98.4	1.5	47.0
1970	116.5	83.6	141.4	45.1	4.9	103.4	12.0	33.5

B. Indices of domestic transportation flows, 1965 = 100

	Cargo				Passenger			
Year	Total	Railway	Motor vehicle	Sea	Total	Railway	Motor vehicle	Bus
1955	31.5	73.1	19.8	35.7	31.0	54.7	16.6	30.9
1960	54.6	95.0	43.3	77.0	49.0	72.7	36.4	58.7
1965	100.0	100.0	100.0	100.0	100.0	100.0	100.0	100.0
1970	262.1	109.4	286.4	179.6	313.7	122.1	396.3	113.1

Sources: Japan. Prime Minister's Office (various dates), *Japan: Statistical Yearbook,* various tables.
Note: [a]N.a. = not available.

Table 8.2

Roads and Cars, Buses, and Trucks in Osaka, 1951–1970

A. Indices for road length and road area and number of bridges (all indices have 1955–1956 = 100).

	Road length and area							
	Osaka City–managed roads		All roads in Osaka City		Paved roads in Osaka City		Bridges	
Period	Length	Area	Length	Area	Length	Area	Number	Area
1951–1955	95.3	98.0	94.7	97.4	96.5	97.3	89.8	96.2
Fourth period in city's history								
1956–1960	100.9	101.9	100.9	102.3	109.4	106.6	111.1	105.3
1961–1965	106.9	108.0	106.5	108.8	133.7	128.5	111.4	104.7
1966–1970	116.6	118.1	117.7	126.5	194.2	195.6	109.2	131.9

B. Indices for cars, buses, and trucks (1955–1956 = 100) and percentage of all vehicles that are trucks (PVTR).

	Indices (1955–1956 = 100) for vehicles registered in city				
Period	Cars	Trucks	Buses	Cars, trucks, & buses	% Vehicles that are trucks
1951–1955	92.5	86.6	85.3	88.8	54.9%
Fourth period in city's history					
1956–1960	105.4	355.3	122.3	246.8	75.4
1961–1965	93.1	1,396.9	400.4	842.2	92.8
1966–1970	108.6	2,528.5	647.9	1497.4	94.9

Sources: Ōsaka-shi Yakusho (various dates), *Ōsaka-shi tōkeisho,* various tables.

network. Indeed, subways are an ideal instrument for transporting large masses of people through the giant conurbations of the Tōkaidō. They do not tie up traffic on the surface. And passengers board them from platforms resting underneath the earth's surface. Moreover, they can be readily linked to intercity railways at huge multilayered transportation nodes like those at Nanba or Umeda.

The new was condemning the old to destruction. As can be seen from Table 8.3, even before the tramway tracks were ripped up, the use of subways—and buses—exploded at a heady pace, far more rapidly than did the other transportation systems managed by the city of Osaka. Once again, old technology was giving away to newer technology. The *jinrikisha* gave way to the tramway, the *niguruma* to the truck. Now the tramway was giving way to the subway. Such is the logic of creative destruction inherent in the innovation waves buttressing the long swings.

Figure 8.1 **Road Area and Cars, Trucks, and Buses in Osaka City: Growth Rates, 1953–1967** (Five-Year Moving Averages)

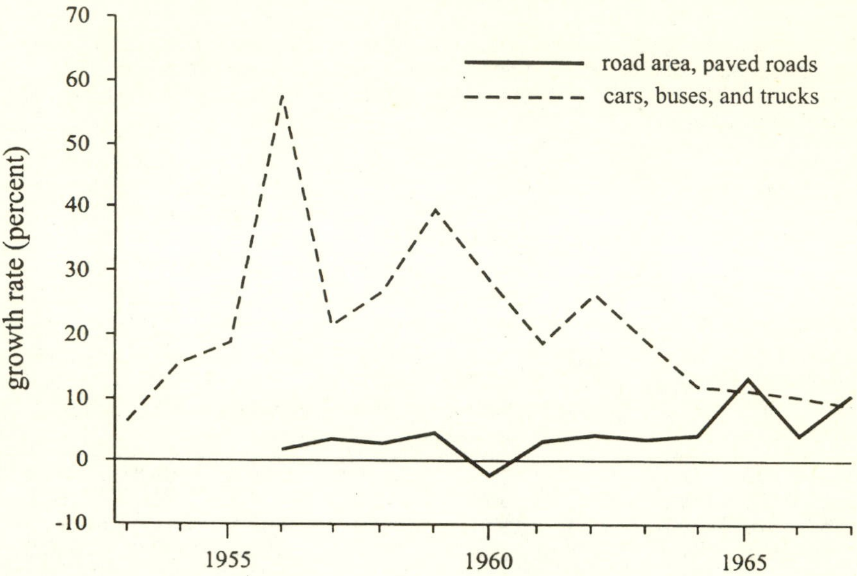

That the great conurbations of the Tōkaidō had the capacity to rebuild, and improve upon, the infrastructure that had been painstakingly erected before 1938 offers a vivid illustration of the thesis that knowing how to do something trumps having the physical capital. Knowledge—in combination with a skilled labor force acting up on that knowledge—is far more powerful than physical structures. Since the technology of modern infrastructure construction had been thoroughly absorbed by Japanese companies and by government organizations prior to the Pacific War, rebuilding in the wake of the defeat in the war proceeded at a brisk pace. By the early 1950s, infrastructure had been sufficiently refurbished within the Tōkaidō core, so that a massive expansion of manufacturing could occur. Infrastructure continued to expand as manufacturing capacity was built up.

As in the prewar period, the key to the buildup of infrastructure and industry in Japan during the convergence long swing was its expansion within the Tōkaidō. For instance, consider the following figures for factories and factory workers in Osaka City and Osaka prefecture.[7] Letting 100 be the base for both the number of workers and the number of factories in Osaka Prefecture for 1950, the index in 1965 for workers is 243.3 and for factories is 172.4. In 1950, 59 percent of the workers and 58 percent of the factories in the prefecture were located in the city. These percentages hardly changed

Table 8.3

City-Managed Transportation in Osaka, 1941–1977

Indices for stock KILO (kilometers of operation) and cars used and indices for daily passenger flow (total and buses and subways only); and percentages of stock measures (PKILO, PCAR) and percentage of total city-managed passenger flow on buses and subways (PDFLOW)[a].

	Indices (1955–1956 = 100)				Percentages in buses and subways: kilometers of operation (PKILO); cars (PCAR); and daily passenger flow (PDFLOW)		
	Stock[b]		Daily passenger flow				
Period	KILO	Cars used	Total	Buses and subways	PKILO	PCAR	PDFLOW
1941–1945	60.4	136.9	85.0	46.9	53.3%	61.6	28.0
1946–1950	59.1	73.4	60.7	43.8	51.6	52.9	35.3
1951–1955	90.6	91.2	88.0	76.0	66.2	60.1	42.1
Fourth period in the city's history							
1956–1960	112.1	126.7	119.4	135.9	68.6	65.8	55.5
1961–1965	133.5	177.9	150.6	212.0	74.2	76.1	68.8
1966–1970	135.7	179.1	137.4	253.9	89.1	91.6	90.5
1971–1977	141.3	160.9	137.4	280.6	100.0	100.0	100.0

Sources: Ōsaka-shi Kōtsū-kyoku (1980b), *Ōsaka-shi kōtsū nanajūgonen shi,* various tables.

Notes: [a]Buses, subways, electric street cars, and trackless electric vehicles (the last category only between 1953 and 1970). [b]Kilometers of distance under operation (stock measure, not flow measure), and cars used in all modes of transportation.

between 1950 and 1965. The percentage of factories in the city rose a bit, and the percentage of workers fell somewhat more. In short, the expansion of manufacturing was massive, and the tendency of the expansion to occur within the great industrial conurbations—a pattern well established before the war—continued after the war.

Indeed, the competitive advantage of the Tōkaidō as a whole was reestablished rapidly. To realize how concentrated Japanese manufacturing was during the upswing between 1953 and 1970, consider the following percentages on proportion of workers within sectors and subsectors located in the Tōkaidō[8]. In 1960, 59.2 percent of manufacturing was concentrated in the Tōkaidō, and in 1970 the percentage was 55.4 percent. Machinery production was far more concentrated in the Tōkadō: 71.7 percent in 1960 and 64.0 percent in 1970. Electrical machinery, one of the key subsectors of machinery, was also highly concentrated in the industrial belt: 76.4 percent in 1960, and 62.4 percent in 1970. Textiles, dependent on low-wage female labor flowing out of farms, was less heavily concentrated in the Tōkaidō, however. For in-

stance, in 1970, only 46.3 percent of workers in textiles were employed in factories located in the industrial belt.

Tokyo/Kantō centrism was also characteristic of the fourth long swing. For instance, in 1960, 25.8 percent of workers in manufacturing were employed in the northern prefectures of the Tōkaidō. In machinery production the percentages were higher: 36.7 percent in total machinery and 48.4 percent in electrical machinery. By contrast, the percentages in Tōkaidō south were lower in 1960. In overall manufacturing, the percentage was 20.4 percent, and in electrical machinery it was 21.4 percent. Furthermore, the percentages in Tōkaidō north fell less sharply between 1960 and 1970, than they did in Tōkaidō south. For instance, in 1970 the percentage of workers in manufacturing employed in Tōkaidō north was 25.5 percent; but in Tōkaidō south it had fallen to 17.6 percent. Strong continuity between the prewar and postwar periods in the evolution of Japan's economic geography is evident from these figures.[9]

Continuity also prevailed in the way the urbanization evolved. Consider the proliferation of satellite cites around the great conurbations of the industrial belt. The buildup of intercity railway lines promoted the creation of bedroom communities along the lines joining the key conurbations. Thus, satellite cities were spun off. For instance, Toyonaka City and Suita City are satellites of Osaka City. Between 1920 and 1924, population growth rates for Osaka City, Toyonaka City, Suita City, and Amagasaki City were 19.6 percent, 46.6 percent, 67.7 percent, and 27.7 percent, respectively.[10] The fact that Osaka City grew more slowly than its satellites continued throughout the 1930s. For instance, in 1935–1940, population growth rates for Osaka City, Toyonaka City, Suita City, and Amagasaki City were 8.8 percent, 22.7 percent, 20.4 percent, and 44.2 percent, respectively.

The push of Osaka's daytime population out to satellite cities encouraged a growing divergence between the size of Osaka's daytime and nighttime populations. The divergence was already quite evident in 1930. But the ratio of daytime to nighttime population continued to increase after the war. This can be seen from Table 8.4. The spilling over of population into the suburban satellites is especially striking after 1965. Indeed, in the decades following that watershed year, net in-migration rates for Osaka City are actually negative.

The hiving off of suburban municipalities around central places in the Tōkaidō was important in the ongoing proliferation of cities in modern Japan, but the number of cities was also increasing in the periphery of the country as well. For instance, consider the following figures on the number of cities in the Tōkaidō and in the non-Tōkaidō periphery of the country. In 1903, of the fifty-eight cities in Japan, ten were in the Tōkaidō (this amounted to 17.2 percent of all cities), and forty-eight were in the periphery. By 1935,

Table 8.4

Osaka: Resident and Daytime Populations, 1945–1985

A. Osaka City: Resident population; and total, natural, and net migration rates of increase (per 1,000 population).

Period	Resident Population	Rates of increase[a]		
		Total	Natural	Net inmigration
1945–1954	1,862,601	+61.8	+16.4	+45.4
1955–1964	2,906,748	+23.7	+11.8	+11.8
1965–1974	2,996,825	–12.5	+12.2	–24.7
1975–1984	2,677,640	–5.8	+5.7	–11.5

B. Indices for daytime and nighttime population (DayPop, NightPop), and ratio of daytime to nighttime population (Dayp/Nigp) with base = 100: Osaka City and city's center zone[b].

	Osaka City (indices with 1955 = 100)			Center (indices with 1955 = 100)		
Date	DayPop	NightPop	Dayp/Nigp	DayPop	NightPop	Dayp/Nigp
1955	100.0	100.0	112.2	100.0	100.0	229.2
1965	135.1	123.9	122.4	146.7	93.7	358.2
1980	127.7	104.0	137.9	159.0	59.7	610.3

C. Indices of resident population for zones; and percentages of population in zones of city[b].

	Indices (1955 = 100)			Percentage of total city population		
Year	Center	Middle ring	Periphery	Center	Middle ring	Periphery
1945	22.7	28.0	50.1	5.6%	11.4%	83.0%
1955	100.0	100.0	100.0	10.6	17.6	71.8
1965	93.7	120.5	129.2	8.0	17.1	74.9
1975	61.0	101.0	118.2	5.9	16.3	77.8
1985	61.0	90.6	112.9	6.3	15.4	78.3

Sources: Shinshū Ōsaka-shi Shi Hensan Iinkai (1994) *Ōsaka-shi shi*, vol. 8, various pages.

Notes: [a]Natural rate of increase = birth rate minus death rate. [b]For the zones of Osaka City, see Table 3.2. The center is the combination of Higashi, Nishi, Kita, and Minami wards.

of the 123 cities in the country, 31 (or 25.2 percent of the total) were in the Tōkaidō core. In 1960, there were 519 cities in Japan, of which 155 (or 29.7 percent) were in the Tōkaidō. By 1990, of the 639 cities in the country, 227 (or 35.5 percent of the total) were in the Tōkaidō. Cities formed in both core and periphery, but they formed more rapidly in the core.[11]

It is apparent from the figures given above that city growth was especially vigorous in the years following the war. Indeed, most of the cities of contemporary Japan were created in the aftermath of the Pacific War. Most of the cities formed right after the war were formed in the periphery and

they were often cities in name only. This point can be seen by comparing the urban percentage with the proportion living in densely inhabited districts.[12] For instance, in 1960, the urban proportion in the Tōkaidō was 82.1 percent and the proportion in densely inhabited districts was 65.0 percent. These figures are fairly close, differing only by about 17 percentage points. By comparison, in the periphery, the proportion urban was 51.7 percent, and the proportion in densely inhabited districts was 26.6 percent. Why were cities that were fundamentally rural in character created in such profusion after the war?

In the postwar period, a strong motivation for creating cities was to promote infrastructure development. For, just as in the cities covered by city planning legislation during the interwar period, the new city administrations of the postwar years actively utilized *kukaku seiri* as a device for creating infrastructure.[13]

For this reason, the logic of postwar growth in the number of municipalities in the core seems to differ somewhat from the logic of growth governing the periphery. In the Tōkaidō core, new cities tended to be formed "organically." They developed around infrastructure that had already been put into place. The city typical of the Tōkaidō core started out as a population concentration built up around a transportation node, like a train station. After it reached a sufficient size, density, and complexity, it applied for incorporation as a city.

But in the periphery, a city was often created by the joining together of sparsely settled rural districts with small population concentrations. The resulting city served as a convenient engine for systematically building infrastructure in the area under its jurisdiction. It was a coordinating agent for carrying out *kukaku seiri*. In this way, schools, public utilities, bus lines, roads, parks, and other amenities were diffused into the less densely settled reaches of the periphery.

To be sure, some postwar city creation within the periphery has been similar to that occurring within the core, in the sense that it has involved satellite cities spinning off from central population concentrations. For instance, as labor-intensive industries like textiles are pushed out into the periphery, factories popped up in cities there. As these cities grew, they hived off suburbs. The location of a number of these major conurbations outside the Tōkaidō is apparent from Map 8.1.

As businesses have sprung up or migrated out of the periphery, the urban hierarchy of the periphery has "built up." Major industrial cities rivaling some of those in the Tōkaidō have emerged there. At the same time, the urban hierarchy of the Tōkaidō has "built down" because satellite cities have been spawned around the giant industrial conurbations. For instance, in 1960, there were

Map 8.1 **Major Cities of Japan in 1960 Inside and Outside of the Tōkaidō**

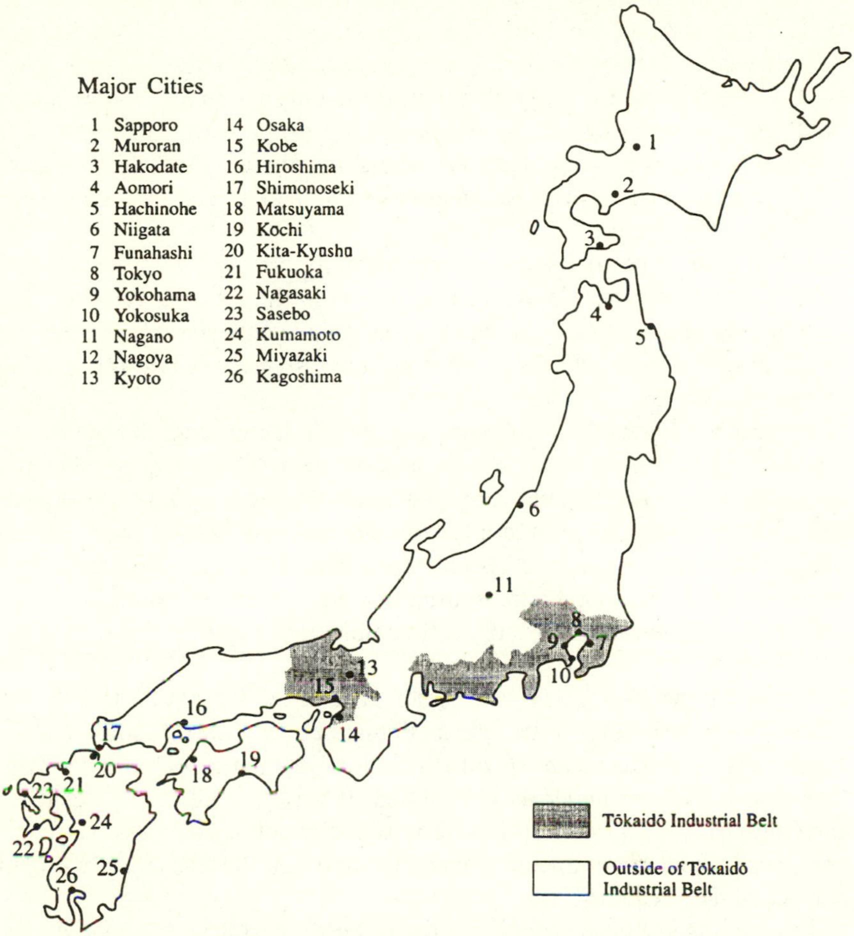

thirty-three Tōkaidō cities that had populations ranging between 100,000 and 500,000 persons. By 1990, there were 101 such cities. Again, in 1960, there were fifty-nine conurbations in the 50,000– to 100,000–person range. By 1990, there were 101 cities of this type. In other words, the periphery has tended to "build up," and the core has tended to "build down." As a result, convergence in the urban hierarchies between core and periphery has occurred.

Convergence can be measured statistically. Let us compare the urban rank order hierarchies for the core and for the periphery for 1990 with those for 1903.[14] In order to do this, compare Figure 8.2, which gives the 1990 hierarchies, with Figure 3.2 in chapter 3, which gives the 1903 hierarchies. The

Figure 8.2 **Rank-Size Distribution for the Cities in Two Regions of Japan, 1990: Tōkaidō Cities and Cities Outside of the Tōkaidō**

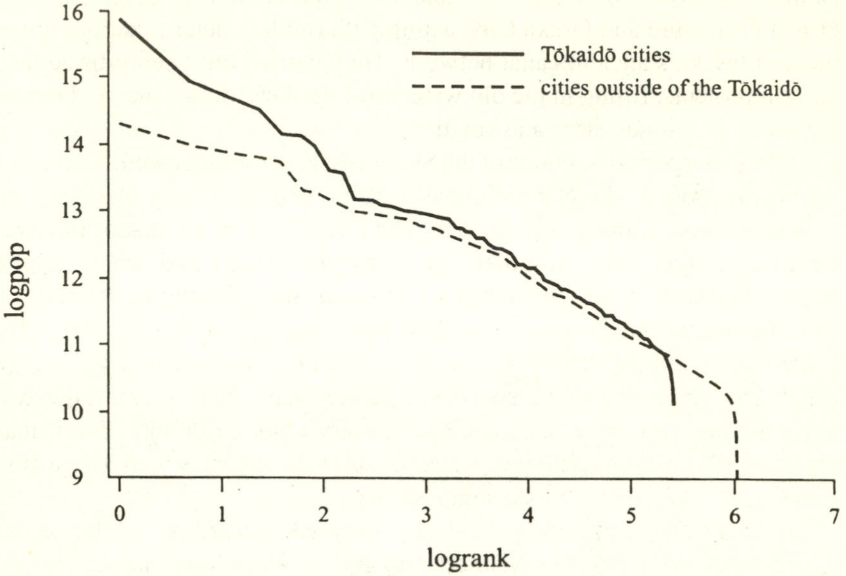

contrast between the two figures is striking. A remarkable convergence has taken place between 1903 and 1990. The hierarchy in the Tōkaidō core became flatter, "building down" from above. And the hierarchy in the periphery became steeper, "building up" from below.[15]

In sum, there are deep-rooted continuities between prewar and postwar periods in the evolution of Japan's urban system. City planning, pioneered first in Tokyo at the end of the nineteenth century and then more broadly developed in the major cities of the country during the interwar period, created a body of practice and doctrine that has been diffused throughout the entire country during the postwar period.

A second continuity involves the close connection between city government and infrastructure improvement and renewal. Of course, all of this has come at a price. In the prewar period, cities had to wrestle with destroying old infrastructure—rooting up irrigation ditches in land being converted from farming to residential purposes, for instance—in order to lay down infrastructure. The postwar fate of Osaka's mighty network of ancient canals offers an excellent illustration of the creative destruction whereby old infrastructure has been swept aside to accommodate new infrastructure. Before the Pacific War, Osaka was famous as a "city of water." Today, Osaka can no longer make this claim. What happened to its waterways?

A Rich Vitality

In the seven years between 1951 and 1958, the combined governments of Osaka Prefecture and Osaka City, acting with ruthless determination, buried most of Osaka's ancient canal network. They carried out twenty-three huge landfill projects, filling in the old waterways. In the process, they pulverized hundreds of bridges large and small.[16]

Within that period, sections of the Shirinashi River disappeared. Also eliminated was most of the Nanbashinbashi River, and the Horie, Nekoma, and Jūsangen rivers. Canals were filled in. The Edo, Kaifu, and Nakanoshimawari canals no longer exist. And there are many other rivers and canals gracing prewar Osaka that no longer can be found in the metropolis. Indeed, the Dōton Canal (Dōtonbori) is one of the few waterways of Osaka left today. More than anything else, it is a forlorn symbol of a bygone era. To be sure, canals and rivers that have been extinguished have been memorialized in street names. Thus, one finds in contemporary Osaka, Edobori (Edo Canal) Street and Nanbashinbashigawa (Nanbashinbashi River) Street. There is no Edo Canal. There is no Nanbashinbashi River.

Osaka's ruthless attitude toward its waterways, toward its reputation as a city of water, is remarkable. Neither Leningrad/St. Petersburg nor Amsterdam nor Venice destroyed their ancient canal networks. However, Osaka did.

History records no groundswell of popular opposition to the filling in of the canals and rivers. No citizens' movement bent on historical preservation seized the initiative and saved the waterways. To be sure, the logic of eliminating them was impeccable. The canals were no longer being used for transport. Many of them had suffered damage in the aerial bombing of the city in 1945. They were filthy and they stank. And they were a potential safety hazard for infants and young children who might fall into them and drown in their fetid waters.

Moreover, think about the positive benefits of filling in the waterways. By burying its ancient network of canals and rivers, Osaka gained land area that could be used for building roads, housing, factories, and parking lots. So the prefecture and city authorities killed two birds with one stone. In an unintended by-product of their efforts, they made a mockery of Osaka's claim to be a water city.

Why should anyone care? After all, Osaka is not a tourist city. Rather it is an industrial and commercial city, and a transportation hub for the Kinai. No one comes to Osaka to see the sights. They come to Osaka to do business. Or they come to Osaka because it is the gateway to the ancient wooden cities of the Kinai, Kyoto, and Nara. Or they hurry through Osaka on their way to the famous hot springs of the Kii peninsula.

This is how it has been for a long time. This was true even when the maze of canals and rivers existed, and one could eat a meal on restaurant boats lining their banks. Consider how the *Official Guide to Eastern Asia* describes the city of Osaka as it was in 1914: "Osaka, though possessing a long history, cannot boast of very many places of interest from either the artistic or antiquarian point of view. For its growth during the last three hundred years has been chiefly industrial and material."[17] Thus, it would seem that Osaka has not been looking back for a longtime. Secure in its faith in the power of technology and the providence of markets, it looks ahead.

Yet, Osaka does look back. For if you pass through Nanba Station and walk on to Umeda, you will see the layering of infrastructure that embodies Japan's modern industrial history. This layered infrastructure harbors a rich vitality. In that vitality, the past informs the present. In that vitality, the past informs the future.

The roots of Osaka run deep. Neither industrialization nor the inexorable logic of the long swings has managed to sever these roots. They are everywhere throughout the great metropolis. I have seen them in the sprawling markets near the Tenpozan docks. I have chanced upon them in a pounding rain in the tiny restaurants huddled among the gigantic manufacturing plants and warehouses of Sakurajima. I have come upon the remnants of Osaka's ancient proto-industrial economy in the narrow winding streets and back alleys of Sumiyoshi Ward.

These are the vital points of modern urban Japan. To these points from all compasses rush past, present, and future. Herein lies our tale. Conceived in water and wood, forged afresh in electricity and steel, reborn triumphant after the rain of fire, flourishes Osaka: a giant arisen upon the plains of the southern Tōkaidō; a giant astride the rushing waters of the Yodo River and the Inland Sea; a giant of the Kinai.

Notes

1. See Willmott (1982, 447 ff.). Willmott is an indispensable English language source concerning the first two years of the war in the Pacific. B-20 raids had been launched from China beginning in June 1944.Tokyo had been bombed in November of that year.

2. For the view that the period 1938–1953 is a downswing of the third long swing, see Minami (1994); Ohkawa and Rosovsky (1973); and Ohkawa and Shinohara with Meissner (1979).

3. See Mosk (1995, chapters 3 and 4). In chapter 4 of that book, an argument is developed that the chief thrust of the American occupation was to upset the balance of power within the bureaucracy. By purging many of the conservative bureaucrats who had promoted ultranationalism during the 1930s, the occupation authorities shifted the weight of policy-making onto the shoulders of liberal bureaucrats who had been intent on promoting land reform and educational reform prior to the war.

4. On the dissolution of the *zaibatsu*, see Bisson (1954), Hadley (1970), and Morikawa (1992).

5. On the ratio of financial assets to income, see Patrick (1984).

6. On convergence and growth in the period 1950–1970, see Abramovitz and David (1996) and Baumol, Nelson, and Wolff (1992). These works take a global perspective. For another type of analysis that also takes a global point of view, see Rostow (1980, 1998).

7. The figures are taken from Ōsaka-shi Kai Jimu-kyoku Chōsa kai (1970, 54). In evaluating the impact of infrastructure investment on the Japanese economy after World War II, it must be kept in mind that infrastructure was less important as a source of innovation after the war than it was before the war. A prominent exception to this proposition is the development of the "bullet train," the *shinkansen*.

8. These figures are taken from Nihon Tsūsan Tōkei Kyōkai (1982, 814 ff.).

9. At present, Kanagawa prefecture—which is located in the Kantō region, situated next to Tokyo prefecture, and is the prefecture where Yokohama and Kawasaki are located—is the highest-ranking prefecture in Japan in terms of industrial output. Aichi prefecture, where Nagoya and Toyota City are located, is ranked second. Osaka prefecture is ranked third.

10. See pages 93–94 in the summary volume of Ōsaka-shi (1969) *Shōwa Ōsaka-shi shi*. Amagasaki City is located in Hyogo prefecture and is a bedroom community for both Osaka and Kobe. To some extent, the other satellite cities of Osaka are also bedroom communities for both conurbations.

11. The figures here are taken from various postwar censuses. The particular source I used is Japan Prime Minister's Office (various dates), Nihon Sōrifu Tōkei-kyoku (various dates), and Zenkoku Shichō Kai (1992).

12. Basically, a DID (densely inhabited district) is a census enumeration district with at least 50,000 or more persons residing within it. In addition, it must have a minimum population density. It must have at least 4,000 persons per square kilometer.

13. For reliance on *kukaku seiri* as a vehicle for city planning in the postwar period, see Honjo (1971), Hebbert (1994), and the chapters written by various authors in Shapira, Masser, and Edgington (1994).

14. For a discussion of urban rank order hierarchy, see chapter 3.

15. The elasticity of city resident population with respect to rank in 1903 is –1.81 in the Tōkaidō, and –0.46 in the periphery. In 1990, the elasticity is –0.92 in the Tōkaidō, and –0.84 in the region outside of the Tōkaidō.

16. See pages 164–65 in Ōsaka-shi (1967), *Shōwa Ōsaka-shi shi. Keizaihen*, vol. 2 (vol. 2 of *Keizaihen*).

17. Quoted from Imperial Japanese Government Railways (1914a, 160). Imperial Japanese Government Railways (1914a, 1914b) provides a useful guide to Japan as it was around World War I.

Bibliography

Abe, Takeshi, and Osamu Saito. 1988. "From Putting-Out to the Factory: A Cotton-weaving District in Late-Meiji Japan." *Textile History* 19, no. 2 (autumn): 143–58.

Abramovitz, Moses, and Paul David. 1996. "Convergence and Deferred Catch-up: Productivity Leadership and the Waning of American Exceptionalism." In Landau, Taylor, and Wright (1996), pp. 21–62.

Adams, T.F.M. 1964. *A Financial History of Modern Japan*. Tokyo: Research (Japan).

Alden, Jeremy, and Hirofumi Abe. 1994. "Some Strengths and Weaknesses of Japanese Urban Planning." In Shapira, Masser and Edgington (1994), pp. 33–58.

Allen, George. 1940. "Japanese Industry: Its Organization and Development." In Schumpeter (1940), pp. 477–787.

Aschauer, David. 1989. "Is Public Sector Expenditure Productive?" *Journal of Monetary Economics* 23: 177–200.

Ashihara, Yoshinobu. 1986. *The Hidden Order: Tokyo through the Twentieth Century*. New York: Kodansha International.

Ausbel, Jesse, and Robert Herman, eds. 1988. *Cities and Their Vital Systems: Infrastructure. Past, Present and Future*. Washington, DC: National Academy Press.

Baumol, William; Richard Nelson; and Edward Wolff, eds. 1992. *Convergence of Productivity: Cross-National Studies and Historical Evidence*. New York: Oxford University Press.

Benevelo, Leonardo. 1971. *The Origins of Modern Town Planning*. Cambridge: MIT Press.

Berry, Mary. 1989. *Hideyoshi*. Cambridge: Harvard University Press.

Bestor, Theodore. 1989. *Neighborhood Tokyo*. Stanford: Stanford University Press.

Bisson, T. A. 1954. *Zaibatsu Dissolution in Japan*. Berkeley: University of California Press.

Boserup, Ester. 1981. *Population and Technological Change: A Study of Long-Term Trends*. Chicago: University of Chicago Press.

Brandstrom, Anders, and Lars-Goran Tederbrand. 2000. *Population Dynamics during Industrialization*. Umea: Umea University.

Cameron, Rondo, with Olga Crisp; Hugh Patrick; and Richard Tilly. 1967. *Banking in the Early Stages of Industrialization: A Study in Comparative Economic History*. New York: Oxford University Press.

Chida, Tomohei, and Peter Davies. 1990. *The Japanese Shipping and Shipbuilding Industries: A History of Their Modern Growth*. London: Athlone Press.

Collcutt, Martin; Marius Jansen; and Isao Kumakura, eds. 1988. *Cultural Atlas of Japan*. Oxford: Phaidon Press.

Dennison, Edward, and William Chung. 1976. "Economic Growth and Its Sources." In Patrick and Rosovsky (1976), pp. 63–151.

de Vries, J. 1984. *European Urbanization, 1500–1800*. Cambridge: Harvard University Press.

de Vries, Jan, and Ad van der Woude. 1997. *The First Modern Economy: Success, Failure and Perseverance of the Dutch Economy, 1500–1815*. New York: Cambridge University Press.

Diederiks, Herman, and Paul Hohenberg. 1992. "The Visible Hand and the Fortune of Cities: A Historiographic Introduction." In Diederiks, Hohenberg, and Wagenaar (1992) , pp.1–16.

Diederiks, Herman; Paul Hohenberg; and Michael Wagenaar, eds. 1992. *Economic Policy in Europe since the Middle Ages: The Visible Hand and the Fortune of Cities*. Leicester: Leicester University Press.

Doane, Donna. 1998. *Cooperation, Technology, and Japanese Development: Indigenous Knowledge, the Power of Networks, and the State*. Boulder, CO: Westview Press.

Dore, Ronald. 1958. *City Life in Japan: A Study of a Tokyo Ward*. Berkeley: University of California Press.

Elvin, Mark. 1974a. "Introduction." In Elvin and Skinner (1974), pp. 1–15.

———. 1974b. "The Administration of Shanghai, 1905–1914." In Elvin and Skinner 1974, pp. 239–62.

Elvin, Mark, and G. William Skinner, eds. 1974. *The Chinese City between Two Worlds*. Stanford: Stanford University Press.

Ericson, Steven. 1996. *The Sound of the Whistle: Railroads and the State in Meiji Japan*. Cambridge: Harvard University Press.

Evans, Alan. 1985. *Urban Economics: An Introduction*. Oxford: Basil Blackwell.

Evans, David, and Mark Peattie. 1997. *Kaigun: Strategy, Tactics and Technology in the Imperial Japanese Navy, 1887–1941*. Annapolis: Naval Institute Press.

Frost, Peter. 1970. *The Bakamatsu Currency Crisis*. Cambridge: Harvard University Press.

Fujimori, S. 1982. *Meiji no Tōkyō keikaku* [Planning Tokyo during the Meiji Era]. Tokyo: Iwanami Shoten.

Fujita, Kuniko, and Richard Hill. 1993. *Japanese Cities in the World Economy*. Philadelphia: Temple University Press.

Fujita, Nobuhisa. 1990. "Ties between Foreign Makers and Zaibatsu Enterprises in Prewar Japan: Case Studies of Mitsubishi Oil Co. and Mitsubishi Electric Manufacturing Co." In Yuzawa and Udagawa (1990), pp. 118–39.

Fujitani, Takashi. 1996. *Splendid Monarchy: Power and Pageantry in Modern Japan*. Berkeley: University of California Press.

Garon, Sheldon. 1997. *Molding Japanese Minds: The State in Everyday Life*. Princeton: Princeton University Press.

Goldsmith, Raymond. 1983. *The Financial Development of Japan, 1868–1977*. New Haven: Yale University Press.

Gordon, Andrew. 1991. *Labor and Imperial Democracy in Prewar Japan*. Berkeley: University of California Press.

Goto, Akira, and Hiroyuki Odagiri. 1997. *Innovation in Japan*. New York: Oxford University Press.

Graham, E., and M. Yoshitomi, eds. 1996. *Foreign Direct Investment in Japan*. Brookfield, VT: Edward Elgar.

Grossman, Gene, and Elhanan Helpman. 1994. "Endogenous Innovation in the Theory of Growth." *Journal of Economic Perspectives* 8, no. 1 (winter): 23–44.

Hadfield, Charles. 1968. *The Canal Age*. Newton Abbot, Devon, England: Latimer Trend.

Hadley, Eleanor. 1970. *Antitrust in Japan*. Princeton: Princeton University Press.

Hall, Peter. 1994. *Innovation, Economics and Evolution: Theoretical Perspectives on Changing Technology in Economic Systems*. New York: Harvester Wheatsheaf.

Hanley, Susan, and Kozo Yamamura. 1977. *Economic and Demographic Change in Preindustrial Japan, 1600–1868*. Princeton: Princeton University Press.

Hanshin Denki Tetsudō Kabushiki Gaisha [Hanshin Electric Railroad Company]. 1985. *Hanshin denki tetsudō hachijū nen shi* [Eighty-Year History of the Hanshin Electric Railroad]. Tokyo: Nihon Keiei Shi Kenkyūsho.

Hanshin Kan Modanizumu Ten Jikkō Iinkai [Realization Committee for the Exhibition "Modernism along the Hanshin (Railroad)]. 1985. *Hanshin kan modanizumu* [Modernism along the Hanshin (Railroad)]. Kyoto: Tankosha.

Harada, Shuichi. [1928]. 1968. *Labor Conditions in Japan*. Reprint, New York: AMS Press.

Hastings, Sally. 1995. *Neighborhood and Nation in Tokyo*. Pittsburgh: University of Pittsburgh Press.

Hatano, Jun. 1994. "Edo's Water Supply." In McClain, Merriman, and Ugawa (1994), pp. 234–50.

Hauser, William. 1974. *Economic Institutional Change in Tokugawa Japan: Ōsaka and the Kinai Cotton Trade*. New York: Cambridge University Press.

Hayami, Akira. 1993. "Overview." In Hayami (1993), pp. 1–15.

———. 1996. "Meiji zenki jinkō tōkei shi nenpyō: Bakufu kunibetsu jinkō hyō." [Chronological Statistical Tables for Pre-Meiji Population History: Tables for Population under the Bakufu and in the Domains of Japan]. Kyoto: International Research Center for Japanese Studies Project on Population and Family Studies Off-Print Series No. 1.

———. 1997. *Rekishi jinkōgaku no sekai* [The World of Historical Demography]. Tokyo: Iwanami Shoten.

Hayami, Akira, ed. 1993. *Population History of East Asia:* Papers Presented to the XXIInd IUSSP General Conference, session 40, August, Montreal, Canada. Montreal: International Union for the Scientific Study of Population.

Hayami, Akira, and Matao Miyamoto, eds. 1997. *Keizai shakai no seiretsu: 17–18 seiki* [The Formation of Economic Society: 17th–18th Centuries]. Tokyo: Iwanami Shoten.

Hayami, Yujiro, in association with Masakatsu Akino; Masahiko Shintani; and Saburo Yamada. 1975. *A Century of Agricultural Growth in Japan: Its Relevance to Asian Development*. Minneapolis: University of Minnesota Press.

Hayashi, Reiko. 1994. "Provisioning Edo in the Early Eighteenth Century: The Pricing Policies of the Shogunate and the Crisis of 1733." In McClain, Merriman, and Ugawa (1994), pp. 211–33.

Hayashi, Takeshi. 1990. *The Japanese Experience in Technology: From Transfer to Self-Reliance*. Tokyo: United Nations University Press.

Hebbert, Michael. 1994. "*Sen-biki* amongst *desakota*: Urban Sprawl and Urban Planning in Japan." In Shapira, Masser, and Edgington (1994), pp. 115–40.

Henderson, Vernon. 2000. "The Effects of Urban Concentration on Economic Growth." National Bureau of Economic Research, working paper no. 7503, Cambridge, MA.

Hidenobu, Jinnai. 1995. *Tokyo: A Spatial Anthropology*. Berkeley: University of California Press.

Higashinari-ku Yakusho [Higashinari Ward Office]. 1957. *Higashinari-ku shi* [History of Higashinari Ward]. Osaka.

Hill, Richard, and Kuniko Fujita. 1993. "Japanese Cities in the World Economy." In Fujita and Hill (1993), pp. 3–25.

Hirakawa, A. 1997. *Kinsei Nihon no kōtsū to chiiki keizai* [Transportation and Regional Economies in Early Modern Japan]. Osaka: Shibunkaku.

Hirsch, Werner. 1993. *Urban Economic Analysis*. New York: McGraw-Hill.

Hirschmeier, James, and Tsunehiko Yui. 1995. *The Development of Japanese Business, 1600–1973*. Cambridge: Harvard University Press.

Hoare, James. 1994. *Japan's Treaty Ports and Foreign Settlements: The Uninvited Guests 1858–1899*. Kent, England: Japan Library.

Hodgson, Geoffrey. 1993. *Economics and Evolution: Bringing Life Back into Economics*. Cambridge, England: Polity Press.

Hohenberg, Paul, and Lynn Lees. 1985. *The Making of Urban Europe, 1000–1950*. Cambridge: Harvard University Press.

Honjo, Masahiko. 1971. *Urban Planning Administration in Japan*. Tokyo: Tokyo University Engineering Department Urban Science Division.

Hotta, A., and T. Nishiguchi. 1995. *Ōsaka Kawaguchi kyoryūchi no kinkyū* [Research about the Osaka Kawaguchi Settlement]. Kyoto: Shibunkaku.

Howe, Christopher, ed. 1981a. *Shanghai: Revolution and Development in an Asian Metropolis*. Cambridge, England: Cambridge University Press.

———. 1981b. "Industrialization under Conditions of Long-run Population Stability." In Howe (1981), pp. 153–87.

Hulten, Charles. 2000. "Total Factor Productivity: A Short Bibliography." National Bureau of Economics, working paper no. 7471, Cambridge, MA.

Imaoka, Junichiro. 1929. "Shipping and Shipbuilding in Japan." In World Engineering Congress, Publications Committee (1929), pp. 353–69.

Imperial Japanese Government Railways. 1914a. *An Official Guide to Eastern Asia:* Vol. 2. *South-Western Japan*. Tokyo: Imperial Japanese Government Railways.

———. 1914b. *An Official Guide to Eastern Asia:* Vol. 3. *North-Eastern Japan*. Tokyo: Imperial Japanese Government Railways.

Ishida, Y. 1987. *Nihon kindai toshi keikaku shi kinkyū* [Research on Japan's Modern City Planning]. Tokyo: Kashiwa Shobosha.

Ito, Takatoshi. 1993. *The Japanese Economy*. Cambridge: MIT Press.

Ito, Y. 1996. *Edo jōsuidō no rekishi* [History of Edo's Waterworks]. Tokyo: Yoshikawa Kōbunkan.

Jansen, Marius, and Gilbert Rozman, eds. 1986. *Japan in Transition: From Tokugawa to Meiji*. Princeton: Princeton University Press.

Japan Department of Communications, Mercantile Merchant Marine Bureau. 1898. *List of Merchant Marine Vessels of Japan, 1898*. Tokyo.

Japan Prime Minister's Office. Various dates. *Japan, Tokyo. Statistical Yearbook.*

Jones, Eric. 1988. *Growth Recurring: Economic Change in World History*. New York: Oxford University Press.

Jones, Hazel. 1980. *Live Machines: Hired Foreigners and Meiji Japan*. Vancouver: University of British Columbia Press.

Kansai Denryoku Kabushiki Gaisha [Kansai Electric Company]. 1978. *Kansai denryoku nijūgonen shi* [Twenty-Five-Year History of Kansai Electric Power Company]. Tokyo: Nihon Keieishi Kenkyūsho.

Kawazoe, N. 1965. *Contemporary Japanese Architecture*. Translated by D. Griffith. Tokyo: Society for International Cultural Relations.

Keihan Shinkyū Kō Denkitetsu Kabushiki Gaisha. 1959. *Keihan Shinkyū Kō denkitetsu gojūnen shi* [Fifty-Year History of the Keihan Shinkyū Electric Railroad]. Osaka: Keihan Shinkyū Kō.

Kelley, William. 1994. "Incendiary Actions: Fires and Firefighting in the Shogun's Capital and the People's City." In McClain, Merriman, and Ugawa (1994), pp. 310–31.

Kikuchi, Toshio. 1966. *Shinden kaihatsu* [Development of New Fields]. Tokyo: Nihon Rekishinsha.

Kinki Nihon Tetsudō Kabushiki Gaisha [Kinki Japan Railroad Company]. 1960. *Kinki Nihon tetsudō: 50 nen no ayumi* [Kinki Japan Railway: 50 Years of Advance]. Osaka: Kinki Nihon Tetsudō Kabushiki Gaisha.

Kiyokawa, Yukihiko. 1995. "Technology Choice in the Cotton-Spinning Industry: The Switch from Mules to Ring Frames." In Minami, Kim, Makino, and Seo (1995), pp. 84–111.

Kujitani, Tsunetaro. 1929. "The Manufacture of Electrical Machines and Materials in Japan." In World Engineering Congress, Publications Committee (1929), pp. 245–53.

Kuroha, H. 1980. *Ōsaka chihō no shiteki kinkyū* [Historical Research about the Osaka Region]. Tokyo: Gotandō Shoten.

Kuznets, Simon. 1971. *Economic Growth of Nations: Total Output and Production Structure*. Cambridge: Harvard University Press.

Ladd, Brian. 1990. *Urban Planning and Civic Order in Germany, 1860–1914*. Princeton: Princeton University Press.

Landau, Ralph; Timothy Taylor; and Gavin Wright, eds. 1996. *The Mosaic of Economic Growth*. Stanford: Stanford University Press.

Leupp, Gary. 1992. *Servants, Shophands, and Laborers in the Cities of Tokugawa Japan*. Princeton: Princeton University Press.

Lewis, Michael. 1990. *Rioters and Citizens: Mass Protest in Imperial Japan*. Berkeley: University of California Press.

Lewis, W. Arthur. 1978. *Growth and Fluctuations, 1870–1913*. London: George Allen and Unwin.

Maddison, Angus. 1987. "Growth and Slowdown in Advanced Capitalist Economies: Techniques of Quantitative Analysis." *Journal of Economic Literature* 25, 649–98.

Mainichi Shinbunsha [Mainichi Newspapers Ltd.]. 1951. *Nihon jishin shiryō* [Historical Data on Japan's Earthquakes]. Tokyo: Meiseki.

Mayehara, Sukeichi. 1929. "The Present State of Electrical Undertakings and Installations in Japan." In World Engineering Congress, Publications Committee (1929), pp. 199–217.

McClain, James. 1994. "Edobashi: Power, Space, and Popular Culture in Edo." In McClain, Merriman, and Ugawa (1994), pp. 105–31.

McClain, James, and John Merriman. 1994. "Edo and Paris: Cities and Power." In McClain, Merriman, and Ugawa (1994), pp. 3–38.

McClain, James, and Osamu Wakita, eds. 1999. *Osaka: The Merchants' Capital of Early Modern Japan*. Ithaca: Cornell University Press.

McClain, James; John Merriman; and Kaoru Ugawa, eds. 1994. *Edo and Paris:*

Urban Life and the State in the Early Modern Era. Ithaca: Cornell University Press.

McKay, John. 1976. *Tramways and Trolleys: The Rise of Urban Mass Transport in Europe*. Princeton: Princeton University Press.

Minami, Ryoshin. 1965. *Tetsudō to denryoku* [Railroads and Electric Utilities]. Tokyo: Toyo Keizai Shinpōsha.

———. 1986 (1st ed.) and 1994 (2d ed.). *Economic Development of Japan: A Quantitative Study*. Houndmills, Basingstoke, Hampshire: Macmillan Press.

———. 1998. "Economic Development and Income Distribution in Japan: An Assessment of the Kuznets Hypothesis." *Cambridge Journal of Economics* 22, no. 1 (January): 39–58.

Minami, Ryoshin, and Fumio Makino. 1995a. "The Development of Appropriate Technologies for Export Promotion in the Silk-Reeling Industry." In Minami, Kim, Makino, and Seo (1995), pp. 31–53.

———. 1995b. "Mechanism of the Diffusion of Technologies: Case Study of the Cotton-Weaving Industry." In Minami, Kim, Makino, and Seo (1995), pp. 54–84.

Minami, Ryoshin; Kwan Kim; Fumio Makino; and Jeong-hae Seo, eds. 1995. *Acquiring, Adapting and Developing Technologies: Lessons from the Japanese Experience*. Houndmills, Basingstoke, Hampshire: Macmillan Press.

Miyamoto, Kenichi. 1993. "Osaka and Tokyo Compared." In Fujita and Hill (1993), pp. 53–81.

Miyamoto, Mataji. 1960. *Senba* [Senba]. Kyoto: Minruba Shobo.

Miyamoto, Mataji, ed. 1968. *Ōsaka no kenkyū. Daini kan: Kindai Ōsaka no keizai shi* [Research on Osaka. Vol. 2, Research about the Modern Economic History of Osaka]. Osaka: Seibundō.

Miyamoto, Matao. 1984. "The Position and Role of Family Business in the Development of the Japanese Company System." In Okochi and Yasuoka (1984), pp. 39–91.

Miyamoto, Matao, and M. Uemura. 1997. "Tokugawa keizai no junkan kōzō" [The Circulation Structure of the Tokugawa Economy]. In Hayami and Miyamoto (1997), pp. 271–324.

Mizutani, F. 1994. *Japanese Urban Railways: A Private-Public Comparison*. Aldershot, England: Avebury.

Mokyr, Joel. 1990. *The Lever of Riches: Technological Creativity and Economic Progress*. New York: Oxford University Press.

Molony, Barbara. 1990. *Technology and Investment: The Prewar Japanese Chemical Industry*. Cambridge: Harvard University Press.

Morikawa, Hidemasa. 1992. *Zaibatsu: The Rise and Fall of Family Enterprise Groups in Japan*. Tokyo: University of Tokyo Press.

Morris, R. 1992. "The State, Elite and the Market: The 'Visible Hand' in the British Industrial City System." In Diederiks, Hohenberg, and Wagenaar (1992), 177–99.

Mosk, Carl. 1983. *Patriarchy and Fertility: Japan and Sweden, 1880–1960*. New York: Academic Press.

———. 1995a. *Competition and Cooperation in Japanese Labour Markets*. Houndmills, Basingstoke, Hampshire: Macmillan Press.

———. 1995b. "Household Structure and Labor Markets in Postwar Japan." *Journal of Family History* 20, no. 1: 103–25.

———. 1996. *Making Health Work: Human Growth in Modern Japan*. Berkeley: University of California Press.

———. 2000a. "Inequality, Ideology, Autarky and Structural Change: The Biological Standard of Living in Japan between the World Wars." Paper read at the meetings of the European Social Sciences History Association, Amsterdam, the Netherlands, April.

———.2000b. "Small-Scale Production and Urban Expansion in Industrializing Japan: Nagoya, 1890–1940." In Brandstrom and Tederbrand (2000), pp. 227–70.

Mosk, Carl, and Shelia Johansson. 1986. "Income and Mortality: Evidence from Modern Japan." *Population and Development Review* 12, no. 3 (September): 415–40.

Murakoshi, Kazunori. 1995. "The Population Demography of the Warrior Class in Tokugawa Japan." Paper read to the 1995 annual meeting of the Social Science History Association, Chicago, Illinois, November.

Murphey, Rhoads. 1974. "The Treaty Ports and China's Modernization." In Elvin and Skinner (1974), pp. 17–71.

Nagamine, H., ed. 1981. *Nation-Building and Regional Development: The Japanese Experience*. Hong Kong: Maruzen Investment.

Nakagawa, Kichizo. 1929. "Civil Engineering in Japan." In World Engineering Congress. Publications Committee (1929), pp. 65–119.

Nakamura, Takafusa. 1983. *Economic Growth in Prewar Japan*. New Haven: Yale University Press.

Naniwa-ku Yakusho [Naniwa Ward Office]. 1957. *Naniwa-ku shi* [History of Naniwa Ward]. Osaka.

Nankai Denki Tetsudō Kabushiki Gaisha [Nankai Electric Railroad Company]. 1978 [1899]. Reprint of the 1899 original. *Nankai tetsudō annai* [Guide to the Nankai Electric Railroad].Osaka: Nankai Tetsudō Sōgō Kenkyūsho.

Nawa, Mitsuo. 1929. "Railways in Japan." In World Engineering Congress, Publications Committee (1929), pp. 137–73.

Nelson, Richard. 1996. *The Sources of Economic Growth*. Cambridge: Harvard University Press.

Nelson, Richard, and Sidney Winter. 1982. *An Evolutionary Theory of Economic Change*. Cambridge: Harvard University Press.

Nihon Jūtaku Kinyū Kōkō [Japan Housing Financial Corporation]. 1970. *Jūtaku kinyū kōkō 20 nenshi* [Twenty-Year History of the Housing Financial Corporation]. Tokyo.

Nihon Naikaku Tōkei-kyoku [Japan Cabinet Bureau of Statistics]. 1882–1940. *Nihon teikoku tōkei nenkan* [Statistical Yearbook of the Japanese Empire]. Tokyo.

———. 1906. *Nihon teikoku seitai tōkei, Meiji sanjūroku nen* [Statistics of the State of Population, 1903]. Tokyo.

———. 1912. *Isshin igo teikoku tōkei sho zairyō, daini shū* [Imperial Statistical Materials after the (Meiji) Restoration, Second Collection]. Tokyo.

———. Various dates. *Taishō kunen kokusei chōsa hōkoku. Fu, ken no bu* [Reports of the 1920 Census. Prefectural Volumes]. Tokyo.

———. Various dates. *Taishō jūyonen kokusei chōsa hōkoku. Fu ken hen* [Reports of the 1925 Census. Prefectural Volumes]. Tokyo.

———. 1930. *Nōgyō chōsa kekka hōkoku, Shōwa yonen* [Report of the Results of the 1929 Agricultural Census]. Tokyo.

———. 1935a. *Shōwa gonen kokusei chōsa hōkoku, shokugyō oyobi sangyō* [Reports of the 1930 Census. Occupation and Industry]. Tokyo.

———. 1935b. *Shōwa gonen kokusei chōsa hōkoku, jūgyō no basho* [Reports of the 1930 Census. Place of Employment]. Tokyo.

———. Various dates. *Shōwa gonen kokusei chōsa hōkoku. Fu ken hen* [Reports of the 1930 Census. Prefectural Volumes]. Tokyo.

———. Various dates. *Shōwa jūnen kokusei chōsa hōkoku. Fu ken hen* [Reports of the 1935 Census. Prefectural Volumes]. Tokyo.

———. 1992. Reprint edited by A. Hayami. *Nihon jinkō tōkei* [Compilation of Japanese Population Statistics]. Tokyo: Harashobo.

Nihon Naimu Daijin Kanbō Toshi Keikaku-ka [Japan Ministry of the Interior. Minister's Secretariat. City Planning Department]. 1988. Reprint of government publications originally issued at various dates. *Toshi keikaku yōkan. Dainikan*; *daisankan*; *daiyonkan*; *dairokukan*; *bekkan* [Basic Summary of City Planning. Vols. 2, 3, 4, and 6; and appended volume]. Tokyo: Kashiwashobō.

Nihon Noriai Jidōsha Kyōkai [Japan Bus Association]. 1957. *Basu jigyō gojūnen shi* [Fifty Years of the Bus Business]. Tokyo: Basu Jigyō Gojūnen Shi Hensan Iinkai.

Nihon Sōrifu Tōkei-kyoku [Japan Prime Minister's Office. Bureau of Statistics]. Various dates. *Japan Statistical Yearbook*. Tokyo.

Nihon Tetsudō-shō. [Japan Ministry of Railroads]. Various dates. *Tetsudō tōkei shiryō to tetsudōin tōkei zuhyō* [Statistical Materials and Statistical Tables Concerning Railroads]. Tokyo.

Nihon Tsūsan Tōkei Kyōkai [Japan Statistics Cooperation Group for the Ministry of International Trade and Industry]. 1982. *Sengo no kōgyō tōkeihyō. Daiichikan. Tōkeihen* [Postwar Factory Statistical Tables. Vol. 1. Statistics]. Tokyo.

Nihon Tsūsanshō Daijin Kanbō Chōsa Tōkei-kyoku [Japan Ministry of International Trade and Industry. Minister's Secretariat. Research and Statistics Division]. 1961. *Kōgyō tōkei 50 nen shi* [Fifty-Year History of the Census of Manufactures]. Tokyo.

Nishinari-gun Yakusho [Nishinari County Office]. 1915. *Nishinari-gun shi* [History of Nishinari County]. Osaka.

Nishiyama, Unotsuke. 1997. *Ajigawa Monogatari* [Aji River Story]. Tokyo: Nihon Keizai Hyōronsha.

Ohkawa, Kazushi, and Katsuo Ōtsuka. 1994. *Technology Diffusion, Productivity Employment, and Phase Shifts in Developing Economies*. Tokyo: University of Tokyo Press.

Ohkawa, Kazushi, and Henry Rosovsky. 1973. *Japanese Economic Growth: Trend Acceleration in the Twentieth Century*. Stanford: Stanford University Press.

Ohkawa, Kazushi, and Miyohei Shinohara, with Larry Meissner, eds. 1979. *Patterns of Japanese Economic Development: A Quantitative Appraisal*. New Haven: Yale University Press.

Ohkawa, Kazushi; Miyohei Shinohara; and Mataji Umemura. 1966. *Zaisei shishutsu* [Government Expenditure].Tokyo: Toyo Keizai Shinposha.

Ōi, T. 1992. *Yomigaeru Meiji no Tōkyō–Tōkyō jūgoku shashin shū* [Photograph Collection for the Fifteen Wards of Tokyo]. Tokyo: Toppan.

Okamoto, R. 1978. *Meiji Taishō zushi, dai 11 kan, Ōsaka* [Meiji/Taishō Illustrated, Vol. 11, Osaka]. Tokyo: Chikuma Shobō.

Okochi, Akio, and Shigeaki Yasuoka, eds. 1984. *Family Business in the Era of Industrial Growth: Its Ownership and Management*. Tokyo: University of Tokyo Press.

Osaka City. Environment and Public Health Bureau. 1996. *Osaka City and the Environment, 1996*. Osaka.

Osaka City Government. City Planning Department. Planning Bureau. 1997. *Planning of Osaka: City Planning in Osaka City*. Osaka.

Ōsaka Kōwan-kyoku [Osaka Harbor Bureau]. 1960. *Edo jidai ni okeru Ōsaka no*

kohai gyōsei ni tsuite [Concerning the Administration of the Rivers and the Bay of Osaka during the Edo Period]. Osaka.

Ōsaka Toshi Kyōkai [Osaka City Cooperation Assembly]. 1933. *Ōsaka toshi keikaku gaiyō* [Summary of the Osaka City Plan]. Osaka.

Ōsaka-fu [Osaka Prefecture]. 1936. *Kōgyō chōsasho, Shōwa kunen* [Factory Survey, 1934]. Osaka.

Ōsaka-fu [Osaka Prefecture]. Various dates. *Ōsaka-fu tōkeisho* [Osaka Prefecture Statistics]. Osaka.

Ōsaka-fu Kenchiku-bu Jūtaku Kaihatsu-ka [Osaka Prefecture. Construction Bureau. Housing Development Department]. 1971. *Ōsaka-fu no jūtaku jijō* [The Housing Situation in Osaka Prefecture]. Osaka.

Ōsaka-shi [Osaka City]. 1951–1969. *Shōwa Ōsaka-shi shi* [Shōwa Osaka City History]. Vols. 1–8 in original series and vols. 1–8 in continuing series. Osaka.

———. 1980. Reprint of the original 1933 edition. *Meiji Taishō Ōsaka-shi shi* [Osaka City History during the Meiji and Taishō Eras]. Osaka.

———. 1989. *Shashin de miru Ōsaka shi 100 nen* [A 100-Year History of Osaka Seen through Photographs]. Osaka: Osaka Toshi Kyōkai.

Ōsaka-shi Chūo Oroshiuri [Osaka Central Wholesale Market]. 1997. *Shijo gaiyo.* [Overview of the Market]. Osaka.

Ōsaka-shi Denki-kyoku [Osaka City Electricity Bureau]. 1990. Reprint of the 1923 original edition. *Kōei kōtsū jigyō enkaku shi* [History of Public Transportation Activities]. Tokyo: Kuresu.

Ōsaka-shi Higashi-ku Yakusho [Osaka City. Higashi Ward Office]. 1892. *Meiji nijūgo tōkeisho* [Statistics for 1892]. Osaka.

Ōsaka-shi Kai Jimū-kyoku Chōsa Kai [Osaka City Assembly Office. Research Committee]. 1970. *Toshi to toshi mondai, 1970* [City and City Problems, 1970]. Osaka.

Ōsaka-shi Kankyō Hoken-kyoku Kankyō Seiri-ka [Osaka City. Environmental Protection Bureau. Environmental Clean-up Section]. 1998. *Ōsaka-shi kinkyō hakusho. Heisei 9 nenpan* [Osaka City Environment White Paper 1997]. Osaka.

Ōsaka-shi Kensetsu-kyoku [Osaka City. Construction Bureau]. 1995. *Ōsaka no machizukuri: tochi kukaku seiri jigyō* [The Making of Osaka's Neighborhoods: The Business of Reorganization and Land Planning]. Osaka.

Ōsaka-shi Konohana-ku Yakusho [Osaka City Konohana Ward Office]. 1955. *Konohana-ku shi* [History of Konohana Ward]. Osaka.

Ōsaka-shi Kōtsū-kyoku [Osaka City Transportation Bureau]. 1980a. *Henden shi* [Electric Power Transmission History]. Osaka.

———. 1980b. *Ōsaka-shi kōtsū nanajūgonen shi* [Seventy-Five-Year History of Osaka City Transportation]. Osaka.

Ōsaka-shi Kōwan-kyoku [Osaka City Harbor Bureau]. 1997. *Port of Osaka*. [Title in English and text in Japanese]. Osaka.

Ōsaka-shi Minato-ku Yakusho. 1956. *Minato shi* [Minato Documents]. Osaka.

Ōsaka-shi Shakai-bu [Osaka City Social Bureau]. 1996. Reprints in sixty volumes with one separate appendix volume of the original reports published between 1927 and 1942. *Ōsaka-shi shakai-bu chōsa hōkokusho* [Reports of the Surveys of the Social Bureau of Osaka City]. Osaka.

Ōsaka-shi Yakusho [Osaka City Hall]. 1935. *Ōsaka-shi kōgyō chōsa. Shōwa hachinen* [Survey of Osaka City Industry. 1933]. Osaka.

———. 1937. *Ōsaka-shi tempo bunpai chōsa hōkoku* [Reports of the Survey of the Main Shopping Arcades within Osaka City]. Osaka.

———. 1940. *Ōsaka-shi kōgyō chōsa. Shōwa jūyonen* [Survey of Osaka City Industry. 1939.] Osaka.

———. Various dates. *Ōsaka-shi tōkeisho* [Osaka City Statistics]. Osaka.

Ōshima, T. 1992. *Jidōsha* [Automobiles]. Tokyo: Nihon Keizai Hyōronsha.

Ostrom, Elinor. 1990. *Governing the Commons: The Evolution of Institutions for Collective Action*. Cambridge: Cambridge University Press.

Ōtsuka, Katsuo. 1995. "Choice of Technologies in the Ship-Building Industry: Modern versus Traditional." In Minami, Kim, Makino, and Seo (1995), pp. 112–37.

Patrick, Hugh. 1967. "Japan, 1868–1914." In Cameron, Crisp, Patrick, and Tilly (1967), pp. 239–89.

———. 1984. "Japanese Financial Development in Historical Perspective." In Ranis, West, Leiserson, and Morris (1984), pp. 302–27.

Patrick, Hugh, and Henry Rosovsky, eds. *Asia's New Giant: How the Japanese Economy Works*. Washington: Brookings Institution.

Ramseyer, J. Mark. 1996. *Odd Markets in Japanese History*. New York: Cambridge University Press.

Ramseyer, J. Mark, and Frances Rosenbluth. 1995. *The Politics of Oligarchy: Institutional Choice in Imperial Japan*. New York: Cambridge University Press.

Ranis, Gustav; Robert West; Mark Leiserson; and Cynthia Taft Morris, eds. 1984. *Comparative Development Perspectives*. Boulder, CO: Westview Press.

Robinson, Olivia. 1992. *Ancient Rome: City Planning and Administration*. London: Routledge.

Romer, Paul M. 1994. "The Origins of Endogenous Growth." *Journal of Economic Perspectives* 8, no. 1 (winter): 3–22.

Rosovsky, Henry. 1961. *Capital Formation in Japan, 1868–1940*. Glencoe, IL: The Free Press.

Rostow, Walt, ed. 1963. *The Economics of Take-Off into Sustained Growth. Proceedings of a Conference Held by the International Economic Association*. New York: St. Martin's Press.

———. 1980. *The World Economy: History and Prospects*. Austin: University of Texas Press.

———. 1998. *The Great Population Spike and After: Reflections of the 21st Century*. New York: Oxford University Press.

Rozman, Gilbert. 1973. *Urban Networks in Ch'ing China and Tokugawa Japan*. Princeton: Princeton University Press.

Rutherford, M. 1994. *Institutions in Economics: The Old and the New Institutionalism*. Cambridge, England: Cambridge University Press.

Saito, Osamu. 1985. *Purotokōgyō no jidai: Seiō to Nihon no hikakushi* [Age of Proto-industrialization: Comparative History of Western Europe and Japan]. Tokyo: Nihonhyōronsha.

———. 1987. *Shōkai no sekai, uradana no sekai: Edo to Osaka no hikaku toshi shi* [The World of Merchant Houses. The World of Back Alleys: The Comparative Urban History of Edo and Osaka]. Tokyo: Riburopooto.

Samuels, Richard. 1983. *The Politics of Regional Policy in Japan: Localities Incorporated?* Princeton: Princeton University Press.

Sawai, Minoru. 1985. "Senzeki Nihon tetsudō sharyō kōgyō no tenkan katei—1890 nendai—1920 nendai" [The Course of Development of the Prewar Japanese Rail-

road Rolling Stock Industry—From the 1890s to the 1920s], *Shakai kagaku kenkyū* [Social Science Research] 37, no. 3: 1–200.

———. 1989. "The Development of Machine Industries and the Evolution of Production and Labor Management." In Yui and Nakagawa (1989), pp. 199–236.

———. 1997. "The Development of Industrial Research in Osaka from 1920s to 1960s." *Osaka Economic Papers* 47, no. 2: 12–45.

———. 1998. "The Japanese National Railways and the Rolling Stock Industry." *Osaka Economic Papers* 47, nos. 2–4: 49–65.

Sawamoto, M. 1981. "One Hundred Years of Public Works in Japan: Lessons of Experience." In Nagamine (1981), pp. 99–135.

Schumpeter, E., ed. 1940. *The Industrialization of Japan and Machukuo 1930–1940: Population, Raw Materials and Industry*. New York: Macmillan.

Schumpeter, Joseph. 1964 [1939]. Abridged by R. Fels. *Business Cycles: A Theoretical, Historical and Statistical Analysis of the Capitalist Process*. New York: McGraw-Hill.

Seki, Keizo. 1956. *The Cotton Industry of Japan*. Tokyo: Japan Society for the Promotion of Science.

Shapira, Philip; Ian Masser; and David Edgington, eds. 1994. *Planning for Cities and Regions in Japan*. Liverpool: Liverpool University Press.

Shibusawa, Motoji. 1929. "The Principal Electrical Transmission Systems in Japan." In World Engineering Congress, Publications Committee (1929), pp. 189–97.

Shimizu, Katsu. 1990. *Ōsaka no rekishi* [History of Osaka (Prefecture)]. Tokyo: Kawabe Shobo Shinshu.

Shinohara, Miyohei. 1979. "Manufacturing." In Ohkawa and Shinohara, with Meissner (1979), pp. 104–21.

Shinpo, Hiroshi, and Osamu Saito, eds. 1997. *Kindai seichō no taidō* [The Quickening of Modern Growth]. Tokyo: Iwanami Shoten.

———. 1997. "Gaisetsu: jūkyū seiki e" [Outline: To the 19th Century]. In Shinpo and Saito (1997), pp. 1–66.

Shinpo, Hiroshi, and Akira Hasegawa.1977. "Shōhin seisan ryūtsū no daiamikkusu" [The Dynamics of Merchandise Production and Distribution]. In Hayami and Miyamoto (1997), pp. 217–70.

Shinshū Ōsaka-shi Shi Hensan Iinkai [Editorial Committee for a New Compilation of Osaka City History]. 1988–1996. *Ōsaka-shi shi* [Osaka City History]. Vols. 1–10. Kyoto: Kahoku.

———. 1996. *Rekishi chizu* [Historical Maps]. Kyoto: Kahoku.

Shobunsha. 1994. *Ōsaka* [Osaka]. Osaka: Shobunsha.

Skinner, G. William. 1964. *Marketing and Social Structure in Rural China*. Ann Arbor: Association for Asian Studies Reprint Series No. 1.

Smith II, Henry. 1986. "The Edo-Tokyo Transition: In Search of Common Ground." In Jansen and Rozman (1986), pp. 347–74.

Smith, Thomas. 1988. *Native Sources of Japanese Industrialization, 1750–1920*. Berkeley: University of California Press.

Smith, W. 1968. *An Historical Introduction to Economic Geography of Great Britain*. London: G. Bell and Sons, Ltd.

Solow, Robert. 1994. "Perspectives on Economic Growth." *Journal of Economic Perspectives* 8, no. 1: 45–54.

Steiner, Kurt. 1965. *Local Government in Japan*. Stanford: Stanford University Press.

Stewart, David. 1987. *The Making of a Modern Japanese Architecture*. New York: Kodansha International.

Sugihara, Kaoru, and Kingo Tamai. 1996. 3d printing. *Taishō Ōsaka Suramu* [Taishō Osaka Slums]. Tokyo: Shinhyōron.

Sumitomo Marine and Fire Insurance Co. 1993. *Sumitomo Marine and Fire Insurance: The First Century 1893–1993*. Tokyo: Sumitomo Marine and Corporate History Office.

Suzuki, Yoshio. 1980. *Money and Banking in Contemporary Japan*. New Haven: Yale University Press.

Szreter, Simon. 1997. "Economic Growth, Disruption, Deprivation, Disease and Death: On the Importance of the Politics of Public Health for Development." *Population and Development Review* 23, no. 4: 693–728.

Takechi, K. 1980. "Kindai Ōsaka ni okeru tetsudō kensetsu no ichi ruikei" [A Type of Railroad Construction in Modern Osaka]. In Kuroha (1980), pp. 261–301.

Takeuchi, Johzen. 1991. *The Role of Labour-Intensive Sectors in Japanese Industrialization*. Tokyo: United Nations University Press.

Taku, K. 1936a. "Ōsaka o chūshin to suru kinkō to gaitetsudō no saikin no gyōseki oyobi naiyo ni tsuite" [Concerning Economic Performance of Osaka Suburban Railroads and of Railroads with Osaka Destination and Their Details]. *Ōsaka-shi Denki-kyoku Gyōmu Kenkyū Shiryō* [Osaka City Electricity Bureau Administration Research Materials] 1, no. 1: 45–60.

———1936b. "Takushii no tōsei sonata ni tsuite" [Concerning Taxi Regulations]. *Ōsaka-shi Denki-kyoku Gyōmu Kenkyū Shiryō* [Osaka City Electricity Bureau Administration Research Materials] 1, no. 2: 19–24.

Tanabe, K. 1982. *Nihon no toshi shisutemu-chirigakuteki kinkyū* [Japan's Urban System-Geography Research]. Tokyo: Kokonshoin.

Thomas, Brinley. 1973. *Migration and Economic Growth: A Study of Great Britain and the Atlantic Economy*. 2d edition. Cambridge: Cambridge University Press.

Tōkyō Hyakunen Henshū Iinkai [Tokyo Hundred-Year History Editorial Committee]. 1979. *Tōkyō hyakunen shi. Daisan kan* [Hundred-Year History of Tokyo. Vol. 3]. Tokyo: Hundred-Year History Editorial Committee.

Tōkyō Kukaisei Gyō Shi [Tokyo Improvement Works Documents]. 1987–1988. Reprints of original documents edited by S. Fujimori. *Tōkyō-shi keikaku shiryō shūsei. Meiji Taishō Hen* [Compilation of Planning Materials of Tokyo for the Meiji and Taishō Eras]. Tokyo: Honnoyūsha.

Tōkyō-shi Yakusho [Tokyo City Hall]. Various dates. *Tōkyō-shi tōkei nenpyō* [Tokyo City Statistical Chronological Tables]. Tokyo.

Tōkyō-to [Tokyo]. 1968. *Me de miru Tōkyō hyakunen* [Looking at One Hundred Years of Tokyo History with One's Eyes]. Tokyo: Toppan.

Totman, Conrad. 1980. *The Collapse of the Tokugawa Bakufu, 1862–1868*. Honolulu: University of Hawaii Press.

———. 1985. *The Origins of Japan's Modern Forests: The Case of Akita*. Honolulu: University of Hawaii Press.

———. 1989. *The Green Archipelago: Forestry in Preindustrial Japan*. Berkeley: University of California Press.

———. 1993. *Early Modern Japan*. Berkeley: University of California Press.

———. 1995. *The Lumber Industry in Early Modern Japan*. Honolulu: University of Hawaii Press.

Tōyō Keizai Shinpōsha. 1935. *Nihon bōeki seiran* [A View of Japan's Trade]. Tokyo: Toyo Keizai Shinpōsha.

Trewartha, Glenn. 1945. *Japan: A Physical, Cultural and Regional Geography*. Madison: University of Wisconsin Press.

Umegaki, Michio. 1986. "From Domain to Prefecture." In Jansen and Rozman (1986), pp. 91–110.

Umemura, Mataji; Keiko Akasaka; Ryoshin Minami; Nobukiyo Takamatsu; Kurotake Arai; and Shigeru Itoh. 1988. *Rōdōryoku* [Manpower]. Tokyo: Toyo Keizai Shinposha.

Umemura, Mataji; Nobukiyo Takamatsu; and Shigeru Itoh. 1983. *Chiiki keizai tōkei* [Regional Economic Statistics]. Tokyo: Toyo Keizai Shinpōsha.

Uriu, Robert. 1996. *Troubled Industries: Confronting Economic Change in Japan*. Ithaca: Cornell University Press.

van Wolferen, Karel. 1989. *The Enigma of Japanese Power: People and Politics in a Stateless Nation*. New York: Alfred A. Knopf.

Vaporis, Constantine. 1994. *Breaking Barriers: Travel and the State in Early Modern Japan*. Cambridge: Harvard University Press.

Wakasugi, Ryuhei. 1997. "Technological Importation in Japan." In Goto and Odagiri 1997, pp. 20–38.

Watanabe, Shunichi. 1993. *"Toshi keikaku" no tanjō: Kokusai hikaku mita Nihon kindai toshi keikaku* ["City Planning": Japan's Modern City Planning in International Perspective]. Tokyo: Kashiwa Shobo.

Westney, D. Eleanor. 1987. *Imitation and Innovation: The Transfer of Western Organizational Patterns to Meiji Japan*. Cambridge: Harvard University Press.

Wigen, Karen. 1995. *The Making of a Japanese Periphery, 1750–1920*. Berkeley: University of California Press.

Williamson, Jeffrey. 1990. *Coping with City Growth during the British Industrial Revolution*. New York: Cambridge University Press.

Willmott, H. P. 1982. *Empires in the Balance: Japanese and Allied Pacific Strategies to April 1942*. Annapolis: Naval Institute Press.

World Engineering Congress, Publications Committee. 1929. *Industrial Japan: A Collection of Papers by Specialists of Industry in Japan*. Tokyo: Kokusai Shuppan Insatsusha.

Wray, William. 1984. *Mitsubishi and the N.Y.K., 1870–1914: Business Strategy in the Japanese Shipping Industry*. Cambridge: Harvard University Press.

———. 1986. "Shipping: From Sail to Steam." In Jansen and Rozman (1986), pp. 248–70.

Wrigley, E. 1987. *People, Cities and Wealth: The Transformation of Traditional Society*. Oxford: Basil Blackwell Ltd.

———. 1988. *Continuity, Chance and Change: The Character of the Industrial Revolution in England*. New York: Cambridge University Press.

Yamamoto, Hirofumi, ed. 1986. *Kōtsū-unyu no hattatsu gijutsu kakushin: Rekishiteki kōsatsu* [The Development of Communication and Transportation and Technological Innovation: A Historical Enquiry]. Tokyo: Tokyo University Publication Association.

———. 1993. *Technological Innovation and the Development of Transportation in Japan*. Tokyo: United Nations University Press.

Yamazawa, Ippei, and Yuzo Yamamoto. 1979. "Trade and the Balance of Payments." In Ohkawa and Shinohara with Meissner 1979, pp. 134–55.

Yui, Tsunehiko. 1985. "Introduction." In Yui and Nakagawa (1985), pp. ix–xviii.
Yui, Tsunehiko, and Keiichiro Nakagawa, eds. 1985. *Japanese Management in Historical Perspective: Proceedings of the Fuji Conference. The International Conference of Business History*. Tokyo: University of Tokyo Press.
Yuzawa, Takeshi, and Masaru Udagawa, eds. 1990. *Foreign Business in Japan before World War II*. Tokyo: University of Tokyo Press.
Zenkoku Shichō Kai [Association of Mayors of All Cities]. 1992. *Nihon toshi nenkan, 1992* [Yearbook of Japanese Cities, 1992]. Tokyo: Daiichi Hoki Shuppan Kabushiki Gaisha.

Index

About the Author

Carl Mosk is professor of economics at the University of Victoria. He has previously taught at Spelman College, the University of California, Berkeley, and Santa Clara University. He has been a visiting professor at the Kyoto Institute of Economic Research, Doshisha University, Nagoya University, the École des Hautes Études en Sciences Sociales, and the International Research Center for Japanese Studies. He is the author of *Patriarchy and Fertility: Japan and Sweden, 1880–1960* (1983, Academic Press), *Competition and Cooperation in Japanese Labour* Markets (1995, Macmillan Press Ltd.), and *Making Health Work: Human Growth in Modern Japan* (1996, University of California Press).

Library of Congress Cataloging-in-Publication Data

Kashiwara, Masaki, 1947–
Foundations of algebraic analysis.
Translation of: Daisū kaisekigaku no kiso.
Bibliography: p.
Includes index.
1. Mathematical analysis. 2. Algebra.
I. Kawai, Takahiro. II. Kimura, Tatsuo, 1947–
III. Title.
QA300.K3713 1986 515 85-43292
ISBN 0-691-08413-0